A GLENCOE PROGRAM

SCIENCE
INTERACTIONS

Bill Aldridge

Russell Aiuto

Jack Ballinger

Anne Barefoot

Linda Crow

Ralph M. Feather, Jr.

Albert Kaskel

Craig Kramer

Edward Ortleb

Susan Snyder

Paul W. Zitzewitz

With Features by:

NATIONAL
GEOGRAPHIC
SOCIETY

GLENCOE

McGraw-Hill

New York, New York   Columbus, Ohio   Mission Hills, California   Peoria, Illinois

A GLENCOE PROGRAM
SCIENCE INTERACTIONS

Student Edition
Teacher Wraparound Edition
Science Discovery Activities
Teacher Classroom Resources
Laboratory Manual
Study Guide
Section Focus Transparencies
Teaching Transparencies
Computer Test Bank

Spanish Resources
Performance Assessment
Performance Assessment in the
 Science Classroom
Science and Technology
 Videodisc Series
Integrated Science Videodisc
 Program
MindJogger Videoquizzes

Glencoe/McGraw-Hill
A Division of The **McGraw·Hill** *Companies*

Send all inquiries to:

Glencoe/McGraw-Hill
936 Eastwind Drive
Westerville, OH 43081
ISBN 0-02-828158-6

Printed in the United States of America.
1 2 3 4 5 6 7 8 9 10 071/043 03 02 01 00 99 98 97

With Features by:

NATIONAL GEOGRAPHIC SOCIETY

Table of Contents

TEACHER GUIDE

Table of Contents

STUDENT EDITION

COURSE 2

Table of Contents

Table of Contents

SCIENCE INTERACTIONS
Contents and Primary Science Emphasis

Course 1		Course 2		Course 3	
Unit 1	Observing the World Around You	Unit 1	Forces in Action	Unit 1	Electricity and Magnetism
Chapter 1	Viewing Earth and Sky	Chapter 1	Forces and Pressure	Chapter 1	Electricity
Chapter 2	Light and Vision	Chapter 2	Forces In Earth	Chapter 2	Magnetism
Chapter 3	Sound and Hearing	Chapter 3	Circulation	Chapter 3	Electromagnetic Waves
Unit 2	Interactions in the Physical World	Unit 2	Energy at Work	Unit 2	Atoms and Molecules
Chapter 4	Describing the Physical World	Chapter 4	Work and Energy	Chapter 4	Structure of the Atom
Chapter 5	Matter in Solution	Chapter 5	Machines	Chapter 5	The Periodic Table
Chapter 6	Acids, Bases, and Salts	Chapter 6	Thermal Energy	Chapter 6	Combining Atoms
Unit 3	Interactions in the Living World	Chapter 7	Moving the Body	Chapter 7	Molecules in Motion
Chapter 7	Describing the Living World	Chapter 8	Controlling the Body Machine	Unit 3	Our Fluid Environment
Chapter 8	Viruses and Simple Organisms	Unit 3	Earth Materials and Resources	Chapter 8	Weather
Chapter 9	Animal Life	Chapter 9	Discovering Elements	Chapter 9	Ocean Water and Life
Chapter 10	Plant Life	Chapter 10	Minerals and Their Uses	Chapter 10	Organic Chemistry
Chapter 11	Ecology	Chapter 11	The Rock Cycle	Chapter 11	Fueling the Body
Unit 4	Changing Systems	Chapter 12	The Ocean Floor and Shore Zones	Chapter 12	Blood: Transport and Protection
Chapter 12	Motion	Chapter 13	Energy Resources	Unit 4	Changes in Life and Earth Over Time
Chapter 13	Motion Near Earth	Unit 4	Air: Molecules in Motion	Chapter 13	Reproduction
Chapter 14	Moving Water	Chapter 14	Gases, Atoms, and Molecules	Chapter 14	Heredity
Chapter 15	Shaping the Land	Chapter 15	The Air Around You	Chapter 15	Moving Continents
Chapter 16	Changing Ecosystems	Chapter 16	Breathing	Chapter 16	Geologic Time
Unit 5	Wave Motion	Unit 5	Life at the Cellular Level	Chapter 17	Evolution of Life
Chapter 17	Waves	Chapter 17	Basic Units of Life	Unit 5	Observing the World Around You
Chapter 18	Earthquakes and Volcanoes	Chapter 18	Chemical Reactions	Chapter 18	Fission and Fusion
Chapter 19	The Earth-Moon System	Chapter 19	How Cells Do Their Jobs	Chapter 19	The Solar System
■ = Physics ■ = Life Science		■ = Earth Science ■ = Chemistry		Chapter 20	Stars and Galaxies

INTRODUCING THE AUTHOR TEAM

Bill Aldridge was Executive Director of the National Science Teachers Association from 1980-1995. After retiring from NSTA, Mr. Aldridge assumed the position of President, Division of Science Education Solutions, of ARS, Inc., Fredericksburg, VA. Prior to becoming NSTA's Executive Director, he served 3 years as a program officer in science education at the National Science Foundation. He received his B.S. and M.S. degrees in physics from the University of Kansas, where he also received an M.S. degree in Educational Evaluation. Mr. Aldridge also received an M.Ed. degree from Harvard University. He taught physics and mathematics at the high school level for 6 years and at the college level for 17 years. He has authored numerous publications, including five textbooks, 13 monographs, and articles in science and science education magazines and journals. As Executive Director of NSTA, Mr. Aldridge worked with the U.S. Congress and with government agencies in designing and producing support programs for science education. He is the recipient of awards and recognition from the National Science Foundation and the American Association of Physics Teachers, and he is a fellow of the American Association for the Advancement of Science. In 1995, he received the Distinguished Public Service Medal from NASA, the highest recognition given to a non-NASA employee.

Russell Aiuto recently retired from the position of Senior Project Officer for the Council of Independent Colleges. Dr. Aiuto is past Director of Research and Development for the National Science Teachers Association in Washington, DC. Throughout his career, Dr. Aiuto has held several prominent positions, including Director of the Division of Teacher Preparation and Enhancement of the National Science Foundation, President of Hiram College in Hiram, Ohio, and Provost of Albion College in Albion, Michigan. He also has 30 years experience teaching biology and genetics at the high school and college levels. Dr. Aiuto received B.A. degrees from Eastern Michigan University and the University of Michigan, and his M.A. and Ph.D. degrees from the University of North Carolina. He has received numerous awards, including the Phi Beta Kappa Faculty Scholar Award, Campus Teaching Award, and Honors Program Faculty Award from Albion College.

Jack Ballinger is a professor in the chemistry department at St. Louis Community College in St. Louis, Missouri, where he has taught for 27 years. He received his B.S. degree in chemistry from Eastern Illinois University, his M.S. degree in organic chemistry from Southern Illinois University, and his Ed.D. in science education from Southern Illinois University. Professor Ballinger has authored numerous articles and several books, including a reference handbook for chemical technicians. Dr. Ballinger has received many awards for his outstanding teaching: the Manufacturing Chemists Association Regional Award for Excellence in Chemistry Teaching, the Emerson Electric Company's "Excellence in Teaching Award," the Missouri Governor's Award for Excellence in Teaching, and the Outstanding Teacher of the Year Award in 1994 at St. Louis Community College at Florissant Valley.

Anne Barefoot is a veteran physics and chemistry teacher with 35 years of teaching experience. Her career also includes teaching biology and physical science, and working with middle school teachers in the summer at Purdue University in the APAST program. A past recipient of the Presidential Award for Outstanding Science Teaching, Ms. Barefoot holds B.S. and M.S. degrees from East Carolina University, and a Specialist Certificate from the University of South Carolina. Other awards include Whiteville City Schools Teacher of the Year, Sigma Xi Award, and the North Carolina Business Award for Science Teaching. Ms. Barefoot is the former District IV Director of the National Science Teachers Association and the former president of the Association of Presidential Awardees in Science Teaching.

Linda Crow is an associate professor in the Department of Natural Sciences at the University of Houston-Downtown. She is the project leader of the Texas Scope, Sequence and Coordination (SS&C) Project. In addition to 24 years as an award-winning science teacher at college and high school levels, Dr. Crow is a recognized speaker at education workshops both in the United States and abroad. Dr. Crow received her BS, M.Ed., and Ed.D. degrees from the University of Houston. She has been twice named the OHAUS Winner for Innovations in College Science Teaching.

Ralph M. Feather, Jr. teaches geology, astronomy, Earth Science, and integrated science, and serves as Science Department Chair in the Derry Area School District in Derry, PA. Mr. Feather has 26 years of teaching experience in secondary education. He holds a B.S. in geology and an M.Ed. in geoscience from Indiana University of Pennsylvania and is currently completing work on his Ph.D. in "Writing Across the Curriculum" at the University of Pittsburgh. Mr. Feather has received the Presidential Award for Excellence in Earth Science Teaching and the Award for Excellence in Earth Science Teaching from the Geological Society of America. Mr. Feather has also received the Outstanding Earth Science Teacher Award from NAGT and the Kevin Burns Citation from the Spectroscopy Society of Pittsburgh.

Albert Kaskel has 32 years experience teaching science, the last 25 at Evanston Township High School, Evanston, IL. His teaching experience includes biology, A.P. biology, physical science, and chemistry. He holds a B.S. in biology from Roosevelt University in Chicago and an M.Ed. degree from DePaul University. Mr. Kaskel has received the Outstanding Biology Teacher Award for the State of Illinois and the Teacher Excellence Award from Evanston Township High School. He is the major author of a leading high school biology textbook and has contributed to more than 30 different science textbook related publications. He was recently a staff member at the Center for Talent Development at Northwestern University.

Craig Kramer has been a physics teacher for 20 years. He is past chairperson for the Science Department at Bexley High School in Bexley, Ohio. Mr. Kramer received a B.A. in physics and a B.S. in science and math education, and an M.A. in outdoor and science education from The Ohio State University. He has received numerous awards, including the Award for Outstanding Teaching in Science from Sigma Xi. In 1987, the National Science Teachers Association awarded Mr. Kramer a certificate for secondary physics, making him the first nationally certified teacher in physics.

Edward Ortleb serves as a science consultant for the St. Louis, Missouri, Public Schools and has 37 years teaching experience. He holds an A.B. in biology education from Harris Teachers College, an M.A. in science education and an Advanced Graduate Certificate from Washington University, St. Louis. Mr. Ortleb is a lifetime member of the National Science Teachers Association, having served as its president in 1978-79. He has also served as Regional Director for the National Science Supervisors Association. Mr. Ortleb is the recipient of several awards, including the Distinguished Service to Science Education Award (NSTA), the Outstanding Service to Science Education Award, and the Outstanding Achievement in Conservation Education Award.

Susan Snyder is a teacher of Earth science at Jones Middle School, Upper Arlington School District, Columbus, OH. Ms. Snyder received a B.S. in comprehensive science from Miami University, Oxford, OH., and an M.S. in entomology from the University of Hawaii. She has 24 years teaching experience and is author of numerous educational materials. Ms. Snyder has been a state recipient of the Presidential Award for Excellence in Science and Math Teaching, a finalist for National Teacher of the Year, and Ohio Teacher of the Year. She also won the Award for Excellence in Earth Science Teaching from the Geological Society of America.

Paul W. Zitzewitz is Professor of Physics at the University of Michigan-Dearborn. He received his B.A. from Carleton College and his M.A. and Ph.D. from Harvard University, all in physics. Dr. Zitzewitz has taught physics to undergraduates for 25 years and is an active experimenter in the field of atomic physics with more than 50 research papers. He has memberships in several professional organizations, including the American Physical Society, American Association of Physics Teachers, and the National Science Teachers Association. Among his awards are the University of Michigan-Dearborn Distinguished Faculty Research Award.

National Geographic Society The National Geographic Society, founded in 1888 for the increase and diffusion of geographic knowledge, is the world's largest nonprofit scientific and educational organization. Since its earliest days, the Society has used sophisticated communication technology, from color photography to holography, to convey geographic knowledge to a worldwide membership. The Education Products Division supports the Society's mission by developing innovative educational programs—ranging from traditional print materials to multimedia programs including CD-ROMs, videodiscs, and software.

Responding to Changes IN SCIENCE EDUCATION

THE NEED FOR NEW DIRECTIONS IN SCIENCE EDUCATION

By today's projections, seven out of every ten American jobs will be related to science, mathematics, or electronics by the year 2000. And according to the experts, if junior high and middle school students haven't grasped the fundamentals, they probably won't go further in science and may not have a future in a global job market. Studies also reveal that high school students are avoiding taking "advanced" science classes.

SCIENCE INTERACTIONS ANSWERS THE CHALLENGE!

At Glencoe Publishing, we believe that *SCIENCE INTERACTIONS* will help you bring science reform to the front lines—the classrooms of America. But more important, we believe it will help students succeed in middle school and junior high science so that they will continue learning science through high school and into adulthood.

When you compare *SCIENCE INTERACTIONS* to a traditional science program, you'll see fewer terms. But you'll also see more questions and more activities to draw your students in. And you'll find broad themes repeated over and over, rather than hundreds of unrelated topics.

SCIENCE INTERACTIONS FITS YOUR CLASSROOM

SCIENCE INTERACTIONS has the right ingredients to help you ensure your students' future. But it also has to work in today's classroom. That's why, on first glance, *SCIENCE INTERACTIONS* may look like a traditional science textbook.

Glencoe knows you have local curriculum requirements. You teach a variety of students with varying ability levels. And you have limited time, space, and support for doing hands-on activities.

No matter. Unlike a purely hands-on program, *SCIENCE INTERACTIONS* lets you offer the perfect balance of content and activities. Your students will be eager to get their hands on science. But *SCIENCE INTERACTIONS* also gives you the flexibility to use only the activities you choose ... without sacrificing anything.

SCIENCE INTERACTIONS IS LOADED WITH ACTIVITIES

You'll choose from hundreds of activities in all three courses. These easy to set-up and manage activities will allow you to teach using a hands-on, inquiry-based approach to learning.

The **Find Out, Explore,** and **Investigate** activities, and **Design Your Own Investigations** are integrated with the text narrative, complete with transitions in and out of the activity. This aids comprehension for students by building continuity between text and activities.

Your teaching methods will include asking questions such as: *How do we know? Why do we believe? What does it mean?* Throughout both the narrative and activities, you'll invite your students to relate what they learn to their own everyday experiences.

SCIENCE INTERACTIONS TEACHES CONCEPTS IN A LOGICAL SEQUENCE
From the Concrete to the Abstract

Research shows that students learn better when they deal with descriptive matters in science for a reasonable portion of their school years before proceeding to the more quantitative, and eventually the more theoretical, parts of science. *SCIENCE INTERACTIONS* helps your students learn in this manner.

Let's look at the way you'll teach the topic of movement within and on the surface of Earth. In Course 1, students learn about the nature of Earth movements by observing that those movements bring about earthquakes and volcanoes. Learning to observe the results of Earth movement gets students to think about it in a concrete way.

In Course 2, students learn more about Earth movements by first observing how simple, everyday objects react to forces. This study of force is extended to the forces inside Earth that result in various observable phenomena—faults, seismic waves, volcanic eruptions.

In Courses 1 and 2, students have learned a lot about Earth movement by observing its results. In Course 3, they gain exposure to theoretical applications of Earth movement when they study plate tectonics.

SCIENCE INTERACTIONS INTEGRATES SCIENCE AND MATH

Mathematics is a tool that all students, regardless of their career goals, will use throughout their lives. *SCIENCE INTERACTIONS* provides opportunities to hone mathematics skills while learning about the natural world. **Skill Builders,** as well as **Find Out, Explore,** and **Investigate** activities and **Design Your Own Investigations** offer numerous options for practicing math, including

making and using tables and graphs, measuring in SI, and calculating. **Across the Curriculum** strategies in the *Teacher Wraparound Edition* provide additional connections between science and mathematics.

SCIENCE INTERACTIONS INTEGRATES TOPICS FOR UNDERSTANDING

According to the experts, by using an integrated approach like *SCIENCE INTERACTIONS*, students will experience dramatic gains in comprehension and retention. For instance, the series helps you teach some of the basic concepts from physical science early on. This, in turn, makes it easier for your students to understand other concepts in life and Earth science.

But you'll be doing more than simply showing how the sciences interconnect. You'll also take numerous "side trips" with your students. Connect one area of science to another. Relate science to technology, society, issues, hobbies, and careers. Show your students again and again how history, the arts, and literature can be part of science. And help your students discover the science behind things they see every day.

IT'S TIME FOR NEW DIRECTIONS IN SCIENCE EDUCATION

The need for new directions in science education has been established by the experts. America's students must prepare themselves for the high-tech jobs of the future.

We at Glencoe believe that *SCIENCE INTERACTIONS* answers the challenge of the 21st century with its new, innovative approach of "connecting" the sciences. We believe *SCIENCE INTERACTIONS* will assist you better in preparing your students for a lifetime of science learning.

Your Questions Answered!

How is *SCIENCE INTERACTIONS* an integrated science program?

Although each chapter has a primary science emphasis, integration of other disciplines occurs throughout the program. Students are more likely to learn and remember a concept because they see it applied to other disciplines. This science integration is evident not only in the narrative of the core part of each chapter, but also in the **Expand Your View** features at the end of each chapter, in the *Teacher Wraparound Edition*, and in the supplements.

How is *SCIENCE INTERACTIONS* different from general science?

There's really no comparing the *SCIENCE INTERACTIONS* program to a traditional general science text. In a general science program, the sequence of the topics and their relationship to one another is of little importance. For example, all the physics chapters in a general science text would be grouped together in a unit at the back of the book and probably have no relationship to the life, Earth, or chemistry units.

SCIENCE INTERACTIONS is different from general science in that chapters from different disciplines are intermixed and sequenced so that what is learned in one discipline can be applied to another.

How is *SCIENCE INTERACTIONS* different from the "layering" or "block" approach?

In a traditional three-year course, students study life science in sixth grade, Earth science in seventh, and physical science in eighth grade. *SCIENCE INTERACTIONS* is different because it contains all of these disciplines in each course. It is true integrated science, where life, Earth, and physical science are integrated throughout the year.

Will *SCIENCE INTERACTIONS* prepare my students for high school science?

The national reform projects agree that the best preparation is a deep understanding of important science concepts. This, rather than requiring students to memorize facts and terms, will keep your students interested in science.

Science will come alive for your students each time they pick up their textbooks. *SCIENCE INTERACTIONS'* visual format lends excitement to the study of integrated science. Students can see fundamental science concepts in living color. Learning concepts by visualizing them also increases cognitive awareness, thus giving students a more solid foundation for future science courses.

SCIENCE INTERACTIONS will help your students frame questions, derive concepts, and obtain evidence. When your students have mastered this language of science, they will be ready for further study.

In addition, *SCIENCE INTERACTIONS* offers your students plenty of reasons to stick with science—including unexpected career choices and examples of women and minorities achieving in science.

Science Interactions Supports
The National Science Education Standards

The *National Science Education Standards*, published by the National Research Council and representing the contribution of thousands of educators and scientists, offer a comprehensive vision of a scientifically literate society. The standards describe not only what students should know but also offer guidelines for science teaching and assessment. Bill Aldridge, a *Science Interactions* author, helped originate this standards initiative and served on its Chair's Advisory Committee. If you are using, or plan to use, the standards to guide changes in your science curriculum, you can be assured that *Science Interactions* aligns with the *National Science Education Standards*.

Science Interactions is an example of how Glencoe's commitment to effective science education is changing the materials used in science classrooms today. More than just a collection of facts in a textbook, *Science Interactions* is a program that provides numerous opportunities for students, teachers, and school districts to meet the *National Science Education Standards*.

Content Standards
The accompanying table shows the close alignment between *Science Interactions* and the content standards. The integrated approach of *Science Interactions* allows students to discover concepts within each of the content standards, giving them opportunities to make connections between the science disciplines. Our hands-on activities and inquiry-based lessons reinforce the science processes emphasized in the standards.

Teaching Standards
Alignment with the *National Science Education Standards* requires much more than alignment with the outcomes in the content standards. The way in which concepts are presented is critical to effective learning. The teaching standards within the *National Science Education Standards* recommend an inquiry-based program facilitated and guided by teachers. *Science Interactions* provides such opportunities through activities and discussions that allow students to discover by inquiry critical concepts and apply the knowledge they've constructed to their own lives. Throughout the program, students are building critical skills that will be available to them for lifelong learning. *The Teacher Wraparound Edition* helps you make the most of every instructional moment. It offers an abundance of effective strategies and suggestions for guiding students as they explore science.

Assessment Standards
The assessment standards are supported by many of the components that make up the *Science Interactions* program. *The Teacher Wraparound Edition* and *Teacher Classroom Resources* provide multiple chances to assess students' understanding of important concepts as well as their ability to perform a wide range of skills. Ideas for portfolios, performance activities, written reports, and other assessment activities accompany every lesson. For more suggestions for assessment and resources, see pages 30T-31T. Rubrics and performance task assessment lists can be found in Glencoe's Professional Series booklet *Performance Assessment in the Science Classroom.*

Program Coordination
The scope of the content standards requires students to meet the outcomes over the course of their education. *Science Interactions Courses 1, 2,* and *3* provide an integrated middle school science curriculum that is aligned not only across a grade level, but also vertically, through several years of instruction. The correlation on the following pages demonstrates the close alignment of this course of *Science Interactions* with the content standards.

Correlation of Science Interactions, Course 2, to the National Science Standards, Content Standards, Grades 5-8

Content Standard	Page Numbers
(UCP) UNIFYING CONCEPTS AND PROCESSESS	
1. Systems, order, and organization	85-90, 100-101, 108, 113, 138, 183, 212-231, 240-258, 270-271, 290, 432, 435, 439, 446-447, 462-465, 484-500, 604-606, 636-641
2. Evidence, models, and explanation	4-17, 24, 26, 29, 32, 36, 39, 41, 44, 47, 54, 58, 60-61, 64, 67, 72, 84, 86, 88, 91-92, 94, 96, 103, 118, 123, 126, 129, 132, 134, 136, 151, 153, 155, 158, 162, 166, 169-170, 182, 184, 187, 190, 194, 200, 202, 212, 214, 216, 221, 225, 228, 230, 240, 242, 244, 248, 252, 254, 271, 274, 282-283, 286, 292, 303, 308, 310, 312, 314, 316, 330-331, 336, 339, 341, 344, 346, 348, 363, 368, 373, 377, 380, 393, 396, 398, 400, 407, 410, 426, 430, 434, 436-437, 440, 442, 457-458, 464, 466, 470, 472, 485, 488, 490, 492, 495, 498, 502, 516, 518-520, 522, 525, 529-530, 532, 536, 540, 552, 554, 557, 563, 568, 571, 584, 586, 588, 590, 596, 598, 642-653
3. Change, constancy, and measurement	27, 32-33, 36, 41, 44, 91-94, 121, 125-127, 129, 131-141, 158-165, 184-185, 187-189, 307-314, 469-475, 552-556, 564-570
4. Evolution and equilibrium	20-51, 70-71, 104, 131-133, 138-141, 186, 220, 318-321, 330-352, 362-367, 372-373, 384-385, 397-401, 536-542, 584-593
5. Form and function	70, 72-73, 101, 143, 150-157, 166-169, 174-175, 178-179, 200-203, 213, 220, 222-224, 226-227, 233, 244-255, 272, 284-285, 296, 302, 332-333, 340-341, 348-349, 407-414, 416-417, 486-493
(A) SCIENCE AS INQUIRY	
1. Abilities necessary to do scientific inquiry	2-17, 21, 24-26, 29, 32-33, 35-36, 39, 41, 43-44, 53-54, 58, 60-61, 64-65, 67, 72-73, 80-81, 83-84, 86, 88, 91-94, 96-97, 103, 108, 111-113, 115, 117-118, 123, 126-129, 132, 134, 136-137, 142, 146-147, 149, 153, 155, 158, 162-163, 166, 169-171, 178-179, 181-182, 184-185, 187, 189-190, 194-195, 200, 202, 206, 208-209, 211-212, 214, 216-219, 221, 225, 228-230, 236-237, 239-240, 242, 244, 248-249, 252-254, 259, 263-265, 267, 271, 274-275, 282, 286-287, 298-299, 303, 308, 310, 312-314, 316-317, 326-327, 330-331, 336-337, 339, 344, 346, 348-349, 358-359, 361, 363, 368-369, 373, 377, 380-381, 388-389, 399-400, 407, 410-411, 419-420, 423, 425-426, 428-431, 434, 437, 440, 442-443, 449-450, 452-453, 455, 457-459, 464, 466-467, 470, 472, 480-481, 485, 488, 490-492, 495-499, 502, 509-511, 513, 516, 518-520, 522-523, 525, 529-530, 532, 536-537, 540-541, 548-549, 551-552, 554-555, 557, 560-561, 563, 568-569, 571, 580-581, 583-584, 586, 588, 590-591, 596, 598-599, 611-613
2. Understandings about scientific inquiry	2-17, 27, 34, 44, 47-48, 109, 144, 183, 188, 201, 246-247, 260-261, 310-311, 432-433, 439, 447, 465, 495, 506, 517, 524, 585, 642-653

Content Standard	Page Numbers
(B) PHYSICAL SCIENCE	
1. Properties and changes of properties in matter	218-219, 266-299, 308-309, 346-349, 422-453, 478, 550-581, 630-631
2. Motions and forces	18-113, 148-179, 340-341, 587-589
3. Transfer of energy	56-57, 114-147, 180-209, 364-365, 390-420, 469-475, 497, 563-570, 592-602
(C) LIFE SCIENCE	
1. Structure and function in living systems	42-43, 82-113, 208-265, 482-549, 578, 582-613, 622-625
2. Reproduction and heredity	514, 536-543
3. Regulation and behavior	8, 42-43, 240-255, 372-373, 375, 482-504, 525-535, 554-555, 572, 578, 584-602
4. Populations and ecosystems	372-373
5. Diversity and adaptations of organisms	46, 78, 350-351, 372-373, 493, 622-625
(D) EARTH AND SPACE SCIENCE	
1. Structure of the Earth system	52-80, 192-193, 284-285, 288, 300-389, 394-406, 414-416, 421, 448-449, 454-481, 559, 582, 626-629, 632-634
2. Earth's history	328, 330-338, 343, 350-351, 382, 394-406
3. Earth in the solar system	2-17, 19, 124-125, 138-139, 196-197, 206, 283, 410-413, 417, 468-477, 562
(E) SCIENCE AND TECHNOLOGY	
1. Abilities of technological design	8-14, 75-77, 142, 173-175, 206, 232, 554-555
2. Understandings about science and technology	46, 48, 89, 174-175, 222-223, 246-247, 324, 376-379, 394-395, 432-433, 506, 524, 536-537, 573, 577, 585

National Standards

Content Standard	Page Numbers
(F) SCIENCE IN PERSONAL AND SOCIAL PERSPECTIVES	
1. Personal health	98-99, 102-106, 109, 168-169, 232, 234, 260-261, 460-461, 500-505, 507, 546, 564-565, 575-576, 608-609
2. Populations, resources, and environments	8-11, 204-205, 281, 288, 293, 320-321, 354-355, 370-375, 396-406, 409, 415, 460-461, 476-477, 573
3. Natural hazards	52-80, 334-335, 362-363, 384-385, 476-477
4. Risks and benefits	107, 204-205, 260-261, 322-323, 354-355, 362-363, 384-385, 402-406, 415-416, 460-461, 607
5. Science and technology in society	46, 48, 143, 176, 246-247, 259, 296, 322-323, 382-386, 390-391, 404-406, 448-449, 506, 543, 575, 578, 607, 609
(G) HISTORY AND NATURE OF SCIENCE	
1. Science as a human endeavor	9, 23, 31, 44, 47, 107, 144, 432, 435, 439, 447, 495
2. Nature of science	47, 517
3. History of science	9, 27, 31, 44, 47, 107, 144, 183, 201, 432, 435, 439, 446-447, 495, 517

Flex Your Brain

A key element in the coverage of problem solving and critical thinking skills in ***SCIENCE INTERACTIONS*** is a critical thinking matrix called **Flex Your Brain.**

Flex Your Brain provides students with an opportunity to explore a topic in an organized, self-checking way, and then identify how they arrived at their responses during each step of their investigation. The activity incorporates many of the skills of critical thinking. It helps students to consider their own thinking and learn about thinking from their peers.

WHERE IS FLEX YOUR BRAIN FOUND?

In the introductory chapter, "Science: A Tool for Solving Problems," is an introduction to the topics of critical thinking and problem solving. **Flex Your Brain** accompanies the text section as an activity in the introductory chapter. Brief student instructions are given, along with the matrix itself. A worksheet for **Flex Your Brain** appears in the ***Critical Thinking/Problem Solving*** book of the ***Teacher Resources.*** This version provides spaces for students to write in their responses.

In the ***Teacher Wraparound Edition,*** suggested topics are given in each chapter for the use of **Flex Your Brain.** You can either refer students to the introductory chapter for the procedure, or photocopy the worksheet master from the ***Teacher Resources.***

USING FLEX YOUR BRAIN

Flex Your Brain can be used as a whole-class activity or in cooperative groups, but is primarily designed to be used by individual students within the class. There are three basic steps.

1. Teachers assign a class topic to be investigated using **Flex Your Brain.**

2. Students use **Flex Your Brain** to guide them in their individual explorations of the topic.

3. After students have completed their explorations, teachers guide them in a discussion of their experiences with **Flex Your Brain,** bridging content and thinking processes.

Flex Your Brain can be used at many different points in the lesson plan.

Introduction: Ideal for introducing a topic, **Flex Your Brain** elicits students' prior knowledge and identifies misconceptions, enabling the teacher to formulate plans specific to student needs.

Development: Flex Your Brain leads students to find out more about a topic on their own, and develops their research skills while increasing their knowledge. Students actually pose their own questions to explore, making their investigations relevant to their personal interests and concerns.

Review and Extension: Flex Your Brain allows teachers to check student understanding while allowing students to explore aspects of the topic that go beyond the material presented in class.

Flex Your Brain

1 Topic: _____
2 ? What do I already know?   1. _____ 2. _____ 3. _____ 4. _____ 5. _____
3 Q: Ask a question
4 A: Guess an answer
5 How sure am I? (circle one)
   Not sure   Very sure
   1   2   3   4   5
6 ? How can I find out?   1. _____ 2. _____ 3. _____ 4. _____ 5. _____
7 EXPLORE
8 Do I think differently?   yes   no
9 ? What do I know now?   1. _____ 2. _____ 3. _____ 4. _____ 5. _____
SHARE
10   1. _____ 2. _____ 3. _____

Using Technology in the Classroom

Technology helps you adapt your teaching methods to the needs of your students. Glencoe classroom technology products provide many pathways to help you match students' different learning styles. To make your lesson planning easier, all of the technology products listed below are correlated to the student text.

Videodiscs

Glencoe's *Integrated Science Videodiscs* are designed to be used interactively in the classroom. Barcodes in this book allow you to step through the programs and pause to discuss and answer on-screen questions. Barcodes for the following videodiscs also appear throughout this teacher edition: *Mr. Wizard's Science and Technology Videodisc, Infinite Voyage, Newton's Apple, and National Geographic Society.*

MindJogger Videoquiz

A videoquiz for each chapter can be used to assess prior knowledge or review content before an exam. Student teams work cooperatively to answer three rounds of questions posed by the video host. Each round requires higher level thinking skills than the previous round.

Glencoe Interactive CD-ROMs

The Glencoe *Life, Earth,* and *Physical Science* CD-ROMs are correlated to *Science Interactions.*

You can:
- use a CD-ROM interaction as a whole class presentation;
- allow 2-3 small groups to use the CD-ROMs at once;
- rotate student groups through a single computer station;
- place the materials in a computer lab; or
- use CD-ROMs as a library resource.

Glencoe also offers three National Geographic Society CD-ROMs. These image-rich, reference CD-ROMs are faithful to the Society's long history of excellence in science teaching and journalism.

Computer Test Bank

Glencoe's Test Generator for Macintosh and for DOS makes creating, editing, and printing tests quick and easy. You also can edit questions or add your own favorite questions and graphics.

Computer Competency Activities

This software is unique to the *Science Interactions* series. Closely related to the content of each chapter, these activities are designed to help students master core computer competencies: word processing, graphing, using spreadsheets, and manipulating databases.

English/Spanish Audiocassettes

Audio chapter summaries in English and in Spanish are a way for auditory learners, lower-level readers, and LEP students to review key chapter concepts. Students can listen individually during class or check out tapes and use them at home. You may find them useful for reviewing the chapter as you plan lessons.

Glencoe Software Support Hotline   1-800-437-3715

Should you encounter any difficulty when setting up or running Glencoe software, contact the Software Support Center at Glencoe Publishing between 8:30 a.m. and 4:30 p.m. Eastern Time.

Using the Internet

If you're already familiar with the Internet, skip to the sites listed at the bottom of this page. If you need some tips on how to get started, keep reading.

The Internet is an enormous reference library and a communication tool. You can use it to quickly retrieve information from computers around the globe. Like any good reference, it has an index so you can locate the right piece of information. An Internet index entry is called a *Universal Resource Locator*, or URL. Here's an example:

http://www.glencoe.com/intro/index.html

The first part of the URL tells the computer how to display the information. The second part, after the double slash, names the organization and the computer where the information is stored. The part after the first single slash tells the computer which directory to search and which file to retrieve. File locations change frequently. If you can't find what you're looking for, use the first part of the address only, for ex-

ample, http://www.glencoe.com/ from the URL shown, and follow links to what you need.

The World Wide Web

The World Wide Web (WWW), a subset of the Internet, began in 1992. Unlike regular text files, web files can have links to other text files, images, and sound files. By clicking on a link, you can see or hear the linked information.

How do I get access?

To use the Internet, you need a computer, a modem, a telephone line, and a connection to the Internet. If your school doesn't have a connection, contact your local public library or a university; they often give free access to students and educators.

CAUTION: Contents may shift!

The sites referenced in Glencoe's Internet Connections are not under Glencoe's control. Therefore,

Glencoe can make no representation concerning the content of these sites. Extreme care has been taken to list only reputable links by using educational and government sites whenever possible. Internet searches have been used that return only sites that contain no content apparently intended for mature audiences.

Where to Start

The brief list of science Internet sites below may prove useful. You can also find Internet addresses throughout this book correlated to selected features.

Useful tools for searching the Internet include:
http://www.yahoo.com/search.html
http://www.altavista.digital.com and
http://www.msn.com/access/
　　allinone/hv1

Science Internet Site	Description
http://ericir.syr.edu/	**Ask ERIC**, an ask-the-expert service for K-12 teachers
http://www.enc.org/	**The Eisenhower National Clearinghouse for Math and Science**, instructional materials
http://medinfo.wustl.edu/~ysp/MSN/	**The Mad Scientist Information Network**, an ask-the-expert service for science students

For more information about Glencoe Technology, call Customer Service at **1-800-334-7344**.

Themes & Scope & Sequence

SCIENCE INTERACTIONS, three science textbooks for middle school, is unique in that it integrates all the natural sciences, presenting them as a single area of study. Our society is becoming more aware of the interrelationship of the disciplines of science. For most people, the ideas that unify the sciences and make connections between them are the most valid.

Themes are the constructs that unify the sciences. Woven throughout *SCIENCE INTERACTIONS*, themes integrate facts and concepts. They are the "big ideas" that link the structures on which the science disciplines are built. While there are many possible themes around which to unify science, we have chosen four: Energy, Systems and Interactions, Scale and Structure, and Stability and Change.

ENERGY

Energy is a central concept of the physical sciences that pervades the biological and geological sciences. In physical terms, energy is the ability of an object to change itself or its surroundings, the capacity to do work. In chemical terms, it forms the basis of reactions between compounds. In biological terms, it gives living systems the ability to maintain themselves, to grow, and to reproduce. Energy sources are crucial in the interactions among science, technology, and society.

SYSTEMS AND INTERACTIONS

A system can be incredibly tiny, such as an atom's nucleus and electrons, or unbelievably large, as the stars in a galaxy. By defining the boundaries of the system, one can study the interactions among its parts. The interactions may be a force of attraction between the positively charged nucleus and negatively charged electron. In an ecosystem, however, the interactions may be between the predator and its prey, or among the plants and animals. Animals in such a system have many subsystems (circulation, respiration, digestion, etc.) with interactions among them.

Course 1

Themes	Chapter																		
	1	2	3	4	5	6	7	8	9	10	11	12	13	14	15	16	17	18	19
Scale and Structure	P			P			S	P	P										S
Energy		S	S												S	P	P		
Stability and Change			P	S	P	S				S	S	P	S	S	P	S		S	
Systems and Interactions	S	P			S	P	P	S	S	P	P	S	P	P		P	S		P

P = PRIMARY THEME   **S** = SECONDARY THEME

SCALE AND STRUCTURE

Used as a theme, "structure" emphasizes the relationship among different structures. "Scale" defines the focus of the relationship. As the focus is shifted from a system to its components, the properties of the structure may remain constant. In other systems, an ecosystem for example, which includes a change in scale from interactions between prey and predator to the interactions among systems inside an animal, the structure changes drastically. In *SCIENCE INTERACTIONS*, the authors have tried to stress how we know what we know and why we believe it to be so. Thus, explanations remain on the macroscopic level until students have the background needed to understand how the microscopic structure was determined.

STABILITY AND CHANGE

A system that is stable is constant. Often the stability is the result of a system being in equilibrium. If a system is not stable, it undergoes change. Changes in an unstable system may be characterized as trends (position of falling objects), cycles (the motion of planets around the sun), or irregular changes (radioactive decay).

THEME DEVELOPMENT

These four major themes, as well as several others, are developed within the student material and discussed throughout the *Teacher Wraparound Edition*. Each chapter of *SCIENCE INTERACTIONS* incorporates a primary and secondary theme. These themes are interwoven throughout each level and are developed as appropriate to the topic presented.

The *Teacher Wraparound Edition* includes a **Theme Development** section for each unit opener and for each chapter opener. These sections discuss upcoming key themes and explain how they are supported. Throughout the chapters, **Theme Connections** show specifically how a topic in the student edition relates to the themes.

Course 2

Themes	Chapter																		
	1	2	3	4	5	6	7	8	9	10	11	12	13	14	15	16	17	18	19
Scale and Structure									P	S		P		P			P		
Energy		P	S	P	S	P	S						P		P	S	S	S	
Stability and Change	S	S						S			P	S							
Systems and Interactions	P		P	S	P	S	P	P	S	P	S			S	S	S	P		P	P

Course 3

Themes	Chapter																			
	1	2	3	4	5	6	7	8	9	10	11	12	13	14	15	16	17	18	19	20
Scale and Structure	S		S	S	P				P	S	P	P	S						P	P
Energy	P	S	P	P			P		P		P			P			P			
Stability and Change		P				S	S	S	S		S		S	S	P		P	P	S	S
Systems and Interactions						P		P	S						S	S	S			

Constructivism

Strategies suggested in *SCIENCE INTERACTIONS* support a constructivist approach to science education. The role of the teacher is to provide an atmosphere where students design and direct activities. To develop the idea that science investigation is not made up of closed-end questions, the teacher should ask guiding questions and be prepared to help his or her students draw meaningful conclusions when their results do not match predictions. Through the numerous activities, cooperative learning opportunities, and a variety of critical thinking exercises in *SCIENCE INTERACTIONS*, you can feel comfortable taking a constructivist approach to science in your classroom.

Activities

A constructivist approach to science is rooted in an activities-based plan. Students must be provided with sensorimotor experiences as a base for developing abstract ideas. *SCIENCE INTERACTIONS* utilizes a variety of learning-by-doing opportunities. **Find Out** and **Explore** activities allow students to consider questions about the concepts to come, make observations, and share prior knowledge. **Find Out** and **Explore** activities require a minimum of equipment, and students may take responsibility for organization and execution.

Investigates and **Design Your Own Investigations** develop and reinforce or restructure concepts as well as develop the ability to use process skills. **Design Your Own Investigation** formats are structured to guide students to make their own discoveries. Students collect real evidence and are encouraged through open-ended questions to reflect and reformulate their ideas based on this evidence.

Cooperative Learning

Cooperative learning adds the element of social interaction to science learning. Group learning allows students to verbalize ideas, and encourages the reflection that leads to active construction of concepts. It allows students to recognize the inconsistencies in their own perspectives and the strengths of others'. By presenting the idea that there is no one, "ready-made" answer, all students may gain the courage to try to find a viable solution. **Cooperative Learning** strategies appear in the *Teacher Wraparound Edition* margins whenever appropriate.

And More …

Flex Your Brain, a self-directed critical thinking matrix, is introduced in the introductory chapter, "Science: A Tool for Solving Problems." This activity, referenced wherever appropriate in the *Teacher Wraparound Edition* margins, assists students in identifying what they already know about a subject, then in developing independent strategies to investigate further. **Uncovering Misconceptions** in the chapter opener suggests strategies the teacher may use to evaluate students' current perspectives.

Students are encouraged to discover the pleasure of solving a problem through a variety of features. **Apply** questions that require higher-level, divergent thinking appear in **Check Your Understanding**. The **Expand Your View** features in each chapter invite students to confront real-life problems. **You Try It** and **What Do You Think?** questions encourage students to reflect on issues related to technology and society. The **Skill Handbook** gives specific examples to guide students through the steps of acquiring thinking and process skills. **Skill Builder** activities give students a chance to assess and reinforce the concepts just learned through practice. **Developing Skills**, **Critical Thinking**, and **Problem Solving** sections of the **Chapter Review** allow the teacher to assess and reward successful thinking skills.

Thinking Processes

Science is not just a collection of facts for students to memorize. Rather it is a process of applying those observations and intuitions to situations and problems, formulating hypotheses and drawing conclusions. This interaction of the thinking process with the content of science is the core of science and should be the focus of study.

THINKING PROCESSES

Observing

The most basic thinking process is observing. Through observation—seeing, hearing, touching, smelling, tasting—the student begins to acquire information about an object or event.

Organizing Information

Students can begin to organize the information acquired through observation. This process of organizing information encompasses *ordering*, *organizing*, and *comparing*.

Communicating

Once all the information is gathered, it is necessary to communicate the findings so that they can be considered and shared by others. Information can be presented in tables, charts, graphs, or models.

Inferring

This leads to another process—*inferring*. Inferences are logical conclusions based on observations and are made after careful evaluation of all the available facts or data. They can be tested and evaluated.

Relating

Relating cause and effect focuses on how events or objects interact with one another. It also involves examining dependencies and relationships between objects and events.

CRITICAL THINKING SKILLS

Making Generalizations

Identifying similarities among events or processes and then applying that knowledge to new events involves *making generalizations*.

Evaluating Information

Developing ability in several categories of information evaluation is important to critical thinking: differentiating fact from opinion, identifying weaknesses in logic or in the interpretation of observations, differentiating between relevant and irrelevant data or ideas.

Applying

Applying is a process that puts scientific information to use. Sometimes the findings can be applied in a practical sense, or they can be used to tie together complex data.

Problem Solving

Using available information to develop an appropriate solution to a complex, integrated question is the essence of *problem solving*.

Decision Making

Decision making involves choosing among alternative properties, issues, or solutions. Making informed decisions is not a random process, but requires knowledge, experience, and good judgment.

Inquiry

The process of *inquiry* involves asking questions or predicting outcomes of future situations. Skills used include the ability to make generalizations, problem solve, and distinguish between relevant and irrelevant information.

INTERACTION OF CONTENT AND PROCESS

SCIENCE INTERACTIONS encourages the interaction between science content and thinking processes by offering hundreds of hands-on activities that are easy to set up and do. In the student text, the **Explore** and **Find Out** activities require students to make observations, and collect and record a variety of data. **Investigates** and **Design Your Own Investigations** connect the activity with the content information.

At the end of each chapter, students use the thinking processes as they complete **Developing Skills**, **Critical Thinking**, **Problem Solving**, and **Connecting Ideas** questions. **Expand Your View** connects the science content to other disciplines.

SKILL HANDBOOK

The **Skill Builder/Skill Handbook** provides the student with another opportunity to practice the thinking processes relevant to the material they are studying. The **Skill Handbook** provides examples of the processes which students may refer to as they do the **Skill Builder** exercises.

Developing Thinking Processes

THINKING PROCESSES	Intro	1	2	3	4	5	6	7	8	9	10	11	12	13	14	15	16	17	18	19
ORGANIZING INFORMATION																				
Classifying		✓	✓	✓	✓	✓		✓	✓	✓	✓	✓	✓	✓		✓	✓	✓	✓	
Sequencing			✓		✓	✓			✓		✓			✓				✓	✓	
Concept Mapping		✓	✓	✓	✓	✓	✓	✓	✓	✓	✓	✓	✓	✓	✓	✓	✓		✓	✓
Making and Using Tables		✓	✓	✓	✓	✓	✓	✓	✓	✓	✓			✓	✓	✓	✓	✓		✓
Making and Using Graphs		✓	✓	✓	✓	✓	✓			✓			✓			✓	✓	✓		✓
THINKING CRITICALLY																				
Observing and Inferring	✓	✓	✓	✓	✓	✓	✓	✓	✓	✓	✓	✓	✓	✓	✓	✓	✓	✓	✓	✓
Comparing and Contrasting	✓	✓	✓	✓	✓	✓	✓	✓	✓	✓	✓	✓	✓	✓	✓	✓	✓	✓	✓	✓
Recognizing Cause and Effect		✓	✓	✓	✓	✓	✓	✓	✓		✓	✓	✓	✓	✓		✓		✓	✓
Forming Operational Definitions		✓			✓		✓		✓	✓	✓	✓		✓			✓	✓	✓	✓
Measuring in SI	✓	✓	✓		✓	✓	✓	✓					✓		✓	✓	✓	✓		✓
PRACTICING SCIENTIFIC PROCESSES																				
Observing	✓	✓	✓				✓	✓		✓		✓	✓	✓	✓		✓	✓		
Forming a Hypothesis	✓	✓	✓	✓	✓	✓	✓	✓	✓	✓	✓	✓	✓	✓	✓	✓	✓	✓	✓	✓
Designing an Experiment to Test a Hypothesis	✓	✓	✓	✓	✓	✓	✓	✓	✓	✓	✓	✓	✓	✓	✓	✓	✓	✓	✓	✓
Separating and Controlling Variables	✓			✓	✓	✓	✓	✓	✓	✓	✓		✓	✓	✓	✓	✓		✓	✓
Interpreting Data	✓	✓	✓	✓	✓	✓	✓	✓	✓		✓		✓	✓	✓	✓	✓	✓	✓	
REPRESENTING AND APPLYING DATA																				
Interpreting Scientific Illustrations		✓	✓	✓	✓	✓	✓	✓	✓	✓	✓	✓	✓	✓	✓	✓	✓	✓	✓	✓
Making Models	✓		✓	✓	✓		✓	✓				✓		✓		✓		✓		
Predicting		✓	✓	✓			✓	✓		✓	✓		✓	✓	✓	✓	✓			✓

Thinking Processes

Multicultural Perspectives

American classrooms reflect the rich and diverse cultural heritages of the American people. Students come from different ethnic backgrounds and different cultural experiences into a common classroom that must assist all of them in learning. The diversity itself is an important focus of the learning experience.

Diversity can be repressed, creating a hostile environment; ignored, creating an indifferent environment; or appreciated, creating a receptive and productive environment. Responding to diversity and approaching it as a part of every curriculum is challenging to a teacher, experienced or not. The goal of science is understanding. The goal of multicultural education is to promote the understanding of how people from different cultures approach and solve the basic problems all humans have in living and learning. *SCIENCE INTERACTIONS* addresses this issue. In the **Multicultural Perspectives** sections of the *Teacher Wraparound Edition*, information is provided about people and groups who have traditionally been misrepresented or omitted. The intent is to build awareness and appreciation for the global community in which we all live.

The *SCIENCE INTERACTIONS Teacher Classroom Resources* also includes a *Multicultural Connections* booklet that offers additional opportunities to integrate multicultural materials into the curriculum. By providing these opportunities, *SCIENCE INTERACTIONS* is helping to meet the four major goals of multicultural education:

1. promoting the strength and value of cultural diversity
2. promoting human rights and respect for those who are different from oneself
3. promoting social justice and equal opportunity for all people
4. promoting equity in the distribution of power among groups

Two books that provide additional information on multicultural education are:

Atwater, Mary, et al. *Multicultural Education: Inclusion of All.* Athens, Georgia: University of Georgia Press, 1994.

Banks, James A. (with Cherry A. McGee Banks) *Multicultural Education: Issues and Perspectives.* Boston: Allyn and Bacon, 1989.

School to Work
Tech-Prep

WHAT IS TECH-PREP?

Tech-prep is a rigorous and focused program of study that aims to create a workforce in the United States that is technically literate. It is designed to prepare students enrolled in a general curriculum for the demands of further education or for employment by providing them with essential academic and technical foundations, along with problem-solving, group-process, and lifelong-learning skills.

Characteristics of the Tech-Prep Curriculum

The Secretary's Commission on Achieving Necessary Skills (SCANS) published a report in June 1991 that outlined several competencies that characterize successful workers. The Tech-Prep curriculum seeks to address these competencies, which include the

- ability to use resources productively.
- ability to use interpersonal skills effectively, including fostering teamwork, teaching others, serving customers, leading, negotiating, and working well with individuals from culturally diverse backgrounds.

- ability to acquire, evaluate, interpret, and communicate data and information.
- ability to understand social, organizational, and technological systems.
- ability to apply technology to specific tasks.

The middle school years provide an opportunity to identify those students who might benefit from a tech-prep curriculum once they reach high school. They also provide an opportunity to introduce unfamiliar students to technological applications leading to career opportunities, or to provide practice for students who already have some knowledge of how technology is used.

Glencoe *Science Interactions* and Tech-Prep Issues

SCIENCE INTERACTIONS helps you develop scientific and technological literacy in your students through a variety of performance-based activities that emphasize problem solving, critical thinking skills, and teamwork. Each chapter contains an activity with a technological application. Look for the Tech-Prep logo to identify these opportunities.

CONCEPT MAPS

In science, concept maps make abstract information concrete and useful, improve retention of information, and show students that thought has shape.

Concept maps are visual representations or graphic organizers of relationships among particular concepts. Concept maps can be generated by individual students, small groups, or an entire class. *SCIENCE INTERACTIONS* develops and reinforces four types of concept maps—the **network tree**, **events chain**, **cycle concept map**, and **spider concept map**—that are most applicable to studying science. Examples of the four types and their applications are shown on this page.

Students can learn how to construct each of these types of concept maps by referring to the **Skill Handbook**. Throughout the course, students will have many opportunities to practice their concept mapping skills through **Skill Builder** activities and **Developing Skills** questions in the **Chapter Review**.

BUILDING CONCEPT MAPPING SKILLS

The **Skill Builders** in each chapter and the **Developing Skills** section of the **Chapter Review** provide opportunities for practicing concept mapping. A variety of concept mapping approaches is used. Students may be directed to make a specific type of concept map and be provided the terms to use. At other times, students may be given only general guidelines. For example, concept terms to be used may be provided and students will be required to select the appropriate model to apply, or vice versa. Finally, students may be asked to provide both the terms and type of concept map to explain relationships among concepts. When students are given this flexibility, it is important for you to recognize that, while sample answers are provided, student responses may vary. Look for the conceptual strength of student responses, not absolute accuracy. You'll notice that most network tree maps provide connecting words that explain the relationships between concepts. We recommend that you not require all students to supply these words, but many students may be challenged by this aspect.

NETWORK TREE

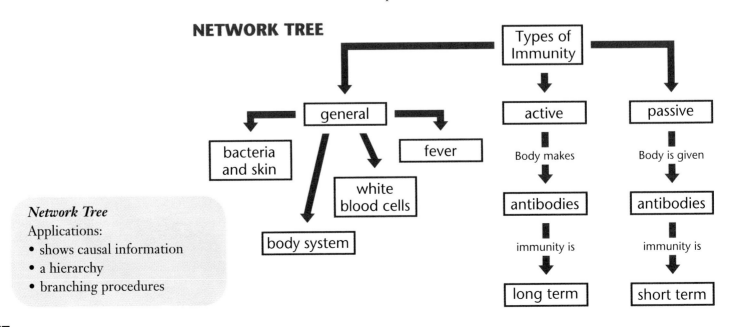

Network Tree
Applications:
• shows causal information
• a hierarchy
• branching procedures

CONCEPT MAPPING BOOKLET

The **Concept Mapping** book of the **Teacher Classroom Resources**, too, provides a developmental approach for students to practice concept mapping.

As a teaching strategy, generating concept maps can be used to preview a chapter's content by visually relating the concepts to be learned and allowing the students to read with purpose. Using concept maps for previewing is especially useful when there are many new key science terms for students to learn. As a review strategy, constructing concept maps reinforces main ideas and clarifies their relationships. Construction of concept maps using cooperative learning strategies as described in this Teacher Guide will allow students to practice both interpersonal and process skills.

CYCLE CONCEPT MAP

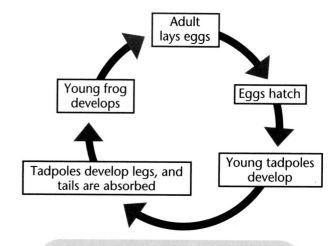

Cycle Concept Map
Application:
• shows how a series of events interact to produce a set of results again and again

EVENTS CHAIN

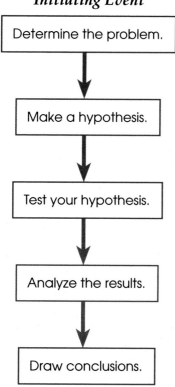

Events Chain
Applications:
• describes the stages of a process
• the steps in a linear procedure
• a sequence of events

SPIDER CONCEPT MAP

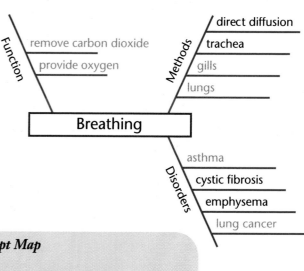

Spider Concept Map
Applications:
• nonhierarchical, except within a category
• unparallel categories

Planning Your Course

SCIENCE INTERACTIONS provides flexibility in the selection of topics and content that allows teachers to adapt the text to the needs of individual students and classes. In this regard, the teacher is in the best position to decide what topics to present, the pace at which to cover the content, and what material to give the most emphasis. To assist the teacher in planning the course, a planning guide has been provided.

SCIENCE INTERACTIONS may be used in a full-year course that is comprised of 180 periods of approximately 45 minutes each. This type of schedule is represented in the table under the heading of Single-class scheduling.

To build flexibility into the curriculum, many schools are introducing a block scheduling approach. In the table shown here, it is assumed that for block scheduling, the course will be taught for 90 periods of approximately 90 minutes each. If you follow a block schedule, you may want to consider either combining lessons or eliminating certain topics and spending more time on the topics you do cover.

Please remember that the planning guide is provided as an aid in planning the best course for your students. You should use the planning guide in relation to your curriculum and the ability levels of the classes you teach, the materials available for activities, and the time allotted for teaching.

Unit	Chapter/Section	Single-class (180 days)	Block (90 days)
Introduction	Science: A Tool for Solving Problems	3	1
UNIT 1	**Forces in Action**	**29**	**15**
1	**Forces and Pressure**	**10**	**5**
1-1	Force and Motion	3	1.5
1-2	How Forces Act on Objects	4	2
1-3	Pressure and Buoyancy	2	1
	Chapter Review and Test	1	0.5
2	**Forces in Earth**	**9**	**5**
2-1	What Causes Earthquakes?	2	1
2-2	Shake and Quake	3	2
2-3	Volcanic Eruptions	3	1.5
	Chapter Review and Test	1	0.5
3	**Circulation**	**10**	**5**
3-1	Circulatory Systems	3	1.5
3-2	A System Under Pressure	4	2
3-3	Circulation	2	1
	Chapter Review and Test	1	0.5
UNIT 2	**Energy at Work**	**46**	**23**
4	**Work and Energy**	**8**	**4**
4-1	Work	2	1
4-2	Forms of Energy	3	1.5
4-3	Conservation of Energy	2	1
	Chapter Review and Test	1	0.5
5	**Machines**	**10**	**5**
5-1	Simple Machines	2	1
5-2	Mechanical Advantage	3	1.5
5-3	Using Machines	4	2
	Chapter Review and Test	1	0.5
6	**Thermal Energy**	**9**	**5**
6-1	Thermal Energy	3	1.5
6-2	Heat and Temperature	3	1.5
6-3	Making Heat Work	2	1
	Chapter Review and Test	1	0.5
7	**Moving the Body**	**10**	**5**
7-1	Living Bones	3	1.5
7-2	Your Body in Motion	2	1
7-3	Muscles	4	2
	Chapter Review and Test	1	0.5
8	**Controlling the Body Machine**	**9**	**4**
8-1	The Nervous System: Master Control	3	1.5
8-2	The Parts of Your Nervous System	3	1
8-3	Your Endocrine System	2	1
	Chapter Review and Test	1	0.5
UNIT 3	**Earth Materials and Resources**	**47**	**23**
9	**Discovering Elements**	**10**	**5**
9-1	Discovering Metals	4	2
9-2	Discovering Nonmetals	4	2
9-3	Understanding Metalloids	1	0.5
	Chapter Review and Test	1	0.5
10	**Minerals and Their Uses**	**8**	**4**
10-1	Minerals and Their Values	2	1
10-2	Identifying Minerals	3	1.5
10-3	Mineral Formation	2	1
	Chapter Review and Test	1	0.5

Unit	Chapter/Section	Single-class (180 days)	Block (90 days)
	11 The Rock Cycle	**10**	**5**
	11-1 Igneous Rocks	3	1.5
	11-2 Metamorphic Rocks	2	1
	11-3 Sedimentary Rocks	4	2
	Chapter Review and Test	1	0.5
	12 The Ocean Floor and Shore Zones	**9**	**4**
	12-1 Shore Zones	3	1.5
	12-2 Humans Affect Shore Zones	2	1
	12-3 The Ocean Floor	3	1
	Chapter Review and Test	1	0.5
	13 Energy Resources	**10**	**5**
	13-1 The Electricity You Use	2	1
	13-2 Fossil Fuels	3	1.5
	13-3 Resources and Pollution	2	1
	13-4 Alternative Energy Resources	2	1
	Chapter Review and Test	1	0.5
UNIT 4	**Air: Molecules in Motion**	**27**	**14**
	14 Gases, Atoms, and Molecules	**9**	**4**
	14-1 How Do Gases Behave?	4	2
	14-2 What Are Gases Made Of?	2	1
	14-3 What Is the Atomic Theory of Matter?	2	1
	Chapter Review and Test	1	0.5
	15 The Air Around You	**9**	**5**
	15-1 So This Is the Atmosphere	3	1.5
	15-2 Structure of the Atmosphere	3	1.5
	15-3 The Air and Sun	2	1.5
	Chapter Review and Test	1	0.5
	16 Breathing	**9**	**5**
	16-1 How Do You Breathe?	3	2
	16-2 The Air You Breathe	3	1.5
	16-3 Disorders of the Respiratory System	2	1
	Chapter Review and Test	1	0.5
UNIT 5	**Life at the Cellular Level**	**28**	**14**
	17 Basic Units of Life	**9**	**5**
	17-1 The World of Cells	3	1.5
	17-2 The Inside Story of Cells	2	1
	17-3 When One Cell Becomes Two	3	1.5
	Chapter Review and Test	1	0.5
	18 Chemical Reactions	**10**	**5**
	18-1 How Does Matter Change Chemically?	1	0.5
	18-2 Word Equations	3	1.5
	18-3 Chemical Reactions and Energy	3	1.5
	18-4 Speeding Up and Slowing Down Reactions	2	1
	Chapter Review and Test	1	0.5
	19 How Cells Do Their Jobs	**9**	**4**
	19-1 Traffic In and Out of Cells	4	1.5
	19-2 Why Cells Need Food	2	1
	19-3 Special Cells with Special Jobs	2	1
	Chapter Review and Test	1	0.5

ASSESSMENT

What criteria do you use to assess your students as they progress through a course? Do you rely on formal tests and quizzes? To assess students' achievement in science, you need to measure not only their knowledge of the subject matter, but also their ability to handle apparatus, to organize, predict, record, and interpret data, to design experiments, and to communicate orally and in writing. *SCIENCE INTERACTIONS* has been designed to provide you with a variety of assessment tools to help you develop a clearer picture of your students' progress.

PERFORMANCE ASSESSMENT

Performance assessment is based on judging the quality of a student's response to a performance task. A performance task is constructed to require the use of important concepts with supporting information, work habits important to science, and one or more of the elements of scientific literacy. The performance task attempts to put the student in a "real world" context so that the class learning can be put to authentic uses.

Performance Assessment in *SCIENCE INTERACTIONS*

Many performance task assessment lists can be found in Glencoe's *Performance Assessment in the Science Classroom*. These lists were developed for the summative and skill performance tasks in the *Performance Assessment* book that accompanies the *SCIENCE INTERACTIONS* program. The Performance Assessment book contains a mix of skill assessments and summative assessments that tie together major concepts of each chapter. Software programs for the assessment lists are also available. The lists can be used to support **Find Out** activities; **Explore** activities; **Investigates; Design Your Own Investigations;** and the **You Try It** and **Going Further** sections of **A Closer Look, Science Connections**, and **Expand Your View**. The assessment lists and rubrics can also be used for the **Project** ideas

described in **Reviewing Main Ideas**. Glencoe's Alternate Assessment in the Science Classroom provides additional background and examples of performance assessment. Activity sheets in the Activity Masters book provide yet another vehicle for formal assessment of student products. The **MindJogger Videoquiz** series offers interactive videos that provide a fun way for your students to review chapter concepts. You can extend the use of the videoquizzes by implementing them in a testing situation. Questions are at three difficulty levels: basic, intermediate, and advanced.

ASSESSING STUDENT WORK WITH RUBRICS

A rubric is a set of descriptions of the quality of a process and/or a product. The set of descriptions includes a continuum of quality from excellent to poor. Rubrics for various types of assessment products are in *Performance Assessment in the Science Classroom.*

GROUP PERFORMANCE ASSESSMENT

Recent research has shown that cooperative learning structures produce improved student learning outcomes for students of all ability levels. *SCIENCE INTERACTIONS* provides many opportunities for cooperative learning and, as a result, many opportunities to observe group work processes and products. *SCIENCE INTERACTIONS: Cooperative Learning Resource Guide* provides strategies and resources for implementing and evaluating group activities. In cooperative group assessment, all members of the group contribute to the work process and the products it produces. An example, along with information about evaluating cooperative work, is provided in the booklet *Alternate Assessment in the Science Classroom.*

SCIENCE JOURNALS AND PORTFOLIOS

A science journal is intended to help the student organize his or her thinking. It is not a lecture or laboratory notebook. It is a place for students to make their thinking explicit in drawings and writing. It is the place to explore what makes science fun and what makes it hard.

The portfolio should help the student see the "big picture" of how he or she is performing in gaining knowledge and skills and how effective his or her work habits are. The portfolio is a way for students to see how individual performance tasks fit into a pattern that reveals the overall quality of their learning. The process of assembling the portfolio should be both integrative (of process and content) and reflective.

OPPORTUNITIES FOR USING SCIENCE JOURNALS AND PORTFOLIOS

SCIENCE INTERACTIONS presents a wealth of opportunities for performance portfolio development. Each chapter in the student text contains projects, enrich-ment activities, investigations, skill builders, and connections with life, society, and literature. Each of the student activities results in a product. A mixture of these products can be used to document student growth during the grading period. Descriptions, examples, and assessment criteria for portfolios are discussed in *Alternate Assessment in the Science Classroom*. Glencoe's *Performance Assessment in the Science Classroom* contains even more information on using science journals and making portfolios. Performance task assessment lists and rubrics for both journals and portfolios are found there.

CONTENT ASSESSMENT

While new and exciting performance skill assessments are emerging, paper-and-pencil tests are still a mainstay of student evaluation. Students must learn to conceptualize, process, and prepare for traditional content assessments. Presently and in the foreseeable future, students will be required to pass pencil-and-paper tests to exit high school, and to enter college, trade schools, and other training programs.

SCIENCE INTERACTIONS contains numerous strategies and formative checkpoints for evaluating student progress toward mastery of science concepts. Throughout the chapters in the student text, **Check Your Understanding** questions and application tasks are presented. This spaced review process helps build learning bridges that allow all students to confidently progress from one lesson to the next.

For formal review that precedes the written content assessment, *SCIENCE INTERACTIONS* presents a **Chapter Review** at the end of each chapter. By evaluating student responses to this extensive review, you can determine whether any substantial reteaching is needed.

For the formal content assessment, a one-page review and a three-page **Chapter Test** are provided for each chapter. Using the review in a whole class session, you can correct any misperceptions and provide closure for the text. If your individual assessment plan requires a test that differs from the **Chapter Test** in the resource package, customized tests can be easily produced using the **Computer Test Bank**.

MANAGING ACTIVITIES
AND PREPARING SOLUTIONS SAFELY

SCIENCE INTERACTIONS engages students in a variety of hands-on experiences to provide all students with an opportunity to learn by doing. The many hands-on activities throughout *SCIENCE INTERACTIONS* require simple, common materials, making them easy to set up and manage in the classroom.

Find Out and **Explore** activities are intended to be short and occur many times throughout the text. The integration of these activities with the core material provides for thorough development and reinforcement of concepts.

SCIENCE INTERACTIONS provides more than the same "cookbook" activities you've seen hundreds of times before. Students work cooperatively to develop their own experimental designs in the **Investigates** and **Design Your Own Investigations.** They discover firsthand that developing procedures for studying a problem is not as hard as they thought it might be. If you want to give your students additional opportunities to experience self-directed activities, the *Science Discovery Activities* in the *Teacher Classroom Resources* will allow you to do just that. The three activities per chapter challenge students to use their critical thinking skills in developing experimental procedures to test hypotheses.

Preparing Students for Open-ended Labs

To prepare students for the investigations, you should follow the guidelines in the *Teacher Wraparound Edition*, especially in the sections titled Possible Procedures and Teaching Strategies. Your introduction to an **Investigate** or **Design Your Own Investigation** will be different from traditional activity introductions in that it will be designed to focus students on the problem without giving them directions for how to set up their experiment. Different groups of students will develop alternative hypotheses and alternative procedures. Check their proce-

dures before they begin. In contrast to some "cookbook" activities, there may not always be just one right answer.

Preparation of Solutions

It is most important to use safe laboratory techniques when handling all chemicals. Many substances may appear harmless but are, in fact, toxic, corrosive, or very reactive. Always check with the manufacturer. Chemicals should never be ingested. Be sure to use proper techniques to smell solutions or other agents. Always wear safety goggles and an apron. The following general cautions should be used.

1. Poisonous/corrosive liquid and/or vapor. Use in the fume hood. Examples: *acetic acid, hydrochloric acid, ammonia hydroxide, nitric acid.*

2. Poisonous and corrosive to eyes, lungs, and skin. Examples: *acids, limewater, iron(III) chloride, bases, silver nitrate, iodine, potassium permanganate.*

3. Poisonous if swallowed, inhaled, or absorbed through the skin. Examples: *glacial acetic acid, copper compounds, barium chloride, lead compounds, chromium compounds, lithium compounds, cobalt(II) chloride, silver compounds.*

4. Always add acids to water, never the reverse.

5. When sulfuric acid or sodium hydroxide is added to water, a large amount of thermal energy is released. Sodium metal reacts violently with water. Use extra care if handling any of these substances.

Unless otherwise specified, solutions are prepared by adding the solid to a small amount of distilled water and then diluting with water to the volume listed. If you use a hydrate that is different from the one specified in a particular preparation, you will need to adjust the amount of the hydrate to obtain the required concentration.

Cooperative Learning

WHAT IS COOPERATIVE LEARNING?

In cooperative learning, students work together in small groups to learn academic material and interpersonal skills. Group members learn that they are responsible for accomplishing an assigned group task as well as for each learning the material. Cooperative learning fosters academic, personal, and social success for all students.

ESTABLISHING A COOPERATIVE CLASSROOM

Cooperative groups in the middle school usually contain from two to five students. Heterogeneous groups that represent a mixture of abilities, genders, and ethnicity expose students to ideas different from their own and help them to learn to work with different people.

Initially, cooperative learning groups should only work together for a day or two. After the students are more experienced, they can work with a group for longer periods of time. Students must understand that they are responsible for group members learning the material.

Before beginning, discuss the basic rules for effective cooperative learning—(1) listen while others are speaking, (2) respect other people and their ideas, (3) stay on tasks, and (4) be responsible for your own actions.

The *Teacher Wraparound Edition* uses the code **COOP LEARN** at the end of activities and teaching ideas where cooperative learning strategies are useful.

USING COOPERATIVE LEARNING STRATEGIES

The *Cooperative Learning Resource Guide* of the *Teacher Classroom Resources* provides help for selecting cooperative learning strategies, as well as methods for troubleshooting and evaluation.

EVALUATING COOPERATIVE LEARNING

At the close of the lesson, have groups share their products or summarize the assignment. You can evaluate group performance during a lesson by frequently asking questions to group members picked at random or having each group take a quiz together. Assess individual learning by your traditional methods.

Meeting Individual Needs

Each student brings his or her own unique set of abilities, perceptions, and needs into the classroom. It is important that the teacher try to make the classroom environment as receptive to these differences as possible.

Recognize that individual learning styles are different and that learning style does not reflect a student's ability level. The chart on pages 34T-35T gives additional tips you may find useful in structuring the learning environment in your classroom to meet students' special needs.

In an effort to provide all students with a positive science experience, this text offers a variety of ways for students to interact with materials so that they can utilize their preferred method of learning. This approach allows students to become familiar with other learning styles as well.

ABILITY LEVELS

The activities are broken down into three levels to accommodate all student ability levels. *SCIENCE INTERACTIONS Teacher Wraparound Edition* designates the activities as follows:

L1 basic activities are designed to be within the ability range of all students.

L2 application activities are designed for students who have mastered the concepts presented.

L3 challenging activities are designed for the students who are able to go beyond the basic concepts presented.

LIMITED ENGLISH PROFICIENCY

In providing for the student with limited English proficiency, the focus needs to be on overcoming a language barrier. Once again it is important not to confuse ability in speaking/reading English with academic ability or "intelligence." Look for this symbol **LEP** in the teacher margin for specific strategies for students with limited English proficiency.

In the options margins of the *Teacher Wraparound Edition* there are two or more **Meeting Individual Needs** strategies for each chapter.

Meeting Individual Needs

	DESCRIPTION	SOURCES OF HELP/INFORMATION
Learning Disabled	All learning disabled students have an academic problem in one or more areas, such as academic learning, language, perception, social-emotional adjustment, memory, or attention.	*Journal of Learning Disabilities* *Learning Disability Quarterly*
Behaviorally Disordered	Children with behavior disorders deviate from standards or expectations of behavior and impair the functioning of others and themselves. These children may also be gifted or learning disabled.	*Exceptional Children* *Journal of Special Education*
Physically Challenged	Children who are physically disabled fall into two categories—those with orthopedic impairments and those with other health impairments. Orthopedically impaired children have the use of one or more limbs severely restricted, so the use of wheelchairs, crutches, or braces may be necessary. Children with other health impairments may require the use of respirators or other medical equipment.	Batshaw, M.L. and M.Y. Perset. *Children with Handicaps: A Medical Primer.* Baltimore: Paul H. Brooks, 1981. Hale, G. (Ed.). *The Source Book for the Disabled.* New York: Holt, Rinehart & Winston, 1982. *Teaching Exceptional Children*
Visually Impaired	Children who are visually disabled have partial or total loss of sight. Individuals with visual impairments are not significantly different from their sighted peers in ability range or personality. However, blindness may affect cognitive, motor, and social development, especially if early intervention is lacking.	*Journal of Visual Impairment and Blindness* *Education of Visually Handicapped* American Foundation for the Blind
Hearing Impaired	Children who are hearing impaired have partial or total loss of hearing. Individuals with hearing impairments are not significantly different from their hearing peers in ability range or personality. However, the chronic condition of deafness may affect cognitive, motor, and social development if early intervention is lacking. Speech development also is often affected.	*American Annals of the Deaf* *Journal of Speech and Hearing Research* *Sign Language Studies*
Limited English Proficiency	Multicultural and/or bilingual children often speak English as a second language or not at all. The customs and behavior of people in the majority culture may be confusing for some of these students. Cultural values may inhibit some of these students from full participation.	*Teaching English as a Second Language Reporter* R.L. Jones (Ed.). *Mainstreaming and the Minority Child.* Reston, VA: Council for Exceptional Children, 1976.
Gifted	Although no formal definition exists, these students can be described as having above-average ability, task commitment, and creativity. Gifted students rank in the top 5% of their class. They usually finish work more quickly than other students, and are capable of divergent thinking.	*Journal for the Education of the Gifted* *Gifted Child Quarterly* *Gifted Creative/Talented*

TIPS FOR INSTRUCTION
With careful planning, the needs of all students can be met in the science classroom.

1. Provide support and structure; clearly specify rules, assignments, and duties.
2. Establish situations that lead to success.
3. Practice skills frequently. Use games and drills to help maintain student interest.
4. Allow students to record answers on tape and allow extra time to complete tests and assignments.
5. Provide outlines or tape lecture material.
6. Pair students with peer helpers, and provide class time for pair interaction.

1. Provide a clearly structured environment with regard to scheduling, rules, room arrangement, and safety.
2. Clearly outline objectives and how you will help students obtain objectives. Seek input from them about their strengths, weaknesses, and goals.
3. Reinforce appropriate behavior and model it for students.
4. Do not expect immediate success. Instead, work for long-term improvement.
5. Balance individual needs with group requirements.

1. Openly discuss with the student any uncertainties you have about when to offer aid.
2. Ask parents or therapists and students what special devices or procedures are needed, and if any special safety precautions need to be taken.
3. Allow physically disabled students to do everything their peers do, including participating in field trips, special events, and projects.
4. Help non-disabled students and adults understand physically disabled students.

1. As with all students, help the student become independent. Some assignments may need to be modified.
2. Teach classmates how to serve as guides.
3. Limit unnecessary noise in the classroom.
4. Encourage students to use their sense of touch. Provide tactile models whenever possible.
5. Describe people and events as they occur in the classroom.
6. Provide taped lectures and reading assignments.
7. Team the student with a sighted peer for laboratory work.

1. Seat students where they can see your lip movements easily, and avoid visual distractions.
2. Avoid standing with your back to the window or light source.
3. Use an overhead projector so you can maintain eye contact while writing.
4. Seat students where they can see speakers.
5. Write all assignments on the board, or hand out written instructions.
6. If the student has a manual interpreter, allow both student and interpreter to select the most favorable seating arrangements.

LS Learning Styles

Kinesthetic
Student learns from touch, movement, and manipulating objects

Visual-Spatial
Student learns by responding to images and illustrations

Logical-Mathematical
Student learns by using numbers and reasoning

Interpersonal
Student learns by interacting with others

Intrapersonal
Student learns by working alone

Linguistic
Student learns by using and understanding words

Auditory-Musical
Student learns by listening to the spoken word and to tones and rhythms

1. Remember, students' ability to speak English does not reflect their academic ability.
2. Try to incorporate the student's cultural experience into your instruction. The help of a bilingual aide may be effective.
3. Include information about different cultures in your curriculum to help build students' self-image. Avoid cultural stereotypes.
4. Encourage students to share their cultures in the classroom.

1. Make arrangements for students to take selected subjects early and to work on independent projects.
2. Let students express themselves in art forms such as drawing, creative writing, or acting.
3. Make public services available through a catalog of resources, such as agencies providing free and inexpensive materials, community services and programs, and people in the community with specific expertise.
4. Ask "what if" questions to develop high-level thinking skills. Establish an environment safe for risk taking.
5. Emphasize concepts, theories, ideas, relationships, and generalizations.

NATIONAL GEOGRAPHIC SOCIETY

and Glencoe Science
We're a Team

Glencoe Science and National Geographic Society

have teamed up to bring exciting new features and technologies to *Science Interactions.* By incorporating National Geographic's world-renowned photographs, illustrations, and content features, *Science Interactions* will engage students as never before.

Engaging new Unit Openers

feature National Geographic photographs that attract students and help them focus on the unit to come.

Fascinating SciFacts

features enrich and extend chapter content with National Geographic illustrations and graphics.

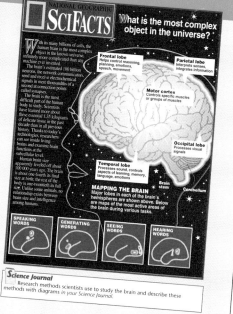

Helpful Teacher's Corner

feature in the Teacher's Wraparound Edition correlates *National Geographic Magazine* articles, technology, and other helpful teaching resources available from National Geographic Society.

Correlated Technology

In addition, **Glencoe** offers a wide variety of National Geographic videodiscs and **CD-ROMs** fully correlated to the textbooks. In fact, **National Geographic Videodisc** barcodes are placed in this Teacher Edition for your convenience at the point of use in each chapter.

LABORATORY SAFETY

Safety is of prime importance in every classroom. However, the need for safety is even greater when science is taught. The activities in **SCIENCE INTERACTIONS** are designed to minimize dangers in the laboratory. Even so, there are no guarantees against accidents. Careful planning and preparation as well as being aware of hazards can keep accidents to a minimum. Numerous books and pamphlets are available on laboratory safety with detailed instructions on preventing accidents. In addition, the **SCIENCE INTERACTIONS** program provides safety guidelines in several forms. The *Lab and Safety Skills* booklet contains detailed guidelines, in addition to masters you can use to test students' lab and safety skills. The *Student Edition* and *Teacher Wraparound Edition* provide safety precautions and symbols designed to alert students to possible dangers. Know the rules of safety and what common violations occur. Know the **Safety Symbols** used in this book. Know where emergency equipment is stored and how to use it. Practice good laboratory housekeeping and management to ensure the safety of your students.

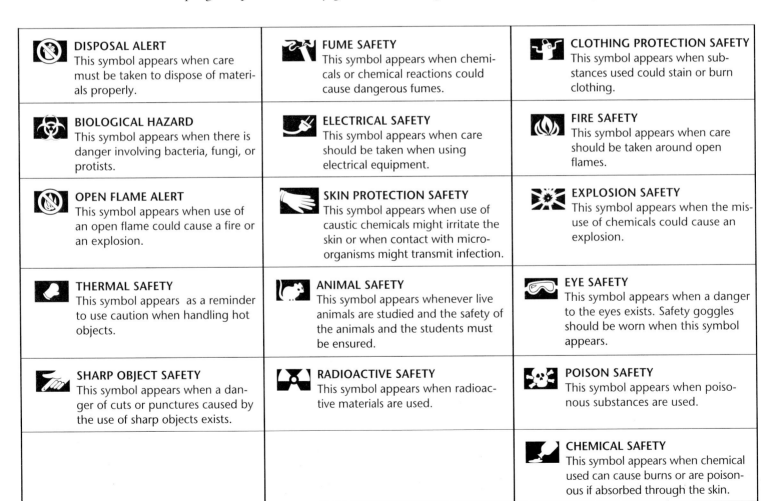

DISPOSAL ALERT This symbol appears when care must be taken to dispose of materials properly.	**FUME SAFETY** This symbol appears when chemicals or chemical reactions could cause dangerous fumes.	**CLOTHING PROTECTION SAFETY** This symbol appears when substances used could stain or burn clothing.
BIOLOGICAL HAZARD This symbol appears when there is danger involving bacteria, fungi, or protists.	**ELECTRICAL SAFETY** This symbol appears when care should be taken when using electrical equipment.	**FIRE SAFETY** This symbol appears when care should be taken around open flames.
OPEN FLAME ALERT This symbol appears when use of an open flame could cause a fire or an explosion.	**SKIN PROTECTION SAFETY** This symbol appears when use of caustic chemicals might irritate the skin or when contact with microorganisms might transmit infection.	**EXPLOSION SAFETY** This symbol appears when the misuse of chemicals could cause an explosion.
THERMAL SAFETY This symbol appears as a reminder to use caution when handling hot objects.	**ANIMAL SAFETY** This symbol appears whenever live animals are studied and the safety of the animals and the students must be ensured.	**EYE SAFETY** This symbol appears when a danger to the eyes exists. Safety goggles should be worn when this symbol appears.
SHARP OBJECT SAFETY This symbol appears when a danger of cuts or punctures caused by the use of sharp objects exists.	**RADIOACTIVE SAFETY** This symbol appears when radioactive materials are used.	**POISON SAFETY** This symbol appears when poisonous substances are used.
		CHEMICAL SAFETY This symbol appears when chemical used can cause burns or are poisonous if absorbed through the skin.

CHEMICAL Storage & DISPOSAL

GENERAL GUIDELINES

Be sure to store all chemicals properly. The following are guidelines commonly used. Your school, city, county, or state may have additional requirements for handling chemicals. It is the responsibility of each teacher to become informed as to what rules or guidelines are in effect in his or her area.

1. Separate chemicals by reaction type. Strong acids should be stored together. Likewise, strong bases should be stored together and should be separated from acids. Oxidants should be stored away from easily oxidized materials, and so on.

2. Be sure all chemicals are stored in labeled containers indicating contents, concentration, source, date purchased (or prepared), any precautions for handling and storage, and expiration date.

3. Dispose of any outdated or waste chemicals properly according to accepted disposal procedures.

4. Do not store chemicals above eye level.

5. Wood shelving is preferable to metal. All shelving should be firmly attached to the wall and should have anti-roll edges.

6. Store only those chemicals that you plan to use.

7. Hazardous chemicals require special storage containers and conditions. Be sure to know what those chemicals are and the accepted practices for your area. Some substances must even be stored outside the building.

8. When working with chemicals or preparing solutions, observe the same general safety precautions that you would expect from students. These include wearing an apron and goggles. Wear gloves and use the fume hood when necessary. Students will want to do as you do whether they admit it or not.

9. If you are a new teacher in a particular laboratory, it is your responsibility to survey the chemicals stored there and to be sure they are stored properly or disposed of. Consult the rules and laws in your area concerning what chemicals can be kept in your classroom. For disposal, consult up-to-date disposal information from the state and federal governments.

DISPOSAL OF CHEMICALS

Local, state, and federal laws regulate the proper disposal of chemicals. These laws should be consulted before chemical disposal is attempted. Although most substances encountered in high school biology can be flushed down the drain with plenty of water, it is not safe to assume that is always true. It is recommended that teachers who use chemicals consult the following book from the National Research Council:

Prudent Practices in the Laboratory: Handling and Disposal of Chemicals. Washington, DC: National Academy Press, 1995.

DISCLAIMER

Glencoe Publishing Company makes no claims to the completeness of this discussion of laboratory safety and chemical storage. The material presented is not all-inclusive, nor does it address all of the hazards associated with handling, storage, and disposal of chemicals, or with laboratory management.

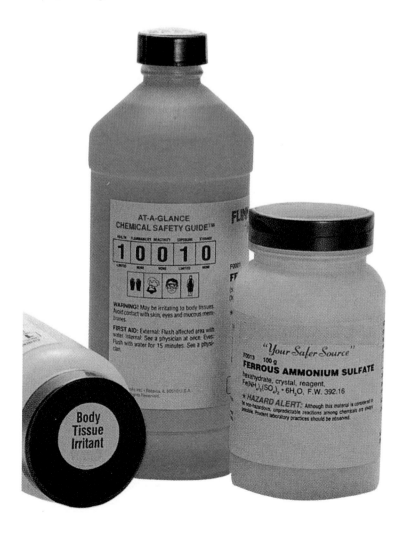

Non-Consumables

Item	INVESTIGATE!	DESIGN YOUR OWN INVESTIGATION	Explore!	Find Out!
Air cylinder (1)	430			
Air pump (5 for class)			202, 455	
Apron (30)	228, 348, 560	274, 312, 568		571
Aquarium w/goldfish (1 for class)			485	
Backpack (15)			118	
Balance, inertia (15)	24			
Balance, pan (15)	162, 286	32, 442, 590	455	169
Balance beam or long board (2 for class)			240	
Basketball or soccer ball, deflated as needed (15)			455	6
Baton				242
Beads, plastic (60)			190	
Beaker, 250-mL (45)	194, 410		557, 583	440, 588
Beaker, 400-mL (30)	458	184, 498	21, 190, 407	44
Beaker, 1000-mL (15)			407	187, 434, 571
Blender (5 to 15 per class)	10			
Blindfold (15)				254
Boards of varying lengths (30)				155, 169, 339
Book (30)	430	32, 126	117, 118, 149, 464	132, 158, 169, 225
Books, small (30)		216		
Bottle, clear plastic (15) w/spray pump		398		
Bottle, plastic squeeze (15)				94
Bottle, glass soda (15)			436	
Bowl (15)	410			344
Box (15)			425	132
Box, clear plastic w/lid (15)	410			470
Box, w/lid (shoe type) (15)			425	377
Brick (15)	430			153
Bucket (15)			123, 407	
Can, coffee, painted black (15)	410			
Can, small coffee (15)		466		
Can openers, various types			166	
Can, small, w/wire handle (15)				44
Can, w/lid, paint or cocoa (15)			151	
Capillary tube (15)				434
Car (1 for class)			200	
Carpet (1 large or 15 squares)			181	
Cart, toy (15)		126		
Chair (15)			426	
Chair, rolling (4 for class)				26
Clamps, alligator (60)				440
Clamps, C type (30)	24			339
Clamp, right angle (30)	136			
Clamp, test tube (15)	458			
Coal (30 pieces)			396	

Non-Consumables

Item	INVESTIGATE!	DESIGN YOUR OWN INVESTIGATION	Explore!	Find Out!
Coffee stirrer		96		
Coins (45 pennies, 15 nickels, 15 dimes, 15 quarters)	162	248, 442	21, 181	
Container, clear plastic (15 large, 15 small)				520
Container, clear food (15)			532	520
Coverslip (120)		522		516, 529, 530
Crutch or cane (15)			39	
Dishpan or sink (15)	10			
Dissecting pan (15)	228			
Drawing compass (15)	522			519
Dropper (30)	348	92, 312	202	103, 529
Electric iron (1 to 5 per class)	10			
Electrodes (30)				440
Eraser		248		
File, steel (15)		312		
Flashlight (15)			437	6, 254
Flask, Erlenmeyer, 500-mL (30)	598			
Forceps (15)	228			
Funnel (15)				88, 584
Glass, clear drinking (15)			21, 586	554
Globe (5 to 10 for class)	64			6
Goggles, safety (30)	348, 560	274, 312, 568, 590	283	571
Graduated cylinder, 10-mL (15)	598			41
Graduated cylinder, 25-mL (15)	560	568		41, 346
Graduated cylinder, 100-mL (15)	286	398, 498, 590		41, 373
Gravel		398		
Hands lens (30)	316, 348	312, 336	396	
Heart, model of (2 for class)				86
Heat lamp (15)				470
Heater, electric immersion (15)			190	
Hole punch (15)				230
Hot plate (15)	316, 410	184	190	187, 331, 393, 434, 571
Human bone and muscle chart (1 for class)	228			
Inflation needle (5 for class)			455	
Inner tube, bicycle (5 for class)			202	
Jar, glass, with lid (30)		590	400	214, 584, 596
Lids, plastic (60)				457
Light (15)			314, 472	8
Magnet (15)		368		
Magnifying glass (15)			329, 330	331, 457
Map, city (15)			83	
Marbles (30)				134
Marker, water-soluble (2 packages)	64			
Marker, permanent (2 packages)	540			187, 457
Masses (15 sets)	24	32, 126	36	29, 41, 44

Equipment List

Non-Consumables

Item	INVESTIGATE!	DESIGN YOUR OWN INVESTIGATION	Explore!	Find Out!
Measuring cup, with mL gradations (15)	458			373
Metal fasteners (75)				230
Meterstick (15)	136	32, 126, 170, 248	129, 472	132, 158, 169
Microscope (30)		522		303, 331, 457, 516, 518, 519, 530, 536
Microscope slides (120)		522		303, 331, 457, 516, 518, 519, 529, 530, 536
Mineral samples (15 sets)	348	312	314, 330	303, 308, 310
Mirror (15)		96		495
Mohs scale of hardness (15 sets)		312		
Object, of unknown mass	24			
Pan—9" x 13" (45)	316		182, 363	44, 346, 373
Paper clips, large (30)	252			54
Pencil sharpener, hand-held types (15)				339
Petri dish, w/cover (15)		92		
Pin				344
Pinwheel			407	393
Power supply, DC (15)				440
Pulley (15)				153
Ring, ring stand (15)	136		472	153, 344
Ramp, 1- to 2-meter (5 to 15 per class)		126		
Research resources			3	
Rocks, various types (as required) (15 sets)	348	336	329, 330, 341, 361	373
Rug (1 for class)				132
Ruler, grooved (15)				134
Ruler, metric (30)	64, 162, 252, 458, 490, 540, 598	522	129, 149, 363	41, 54, 187, 230, 377, 434, 440
Scale, bathroom (2 for class)		170	39	
Scale, spring (15)		126	36, 129	41, 44, 132, 153
Scalpel (15)	228			
Scissors (30)	228, 458	466, 522	583	230
Screwdrivers (15)				155
Shoes, flat-heeled (15)			39	
Sink			407	
Small object		248		
Softball (15)			123	6
Spatula (15)	286			571
Spoon (15 wooden, 15 plastic, 15 metal)			84, 151, 190	571
Spring cart (15)		32		
Spring, coiled, slinky type (15)			60	
Stereomicroscope (15)		368		
Stirring rod (45)	194	184		187, 584
Stopper, 1-hole rubber (30)	598			94

Non-Consumables

Item	INVESTIGATE!	DESIGN YOUR OWN INVESTIGATION	Explore!	Find Out!
Stopper, 2-hole medium rubber (15)	136			
Stopper, various sizes (60)	286		283	
Stopper, rubber, w/tube assembly (30)	598			94
Stopwatch (15)		126, 184		169
Streak plate (30)		312		308
Support rod and clamp (15)	136			
Tape measure, metric			211, 492	
Teakettle (5 for class)				393
Teaspoon				346
Tent stake (15)			123	
Test tubes (120)	286, 316, 458, 560, 598	274, 568	283	440, 571
Test-tube holder (15)	286, 316, 458, 560, 598		283	440, 571
Test-tube rack (15)	286, 316, 458, 560	274, 568		571
Thermal mitt (15)	316	184, 274		331, 393
Thermometer (30)	194, 410	184, 568		434, 470, 596
Thermometer, unmarked, alcohol (15)				187
Timer, with second hand (15)	24	92, 170, 498, 568	91, 118, 182, 483	
Tongs (15)	194, 458			
Tongs, beaker (15)				434
Towel (15)				495
Toy car, wind-up (15)				169
Tube, Y-shaped (15)				88
Tubing, clear plastic and rubber, 2-foot lengths (45)	598	398		88, 94, 103
Tweezers (15)		274		103
Vise (15)				339
Watch (15)	598			
Wood, 2" x 4" 1-foot lengths (30)			58	
Wood screw (30)				155

Living Organisms

Item	INVESTIGATE!	DESIGN YOUR OWN INVESTIGATION	Explore!	Find Out!
Earthworms (15)		92		
Elodea (10 sprigs)				516, 519
Flower petals (30)		522		
Goldfish or guppy (5 for class)			485	
Leaves (for guard cells) (30)				518
Plants, same size and type (45)				8
Seedlings, corn (75)	540			
Yeast (1 package dry)				518

Equipment List

Consumables

Item	INVESTIGATE!	DESIGN YOUR OWN INVESTIGATION	Explore!	Find Out!
Aluminum foil (2 rolls)	410			339
Bag, paper lunch (60)			472	8
Bag, plastic quart (75)	540			
Bag, plastic resealable sandwich (30)			525, 532	
Balloon (120)	490	466	425, 436	54, 134
Banana (15)				552
Butter or margarine (1 lb)			190	
Candle or burner (15)		274		
Cardboard	316			470
Cheesecloth (10" x 10" squares) (30)				584
Chicken bones (30)				214
Chicken, boiled (5)	228			
Clay, modeling 3 different colors (7 packages)		96	53, 212, 464	377
Clothes hanger (15)	10			
Cotton balls (1 box)				103, 373
Cotton batting (1 box)				596
Crayons (4 boxes of 96)				339
Cup, foam (30)				134
Cup, paper (60)				344
Detergent (1 small box)				373
Detergent, liquid (1 32-oz. bottle)		398		
Dialysis membrane tubing (10-foot, 1 3/4" roll)				588
Egg, raw (30)		590		
Feathers (30)				373
Gelatin (10 packages)			525, 532	
Glue, white (15 bottles)	10, 410		58	344
Gravel (1 bag)				584
Honey (1 pint)			84	
Ice cubes (8 trays)	194	184	437	
Ice, crushed (1 bag each time)			182	187, 434
Index cards (3" x 5") (300)			21, 563	230
Kidney beans, dry (4 pounds)				596
Labels, self-stick (several packs)	540			6, 124
Liver, raw (1/2 pound for class)		568		
Marshmallows (1 bag large)				552
Matches, safety (1 box)	286	274	283	552, 571
Mineral supplement labels (15)			292	
Newspaper (15)	10, 410	522		
Note cards (4" x 6") (300)		216		
Object with an odor or fragrance (1 for class)			426	
Onion, white (1)				530
Onion, green (1)			583	
Onion, red (1)				529
Onion skin		522		
Pantyhose, old (15 pairs)	10			

Consumables

Item	INVESTIGATE!	DESIGN YOUR OWN INVESTIGATION	Explore!	Find Out!
Paper, carbon (15 sheets)			39	
Paper, construction		466		
Paper graph (250 sheets)	24, 380, 458		39	61, 377
Paper, heavy white (30 sheets)	162			
Paper, notebook (500 sheets)	286		239	54
Paper, plain white (210 sheets)	64	72, 522	21, 244	29, 552
Paper, waxed (1 roll)			464	
Pencil		72		
Pencil, wax (15)	598	590		457
Pencils, colored (15 sets)	380			
Pin (15)				344
Plastic, clear and black (15 sheets each)	410			
Plastic foam sheets, thin (15)	410			
Plate, paper (30)			84	
Poster-making supplies (15)			3	14
Potatoes, raw (1 large)		568		588
Rice, cooked (4 cups)				588
Rubber bands (90)	24, 410, 458	466	563	377
Rubber cement (15 small bottles)		442		
Sand (large bag)		398	123	
Sand, different type (5 lb each)		368	361, 363	344, 584
Sandpaper, coarse (90 sheets)			58	29
Soft drink, carbonated (2-L bottles) (2)				554
Soil (1 large bag)				8, 470
Steel wool (1 bag)	458			
Stick, relay (12 inches long) (1)				242
Sticks, wooden craft (500)			212	
Straw, drinking (200)		466, 498	436	377, 554
String (1 ball)	64, 136		472	153, 230
Sugar (5-pound bag) (1)	316			
Syrup (15 bottles)		590	67	
Tape (15 rolls)	410	216, 466		134, 230
Tape, masking	136	32, 126	221	
Tea bag (30)			586	
Thread, cotton (1 spool)	316			
Tissue paper (3 boxes), or fabric			212	
Tomato skin		522		
Tongue depressor (60)				134, 373
Toothpicks, flat (1 box)	316			
Towel, paper	228, 348, 458, 540	92		373, 530
Vegetable oil (1 gallon)		398	400	373
Wood splints (100)	286, 560	274	283	134, 571
Yeast, cubes (30)	598			

Chemical Supplies

Item	INVESTIGATE!	DESIGN YOUR OWN INVESTIGATION	Explore!	Find Out!
Alcohol, rubbing (1 pint)			202	
Alum				346
Bleach, laundry (1 gallon)	286			
Bromothymol blue solution (3 L)		498		
Cobalt chloride (8 g)	286			
Element samples (15 sets)			271	
Food coloring (1 bottle)			84	
Hydrochloric acid (500 mL)	348, 560	312		
Hydrogen peroxide solution, 3% (500 mL)		568		571
Iodine, solid (25 g)				530, 588
Limewater, colorless (1 L)				554
Magnesium ribbon (30 small pieces)			283, 557	
Manganese dioxide (10 g)				571
Metal salt solutions (75 mL each)		274		
Metal samples (15 each of copper, zinc, and magnesium)	560			
Mineral oil (2 L)				103
Petroleum jelly (1 jar)	430			457
Potassium iodide (50 g)				530, 588
Potassium permanganate crystals (1 bottle)				520
Salol (phenol salicylate, 1 bottle)				331
Salt, table (1 large container)	194, 316		583	303, 516
Sucrose solution (1L)	598			
Talcum powder (1 container)				516
Vial of helium or neon			271	
Vinegar (5 gallons)	458	590	283, 557	214
Washing soda solution (3 L)				440
Water (as required throughout)				
Water, distilled (4L)		590	583	

Preserved Specimens

Item	INVESTIGATE!	DESIGN YOUR OWN INVESTIGATION	Explore!	Find Out!
Blood, frog, prepared slide (30)				516
Frog skin cells, prepared slide of (30)		522		
Human cheek cells, prepared slide of (30)				518, 519
Mitosis in onion root tips (30)				530, 536
Leaf, prepared slide of (30)				518
Yeast cells (30)				518

BIBLIOGRAPHY

GENERAL SCIENCE CONTENT

Cash, Terry. 175 *More Science Experiments To Amuse and Amaze Your Friends: Experiments! Tricks! Things to Make!* New York: Random House, 1991.

Churchill, E. Richard. *Amazing Science Experiments with Everyday Materials.* New York: Sterling Publishing Co., Inc., 1991.

Lewis, James. *Hocus Pocus Stir and Cook, The Kitchen Science-Magic Book.* New York: Meadowbrook Press, Division of Simon and Shuster, Inc., 1991.

Mandell, Muriel. *Simple Science Experiments with Everyday Materials.* New York: Sterling Publishing Co., Inc., 1989.

Roberts, Royston. *Serendipity: Accidental Discoveries in Science.* New York: John Wiley and Sons, Inc., 1989.

Schultz, Robert F. *Selected Experiments and Projects.* Washington, DC: Thomas Alva Edison Foundation, 1988.

Strongin, Herb. *Science on a Shoestring.* Menlo Park, CA: Addison-Wesley Publishing Co., 1985.

PHYSICS

Arons, A.B. *A Guide to Introductory Physics Teaching.* New York: John Wiley and Sons, 1990.

Aronson, Billy. "Water Ride Designers Are Making Waves." *3-2-1 Contact.* August, 1991, pp. 14-16.

Cash, Terry. *Sound.* New York: Warwick Press, 1989.

Hajda, Joey and Lisa B. Hajda. "Sparking Interest in Electricity." *Science Scope,* Nov./Dec., 1994, pp. 36-39.

Hardy, John W. "Adaptive Optics." *Scientific American,* June 1994, pp. 60-65.

Heiligman, Deborah. "There's a Lot More to Color Than Meets the Eye." *3-2-1 Contact.* November, 1991, pp. 16-20.

McGrath, Susan. *Fun with Physics.* Washington, D.C.: National Geographic Society, 1986.

Taylor, Barbara. *Sound and Music.* New York: Warwick Press, 1990.

Terres, John K. *How Birds Fly.* Mechanicsburg, PA: Stackpole Books, 1994.

Ward, Allen. *Experimenting with Batteries, Bulbs, and Wires.* New York: Chelsea House, 1991.

CHEMISTRY

Barber, Jacqueline. *Of Cabbage and Chemistry.* Washington, DC: Lawrence Hall of Science, NSTA, 1989.

Barber, Jacqueline, *Chemical Reactions.* Washington, DC: Lawrence Hall of Science, NSTA, 1986.

Cornell, John. *Experiments with Mixtures.* New York: Wiley, John and Sons, Inc., 1990.

Joesten, Melvin. *World of Chemistry.* Philadelphia, PA: Saunders College Publishing, 1991.

Laidler, Keith J. *The World of Physical Chemistry.* New York: Oxford University Press, 1993.

Mitchell, Sharon and Juergens, Frederick. *Laboratory Solutions for the Science Classroom.* Batavia, IL: Flinn Scientific, Inc., 1991.

Snyder, Carl H. *The Extraordinary Chemistry of Ordinary Things,* 2nd ed. New York: Wiley, 1995.

LIFE SCIENCE

Children's Atlas of the Environment. Chicago: Rand McNally, 1991.

Dewey, Jennifer Owings. *A Day and Night In the Desert.* Boston, MA: Little Brown, 1991.

Hancock, Judith M. *Variety of Life: A Biology Teacher's Sourcebook.* Portland, OR: J. Weston Walch, 1987.

Johnson, Cathy. *Local Wilderness.* New York: Prentice Hall, 1987.

McGrath, Susan. *The Amazing Things Animals Do.* Washington, DC: National Geographic Society, 1989.

Markmann, Erika. *Grow It! An Indoor/Outdoor Gardening Guide for Kids.* New York: Random House, 1991.

VanCleave, Janice Pratt. *Biology for Every Kid: 101 Easy Experiments that Really Work.* New York: Wiley, 1990.

Wilson, Edward O. *The Diversity of Life.* New York: Norton, 1993.

EARTH SCIENCE

Ardley, Neil. *The Science Book of Air.* New York: Gulliver Books, Harcourt, Brace, Jovanovich, Publishers, 1991.

Barrow, Lloyd H. *Adventures with Rocks and Minerals: Geology Experiments for Young People.* Hillsdale, NJ: Enslow, 1991.

Booth, Basil. *Volcanoes and Earthquakes.* Englewood Cliffs, NJ: Silver Burdett Press, 1991.

Javna, John. *50 Simple Things Kid Can Do to Save the Earth.* Kansas City: The Earth Works Group, Andrews and McMeel, a Universal Press Syndicate Co., 1990.

Norman, David. *Dinosaur!* London: Boxtree Limited, 1991.

Robinson, Andrew. *Earth Shock: Hurricanes, Volcanoes, Earthquakes, Tornadoes and Other Forces of Nature.* Thames and Hudson, 1993.

Seeds, Michael A., *Horizons, Exploring the Universe.* Belmont, CA: Wadsworth Publishing Company, 1995.

VanCleave, Janice. *Earth Science for Every Kid.* New York: John Wiley and Sons, Inc., 1991.

Wood, Robert W. *Science for Kids: 39 Easy Geology Activities.* Blue Ridge Summit, PA: Tab Books, 1992.

References

SUPPLIER ADDRESSES

SCIENTIFIC SUPPLIERS

Science Kit & Boreal Laboratories
777 East Park Drive
Tonawanda, NY 14150-6748

Carolina Biological Supply Co.
2700 York Road
Burlington, NC 27215

Fisher Scientific Co.
1600 W. Glenlake
Itasca, IL 60143

Flinn Scientific Co.
P.O. Box 219
Batavia, IL 60510

Frey Scientific
100 Paragon Parkway
Mansfield, OH 44903

Kemtec Educational Corp.
9889 Cresent Drive
West Chester, OH 45069

Sargent-Welch Scientific Co.
P.O. Box 5229
Buffalo Grove, IL 60089

Ward's Natural Science
Establishment, Inc.
P.O. Box 92912
Rochester, NY 14692

SOFTWARE DISTRIBUTORS

(AIT) Agency for Instructional
Technology
Box A
Bloomington, IN 47402-0120

Cambridge Development Lab (CDL)
1696 Massachusetts Avenue
Cambridge, MA 02138

COMPress
P.O. Box 102
Wentworth, NH 03282

Earthware Computer Services
P.O. Box 30039
Eugene, OR 97403

Educational Activities, Inc.
1937 Grand Avenue
Baldwin, NY 11510

Educational Materials and Equipment
Company (EME)
P.O. Box 2805
Danbury, CT 06813-2805

Gemstar (Classroom Consortia
Media, Inc.)
P.O. Box 050228
Staten Island, NY 10305

IBM Educational Systems
Department PC
4111 Northside Parkway
Atlanta, GA 30327

McGraw-Hill Webster Division
1221 Avenue of the Americas
New York, NY 10020

Microphys
1737 W. Second Street
Brooklyn, NY 11223

Minnesota Educational Computing
Corporation (MECC)
3490 Lexington Avenue N.
Saint Paul, MN 55126

Queue, Inc.
562 Boston Avenue
Bridgeport, CT 06610

Texas Instruments, Data Systems Group
P.O. Box 1444
Houston, TX 77251

Ventura Educational System
3440 Brokenhill Street
Newbury Park, CA 91320

AUDIOVISUAL DISTRIBUTORS

Aims Media
9710 Desoto Avenue
Chatsworth, CA 91311-4409

BFA Educational Media
468 Park Avenue S.
New York, NY 10016

Churchill Films
662 N. Robertson Blvd.
Los Angeles, CA 90069

Coronet/MTI Film and Video
Distributors of LCA
108 Wilmot Road
Deerfield, IL 60015

CRM Films
2233 Faraday Avenue
Suite F
Carlsbad, CA 92008

Diversified Education Enterprise
725 Main Street
Lafayette, IN 47901

Encyclopaedia Britannica Educational
Corp. (EBEC)
310 S. Michigan Avenue
Chicago, IL 60604

Focus Media, Inc.
839 Stewart Avenue
P.O. Box 865
Garden City, NY 11530

Hawkill Associates, Inc.
125 E. Gilman Street
Madison, WI 53703

Journal Films, Inc.
930 Pitner Avenue
Evanston, IL 60202

Lumivision
1490 Lafayette
Suite 305
Denver, CO 80218

National Earth Science Teachers
c/o Art Weinle
733 Loraine
Grosse Point, MI 48230

National Geographic Society
Education Products Division
17th and "M" Streets, NW
Washington, DC 20036

Science Software Systems
11890 W. Pico Blvd.
Los Angeles, CA 90064

Time-Life Videos
Time and Life Building
1271 Avenue of the Americas
New York, NY 10020

Universal Education & Visual Arts
(UEVA)
100 Universal City Plaza
Universal City, CA 91608

Video Discovery
1515 Dexter Avenue N.
Suite 400
Seattle, WA 98109

Glencoe

SCIENCE INTERACTIONS

Course 2

With Features by:

NATIONAL
GEOGRAPHIC
SOCIETY

GLENCOE

McGraw-Hill

New York, New York   Columbus, Ohio   Mission Hills, California   Peoria, Illinois

Science Interactions

Student Edition

Teacher Wraparound Edition

Science Discovery Activities

Teacher Classroom Resources

Laboratory Manual

Study Guide

Section Focus Transparencies

Teaching Transparencies

Performance Assessment

Performance Assessment in the Science Classroom

Computer Test Bank: IBM and Macintosh Versions

Spanish Resources

English/Spanish Audiocassettes

Science and Technology Videodisc Series

Integrated Science Videodisc Program

MindJogger Videoquizzes

Glencoe/McGraw-Hill

A Division of The McGraw·Hill Companies

Copyright © 1998 by The McGraw-Hill Companies, Inc. All rights reserved.
Except as permitted under the United States Copyright Act, no part of this publication may be reproduced or distributed in any form or by any means, or stored in a database or retrieval system, without prior written permission of the publisher.

Series and Cover Design: DECODE, Inc.

Send all inquiries to:
Glencoe/McGraw-Hill
936 Eastwind Drive
Westerville, OH 43081

ISBN 0-02-828157-8

Printed in the United States of America

1 2 3 4 5 6 7 8 9 10 071/043 06 05 04 03 02 01 00 99 98 97

With Features by:

NATIONAL GEOGRAPHIC SOCIETY

Authors

Bill Aldridge, M.S.
Director-Division of Science Education Solutions
Airborne Research and Services, Inc.
Fredericksburg, Virginia

Russell Aiuto, Ph.D.
Education Consultant
Frederick, Maryland

Albert Kaskel, M.Ed.
Biology Teacher, Emeritus
Evanston Township High School
Evanston, Illinois

Jack Ballinger, Ed.D.
Professor of Chemistry
St. Louis Community College at Florissant Valley
St. Louis, Missouri

Craig Kramer, M.A.
Physics Teacher
Bexley High School
Bexley, Ohio

Anne Barefoot, A.G.C.
Physics and Chemistry Teacher, Emeritus
Whiteville High School
Whiteville, North Carolina

Edward Ortleb, A.G.C.
Science Consultant
St. Louis Board of Education
St. Louis, Missouri

Linda Crow, Ed.D.
Associate Professor
University of Houston-Downtown
Houston, Texas

Susan Snyder, M.S.
Earth Science Teacher
Jones Middle School
Upper Arlington, Ohio

Ralph M. Feather, Jr., M.Ed.
Science Department Chair
Derry Area School District
Derry, Pennsylvania

Paul W. Zitzewitz, Ph.D.
Professor of Physics
University of Michigan-Dearborn
Dearborn, Michigan

With Features by:

NATIONAL
GEOGRAPHIC
SOCIETY

The National Geographic Society, founded in 1888 for the increase and diffusion of geographic knowledge, is the world's largest nonprofit scientific and educational organization. Since its earliest days, the Society has used sophisticated communication technologies, from color photography to holography, to convey geographic knowledge to a world-wide membership. The Education Products Division supports the Society's mission by developing innovative educational programs—ranging from traditional print materials to multimedia programs including CD-ROMs, videodiscs, and software.

Consultants

Chemistry

Richard J. Merrill
Director,
Project Physical Science
Associate Director, Institute
for Chemical Education
University of California
Berkeley, California

Robert W. Parry, Ph.D.
Dist. Professor of Chemistry
University of Utah
Salt Lake City, Utah

Earth Science

Allan A. Ekdale, Ph.D.
Professor of Geology
University of Utah
Salt Lake City, Utah

Janifer Mayden
Aerospace Education Specialist
NASA
Washington, DC

James B. Phipps, Ph.D.
Professor of Geology
and Oceanography
Gray's Harbor College
Aberdeen, Washington

Life Science

Mary D. Coyne, Ph.D.
Professor of Biological Sciences
Wellesley College
Wellesley, Massachusetts

Joe W. Crim, Ph.D.
Associate Professor of Zoology
University of Georgia
Athens, Georgia

Richard D. Storey, Ph.D.
Associate Professor of Biology
Colorado College
Colorado Springs, Colorado

Physics

David Haase, Ph.D.
Professor of Physics
North Carolina State University
North Carolina

Patrick Hamill, Ph.D.
Professor of Physics
San Jose State University
San Jose, California

Middle School Science

Garland E. Johnson
Science and Education Consultant
Fresno, California

Barbara Sitzman
Chatsworth High School
Tarzana, California

Multicultural

Thomas Custer
Coordinator of Science
Anne Arundel County Schools
Annapolis, Maryland

Francisco Hernandez
Science Department Chair
John B. Hood Middle School
Dallas, Texas

Carol T. Mitchell
Instructor
Elementary Science Methods
College of Teacher Education
University of Omaha at Omaha
Omaha, Nebraska

Karen Muir, Ph.D.
Lead Instructor
Department of Social and
Behavioral Sciences
Columbus State
Community College
Columbus, Ohio

Reading

Elizabeth Gray, Ph.D.
Reading Specialist
Heath City Schools
Heath, Ohio
Adjunct Professor
Otterbein College
Westerville, Ohio

Timothy Heron, Ph.D.
Professor, Department
of Educational
Services & Research
The Ohio State University
Columbus, Ohio

Barbara Pettegrew, Ph.D.
Director of Reading
Study Center
Assistant Professor of Education
Otterbein College
Westerville, Ohio

LEP

Ross M. Arnold
Magnet School Coordinator
Van Nuys Junior High
Van Nuys, California

Linda E. Heckenberg
Director
Eisenhower Program
Van Nuys, California

**Harold Frederick
Robertson, Jr.**
Science Resource Teacher
LAUSD Science Materials Center
Van Nuys, California

Safety

Robert Tatz, Ph.D.
Instructional Lab Supervisor
Department of Chemistry
The Ohio State University
Columbus, Ohio

Reviewers

Lillian Valeria Jordan Alston
Science Consultant
Institute of Government
University of North Carolina
Chapel Hill, North Carolina

Janet P. Bailey
Science Teacher
East Wake Middle School
Youngsville, North Carolina

Jamye Barnes
Science Teacher
Prescott Middle School
Prescott, Arkansas

Betty Bordelon
Science Teacher
Haynes Middle School
Metairie, Louisiana

James Carbaugh
Science Teacher
Greencastle Middle School
Greencastle, Pennsylvania

Elberta Casey
8th Grade Earth Science Teacher
Crawford Middle School
Lexington, Kentucky

Linda Culpeper
Science Department Chairperson
Piedmont Open Middle School
Charlotte, North Carolina

Nancy Donohue
General Science Teacher
Emerson Junior High School
Yonkers, New York

Susan Duhaime
5th/6th Grade Science Teacher
Assistant Principal
St. Anthony School
Manchester, New Hampshire

Ken Eiseman
Science Supervisor
West Chester Area School District
West Chester, Pennsylvania

Mel Fuller
Professor of Science Education
Department of Teacher Education
University of Arkansas
at Little Rock
Little Rock, Arkansas

Janet Grush
7th Grade Science/Math Teacher
Wirth Middle School
Cahokia, Illinois

Joanne Hardy
7th Grade Science/Social
Studies/Language Arts Teacher
Memorial Middle School
Conyers, Georgia

Nancy J. Hopkins
Gifted/Talented Coordinating
Teacher for Middle School Science
Morrill Elementary
San Antonio, Texas

Amy Jacobs
7th Grade Life Science Teacher
Morton Middle School
Lexington, Kentucky

Rebecca King
Chemistry Teacher
New Hanover High School
Wilmington, North Carolina

Ken Krause
Science Teacher
Harriet Tubman Middle School
Portland, Oregon

Martha Sculley Lai
Science Teacher
Department Chairperson
Highland High School
Medina, Ohio

William Lavinghousez
MAGNET Program Director
Ronald McNair MAGNET School
Cocoa, Florida

Norman Mankins
Science Specialist (Curriculum)
Canton City Schools
Canton, Ohio

John Maxwell
7th Grade Life Science Teacher
Claremont Middle School
Claremont, New Hampshire

Fred J. Mayberry
Earth/Life Science Teacher
Department Chairperson
Vernon Junior High School
Harlingen, Texas

Michael Parry
Science Supervisor
Boyerstown Area School District
Boyerstown, Pennsylvania

Lola Perritt
Science Specialist
Instructional Resource Center
Little Rock, Arkansas

Chuck Porrazzo
Science Department Chairperson
Bronx Career Technical
Assistance Center
Junior High 145
New York, New York

James Stewart
Life Science Teacher
W. E. Greiner Middle School
Dallas, Texas

James Todd
7th/8th Grade Science Teacher
East Hardin Middle School
Glendale, Kentucky

Deborah Tully
8th Grade Earth Science Teacher
Department Chairperson
Winburn Middle School
Lexington, Kentucky

Marianne Wilson
Science-Health-Drug
Coordinator
Pulaski County Special Schools
Sherwood, Arkansas

UNIT 1 — Forces in Action   18

Chapter 2   Forces in Earth   52

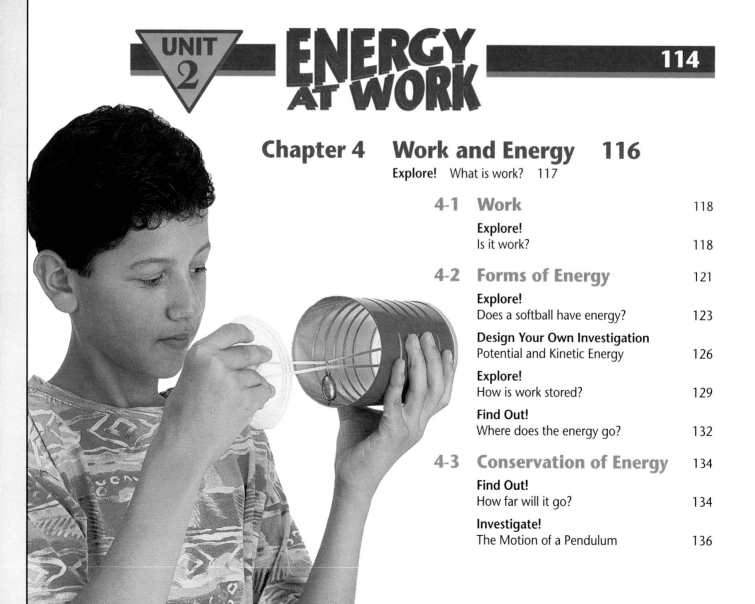

UNIT 2 ENERGY AT WORK 114

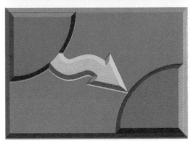

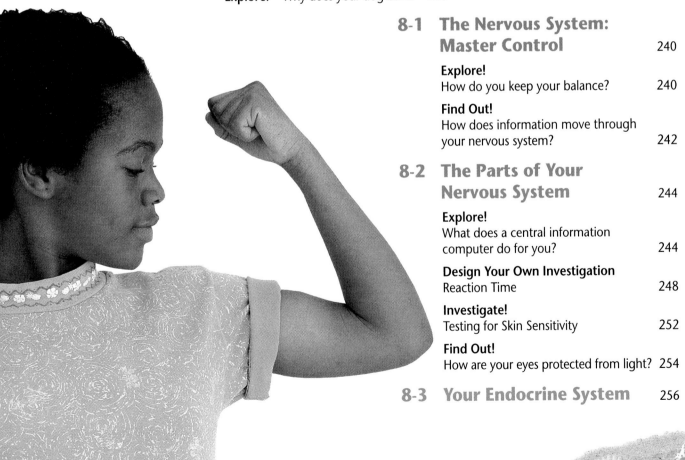

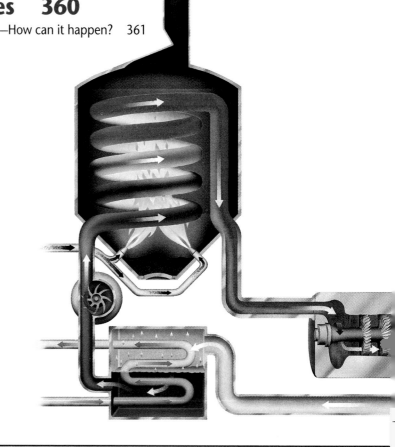

UNIT 4   Air: Molecules in Motion   422

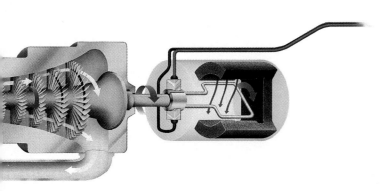

Chapter 15 The Air Around You   454

Chapter 16 Breathing   482

UNIT 5 Life at the Cellular Level | **512**

Chapter 18   Chemical Reactions   550

SCIENCE CONNECTIONS

Have you ever noticed that you really can't talk about earthquakes without mentioning forces? How is one science related to another? Expand your view of science through A CLOSER LOOK and Science Connections features in each chapter.

Earth Science

Life Science

Physics and Chemistry

A CLOSER LOOK

SCIENCE CONNECTIONS

Science is something that refuses to stay locked away in a laboratory. In both the Science and Society and the Technology features, you'll learn how science impacts the world you live in today. You may also be asked to think about science-related questions that will affect your life fifty years from now.

Science and Society

Technology Connection

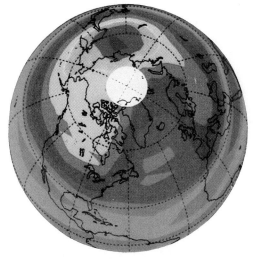

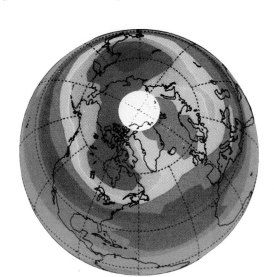

CROSS-CURRICULUM
CONNECTIONS

With the EXPAND YOUR VIEW features at the end of each chapter, you'll quickly become aware that science is an important part of every subject you'll ever encounter in school. Read these features to learn how science has affected history, health, and your buying power.

As you begin each unit of Science Interactions, start by envisioning the big picture with the help of an exciting National Geographic Society photograph. Then look for the National Geographic SciFacts article in each unit to enrich and extend your understanding of science in the real world.

NATIONAL GEOGRAPHIC SOCIETY

Science:
A Tool for Solving Problems

National Content Standards: (5-8) UCP2, A1-2, C3, D3, E1, F2, G1, G3

THEME DEVELOPMENT

The primary theme of this chapter is stability and change. Scientific methods help us understand stability and change in the natural world. When we understand a process, we can improve upon it, leading to technological advances.

The chapter's secondary theme of systems and interactions is illustrated as scientific methods help us define and predict the interactions of elements within a system and of systems with other systems. Scientific methods help us understand how systems interact to enable us to live on Earth and some day, perhaps, on the moon.

CHAPTER OVERVIEW

In this chapter, a science class learns the purpose of scientific methods by using them to plan a city on the moon. Students practice testing hypotheses by making a model and setting up controlled experiments. They also propose scientific methods to help them answer questions about living on the moon.

Consider not only having students complete the Explore and other activities in the chapter, but also having them work in groups to plan a city on the moon. Through hands-on application, your students, like the class in this chapter, will begin to appreciate the practical uses of science.

SCIENCE:
A Tool for
Solving Problems

What good is science, anyway? Can everyone use science, or is it reserved for trained scientists? How can it help in your life today? How can scientific methods help you plan ahead or make tough decisions?

Follow Philip, Sachi, Lena, and the rest of their science class as they explore building a city on the moon and learn how to put science to work in their lives. The tools and methods they use can help you answer important questions in your own life, in school and out of school—even if you don't consider yourself a "real scientist."

Who Can Use Science?

"I don't have time to help plan a city on the moon!" Philip told Lena and Sachi. They were standing in the hallway after their science class. "I need every spare second to practice for the track meet this Friday!"

Sachi frowned. "I'd rather finish writing my story. This girl is adopted, and she thinks that her gym teacher is her real mother ..."

"What if we end up living on the moon?" Lena interrupted. "Remember what Ms. Howard said? The first people to live on the moon might be in middle school or junior high school right now."

"She didn't mean me," Sachi said. "I'm a writer, not a scientist!"

Lena nodded. "You know, Rachel Carson didn't want to take a science course. She started out as a writer."

"You mean Rachel Carson, the person who wrote *Silent Spring*? She was always a writer!" Sachi said.

"Yes, she always was a writer," Lena agreed, "but she was a scientist, too. Because she was such a good writer and she understood science so well, she helped millions of people learn about pesticides and pollution."

Philip thought for a minute. "My track coach said he studied physics in college to figure out how we could cut down wind resistance when we run. So I guess he uses science, too. I wonder how many other people use science in their work."

La Bec du Hoc, *a landscape painting by Georges Seurat*

Explore! ACTIVITY

How does science contribute to other studies?

Georges Seurat and Rachel Carson, whose works are shown on this page, are just two people who combined science with other fields to produce beautiful paintings and writings. How does science contribute to other parts of life?

What To Do

1. Research a person, living or in the past, who combined science with another field, such as cooking, playing a sport, farming, writing, or another of your own choosing.

2. Make a poster or write a short biography or skit that tells how that person used science to help improve other areas of life. When you're through, display your work.

3. Observe your classmates' posters and writings. *In your Journal*, describe how science contributed to improving other fields.

Science: A Tool for Solving Problems   **3**

PREPARATION

Concepts Developed

The goal throughout this chapter is not for students to plan a city on the moon, but to gain experience in using scientific methods. The answers they obtain in the activities are not nearly so important as the methods they use to obtain them.

1 MOTIVATE

Concept Development

Activity Students can pretend they are planning a different project, such as selecting a piece of playground equipment for an elementary school. Have them list several things they might already know about the situation and five or more questions they would need to answer before they start looking through equipment catalogs. Discuss the reasons for using this Flex Your Brain approach. *It helps you think about a situation and clearly define the questions before starting to look for answers.*

2 TEACH

Tying to Previous Knowledge

Students are familiar with activities that require planning and research. Ask questions about such activities after having students identify them.

How Do We Find Out?

The next day at school, the science class began to plan its city on the moon. "Cities don't just happen," Ms. Howard pointed out, "especially on the moon. What are some things a city needs?"

"A swimming pool and a video store!" offered Philip.

Ms. Howard smiled. "Let's call that 'recreation.'" After everyone called out ideas, the class narrowed the list to these areas: food, water, air, clothing, housing, transportation, health care, recreation, and waste control/removal.

"Our city will be expensive to build," Ms. Howard pointed out. "We'll also have to get supplies from Earth to keep it going after it's built. How are we going to pay for this? Hold a bake sale?"

Alberto said, "We need to find or make something on the moon that people on Earth will buy so we can get money to pay for supplies."

"Or maybe we could offer some special service people want, but can't get anywhere else," Sachi suggested.

Ms. Howard nodded and added "product/service" to the list on the chalkboard. Lena couldn't help wondering what people living on the moon could make and sell. Wasn't the moon just dust and craters?

"Let's recall what we KNOW about the moon." Ms. Howard tacked up a list of statistics.

> *Facts About the Moon*
> Distance from Earth: 384,403 km
> Diameter: 3,476 km (.27 times Earth)
> Revolution around Earth: 29 days, 12 hours, 44 minutes
> Rotation on axis: 27 days, 7 hours, 43 minutes
> Temperature: 127° C day; -173° C night
> Atmosphere: Ultra thin: neon, hydrogen, helium, argon
> Surface gravity: 0.17 times Earth (1/6 of Earth's)
> Surface: silicate crust

■ Asking a Question

Ms. Howard showed them a form. "This Flex Your Brain chart will help us use what we know to find out what we want to know. What else do we need to know about the moon before we can start planning?" she asked.

Hands went up. As the class suggested questions to be answered, Ms. Howard wrote the questions down.

"If the atmosphere is thin, does that mean there's no ozone layer?" Michelle asked.

"Is there anything on the moon we could use as a building material?" Mieko asked. "Or does everything have to be shipped from Earth?"

Then Lena asked, "Is it always night on one side of the moon and day on the other? I can't remember."

"It's always dark on one side," Philip told her. "So we have to put our city on the 'day' side."

Ms. Howard stopped writing. "I see some puzzled faces. Does everyone agree with Philip? How can we find out if the moon has a dark side? We can't experiment with the sun and moon, so what else could we do? Let's use the Flex Your Brain chart to develop some ideas." The class completed the form and developed one way to investigate whether or not the moon has a dark side.

Flex Your Brain

1 Topic: _____

2 ? **What do I already know?** 1. ___ 2. ___ 3. ___ 4. ___ 5. ___

3 Q: Ask a question _____

4 A: Guess an answer _____

5 How sure am I? (circle one)

Not sure   Very sure
1   2   3   4   5

6 ? **How can I find out?** 1. ___ 2. ___ 3. ___ 4. ___ 5. ___

7 EXPLORE

8 Do I think differently? →   yes   no

9 ? **What do I know now?** 1. ___ 2. ___ 3. ___ 4. ___ 5. ___

10 SHARE → 1. ___ 2. ___ 3. ___

Flex Your Brain The Flex Your Brain strategy helps students think about what they already know and define the question they want to answer. Here is how students might complete the form for the Find Out activity on page 6.

1. Topic: The moon

2. In listing what they already know, students might include the statistics from page 4.

3. Students form a question: Does the moon have a "dark side"?

4. Students may guess that the answer is "yes."

5. Students indicate their level of confidence that their answer is correct.

6. Students list ways to test their answer, such as doing library research and making a model.

7. In this case, guide students to choose making a model as the way to test their answer. Involve them in selecting objects to represent the sun, moon, and Earth and in planning how to show the relationships among them. Ask them to carry out their plans.

8. Students check the accuracy of their answer in Step 4.

9. Students should now know that the moon does not have a dark side. Urge them to consider other things they have learned by testing this answer with a model.

10. Have students share what they considered as they completed each step. Discuss what they have learned about using the Flex Your Brain strategy to ask and answer a question.

Science Journal Students can use their journals to record their observations, charts and results of activities, and impressions. You may wish to review these journals for evaluation and assessment.

Encourage students to keep a record of their Flex Your Brain exercises in their journals and later add them to their portfolios.

A Look at the "Dark Side"

Time Needed 10 minutes

Materials globe, basketball, or soccer ball; small ball; sticker; flashlight

Thinking Processes observing, inferring

Purpose To model the rotation and revolution of the moon.

Expected Outcomes

Students will discover that the moon has no dark side.

Teaching the Activity

Science Journal Have students record their observations in their journals.

✔ Assessment

Oral Students should create a diagram illustrating their experiment and then use the poster to explain their observations to their classmates. Use the Performance Task Assessment List for Oral Presentation in **PASC**, p. 71.

COOP LEARN

Across the Curriculum

Math

Divide the class into groups. Ask half of the groups to choose a large object, record its actual dimensions, and then figure out the dimensions of a model one-tenth its size. Ask the other groups to choose a very small object, record its actual dimensions, and then figure out the dimensions of a model ten times larger. Have groups exchange their calculations and check each other for accuracy. Discuss the value of making models that are much smaller or larger than the original objects.

A Look at the "Dark Side"

Construct a model to find out if the moon has a dark side. You'll need a globe or large ball (Earth), a small ball (moon), a flashlight (sun), and a sticker.

What To Do

1. Put the sticker on the "equator" of the moon ball.
2. Shine the flashlight on Earth.

3. Then, move the moon slowly around Earth to represent its one-month orbit. Remember, as the moon orbits, it must also rotate on its axis. The moon's orbit around Earth and one rotation take about the same length of time.
4. Watch how long the "sun" shines on the sticker as the moon orbits Earth and rotates on its axis.
5. Record your observations and the answers to the following questions *in your Journal.*

Conclude and Apply

1. Does the moon have a "dark side"?
2. How long is one "day" on the moon, according to your model?
3. How did constructing this model help you answer the question "Does the moon have a dark side?"

■ **Planning an Approach**

After the class used a model to figure out whether the moon had a dark side or not, they listed more questions. Then, everyone divided into groups to start planning.

Philip thought about joining the Recreation Group. That group would examine how the moon's gravity and other conditions would affect the games we play on Earth. For example, what might happen if you bunted a baseball on the moon? The group could rewrite the rules for a game—or invent a new game, taking into account the conditions on the moon.

But Philip, who was always hungry, decided to volunteer for the Food Group instead. This group would investigate the kinds of food that could be grown on the moon. It would study how conditions on the moon might affect the growth of plants. The group would also figure out how much it would cost to bring food from Earth.

Sachi chose the Waste Control/Removal Group. This group would explore which materials could be reused, which could be recycled, and which had to be thrown away. The group's aim was to conserve materials

6 Science: A Tool for Solving Problems

Meeting Individual Needs

Learning Disabled Invite students to repeat the "Dark Side" Find Out! activity in groups of three so they can experience the movement of the moon and its relationship to Earth and the sun. Ask them to use the model to show a full moon or a moon in the first or third quarter.

and avoid polluting their new city.

Lena wanted to be in the Product/Service Group. She was curious about what the moon could have that people on Earth would want. After all, how many samples of moon rock would anyone buy?

Each group began by brainstorming a list of questions. They chose some of the questions from the class list. Then, they added more things they needed to know in order to plan their own aspect of the city. Each group had a week to select a question from its list and try to find an answer, using the "Flex Your Brain" chart to plan a scientific approach.

■ Experimenting

At the end of the week, the groups shared what they had learned. The Food Group shared first. Philip reported that the group members— Alberto, Michelle, and himself— already were fairly sure seeds could be transported to the moon safely. In fact, a science class at the school had grown great-tasting tomatoes from seeds that had spent six years in space aboard NASA's Long Duration Exposure Facility (LDEF).

"But," Michelle told the class, "even starting with healthy seeds, we thought the moon's long night might kill most plants. So we thought that plants on the moon would probably be grown in artificially controlled light. We wondered how many hours of light would produce the best plants."

"So we set up an experiment," Alberto explained. "We gave some plants a short day, some a 14-hour 'summer' day, and some an extra-long day. We also left some plants in the light all the time, to see how they grew."

After the Food Group explained the results of its experiment, Michelle laughed. "Now we know how many hours of light make the plants grow fastest, but there are a lot of other questions we need to answer before we're ready to feed a city!"

How Will Your Garden Grow?

Time Needed 20–30 minutes plus approximately 5 minutes each day to take care of the plants and record observations

Materials plants in planting medium, artificial lights, light-blocking materials such as paper bags

Thinking Processes designing an experiment, setting up experimental controls, carrying out a strategy, measuring growth, collecting and evaluating data

Purpose To design and carry out an experiment.

Teaching the Activity

Discussion Remind students that this chapter's activities are helping them learn scientific ways to solve problems. Learning the best light schedule for plant growth is not nearly so important as learning a way to find out the best schedule.

Science Journal Have students record their experimental design and their observations in their journals. L1

Expected Outcomes

Students should design and execute an experiment to find optimum conditions for food-plant growth.

Conclude and Apply

2. A control is important in an experiment as a standard against which to compare experimental results.

✔ Assessment

Process Students should evaluate the results of their experiments and decide what conditions are best and worst for plant growth. Use the Performance Task Assessment List for Assessing the Whole Experiment in **PASC,** p. 33.
COOP LEARN

How will your garden grow?

Try developing your own experiment to test the same problem the food group faced.

What To Do

1. First, set up three different schedules for the amount of light each group of plants will receive. One group of plants should be a "control" and follow a normal Earth daylight schedule.

2. Predict which schedule will result in best plant growth. Your prediction will be the hypothesis, or suggested solution, for this experiment.

3. Now, design an experiment to test your hypothesis. Remember, you are only testing the effects of light, so be certain that all the other conditions, or variables, of the experiment are the same for all groups.

4. Get your design approved by your teacher and conduct your experiment.

Conclude and Apply

1. *In your Journal*, write a summary of your experiment. Include your hypothesis, which is your suggested answer to the problem, how you tested your hypothesis and the results of your experiment. Also tell how you evaluated your results. For example, how did you define and evaluate "best growth"?

2. Why is a control important in an experiment?

■ More Experimenting

Sachi told the class that the Waste Control/Removal Group—Kareem, Jeff, and herself—had begun by discussing its goals. "Pollution is the main thing," Kareem said. "We sure don't want to turn the moon into a garbage pit. We thought about setting up an ecosystem that would recycle everything."

"We didn't know much about recycling," Jeff said. "So we became scientists." He smiled, "We also set up an experiment!"

"We did the experiment with paper," Sachi explained. "We wanted to see whether it was easier to reuse paper, recycle it, or make new paper.

We thought we could use what we learned about paper to control other waste materials on the moon, too.

"Our hypothesis (the answer we expected) was that reusing paper would be the easiest, recycling it would be harder, and making new paper would be the hardest. We already knew how to reuse paper— just turn it over and write on the other side.

"To make

8 Science: A Tool for Solving Problems

Two "Amateur" Scientists at Work

Gregor Mendel, who defined the laws of heredity, was an amateur scientist. He studied science on his own, took college science courses, and taught science in a high school.

Mendel began to experiment on plants in his own garden. His careful experimentation led him to understand how characteristics such as tallness and color are passed from one plant to another.

Valentina Tereshkova spent her early years working in a tire factory and in a cotton mill in the Soviet Union. Inspired by cosmonaut Yuri Gagarin, Valentina Tereshkova wrote to the Soviet government and asked to be part of its space program.

Tereshkova was selected and received one year of training. (When she started, her only qualifications were her great interest in the program and her skill in parachuting.) However, her interest and dedication led her to learn all of the mechanics and physics that went into piloting a space capsule.

On June 16, 1963, she became the sixth cosmonaut and the first woman in space, orbiting Earth for three days. That was longer than any of the six U.S. astronauts who had flown, up to that time.

Across the Curriculum

Daily Life

Ask students to describe scientific methods farmers use to increase their harvests each year. *Farmers often use observation and controlled experiments to see which crops grow best under which conditions.*

Inquiry Question Why did the Waste Control/Removal Group decide to try recycling paper? *They were testing their hypothesis that it was easier to reuse paper than to recycle it.*

new paper, you have to cut down trees in the forest. That's much harder than reusing paper and wouldn't work on the moon, for sure. Our group didn't know much about recycling paper, the middle part of our hypothesis, so we decided to test that part by finding out how hard it is to recycle paper."

ENRICHMENT

Activity Students can design a way to find out the best method to use in studying for a test. Ask them to use Steps 1 to 6 of the Flex Your Brain strategy.

Paper, the Second Time Around

Planning the Activity

Time needed 20 minutes plus 10 minutes after the paper has dried

Purpose To compare and contrast reusing, recycling, and making new paper.

Process Skills comparing and contrasting, analyzing observations

Materials See student text activity.

Teaching the Activity

Process Reinforcement Be sure that students clearly describe or illustrate the procedure. L1

Safety If the blender blades jam, unplug the blender before trying to free them. Students should be careful, even when the blender is unplugged, since the blades are sharp and can cut skin.

Troubleshooting Students should add paper to the blender in small batches. Adding too much paper at once could jam the blades.

Science Journal Have students record their observations in their journals. They should include a description of the paper they made.

Paper, the Second Time Around

Follow these directions to make one sheet of recycled paper by hand. Machines recycle paper in factories, of course. This activity will show you what's involved in the process.

Problem

Is it easier to reuse, recycle, or make new paper?

Materials

2 pages of newspaper, torn into small squares	2 tablespoons of white glue
	dishpan
	one leg of an old pair of pantyhose
2 to 3 cups of water	electric iron
blender	clothes hanger

Safety Precautions

Do not take the lid off of the blender while it is operating. Do not get your hands near the blender blades.

 **What To Do**

1 Carefully untwist the clothes hanger and form it into a 6-inch square (see photo **A**).

2 Carefully slip the wire square inside the pantyhose, trying not to snag the hose (see photo **B**). Make sure the hose is tight and flat. Tie each end of the hose into a knot.

3 Put some torn paper and water into the blender. Close the lid and turn it on high. Add more paper and water until the paper disappears and the mixture turns into a large ball of pulp. Then, let the blender run for two more minutes.

4 Put about 4 inches of water in the dishpan and add the glue.

5 Add the pulp to the water and mix well.

A   B   C

6 Stir the water. Quickly slip the wire frame under the pulp and rest it at the bottom of the dishpan. Then lift the frame slowly as you count to 20.

7 Let the paper on the frame dry completely. (You might put the frame in the sun.) When the paper is totally dry, gently peel it off the frame (see photo C).

8 Use the iron on the hottest setting to steam your paper flat. When the paper dries again, it's ready to use!

Analyzing

1. Compare the amount of energy that goes into recycling paper to reusing one sheet of paper.

2. Which process would be more energy efficient?

You can create handmade paper in a variety of textures and colors using not only newspaper, but materials such as flowers, grasses, food coloring, even lint from the clothes dryer!

Concluding and Applying

3. If you did this experiment again, do you think you would come to the same conclusions?

4. When would you choose to use an experiment rather than a model?

5. Why is an experimental method sometimes used in science?

6. Did this experiment raise any questions that were not answered? What questions? How would you go about finding an answer to those questions?

7. **Going Further** Redesign the experiment to make it more accurate or more measurable. For instance, can you figure out a way to measure how much effort each of the three different paper treatments used?

Answers to Analyzing/ Concluding and Applying

1. Student answers may vary but should indicate that much more energy is needed to recycle paper than to reuse it.

2. Reusing paper is more energy efficient.

3. Yes.

4. Use an experiment when actual conditions can be duplicated, even if only on a small scale.

5. An experimental method is used to investigate and explore a hypothesis, to investigate new procedures on a small scale, and to simulate real-life situations on a small scale.

6. Answers may vary. If students have unanswered questions, they should also suggest how to get answers to the questions.

7. Answers may vary. Students may suggest timing how long the blender runs, then finding out how much electricity the blender uses.

✔ Assessment

Process Have students repeat the experiment, starting with other kinds of paper, including waxed paper and cardboard. Then they can compare the energy needed to recycle different kinds of paper. Students could then write a newspaper article about recycling and reusing paper. Use the Performance Task Assessment List for Newspaper Article in **PASC,** page 69.

`COOP LEARN` `P`

Program Resources

Activity Masters, pp. 5-6
Study Guide, pp. 5-6
Performance Assessment, Introductory Chapter

Content Background

Lunar soil is 40 percent oxygen by weight. This oxygen might be usable in a colony's air supply and in rocket fuel. Today's rockets burn 8 kilograms of oxygen for every 1 kilogram of hydrogen. Lunox, oxygen taken from lunar soil, would drastically reduce the need to ship rocket fuel to the moon. It also could allow the moon to serve as a "pit stop" for outward-bound spacecraft.

■ Analyzing Data

Sachi, Kareem, and Jeff held up the sheets of recycled paper they had made. "So you see," Sachi said, "our hypothesis was right. It's easy to reuse paper, and it's harder to recycle it. But it would be hardest to make new paper, especially when you'd have to transport trees to the moon!" she added.

"So we recommend that all paper on the moon be reused and then recycled," Jeff told the class. "But our Waste Control/ Removal Group still has a lot more materials to experiment with before we're ready to help set up an ecosystem in our city!"

The Product/Service Group shared next. "Lewis, Mieko,

and I had a little trouble getting started," Lena told the class. "The first time we met, we just stared at one another. No one could think of anything to make on the moon and sell on Earth. We even started to think that maybe a bake sale was a good idea!"

Everyone laughed. Then Mieko said, "We finally realized we had to

find some way to take advantage of the conditions on the moon, so we listed ways the moon is different from Earth. Here are some ways we thought of, along with questions we needed to answer." She taped a newsprint list on the wall. The list is shown on the right.

"Then," Lewis explained, "we thought about the problem a different way. We listed the different things we value here on Earth and tried to pick ones that might be produced more cheaply or easily on the moon."

"What we finally realized," Lena told the class, "was that we had a lot more questions than answers. We needed to choose a question or two and figure out some way to get the answers. So each of us picked one possible way to make money on the moon and wrote a proposal explaining how we could find out if our idea was practical."

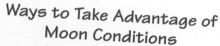

Ways to Take Advantage of Moon Conditions

1. **The moon has lower gravity than Earth.**
   What does that mean for space vehicles taking off from the moon's surface? What products could be produced more easily if gravity weren't pulling on them? Heavy things? Delicate things? Things that must be a certain shape?

2. **The moon has almost no atmosphere.**
   How could we take advantage of that? No atmosphere means no humidity. What about making computer parts and other things that are affected by dampness in the air? (The moon also has a dusty surface that might wreck computer parts!)

3. **The moon might have different minerals than Earth has.**
   Maybe the moon has valuable minerals that could be mined and sold. Or maybe certain minerals in the moon's crust would make plants grow super-fast. (The minerals might also be useful as building materials on the moon.)

4. **The moon and Earth are in different positions in space.**
   How could we use the moon's position to offer a service to space vehicles or astronomers?

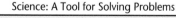

Science: A Tool for Solving Problems **13**

Paying for the Moon

Time Needed One hour plus one class period for presentations.

Materials Poster materials

Thinking Processes Collecting and Evaluating Data, Designing an experiment

Purpose To design a research investigation.

Teaching the Activity

Discussion Encourage students to discuss possible strategies and investigate the pros and cons of each before they start to finalize their plans. Emphasize to students that they do not need to be able to carry out these experiments with materials in the classroom. They will not be doing the investigation, only planning the best way to find out what they need to know.

Science Journal Have students record not only the plan they finally decide on, but other plans suggested by group members that were not developed. L1

Expected Outcomes

Each student group should design some investigation to determine what resources could be used to pay for a lunar colony.

✔ Assessment

Process Students should evaluate each plan and determine which plans seem most feasible. Use the Performance Task Assessment List for Designing an Experiment in **PASC,** p. 23. COOP LEARN

Paying for the Moon

You need to pay for your moon city. You need a product—but what? How could you find a product or service to sell on the moon?

What To Do

1. With a partner, brainstorm some ways a moon colony might make money.

2. Choose one of your ideas to be a hypothesis. A hypothesis is a suggested or possible solution to a problem. In this case the problem is paying for a space colony.

3. Design a way to test your hypothesis and see whether your solution is possible and practical. You won't perform your test, so don't worry about what equipment you'll need to perform it.

4. Here are some of the ways you could test your hypothesis:

   Make a model Create a scale model of what you're proposing and see if it works. Or make a computer model of a process, taking into account the conditions on the moon.

   Experiment Try what you're proposing under different conditions and see which condition works best.

   Observe Go to the moon and look for valuable minerals. Or record the conditions and process needed to manufacture something here on Earth and consider whether they are practical for the moon.

5. Now, write a proposal for your research plan. Start with the problem and your suggested solution. Then explain how you'll test your hypothesis and determine whether your solution makes sense. Create a poster to convince your classmates that your research is reasonable and should be funded.

6. After each group presents its poster and answers questions from the class, vote for the idea that seems most likely to lead to a profitable product. Summarize your classmates' ideas and the reasons for your own vote *in your Journal.* Did some ideas seem more "scientific" than others? What do you think makes good scientific research?

14

Space Research

Every year, a group of 66 Japanese companies known as *keidanren* spends twice as much as the Japanese government on space research and development. This group is looking for ways to profit from space-age technology.

At the same time, some Japanese construction companies are designing space hotels for the tourists they expect only 25 or 30 years from now.

Have students research the space programs of Japan and other nations and report on the goals of these programs.

■ Using Science Processes Every Day

After the groups shared, Ms. Howard asked the class for their impressions of the moon project so far.

Kareem laughed. "I think we need a lot more planning before we can start packing for the moon! Our groups have more questions than answers!"

Ms. Howard smiled. "Often good scientific research will answer one question only to come up with several new, unanswered questions."

Sachi said, "But I've thought of a way to use a scientific approach in my writing. See, in this story an adopted girl is trying to find her mother. She carefully observes everyone around her and starts gathering information. . . ."

I n this chapter, you've learned some of the ways scientists try to solve problems. You'll use these processes as you study this book and continue to discover what science is about. But you can also use scientific approaches as you try to solve problems in your everyday life.

3 ASSESS

Check for Understanding

Ask students whether they would use a model, an experiment, or observation to answer these questions:

1. What is the best location for a moon colony? *Observation*
2. How does weightlessness affect muscle strength? *Experiment*
3. What is the best design for a moon colony? *Model*

Use the Performance Task Assessment List for Group Work in **PASC**, p. 97. **COOP LEARN**

Reteach

Discussion To help students understand the difference between a model and an experiment, discuss why the "Dark Side" Find Out activity is not an experiment. *We cannot experiment with the relationships between the sun, moon, and Earth. We can gather information about them through this model.*

Extension

Activity Working in small groups, students can demonstrate a scientific approach by creating a model, performing a controlled experiment, or sharing a question they have answered by using observation. Encourage students to demonstrate and explain what they have learned about scientific tools for solving problems.

4 CLOSE

Activity

Ask students to plan how to help a younger class learn about scientific tools by performing an experiment or demonstrating how a model works. Have them display a poster of the Flex Your Brain strategy and explain to the younger students how they used this approach to ask and answer a question scientifically.

Have students work in small groups to illustrate the main ideas of the chapter.

Teaching Strategies

Divide the class into three groups and assign one of the main ideas to each group. Have each group make a collection of pictures that illustrate their topic.

Answers to Questions

1. The "Flex Your Brain" approach gives a sequence of steps to guide the approach to an investigation. The questions and step-by-step outline provide a framework in which to work.

2. It is most appropriate to use a model to help find the answer to a question when duplicating the real situation is impossible, as in observing the dark side of the moon, or too expensive, as when working with a model of a new building or when comparing models of several new pieces of machinery.

3. Some examples of when an artist can use science include choosing the most appropriate paper or surface for painting, choosing a material for a sculpture, choosing a glue or paste for a collage, choosing a paint medium, choosing a drying agent, and choosing a cleaner to remove paint from hands or clothing. Some examples of when a musician can use science include choosing electronic amplification and recording apparatus, choosing the material from which an instrument is made, and choosing how loud to play so that the audience hears the correct level.

Science Journal
Review the statements below about the big ideas presented in this chapter, and answer the questions *in your Science Journal.*

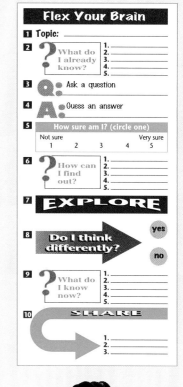

Flex Your Brain

1 Topic: _____
2 **What do I already know?** 1.___ 2.___ 3.___ 4.___ 5.___
3 Q: Ask a question
4 A: Guess an answer
5 **How sure am I? (circle one)**
Not sure   Very sure
1   2   3   4   5
6 **How can I find out?** 1.___ 2.___ 3.___ 4.___ 5.___
7 **EXPLORE**
8 **Do I think differently?** → yes / no
9 **What do I know now?** 1.___ 2.___ 3.___ 4.___ 5.___
10 **SHARE** 1.___ 2.___ 3.___

❶ There are many ways to approach a problem scientifically and the first important step in each is to define the question you are investigating. *How does the "Flex Your Brain" approach help you define the question you wish to research?*

❷ Scientists use observation, models, and controlled experiments, among other methods to help find the answers to scientific questions. *When is it most appropriate to use a model to help you find the answer to a question?*

❸ Ordinary people use science and scientific methods in their everyday work. *List three ways in which an artist or a musician can use science or scientific methods in their work.*

16

Understanding Ideas

Answer the following questions in your Journal using complete sentences.

1. List three activities you do every day that you could do better if you had a scientific understanding of how that activity was done. Explain
2. Describe one way in which to develop a scientific question to investigate.
3. Why is saying what you know about a problem helpful in asking new questions about the problem?

Critical Thinking

Use your understanding of the concepts developed in the chapter to answer each of the following questions.

1. If you wished to study the motion of atoms, which scientific method would you use? Explain.
2. Grena Phacops is a scientist who wants to study communities of clams. What scientific method might she employ? Why?
3. People in the town of Persimmon Gap are concerned about possible pollution from farm water runoff. They want to know the effects of runoff on algae growth in Persimmon Pond. What scientific method should they use? Why?
4. How does using a model differ from designing a controlled experiment? How are they similar?
5. Why is asking the right question such an important part of any scientific method?

Problem Solving

Read the following problem and discuss your answer in a brief paragraph.

Rachelle knows that long-term space travel causes calcium-loss from bone tissue in astronauts. She also knows that this is related to very low gravity conditions that astronauts experience in space. She is concerned that long-term exposure to the low gravity of the moon might have a similar effect. She comes to you and asks you for help in finding out whether or not this is true. Help Rachelle choose an appropriate scientific method and plan an investigation that will help her find the answer.

swered until someone goes to the moon and spends time there.

Understanding Ideas

1. Possible answers include cooking, photography, gardening, auto or bicycle repair.
2. Sample answer: Examine a common daily activity, then think about how it could be improved. Come up with several alternatives to test, using the usual way as a control.
3. Saying what you know can make you aware of what you don't know and need to find out.

Critical Thinking

1. To study the motion of atoms, you would probably use a model such as shining a flashlight on air and watching the dust movements.
2. She might observe real-life communities of clams at the sea shore or she might set up an artificial environment that duplicates the real environment and study the clams in a laboratory.
3. They could do an experiment that duplicates real conditions in a laboratory on a small scale.
4. A model differs from a controlled experiment in that it mimics but doesn't duplicate actual materials and/or conditions. A model is similar to a controlled experiment in that both allow scientists to perform investigations with controlled variables.
5. Asking the right question helps define what you want to learn and guides the experiment.

Problem Solving

Have students refer to the Find Out activity on page 8 in which they developed a controlled experiment. Students may suggest that Rachelle might do library research to compare the conditions on the moon to the conditions astronauts experience in space. Then she might devise a model. However, this is the type of scientific question that cannot be accurately an-

Forces in Action

UNIT OVERVIEW

UNIT FOCUS

In Unit 1, students will learn about forces. Students will explore the physical forces that act on moving objects. They will learn about the forces in Earth that cause earthquakes and volcanoes. Students will discover how forces inside the bodies of animals that cause blood to flow.

THEME DEVELOPMENT

Two themes developed in Unit 1 are systems and interactions and stability and change. The interaction of natural forces with matter, on the surface of Earth as well as inside, causes motion and change in motion. Examples of this are the forces that cause earthquakes and volcanoes. Similar forces interact with the blood in the body and are vital to the efficient function of the body's circulatory system.

Connections to Other Units

The concepts in Unit 1 provide a foundation for understanding events that occur in the physical world, in Earth, and in the bodies of living things. Students will use what they learn in this unit about force and pressure to explore how people use force to do work in Chapters 4 and 5 of Unit 2. In addition, students will link the information in Chapter 3, *Circulation,* to discover how the body moves and is controlled, topics of Chapters 7 and 8 in Unit 2.

UNIT 1

Forces in Action

A molten river of fire courses through the night from a volcano on the Pacific Rim. Everywhere—around you, within you, within Earth itself—forces make things move. Though you may not be aware of them, you depend on these forces to get things done and to keep you alive. Sometimes you can observe the disastrous effects of forces unleashed within Earth. Plunge in and learn how forces shape your life.

18

GLENCOE TECHNOLOGY

 Videodisc

Use the *Science Interactions, Course 2* **Integrated Science Videodisc** lesson, *Buoyancy: Camcorder Carl Goes Fishing,* after Chapter 1 Section 3 of this unit.

NATIONAL GEOGRAPHIC
try it!

The weight of air in our atmosphere produces a force on all things on the surface of Earth. How can you observe this force?

What To Do

1. Lay a meterstick on the table with 45 cm extending over the edge.

2. Place a large sheet of newspaper over the part of the meterstick on the table.

3. Now push quickly down on the free end of the meterstick extending over the edge of the table. What do you feel? How might air pressure be related to what you feel?

19

GETTING STARTED

Discussion Some questions you may want to ask your students are

1. Why doesn't Earth fly off into space? Students may suggest a wide variety of causes before they realize that Earth is held in orbit by the gravitational force of the sun.

2. Can an earthquake cause a volcano? Some students may have the cause-and-effect relationship reversed. Volcanic activity can cause earthquakes but usually not vice versa.

3. How might forces be involved in the circulation of your blood? Students may have heard of blood pressure, especially high blood pressure, but not know exactly what the term means. The peak pressure of blood flow is caused by the contraction of the heart's ventricles and is known as systolic pressure. This pressure is a force that pushes blood through the blood vessels.

Responses to these questions will help you determine misconceptions that students have.

✔ Assessment

Performance Ask students to form hypotheses about the effects of using a smaller or larger sheet of paper on the meterstick and the effects of covering more or less of the meterstick with the paper. Then have students repeat the experiment, changing the variables as needed to test their hypotheses. Use the Performance Task Assessment List for Evaluating a Hypothesis in **PASC,** p. 31.

Try It!

Purpose Students will observe the resistance of objects under pressure.

Background Information The force of air resistance on the newspaper increases the effort needed to lift the meterstick. Engineers who design automobiles and other vehicles employ knowledge of this principle when they create aerodynamic designs that will decrease the resistance of air against the moving vehicle.

Materials meterstick, large sheet of newspaper, table

Troubleshooting When placing the newspaper over the meterstick, have students smooth it out to remove any air pockets beneath it.

Answers to Questions

Students should feel the force of the air resistance on the newspaper. This increases the effort needed to lift the meterstick.

Chapter Organizer

SECTION	OBJECTIVES	ACTIVITIES & FEATURES
Chapter Opener		**Explore!**, p. 21
1-1 Force and Motion (3 sessions, 1.5 blocks)	1. **Relate** inertia to mass. 2. **Identify** the forces acting when objects interact. 3. **Describe and use** Newton's First Law of Motion. **National Content Standards: (5-8) UCP2-4, A1-2, B2, D3, G1, G3**	**Investigate 1-1:** pp. 24–25 **Find Out!**, p. 26 **Find Out!**, p. 29
1-2 How Forces Act on Objects (4 sessions, 2 blocks)	1. **Describe** the relationships between force, mass, and acceleration. 2. **Use** Newton's second law to predict acceleration. 3. **Identify** pairs of forces. 4. **Describe** the difference between balanced forces and action-reaction forces. **National Content Standards: (5-8) UCP2-4, A1-2, B2, G1, G3**	**Design Your Own Investigation 1-2:** pp. 32–33 **Explore!**, p. 36 **Skillbuilder:** p. 37 **A Closer Look**, pp. 34-35
1-3 Pressure and Buoyancy (2 sessions, 1 block)	1. **Calculate** pressure from a given force and surface area. 2. **Explain** how objects can float in a liquid. 3. **Interpret and use** Archimedes' Principle. **National Content Standards: (5-8) UCP2-4, A1-2, B2, C1, C3, C5, E2, F5, G1-3**	**Explore!**, p. 39 **Find Out!**, p. 41 **Find Out!**, p. 44 **History Connection**, p. 47 **Technology Connection**, p. 48 **Life Science Connection**, pp. 42-43 **Science and Society**, p. 46

ACTIVITY MATERIALS

EXPLORE!
p. 21* glass or beaker, index card, penny, small piece of paper
p. 36* 1-kg mass, spring scale calibrated in newtons
p. 39* graph paper, sheet of carbon paper, crutch, shoes, scale

INVESTIGATE!
pp. 24–25* inertial balance, pan balance, clamp, assorted marked masses, second timer, heavy rubber bands, object of unknown mass, graph paper

DESIGN YOUR OWN INVESTIGATION
pp. 32–33* balance with 5-kg capacity, masses of 1 kg and less, spring carts, 2 books, masking tape, meterstick

FIND OUT!
p. 26* 2 rolling chairs
p. 29* plain white paper, 20-g mass, coarse sandpaper
p. 41* scale, 3 or 4 graduated cylinders of different diameters, ruler, water
p. 44* beaker, pan, spring scale, water, small can with wire handle, 50-g mass

KEY TO TEACHING STRATEGIES

The following designations will help you decide which activities are appropriate for your students.

L1	Basic activities for all students
L2	Activities for average to above-average students
L3	Challenging activities for above-average students
LEP	Limited English Proficiency activities
COOP LEARN	Cooperative Learning activities for small group work
P	Student products that can be placed into a best-work portfolio
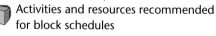	Activities and resources recommended for block schedules

Need Materials? Call Science Kit (1-800-828-7777).

⏲ OUT OF TIME? We recommend that students do the activities with an asterisk.

Chapter 1 Forces and Pressure

TEACHER CLASSROOM RESOURCES

Student Masters	Transparencies
Study Guide, p. 7 **Take Home Activities**, p. 6 **Critical Thinking/Problem Solving**, p. 9 **Activity Masters**, Investigate 1-1, pp. 7–8 **Making Connections: Across the Curriculum**, p. 5 **Science Discovery Activities**, **1–1** **Laboratory Manual**, pp. 1–2, Newton's Laws of Motion	**Section Focus Transparency 1**
Study Guide, p. 8 **Concept Mapping**, p. 9 **Multicultural Connections**, p. 5 **Activity Masters**, Design Your Own Investigation 1–2, pp. 9–10 **Science Discovery Activities**, **1–2** **Making Connections: Integrating Sciences**, p. 5 **Making Connections: Technology and Society**, p. 5	**Teaching Transparency 1**, Newton's Third Law **Section Focus Transparency 2**
Study Guide, p. 9 **How It Works**, p. 5 **Science Discovery Activities**, **1–3** **Laboratory Manual**, pp. 3–4, Density and Buoyancy	**Teaching Transparency 2**, Archimedes' Principle **Section Focus Transparency 3**

ASSESSMENT RESOURCES	TEACHING & TECHNOLOGY
Review and Assessment, pp. 5–10 **Performance Assessment**, Ch. 1 **PASC*** **MindJogger Videoquiz** **Alternate Assessment in the Science Classroom** **Computer Test Bank**	**Spanish Resources** **Cooperative Learning Resource Guide** **Lab and Safety Skills** **Science Interactions, Course 2, CD-ROM** **Computer Competency Activities**

*Performance Assessment in the Science Classroom

NATIONAL GEOGRAPHIC TEACHER'S CORNER

Index to National Geographic Magazine	National Geographic Society Products Available From Glencoe	Additional National Geographic Society Products
The following articles may be used for research relating to this chapter: • "Searching for the Secrets of Gravity," by John Boslough, May 1989.	To order the following products for use with this chapter, contact your local Glencoe sales representative or call Glencoe at 1-800-334-7344: • *Newton's Apple Physical Sciences* (Videodisc)	To order the following products for use with this chapter, call the National Geographic Society at 1-800-368-2728: • *Everyday Science Explained* (Book)

Teacher Classroom Resources

These are key components of the classroom resources package.

TEACHING AIDS

Section Focus Transparencies

1 SECTION FOCUS TRANSPARENCY Section 1-1

A ROLLER-COASTER RIDE

Imagine that you are on this roller coaster. What makes the ride exciting? Think about how you move and feel as the roller coaster speeds up or shoots around a curve.

1. How do you feel when the roller coaster plunges down a hill? Which way do you move? What do you feel?

2. Which way do you move when the roller coaster speeds around a curve to the right? To the left? Why?

3. Which way do you move when the roller coaster suddenly slows down? Why?

L1

2 SECTION FOCUS TRANSPARENCY Section 1-2

LIFTOFF

Imagine the time and effort it took to develop something as complex as a rocket. Think of the power it takes to send this massive structure rocketing through the atmosphere.

1. Knowing what you do about gravity, how do you think the rocket manages to lift off?

2. Can you think of some other forces the rocket must overcome before it can be launched?

3. Name some ways that gravity affects you.

L1

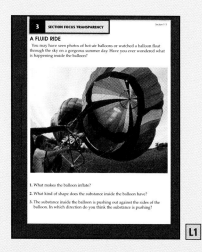

3 SECTION FOCUS TRANSPARENCY Section 1-3

A FLUID RIDE

You may have seen photos of hot-air balloons or watched a balloon float through the sky on a gorgeous summer day. Have you ever wondered what is happening inside the balloon?

1. What makes the balloon inflate?

2. What kind of shape does the substance inside the balloon have?

3. The substance inside the balloon is pushing out against the sides of the balloon. In which direction do you think the substance is pushing?

L1

Teaching Transparencies

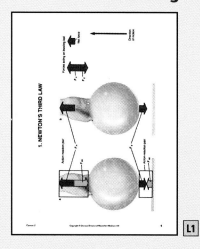

1. NEWTON'S THIRD LAW

L1

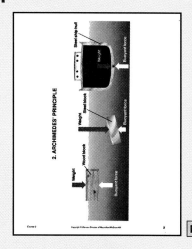

2. ARCHIMEDES' PRINCIPLE

L2

HANDS-ON LEARNING

Science Discovery Activity*

ACTIVITY 1-1

Slip Sliding Away

Everyone seems to like sliding down hills. You can see people trying it in cold weather with skis and sleds. In hot weather, you see people having fun on slick water slides. Did you ever watch people on a water slide? What are some factors that affect sliding?

Getting Started

It's difficult to slip on a level surface unless it's very smooth or icy. You slide faster and farther if the surface is sloped. That's why ski runs and sliding boards are sloped. Think about what will happen if you put some boxes on a ramp and then begin to raise one end of the ramp. Will all the boxes start to slip at the same time? Will a lightweight box slip before a heavier one? Will a rough box slip less than a smooth one?

Hypothesize

An object's weight and surface texture seem to be important factors in determining if it slips on a raised ramp. Write a hypothesis about how you think the steepness of a ramp at which a box just begins to slip down the ramp is related to the box's weight. Then write one relating the ramp's steepness to the box's surface texture.

Try It!

You may want to use the following materials:
- ramp
- books
- sand
- glue
- metric ruler
- small cardboard box with cover
- sandpaper

1. What variable is the independent variable in your first hypothesis? What is the independent variable in the second? How are you going to measure them? How will you change them?

5

L1

Laboratory Manual*

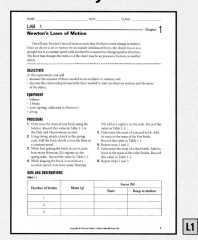

NAME _____ DATE _____ CLASS _____

LAB 1 Chapter **1**

Newton's Laws of Motion

One of Isaac Newton's laws of motion states that all objects resist change in motion. Once an object is set in motion by an outside unbalanced force, the object moves in a straight line at a constant speed until another force causes it to change speed or direction. The force that changes the motion of the object may be air pressure, friction, or another object.

OBJECTIVES

In this experiment, you will
- measure the amount of force needed to set an object in motion, and
- discover the relationship between the force needed to start an object in motion and the mass of the object.

EQUIPMENT
- balance
- 3 bricks
- scale (spring, calibrated in Newtons)
- string

PROCEDURE

1. Determine the mass of one brick using the balance. Record this value in Table 1-1 in the Data and Observations section.
2. Using string, attach a brick to the spring scale. Pull the brick slowly across the floor at a constant speed.
3. While first getting the bricks to move, note how many Newtons (N) register on the spring scale. Record the value in Table 1-1.
4. While keeping the brick in motion at a constant speed, note how many Newtons

(N) of force register on the scale. Record the value in Table 1-1.
5. Determine the mass of a second brick. Add its mass to the mass of the first brick. Record this value in Table 1-1.
6. Repeat steps 2 and 3.
7. Determine the mass of a third brick. Add its mass to the mass of the other bricks. Record this value in Table 1-1.
8. Repeat steps 2 and 3.

DATA AND OBSERVATIONS

TABLE 1-1

Number of bricks	Mass (g)	Force (N)	
		Start	Keep in motion
1			
2			
3			

L1

Take Home Activity

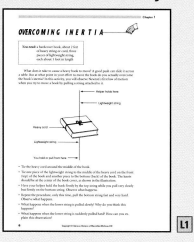

Chapter 1

OVERCOMING INERTIA

You need: a hardcover book, about 2 feet of heavy string or cord, three pieces of lightweight string, each about 1 foot in length

What does it take to cause a heavy book to move? A good push can slide it across a table. But at what point in your effort to move the book do you actually overcome the book's inertia? In this activity, you will observe Newton's first law of motion when you try to move a book by pulling a string attached to it.

- Helper holds here
- Lightweight string
- Heavy cord
- Lightweight string
- You hold or pull from here

- Tie the heavy cord around the middle of the book.
- Tie one piece of the lightweight string to the middle of the heavy cord on the front (top) of the book and another piece to the bottom (back) of the book. The knots should be at the center of the book cover, as shown in the illustration.
- Have your helper hold the book firmly by the top string while you pull very slowly but firmly on the bottom string. Observe what happens.
- Repeat the procedure, only this time, pull the bottom string fast and very hard. Observe what happens.
- What happens when the lower string is pulled slowly? Why do you think this happens?
- What happens when the lower string is suddenly pulled hard? How can you explain this observation?

6

L1

*There may be more than one activity for this chapter.

Chapter 1 Forces and Pressure

Study Guide*

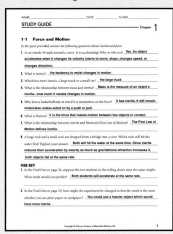

Concept Mapping

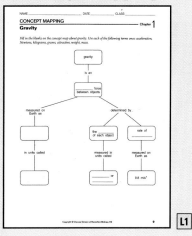

Critical Thinking/ Problem Solving

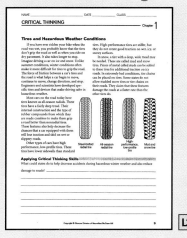

Integrating Sciences

Across the Curriculum

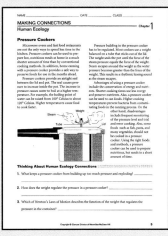

Technology and Society

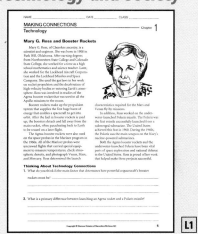

Multicultural Connection**

Performance Assessment

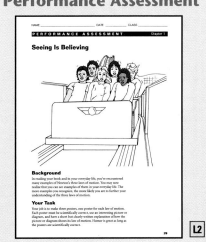

Review and Assessment

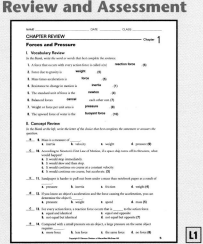

*One per section **Two per chapter

Forces and Pressure

THEME DEVELOPMENT

The themes that this chapter supports are systems and interactions and stability and change. Newton's Laws of Motion deal with the interactions of forces and objects, which cause motion. The first law relates to inertia. The second law explains the relationship among mass, acceleration, and force. The third law involves pairs of forces.

CHAPTER OVERVIEW

The first section of the chapter explains the relationship between inertia and mass using Newton's First Law of Motion.

The chapter also covers Newton's Second and Third Laws of Motion.

The chapter concludes with a discussion of pressure and buoyancy.

Tying to Previous Knowledge

Hold a pencil between two fingers, then drop it. Ask volunteers to tell what forces held the pencil up and what forces caused it to drop. *Students should be able to recognize gravity as the force that made the pencil fall.* Explain to students that they will be studying forces and their effects on motion.

INTRODUCING THE CHAPTER

Ask students if they have ever watched a juggler or a magician. Students can describe any motion they observed.

`00:00` OUT OF TIME?

If time does not permit teaching the entire chapter, use the Chapter Overview on this page, Reviewing Main Ideas at the end of the chapter, and the Chapter 1 audiocassette to point out the main ideas of the chapter.

FORCES & PRESSURE

Did you ever wonder...

✓ **Why a revolving door continues to turn after the last person has left?**
✓ **Why snowshoes help you walk on top of snow?**
✓ **Why you float in water?**

Science Journal

Before you begin to study forces, think about these questions and answer them *in your Science Journal*. When you finish the chapter, compare your journal write-up with what you have learned.

Answers are on page 49.

A tightrope walker moves across a rope high overhead. Dogs stand on trotting horses. Elephants perform amazing feats of balance. Jugglers spin their rings, and clowns keep everyone laughing. You can't decide where to look first. You're afraid you'll miss something if you look away for a second.

As amazing as these circus acts are, they're all based on basic principles of motion. These same principles apply to things that go on around you each day. For example, the grocery cart you push continues to roll down the aisle by itself. You slam your locker door, and it flies back at you.

▶ *In this chapter, you'll learn the laws of motion. You'll also see what makes objects move, and what makes them change their motion. Start your exploration on the next page.*

20

Learning Styles	**Kinesthetic**	Explore, p. 21; Find Out, p. 26; Activity, pp. 30, 36, 45; Design Your Own Investigation, pp. 32-33; Visual Learning, p. 37
	Visual-Spatial	Multicultural Perspectives, p. 27; Visual Learning, pp. 27, 31, 42; Activity, pp. 31, 37, 38, 45; Demonstration, pp. 34, 37, 42; Find Out, p. 44; Research, p. 45
	Logical-Mathematical	Investigate, pp. 24-25; Find Out, pp. 29, 41; A Closer Look, pp. 34-35; Explore, pp. 36, 39; Discussion, p. 38; Visual Learning, p. 40
LS	**Linguistic**	Activity, p. 22; Across the Curriculum, p. 23; Multicultural Perspectives, pp. 23, 40; Visual Learning, p. 28; Debate, p. 28; Research, p. 30; Discussion, pp. 37, 39; Life Science Connection, pp. 42-43; Project, p. 49

Explore! ACTIVITY

Why doesn't the penny move?

Have you ever seen a magician pull a tablecloth from a table and leave all the items on the table standing? Here's your chance to do a similar sort of trick.

What To Do

1. Lay an index card over the top of a glass or beaker.

2. Place a penny on the card, centered over the glass.

3. With a flick of your finger, give the card a quick horizontal push. What happens?

4. *In your Science Journal*, record your observations and your explanation for what you saw.

21

PREPARATION

Planning the Lesson

Refer to the Chapter Organizer on pages 20 A–D.

Concepts Developed

This section explores Newton's First Law of Motion, which states that a body at rest or in uniform motion tends to remain so unless it is acted on by a force. The section also deals with the forces of friction and gravity.

1 MOTIVATE

Bellringer

Before presenting the lesson, display **Section Focus Transparency 1** on an overhead projector. Assign the accompanying **Focus Activity** worksheet. **L1**

LEP

Activity Ask students to recall the sensation of rides in amusement parks. Have them describe the motion of some of the rides. Explain that the effects of motion they may have experienced, such as pushes they felt during rapid turning or dropping from a height, are related to inertia, which they will study in this section. **L1**

2 TEACH

Tying to Previous Knowledge

Ask students to recall the common sensation of moving forward in a car when the brakes are applied quickly. Their bodies continue to move forward due to inertia.

Student Text Question

What's happening to make you feel as if something is pushing you? *As the car (and the seat) speed up, you tend to remain at rest and the back of the seat pushes against your back.*

1-1 Force and Motion

Section Objectives

- Relate inertia to mass.
- Identify the forces acting when objects interact.
- Describe and use Newton's First Law of Motion.

Key Terms

inertia, force

Keep on Moving

Picture yourself in a car on your way to the grocery store. You're waiting at a stoplight. When the light turns green, the car moves forward. You feel like you're being pushed back into the seat. When the car slows to a stop, you feel pushed forward. And when the car turns a corner, you feel as if you're being pushed outward. What's happening to make you feel as if something is pushing you?

Velocity, Acceleration, and Motion

You may have already learned about velocity and acceleration. Velocity is how fast an object is going in a given direction. An acceleration results in a change in either the speed of an object or its direction of travel.

When you travel in a car, you may feel pushes and pulls as a result of speeding up, slowing down, or turning a sharp curve. Where do these

Figure 1-1

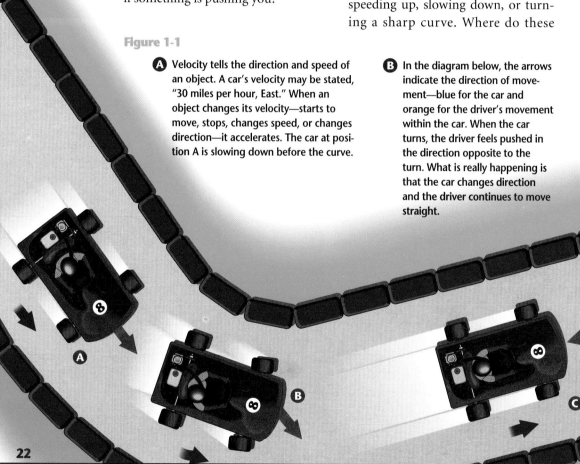

A Velocity tells the direction and speed of an object. A car's velocity may be stated, "30 miles per hour, East." When an object changes its velocity—starts to move, stops, changes speed, or changes direction—it accelerates. The car at position A is slowing down before the curve.

B In the diagram below, the arrows indicate the direction of movement—blue for the car and orange for the driver's movement within the car. When the car turns, the driver feels pushed in the direction opposite to the turn. What is really happening is that the car changes direction and the driver continues to move straight.

22

Program Resources

Study Guide, p. 7
Take Home Activities, p. 6, Overcoming Inertia **L1**
Science Discovery Activities, 1-1
Section Focus Transparency 1

feelings come from? Examine **Figure 1-1** to discover more about your motion in a car.

■ Inertia

When you experience pushes and pulls associated with acceleration, something very simple is happening. You continue the motion, or lack of motion, that you had before the acceleration began.

The tendency to resist changes in motion is **inertia**. The Explore activity at the beginning of the chapter showed an example of inertia. Both the card and the coin were at rest until you flicked the card and caused it to accelerate horizontally. Why didn't the coin accelerate with the card?

The property of inertia can give you useful information about objects. Investigate how on the following pages!

The word *inertia* **comes from a Latin word that means "idle or lazy." Objects seem lazy because they don't easily change the way they move.**

C As the car speeds up, the driver tends to keep moving at the previous speed. The driver feels pushed back into the seat. Actually, the back of the seat catches up and pushes against the slower-moving driver.

D Once the velocities of car and driver are the same, the driver feels no push or pull until the car again changes speed or direction.

E If the car stops suddenly, the driver feels pushed forward. Even though the forward motion of the car stops, the forward motion of the passenger continues. What might happen if you were on a slippery seat with no back and your car suddenly accelerated forward?

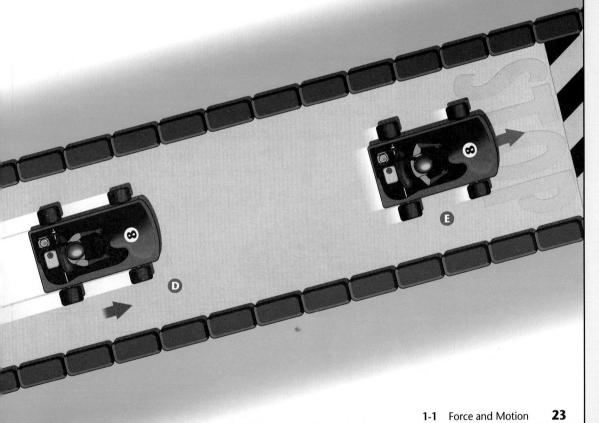

Language Arts

Have students keep logs in their journals of how terms such as *force*, *pressure*, and *motion* are used in other classes and in conversation. At the end of the chapter, have them compare and contrast the use of these words in science with their use in other contexts. **L1**

Uncovering Preconceptions

Because they have felt the tendency to slide sideways in the seat when the car rounds a curve, many people think that inertia would put them on a path perpendicular to the curve. Have students look at Figure 1-1B. Point out that inertia would cause the passenger to continue moving in the direction he was moving prior to the curve.

Student Text Questions

Why didn't the coin accelerate with the card? *The force was applied to the card only. Inertia kept the coin from moving.* **What might happen if you were on a slippery seat with no back and you suddenly accelerated forward?** *You remain at rest and the car drives out from under you—you'd fall off the seat.*

GLENCOE TECHNOLOGY

 Videodisc

STVS: Physics

Disc 1, Side 1
Science of Bowling (Ch. 2)

Safer Roads (Ch. 3)

New Skid Control (Ch. 4)

Multicultural Perspectives

Timely Measures

Until electric or battery-operated clocks were available, time was often measured by mechanical means. Early Chinese clocks (1200s) were driven by fluids flowing back and forth between vessels. Early European clocks (1300s) were driven by weights that dropped by the force of gravity. Later, in the 1600s, grandfather clocks with pendulums and counterweights were used. Students can research how one of these early mechanical timepieces worked and explain it to the class.

1-1 Measuring Inertial Mass

Planning the Activity

Time needed 40 minutes

Purpose To observe the relationship between mass and acceleration.

Materials See student activity.

Process Skills making and using tables, predicting, measuring in SI, making and using graphs, inferring

Preparation You may wish to set up the balance and materials before class.

Teaching the Activity

Process Reinforcement To reinforce observation and recording skills, have students carefully read all the steps of the procedure before they begin the activity. Have students define acceleration. *Acceleration is any change in motion of an object.*

Students should label their data tables and graph axes correctly. If students have any difficulty, go over the mechanics of using tables and graphs in the Skill Handbook. L1

Troubleshooting Check to make sure that masses are securely attached to the balances before the platform is released.

• Remind students to start their timer at the instant the platform is released, to count the number of cycles carefully, and to stop their timers at the end of 10 complete periods.

INVESTIGATE!

Measuring Inertial Mass

You can find the mass of an object by measuring the way an object speeds up or slows down, and comparing that with the way known masses speed up or slow down.

Problem
How can you find the mass of an object anywhere in the universe without a laboratory balance or a force of gravity?

Materials
inertial balance	pan balance
clamp	second timer
heavy rubber bands	graph paper
object of unknown mass	assorted marked masses

Safety Precautions

Be careful handling the inertial balance. Metal may have sharp edges.

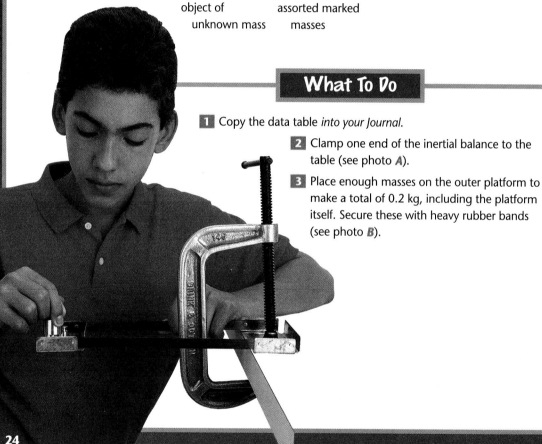

What To Do

1 Copy the data table *into your Journal.*

2 Clamp one end of the inertial balance to the table (see photo **A**).

3 Place enough masses on the outer platform to make a total of 0.2 kg, including the platform itself. Secure these with heavy rubber bands (see photo **B**).

24

Program Resources

Activity Masters, pp. 7–8, Investigate 1-1

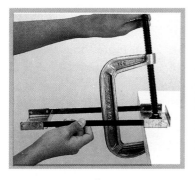
A

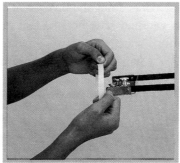
B

4 Pull the outer platform sideways and let go. What happens? Measure the time for 10 complete back-and-forth cycles. Record your data and observations *in your Journal.*

5 Repeat Steps 3 and 4 two more times, with total masses of 0.4 kg and 0.6 kg.

6 Attach the unknown mass to the platform and measure the time for 10 complete cycles. Record.

Sample data

Data and Observations		
Mass	Time For 10 Cycles	Period
0.2 kg	Results obtained will	
0.4 kg	depend on	
0.6 kg	equipment used.	
Unknown mass		

Analyzing

1. The period of a cycle is the time it takes for one complete back-and-forth motion. *Calculate* the period for each of your four masses.

2. *Plot a graph* of the results with three known masses. Plot period on the horizontal axis and mass on the vertical axis. Connect the points with a smooth curve.

3. Find the period of the unknown mass on the graph. *Predict* what the mass should be.

4. Subtract the mass of the platform from the mass you predicted. This will give you the mass of the unknown object.

5. *Measure* the unknown mass on a pan balance to check your results.

Concluding and Applying

6. What trends do you observe in the relationship between the mass and the period?

7. How did your prediction compare with the actual mass?

8. A period of 0.75 seconds on this inertial balance would indicate how much mass?

9. **Going Further** What would be the period of an object with a mass of 5 kg?

Expected Outcomes

Students should observe that the period of the inertial balance is directly related to the mass of the object. This knowledge can be used along with careful graphing and recording to predict the periods of unknown masses and to find the mass of an object.

Answers to Analyzing/ Concluding and Applying

1.–5. Results will vary depending on students' observations and recorded data.

6. The greater the mass, the longer the period.

7. Answers will vary depending on students' predictions.

8.–9. Answers depend on the specific balance used. Since students used the balance to determine the information plotted on their graphs, they can determine the answers to these questions from their graphs.

✔ Assessment

Process Have students work in cooperative groups to design an investigation in which a spring or an elastic band is used in place of the inertial balance. Once the design has been completed and approved, have them carry out the investigation and compare their results with those achieved with the inertial balance. Use the Performance Task Assessment List for Designing an Experiment in **PASC,** p. 23.

What factors affect the acceleration of objects?

Time needed 15 minutes

TECH PREP

Materials two rolling chairs

Thinking Processes observing and inferring, comparing and contrasting

Purpose To discover the relationship among mass, inertia, and acceleration.

Teaching the Activity

The rolling chairs used in this activity should be identical. The directions tell students to form teams based on their relative weights. Because inertia depends on mass, students are actually comparing the effect of mass on acceleration. However, students at this level have little experience comparing how "massive" two objects are, so "weight" is used instead.

Science Journal Have students compare and contrast results they achieved throughout the activity. **L1**

Expected Outcome

Students should observe that the lighter student accelerates faster and travels farther than the heavier student.

Conclude and Apply

1. no

2. The heavier person will accelerate the least.

3. There is a feeling of being pushed backward into the chair.

4. Your partner pushes the chair and it pushes you.

✔ Assessment

Performance Have students perform a similar activity in which a single student sits in a rolling chair and pushes against a wall. Have students compare and contrast their results with those attained during the regular activity. Use the Performance Task Assessment List for Carrying Out a Strategy and Collecting Data in **PASC**, p. 25.

Mass and Inertia

While doing the Investigate activity you may have noticed that mass is more than how much matter an object has. Mass is also a measure of an object's inertia—how much it resists changes in motion. Let's find out more about the inertia of objects as they interact.

Find Out! ACTIVITY

What factors affect the acceleration of objects?

You know that mass and inertia are related. How does mass relate to acceleration?

What To Do

1. Team up with another student who is either much lighter or much heavier than you are. Find two rolling chairs.

2. You sit in one and have your partner sit in the other.

3. Bring your chairs close together. Place your palms against those of your partner. Then gently push each other away.

4. Which chair moves away with the greater final velocity? Which chair is given the greater acceleration? How does the direction of your acceleration compare with that of your partner?

5. Record your observations and the answers to these questions *in your Journal.*

6. Repeat this activity while both of you are pushing and then when only one of you is pushing.

Conclude and Apply

1. Are there any differences in the resulting accelerations for these different ways of pushing each other?

2. How does the acceleration of each chair compare with how heavy each of you is?

3. Turn your chair around and let your partner push on the back of your chair, rather than pushing against your hands. How is what you experience different from when you pushed hand to hand?

4. What causes you to accelerate in this case?

Program Resources

Laboratory Manual, pp. 1–2, Newton's Laws of Motion **L1**

Critical Thinking/Problem Solving, p. 9, Tires and Hazardous Weather Conditions **L2**

Making Connections: Across the Curriculum, pp. 5–6, Pressure Cookers **L1**

Meeting Individual Needs

Learning Disabled Have students do this activity to see whether they can detect evidence of movement without seeing motion. Have students close their eyes while one student moves something in the classroom. Then have students open their eyes and look for a change. Have the group focus on the senses they are using. Are they using just their eyes, or other senses, such as hearing, as well?

You've already learned that objects with large mass have greater inertia than objects with small mass. In the Find Out activity you discovered that objects with large mass accelerate less than objects with small mass when acted on by the same force. From this information you can conclude that objects with great inertia (large mass) have a great resistance to acceleration.

■ Newton's First Law

If you had done the Find Out activity on a surface that had no resistance to motion, how would your motion have been affected? Would you have gone farther? As it turns out, under those conditions, you might go on forever.

As a result of a thought experiment, Galileo concluded that objects at rest remain at rest and objects in motion remain in motion. Sir Isaac Newton agreed, except when objects are acted on by a push or pull. Newton's First Law of Motion is that an object remains in motion or at rest unless acted on by a push or pull. One example of this law is that a soccer ball sitting on flat ground will not begin to roll until you apply a push by kicking it.

Figure 1-2

You need a lot of force to start a big truck moving, but less force to start to move a small car. This is due to the relationship between the mass of an object and its inertia.

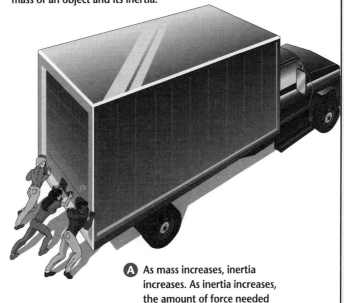

A As mass increases, inertia increases. As inertia increases, the amount of force needed to accelerate the object also increases.

B The opposite is also true. As an object's mass decreases, its inertia decreases, and the force needed to cause it to accelerate decreases.

C When an object accelerates, its velocity changes. To change velocity, a force must act on an object. Which of the three vehicles pictured will require the most force to start moving? To stop moving? Why?

Visual Learning

Figure 1-2 Ask students to compare and contrast the three situations shown in the illustration. Help students use prior experiences to understand the relationship among mass, inertia, and force.
Which of the three vehicles pictured will require the most force to start moving? To stop moving? Why? *The truck in both cases; it has the greatest mass and therefore the greatest inertia.*

NATIONAL GEOGRAPHIC SOCIETY

Videodisc

Newton's Apple: Physical Sciences

Newton's Laws
Chapter 3, Side B
Inertia: Newton's first law of motion

26249-29476

Gravity acts on moving objects

29506-33797

Acceleration: Newton's second law of motion

33797-36846

Action and reaction: Newton's third law of motion

37320-39243

Reviewing Newton's laws of motion

39339-41694

Moving Heads

Multicultural Perspectives

Show students a photograph of the colossal heads at Easter Island. The Dutch explorers who first saw the island on Easter Sunday 1722 gave the island this name. Its Polynesian inhabitants call the island Rapa Nui. The stone heads were carved by prehistoric inhabitants. No one knows how the heads were moved into place at a time when no machines existed that could have lifted them. Students can brainstorm ways these massive heads might have been moved. According to local legend, small stones were used as rollers beneath the stones, which were then dragged into place with ropes.

Interactions That Push or Pull

When you sat in the rolling chair and pushed on your partner's hands, each of you pushed the other. You both accelerated in opposite directions. In each case the person with less mass (less inertia) experienced greater acceleration. A push or pull is commonly referred to as a **force**. Let's see how forces are involved in motion.

Figure 1-3

Friction is a force that resists motion between two objects in contact, such as the soles of a runner's shoes and the running surface.

A Ice does not produce enough friction between itself and the shoe soles for the runner to get a firm, controlled push-off.

B A sticky surface produces too much friction for efficient running.

C A dry surface produces the right amount of friction between shoe soles and surface for efficient running.

When you pushed your classmate in the rolling chair, what caused you to accelerate—the force you used to push, or the force your classmate exerted on you? Can you push on yourself and make your body speed up? No, something else must push on you. When you walk on the floor, it is the floor pushing back on your feet that makes you move. You must have a force pushing on you, and then you will accelerate in the direction of that force. So, when you pushed your classmate and he or she pushed back, it was the force he or she exerted on you that caused your acceleration.

■ Friction

Earlier in the chapter, you thought about what would happen if you were in a car with a slippery plastic seat and no seat back. Let's take that one step further. Imagine a block of ice on a flatbed truck.

The truck is at rest and the block of ice is right behind the driver. What will happen when the truck starts up? What force is needed for the ice to accelerate with the truck? There must be an interaction between the ice and the truck bed. That interaction is friction. Friction is necessary when we want one object to accelerate with another one, as with the ice on the flatbed truck. At other times, when we want something to keep moving, friction slows it down. But is friction a force—a push or a pull? Let's find out.

Is friction a force?

Friction is an important aspect of everyday life. What might be some important aspects of friction?

What To Do

1. Place a sheet of plain white paper on a flat surface.

2. Set a 20-g mass on the paper about 7 cm from one end.

3. Grip the other end of the paper and give it a smooth, quick pull.

4. Now repeat the procedure replacing the white paper with a sheet of coarse sandpaper, rough side up.

5. Record your observations *in your Journal.*

Conclude and Apply

1. Compare and contrast what happens to the 20-g mass with each piece of paper.

2. Use Newton's First Law of Motion to explain your observations.

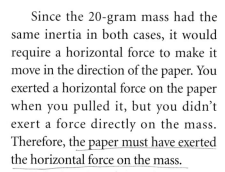

Since the 20-gram mass had the same inertia in both cases, it would require a horizontal force to make it move in the direction of the paper. You exerted a horizontal force on the paper when you pulled it, but you didn't exert a force directly on the mass. Therefore, the paper must have exerted the horizontal force on the mass.

Friction is a force that resists motion between two objects in contact. The force of friction between the paper and the weight was not large, so the weight slid off the paper. However, the force of friction resisted motion between the sandpaper and the weight. As a result, the weight moved with the sandpaper.

What if there were no friction? Your feet wouldn't be able to grip the ground. You wouldn't accelerate unless something bumped into you

and started you going. Once you were moving, you'd keep moving at a constant speed in the same direction until you interacted with something else.

■ Gravity

Another force that results from the interactions of two objects is gravity. Like friction, gravitational interaction is so common that you usually ignore it. You know that free-falling objects accelerate downward at about 9.8 meters per second per second (9.8 m/s^2). This acceleration is due to gravitational force, which pulls you toward Earth's center.

When a meteor is falling, Earth exerts a force that accelerates the meteor downward. Does the meteor cause Earth to accelerate upward, too?

You can see that if the mass of the object you are interacting with is

DID YOU KNOW?

In the English system, the pound is the unit of force and the unit of acceleration is feet/second². What is the unit of mass? Believe it or not, it's called a slug! One slug = 14.6 kg.

Meeting Individual Needs

Learning Disabled A falling object accelerates at 9.8 m/s² unless a force opposes the downward motion. Have a parachute-making contest. Students can experiment with a variety of materials, sizes, and shapes. Students should use similar objects to weight the parachutes. Have them determine which parachutes kept the weight aloft the longest by causing the greatest air resistance. L1

Is friction a force?

Time needed 10 minutes

Materials sheet of white paper, 20-g mass, sheet of coarse sandpaper

Thinking Processes observing and inferring, comparing and contrasting, forming operational definitions

Purpose To observe that friction is a force that resists motion between two surfaces in contact with each other and its size (magnitude) depends on the nature of the surfaces.

Teaching the Activity

Science Journal Students can compare and contrast the surfaces of the white paper and the sandpaper in a paragraph in their journals. Have them hypothesize how the differences in the two surfaces affect the force of friction between each type of paper and the 20-g mass. L1

Conclude and Apply

1. With the smooth paper, the mass stays in place. With the sandpaper, the mass moves in the direction that the paper moves.

2. The 20-g mass had the same inertia in each case. Therefore, it should have remained at rest. To make it move horizontally, there must have been a horizontal force on it. That force came from the friction between the mass and the sandpaper.

✔ Assessment

Performance Have students extend the activity by testing different kinds of paper, such as newspaper, waxed paper, paper toweling, and different grades of sandpaper. Use the Performance Task Assessment List for Making Observations in **PASC**, p. 17.

Check for Understanding

Ask students questions 1–3 and the Apply question in Check Your Understanding. Discuss students' answers to the Apply question. If they have difficulty responding, suggest that they consider whether speed or safety is more desirable in different activities. Use the Performance Task Assessment List for Group Work and the Individual in **PASC**, p. 97.

Reteach

Activity Use two identical toy cars to demonstrate Newton's First Law of Motion. Tape a weight to one of the cars. Students can gently push both cars from a standstill with equal force. Ask which car is more difficult to move. Students can discuss how the motion of the cars represents the principle of inertia discussed in this section. L1

Extension

Research Ask students to explore substances that are used to increase friction to provide greater force. Some possibilities are rosin that dancers use on their shoes and rosin used by violinists to increase friction between the bow and strings. L3

4 CLOSE

Students can explain why some people have the feeling that they are falling as they step off at the top of an escalator. *The person was moving steadily upward on a diagonal path. When the person steps off, the path of motion becomes horizontal. The loss of upward motion feels like downward motion.* L1

Figure 1-4

A All objects exert a gravitational force that pulls on surrounding objects. For example, as Earth pulls a 1 kilogram meteor toward its center, the meteor, at the same instant, pulls Earth toward the meteor's center.

1 kg

Acceleration
9.8 m/s²

Acceleration
0.000 000 000 000 000 000 000 001 6 m/s²

6,000,000,000,000,000,000,000,000 kg

much larger than your own mass, the object's acceleration would be so small you couldn't even measure it. In the same way, the gravitational interaction between the meteor and Earth, shown in **Figure 1-4**, is such that only the meteor appears to accelerate.

As objects fall, more massive objects are more strongly attracted to Earth, and for that reason you would expect them to fall more quickly. On the other hand, they are also harder to accelerate. Their inertia reduces their acceleration by exactly as much as their greater attraction increases it. As a result objects fall at the same rate unless air resistance slows them down, as in the case of a feather or a single flat sheet of paper.

In the next section, we'll discover how mass, force, and acceleration are related. They are a part of everything you do and everything that happens around you.

B Earth's mass causes objects, including the meteor to accelerate toward Earth's center at about 9.8 meters per second per second. The meteor causes Earth to accelerate toward it at about 0.000 000 000 000 000 000 000 001 6 meters per second per second. Because the accelerations are so different, only the acceleration of the meteor can be observed.

check your UNDERSTANDING

1. What determines an object's inertia?
2. Describe two forces that don't depend upon physical contact.
3. Briefly state and give an example of Newton's First Law of Motion.
4. **Apply** Describe two activities where you need frictional forces and two where you would like smaller frictional forces.

check your UNDERSTANDING

1. Its mass—greater mass, greater inertia.
2. Magnetic attraction between opposite poles and electric attraction between opposite charges; also, gravitational attraction
3. An object tends to continue in its state of rest or uniform motion unless it interacts with another force. One example is a book lying on a table. It lies still unless a force sets it in motion. Then it continues in motion until stopped by some other force.
4. **Apply** You need friction to ride a bicycle and to drive a car. You want less friction to ice-skate or roller-skate.

1-2 How Forces Act on Objects

Acceleration, Mass, and Force

You know that if you throw a ball harder it moves faster. From your experience, you also know that a car is more difficult to push than a bicycle. How is the force you use related to an object's acceleration? Suppose you have two objects and each object can move. What kind of relationship might exist between the acceleration and forces of the two objects?

Isaac Newton was one of the first to describe it. His observations led him to develop his second law of motion. You can investigate this law for yourself in the activity that follows.

Section Objectives

■ Describe the relationships between force, mass, and acceleration.

■ Use Newton's second law to predict acceleration.

■ Identify pairs of forces.

■ Describe the difference between balanced forces and action-reaction forces.

Key Terms

newton, balanced forces, weight, action force, reaction force

Figure 1-5

A At any given time several forces may act upon an object. When there is an unbalanced force in one direction, the object accelerates.

B In these pictures the same soccer ball is kicked with different forces. Which of the following is the same in both pictures: the mass, the force, the acceleration? Which is different? What is the result?

⌐Program Resources⌐

Study Guide, p. 8
Concept Mapping, p. 9, Gravity L1
Multicultural Connections, p. 5, Frozen Domes; p. 6, Ancient Builders L1
Science Discovery Activities, 1-2
Section Focus Transparency 2
Teaching Transparency 1

Visual Learning

Figure 1-5 **Which of the following is the same in both pictures: the mass, the force, the acceleration? Which is different? What is the result?** *The mass is the same; the force in B is greater, resulting in a greater acceleration of the ball.*

Planning the Lesson

Refer to the Chapter Organizer on pages 20A–D.

Concepts Developed

This section describes the relationships among force, mass, and acceleration; identifies pairs of forces; and describes the differences between balanced forces and action-reaction forces.

1 MOTIVATE

Bellringer

🔦 Before presenting the lesson, display **Section Focus Transparency 2** on an overhead projector. Assign the accompanying **Focus Activity** worksheet. L1 LEP

Activity   Take a walk outside. Point out equipment such as the swings, seesaw, and glider that take advantage of the First and Second Laws of Motion in their design. Explain that you will be investigating some of the principles that make the equipment work. L1

2 TEACH

Tying to Previous Knowledge

Newton's Second Law of Motion builds on the First Law of Motion, that objects at rest tend to remain at rest and objects in motion remain in motion unless acted on by a force.

1-2 Acceleration and Mass

Preparation

Purpose To determine the relationship between the mass and acceleration of an object through predicting and verifying that prediction in an experiment.

Process Skills predicting, making and using tables, observing, designing an experiment, forming a hypothesis, separating and controlling variables, interpreting data

Time Required One class period

Materials See reduced student text. If you do not have spring carts, try to borrow them from a high school or rig them from skateboards, a compression spring, and string.

Possible Hypotheses Students should hypothesize that the same force accelerates a small mass more than a heavier mass.

The Experiment

Process Reinforcement In this experiment, it is important to measure variables and results accurately. Carts should be weighed on the pan balance more than once for accuracy. Before the experiment begins, students should try a few trial runs with the carts to find out how to work them, how best to measure the distance they go, and how to coordinate keeping time constant.

Possible Procedures Measure and record the masses of Carts 1 and 2. Run a trial with the carts to see how far they go. Double the mass on one of the carts and run another trial. Observe and record the position of the slower cart when the faster one hits a book. Then predict where to start the carts so they will hit the book at the same time. Discuss how students will measure distance in each trial of the experiment.

Acceleration and Mass

When a car speeds up, we say it accelerates. That means its velocity changes. The car moves through a greater distance during each second. In this Investigation, you will find out how objects of different masses accelerate.

Preparation

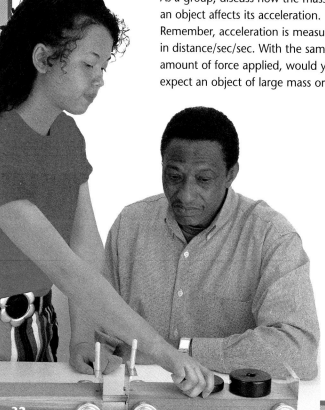

Problem
What is the relationship between the mass and acceleration of an object?

Form a Hypothesis
As a group, discuss how the mass of an object affects its acceleration. Remember, acceleration is measured in distance/sec/sec. With the same amount of force applied, would you expect an object of large mass or small mass to move farther? As you discuss this, think about pushing a car or a bicycle. Which one can you move farther?

Objectives
• Measure distance traveled in the same period of time by objects of different masses.

• Relate mass to acceleration.

Materials
spring carts
2 books
meterstick
masking tape
balance pan with 5 kg capacity
masses of 1 kg and less

Program Resources

Activity Masters, pp. 9–10, Design Your Own Investigation 1-2

Making Connections: Integrating Sciences, p. 5

Making Connections: Technology and Society, p. 5

Meeting Individual Needs

Learning Disabled Have students use a rubber band to accelerate a Ping-Pong ball across a tabletop. Students can predict how the motion of the Ping-Pong ball would change if a longer or thicker rubber band were used. Students can test their predictions and draw conclusions. *They should observe that increased force increases the acceleration of the same mass.* L1 LEP

Plan the Experiment

1 Before planning the experiment, try out the spring carts and see how they work. What force provides the acceleration?

2 How can you design an experiment to test your hypothesis? You need to plan your experiment so that you have only one variable in each test. How will you keep time constant? Is the force constant? What will you change in each test? How will the results of each test change?

3 After you have taken all these points into consideration, write a procedure for testing your hypothesis.

4 Design a data table *in your*

Science Journal. What will you measure? What data will you put in the table?

Check the Plan

1 As you can see from the photo on this page, it takes a few people to run the experiment. Does everyone know what they are to do?

2 How do you plan to measure the distance the cars travel?

3 Before you proceed with your experiment, make certain your teacher has approved your plan.

4 Carry out your investigation and record your measurements in your data table.

Analyze and Conclude

1. Analyze Was your hypothesis supported? Explain.

2. Compare and Contrast Compare the mass of each cart to the distance it traveled. State the relationship.

3. Analyze What force accelerated each cart? Was it always the same?

4. Analyze What would change in your investigation if the carts traveled the distances in different amounts of

time? Would the experiment work?

5. Predict From your data, predict how far a cart twice as heavy as the heaviest cart would go in the same period of time.

How have the students kept time constant? Use masking tape to mark the starting point of each experiment and where to place the books.

Teaching Strategies

Science Journal Students need to prepare tables in their Science Journals. Trials should run downward and mass of cart 1, distance cart 1, mass of cart 2, distance cart 2 and comparison should run across the top.

Expected Outcome

Students will observe that with a constant force, acceleration is inversely related to mass, or stated more simply, a constant force on an object of small mass produces greater acceleration than the same force on an object of greater mass.

Analyze and Conclude

1. Answers may vary, but students should come up with the expected outcome.

2. Mass and acceleration are inversely related. When mass goes up, acceleration goes down.

3. The force of the spring pushes the carts apart. It is the same for both carts.

4. You would have two variables—both distance and time. An experiment with two variables does not work.

5. It would go half as far.

Going Further

Going Further

If you rolled two miniature cars down a ramp, which would be faster, the lighter car or the heavier? What provides the constant force on the cars? Explain your answers.

Both cars should accelerate at the same rate because the acceleration due to gravity is the same for both cars.

✔ Assessment

Process Have students conduct a similar investigation in which the force—the size of the spring—is varied while the mass remains constant. Ask students to predict the outcome, then experiment to check their predictions. Use the Performance Task Assessment List for Formulating a Hypothesis in **PASC**, p. 21.

Demonstration Show the effect of mass on acceleration with a constant force.

Materials needed are a flexible plastic ruler, a golf ball, and another small ball of a different mass.

Bend a plastic ruler and hold it next to a golf ball. Release the ruler. Have students observe the motion of the ball. Repeat the demonstration using the second ball. Be certain to bend the ruler with the same amount of force used on the golf ball. Have students observe and compare the accelerations and the forces.

Teacher F.Y.I.

Vehicles traveling on an icy roadway need three to seven times more room to stop than they would on a dry road—a good reason to keep plenty of distance between your car and other vehicles.

Figure 1-6

1 newton of force = 1 kg mass × 1 m/s² acceleration

1 kg

1 m/s²

The standard unit of force is the newton. One newton is equal to the amount of force required to accelerate a mass of 1 kilogram at 1 meter per second per second.

■ Force and Newton's Second Law of Motion

If you had been able to measure the acceleration, the mass, and the force applied to each of the two carts in the Investigate activity, you'd have seen that the initial force applied to each cart was the same and the product of the mass and the acceleration was the same. Newton made similar observations and concluded that the force on an object equals the mass of the object times the acceleration of the object. This is Newton's Second Law of Motion. It can be written as an equation:

Force = Mass × Acceleration.

Force is mass times acceleration. In other words, a force causes a mass to accelerate. In what unit do we measure force? You know that mass is measured in kilograms and acceleration in meters per second squared

Centripetal Force

When you swing a mass on the end of a string, the mass whirls around with uniform circular motion. The speed of the mass may be constant, but the velocity is always changing because the direction is changing. Therefore, you can say that the mass is accelerating. Newton's second law tells you that in order for a mass to accelerate, a force must act upon it. What forces act upon an object in uniform circular motion to keep it accelerating?

Look at the print on the next page that shows an example of uniform circular motion. When the gaucho lets go, the bola moves away in a straight line.

The only thing that keeps it moving in a circle is the force acting toward the center of the circle—the pull that the person exerts on the bola.

Satellites and Centripetal Force

If a satellite were in orbit and Earth's gravity were suddenly shut off, what would happen to the satellite? Because of its inertia, it would move off in a straight line. The force of gravity keeps pulling it back in. At any point on the circle, you could think of the path of the satellite as a tiny straight line. To change the direction, there must be a force acting toward

Purpose

A Closer Look reinforces Section 1-2 by explaining the relationship between force and motion. Because of an object's inertia, its motion is constant unless a force acts on the object. In this example, the force is centripetal force.

Content Background

Centripetal force keeps a body moving in a circular path. The force can be calculated using the following formula:

$$F = \frac{mv^2}{r}$$

In this formula, the centripetal force equals the product of the object's mass, and the square of its velocity, divided by the radius of the circle in which it travels.

Teaching Strategies

Take the class outdoors to demonstrate the effect of centripetal force. Swing a pail filled with water or confetti quickly around

(m/s²). The unit of force is therefore,

kilogram × meter/second²

The unit of force is also called a **newton**, abbreviated N. In the English system, the unit of force is the pound.

■ Balanced Forces

When does an object accelerate? Hold a book steady on the palm of your hand. What forces are acting? Earth exerts a downward force on the book, and your hand exerts an upward force on the book. Both forces act on the same object, but in opposite directions. Are the forces equal? If they weren't, the book would be accelerating.

But the forces are balanced. **Balanced forces** means that the forces acting on an object cancel one another. The effect is as if no force is acting.

■ Unbalanced Forces and Acceleration

What happens when forces are not balanced? If you increase your upward force against the book, it moves. Objects accelerate only when they are acted upon by unbalanced forces, that is, when there is a greater force acting on an object in some direction. An unbalanced force produces a change in motion.

Content Background

The general study of motion in physics is *mechanics*. Mechanics is subdivided into two areas of study. *Kinematics* is the study of the relationships among the quantities that describe motion, such as position, time, velocity, and acceleration. The equations of average speed and acceleration are examples of kinematic equations. *Dynamics* is the quantitative study of the forces that cause motion.

GLENCOE TECHNOLOGY

 Videodisc

STVS: Physics
Disc 1, Side 2
Prop-Fan Propellers (Ch. 4)

VTOL Airliners (Ch. 5)

Laminar Flow Over Airplane Wings (Ch. 6)

Cutting with Water (Ch. 9)

The Infinite Voyage: The Search for Ancient Americans

Chapter 3
Ancient Weapons for Hunters

the center of the circle—in this case, toward Earth. This is centripetal force.

Centripetal force is the force exerted to keep an object moving in a circle. An object traveling in a circle is constantly accelerating toward the center of the circle.

Science Journal
In your Science Journal, explain how gravity is a centripetal force for objects in orbit around a planet. Think about the direction that gravity pulls with respect to the center of a planet.

South American gauchos used bolas as a tool to hunt animals.

in a circle, perpendicular with the ground. Have students observe that the substance in the pail remains in the pail, even when the container is upside-down. Ask what force the pail exerts on its contents. *centripetal force.*

Discussion

Students can discuss other examples of centripetal force they have experienced, such as the amusement park ride in which they can stay inside the perimeter of a spinning drum. At a certain speed, the floor is dropped. Riders are kept from falling through the hole by friction, their inertia and the wall's centripetal force which holds them in place on the wall.

Science Journal
Gravity is the force that keeps objects around a planet. Gravity pulls the object in and keeps it in circular motion around the planet.

Going Further ⬛⬛⬛⬛➤

Invite students to explore the effect of speed on the relationship between gravity and centripetal force. Let them experiment by speeding up and slowing down circular motion. They might repeat the experiment with confetti in a pail, or place small objects on a record player and vary the speed. Use the Classroom Assessment List for Group Work and the Individual in **PASC,** p. 97.

Weight As Force

Up until now you've studied motion and some of the effects of forces on motions. You've seen that if you know the mass of an object and the acceleration of the object, you can calculate the force on the object. But is there any way to directly measure a force? Let's explore.

Explore! ACTIVITY

How can forces be measured?

What To Do

1. Hang a 1-kg mass from the hook of a spring balance that is calibrated in newtons.
2. What causes the spring to stretch? How much force does the mass exert on the spring?
3. Record your observations and answers *in your Journal.*

You discovered that a 1-kg mass exerts a downward force of 9.8 N. How does this happen? The force due to gravity accelerates the mass downward at 9.8 m/s² and the mass, in turn, pulls the spring downward. If the spring did not exert an equal opposing force, the mass would continue to accelerate downward.

When you jump off a step onto the floor, you accelerate downward at the rate of 9.8 m/s². When you reach the ground you stop because the floor exerts an equal force upward on your feet. Earth still pulls on you with the same force. The gravitational force on you or any object is called **weight**.

Weight is calculated by multiplying the acceleration due to gravity by the mass of the object. In the Explore you used a one-kilogram mass. Its weight is:

Force (N) = mass (kg) × acceleration (m/s²)
9.8 N = 1 kg × 9.8 m/s²

The force exerted on the 1-kg mass is actually its weight, measured in newtons. When the force being measured is gravitational force, force and weight are equivalent.

Action and Reaction Forces

In all of the interactions we have seen so far, there are always two equal and opposite forces involved. The forces are always equal in size but opposite in direction.

Whenever you push on something, that force is called an **action force**. The force that pushes back on you occurs at exactly the same moment and it is called the **reaction force**. Action and reaction forces always occur in pairs and they always act on different objects. Newton's Third Law of Motion states that for every action force there is an equal and opposite reaction force. These forces occur at exactly the same time. An action force does not produce a reaction force a split second later; they always occur together.

Every day, we see hundreds of examples of the third law in action. When you press lightly on a wall, the wall presses lightly back. A car is set in motion by the push of the ground on the tires as the tires push back on the ground. As you walk, you push on the ground and the ground pushes back. See **Figure 1-7** for another example.

SKILLBUILDER

Recognizing Cause and Effect

When using a high-pressure hose, why is it necessary for firefighters to grip the hose strongly and plant their feet firmly? If you need help, refer to the **Skill Handbook** on page 643.

Figure 1-7

Forces always occur in pairs. Whenever a force is applied to an object there is an equal and opposite force applied to another object.

A The equal and opposite actions or forces happen exactly at the same moment. There is no time lag between them.

B The woman's head exerts an upward force on the bundle that is exactly equal to the downward force the bundle exerts on the woman's head.

37

SKILLBUILDER

Firefighters need to brace themselves against the reaction. The force on the water by the hose causes an equal force of the water pushing back on the hose. L1

Discussion Ask students if they have ever watched someone in a movie or television show shoot a pistol or rifle. Have a volunteer describe the "kick" as the force of the bullet leaving the gun produces a reaction felt by the person shooting.

Demonstration Newton's Third Law of Motion can explain how gases can propel an object.

Inflate a small balloon and then let it go. Students will see the balloon fly around the room. Have students relate the flow of air from the balloon to the movement of the balloon. L1

Visual Learning

Figure 1-7 To help students practice formulating models of equal and opposite forces, ask them to rest their hands on top of their heads. The downward force of their hands is equal to the upward force of their heads.

Figure 1-8 (p. 38) Be sure students understand that the pairs of **balanced** action-reaction forces include: hand on the ball/ball on the hand **and** Earth on the ball/ball on Earth. The **unbalanced** forces that cause the ball to move are both acting on the same object—the ball.

ENRICHMENT

Activity Have students bring in several action photographs of various sporting events. Photographs can come from newspapers or magazines. Students can identify action-reaction pairs and balanced forces and explain their choices to the class. L1

Figure 1-8

3 ASSESS

Check for Understanding

Ask students questions 1–3 and the Apply question in Check Your Understanding. Ask if students have seen a stuntperson fall into the air bag. Have them describe what they saw. Ask what similar device is available on some cars. *the air bag.* Use the Performance Task Assessment List for Group Work in **PASC,** p. 97.

Reteach

Activity Tie a piece of string around a chalkboard eraser and suspend it. Then cut the string. Point out that before the string was cut, the eraser was at rest, indicating that the forces holding it up and pulling it down toward Earth were in balance. After the string was cut, the eraser accelerated, indicating that there was a net force acting on it. L1

Extension

Discussion Imagine that you are carrying ice skates and hockey equipment as you walk over a frozen pond. You slip and fall on your back. Your friend challenges you to move to the edge of the pond without turning over or standing up. You try pushing on the ice with your hands and feet, but it is so slippery that you cannot move. How can you get to the edge of the pond using action-reaction forces? *One possible solution is to throw the equipment, one piece at a time, toward the center of the pond. The reaction will cause you to slide partway in the opposite direction toward the edge of the pond.* L2

4 CLOSE

Discuss how you could propel yourself in a space that had zero gravity. Discuss how Newton's Third Law would account for the way things move in a place with zero gravity. L2

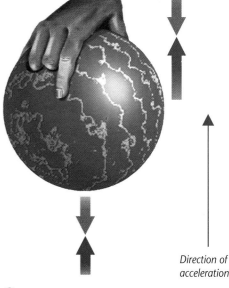

If every action force produces an equal and opposite reaction force, how do objects ever get moving?

A Think about the last time you went bowling. You picked up the ball from the return rack and carried it to the starting point.

Direction of acceleration

B When the ball was hanging beside you, motionless, there were two sets of action-reaction forces acting on it. Your hand pulled on the ball exactly as much as the ball pulled on your hand. The ball pulled on the Earth exactly as much as Earth pulled on the ball. The force of your hand on the ball and Earth on the ball were balanced so there was no acceleration.

C So, how could you raise the ball to get ready to bowl? When you raised the ball, the same two pairs of action-reaction forces were acting on the ball, but the action-reaction pair between your hand and the ball was much greater than the pair between Earth and the ball. The forces acting on the ball were unbalanced and the ball accelerated upward.

■ Action and Reaction: An Example

You've already seen that an object must be acted on by unbalanced forces in order to accelerate. If action-reaction forces are always balanced, how can anything ever move? Look at

Figure 1-8, which describes the forces acting on a bowling ball.

Forces always come in pairs. When one object exerts a force on a second, the second exerts an equal and opposite force on the first. This is Newton's Third Law of Motion.

check your UNDERSTANDING

1. Write a short paragraph explaining how the following words are related: force, weight, newton, pound.
2. A 10-kg object collides with a 30-kg object. As a result, the 10-kg object accelerates at 60 m/s². What would be the acceleration of the 30-kg object?

   $F = M \times A$

   $\times$?

3. When you push a door closed, you're exerting a force on the door. Where is the equal and opposite force and what is it pushing on?
4. **Apply** When a stuntperson jumps from a height, he or she lands on a large bag filled with air. What is the purpose of this bag in terms of what you learned in this section?

check your UNDERSTANDING

1. Forces are measured in newtons in the SI system and pounds in the English system. Weight is a force—the force of gravity between two objects.
2. 20 m/s²
3. The equal and opposite force is exerted on you by the door.
4. **Apply** Since you want as little force as possible acting on the person and you cannot change the person's mass, you must decrease acceleration. The velocity is fixed by the height he or she falls, so the only thing to change is the time it takes him or her to stop. This is greatly increased by the bag.

Pressure and Buoyancy

Area and Force

In a previous section, you measured the force exerted by a 1-kg mass and found that it was 9.8 N. If you set the mass on the table, does it still exert the same force? What if you hammered the mass into a flat plate—would the force change? You can start to explore the relationship of the force exerted to the area that the force acts on by doing the activity below.

Section Objectives

- Calculate pressure from a given force and surface area.
- Explain how objects can float in a liquid.
- Interpret and use Archimedes' Principle.

Key Terms

pressure
buoyant force

 **Explore!** **ACTIVITY**

How are weight and pressure related?

What To Do

1. Place a sheet of graph paper lined 1 inch by 1 inch on the floor and cover it with a sheet of carbon paper.

2. Place the tip of a crutch on the carbon paper and lean your full weight on the crutch. This should make an impression showing the surface area of the crutch tip.

3. Find your weight.

4. From the impression, estimate what area the crutch covered.

5. Divide your weight by this surface area to find the force per square inch.

6. Have a much heavier person stand on one foot on the carbon paper and graph paper underneath. Trace the outline of the shoe.

7. Use this person's weight and divide it by this area.

8. How do the numbers you calculated differ? Which push had greater force per square inch? Which push had greater pressure? Answer the question *in your Journal.*

1-3 Pressure and Buoyancy **39**

PREPARATION

Planning the Lesson

Refer to the Chapter Organizer on pages 20A–D.

Concepts Developed

This section examines the effects of pressure on solids and liquids. The force of buoyancy is also presented.

1 MOTIVATE

Bellringer

Before presenting the lesson, display **Section Focus Transparency 3** on an overhead projector. Assign the accompanying **Focus Activity** worksheet. L1 LEP

Discussion Bring in a high-heeled shoe and a flat-heeled shoe. Hold them up and ask if anyone has had his or her foot stepped on by either type of shoe. Ask the class which would hurt more. Tell students that this section will teach them the reason why the high-heeled shoe hurts more. L1

greater than that produced by the same force over a greater area.

Answers to Questions

The heavier the person, the greater the result of the division of weight by area. The smaller the area, the greater the pressure for a given weight.

✓ Assessment

Performance Have students repeat the activity using different forces and compare their results with those attained in the Explore! Activity. Use the Performance Task Assessment List for Analyzing the Data in **PASC**, p. 27.

 Explore!

How are weight and pressure related?

Time needed 30 minutes

Materials crutch (or cane), flat-heeled shoe, graph paper, carbon paper, scale

Thinking Processes observing and inferring, comparing and contrasting

Purpose To define operationally how weight and pressure are related.

Teaching the Activity

Science Journal Have students record their results and calculations in their journals. L1
COOP LEARN

Expected Outcome

Students will observe that the pressure produced by a force exerted over a small area is

Tying to Previous Knowledge
Review with students the Third Law of Motion discussed in the last section. Explain that in this section they will learn about how force relates to the area it acts on. Remind students that when a fluid, such as milk, pushes on its container, the container pushes back on the fluid.

Visual Learning

Figure 1-9 Guide students to understand that both walkers exert the same amount of force on the snow—100 pounds.
Answer to part C:
Boots: 100 lb ÷ 60 in.² = 1.67 lb/in.²
Snowshoes: 100 lb ÷ 1296 in.² = 0.077 lb/in.²
Difference: 1.67 − 0.077 = 1.59 lb/in.²

GLENCOE TECHNOLOGY

 Software
Computer Competency Activities
Chapter 2

 Videodisc
Science Interactions, Course 2 Integrated Science Videodisc
Lesson 1

■ Pressure

In the activity, when you divided the weight of an object by the surface area the object occupied, you were finding pressure. **Pressure** is defined as the weight or force acting on each unit of area. The equation is:

(Pressure) = (Force/Area on which the force acts)

Look at **Figure 1-9** to get an idea of practical applications of the concept of pressure.

When you reduce the area on which a force acts, you can get a lot of pressure from a small force. Think about a needle. Suppose its point has a surface of 0.001 square inches. How much pressure will that needle exert if you push on it with just 1 pound of pressure? Divide 1 by 0.001 and you get 1000 pounds per square inch. You can see, then, why doctors and people who sew would want very sharp needles for stitching up wounds and clothes.

Figure 1-9

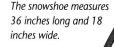
The snowshoe measures 36 inches long and 18 inches wide.

The boot measures 10 inches long and 3 inches wide.

A The surface of snow can support a certain amount of weight without cracking, breaking, or crushing. This ability to resist pressure is measured in number of pounds per square inch.

The two walkers shown both weigh 100 pounds. The one wearing boots crushes the snow and sinks with each step. The one with snowshoes walks along on top of the snow.

B By wearing snowshoes, the girl has spread her weight over a larger area of the snow. By spreading her weight over a larger area, she has limited the force she exerts on the snow in any square inch. The snow supports her weight, and she glides along without sinking.

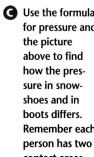 **C** Use the formula for pressure and the picture above to find how the pressure in snowshoes and in boots differs. Remember each person has two contact areas.

Program Resources

Study Guide,   p. 9
Laboratory Manual,   pp. 3–4, Density and Buoyancy ☐L2
Teaching Transparency 2 ☐L2
How It Works,   p. 5. Why do Steel Ships Float? ☐L3
Section Focus Transparency 3
Science Discovery Activities, 1-3

Multicultural Perspectives

Even Pressure
Yoga is a physical discipline that is part of Hinduism. Explain to students that some advanced practitioners of yoga lie on a bed of nails or spikes. Ask students how this can be accomplished without injury. Point out that the secret to this is concentration and relaxation. If the body is relaxed, the pressure is distributed across the person's body.

It's Different with Fluids

What do you think exerts more pressure—10 cm of water on the bottom of a coffee can, or the same depth of water on the bottom of a wading pool? Or does it make any difference?

Find Out! ACTIVITY

What is the relationship of fluid depth to pressure?

If you've ever been in a pool and gone to the bottom, you've experienced one effect of water pressure. How does the pressure relate to the depth of the water? You probably know from experience, but use this activity to confirm your observations.

What To Do

1. Weigh three or four graduated cylinders of different diameters. If you don't have graduated cylinders, you may use cans or glasses, but they must have vertical sides.

2. Record the weights.

3. Fill each container with water to the same height from the base.

4. Weigh the containers again.

5. Subtract the weights of the empty containers from the weights of the filled containers. This gives you the weight of each column of water.

6. If you are using a circular container, measure the inside diameter of each container. Divide each diameter by 2 to get the radius. Multiply the radius by itself, and then multiply that number by 3.14. That gives you the cross-sectional area of each cylinder. Otherwise calculate the rectangular or square cross-sectional area.

7. Divide the weight of each column of water by its container's cross-sectional area. This tells you how much pressure the water exerts on the bottom of its cylinder.

Conclude and Apply

1. How do the pressures compare?

2. If you had a 30-foot vertical pipe filled with water, how do you think the pressure on the bottom of that pipe would compare with the pressure at the bottom of a 30-foot deep lake?

What is the relationship of fluid depth to pressure?

Time needed 30 minutes

Materials three or four graduated cylinders of different diameters, metric ruler, water, scale

Thinking Processes observing and inferring, comparing and contrasting

Purpose To measure and define the relationship between the height of a column of water and the pressure the water exerts on the bottom of its container.

Teaching the Activity

Have students work in groups of three or four. If each group fills its containers to a different depth, the class will be able to generalize results more easily. **COOP LEARN** L1

Troubleshooting If students use containers of various sizes as shown in the photograph, make sure that students use only the vertical portions of each container.

Science Journal Have students record their data and calculations in their journals.

Expected Outcome

Students will observe that water exerts the same pressure at the same depth, regardless of the cross-sectional area of the container.

Conclude and Apply

1. They should be the same within experimental error.

2. They would be the same.

✔ Assessment

Performance Have students sketch the containers of water in their journals. Have them predict how the pressures would compare at different depths and ask them to hypothesize what causes pressure to change with depth. Use the Performance Task Assessment List for Scientific Drawing in **PASC**, p. 55. L1

Meeting Individual Needs

Behaviorally Disordered, Learning Disabled For students who need more structure when presented with many numbers and calculations, prepare data tables for the Find Out activity. Column heads should be the numbers of the graduated cylinders. Rows should list the data to be filled in. Be sure items in the rows are listed in the order that students will follow in the activity. **LEP**

Figure 1-10

The column of water above the top hole is short, so the water is pushed out with a small force. The column of water above the bottom hole is taller and therefore heavier, so the water is pushed out with greater force.

When you did the Find Out activity, you saw that the pressure at the bottom of a container of water depends only on the height of the water. For liquids other than water the same rule holds true; however, the pressures in other liquids would be different from the pressure in water.

But what about pressure other than at the bottom of a column of liquid? A fluid is any substance that does not have a definite shape. Liquids and gases are fluids. Does fluid exert pressure in any direction other than down? Your experience tells you the answer. If you fill a paper cup with water and poke a hole halfway down the cup's side, what happens? Water flows from the hole. So there must be sideways pressure forcing the water out.

If you could move a pressure detector around in any container of fluid, you'd find that the fluid exerts pressure in all directions.

Life Science CONNECTION

Buoyancy in Fishes

For fish that stay more or less at a certain level, the pressure on their bodies is always about the same. Some fish, however, migrate upward or

downward in the water. They may, for example, remain on the bottom during the day but swim upward at night to feed. They have to be able to compensate for the change in pressure on their bodies. Through time, bony fish evolved an adaptation that made compensating for pressure possible. The same adaptation makes it possible for the fish to adjust its buoyancy.

An Adaptation for Buoyancy

The swim bladder is a sac-like structure that holds air. This organ helps a fish adjust its buoyancy by changing the

42 Chapter 1

Life Science CONNECTION

Purpose

The Life Science Connection extends the concepts of pressure and buoyancy in Section 1-3 by relating those concepts to an adaptation in fish that change levels, namely, the swim bladder.

Content Background

Fish that do not remain at the same level in water often compensate for pressure changes through their swim bladders. However, swim bladders sometimes have

other functions. The noise produced by the swim bladder is often a form of communication used in courtship and mating.

Teaching Strategy

Have students work in groups to make models of "swim bladders" that can be used to float or submerge small objects. Model materials should include pieces of plastic foam, balloons, corks, washers or coins, and materials for joining such as pins, tape, glue, or paper clips. Students

Why You Float

So far you've seen how water pressure depends on the depth of the water. You've also seen that water exerts pressure not only on the bottom of a container but in all directions. Now think about another more familiar occurrence that involves water. Why do you float in a body of water?

Imagine yourself in a swimming pool, pond, lake, or the ocean. You know that something about the water allows you to float. You know that you can easily lift friends in the water, even if you could barely budge them on land. Why is that?

Let's first examine the question in a smaller body of water. We'll move from the swimming pool to the bathtub. When you sit in your tub, the water level rises. That's no surprise. After all, you added something to the water: yourself. How much do you think the water rises?

When you float in the pool, the water doesn't rise by an amount equal to the volume of your entire body. Your entire body isn't under the water. How much does it rise, then? What determines just how far your body sinks into the water? Let's find out.

Connect to...
Life Science

Divers need to understand pressure. If they return to the surface too quickly, they can suffer from decompression sickness or "the bends." Research the bends and write some suggestions for avoiding the problem.

Connect to . . .
Life Science

Some people are more susceptible to the bends and other pressure-related diseases. A test in low-pressure chambers will allow a person to decide whether diving is a profession or sport he or she should pursue. Air locks and the breathing of pure oxygen are preventative measures for the bends.

NATIONAL GEOGRAPHIC SOCIETY

⊙ Videodisc

Newton's Apple: Physical Sciences

Buoyancy
Chapter 1, Side A
Buoyancy principle

4355-5410
Why huge ships float?

8175-9712
Fascinating fact

13648
Buoyancy
Chapter 1, Side A
The icebreaker out of water

8660
The icebreaker displacing water

8785

---

amount of gas in the bladder. More gas in the bladder increases the volume of the fish. The fish then displaces just the amount of water needed to keep it buoyant.

Coping with Pressure Changes

As a fish swims upward or downward in the water, the pressure on it changes. The swim bladder helps it react to the changes. The bladder is equipped with two separate organs. One organ allows additional gas from the blood to seep into the swim bladder as it is needed. The other organ removes excess gas from the bladder and returns it to the blood. These organs help a fish

Swim bladder

maintain buoyancy as depth increases.

The swim bladder makes it possible for a fish to swim at different depths in the water. The ability to swim at different depths allows fish to search for food at any depth.

What Do You Think?

How does the swim bladder help fish survive? A diver does not have a swim bladder. How does a diver overcome the buoyant force on the body at different depths?

---

can attach their bladders to small classroom objects to help them rise or fall in a basin of water. L1 **COOP LEARN**

Answers to
What Do You Think?

The swim bladders enable fish to change depth in search of food or to escape from predators. A diver must wear a weight belt or use additional weights to counter the buoyant force of the water.

Going Further ▸▸▸▸▸➤

Students may be interested in more information about how submarines or divers accommodate to extreme pressure underwater. Encourage them to research topics such as submarine engineering or the use of decompression chambers for divers who get the "bends" after experiencing rapid changes in water pressure. Use the Performance Task Assessment List for Writing in Science in **PASC**, p. 87. L1 **P**

Find Out! ACTIVITY

How much water does an object displace?

You know that in order for an object to float, some force must be acting upon it. But what? As you do the activity, remember that weight is a measure of force.

What To Do

1. Fill a beaker to the brim.
2. Put a pan underneath it so that you can catch any water that overflows.
3. Using a spring scale, weigh an object dense enough to sink in the beaker. Record its weight.

4. Leaving the object attached to the spring scale, suspend the object in the water so that it is completely beneath the surface, but not touching the bottom. Use the figure as a guide. Some water will overflow into the pan below.
5. Use the spring scale to determine the object's weight while in the water.
6. Weigh a small can with a wire handle.
7. Transfer as much of the water that overflowed from the beaker into the can as possible. Weigh the can again. Subtract to find the weight of the water.

Conclude and Apply

1. How much weight did the object lose when you submerged it in water?
2. What is the relationship between the weight of the water that overflowed and the weight the object lost?

■ Archimedes' Principle

As you saw in the Find Out activity, when an object is immersed in water, it pushes aside some amount of water. The weight of the object is reduced by the weight of the water that is pushed aside. This relationship was discovered by the Greek mathematician Archimedes in the third century B.C.E. Archimedes' Principle says that the weight of water displaced by an object is equal to the amount of weight lost by the object. The greater the weight of water you displace, the

more your weight is reduced.

But hold on. What's this about losing weight? Water doesn't get rid of gravity, does it? No, but it can lift you up as much as gravity pulls you down. If you lose 40 pounds while standing in a pool, the water is exerting an upward force of 40 pounds on you. This force is called the **buoyant force.** When something is pushed upward by fluid, that's known as buoyancy. All objects experience buoyancy, whether they're more or less dense than the fluid they're in.

Figure 1-11

A According to Archimedes' Principle, when an object is immersed in water, its weight, while in the water, is reduced by an amount equal to the weight of the water the object is displacing, or moving aside.

B The reduction in weight during immersion is due to the upward force of water, called buoyant force. Just as gravity pulls you down, water pushes you up. Water pushes you up with a force equal to the weight of the water you displace.

Because your weight is a measure of the force with which gravity is pulling you downward, your apparent weight in water is reduced by the amount of the buoyant force, which pushes you upward.

C When buoyant force pushes up on an object with force that is greater than or equal to the weight of the object, the object floats.

We began this chapter by considering the many different kinds of motion and force present in a variety of circus acts. Can you understand now how, regardless of how amazing these acts seem, they are all based on a few basic laws developed by Isaac Newton several hundred years ago?

check your UNDERSTANDING

1. Describe what happens to the pressure exerted by a force as the area that the force acts on increases.

2. How does the diameter of a column of water affect the pressure at the bottom of the column?

3. Explain why a block of wood can float in water and a block of metal of equal volume cannot.

4. **Apply** If you step into a pool and displace 50 pounds of water, how much weight have you lost?

check your UNDERSTANDING

1. As the area increases, the pressure exerted decreases.

2. The diameter of the column of water has no effect on the pressure at the bottom of the column.

3. The buoyant force can balance the weight of the wood but not the weight of heavier metal.

4. **Apply** You "lose" 50 pounds. In reality, the water exerts 50 pounds of buoyant force on you.

Check for Understanding

Ask students questions 1–3 and the Apply question in Check Your Understanding. Discuss students' answers to the Apply question. Be sure students understand that you do not actually lose mass; rather, there is an added buoyant force that lessens the force of gravity. Use the Classroom Assessment List for Group Work in **PASC**, p. 97.

Reteach

Activity Bring in a ski or a snowshoe and ask students why they could ski or snowshoe over snow that they would sink into if they wore boots or shoes. They should understand that the skis or snowshoes distribute their weight over a greater surface, thereby exerting less pressure on the surface of the snow. L1

Extension

Research Encourage students to investigate what makes different materials such as wood, steel, and fiberglass suitable materials for boatbuilding. Challenge students to make a diagram showing how a ship floats if it is made of materials denser than water. L3

4 CLOSE

Activity

Bring in a bicycle pump. Have students use it to inflate a basketball and observe how the pump works. *As the ball fills, they will observe its volume change.* Once the ball appears full, ask what happens when more air is added. *As the ball fills, the air pressure inside exceeds the pressure outside the ball, expanding the wall of the ball. As more air is added to the ball, the pressure inside increases. Eventually, if too much air is added, the ball will burst to relieve the extreme pressure.* L1

Science and Society

Science and Society

Purpose

Science and Society reinforces Section 1-3 by explaining the workings of deep-sea submersibles. This excursion also extends the concept of buoyancy, as changes in water displacement are described.

Content Background

Deep-sea submersibles were originally developed to allow underwater exploration despite the extreme pressure at lower ocean depths. The Alvin was developed by Woods Hole Oceanographic Institute and has made over 2400 dives. It is capable of making dives for a full workday of eight or nine hours, but can remain submerged for as long as 72 hours. This submersible was used by Dr. Robert Ballard to find and explore the Titanic. Today, some of its technical capabilities have been surpassed by more modern submersibles such as the Mir 1 and Mir 2, a pair of Russian submersibles, whose names mean peace.

Teaching Strategies

Have students choose an area of the world where Alvin has been used for research, such as the North Atlantic search for the Titanic, archaeological explorations in the Mediterranean, or marine biology research in the Galápagos Islands. Have them write one or two journal entries as if they were researchers on the Alvin making underwater observations. L2

Deep-Sea Submersibles

Alvin is a lightweight, self-propelled vehicle that travels underwater with a crew of three to five persons.

Alvin is lowered into the water, carrying a weight that will help it descend to the bottom. This saves the pilot from turning on the vertical thrusters, which could be used to maneuver the vessel downward. The thrusters run on batteries, and batteries must be conserved for horizontal travel when the vessel is in deep waters near the ocean floor.

Submersibles and Buoyancy

To make small changes in buoyancy while under water, some submersibles have spheres filled with air and oil. The oil can be pumped into flexible bags on the outside of the vessel. The bags make the vessel as buoyant as the pilot wants, depending on how much oil is pumped into the bags. The more oil in the bags, the greater upward push the vessel receives from the water displaced. When a submersible is ready to ascend, however, using the oil bags is not enough to get it to the top. The pilot must also drop the weight that helped it descend. Then the submersible becomes positively buoyant and rises to the surface.

Discoveries in the Galápagos Rift

In 1977, the crew aboard Alvin descended to the ocean floor to study the hot springs at the Galápagos Rift in the Pacific Ocean. Imagine their surprise when they found a complex community of living things two kilometers below the surface, where not even the faintest light could reach! They had come across a whole new food web on the ocean floor—one that did not depend on plants that convert energy from the sun into chemical energy!

Submersibles like Alvin have made it possible for us to increase our knowledge and understanding of both the ocean and life at great ocean depths. Because of Alvin, we are now aware of an entire ocean community that can survive without light!

Science Journal

Think about how submersibles may affect people's lives in the future. It may be possible that at an underwater resort, you could hire a submersible.

In your Science Journal, write about how widespread use of submersibles might affect wildlife in deep-sea habitats. In what way might people enjoy this experience without harming fragile habitats?

Science Journal

Accept all logical responses from students. Widespread use of submersibles may lead to several problems, including introduction of organisms that wouldn't otherwise be present in those environments. Inexperienced submersible pilots may bump into fragile habitats or stir up sediment with the propeller wash. The increased traffic may disturb shy species, causing them to emigrate or even interrupt reproduction.

Going Further ▸

Have students find out more about specific research projects in which Alvin has been used. Some possible resources for further study on submersibles are:

Ballard, R.D. "Window on Earth's Interior," *National Geographic,* (August 1976)

Ballard, R.D. "How We Found the Titanic," *National Geographic,* (December 1985)

Use the Performance Task Assessment List for Group Work in **PASC,** p. 97. L2

HISTORY CONNECTION

A Royal Solution

Have you ever tried to figure out a difficult puzzle? If so, you know how it holds your attention until you think of a solution. That's what happened to Archimedes, the Greek mathematician you read about in connection with buoyancy. Legend has it that King Hiero II of Sicily ordered a gold crown to be made. When the crown was delivered, the king suspected that the gold had been mixed with silver. Archimedes was asked to find out whether the crown was made of pure gold.

A Volume Puzzle

Archimedes was puzzled about how he could do this. He could weigh the crown and then find a gold piece that weighed the same as the crown. If both the gold piece and the crown had the same volume, he could be certain that the crown was pure gold. If the crown had a larger volume than the gold piece, he could be sure that silver had been mixed with the gold. He based his strategy on the fact that, for equal volumes, silver weighs less than gold. But, there was a problem. Even though Archimedes was a mathematician, he had no idea how to find the volume of the crown. How could he measure such an irregular object?

Discovering Displacement

One day, as he sank into a bathtub filled with water, Archimedes noticed that water

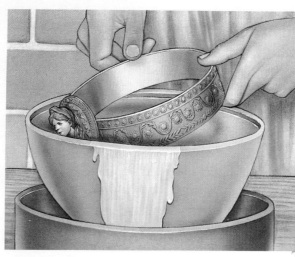

spilled over the top of the tub. He realized that his body had displaced some of the water in the tub. He reasoned that the volume of the water that spilled over was equal to the space that his body took up.

Archimedes submerged the crown and collected the water the crown displaced. He measured the volume of the displaced water because he knew it equaled the volume of the crown. Archimedes had the solution to his problem. He proved that the crown was made of a mixture of gold and silver. He concluded that the man who had made the crown had tried to cheat the king.

You Try It!

Find and compare the volume of five objects of irregular shape by measuring the volume of water each displaces when submerged.

Purpose
History Connection reinforces Section 1-3 by explaining how Archimedes used his understanding of buoyancy to solve a problem. This excursion discusses the relationship between volume, buoyancy, and displacement.

Content Background
Archimedes' Principle demonstrates that a body displaces an amount of water that equals its own volume. A related factor is density, which is the ratio of mass to volume.

Archimedes took advantage of other principles of physics in some of his inventions, notably the laws governing levers and pulleys. He is also believed to have developed the Archimedean screw, a device that was used for thousands of years to raise water from the Nile River for use in crop irrigation.

Teaching Strategy Demonstrate how objects of varying densities displace different amounts of water. Place a pan filled to the brim with water inside an empty dishpan. Use two objects of identical size and shape, but made of different materials in this demonstration. One possibility is two cans of the same brand of soda: one regular, and one sugar free. The sugar in the regular soda makes it more dense than the diet soda. Measure and compare the amount of water displaced by each object. Then use a balance to compare their masses. Although volumes are equal, the regular soda will have a greater mass.

Answers to You Try It!
Results of this activity will vary, depending on the objects chosen.

Going Further ⁞⁞⁞⁞➡
Have students research other topics related to the work of Archimedes. They might then make a presentation of their findings in the form of models or oral reports. Possible topics include how submarines vary the depth at which they stay in the water, or why ballast is sometimes added to ships. Use the Performance Task Assessment List for Oral Report in **PASC**, p. 71. L3

Technology Connection

Purpose

Technology Connection reinforces Section 1-3 by explaining how transducers are used. A force causes motion. In a transducer, this motion generates a signal of some form. Section 1-3, which discusses pressure, is extended by this discussion of blood pressure.

Content Background

Transducers work by converting electrical waves to mechanical vibrations or vice versa. Sonar transducers enable people to do what certain marine mammals, in particular dolphins, can do anatomically—send and receive sound waves in water.

Transducers are widely used in oceanographic research. Sonar waves sent from a ship at a particular frequency strike the ocean floor, layers of water with different temperatures, and large objects in the water such as schools of fish. The sonar waves are reflected back to the ship and recorded by machines. The resulting graphs provide extensive information about ocean depths, temperature, and sea life.

Teaching Strategy

If possible, bring in one or more examples of transducers for students to examine firsthand. Have them trace the conversion by identifying the original force and tracking the transfer of energy until the final signal is produced.

Answers to
What Do You Think?

An electrical switch changes the form of a signal from a mechanical signal (such as a toggle switch) to an electrical signal. Once contact is made, a circuit is completed and electricity is available for use in some form, such as a lightbulb or buzzer. Transducers may be used to cut

---

Technology Connection

Transducers in Your Life

You may not have heard of the word transducer before, but you've certainly used gadgets that contain them. A transducer is a device that converts one form of signal into another form. When you speak into a telephone, a transducer converts one kind of information—sound, or variations in air pressure—into another kind of information—electrical signals. The electrical signals are carried over wires to another place, where they are then converted back to sound.

Motion into Signals

A transducer can convert motion to electrical signals, or vice versa. The fuel gauge in an automobile is an example. The transducer is mounted on the fuel tank. A float moves up and down as the amount of fuel changes. The motion of the float is converted to an electrical signal that is transmitted by a wire to the gauge on the dashboard.

Many of the instruments used in laboratories have transducers. In this chapter, you have measured force, pressure, and acceleration. Some instruments can convert a measurement of a mechanical quantity, such as force, to an equivalent electrical signal. This happens when you weigh something on an electronic scale.

Useful Tools From Transducers

When machine parts are cast in metal, it's important that the mold be held together by a large enough force. If the force is less than needed, the part that is cast will be defective. A force transducer indicates on a gauge how much force is being exerted.

Your blood pressure can be measured using a blood-pressure transducer, such as the one shown on this page. Because of the pumping action of the heart, blood pressure pulsates. The transducer keeps track of these pulsations and converts them into an accurate and fast record of the blood pressure.

A transducer that measures acceleration is used in designing safe containers for the shipment of fragile objects. A table to which a delicate instrument is attached is raised or lowered abruptly to determine the amount of acceleration the instrument can tolerate before it malfunctions.

What Do You Think?

Transducers work by changing signals from one form to another. How is an electric switch a transducer? In what ways, other than those mentioned, might transducers be used?

---

extremely hard objects, clean fine instruments, and drill oil wells. They are also used in phonographs and sonar equipment.

Going Further ||||||▶

Encourage students to research further the use of transducers in sonar. Some resources are:

Macaulay, D. *The Way Things Work*, Boston: Houghton Mifflin, 1988.

The Ocean World of Jacques Cousteau. vol 7; Invisible Messages. Danbury; The Danbury Press.

Science Journal

Review the statements below about the big ideas presented in this chapter, and answer the questions. Then, re-read your answers to the Did You Ever Wonder questions at the beginning of the chapter. *In your Science Journal*, write a paragraph about how your understanding of the big ideas in the chapter has changed.

1 Objects in motion continue moving in a straight line at the same speed and objects at rest remain at rest unless the objects are acted upon by an unbalanced force. This tendency is called inertia. *Use inertia to explain why protection, such as a seat belt or a helmet, is important in situations where sudden stops may occur.*

2 When objects interact, the mass times the acceleration of the objects is equal to the force of the interaction (F=ma). *Why is the acceleration of a tennis ball toward Earth so much greater than the acceleration of Earth toward the tennis ball?*

4 Pressure is equal to force divided by the surface area over which the force acts. *You have a cube of metal that keeps falling through the snow. What would you do to the metal to get it to rest on top of the snow?*

3 Forces always occur in action-reaction pairs, and these forces are always equal in strength and opposite in direction. *List a pair of action-reaction forces that involve the rope during a tug-of-war with a friend. List a pair of forces also involving the rope that are not action-reaction.*

5 The buoyant force (upward force fluids exert on all objects) is equal to the weight of the fluid displaced by an object. *If the buoyant force is equal to the weight of the fluid displaced by an object, why don't rocks float?*

Science Journal

Did you ever wonder...
• A revolving door illustrates Newton's First Law of Motion: an object moving remains in motion until it interacts with another force. (p. 27)
• Snowshoes spread the weight over a greater surface area than an ordinary shoe, decreasing the pressure on the snow at any one area. (p. 40)
• Buoyant force opposes gravity when a person is submerged. A person floats at a level such that the buoyant and gravitational forces are balanced. (pp. 43-45)

Project

Have the class make a handbook illustrating Newton's First, Second, and Third Laws of Motion. The handbook may contain illustrations, photographs, or written descriptions of examples of the laws that students observe around them. This can also be prepared as an exhibit for a school display case. Use the Classroom Assessment List for Booklet or Pamphlet in **PASC**, p. 57. **L2**

Have students examine the five statements on this page to review the main ideas from the chapter.

Teaching Strategies

Divide the class into five groups. Assign each group one of the main ideas statements. Then have members of each group work together to prepare a brief skit that demonstrates the principle in the statement. Encourage them to use humor, so long as the principle is not obscured. Then have each group present its skit, ending by repeating the law or physical principle. Possible skit content is listed below. **COOP LEARN**

Answers to Questions

1. Students could act out being on a bicycle or car that stops suddenly. They will continue moving because of their bodies' inertia. A seat belt restrains them and prevents them from hitting the windshield; a helmet protects their head in the event they fly over the handlebars.

2. Earth's mass is so much greater than that of a tennis ball.

3. The force of a student's hands on the rope and the force that the rope exerts on the hands are action-reaction forces. The force a student exerts on the rope and the force the rope exerts on the student's friend are not balanced forces.

4. Students could hammer it into a flat disk.

5. The force of gravity on most rocks is much greater than the buoyancy force on the rocks. Pumice is an exception to this.

GLENCOE TECHNOLOGY

 MindJogger Videoquiz

Chapter 1 Have students work in groups as they play the Videoquiz game to review key chapter concepts.

Using Key Science Terms

1. Weight is the force on an object due to gravity; mass is the amount of matter in an object.

2. The more mass an object has, the greater its inertia.

3. Responses may vary. Possible response may be the action of a person's shoe pushing the ground and the reaction of the ground pushing against the shoe.

4. Balanced forces are equal in size and opposite in direction. When balanced forces act on an object, they result in no change in the object's motion. When unbalanced forces act on an object, the object moves in the direction of the greater force.

5. Force is a push or a pull; pressure is the amount of force divided by the area it acts on.

6. Force—newtons (N); mass—kilograms (kg); acceleration—meters per second squared (m/s^2); weight—newtons; pressure—N/m^2 or (lb./in.²)

7. Buoyant force on an object is equal to the amount of weight the object loses when submerged.

Understanding Ideas

1. An unbalanced force—gravity—acts on the picture and causes it to accelerate down.

2. $F = ma$; $F = 10 \text{ kg} \times 40 \text{ m/s}^2 = 400$ N. 400 newtons must be applied to a 10-kg mass to accelerate it at 40 m/s^2.

3. The heavier bat will exert a greater force on the ball, causing it to travel farther.

4. Your mass would be the same on Earth and on the moon. You would weigh less on the moon than on Earth.

Developing Skills

1. **Concept Mapping** See student page for answers.

2. **Predicting** Answers will vary depending on the student's results for the experiment. Generally, the greater the mass, the longer the period.

3. **Comparing and Contrasting**

Using Key Science Terms

action force	newton
balanced force	pressure
buoyant force	reaction force
force	weight
inertia	

Answer the following questions using what you know about the science terms.

1. Describe the difference between weight and mass.
2. How is inertia related to mass?
3. Give an example of something you do every day that demonstrates action and reaction forces.
4. Contrast balanced and unbalanced forces.
5. Distinguish between force and pressure.
6. Write the appropriate unit each quantity is measured in: force, mass, acceleration, weight, and pressure.
7. How is buoyant force related to weight?

Understanding Ideas

Answer the following questions in your Journal using complete sentences.

1. What forces act on a picture hanging on a wall to cause it to fall?
2. What amount of force would have to be applied to a 10-kg object to make it accelerate 40 m/s²?
3. If two bats are swung with equal acceleration by two batters, and bat A is 1 kg heavier than bat B, which bat will make a tossed ball travel further? Why?

The heel or cleat with smallest area will exert the greatest pressure.

Critical Thinking

1. Once an object is moving, it continues at constant velocity until some other unbalanced force acts on it.

2. Weight is a measure of the force of gravity on one mass caused by the other. Therefore, the weight of the masses will change any time the gravitational force between them changes.

---

4. Which measurement, your weight or your mass, would be the same when taken on Earth and on the moon?

Developing Skills

Use your understanding of the concepts developed in this chapter to answer each of the following questions.

1. **Concept Mapping** Using the following terms and phrases, complete the concept map about Newton's Laws of Motion: *force, tendency to remain in motion or still, mass × acceleration, equal and opposite reaction forces.*

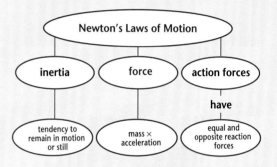

2. **Predicting** Use the information acquired by doing the Investigate activity on page 24 to predict what the period would be if a 0.5 kg mass were placed on the outer platform.

3. **Comparing and Contrasting** Repeat the Explore activity on page 39 using different types of shoes. Use football cleats, clogs, tennis shoes, mid-size heels, and western boots. Find the pressure exerted by each shoe. Compare the results of the different shoes. What can you conclude from the comparison of different shoes?

Critical Thinking

Use your understanding of the concepts developed in the chapter to answer each of the following questions.

1. How can an object be moving if there is no unbalanced force acting on it?

2. How can an object's weight change even though it contains the same mass?

3. The photographs show pictures of a cargo ship taken at two different times. What can you tell about the cargo ship from the pictures? Explain your reasoning.

Problem Solving

Read the following problem and discuss your answers in a brief paragraph.

"Newton was wrong and I can prove it!" declared Denton McQuarrel. He took two magnets—one big one and one little one—and he pushed their north poles together. "According to Newton," he said, "each magnet experiences equal force. Watch what happens." He held one of the magnets in place on the table as he let go of the other. "All I had to do was hold one magnet in place, and Newton's law doesn't work any more. The magnet I held in place didn't move at all." What's wrong with McQuarrel's argument?

CONNECTING IDEAS

Discuss each of the following in a brief paragraph.

1. **Theme—Systems and Interactions** Describe the two pairs of forces that are acting when you pick up your books. Why do the books move in an upward direction?

2. **Theme—Stability and Change** Use tug-of-war to explain how forces can be balanced and unbalanced.

3. **History Connection** Describe how you would show that an irregularly shaped piece of metal was pure silver.

4. **Life Science Connection** Sharks must swim constantly to maintain their position in the water. If sharks had swim bladders in their bodies, how would this affect their daily lives?

5. **Science and Society** How are submersibles used to help scientists describe and collect information about ocean depths?

✔ Assessment

Portfolio Review the portfolio options that are provided throughout the chapter. Encourage students to select one product that demonstrates their best work for the chapter. Have students explain what they learned and why they chose this example for placement into their portfolios.

Additional portfolio options can be found in the following **Teacher Classroom Resources:**

Making Connections: Integrating Sciences, p. 5
Multicultural Connections, p. 5
Making Connections: Across the Curriculum, p. 15
Concept Mapping, p. 9
Critical Thinking/Problem Solving, p. 9
Take Home Activities, p. 6
Laboratory Manual, pp. 1–2; 3–4
Performance Assessment P

3. The tanker in the photograph at the left is heavier than the tanker on the right. You can tell the tanker is heavier because it displaces more water.

Problem Solving

The magnet he held in place became part of a larger mass that included Denton and the table. This large mass does accelerate away from the magnet, but at an unnoticeable rate.

Connecting Ideas

1. **Theme—Systems and Interactions** The balanced pairs are: (1) your pull on the books and the pull of the books on you, and (2) Earth's pull on the books and the pull of the books on Earth. The unbalanced forces are your pull and Earth's pull. The books move up because your upward pull is greater than Earth's downward pull.

2. **Theme—Stability and Change** When each team pulls with equal force on the rope, the forces are balanced and no motion takes place. When one team pulls with greater force than the other, the forces are unbalanced, and the rope moves.

3. **History Connection** Find the mass of the metal. Find out how much water it displaces. The volume of displaced water equals the volume of the metal. Find the density of the metal by dividing its mass by its volume. Finally, compare the result with silver's density.

4. **Life Science Connection** Responses may include changes in the shark's diet, behavior, movement, and feeding habits. Also, they would not need to keep swimming to maintain buoyancy.

5. **Science and Society** Responses may include the use of remote controlled submersibles that measure and record conditions with minimal disturbance of deep-sea habitats.

Chapter Organizer

SECTION	OBJECTIVES	ACTIVITIES & FEATURES
Chapter Opener		**Explore!**, p. 53
2-1 What Causes Earthquakes? (2 sessions; 1 block)	1. **Explain** how earthquakes result from the buildup of pressure inside Earth. 2. **Describe** the forces inside Earth that result in faults. 3. **Compare and contrast** normal, reverse, and strike-slip faults. **National Content Standards: (5-8) UCP2, A1, B2-3, D1, F3**	**Find Out!**, p. 54 **Skillbuilder:** p. 56 **Explore!**, p. 58 **Physics Connection**, pp. 56–57
2-2 Shake and Quake (3 sessions; 2 blocks)	1. **Compare and contrast** primary, secondary, and surface waves. 2. **Explain** how an earthquake's epicenter is located by using seismic wave information. **National Content Standards: (5-8) UCP2, A1, B2, D1, F3**	**Explore!**, p. 60 **Find Out!**, p. 61 **Investigate 2-1:** pp. 64–65 **Science and Society**, pp. 75–76 **Technology Connection**, p. 77
2-3 Volcanic Eruptions (3 sessions; 1.5 blocks)	1. **Explain** what causes volcanoes to erupt. 2. **Explain** the difference between a quiet and an explosive eruption. **National Content Standards: (5-8) UCP2, UCP4-5, A1, B2, C5, D1, E1, F3**	**Explore!**, p. 67 **Design Your Own Investigation 2-2:** pp. 72–73 **A Closer Look**, pp. 68–69 **SciFacts**, p. 78

ACTIVITY MATERIALS

EXPLORE!

p. 53 3 different colored clays
p. 58 2 pieces of 2 inch × 4 inch wood, sandpaper, glue
p. 60* coiled spring toy
p. 67 bottle of cold syrup (maple or corn)

INVESTIGATE!

pp. 64–65* paper, string, metric ruler, globe, water-soluble marker or chalk

DESIGN YOUR OWN INVESTIGATION

pp. 72–73* pencil, paper

FIND OUT!

p. 54* large paper clip, long balloon, 50 sheets of notebook paper, metric ruler, pencil
p. 61* graph paper

KEY TO TEACHING STRATEGIES

The following designations will help you decide which activities are appropriate for your students.

- **L1** Basic activities for all students
- **L2** Activities for average to above-average students
- **L3** Challenging activities for above-average students
- **LEP** Limited English Proficiency activities
- **COOP LEARN** Cooperative Learning activities for small group work
- **P** Student products that can be placed into a best-work portfolio
- Activities and resources recommended for block schedules

Need Materials? Call Science Kit (1-800-828-7777).

⏱ OUT OF TIME? We recommend that students do the activities with an asterisk.

Chapter 2 Forces In Earth

TEACHER CLASSROOM RESOURCES

Student Masters	Transparencies
Study Guide, p. 10 **Critical Thinking/Problem Solving,** p. 10 **Concept Mapping,** p. 10 **Science Discovery Activities,** 2-1	**Teaching Transparency 3,** Types of Faults **Section Focus Transparency 4**
Study Guide, p. 11 **Multicultural Connections,** p. 7 **Making Connections: Across the Curriculum,** p. 7 **Activity Masters,** Investigate 2-1, pp. 11–12 **Science Discovery Activities,** 2-2 **Laboratory Manual,** pp. 5–8, Locating an Earthquake	**Section Focus Transparency 5**
Study Guide, p. 12 **Making Connections: Integrating Sciences,** p. 7 **Making Connections: Technology & Society,** p. 7 **Critical Thinking/Problem Solving,** p. 5 **Multicultural Connections,** p. 7 **Activity Masters,** Design Your Own Investigation 2-2, pp. 13–14 **Take Home Activities,** p. 7 **Science Discovery Activities,** 2-3	**Teaching Transparency 4,** Volcanic Activity **Section Focus Transparency 6**

ASSESSMENT RESOURCES	TEACHING & TECHNOLOGY
Review and Assessment, pp. 11–16 **Performance Assessment,** Ch. 2 **PASC*** **MindJogger Videoquiz** **Alternate Assessment in the Science Classroom** **Computer Test Bank**	**Spanish Resources** **Cooperative Learning Resource Guide** **Lab and Safety Skills** **Science Interactions, Course 2, CD-ROM** **Computer Competency Activities**

*Performance Assessment in the Science Classroom

NATIONAL GEOGRAPHIC TEACHER'S CORNER

Index to National Geographic Magazine

The following articles may be used for research relating to this chapter:

- "Living with California's Faults," by Rick Gore, April 1995.
- "Volcanoes: Crucibles of Creation," by Noel Grove, December 1992.
- "Earthquake—Prelude to The Big One?" by Thomas Y. Canby, May 1990.

National Geographic Society Products Available From Glencoe

To order the following products for use with this chapter, contact your local Glencoe sales representative or call Glencoe at 1-800-334-7344:

- *STV: Restless Earth* (Videodisc)

Additional National Geographic Society Products

To order the following products for use with this chapter, call the National Geographic Society at 1-800-368-2728:

- *Raging Forces: Earth in Upheaval* (Book)
- *Nature's Fury* (Video)
- *Our Dynamic Earth* (Video)
- *Volcano!* (Video)

Teacher Classroom Resources

These are key components of the classroom resources package.

TEACHING AIDS

Section Focus Transparencies

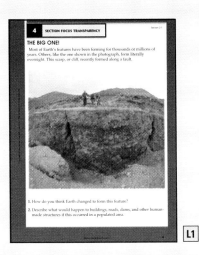

4 SECTION FOCUS TRANSPARENCY

THE BIG ONE!

Most of Earth's features have been forming for thousands or millions of years. Others, like the one shown in the photograph, form literally overnight. This scarp, or cliff, recently formed along a fault.

1. How do you think Earth changed to form this feature?

2. Describe what would happen to buildings, roads, dams, and other human-made structures if this occurred in a populated area.

L1

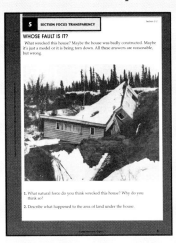

5 SECTION FOCUS TRANSPARENCY

WHOSE FAULT IS IT?

What wrecked this house? Maybe the house was badly constructed. Maybe it's just a model or it is being torn down. All these answers are reasonable, but wrong.

1. What natural force do you think wrecked this house? Why do you think so?

2. Describe what happened to the area of land under the house.

L1

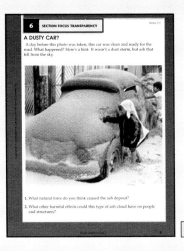

6 SECTION FOCUS TRANSPARENCY

A DUSTY CAR?

A day before this photo was taken, this car was clean and ready for the road. What happened? Here's a hint: It wasn't a dust storm, but ash that fell from the sky.

1. What natural force do you think caused the ash deposit?

2. What other harmful effects could this type of ash cloud have on people and structures?

L1

Teaching Transparencies

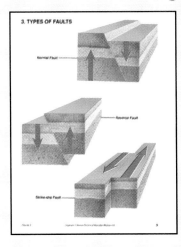

3. TYPES OF FAULTS

L1

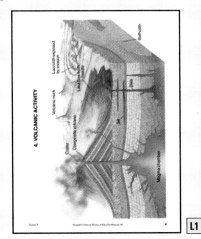

4. VOLCANIC ACTIVITY

L1

HANDS-ON LEARNING

Science Discovery Activity*

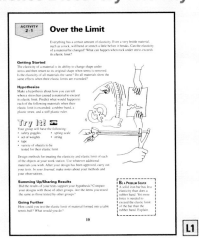

ACTIVITY 2-1 Over the Limit

L1

Laboratory Manual*

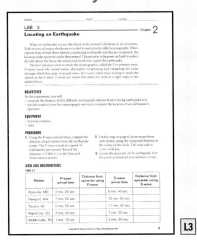

LAB 3
Locating an Earthquake

L3

Take Home Activity

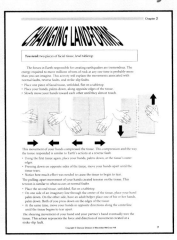

CHANGING LANDFORMS

L1

52C   Chapter 2   Forces in Earth   *There may be more than one activity for this chapter.

Chapter 2 Forces in Earth

Study Guide*

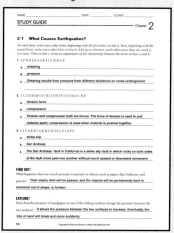

Concept Mapping

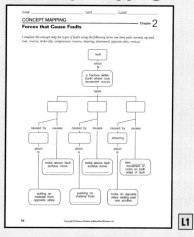

Critical Thinking/ Problem Solving

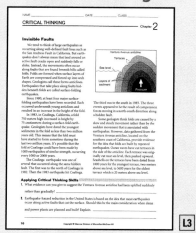

Integrating Sciences

Across the Curriculum

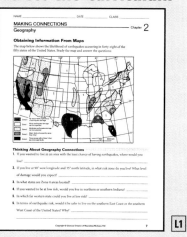

Technology and Society

Multicultural Connection**

Performance Assessment

Review and Assessment

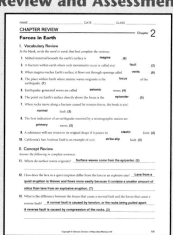

*One per section **Two per chapter

Chapter 2 Forces in Earth **52D**

Forces In Earth

THEME DEVELOPMENT

The themes that this chapter supports are stability and change, and energy. Energy is stored and released by Earth's rocks as a result of forces within our planet. Earthquakes and volcanoes release enormous amounts of energy and cause changes in Earth's surface.

CHAPTER OVERVIEW

In this chapter, students study earthquakes and volcanoes and the forces inside Earth that cause them. Pressures inside Earth can cause layers of rock to move and then break, causing an earthquake. Vibrations from an earthquake travel through Earth as waves.

When magma, or liquid rock, rises to the surface of Earth and erupts, volcanoes result.

Tying to Previous Knowledge

Have students recall the definition of pressure from Chapter 1. Ask students to brainstorm a list of ways they use or are affected by pressure in their daily lives, for example, the air in bicycle tires. Then explain that forces within Earth exert pressure on Earth's rocks to produce earthquakes and volcanoes.

INTRODUCING THE CHAPTER

Ask students to describe what is shown in the photograph. Students may have heard of the San Andreas Fault, in California. Ask

⏱ 00:00 OUT OF TIME?

If time does not permit teaching the entire chapter, use the Chapter Overview on this page, Reviewing Main Ideas at the end of the chapter, and the Chapter 2 audiocassette to point out the main ideas of the chapter.

CHAPTER 2

FORCES IN EARTH

Did you ever wonder...

✓ What causes the ground to shake during an earthquake?

✓ How scientists know the exact spot where an earthquake begins?

✓ Why there are different shapes of volcanoes?

📝 *Science Journal*
Before you begin to study forces in Earth, think about these questions and answer them *in your Science Journal.* When you finish the chapter, compare your journal write-up with what you have learned.

Answers are on page 79.

You've probably heard about the San Andreas Fault, pictured below. Over time, tremendous forces inside Earth exert so much pressure on the rocks along the San Andreas Fault that they move.

Sometimes this movement is gradual. Rocks on both sides of the fault may slowly move past each other as these forces are exerted on them. But if the rocks aren't free to move, pressure builds up. When the rocks finally break free, the movement may be sudden and violent—an earthquake! You can feel the earthquake if you are not too far away from the moving rocks and if the vibrations they produce are strong enough.

▶ **In the activity on the next page, explore how forces inside Earth can affect the rocks within Earth.**

San Andreas Fault

52

Learning Styles	**Kinesthetic**	Explore, pp. 58, 60; Demonstration, pp. 59, 60; Activity, p. 73; Science at Home, p. 79
	Visual-Spatial	Explore, pp. 53, 67; Find Out, p. 54; Visual Learning, pp. 55, 56, 58, 62, 68, 70, 71; Physics Connection, pp. 56-57; Demonstration, pp. 57, 59, 67, 70, 74; Across the Curriculum, p. 63; Research, pp. 64, 74; A Closer Look, pp. 68-69
	Interpersonal	Activity, p. 62
	Logical-Mathematical	Across the Curriculum, p. 57; Find Out, p. 61; Visual Learning, p. 63; Investigate, pp. 64-65
LS	**Linguistic**	Discussion, pp. 55, 66, 69, 71; Multicultural Perspectives, pp. 55, 65; Across the Curriculum, pp. 62, 69, 71; Activity, pp. 70, 74; Investigate, pp. 72-73

How do forces inside Earth affect rock layers?

What To Do

1. Place three rectangular layers of different-colored clay on top of one another.

2. Now place your hands on opposite ends of the clay. Slowly push your hands together, compressing the clay.

3. What happens? Draw a picture to show what happened to the clay.

4. Now place three layers of clay on top of one another as before.

5. With your hands on the opposite ends of the clay, gradually pull the clay apart. What happens?

6. Finally, *in your Journal*, draw a picture to show what you observed.

students what kind of movement has occurred here in the past. Students may guess that the movement was lateral, since the two sides of the fault are at the same level. Ask what the consequences of movement would be. Students may realize that the motion causes earthquakes.

Uncovering Preconceptions

After they study the photo ask students to explain what happens at a fault when an earthquake occurs. Students may think that the rock on either side of the fault separates and creates a big crack. Explain that as masses of rock on either side of the fault push against each other, waves travel through the surrounding Earth.

Explore!

How do forces inside Earth affect rock layers?

Time needed 15 minutes

Materials small slabs of three different colors of clay

Thinking Processes observing, making models

Purpose To formulate a model of the effect of forces inside Earth on layers of rock.

Teaching the Activity

Science Journal Have students record their sketches and responses in their journals. [L1]

Answers to Questions

3. the clay folds

5. the clay stretches and tears

✔ Assessment

Process To compare and contrast the results of the pushing and pulling actions, ask students to write descriptions of the clay after each action. Use the Performance Task Assessment List for Writing in Science in **PASC**, p. 87.

ASSESSMENT PLANNER

PORTFOLIO
Refer to page 81 for suggested items that students might select for their portfolios.

PERFORMANCE ASSESSMENT
Process, pp. 53, 54, 65, 73
Skillbuilder, p. 56
Explore! Activities, pp. 53, 58, 60, 67
Find Out! Activities, pp. 54, 61
Investigate, pp. 64–65, 72–73

CONTENT ASSESSMENT
Oral, pp. 58, 67
Check Your Understanding, pp. 59, 66, 74
Reviewing Main Ideas, p. 79
Chapter Review, pp. 80–81

GROUP ASSESSMENT
Opportunities for group assessment occur with Cooperative Learning Strategies.

PREPARATION

Planning the Lesson

Refer to the Chapter Organizer on pages 52A–D.

Concepts Developed

Students discover what causes earthquakes. A break or fracture where rock movement occurs is called a fault. Normal, reverse, and strike-slip faults are produced by different types of force.

1 MOTIVATE

Bellringer

Before presenting the lesson, display **Section Focus Transparency 4** on an overhead projector. Assign the accompanying **Focus Activity** worksheet. L1 LEP

Find Out!

How do different objects react when they are bent or stretched?

Time needed 25 minutes

Materials large paper clips, sheets of paper, metric ruler, long balloons, pencils

Thinking Processes observing and inferring, comparing and contrasting, making models, predicting

Purpose To observe and compare the effects of bending and stretching on objects.

Teaching the Activity

Science Journal In their journals, have students record their observations of the effects of bending and stretching on objects. L1 LEP

Expected Outcomes

Students will observe that objects that have a slight force

What Causes Earthquakes?

Section Objectives

- Explain how earthquakes result from the buildup of pressure inside Earth.
- Describe the forces inside Earth that result in faults.
- Compare and contrast normal, reverse, and strike-slip faults.

Key Terms
fault

Pressure

Have you ever felt an earthquake? Feeling the ground move and seeing things fall off shelves can be scary. A great deal of pressure inside Earth is released as a result of an earthquake. This pressure built up as a result of force on the rocks underground. You learned in Chapter 1 that pressure is force acting on an area. In the following activity, you'll apply pressure to different objects by bending and stretching them.

Find Out! ACTIVITY

How do different objects react when they are bent or stretched?

How does pressure affect everyday items? What can this tell you about how pressure affects rocks in Earth?

What To Do

1. Use a paper clip to hold two sheets of paper together. Remove the paper clip. Has it changed? Repeat this procedure with 5, 20, and 50 sheets of paper. Did the clip change its shape at any point?

2. Now measure the length of a balloon. Stretch the balloon so that it's 1.5 times its original length.

Stop stretching and measure it. Now stretch the balloon farther, so that after releasing it, it won't return to its original length. Measure the length.

3. Without breaking a pencil, can you make it flex by pushing up on it with your thumbs?

Conclude and Apply

1. What happened to the paper clip and the balloon when they were bent or stretched too far? Why did the objects react this way?

2. What would happen if the pencil were bent too far? Record your answers *in your Journal.*

As you learned in the Find Out activity, objects will stretch and bend only so much. Bending and stretching are ways to apply pressure to objects. There is a limit to how much pressure a paper clip, a pencil, a balloon, or some other object can withstand and still return to its original shape. This limit is called the elastic limit. Once the elastic limit is passed, a substance will remain bent or stretched out of shape, or it will break.

---

applied to them will return to their original shape when the force is released, but that objects that have a greater force applied to them may be permanently altered. Students may infer that lesser and greater amounts of force have a similar effect on rocks.

Conclude and Apply

1. They did not return to their original shapes. Too much pressure was applied.

2. It would break.

✔ Assessment

Process Ask students to observe and compare how other objects react when they are bent or stretched. Possible objects include tissue paper, cardboard, or plant leaves. Students could then make a table from their findings. Use the Performance Task Assessment List for Data Table in **PASC**, p. 37.

Pressure in Rock Layers

Layers of rock behave in much the same way when pressure inside Earth bends or stretches them too far. Just as with the paper clip, balloon, and pencil, forces applied to rocks can cause them to fold or stretch without permanent change, but only up to a point. The rocks will remain folded or will break once their elastic limit is passed. Rocks that have been folded look like those shown in **Figure 2-1B**. When rocks break, they produce vibrations that travel throughout Earth. These vibrations are called earthquakes.

Figure 2-1

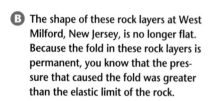

Ⓐ The shape and position of rock layers indicate the pressure the rock has experienced. These rocks at Glen Canyon in Arizona are still flat and horizontal. Therefore, you know that pressures on the rock have been less than the elastic limit of the rock.

Ⓑ The shape of these rock layers at West Milford, New Jersey, is no longer flat. Because the fold in these rock layers is permanent, you know that the pressure that caused the fold was greater than the elastic limit of the rock.

Ⓒ Pressures beneath Earth's surface can exceed the elastic limit of rock so that the rock breaks, causing powerful vibrations which can crack or shatter rock. Such vibrations damaged the rock layers shown here.

Multicultural Perspectives

Tales about Earthquakes

Have students research myths, folktales, and legends about earthquakes. Ask students to find out how people explained these events within the context of their cultures and belief systems. Students can share their research with classmates.

Program Resources

Study Guide, p. 10
Concept Mapping, p. 10, Forces that Cause Faults L1
Critical Thinking/Problem Solving, p. 10, *Invisible Faults* L3
Teaching Transparency 3 L1
Science Discovery Activities, 2-1
Section Focus Transparency 4

Discussion To explore cause and effect, ask students to share what they know about the cause of the California earthquake of 1989 and the damage it caused. Have students identify the fault that caused the earthquake. *San Andreas Fault* L1

2 TEACH

Tying to Previous Knowledge

Ask students to recall the definition of pressure from Chapter 1. *Pressure is the weight or force acting on each unit of area.* Then have students list examples of objects that break, bend, or tear under the application of too much pressure. *Answers might include rubber bands, metal nails, and pencil lead, among others.*

Theme Connection As students do the Find Out activity and read about forces inside Earth, they will see evidence of the theme of stability and change. When force is applied to an object such as a paper clip or a rock, the object changes shape, but usually returns to its original shape when the force is removed. If the force is great enough, this stability is overcome and the object is permanently changed.

Visual Learning

Figure 2-1 As students observe the three photographs, ask them to make a hypothesis, based on the photos, about which area is most likely to have had an earthquake in the past. *The location shown in Figure 2-1C is most likely to have had an earthquake, since rocks in that location were broken.*

Finding Fault

When rocks break under pressure, they may move. A fracture within Earth where rock movement occurs is called a **fault**. At the beginning of this chapter, you read about the San Andreas Fault. The rocks on one side of this fault move at a different speed from the rocks on the other side of the fault.

Faults can be found very near the surface, as in the case of the San Andreas Fault, or they can be found deep beneath Earth's surface. Regardless of where faults are found, they are always caused by forces within Earth that push the rocks together (compression), pull the rocks apart (tension), or cause the rocks to slide past each other (shearing).

Movement along faults can result in dramatic changes on Earth's surface. **Figure 2-3** on pages 58-59 illustrates the different types of faults and some of the land features that result from these faults.

SKILLBUILDER

Sequencing

Arrange these events that lead up to an earthquake in correct order. If you need help, refer to the **Skill Handbook** on page 636.

- 1 rocks undergo pressure
- 5 earthquake
- 3 elastic limit exceeded
- 2 rocks bend and stretch
- 4 rocks break and move

Physics CONNECTION

Passing the Limit

Elasticity exists in both natural objects and those made by humans. Trees, for example, may look straight and stiff, but what happens when strong winds blow through a forest? Do all the trees snap and fall over? Of course not. Trees change their shape, bending when the force of the wind strikes them, then returning to normal when the wind stops.

Hurricane Force

But what happens during a hurricane? If the wind is too strong, the force becomes too great for tree trunks and branches, and they pass their elastic limit and break. The same thing happens when you break a stick. If you bend it gently and then let go, it springs back to its original shape. But if you exert too much pressure, you force the stick beyond its elastic limit, and it snaps in two.

What Happens?

What actually happens when objects pass their elastic limit? As you bend a stick, the energy you exert is stored inside. When you let go, that energy is

Physics CONNECTION

Purpose

All materials, including Earth's rocks, have unique elastic limits. Once a material exceeds its elastic limit it will not return to its original shape when the applied force is released. The Physics Connection explains the property of elastic limit, in terms of the energy stored and released by a material.

Content Background

The energy an object has because of its position or condition is called its potential energy. The energy of motion is called kinetic energy.

Stress is a force divided by the area over which it is exerted. Strain is the deformation caused by stress. Initial deformation of rocks is reversible, but if stress is exerted beyond the elastic limit, rocks will experience permanent deformation. Some rocks change shape as they are deformed; brittle rocks break if stress is applied beyond the elastic limit.

Teaching Strategy

Use a rubber band to demonstrate how

Figure 2-2

Faults along the surface are easily-observed evidence of the powerful forces at work deep within Earth.

Uncovering Preconceptions
Explain that mountains rise at relatively slow rates.

Demonstration Use a slender, flexible stick to demonstrate that the elastic limit of a material is related to the force exerted on it. Hold the stick at both ends and gently bend it. Have students observe that when you release the force, the stick returns to its original shape. Now, ask students to predict what will happen if you apply a stronger force to the stick. Bend the stick beyond its elastic limit so that it breaks in two. Have students relate this demonstration to forces exerted on Earth's rocks.

released as the stick springs back into its original shape. Push too hard, and the energy is released as the stick breaks. This release of energy is called elastic rebound.

If you blow into a balloon, you exert a force that pushes the walls of the balloon out. If you continue to inflate the balloon, it will explode when the balloon wall becomes too weak to store the energy. It will reach its elastic limit and release the stored energy, resulting in a pop. The release of energy in the pop is the elastic rebound.

If a car tire were inflated beyond its elastic limit, the resulting explosion would be much greater than that of a balloon bursting. That's because the tire could store much more energy before reaching its elastic limit.

Earthquake!

When different forces are exerted on rocks inside Earth, the rocks, like the balloon, can store energy—up to a point. But eventually the forces are too great, and the rocks reach their elastic limit. They break. And what happens next? An earthquake! As the rocks pass their elastic limit, energy is released, causing seismic waves to travel out in all directions from the focus of the earthquake.

What Do You Think?

Can you think of examples of objects around you in which energy is stored? Look for objects in your classroom that have not yet reached their elastic limit.

Across the Curriculum

Math

The January, 1994, Northridge earthquake registered 6.5 on the Richter scale. The 1989 Loma Prieta earthquake was 7.1. For each full unit in the scale, the energy release increases 30 fold. The amplitude of each surface wave increases 10 fold. Have students calculate whether there was a bigger increase in energy released or amplitude of surface wave for the two earthquakes.

energy can be stored in an object. Pull the rubber band to show how it gains potential energy because of its stretched condition. Release one end and help students conclude that the potential energy could be transferred to another object such as a paper clip.

Answers to
What Do You Think?
Students themselves have energy because of the foods they eat. The potential energy stored in foods is chemical potential energy. Most objects in the classroom will not have reached their elastic limits, unless they have been bent, stretched, or compressed beyond their limit.

Going Further ⅢⅢ➡
Have groups of students use chalk and an eraser to demonstrate the concepts of potential and kinetic energy. Have one student lift an eraser. Ask the group to discuss if there is a change in energy. *The eraser has gained potential energy.* Now ask the student to drop the eraser and ask what happened to the eraser. *It loses potential energy but gains kinetic energy when it falls.* Then have another student break a piece of chalk and have the group reach a consensus as to what happened using the terms *potential energy, kinetic energy,* and *elastic limit.* **COOP LEARN**

Figure 2-3

In the Explore activity at the beginning of this chapter, you used tension to pull clay apart and compression to push clay together. Rock layers experience tension, compression, and a third type of force called shearing. You can start to explore movement along a fault in the activity below.

Normal fault

Force

A When rock moves along a fracture caused by tension forces, the break is called a normal fault. Rock above the fault moves downward in relation to the rock below the fault surface. Normal faults can form mountains such as the Sierra Nevada which borders California on the east.

Explore! ACTIVITY

How do rocks at a strike-slip fault move past each other?

What To Do

1. Glue sandpaper to the narrow side of two 2 inch × 4 inch blocks of wood. Use just enough glue to make it stick. (Don't use too much glue.) Press the sandpaper covered edges together to help set the glue and hold the sandpaper in place.

2. Now flip the blocks over so that the sides without sandpaper are together on a desk or table. Push them past each other in different directions so that the side of one block rubs against the side of the other.

3. After the glue has dried, place the sandpaper sides together and push them past each other in different directions so they rub together.

4. *In your Journal*, describe the differences between pushing the smooth surfaces of the blocks past each other and pushing the rough surfaces of the blocks past each other.

How do rocks at a strike-slip fault move past each other?

Time needed 15–20 minutes

Materials two 2 inch × 4 inch blocks of wood per student, sandpaper, glue

Thinking Processes observing and inferring, recognizing cause and effect, making models

Purpose To model the way rocks at a strike-slip zone move past each other.

Teaching the Activity

The sandpaper need not cover the entire side of each block to produce the desired effect. **L1**

Science Journal
Have students record their observations in their journals. **L1**

Expected Outcome

Students should compare and contrast the movements of the sandpapered and bare blocks and conclude that the sandpapered blocks are more difficult to move against each other.

✔ Assessment

Oral Have students relate their observations to the movements of rocks at a strike-slip fault. Use the Performance Task Assessment List for Oral Presentations in **PASC**, p. 71.

Reverse fault

Force ← →

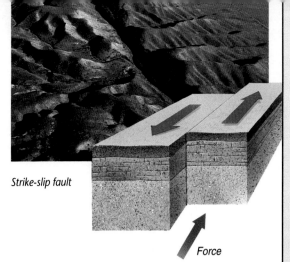

Strike-slip fault

Force

B When compression forces break rock, the rock above the fault surface moves upward in relation to the rock below the fault surface. The mountains shown make up part of the Himalayas separating India and China, and contain many reverse faults.

C Shearing forces push on rock in opposite, but not directly opposite, horizontal directions. When strong enough, these forces split rock and create strike-slip faults such as California's San Andreas Fault. Movement along faults is usually short and sudden but can be slow and steady.

Why is it important to know about faults? Think back to the Explore activity in which you studied how rocks move past each other at strike-slip faults. Suppose the blocks you used were really rocks on either side of the San Andreas Fault. Do you think the rocks would slide past each other at a constant rate? What would happen if the irregular surfaces of the rocks snagged, but shearing forces inside Earth continued to push them? Pressure would build up, and the elastic limit of the rocks would be exceeded. Then rocks along the fault surface would break and move. An earthquake might result.

Most earthquakes are the result of faulting in Earth's rocks. Earthquakes occur as a result of all three types of faults. In the next section you'll find out how scientists locate earthquakes.

check your UNDERSTANDING

1. Explain how earthquakes can result from the buildup of pressure inside Earth.
2. What happens when compression forces are exerted on rock layers? Tension forces? Shearing forces?
3. How do strike-slip faults differ from reverse and normal faults?

4. **Apply** A stream flows through an area where there are many faults. The stream takes a sharp turn to the left, flows for a short distance, and then takes a sharp turn to the right. It flows straight beyond that. How can you explain the stream's sharp turns? Draw diagrams to explain your answers.

check your UNDERSTANDING

1. Pressure builds up because of forces acting on rocks. When the rocks under pressure break and move, vibrations are sent out and an earthquake may occur.
2. Compression can cause rocks to bend and reverse faults to form. Tension causes normal faulting. Shearing causes strike-slip faults to form.

3. Rocks along a strike-slip fault move past each other without much vertical movement. Rocks at normal and reverse faults move vertically.
4. Strike-slip faults could cause the stream to make sudden sharp turns as they follow the direction of the fault.

Check for Understanding

Have students use their hands to demonstrate the movements represented by the terms *compression, tension, shearing, normal fault, reverse fault,* and *strike-slip fault.*

Reteach

Demonstration To differentiate between motions of rock masses at faults, use two of the sandpaper-covered blocks from the activity on p. 58. Move the sandpapered surfaces together to show a strike-slip fault and use two edges without sandpaper to show movement at normal and reverse faults. L1

Extension

Activity To determine how many people live near the San Andreas Fault, have students use maps to locate the fault and make a list of cities or towns located along it. L3

4 CLOSE

Demonstration

To contrast the movements along the three types of faults, have students build a six-layer sandwich. Materials needed are bread, jelly, and peanut butter.

Alternate jelly and peanut butter. Cut off crusts so layers can be seen. Construct a "fault" at about 30° by cutting through the sandwich. Move the halves of the sandwich to demonstrate normal, reverse, and strike-slip faults. L1

PREPARATION

Planning the Lesson

Refer to the Chapter Organizer on pages 52A–D.

Concepts Developed

Seismic waves and how they are used to locate earthquakes are discussed. Primary, secondary and surface waves are differentiated.

1 MOTIVATE

Bellringer

Before presenting the lesson, display **Section Focus Transparency 5** on an overhead projector. Assign the accompanying **Focus Activity** worksheet. L1

LEP

Demonstration
To simulate how surface waves move, use a rock and a shallow pan of water. Fill the pan three-fourths full. Ask a volunteer to drop a small rock into the water. Have students observe the waves generated. Explain that the wave movement in the water is similar to the movement of certain types of seismic waves. L1

 2-2 **Shake and Quake**

Section Objectives

- Compare and contrast primary, secondary, and surface waves.
- Explain how an earthquake's epicenter is located by using seismic wave information.

Key Terms
seismic waves
focus
epicenter

Seismic Waves

Have you ever played with a coiled-spring toy? If so, you probably know how it behaves going down stairs. You may also have used a coiled spring to study how waves travel through matter. Such a coil can help you get an idea of how vibrations travel as waves through Earth after an earthquake has occurred at a fault. Do the following activity to help you understand more about earthquake waves.

Explore! ACTIVITY

How can a coiled spring be used to demonstrate two types of earthquake waves?

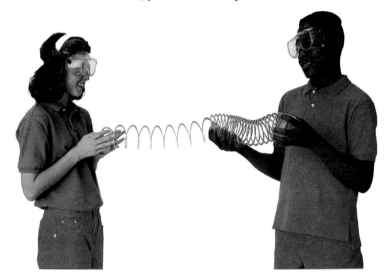

What To Do

1. Stretch a coiled spring between another person and yourself.

2. Squeeze 4 or 5 coils together and then release the squeezed portion. What happens?

3. Now move one end of the spring up and down quickly. *In your Journal*, record your observation of how the wire in the spring moves.

When you let go of the squeezed spring, you created a compression wave. Matter that is squeezed and stretched has a compression wave traveling through it.

The second wave you created was a transverse wave. A transverse wave causes matter to move at right angles to the direction that the wave is moving.

60 Chapter 2 Forces in Earth

Explore!

How can a coiled spring be used to demonstrate two types of earthquake waves?

Time needed 15 minutes

Materials coiled spring toy

Thinking Processes observing, making models

Purpose To model two types of seismic waves. LEP

Teaching the Activity

Form groups of two to three students. COOP LEARN

Science Journal Have students compare and contrast how the wave movements differed in each part of the experiment. L1

Expected Outcome

Students should observe horizontal and vertical movements.

✔ Assessment

Content Have students make bulletin board displays to show how the coils moved in step 2 and step 3 of the Explore activity. Use Performance Task Assessment List for Bulletin Board in **PASC**, p. 59.

Earthquakes generate waves that are similar to the waves you made with the coiled spring. Such waves are called **seismic waves**. The point in Earth's interior where seismic waves originate is the **focus** of the earthquake. The focus can be between 5 and 700 kilometers below the surface. Seismic waves travel outward through Earth in all directions from the focus. Scientists use an instrument called a seismograph to detect the waves. Seismographs can detect seismic waves that originated as far away as the other side of Earth. In this Find Out activity, you will graph two seismic waves to compare how fast they travel.

Find Out! ACTIVITY

Which is faster, a compression wave or a transverse wave?

The table shows when either a compression wave or a transverse wave arrived at a seismograph station at some distance from an earthquake. Assume that the earthquake occurred at precisely 4:00 P.M. The times given in the table are hour, minutes, and seconds. So a time of 4:06:30 indicates that the wave arrived at a station 6 minutes and 30 seconds after the earthquake occurred.

What To Do

1. Make a graph comparing the distance from the earthquake focus and the time it took for the waves to travel there.

2. How many lines will you plot on your graph? Answer the questions below *in your Journal*.

Conclude and Apply

1. Based on your graph, which is faster—a compression wave or a transverse wave?

2. Do the waves remain the same distance apart as they travel?

3. Do they grow closer together or farther apart over time?

4. How could you use your graph to fill in the missing arrival times?

5. Use your observations to tell why compression waves are called primary waves and transverse waves are called secondary waves.

Compression and Transverse Waves		
Distance from focus (km)	Time of Arrival (Hr:Min:Sec)	
	Compression wave	Transverse wave
250	4:01:00	—
500	—	4:03:00
1000	4:02:30	—
1750	4:04:00	—
2000	—	4:07:30
3000	—	4:10:15
4500	4:08:00	—
5000	—	4:15:00
6000	4:09:30	—
6500	—	4:18:00
7250	4:11:00	—
8000	4:11:45	—
8250	—	4:21:15
9000	4:12:30	—
9500	—	4:22:45

2 TEACH

Tying to Previous Knowledge
Have students list other waves with which they are familiar. *Answers might include water waves, sound waves, and/or light waves.*

Find Out!

Which is faster, a compression wave or a transverse wave?

Time needed   20 minutes

Materials   graph paper

Thinking Processes measuring in SI, making and using graphs

Purpose To make a graph to compare the speeds of seismic waves.

Teaching the Activity
The scale for the distance axis can use units of 250 or 500 km and must extend to 9500 km. The time axis must extend from 4:00 to 4:22:45. The scale for this axis can use units of 15 or 30 s. L1

Science Journal Have students make their graphs on graph paper and paste them into their journals. L1

Expected Outcome

Students should plot two lines—compression wave and transverse wave.

Conclude and Apply

1. Compression wave

2. No

3. The waves are farther apart the farther they travel.

4. The table can be completed by finding the distance of the station in the table, and then reading the time difference on the graph.

5. Compression waves travel faster, and arrive first.

✔ Assessment

Performance Have students draw freehand graphs that demonstrate the difference in speed of compression waves and transverse waves. Use the Performance Task Assessment List for Graphs in **PASC**, p. 39.

Figure 2-4

Primary and secondary waves can be detected at Earth's surface by a seismograph. But remember that primary and secondary waves originate from an earthquake's focus and generally travel through Earth's interior. The point on Earth's surface directly above the focus is the **epicenter**. When seismic waves from the focus of an earthquake reach the epicenter, they generate surface waves. These surface waves travel outward from the epicenter the way ripples on a pond's surface travel outward when you throw a stone into the water. Surface waves travel more slowly than secondary and primary waves and they cause the greatest damage. You can see the directions of motion of all three kinds of seismic waves in **Figure 2-4**.

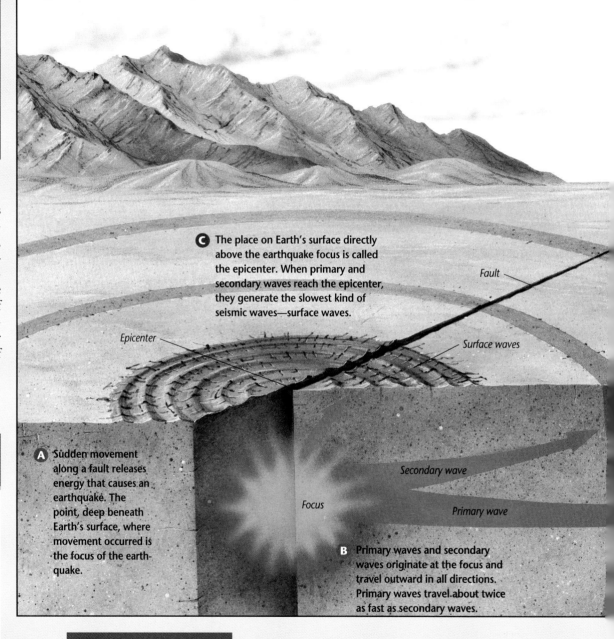

C The place on Earth's surface directly above the earthquake focus is called the epicenter. When primary and secondary waves reach the epicenter, they generate the slowest kind of seismic waves—surface waves.

Epicenter

Fault

Surface waves

A Sudden movement along a fault releases energy that causes an earthquake. The point, deep beneath Earth's surface, where movement occurred is the focus of the earthquake.

Secondary wave

Focus

Primary wave

B Primary waves and secondary waves originate at the focus and travel outward in all directions. Primary waves travel about twice as fast as secondary waves.

Locating an Epicenter

As you discovered, seismic waves don't travel through Earth at the same speed. **Figure 2-5** shows how far primary and secondary waves travel in a certain amount of time. You can use the graph to determine how far away an epicenter is from a seismograph station. Say your seismograph detects a primary seismic wave at precisely 3:00 P.M. At 3:06 and 30 seconds, the secondary wave arrives. Find the place on the graph where the two curved lines are separated by 6 minutes and 30 seconds. Where on the graph does this occur? It occurs at the 5000-kilometer mark. How far away is your seismograph station from the earthquake epicenter?

Based on your calculations, you can say that an earthquake occurred about 5000 kilometers from your seismograph station. But can you say in which direction? So far, you don't have enough information to answer that question. Can you think of a way to determine the exact location of the epicenter? Do the following Investigate activity to see if you're right.

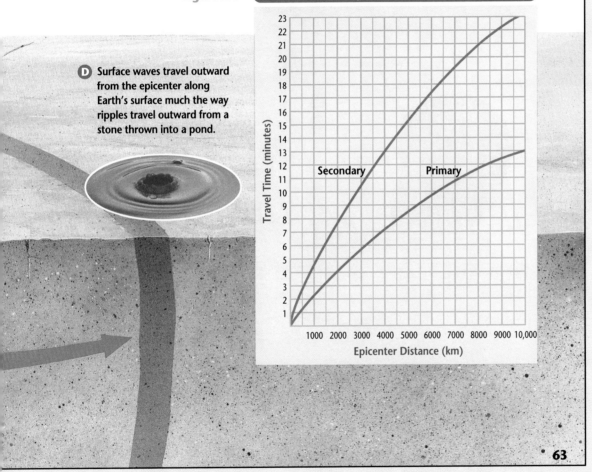

Figure 2-5

D Surface waves travel outward from the epicenter along Earth's surface much the way ripples travel outward from a stone thrown into a pond.

Primary and Secondary Waves

(Graph: Travel Time (minutes) vs. Epicenter Distance (km), showing Secondary and Primary wave curves)

INVESTIGATE!

2-1 Where's the Epicenter?

Planning the Activity

Time needed 40 minutes

Purpose To interpret arrival time data to locate an earthquake epicenter.

Process Skills making and using tables, measuring in SI, interpreting data

Preparation Gather enough equipment so that students can work in groups of three.

Teaching the Activity

Have students make their measurements of the circumference of the globe at the equator. If globes are mounted on stands so that they rotate, have students take turns holding the globe steady and measuring its circumference. `L1` **COOP LEARN**

Process Reinforcement Remind students that a scale is a way of converting real-world distances to distances on a map or globe. Ask students which map would record a larger area of Earth's surface, a 20 cm by 30 cm map with a scale of 1 centimeter to 2 kilometers, or a 20 cm by 30 cm map with a scale of 2 centimeters to 50 kilometers? *The second map would record a larger area. The first scale might be used for a detailed street map of a city or town; the second scale might be used for a road map of a state or region.*

Science Journal Have students record their data, make their sketches, and write their answers in their journals.

Where's the Epicenter?

You know that primary waves travel faster than secondary waves and therefore arrive at a seismograph station first. In the following activity, you'll locate epicenters using the arrival times of these waves.

Problem
How are epicenters located?

Materials
paper	Figure 2-5
string	metric ruler
globe	water-soluble marker or chalk

What To Do

1 Copy the data table on the next page *into your Journal.*

2 Determine the difference in arrival times between the primary and secondary waves at each station for each earthquake in the table on page 65.

3 *Interpret the graph* in Figure 2-5 to determine the distance in kilometers of each seismograph from the epicenter of each earthquake. Record these data. An example has been done for you.

64 **Chapter 2** Forces In Earth

Program Resources

Activity Masters, pp. 11–12, Investigate 2-1

ENRICHMENT

Research Have students conduct research about major earthquakes that have occurred during recent history. Then, ask them to make time lines of these earthquakes from the oldest to the most recent. The magnitude of each quake should also be included. Students can then mark the locations on a world map. Display time lines in the classroom. `L2`

Data and Observations

Sample data

Earthquake	Calculated distance from epicenter (km) for each seismograph location				
	(1)	(2)	(3)	(4)	(5)
A	3750	4250	8000	9750	
B	4500	1250		9750	8500

A

4 Using the string, *measure* the circumference of the globe by using the ruler to measure the length of string needed to circle the globe (see photo **A**). Determine a scale of centimeters of string to kilometers on the surface of Earth (Earth's circumference = 40 000 km).

5 For each earthquake, A and B, place one end of the string at each seismic station location. Use the marker or chalk to draw a circle with a radius equal to the distance to the epicenter of the earthquake.

Location of Seismograph	Wave	Wave Arrival Times	
		Earthquake A	Earthquake B
(1) New York	P	2:24:05 PM	1:19:00 PM
	S	2:28:55 PM	1:24:40 PM
(2) Seattle	P	2:24:40 PM	1:14:37 PM
	S	2:30:00 PM	1:16:52 PM
(3) Rio de Janeiro	P	2:29:00 PM	——
	S	2:38:05 PM	——
(4) Paris	P	2:30:15 PM	1:24:05 PM
	S	2:40:15 PM	1:34:05 PM
(5) Tokyo	P	——	1:23:30 PM
	S	——	1:33:05 PM

Analyzing

1. The epicenter is the point at which all the circles intersect. What is the location of the epicenter of each earthquake?

2. Compare the distance of a seismograph from the earthquake with the difference in arrival times of the waves. How are these data related?

Concluding and Applying

3. What is the minimum number of seismograph stations needed to locate an epicenter accurately?

4. Going Further What information would only two seismograph stations give you in regard to the location of an epicenter? Make a drawing to show what information two stations would provide about the epicenter.

Expected Outcome

Using a table of arrival times of seismic waves, students should be able to make measurements to locate earthquake epicenters.

Answers to Analyzing/ Concluding and Applying

1. A: Mexico City, Mexico; B: San Francisco, California

2. The difference in arrival times between primary waves and secondary waves increases as the distance of the seismic station to the earthquake increases.

3. three

4. Data from two seismic stations would yield two possible locations for the epicenter. (Two circles would intersect at 2 points.)

✍ **Assessment**

Process Have students use the table of differences in the arrival of primary and secondary waves and Figure 2-5 to find the distance the epicenters of earthquakes A and B are from the seismographic station C.

Readings at Station C

A P 2:37:08   **B** P 1:28:17
  S 2:43:17   S 1:35:53
time difference A about 6 minutes; time difference B about 7 minutes and 30 seconds; distance from epicenter A to C about 4500 km; distance epicenter B to C about 6000 km Use the Performance Task Assessment List for Using Math in Science in **PASC**, p. 29.

Multicultural Perspectives

Earthquake Detection

Have students find out about ancient devices made by the Chinese to detect Earth movements. Many of these earthquake-detection devices included statues of mythical creatures with metal balls balanced in their mouths. The metal balls fell in response to certain Earth movements. Have interested students contrast these detectors with modern seismographs. Students can record their findings as a written report. Some students may wish to draw a picture of a device of their own invention to detect Earth movements. **P**

Check for Understanding

Have students work in small groups to review what they learned about seismic waves. Orally quiz groups to assess their understanding. For example, ask students to distinguish between primary and secondary waves. [L1] **COOP LEARN**

Reteach

Demonstration Use Figure 2-6 to help students differentiate primary, secondary, and surface waves and their motion. [L1]

Extension

Discussion Ask students who understand the concepts to think about factors that influence the amount of damage caused by an earthquake. Students should be able to deduce that strength, the types of structures present, population density, preparedness, and the type of ground all affect the amount of damage done. [L3]

4 CLOSE

Using the graph on page 63, have students identify the primary and secondary waves and determine the time lag between the arrival of primary waves and the arrival of secondary waves at a spot 1500 km away, 2250 km, 4000 km, and 7000 km. *2:45, 3:45, 5:30, 8:30* Have students use the graph to determine the distance from the epicenter when the difference in primary and secondary wave arrival time is 6 hr. 30 min. *5000 km* [L1]

Visualizing Waves

Figure 2-6

A Primary waves cause alternating compression (pushing together) and stretching (pulling apart) in rock, in the direction of the wave. Primary waves cause back-forth, rocking movements on Earth's surface.

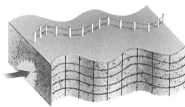

B Secondary waves cause rock to vibrate at right angles to the direction of the wave. This motion causes side-to-side movements on Earth's surface.

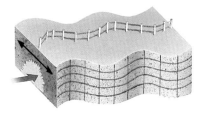

C One type of surface wave creates side-to-side, rocking motions on Earth's surface.

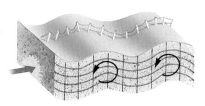

D A second type of surface wave creates up-down, and rolling motions on the surface.

Primary and secondary waves are used to determine the location of an earthquake's epicenter. However, these waves don't usually cause major damage at the surface. When you think of an earthquake, you probably think of shaking ground and crumbling buildings. Surface waves are responsible for most of this destruction.

Figure 2-6 shows the effects of earthquake waves. You can see why surface waves cause the most damage. The waves cause one part of a building to move up, while another part moves down. At the same time, the building moves from side to side. In some areas of the United States, Japan, and elsewhere in the world, many buildings are constructed so that they can better withstand vibrations caused by surface waves.

Thus far, you've learned how forces inside Earth can send seismic waves through Earth and along its surface, and how seismologists use these waves to locate the epicenter of an earthquake. In the next section, you will find out how forces inside Earth can cause volcanoes to erupt.

check your UNDERSTANDING

1. Compare the origins of the three kinds of seismic waves. Which originate at an earthquake's focus? Which originate at an earthquake's epicenter?
2. What do you think would happen to a row of evenly spaced utility poles as a surface wave travels along the row?
3. **Apply** If all seismic waves traveled at the same speed, could the epicenter be located? Explain.

check your UNDERSTANDING

1. Primary and secondary waves originate at the focus of an earthquake. Surface waves originate at the epicenter of an earthquake.
2. The poles would move vertically and from side to side as the waves passed by.
3. No; the difference in the speeds of the waves is used to determine the distance from a seismic station to an earthquake's epicenter. If there were no difference in arrival time, it would not be possible to calculate the distance from the epicenter. However, if you knew when the earthquake occurred, you could use the time and speed to determine the distance.

2-3 Volcanic Eruptions

What Causes Volcanoes to Erupt?

You've learned that earthquakes occur along faults such as the San Andreas Fault shown in the picture at the beginning of this chapter. Volcanoes and earthquakes often occur in the same regions. In fact, the movement of magma and volcanic eruptions may trigger some earthquakes.

Although heat and pressure within Earth can cause rocks to melt and form magma, some rocks deep inside Earth are already melted. Others are so hot that only a small rise in temperature or slight change in pressure is needed to melt them and form magma.

What causes magma to rise toward the surface and erupt to form a volcano? You can see how this process takes place if you do the following activity.

Section Objectives

- Explain what causes volcanoes to erupt.
- Explain the difference between a quiet and an explosive eruption.

Key Terms
vent

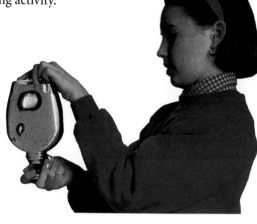

Explore! ACTIVITY

How does magma move?

What To Do

1. Turn a closed bottle of cold syrup upside down.
2. Observe what happens. Record your observations *in your Journal.*
3. What causes the bubbles to rise? Which is less dense—the syrup or the air?

Magma rises toward Earth's surface for the same reason that air bubbles in syrup rise. Less-dense materials are pushed upward by denser materials. Magma is less dense than the rocks around it, so it's pushed up by the denser material and rises toward the surface.

When magma reaches Earth's surface, it flows out through openings called **vents**. This event is called an eruption, and at this point, magma is called lava. Lava flows out, cools, and forms layers of volcanic rock around the vent. Volcanic material may pile up around the vent to form a cone. How does the type of magma determine the type of eruption that occurs and the kind of volcanic cone that forms?

Explore!

How does magma move?

Time needed   5 minutes

Materials   bottles of cold maple or corn syrup

Thinking Processes   observing and inferring

Purpose   To model the action of the forces that cause magma to rise

Teaching the Activity

Safety   Be sure bottles are tightly closed. Tell students to hold one hand under the bottle cap.

Troubleshooting   Use very cold syrup.

Science Journal   Have students describe their observations and answer the questions in their journals. L1

Answers to Questions

3. They are less dense than the surrounding syrup, therefore the syrup pushes the air bubbles up. The air is less dense.

PREPARATION

Planning the Lesson

Refer to the Chapter Organizer on pages 52A–D.

Concepts Developed

Volcanoes can erupt explosively or quietly. The nature of the volcanic eruption determines the kind of volcano formed.

1 MOTIVATE

Bellringer

Before presenting the lesson, display **Section Focus Transparency 6** on an overhead projector. Assign the accompanying **Focus Activity** worksheet. L1
LEP

Demonstration   To simulate a volcanic eruption, you can make a model. Materials needed are modeling clay, ¼ cup baking soda, red food coloring, vinegar, a small metal cup, and safety goggles. Beforehand, make a hollow clay model of a volcano with a crater at the top. Have students put on their safety goggles. Put the baking soda and a few drops of red food coloring into the metal cup and put the cup inside the volcano. Add about 20 mL of vinegar to the baking soda and have students observe. L1

✔ Assessment

Oral   Have students name another liquid that is more dense than air and describe how the experiment would work with this liquid. Use the Performance Task Assessment List for Oral Presentations in **PASC**, p. 71.

2 TEACH

Tying to Previous Knowledge

Remind students that in Chapter 1, they learned that fluid in a container exerts pressure in all directions. Ask students how this principle would relate to a fluid beneath Earth's surface. *Magma, a fluid, exerts pressure in all directions.*

Theme Connection The theme of energy can be seen as students learn about volcanic eruptions. Thermal energy is involved in the formation of magma and the buildup of pressure that leads to volcanic eruptions.

Visual Learning

Figure 2-7 To compare the results of different types of volcanic eruptions, have students look at the photographs and read the captions for Figures 2-7A and 2-7B. Ask them what they think determines how explosively a volcano will erupt.

GLENCOE TECHNOLOGY

Software
Computer Competency Activities
Chapter 2

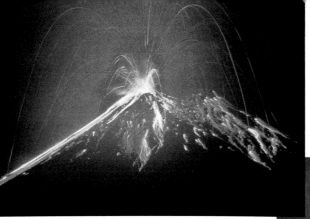

Figure 2-7
Ⓐ Some volcanoes erupt with such violent explosions that the sound is heard thousands of miles away and the force of the blast blows away great chunks of the volcano. The Arenal volcano in Costa Rica erupted with violence in July 1991.

Ⓑ By comparison, some other volcanoes erupt rather quietly. Melted rock boils out of one or more openings in Earth and flows steadily outward, until it cools. Mount Kilauea, in Hawaii, had such a history of frequent but relatively quiet eruptions that it was chosen as the site of a permanent volcanic observatory in 1912.

Plate Tectonics

Are there earthquakes and volcanic eruptions near where you live? Scientists have developed a theory called *plate tectonics* that explains how and where earthquakes and volcanic eruptions occur.

According to the theory of plate tectonics, Earth's outer layer is made of many rigid pieces, called plates, that move around on the denser plastic-like rock forming the layer beneath.

When Plates Collide

When plates move toward each other, they collide and the edge of one plate can slip underneath the edge of the other. The collision can cause earthquakes along the meeting edges. The portion of the plate that slips underneath begins to melt in the great heat of the mantle. This newly melted material is less dense than the rock of the mantle, so the newly melted material is forced upward and may flow through the crust in places, causing volcanic eruptions.

What kind of volcanic eruptions occur where plates are colliding? You already know that the newly melted material is less dense than mantle material. So you might expect the newly melted material to contain lots of dissolved gases—and you'd be right. Magma

---

Purpose
A Closer Look extends the discussion describing how scientists apply the theory of plate tectonics to explain earthquake and volcanic activity.

Teaching Strategies
Show students a map of Earth's tectonic plates. Have them locate the volcanoes listed on page 73 on the map. Ask: **Why are earthquakes and volcanoes likely to occur in the same areas?** *Because these* are the areas where movement of tectonic plates allows magma to reach Earth's surface.

*inter*NET CONNECTION
The World Wide Web site for the Hawaii Volcano Observatory can be found at (http://www.soest.hawaii.edu/hvo).

Types of Eruptions

Gases such as water vapor and carbon dioxide are trapped in magma. Surrounding rock puts pressure on magma when it is deep underground. Under pressure, magma can contain many dissolved gases. But when the magma is pushed toward the surface, the pressure on it decreases. Magma under low pressure cannot hold as much gas. As a result, the gas begins to escape as the magma nears Earth's surface in an explosive blast. Gases in thin, fluid magma escape in a quiet way because they are released gradually before pressure builds up.

Some magma contains a lot of the compound silica. Silica-rich magma tends to be very thick. In fact, it can be so thick that it clogs a volcanic vent. Magma that is trapped below the clogged vent and the gases within the magma are under even greater pressure than before. Eruptions of volcanoes with silica-rich magma are usually very explosive. The magma itself often explodes forming dust, ash, and rock fragments as it cools. When low-silica magma reaches the surface, it tends to flow from a vent in a much less explosive manner.

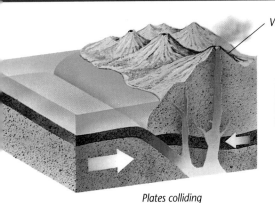

Plates colliding

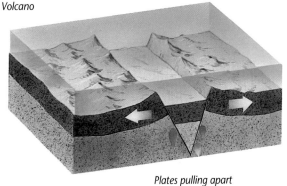
Volcano

Plates pulling apart

containing much gas under pressure causes explosive eruptions that hurl dust, ash, and rock fragments.

When Plates Pull Apart

As plates move away from each other, magma located a few kilometers below the surface rises to fill the gap. There isn't much stress on the plates here, so earthquakes are not as violent as they are where plates come together. The volcanic eruptions here tend to flow easily rather than exploding, partly because the magma from the deep mantle is very fluid and contains few dissolved gases.

inter NET CONNECTION

Use the World Wide Web to find out more about volcanoes on Hawaii. When and where was the most recent lava flow from Mt. Kilauea? How are the volcanoes on Hawaii different from those discussed in the article? How are they similar?

2-3 Volcanic Eruptions **69**

Going Further ▸
Have students trace the major tectonic plates on outline maps of the world. Have them use data from the table on page 73 to determine if there is a relationship between location and type and force of the volcano. L3 P

Program Resources

Study Guide, p. 12.
Making Connections: Integrating Sciences, p. 7, *Are There Volcanoes on Other Planets?* L2
Making Connections: Technology & Society, page 7, *Deep Heat* L2
Section Focus Transparency 6

Across the Curriculum

Writing

Have students read back issues of news weeklies and general interest magazines for articles that include first person accounts of the eruption of Mt. St. Helens. Then ask them to imagine that they witnessed a volcanic eruption and write a first-person fictionalized account.

Discussion From 1986 to 1989, Kilauea produced enough lava to cover a four-lane highway from New York to San Francisco nine meters deep! Ask students how would the impact of Kilauea be different if the volcano erupted explosively?

Inquiry Questions Why does silica-rich magma erupt so explosively? *The silica-rich magma tends to trap water vapor and other gases. Pressure then builds within the magma, and when it is finally released, an explosive eruption occurs.* What does magma tell us about Earth's interior? *Because magma is melted rock material, we can assume that Earth's interior is hot.*

GLENCOE TECHNOLOGY

Videodisc

The Infinite Voyage: Living With Disaster

Chapter 8
Nevada del Ruiz: A Volcanic Disaster

Chapter 9
The Nature of Volcanoes: Nevada del Ruiz

Chapter 10
Adapting to the Nature of Volcanoes: Sakurajima

The Infinite Voyage: To The Edge Of The Earth

Chapter 3
Exploring Volcanoes: To the Center of the Earth

Forms of Volcanoes

Figure 2-8

Magma   *Shield volcano*

Ⓐ When hot, thin lava flows without violent explosions out of one or more vents and then cools and hardens, it builds into a gentle slope. The result is a shield volcano, such as this one at Mauna Loa in Hawaii.

Ⓑ When volcanic ash and slightly cooled lava are forced out of a single vent during an explosive eruption, they fall back to Earth. The ash and cooled lava form the steep slope of a cinder cone around the vent. Parícutin in Mexico, pictured in the photograph, is a cinder cone.

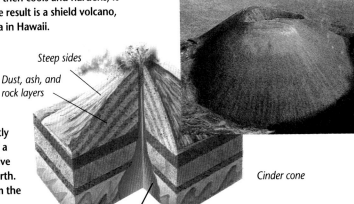

Steep sides
Dust, ash, and rock layers
Magma
Cinder cone

Ⓒ If a volcano throws out flowing lava, hardened lava, and chunks of ash, a composite cone, such as Mt. Shasta in California shown here, is formed. Because the layers of lava cover and protect the layers of loose materials from erosion, composite cones build into steep-sided towering mountains.

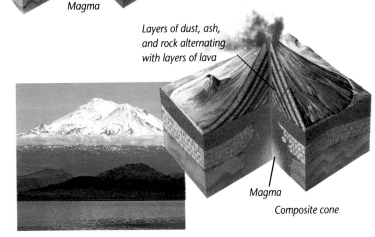

Layers of dust, ash, and rock alternating with layers of lava
Magma
Composite cone

The form of volcano that is produced is determined by the nature of its eruption. If the eruption is quiet, a gently sloping shield volcano is produced, like the one shown in **Figure 2-8A**. If the eruption is explosive, a steep-sloped cinder cone is produced, like the one shown in **Figure 2-8B**. Sometimes, however, eruptions alternate between quiet and explosive. Then a composite cone volcano is produced, like the one shown in **Figure 2-8C**.

 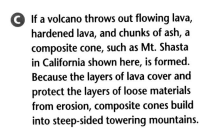

Volcanoes and Humans

Volcanoes affect people's lives all over the world. On November 13, 1985, Nevado del Ruiz erupted explosively, killing nearly 23,000 people in Armero, Colombia. On May 26, 1991, Mount Unzen in Japan erupted explosively after having been dormant for about 200 years. Soon after, the eruptions of Mount Pinatubo in the Philippines killed nearly 900 people during the month of June 1991.

Figure 2-9

A Before its eruption in the spring of 1980, Mount St. Helens's snow-capped cone was a magnificent sight and a favorite of many people.

B In March of 1980, Mount St. Helens began a series of explosive eruptions that triggered fires and flooding. Hundreds of tons of volcanic ash fell. More than 60 people lost their lives. Thousands of large animals including deer, elk, and bear, and all the birds and small mammals in the area also died. Crops were ruined. The local timber industry was wiped out.

C Within 10 days of the eruption, fresh deer tracks were found in the ash on Mt. St. Helens. Within a month, fireweed was blooming at a nearby lake. Today, Mount St. Helens continues its comeback.

Discussion Ask students why a volcanic eruption is often a traumatic event. *Because eruptions can destroy croplands and forests and may cause fires and floods that destroy homes and buildings.* Point out that volcanic eruptions can have effects far beyond the immediate area. For example, after the 1980 explosion of Mt. St. Helens in Washington state, winds deposited volcanic debris over most of the United States.

Visual Learning

Figure 2-9 To reinforce the sequence of events have students discuss the differences among the three photographs of Mt. St. Helens. Ask them to compare the renewal of the area around the volcano with what they know about the aftermath of a forest fire.

Across the Curriculum

Reading

Many famous volcanic eruptions around the world have been described in detail in books. Have students go to the library and work in pairs to find and read books about famous eruptions, such as the 1902 eruption of Mount Pelée, and how they affected the people nearby. **L1** **COOP LEARN**

NATIONAL GEOGRAPHIC SOCIETY

 Videodisc

STV: Restless Earth
Volcanoes in Iceland
Unit 2, Side 1
Volcanoes in Iceland

14560-37468

Kauai, Hawaii

52231

Multicultural Perspectives

A Different View

In 1793 near the town of San Andre's de Tuxtla, Vera Cruz, Mexico, a long dormant volcano erupted violently and repeatedly. Its explosions could be heard 200 miles away. A daring 36-year-old scientist named José Mariano Moziño climbed to the edge of the crater to observe the lava, putting himself in great danger to gain scientific knowledge. Today researchers can survey eruptions from the relative safety of helicopters or airplanes.

2-2 Identifying Types of Volcanoes

Preparation

Purpose To interpret data to make generalizations about factors that influence types of volcanic eruptions.

Process Skills classifying, making and using tables, recognizing cause and effect, interpreting data

Time Required 40 minutes

Materials paper, pencil, table on page 73

Possible Hypotheses Before beginning the activity, ask students to pose some possible hypotheses about which factors determine the form of a volcano. Possible hypotheses may include: Volcanoes that erupt with great force usually produce a cinder cone, while volcanoes that erupt with low force produce a shield cone. After reviewing the type of data contained in the table on page 73, ask students what kind of information they will look for to prove or disprove each of the hypotheses that have been suggested.

Plan the Experiment

Process Reinforcement Have students compare the type of volcano and the eruptive force that causes each type. Is any type of cone associated with all three levels of eruptive force? *no* What type of cone is associated with a high ability of lava to flow? *shield* What other characteristics are shared by volcanoes that have magma with a high ability to flow? *The magma has low silica and water vapor content.*

Possible Procedures Design a chart to record your results. Label the vertical axis "Silica Content of Magma." Label the horizontal axis "Water Content of Magma." Divide the vertical axis into two equal parts, labeled "High" and "Low." Do the same for the horizontal axis. The graph should now contain four equal squares. Using the information provided in the table on page 73, plot the magma

Identifying Types of Volcanoes

Some volcanoes erupt explosively with little or no lava. Others erupt quietly with thin, runny lava. As you have learned, certain properties of magma are related to the type of volcanic eruption that occurs and the form that the volcano will develop. Try this activity to see how magma is related to volcanic eruptions and volcanic forms.

Preparation

Problem

How could you interpret data to show the relationship between the properties of magma and the properties of volcanoes?

Form a Hypothesis

As a group, discuss the factors that determine different volcanic forms and different volcanic eruptions. Then form a hypothesis about the relationship between magma type and volcano type.

Objectives

- Infer which properties of magma are related to volcanic eruptions and volcanic forms.
- Classify different types of volcanoes according to magma content.
- Recognize the effect magma content has on eruptive force and volcanic form.

Materials

paper
pencil

Program Resources

Activity Masters, pp. 13–14, Design Your Own Investigation 2-2 L2

Plan the Experiment

1 As a group, agree upon a way to test your hypothesis. Write down what you will do at each step of your test.

2 Examine the chart of selected volcanoes at right. Then make a list of the properties of magma that you will use to classify the volcanoes according to form, eruptive force, and ability of magma to flow.

3 With your teacher's help, design a way to summarize your data *in your Science Journal.*

Check the Plan

Discuss and decide upon the following points and write them down.

1 Which properties of magma will you use to classify the volcanoes?

If you're unsure, examine the chart below again.

2 How many data tables will you construct? How will you identify the volcanoes in each of your tables?

3 Make sure your teacher approves your experiment before you proceed.

4 Carry out your experiment. Record your observations.

Selected Volcanoes					
Volcano and Location	Magma Content		Type of Volcano	Eruptive Force	Ability of Magma to Flow
	Silica	Water Vapor			
(1) Etna, Italy	high	low	composite	moderate	medium
(2) Tambora, Indonesia	high	high	cinder	high	low
(3) Lassen, California	high	low	composite	moderate	low
(4) Mauna Loa, Hawaii	low	low	shield	low	high
(5) Parícutin, Mexico	high	low	cinder	moderate	medium
(6) Kelut, Indonesia	high	high	cinder	high	low
(7) Helgafell, Iceland	low	high	shield	moderate	medium
(8) Saint Helens, WA	high	high	composite	high	medium
(9) Laki, Iceland	low	low	shield	moderate	medium
(10) Kilauea Iki, Hawaii	low	low	shield	low	high

Analyze and Conclude

1. Compare and Contrast Which eruptions are the most explosive— eruptions with high silica content or low silica content?

2. Interpret Data What effect does the silica content and water-vapor content of the magma have on its ability to flow?

3. Recognize Cause and Effect Infer which of the two variables (silica or water vapor) has the greater effect on the eruptive force of a volcano.

Going Further

Summarize the relationship between volcanic form, the silica content, and the water-vapor content of the magma.

ENRICHMENT

Activity Students will use sand, gravel, and a protractor to determine the angle of repose of a slope of a volcano. Have students make small piles of each material, then measure the angle each slope makes with the horizontal. Students will find that the steepness of a slope depends on the size of the particles that make up the slope. Have students relate this to the kinds of solid ejecta thrown from volcanoes and the resulting volcano form. **COOP LEARN**

Meeting Individual Needs

Learning Disabled, Visually Impaired Learning-disabled and visually impaired students may enjoy working in small groups to make clay models of shield volcanoes, cinder cones, and composite volcanoes. Lava can be represented by red clay, and solid volcanic debris can be represented by sand and gravel. L1
LEP

content data for each of the volcanoes listed by writing the name of the basic type of volcano (composite, cinder, or shield) in the appropriate square in the chart. When all the volcanoes have been plotted, analyze the patterns of volcanic types on the diagram to answer the questions.

Teaching Strategies

Troubleshooting Assist students with plotting the first few volcanoes so that they are able to complete the table correctly. L1

Student Journal Have students make their charts and write their answers to the questions in their journals. L1

Expected Outcome

Students will recognize a relationship between the properties of magma and volcano type.

Analyze and Conclude

1. those with high silica content

2. The ability of magma or lava to flow is greater when the silica and water-vapor contents are low.

3. Silica content seems to have a greater effect than water-vapor content.

✔ Assessment

Process Have students make a table comparing the properties of volcanoes in Indonesia and Iceland. Ask students what they can infer about volcanic activity in these two parts of the world. Use the Performance Task Assessment List for Data Table in **PASC**, p. 37. **P**

Going Further

Shield volcanoes seem to result from magma with relatively low amounts of silica and water vapor. Cinder cones seem to be produced from materials high in silica and water vapor. Composite volcanoes appear to result from magma with high silica content and generally low water-vapor content.

Life Science

Student posters should show steps in succession as ground cover is gradually replaced by pines and deciduous trees.

3 ASSESS

Check for Understanding

Have volunteers make drawings on the chalkboard of their answers to the Apply question.

Reteach

Activity Use the table on page 73 to review the relationship among magma type, the force of the volcanic eruption, and the type of volcanic cone formed. For example, call out a volcano name and ask for the type of cone it is. Then have students look for generalizations. For example, most cinder cone volcanoes are associated with high eruptive force. **L1**

Extension

Research For students who understand the concepts presented in this section, show the laserdisc *Born of Fire* to enrich and extend what has been learned about volcanoes. **L2**

4 CLOSE

Demonstration

Use a full tube of toothpaste with the lid securely on and a straight pin to illustrate how magma flows through vents to form volcanoes. Prick a small hole into the center of the tube. The paste is under pressure and will flow without any additional pressure. Allow students to observe the tube after you make the hole. Students should be able to conclude that the paste represents lava and the hole a volcanic vent. **L1**

Figure 2-10

A The eruption of Mount Pinatubo in June 1991 caused destruction and the death of nearly 900 people in the Philippines. The eruption released a cloud of ash and millions of tons of sulfur dioxide gas. The immediate effects of volcanoes are always negative and often devastating. But when viewed over a long span of years, volcanoes also have a positive side.

B The long range positive effects of volcanoes include the production of sulfur and other mineral deposits. Rocks formed by lava are used in construction of roads. Volcanic ash, once it has time to break down, increases soil fertility. The Canary Islanders pictured here are harvesting onions grown in fertile lava soil.

Connect to...

Life Science

A volcanic eruption can cause a lot of destruction to trees, plants and animals. Within a short time, however, new plants are seen in the area. How is a stable ecosystem formed after a volcanic eruption? Make a poster that shows the steps involved in rebuilding the ecosystem.

In this chapter, you have explored the causes of earthquakes and the different types of volcanoes. You will probably read and hear in the news about earthquakes and volcanoes throughout your life. These naturally occurring events are interesting to everyone!

check your UNDERSTANDING

1. What characteristic of magma causes it to be forced upward to Earth's surface and eventually erupt?
2. Some volcanic eruptions are quiet, yet others are explosive. What causes this difference?

3. **Apply** A large body of magma is forced upward close to the surface under what has been flat land, but the magma has no vent to reach the surface. How do you think the magma would affect the land? Why?

check your UNDERSTANDING

1. Magma is less dense than surrounding rocks and thus is forced upward toward Earth's surface.
2. The amount of water vapor and other gases present in the magma and the composition of the magma determine whether eruptions are quiet or explosive.
3. The magma would tend to push the land upward to form a dome because the land is under pressure from the magma below it. If there is a break in the surface, lava may flow out onto the surface.

Science and Society

Preparing Buildings for Earthquakes

Experts know where earthquakes are most likely to occur, but they don't know when the quakes will happen. If they could predict earthquakes, buildings could be evacuated and lives saved.

While experts are trying to better predict the *when* of earthquakes, others are building earthquake-resistant buildings or modifying older buildings to help them survive earthquakes.

Constructing for Safety

For example, an earthquake safety commission in Japan is working closely with construction companies to determine what features help make a building earthquake safe. They've designed and built the small-scale structures you see in the picture to the right for use in earthquake trials. At each step, the building is fitted with a design feature, then vibrated with the force of a powerful earthquake to test its effectiveness. Note the second structure from the left.

The National Center for Earthquake Engineering Research (NCEER) in Buffalo, New York has undertaken a similar project.

Active Controls

NCEER has also worked with Japanese engineers to build an experimental earthquake-resistant building in Tokyo. The six-story building contains an active system for controlling a building's response to tremors. A computer system in the building tells giant pistons, which move up and down, to shift the level of different areas of the building to counterbalance earthquake waves.

This building is considered strong enough to withstand earthquakes common to Tokyo. Researchers study the effectiveness of the computerized active system during the small earthquakes that occur there every year.

ers, earthquakes can occur anywhere. Have students obtain information on what to do should an earthquake occur in your area, and conduct mock drills to ensure that all students are able to respond safely and efficiently.

Science and Society

Purpose

Science and Society is an extension of the material presented in Section 2-2 regarding the destructiveness of seismic waves.

Content Background

In addition to a building's structure, one of the most critical factors that affect the stability of a building in earthquake-prone areas is the nature of the material on which the structure is erected. Bedrock has proven to be the most stable material in terms of the least amount of structural damage incurred as the result of an earthquake. Unconsolidated sediments, on the other hand, amplify, rather than dampen, seismic waves.

Single-story structures made from wood generally suffer the least amount of damage during an earthquake. Masonry buildings with no type of reinforcement are most prone to earthquake damage.

In the active system described in this feature, cables, which are operated by computers, are attached to the hydraulic pistons. The pistons respond to external pressure due to the earthquake.

Teaching Strategies
After students have read about both passive and active systems, have them debate the advantages and disadvantages of each. Students should realize that although active systems may be more effective in preventing damage to structures and in reducing the number of lives lost, these systems are very costly. Passive systems, in most cases, are effective in many earthquake-prone areas and are much cheaper.

It is essential that everyone know how to respond in the event of an earthquake because although some areas are more prone to earthquakes than oth-

Discussion

Have an interested student find out about passive systems and report his or her findings to the class. The student should be able to answer most of the questions posed by classmates. Most passive systems consist of flexible moorings made of alternating metal plates and cured rubber. The rubber absorbs most of the energy released during an earthquake while the metal provides the needed support. The result is that buildings with these flexible moorings tend to sway gently as energy is transferred through them.

Answers to

What Do You Think?

Students can obtain information from your municipal Chamber of Commerce that will enable them to answer these questions. If your area has no earthquake codes, have students obtain information from the National Earthquake Information Service in Golden, Colorado. Items in the kit could include canned food, bottled water, flashlight, blankets, battery-powered radio, and a first aid kit. Accept all reasonable answers.

Going Further ▐▐▐▐▶

Have students work in small groups to design posters that will instruct others how to respond in the event of an earthquake. Each group should be assigned a specific topic, such as where to stand if you are inside a sturdy structure, what to do prior to a quake about securing items such as water heaters and gas appliances, which can cause fires, how to place heavy items on shelves to prevent injury should toppling occur, and what to do if you happen to be on a crowded city street or highway when an earthquake occurs. Use the Performance Task Assessment List for Poster in **PASC**, p. 73. **COOP LEARN** **P**

Passive Controls

A much more common approach to stabilizing buildings during earthquakes is through passive control systems. For example, a building sitting on a huge cushion-like foundation might ride out an earthquake relatively well. This is called a passive system because the cushion is always responding, not waiting like an active system.

"A lot of people feel that a dumb (passive) building that performs well is much preferable to a smart (active) building that may have something go wrong with it," says Professor James D. Jirsa, a structural engineer specializing in earthquake-resistant buildings.

Dr. Jirsa explains that passive and active systems are usually added to already existing buildings to prepare them for future earthquakes. New buildings can be built more earthquake resistant in the first place.

Construction Problems

One of the biggest problems in constructing and modifying buildings for earthquake resistance is that each earthquake is different and experts can't always predict the earthquake's strength or its duration. So experts can't be absolutely certain how a building will respond to the next earthquake.

Engineers also have to decide whether to construct buildings that only protect people from injuries or that protect people and contents. Buildings designed to protect only people may move during an earthquake, but they won't collapse. The movement of the building, however, might cause the contents of the building to fall or break. Buildings that protect people and contents are more expensive but may be

necessary when computers, medical equipment, and communications are involved.

Earthquake Education

In addition to efforts to prevent human injury and building damage through earthquake-resistant construction, engineers agree there is another extremely important way people can try to protect themselves—education.

"A very important part of this whole thing is education," says Dr. T.T. Soong of NCEER. "People should take precautions so that lives can be saved. One can have earthquake drills, very much like fire drills, so that people can be in safe places when the earthquake strikes."

What Do You Think?

Does your city have earthquake codes for new buildings? Does it have special requirements for existing buildings? Make a list of items that should be included in a disaster readiness kit.

Technology Connection

Seismic Waves and the Search for Oil

As you learned in this chapter, earthquakes generate seismic waves, which help determine exactly where an earthquake took place.

With explosive devices, geologists can create their own seismic waves and use them in the difficult search for oil beneath the ocean floor. Oil formed millions of years ago from the remains of marine plants and animals buried deep inside Earth. Today, much of that oil is contained in reservoir rocks—porous rocks with many tiny holes that trap oil. The reservoir rocks themselves are concealed by faults, folds, and other rock formations that further trap the oil.

beneath the sea. The waves travel down, then bounce back off the different underground rock layers.

Seismic waves travel at different speeds, depending on the kinds of rock through which they're passing. By measuring how long it takes the waves to return to the surface, geologists can map out a cross section of the rock layers.

Using these maps, as well as their knowledge of the area and other drilling sites, seismologists can then estimate where oil might be found. As you may have guessed, these maps help limit the possibilities and save money in the search for oil.

Using Waves To Find Oil

Geologists look beneath the ocean for the right combination of reservoir rocks and protective layers, which point to the possibility of trapped oil. To do this, they cause explosions, which send seismic waves into the rocks

Science Journal

Look up the words *seismology* and *geology* in the dictionary. Discuss *in your Science Journal* why you think the search for oil in the rocks beneath the sea is being done by seismologists and geologists.

Going Further ▮▮▮▮➤

Have students write to the public relations department of a major oil company such as Exxon, Shell, Amoco, or Chevron and request copies of published seismic lines that depict an oil and gas reservoir. Have students also request information on how to read the seismic data and how the data are used in the oil and gas industry. Use the Performance Task Assessment List for Letter in **PASC,** p. 67. [P]

Technology Connection

Purpose

In Section 2-2, students learned that earthquakes generate waves that travel outward from a focus through Earth's interior. Technology Connection explains how artificial seismic waves can be used to locate hydrocarbons buried deep within Earth's crust.

Content Background

Seismic data are used in the oil industry to measure the differences in densities among encountered materials, which are a function of the travel times of the waves. The actual seismic reflections and refractions primarily designate differences in densities. Fluids—oil, natural gas, and water—trapped in the rocks influence the travel times and thus assist seismologists and petroleum geologists in locating hydrocarbon reservoirs.

Teaching Strategy

Although the feature discusses only offshore oil and gas reservoirs, have students think about where most of the oil and gas fields are in the United States. You may have to help them determine that most of the known reserves are in onshore deposits in California, Texas, Alaska, and Louisiana. Ask students to hypothesize how these deposits may have formed.

Science Journal

Seismology is the study of earthquakes. Geology is the study of Earth, its materials, the processes that affect it, as well as the history of the planet and its inhabitants. Seismologists and geologists are both knowledgeable about where oil is likely to be found.

Purpose

The extensive property damage and loss of life that can be caused by volcanic activity could be minimized if the affected areas had advance warning of an impending eruption. These Sci-Facts expand the information about volcanoes found in Section 2-3 by discussing recent advances in predicting eruptions.

Content Background

Measuring tremor frequency and intensity, determining the amount of sulfurous gases vented, and detecting formation of bulges on volcanic mountains are not foolproof methods of predicting eruptions. However, these factors were used in 1991 to predict the eruptions of Mount Unzen in Japan and Mount Pinatubo in the Philippines.

Discussion

Students may wonder why humans don't avoid volcanic areas if they are potentially dangerous. Emphasize to them that volcanic activity has produced many products that are of great economic value. For example, volcanic rocks such as pumice are used as polishing compounds and abrasives. Useful elements such as sulfur, zinc, lead, and copper are found in relatively high concentrations in volcanic deposits. Have students discuss whether the economic advantages of obtaining these materials is worth possible loss of life and property from an eruption.

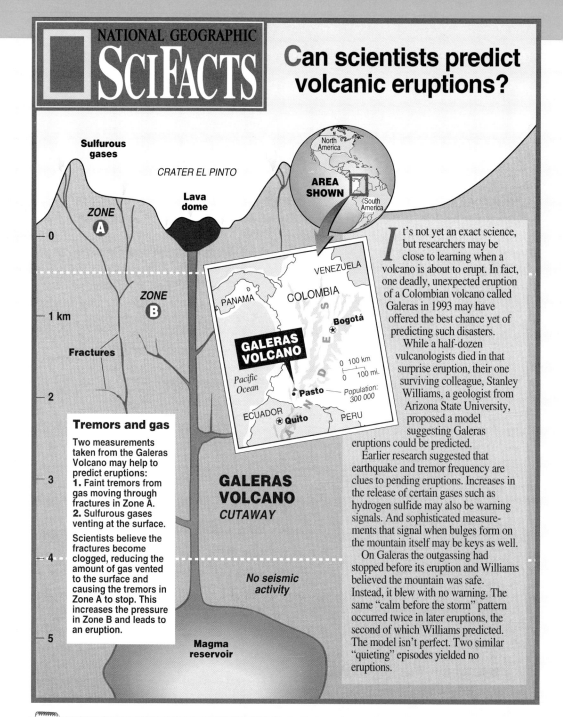

NATIONAL GEOGRAPHIC
SciFacts

Can scientists predict volcanic eruptions?

It's not yet an exact science, but researchers may be close to learning when a volcano is about to erupt. In fact, one deadly, unexpected eruption of a Colombian volcano called Galeras in 1993 may have offered the best chance yet of predicting such disasters.

While a half-dozen vulcanologists died in that surprise eruption, their one surviving colleague, Stanley Williams, a geologist from Arizona State University, proposed a model suggesting Galeras eruptions could be predicted.

Earlier research suggested that earthquake and tremor frequency are clues to pending eruptions. Increases in the release of certain gases such as hydrogen sulfide may also be warning signals. And sophisticated measurements that signal when bulges form on the mountain itself may be keys as well.

On Galeras the outgassing had stopped before its eruption and Williams believed the mountain was safe. Instead, it blew with no warning. The same "calm before the storm" pattern occurred twice in later eruptions, the second of which Williams predicted. The model isn't perfect. Two similar "quieting" episodes yielded no eruptions.

Tremors and gas

Two measurements taken from the Galeras Volcano may help to predict eruptions:
1. Faint tremors from gas moving through fractures in Zone A.
2. Sulfurous gases venting at the surface.

Scientists believe the fractures become clogged, reducing the amount of gas vented to the surface and causing the tremors in Zone A to stop. This increases the pressure in Zone B and leads to an eruption.

Science Journal

In your Science Journal, research further problems involved in predicting volcanic eruptions

Science Journal

Students may find that many volcanoes are in isolated areas that are difficult to monitor. Another problem may be absence of baseline information that can be used for comparison. Also, underground factors that affect eruptions are difficult or impossible to observe directly.

Science Journal

Review the statements below about the big ideas presented in this chapter, and answer the questions. Then, re-read your answers to the Did You Ever Wonder questions at the beginning of the chapter. *In your Science Journal*, write a paragraph about how your understanding of the big ideas in the chapter has changed.

1 Forces within Earth cause faults and earthquakes. *Describe three types of faults and the forces they are associated with.*

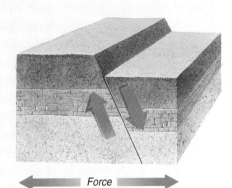

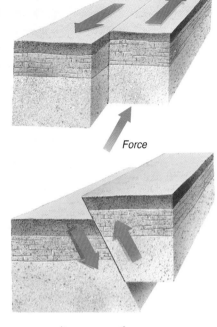

Force

Force

Force

Force

2 Earthquakes generate seismic waves—primary, secondary, and surface waves—that travel outward from the epicenter. *Which seismic waves cause the greatest destruction? Why?*

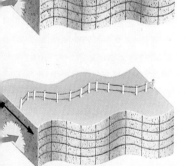

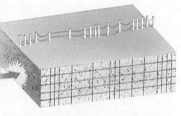

3 Volcanoes form from the eruption of magma onto Earth's surface. *Why does magma rise to Earth's surface?*

Have students work in groups to answer the questions.

Teaching Strategies

Form small groups, each of which will be responsible for illustrating the answer to one of the Reviewing Main Ideas questions. Students may choose any method of illustration, including making models, drawing diagrams or pictures, making tables or graphs. Later, have groups share their illustrations.

Answers to Questions

1. Normal (Tension)
Reverse (Compression)
Strike-slip (shear)

2. Surface waves; because they cause Earth's surface to move like waves, which cause buildings to fall.

3. Because it is less dense than the surrounding rock, it is forced toward the surface.

GLENCOE TECHNOLOGY

MindJogger Videoquiz

Chapter 2 Have students work in groups as they play the Video-quiz game to review key chapter concepts.

Science Journal

Did you ever wonder...

• Rocks under pressure break and move producing vibrations that make Earth move. (pp. 52–56)

• Scientists locate an epicenter by timing the movements of seismic waves and calculating the point where they originated. (pp. 63–65)

• The shapes of volcanoes are determined by the type of eruption and the composition of the magma. (p. 70)

Science at Home

Have students use two phone books to simulate movements that occur along faults. Instruct students to place the unbound edges together and push the books together. Students should observe what happens to the edges of the pages. The bending, folding, and crumpling of the pages simulate what happens to rocks along faults. Point out that other types of movements occur when rocks are pulled apart instead of pushed together.

Using Key Science Terms

1. The focus of an earthquake is the point inside Earth where seismic waves originate, and the epicenter is the place directly above the focus on Earth's surface.

2. Magma may flow out through the vent and become lava. The lava cools and forms layers of volcanic rock. If enough volcanic material piles up around the vent, a cone may form.

3. A fault is a break in the rock layers of Earth's crust where movement occurs generating seismic waves.

Understanding Ideas

1. Magma is called lava when magma reaches Earth's surface.

2. Magma of an explosive volcano is thick, and silica- and water vapor-rich. Magma of a nonexplosive volcano is thin and does not contain much gas or water vapor.

3. Since the rocks have not reached their elastic limit they will return to their original form.

4. Surface waves originate from the earthquake's epicenter and travel outward. Primary and secondary waves originate from an earthquake's focus. Primary waves move in compressional movement, whereas secondary waves move transversely. Surface waves move in an elliptical manner with additional side-to-side motion.

5. Earthquakes and volcanoes are examples of how forces inside Earth can surface and cause destruction.

6. Volcanoes and earthquakes appear where there is a great amount of heat, pressure, and movement within or beneath Earth's crust.

Developing Skills

1. Heat and pressure build up; rocks melt and form magma; magma is pushed to Earth's surface; volcano erupts.

Using Key Science Terms

epicenter   seismic waves

fault   vent

focus

Answer the following questions using what you know about the science terms.

1. Distinguish between the focus and the epicenter of an earthquake.

2. Describe what you might find in the area of a vent.

3. What is the relationship between seismic waves and faults?

Understanding Ideas

Answer the following questions in your Journal using complete sentences.

1. At what point does magma turn into lava?

2. Describe what type of magma makes a volcano erupt nonexplosively, and what type makes an explosive eruption.

3. When rocks are compressed and released before their elastic limit is reached, how will the shape of the rocks be affected?

4. What differentiates the three types of seismic waves?

5. Describe how volcanoes and earthquakes are alike.

6. Why do many earthquakes occur in the same regions as many volcanoes?

Developing Skills

Use your understanding of the concepts developed in each section to answer each of the following questions.

1. Sequencing Outline the process of a volcanic eruption.

2. Observing and Inferring Repeat the Explore activity on page 53 to model the clay by compression and tension so that the bottom layer of clay is seen at the surface. Infer how this could happen in rocks on Earth.

3. Concept Map Complete the concept map about earthquakes using the following terms: *earthquakes, faults, normal, strike-slip, tremendous pressure.*

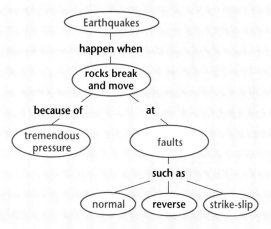

4. Forming a Hypothesis Refer to the Explore activity on page 67. Hypothesize what effect thinner syrup would have on the air bubbles. Add water to the syrup and mix. What do you observe? What hypothesis would you make about thinner magma of a volcano?

2. Reverse faults result in rocks from under Earth's surface moving above the surface, because of compression of rocks against one another.

3. See reduced student page for answers.

4. The air bubbles would contain less gas and be of less pressure. Thinner magma would flow more quickly, since the magma was less dense.

Program Resources

Review and Assessment, pp. 11–16 [L1]

Performance Assessment, Ch. 2 [L2]

PASC

Alternate Assessment in the Science Classroom

Computer Test Bank [L1]

Critical Thinking

Use your understanding of the concepts developed in this chapter to answer each of the following questions.

1. The data table shows some travel times of two waves from an earthquake. Classify each time as belonging to the primary or the secondary wave. How do you know?

Distance from Earthquake (km)	Time (minutes)
1500	5.0
2000	2.5
5000	14.0
5500	7.0
8600	11.0
10000	23.5

2. In an area around an active volcano, you discover piles of ash, cinders, and rocks. A geologist tells you that the rocks are volcanic rocks. What can you infer about the form of the volcano and the kind of eruptions it undergoes?

3. Do you think it is possible for tension forces and compression forces to be acting on the same rock layer? Explain.

Problem Solving

Read the following problem and discuss your answers in a brief paragraph.

Scientists want a reliable approach to predicting earthquakes. Both seismographs and satellites detect motion. Seismographs detect seismic waves. Satellites send signals that are picked up by radio receivers on Earth. When a radio receiver moves just a few centimeters, it can be detected.

1. Would seismographs be useful for predicting earthquakes? Why or why not?

2. Where could scientists place radio receivers to help detect rock movement that may lead to an earthquake?

CONNECTING IDEAS

Discuss each of the following in a brief paragraph.

1. **Theme—Systems and Interactions** Explain the force behind each type of fault.

2. **Theme—Scale and Structure** In what ways do layers of rock behave when forces inside Earth bend or stretch them too far?

3. **Theme—Systems and Interactions** How do less-dense materials react when they are trapped within denser materials? How is this principle related to volcanic activity?

4. **Technology Connection** Different seismic waves travel at different speeds. What other factor affects the speed at which a wave travels? How do geologists use this information to locate oil?

5. **Physics Connection** What do rocks store and release as forces act upon them? Explain.

Critical Thinking

1. Primary waves arrive at 2.5 min, 7.0 min, and 11.0 min. Secondary waves arrive at 5.0 min, 14.0 min, and 23.5 min. Secondary waves are slower than primary waves.

2. The volcano could be either a cinder cone or a composite, both of which produce much ash and rock particles.

3. Rock layers can extend for hundreds or thousands of kilometers. Tension forces may be acting on one part of a rock layer while compression forces are acting on another part.

Problem Solving

1. Yes; they may detect small vibrations that could indicate that a larger, more powerful earthquake is on the way.

2. on both sides of the fault

Connecting Ideas

1. The force in a normal fault is tension. In reverse faults the force is compression. The force behind strike-slip faults is shearing action.

2. If bent or stretched too far, layers of rock will break or be bent out of shape.

3. Less dense materials tend to be forced up through the denser materials. Because magma is less dense than the material that surrounds it, it is forced toward Earth's surface where it may erupt.

4. The type of material through which the wave travels affects its speed. By knowing how fast seismic waves travel through oil in comparison with rock, geologists can evaluate wave speeds to determine areas beneath Earth's surface where oil may be found.

5. Energy is stored and released as forces act on rocks.

Assessment

Portfolio Review the portfolio options that are provided throughout the chapter. Encourage students to select one product that demonstrates their best work for the chapter. Have students explain what they learned and why they chose this example for placement into their portfolios.

Additional portfolio options can be found in the following **Teacher Classroom Resources: Making Connections: Integrating Sciences,** p. 7
Multicultural Connections, pp. 7, 8

Making Connections: Across the Curriculum, p. 7
Concept Mapping, p. 10
Critical Thinking/Problem Solving, p. 10
Take Home Activities, p. 7
Laboratory Manual, pp. 5–8
Performance Assessment P

Chapter Organizer

SECTION	OBJECTIVES	ACTIVITIES & FEATURES
Chapter Opener		**Explore!**, p. 83
3-1 Circulatory Systems (3 sessions; 1.5 blocks)	1. **Explain** the role of a circulatory system in animals. 2. **Compare and contrast** open and closed circulatory systems. 3. **Describe** the path of blood through the heart, lungs, and body. 4. **Compare and contrast** arteries, veins, and capillaries. **National Content Standards: (5-8) UCP1-2, A1, B2, C1, E2**	**Explore!**, p. 84 **Find Out!**, p. 86 **Find Out!**, p. 88 **A Closer Look**, pp. 88–89 **History Connection**, p. 107
3-2 A System Under Pressure (4 sessions; 2 blocks)	1. **Explain** what causes pulse. 2. **Explain** how blood moves through your body under pressure. 3. **Compare** the structural adaptations of blood vessels. 4. **Compare** the circulatory systems of fish, birds, amphibians, and mammals. **National Content Standards: (5-8) UCP1-3, UCP5, A1, B2, C1, F1**	**Explore!**, p. 91 **Design Your Own Investigation 3-1:** pp. 92–93 **Find Out!**, p. 94 **Investigate 3-2:** pp. 96–97 **Skillbuilder:** p. 98 **Physics Connection**, pp. 98–99 **Teens in Science**, p. 109
3-3 Disorders in Circulation (2 sessions; 1 block)	1. **Describe** the role of fatty deposits in heart disease. 2. **Relate** lifestyles to high blood pressure. **National Content Standards: (5-8) UCP1-2, UCP4, A1-2, B2, C1, F1, F4, G1, G3**	**Find Out!**, p. 103 **Science and Society**, p. 106 **How It Works**, p. 108

ACTIVITY MATERIALS

EXPLORE!

p. 83* city map showing interstate
p. 84 spoonful of honey, food coloring, paper plate
p. 91* watch with second hand

INVESTIGATE!

pp. 96–97 clay or putty, coffee stirrer, small mirror

DESIGN YOUR OWN INVESTIGATION

pp. 92–93* live earthworm, petri dish with cover, paper towels, water at room temperature, medicine dropper, clock or watch with second hand

FIND OUT!

p. 86* model of a heart showing the chambers
p. 88* funnel, Y-shaped tube, 3 pieces of flexible tubing
p. 94 plastic squeeze bottle, stopper with plastic or rubber tubing, water
p. 103* dropper, mineral oil, two 10-cm pieces of plastic tubing, cotton, tweezers

KEY TO TEACHING STRATEGIES

The following designations will help you decide which activities are appropriate for your students.

L1 Basic activities for all students
L2 Activities for average to above-average students
L3 Challenging activities for above-average students
LEP Limited English Proficiency activities
COOP LEARN Cooperative Learning activities for small group work
P Student products that can be placed into a best-work portfolio
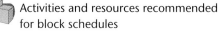 Activities and resources recommended for block schedules

Need Materials? Call Science Kit (1-800-828-7777).

[00:00] OUT OF TIME? We recommend that students do the activities with an asterisk.

Chapter 3 Circulation

TEACHER CLASSROOM RESOURCES

Student Masters	Transparencies
Study Guide, p. 13 **Concept Mapping,** p. 11 **Making Connections: Across the Curriculum,** p. 9 **Take Home Activities,** p. 8 **Making Connections: Technology and Society,** p. 9 **Science Discovery Activities, 3-1, 3-2**	**Teaching Transparency 5,** The Heart **Section Focus Transparency 7**
Study Guide, p. 14 **Making Connections: Integrating Sciences,** p. 9 **Activity Masters,** Design Your Own Investigation 3-1, pp. 15–16 **Activity Masters,** Investigate 3-2, pp. 17–18 **Science Discovery Activities, 3-2, 3-3** **Laboratory Manual,** pp. 9–12, Blood Pressure	**Teaching Transparency 6,** Comparing Hearts **Section Focus Transparency 8**
Study Guide, p. 15 **Critical Thinking/Problem Solving,** p. 11 **Multicultural Connections,** pp. 9–10 **How It Works,** p. 6	**Section Focus Transparency 9**

ASSESSMENT RESOURCES	TEACHING & TECHNOLOGY
Review and Assessment, pp. 17–22 **Performance Assessment,** Ch. 3 **PASC*** **MindJogger Videoquiz** **Alternate Assessment in the Science Classroom** **Computer Test Bank**	**Spanish Resources** **Cooperative Learning Resource Guide** **Lab and Safety Skills** **Science Interactions, Course 2, CD-ROM** **Computer Competency Activities**

*Performance Assessment in the Science Classroom

NATIONAL GEOGRAPHIC TEACHER'S CORNER

National Geographic Society Products Available From Glencoe	Additional National Geographic Society Products
To order the following products for use with this chapter, contact your local Glencoe sales representative or call Glencoe at 1-800-334-7344: • STV: Human Body Series, "Respiratory, Circulatory, Digestive Systems." (Videodisc)	To order the following products for use with this chapter, call the National Geographic Society at 1-800-368-2728: • Everyday Science Explained (Book) • The Incredible Machine (Book) • The Incredible Human Machine (Video) • Man: The Incredible Machine (Video) • Your Body Series, "Circulatory and Respiratory Systems." (Video)

GLENCOE TECHNOLOGY

The following multimedia resources are available from Glencoe.

The Infinite Voyage Series
The Champion Within
A Taste of Health

Glencoe Life Science Interactive Videodisc
Photosynthesis and Cellular Respiration

Life Science CD-ROM

Teacher Classroom Resources

These are key components of the classroom resources package.

TEACHING AIDS

Section Focus Transparencies

7 SECTION FOCUS TRANSPARENCY · Section 3-1

CLEAN WATER IN, DIRTY WATER OUT

Each day, you use water for a variety of purposes in your home. This water is carried into your home by way of pipes that make up the plumbing system of the house. Similarly, the plumbing system carries wastewater into the sewage system.

1. Imagine you are going to shower in the home shown. Describe the path water must follow to allow you to take your shower. What happens to the dirty water after you have showered?

2. In what way is the job carried out by the plumbing system of a house similar to the functions of your circulatory system? How is it different?

L1

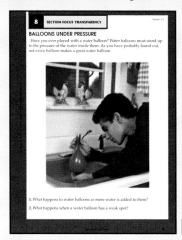

8 SECTION FOCUS TRANSPARENCY · Section 3-2

BALLOONS UNDER PRESSURE

Have ever played with a water balloon? Water balloons must stand up to the pressure of the water inside them. As you have probably found out, not every balloon makes a great water balloon.

1. What happens to water balloons as more water is added to them?

2. What happens when a water balloon has a weak spot?

L1

9 SECTION FOCUS TRANSPARENCY · Section 3-3

SKIING FOR LIFE

Skiing—both downhill and cross-country—is a popular pastime for many individuals in countries with snowy winters. Families and schools often take skiing trips during their winter holidays. Such activities are fun when proper safety precautions are taken.

1. What do you think may have happened prior to this picture being taken?

2. In addition to caring for other injured skiers, how does the work of ski patrol members help them stay healthy?

L1

Teaching Transparencies

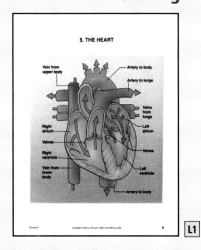

5. THE HEART

L1

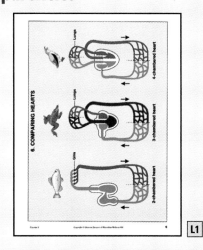

6. COMPARING HEARTS

L1

HANDS-ON LEARNING

Science Discovery Activity*

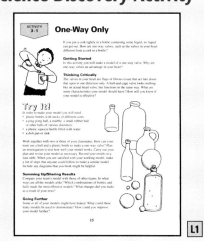

ACTIVITY 3-1

One-Way Only

If you put a cork tightly in a bottle containing some liquid, no liquid can get out. How are one-way valves, such as the valves in your heart different from a cork in a bottle?

Getting Started
In this activity you will make a model of a one-way valve. Why are one-way valves an advantage in your heart?

Thinking Critically
The valves in your heart are flaps of fibrous tissue that act like doors that open in one direction only. A ball-and-cage valve looks nothing like an actual heart valve, but functions in the same way. What are some characteristics your model should have? How will you know if your model is effective?

Try It!
In order to make your model you will need:
• plastic bottles with necks of different sizes
• a ping-pong ball, a marble, a small rubber ball or other balls of various diameters
• a plastic squeeze bottle filled with water
• a dish pan or sink

Work together with two or three of your classmates. How can you team use a ball and a plastic bottle to make a one-way valve? Plan an investigation to test how well your model works. Carry out your plan and revise your model as necessary. Record your results in a data table. When you are satisfied with your working model, make a list of steps that anyone could follow to make a similar model. Include any diagrams that you think might be helpful.

Summing Up/Sharing Results
Compare your team's model with those of other teams. In what ways are all the models alike? Which combinations of bottles and balls made the most effective models? What changes did you make as a result of your tests?

Going Further
Some or all of your models might have leaked. What could those leaky models be used to demonstrate? How could you improve your model further?

15

L1

Laboratory Manual*

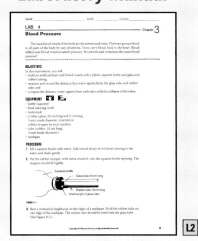

NAME _____ DATE _____ CLASS _____

LAB 4 — Chapter 3

Blood Pressure

The main blood vessels of the body are the arteries and veins. The heart pumps blood to all parts of the body by way of arteries. Veins carry blood back to the heart. Blood within your blood vessels is under pressure. Do arteries and veins have the same blood pressure?

OBJECTIVES
In this experiment, you will
• build an artificial heart and blood vessels with a plastic squeeze bottle and glass and rubber tubing,
• measure and record the distance that water squirts from the glass tube and rubber tube, and
• compare the distance water squirts from each tube with the softness of the tubes.

EQUIPMENT
• bottle (squeeze)
• food coloring (red)
• meterstick
• 2 tubes (glass, 20 cm long and 5 cm long, 5 mm inside diameter, inserted in rubber stopper by your teacher)
• tube (rubber, 18 cm long,
5 mm inside diameter)
• washpan

PROCEDURE
1. Fill a squeeze bottle with water. Add several drops of red food coloring to the water and shake gently.

2. Put the rubber stopper, with tubes attached, into the squeeze bottle opening. The stopper should fit tightly.

3. Rest a meterstick lengthwise on the edges of a washpan. Hold the rubber tube on one edge of the washpan. The rubber tube should be level with the glass tube. (See Figure 4-2.)

L2

Take Home Activity

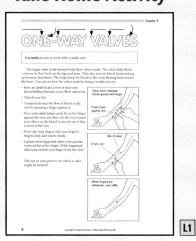

Chapter 3

ONE-WAY VALVES

You need: an arm or wrist with a visible vein

The larger veins in the human body have valves inside. The valves help blood continue its flow back up the legs and arms. They also prevent blood from backing up between heartbeats. This helps keep the blood in the veins flowing back toward the heart. You can see how the valves keep blood flowing back toward

• Have an adult locate a vein in your arm about halfway between your elbow and wrist.
• Clench your fist.
• Temporarily stop the flow of blood in the vein by pressing a finger against it.
• Have your adult helper push his or her finger against the vein and then rub the vein toward your elbow as the blood is moved out of that section of the vein.
• Now take your fingers and your helper's fingers away and watch closely.
• Explain what happened when your partner removed his or her finger. What happened when you moved your finger from the vein?

Did you see your partner see where a valve might be located?

L1

82C   Chapter 3   Circulation   *There may be more than one activity for this chapter.

Chapter 3 Circulation

REVIEW AND REINFORCEMENT

Study Guide*

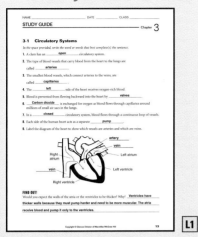

L1

Concept Mapping

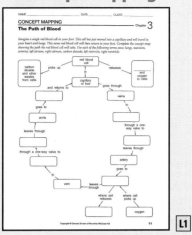

L1

Critical Thinking/ Problem Solving

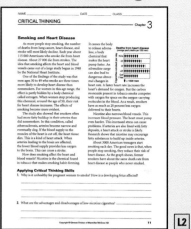

L2

ENRICHMENT AND APPLICATION

Integrating Sciences

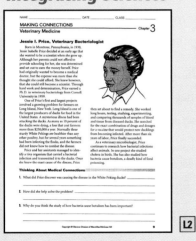

L2

Across the Curriculum

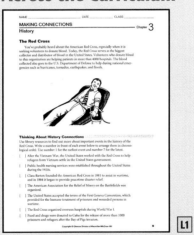

L1

Technology and Society

L1

Multicultural Connection**

L1

ASSESSMENT

Performance Assessment

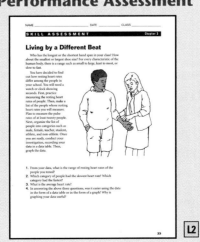

L2

Review and Assessment

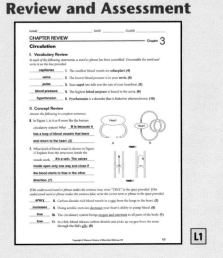

L1

*One per section **Two per chapter

CHAPTER 3

Circulation

THEME DEVELOPMENT

The themes that this chapter supports are systems and interactions and energy. The parts of the circulatory system (the heart, the blood, and the blood vessels) interact to accomplish the tasks of bringing oxygen and nutrients to and removing wastes from cells.

CHAPTER OVERVIEW

Students will compare and contrast open and closed circulatory systems in animals, study the structure of the heart and blood vessels, and learn how blood travels through the body under pressure. Students will learn how a healthy lifestyle can help prevent circulation-related diseases.

Tying to Previous Knowledge

Ask students to list evidence that their circulatory systems exist. They should list such things as heartbeat or the fact that people bleed when cut.

INTRODUCING THE CHAPTER

Give students maps and have them brainstorm the similarities between a circulatory system and a highway system.

Uncovering Preconceptions

Students may think that their bodies are filled with uncontained blood because they bleed anywhere they are cut. Explain that all blood is contained in blood vessels, such as those they can see in their wrists.

Did you ever wonder...

✓ **What keeps blood moving through your body?**

✓ **Whether all animals have the same kind of heart and circulation?**

✓ **How blood pressure affects blood vessels?**

Science Journal

Before you begin to study about circulation, think about these questions and answer them *in your Science Journal.* When you finish the chapter, compare your first journal write-up with what you have learned.

Answers are on page 110.

*T*aking a trip to an unfamiliar city requires a good road map to guide you along endless miles of intersecting highways, like those in the photograph below. There are routes or "arteries" that take you into the "heart" of the city or out to specific streets in the suburbs. There are one-way streets, traffic jams, slow ups, and even the occasional accident. Living organisms have something like a highway system through which needed materials move.

Are there things like one-way streets in living organisms? Are there ever any slow ups? Are there ever accidents as materials are transported in organisms?

▶ *In the following activity, begin to develop an idea of what circulation is like by using a city map.*

82

Learning Styles	**Kinesthetic**	Explore, p. 84; Find Out, p. 94; Investigate, pp. 96-97
	Visual-Spatial	Explore, p. 83; Demonstration, pp. 85, 91, 104; Visual Learning, pp. 85, 87, 89, 90, 95; Find Out, pp. 86, 103
	Interpersonal	Enrichment, pp. 87, 95
	Linguistic	Inquiry Question, p. 85; A Closer Look, pp. 88-89; Multicultural Perspectives, p. 92; Discussion, pp. 99, 100, 102; Enrichment, p. 102; Across the Curriculum, p. 104
LS	**Logical-Mathematical**	Explore, p. 91; Design Your Own Investigation, pp. 92-93; Physics Connection, pp. 98-99; Discussion, p. 100; Visual Learning, pp. 100, 105
	Auditory-Musical	Find Out, p. 88

Explore! ACTIVITY

How many ways can you get from one place to another in a city?

Many big cities now have a beltway around them and usually have one or more interstate highways.

What To Do

1. Obtain a map that shows the streets, interstate roads, and a beltway around a large city.

2. Study the map to find the center or "heart" of the city. Use the map key to identify roads that are interstates and roads that are state and county routes.

3. *In your Journal,* plan a route from the heart of your city to a street out in the suburbs. Describe the different types of streets you would take. Are they all the same size? How does the final street compare in size with the one you started out on downtown?

4. *In your Journal,* describe two ways to get from the east side to the west side of the city. State the advantages and disadvantages of both routes.

5. If the city represented a human body, what would the center of town represent? What would the suburbs represent?

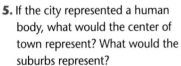

Explore!

How many ways can you get from one place to another in a city?

Time needed 30 minutes

Materials city map showing beltway or Interstate

Thinking Processes comparing and contrasting, observing and inferring

Purpose To infer the features of a circulatory system by comparing it to a road map.

Teaching the Activity

Troubleshooting Make sure students can identify and use the map key.

Science Journal Routes will vary. Streets usually become smaller in the suburbs. Advantages of interstate roads and beltways include higher speed limits, more lanes, and fewer stops. An advantage of smaller streets is the possibility of less traffic congestion. Disadvantages include lower speed limits, fewer lanes, and more stops. **L1**

Expected Outcome

Students will observe that a large highway or beltway connects the downtown with the suburbs.

Answers to Questions

3. See Science Journal above.

4. Plans may go through town or around a beltway. The beltway is longer at higher speeds; through town is shorter at slower speeds.

5. Center—heart; the suburbs—arms and legs

✔ Assessment

Performance Have students make a display showing an outline of the human body indicating the areas that correspond to the center city district (heart); university (brain); suburbs (arms and legs). Use the Performance Task Assessment List for Display in **PASC**, p. 63. **P**

ASSESSMENT PLANNER

PORTFOLIO
Refer to page 112 for suggested items that students might select for their portfolios.

PERFORMANCE ASSESSMENT
Process, pp. 91, 93, 97
Skillbuilder, p. 98
Explore! Activities, pp. 83, 84, 91
Find Out! Activities, pp. 86, 88, 94, 103
Investigate, pp. 92–93, 96–97

CONTENT ASSESSMENT
Oral, pp. 86, 88, 94
Check Your Understanding, pp. 90, 101, 105
Reviewing Main Ideas, p. 110
Chapter Review, pp. 111-112

GROUP ASSESSMENT
Opportunities for group assessment occur with Cooperative Learning Strategies.

PREPARATION

Planning the Lesson

Refer to the Chapter Organizer on pages 82A–D.

Concepts Developed

In this section, students will compare open and closed circulatory systems; examine the pathway of blood through the heart, lungs, and body; and compare and contrast arteries, veins, and capillaries.

1 MOTIVATE

Bellringer

Before presenting the lesson, display **Section Focus Transparency 7** on an overhead projector. Assign the accompanying **Focus Activity** worksheet.

L1 LEP

Explore!

How can you make a model of circulation in a one-celled organism?

Time needed 10 minutes

Materials paper plate, honey, food coloring

Thinking Processes observing and inferring, comparing and contrasting

Purpose To observe food coloring diffusing through honey and compare this process to nutrient circulation in a one-celled organism.

Teaching the Activity

Science Journal Have students predict what will occur when they tilt the plate and write their predictions in their journals. L1

84 Chapter 3 Circulation

3-1 Circulatory Systems

Section Objectives

In this section, you will

- Explain the role of a circulatory system in animals.
- Compare and contrast open and closed circulatory systems.
- Describe the path of blood through the heart, lungs, and body.
- Compare and contrast arteries, veins, and capillaries.

Key Terms

arteries
veins
capillaries

Getting What You Need to Live

In the living world, there are organisms of all different sizes and shapes. But whether they are one cell in size or trillions of cells in size, whether they live in a puddle or an apartment building, living organisms all circulate similar things throughout their bodies to stay alive.

You may never have thought about the differences between small and large organisms. However, size makes a big difference in how different organisms have their needs supplied. Those organisms that are one cell in size or only a few cells thick have very different kinds of problems from larger organisms like yourself. Try the following activity to learn about what might take place as materials circulate in a one-celled organism.

Explore! ACTIVITY

How can you make a model of circulation in a one-celled organism?

Even one-celled organisms depend on having nutrients available to all the parts of their bodies.

What To Do

1. Place a spoonful of honey on a paper plate.
2. Add one drop of food coloring at the edge of the blob of honey.
3. Gently tilt the paper plate to make the honey flow in different directions. Observe what happens to the food coloring.
4. *In your Journal*, describe what happens to the drop of food coloring as the honey flows in different directions.
5. Think of the blob of honey as a one-celled organism. *In your Journal*, describe how materials are distributed throughout the organism.
6. How efficient do you think this method is for distributing materials in an organism's body?

84 Chapter 3 Circulation

Expected Outcome

Students will observe that the honey swirls and flows in the direction in which it is moved.

Answers to Questions

4. The color swirls and flows in the same direction as the honey.
5. It moves as fluid in the organism moves.
6. Not very efficient; seems to depend on how much the organism moves.

✔ Assessment

Performance Have students working in groups demonstrate and observe how the membrane of a one-celled organism works. Have each group stretch a piece of closely woven cloth over a basin, place small objects on top of the cloth, and then pour sugar water slowly onto the cloth. Ask students to describe what they observe. Use the Performance Task Assessment List for Group Work in **PASC**, p. 97. COOP LEARN

Patterns of Circulation

A little one-celled organism, such as the blob of honey represented, has only a thin barrier between the inside and outside of its body. One-celled organisms are usually found completely immersed in water. Nutrients and oxygen are right there to move into the body. Once inside, these substances flow or stream through the liquid that makes up most of the organism's body. As the organism uses up these substances, wastes are produced and move out.

■ Open and Closed Systems

How do nutrients and oxygen reach all the body parts in larger organisms? In the chapter opening exercise, you plotted a route from the center of a city to the suburbs. You may have first moved along a four-lane highway, then onto a county road, and finally along a narrow street. Your body is similar in that it has blood vessels of various sizes going to all body parts.

The circulatory systems of complex animals are of two types, open and closed. Learn about open and closed systems in **Figure 3-1.**

Figure 3-1

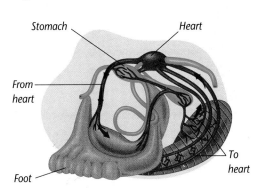

Stomach *Heart*

From heart

Foot

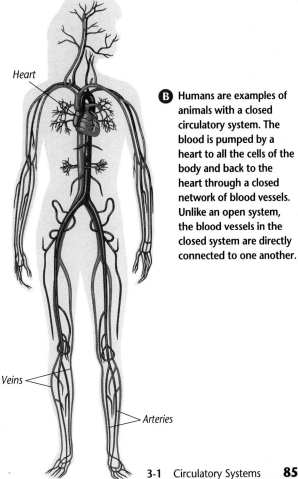

Heart

 B Humans are examples of animals with a closed circulatory system. The blood is pumped by a heart to all the cells of the body and back to the heart through a closed network of blood vessels. Unlike an open system, the blood vessels in the closed system are directly connected to one another.

To heart

Veins

Arteries

 A Clams are examples of animals that have an open circulatory system. Clam hearts pump blood through blood vessels that lead to open spaces within the body. Blood washes through these spaces, and supplies body organs with a bath of nutrients and oxygen. The blood then collects in larger vessels. These vessels are squeezed by the movement of the animal, thereby moving blood back toward the heart.

3-1 Circulatory Systems **85**

Content Background

Blood helps people maintain an even body temperature by carrying heat from one part of the body to another. It carries hormones that help regulate growth and other body functions to organs. The blood also helps fight disease. It carries antibodies and certain cells that help fight and prevent infection. The blood contains clotting factors, which help stop bleeding. Ask students what might happen if one part of the body, such as the foot, did not get an adequate supply of blood. *The body part may feel cold, become infected, and suffer tissue damage and cell death due to lack of nutrients.*

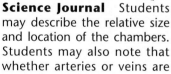

Find Out!

What are the parts of the heart?

Time needed 15 minutes

Materials model of the heart showing the chambers

Thinking Processes observing and inferring, comparing and contrasting

Purpose To observe the structures of the heart, compare the functions of the heart chambers, and infer each chamber's role from its structure.

Teaching the Activity

Science Journal Students may describe the relative size and location of the chambers. Students may also note that whether arteries or veins are connected to each chamber tells something of the chamber's role. L1

Expected Outcome

Students will observe that the atria are smaller than the ventricles and have thinner walls. Students should note that vessels carry blood out of the ventricles whereas vessels carry blood into the atria.

Conclude and Apply

1. The thicker walls of the ventricles and their larger size may indicate that ventricles pump harder than atria.

In the early 1600s, Dr. William Harvey confirmed that blood flowed in one direction in the human body. Centuries before, the Chinese recorded similar conclusions.

The Heart and Blood Vessels

What is a heart? How important is your heart for circulation? If blood moved through your body just by the action of some body muscles, how fast do you think it would move? Your heart is a two-pump system made up of a special type of muscle found only in your heart. This muscle started to contract about 45 days after you were conceived. Now it contracts at a rate of about 72 beats per minute. Each beat exerts pressure on a fluid enclosed in blood vessels. Use a model in the following activity to see how the two pumps of the human heart compare.

Find Out! ACTIVITY

What are the parts of the heart?

S ometimes, it is easier to learn about some structures, such as hearts, by looking at a model.

What To Do

1. Obtain a model of the human heart that you can open.
2. Look at all the external and internal features shown on the model.
3. Find the two upper chambers. These are the atria. Examine them from the outside and the inside.
4. Next, locate the two lower chambers, the ventricles, and examine them.

Conclude and Apply

1. *In your Journal*, describe any differences you observed between the atria and the ventricles. What hint does the structure of each side give you about the work each side does?
2. Each side of the human heart is a pump. Explain what parts make up each pump.

■ Blood Vessels

Did you notice what looked like large and small tubes attached to the model heart? These tubes represent blood vessels. There are three types of blood vessels in the body. **Arteries** are blood vessels that carry blood away from the heart—to the lungs or to the body. **Veins** are vessels that transport blood back to the heart from the lungs or body. The smallest vessels, called **capillaries**, form an extensive network of vessels in the body organs, connecting arteries to veins.

2. The right atrium and right ventricle make up one pump; the left atrium and left ventricle make up the second.

✔ Assessment

Oral Display the model of a heart. Have one group of students name each of the heart's chambers. Have a second group explain the function of each chamber. Use the Performance Task Assessment List for Oral Presentation in **PASC**, p. 71.

Blood Flow Through the Heart

As you observed on the model, each side of your heart has an atrium and a ventricle. In an office building, an atrium is an area that people enter before going off to individual offices. Each atrium in the heart is an area where blood first enters the heart before being sent to body parts. The job of each ventricle is to pump blood out of the heart. Just as you enter some buildings through one particular door and leave by another, blood always enters the heart through an atrium and leaves through a ventricle. Follow the pathway of blood through the heart shown in **Figure 3-2.**

Figure 3-2

During a single heartbeat, both atria contract at the same time, then relax. Then both ventricles contract at the same time and then relax.

A The right side of the heart receives oxygen-poor blood (represented by blue arrows) from the body and pumps it to the lungs, where it picks up oxygen.

B The left side receives oxygen-rich blood (represented by dark red arrows) from the lungs and pumps it to the body cells.

C On the diagram, find the valves that separate the atria from the ventricles. These valves open in only one direction. What can you say about the direction of blood flow between an atrium and a ventricle?

D Valves in each ventricle control the flow of blood out of the heart. If blood starts to flow back into the heart, the valves are closed by the pressure of the back-flowing blood.

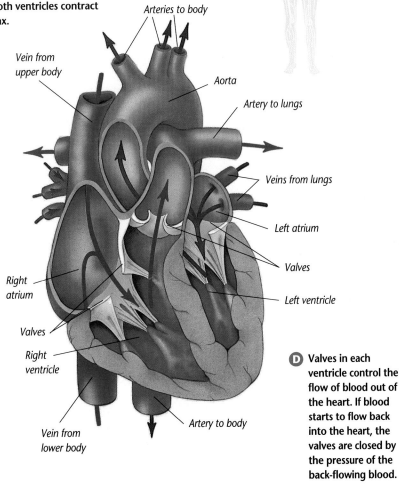

Heart

Arteries to body
Vein from upper body
Aorta
Artery to lungs
Veins from lungs
Left atrium
Valves
Left ventricle
Right atrium
Valves
Right ventricle
Vein from lower body
Artery to body

Heart Valves

As you noticed in **Figure** 3-2 valves keep blood flowing only in one direction through the heart. But valves can tell you more about the heart. Do the following activity to find out more about heart valves.

Find Out! ACTIVITY

What makes heart sounds?

What To Do

1. Put together a simple stethoscope using a y-shaped tube, a funnel, and three pieces of flexible tubing.
2. Using your stethoscope, listen to your heart for about 30 seconds.

Conclude and Apply

1. *In your Journal*, describe the sounds your heart valves make.
2. What might be the disadvantage of faulty valves?

A Valve Job

Because heart disease remains the number one killer of people in the United States, much research goes on to find ways to prevent heart attack, high blood pressure, and coronary artery problems. Each part of the heart is subject to some sort of problem, including the valves of the heart.

Valve Disorders

The purpose of valves in the heart is to prevent backward flow of blood. Healthy valves make specific sounds when they close. When something happens to damage a valve, a murmur is heard. A murmur is often a swishing sound that is heard as blood slips back through a faulty valve. The sound is different from the normal snap-shut sound "lub-dup," of a healthy valve. Blood flow is slowed from the heart. This means that the heart muscle has to work harder. It often becomes larger to accomplish that work.

There are several diseases that cause heart valve problems. One of the best documented causes of valve problems is rheumatic heart disease in which the valves become scarred as a result of inflammation.

Purpose

A Closer Look extends this section's discussion of valves by explaining valve disorders and describing a mechanical device used to replace a faulty valve.

Teaching Strategies

Obtain a transparent check valve and tubing from a chemistry lab. Connect it to a water line, and demonstrate how the valve works. With the valve positioned so the water flows straight up, students can see the ball jiggling freely as water passes in the desired direction (upward), but watch the ball seat itself firmly when the water is shut off, preventing backflow in the tubing. Explain that this is how the one-way valves operate in the heart and veins.

Answers to You Try It!

Student research will describe a variety of artificial valves and some pig valves. Congenital problems include missing or

Blood Flow Through the Body

When blood leaves the left ventricle, it travels first through arteries, then capillaries, and finally veins as in **Figure 3-3**, before it returns to your heart.

Figure 3-3

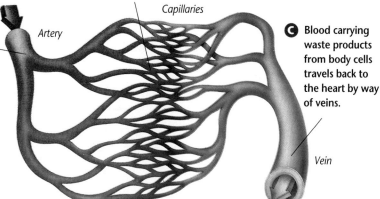

A After leaving the left ventricle, the aorta branches into smaller and smaller arteries that carry blood to every part of the body.

B Eventually, in body tissues, these arteries become smaller and smaller until they are microscopic, thin-walled capillaries through which oxygen and waste products pass easily.

C Blood carrying waste products from body cells travels back to the heart by way of veins.

Artery

Capillaries

Vein

Visual Learning

Figure 3-3 To help students comprehend and compare the different blood vessels, demonstrate taking apart a thick piece of rope. The whole can be thought of as the aorta; the separate pieces twisted to make up the "aorta" can be visualized as arteries; the wispy single strands can serve as a model for an extensive capillary network. However, students should be informed that blood vessels are not twisted or intertwined.

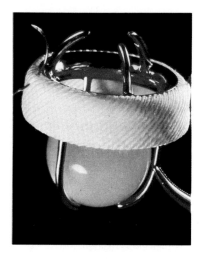

The cage-ball valve above consists of a tiny stainless steel cage that encloses a heat-treated carbon ball. Once sewn into the heart, the valve allows blood to pass in only one direction.

Some damaged valves, such as the one leading out to the aorta, can be replaced by valves from pig hearts. Others are replaced with a mechanical model.

Ball in the Cage

Does the above title sound like some kind of game? It's actually a description of a heart valve—not a natural valve, but a mechanical replacement valve.

The ball-and-cage valve is a perfect name for this replacement valve. Look at the photograph. In it, you'll see that the valve consists of a small cage with a ball inside. When the ventricle contracts, the ball moves up within the cage. This action closes the valve and blood leaves the heart. Once all the blood is out, the ventricle relaxes, and the ball falls back to its open position, allowing blood to flow into the ventricle from the atrium.

Because the valve is ball-shaped, fluids flow past it easily resulting in little wear on moving parts.

You Try It!

The ball-and-cage valve was the first replacement valve invented. Find out more about open heart surgery and different types of replacement valves.

3-1 Circulatory Systems **89**

GLENCOE TECHNOLOGY

 Videodisc
STVS: Human Biology
Disc 7, Side 1
Modeling Blood Flow (Ch. 2)

Testing Heart Valves (Ch. 4)

NATIONAL GEOGRAPHIC SOCIETY

 Videodisc
STV: Human Body Vol. 1
Circulatory and Respiratory Systems
Unit 1, Side 1
Need for Oxygen

7795-9506
Blood and Circulation

9508-12959
Systemic Circulation

20314-23616

---

deformed valves. Usually, these can be surgically corrected or replaced. Acquired problems result from diseases such as rheumatic fever, which can cause lesions on valves that prevent them from working properly. This also creates a heart "murmur" which is audible through a stethoscope.

Going Further

Artificial heart valves are in common use. But what about use of a heart from another animal to replace a badly damaged human heart? Have students form small groups to debate this question. Will we get to the point where doctors can routinely replace hearts, kidneys, and other vital organs with organs from other animals? Use the Performance Task Assessment List for Investigating an Issue Controversy in **PASC**, p. 65. L2

3 ASSESS

Check for Understanding

Assign the Check Your Understanding questions to small groups or pairs of students. Suggest that students make a table as they answer question 2.

Reteach

Sequencing Reinforce students' ability to sequence the flow of blood through the circulatory system, by obtaining a large diagram of the circulatory system and having students trace the path of the blood with their fingers. L1

Extension

Acquiring Information Have students who have mastered the concepts of this section investigate blue babies and the work of Dr. Helen Taussig. L3

4 CLOSE

Discussion

Show students a road map of your state. Ask why traffic reporters often refer to major highways as "arteries." *They are large and many cars travel on them.* If highways are like large arteries or veins, what roads are like capillaries? *local streets* L2

Blood and Breathing

Why does blood travel to your lungs? Just as you wash clothes before you wear them again, your blood isn't sent on another trip through your body until the carbon dioxide from your cells has been removed and oxygen has been resupplied. How is this accomplished?

Blood rich with carbon dioxide flows from the right atrium to the right ventricle. From there, it is pumped to your lungs. There it is exposed to oxygen as it flows through capillaries around millions of small thin-walled sacs containing air that you've breathed in. Follow this pathway in **Figure 3-4**.

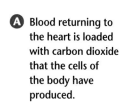

Capillaries in lungs
Heart

Figure 3-4

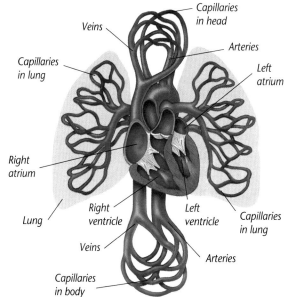

Capillaries in head
Veins
Arteries
Capillaries in lung
Left atrium
Right atrium
Right ventricle
Lung
Left ventricle
Capillaries in lung
Veins
Arteries
Capillaries in body

A Blood returning to the heart is loaded with carbon dioxide that the cells of the body have produced.

B When the right ventricle contracts, it forces the blood, rich with carbon dioxide, through an artery to the lungs. In the lungs, the carbon dioxide is exchanged for oxygen, and the carbon dioxide is then exhaled.

C Replenished with oxygen, the blood returns to the left atrium, then into the left ventricle. When the heart contracts, it pumps the now oxygen-rich blood out to cells throughout the body.

check your UNDERSTANDING

1. Diagram the pathway of blood through the human heart.
2. Compare the three types of blood vessels.
3. Describe the structural similarities and differences between an open and a closed circulatory system.
4. Discuss whether a closed circulatory system is more efficient at delivering oxygen and nutrients to cells than an open circulatory system.
5. **Apply** Fish have gills lined with capillaries. Water contains dissolved oxygen. Explain what you think happens when water flows over the blood supply in fish gills.

check your UNDERSTANDING

1. Diagrams should resemble Figure 3-2.
2. Arteries carry blood away from the heart. Capillaries bring oxygen and nutrients to cells. Veins carry blood back to the heart.
3. Open systems pump blood through small vessels into open cavities. A closed system has a heart with a loop of closed vessels.
4. A closed system is more efficient. In a closed circulatory system, capillaries ensure all cells get the oxygen and nutrients they need. In an open system, there is a chance that not all cells will receive the same amounts of oxygen and nutrients.
5. Oxygen moves in through gills.

3-2 A System Under Pressure

With Every Beat of Your Heart

What happens when you put air in a bike tire? Does the air pour in as a continuous stream? Or do you have to apply pressure, let up, then apply pressure again? Is blood in your body pumped by your heart in the same way you pump air into a bicycle tire or is the force continuous? Think about this question as you do the following Explore activity.

 ACTIVITY

How can you feel and measure heart rate?

What To Do

1. Place your fingers on your neck between your ear and your Adam's apple.

2. Push gently and move your fingers around until you feel a strong beat. What you are feeling is blood as it pulses through your carotid artery.

3. Count the number of beats you feel for 15 seconds. Then multiply that number by four. This number is your heart rate for one minute.

4. *In your Journal*, record your heart rate and complete the following statement: If the left ventricle of my heart exerts pressure on the blood each time my heart beats, then I feel my heartbeat in my neck because....

Both air in a tire and blood are classified as fluids. Both air in a tire and blood in vessels are in confined spaces. When pressure is placed on or released from a fluid in a closed space, then all the fluid in the system feels the pressure.

In the Explore activity, you felt the walls of an artery expand each time the left ventricle contracted. You felt your blood apply more, then less pressure on the walls of that artery. The rhythmic expansion and contraction of an artery is your **pulse**. Pulse is also your heartbeat. Find out if you are the only organism with a pulse.

Section Objectives
In this section, you will
- Explain what causes pulse.
- Explain how blood moves through your body under pressure.
- Compare the structural adaptations of blood vessels.
- Compare the circulatory systems of fish, birds, amphibians, and mammals.

Key Terms
pulse
blood pressure

PREPARATION

Planning the Lesson
Refer to the Chapter Organizer on pages 82A–D.

Concepts Developed
In this section, students learn how the heart produces a pulse and how it causes blood to move through the body under pressure. Students compare the circulatory systems of fish, birds, amphibians, and mammals.

1 MOTIVATE

Bellringer
Before presenting the lesson, display **Section Focus Transparency 8** on an overhead projector. Assign the accompanying **Focus Activity** worksheet. L1 LEP

Demonstration Fill a container with water and use a baster to transfer water to an empty container. Ask students to suggest how this procedure resembles the action of the heart. *They should note that both involve pumping.* L1

2 TEACH

Tying to Previous Knowledge
Review Pascal's Principle. Blood in a closed circulatory system is subject to this same law.

 Explore!

How can you feel and measure heart rate?

Time needed 5 minutes

Materials watch with second hand

Thinking Processes observing and inferring, recognizing cause and effect

Purpose To measure heart rate and recognize the effect of blood moving through vessels under pressure.

Teaching the Activity

Troubleshooting Caution students not to use their thumb—it has a pulse.

Science Journal Students' heart rates will vary. L1 LEP

Expected Outcome
Students will observe a regular pulse in their necks and relate it to blood pressure.

Answer to Question
4. . . . the pressure is felt throughout the arteries of the body.

✔ Assessment
Process Have students make a graph showing the effects of exercise on heart rate. Use the Performance Task Assessment List for Graph from Data in **PASC**, p. 39. P

3-1 Pulse! Pulse! Who Has a Pulse?

Preparation

Purpose To measure and compare an earthworm pulse to a human pulse.

Process Skills observing and inferring, comparing and contrasting, making and using tables, interpreting data

Time Required 30 minutes—20 minutes for finding human and worm pulse and 10 minutes for calculating averages and answering questions

Materials See reduced student text.

Possible Hypotheses Students may hypothesize that an earthworm has a measurable pulse rate that is faster or slower than a human pulse rate.

The Experiment

Process Reinforcement To reinforce student data interpretation skills, have them make tables to record their own pulse rate. Students can measure their pulse rate for 15 seconds and then multiply that number by 4 to obtain their pulse rate per minute. For accuracy, they should repeat the procedure three times. Use this data to calculate the average pulse rate per minute. Ask students to calculate how many times their heart beats in one hour, one day, one month, one year, and in an average lifetime of 70 years.

Possible Procedures Put the earthworm on a damp towel in a petri dish, and obtain a pulse by placing the index finger on the blood vessel along the top surface of the earthworm. For accuracy, take the worm pulse three times.

Teaching Strategies

Troubleshooting Remind students to use water at room temperature to keep the worm damp. If students have trouble seeing the dorsal blood vessel, have them use a hand lens.

Pulse! Pulse! Who Has a Pulse?

You've learned that you have a closed circulatory system in which you can measure a pulse. In this activity, investigate how to measure pulse in another animal with a closed system.

Preparation

Problem

How can you measure an earthworm's pulse rate, and how does it compare with your own?

Form a Hypothesis

By observing the earthworm, form a hypothesis about its pulse rate. Do you think the earthworm has a measurable pulse rate faster or slower than a human's?

Objectives

• *Measure* the pulse of an earthworm.
• *Compare* the earthworm pulse to a human pulse.

Materials

live earthworm
paper towels
stopwatch
Petri dish with cover
medicine dropper
water at room temperature

Safety Precautions

Earthworms are living organisms. They need to be handled so that they are not harmed or allowed to dry out.

Meeting Individual Needs

Learning Disabled, Behaviorally Disordered To get learning-disabled or behaviorally disordered students actively involved in the Investigate activity, ask one student to count aloud the pulses of the earthworm's heart, another to serve as a timekeeper, and another to record the data. Provide students with prepared data sheets to help them record and study their data. **L1** **COOP LEARN**

Multicultural Perspectives

Egyptian Embalming
Egyptians learned much about anatomy from their practice of embalming corpses, but believed the heart to be the center of human intelligence. Ask students to find out more about embalming methods and prepare an oral presentation for the class.

DESIGN YOUR OWN
INVESTIGATION

Plan the Experiment

1 Place the live earthworm in a Petri dish that has been lined with a moist paper towel. Keep the earthworm moist with water, but NEVER flood the worm.

2 The earthworm has a blood vessel as shown in the illustration.

3 As a group, decide how to measure the pulse of the earthworm safely. *In your Science Journal,* make a table to record your data.

4 Determine how to measure a human pulse and record this data in your table.

Check the Plan

1 What are you comparing?

2 How many trials do you need to do to obtain accurate information? Do you know how to find an average of the trials?

3 Make sure your worm is still moist.

4 Have your teacher review your plan before you proceed.

5 Carry out the experiment, observe, and record your data.

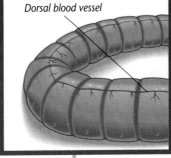
Dorsal blood vessel

Analyze and Conclude

1. Calculate What was the average pulse rate for one minute for your earthworm? What was your pulse rate for one minute?

2. Compare and Contrast How does the earthworm's pulse rate compare with yours?

3. Analyze Why was it important to repeat your measurements and determine an average?

4. Draw a Conclusion What conclusion might you draw about animal activity and pulse rate?

Going Further

Testing a Hypothesis

Is the range of pulse rates of male students in the class the same as that for female students? Hypothesize whether one or the other might be faster and design an experiment to test your hypothesis.

Program Resources

Activity Masters, pp. 15–16, Design Your Own Investigation 3-1 L2

Study Guide, p. 14

Making Connections: Integrating Sciences, p. 9, Jessie I. Price, Veterinary Bacteriologist L2

Section Focus Transparency 8

Going Further ▐▐▐▐▐►

Students may hypothesize that there are or are not differences in the pulse rates between the sexes. Their experiments should maintain identical conditions. Male pulse rates may be lower.

Discussion Discuss with students whether the size of an organism affects the pulse rate. *It doesn't.* Then ask if they think activity level of an animal is related to pulse rate. This is part of their hypotheses, so let them discuss without giving a definite answer. To further discussion, mention the lifetime pulse of a human. A newborn may have a pulse rate as high as 140. A child may have a pulse rate of 90 beats. Adults have a pulse of around 72-76. Elderly people have a slower pulse rate.

Science Journal Have students record their hypothesis and their data in their Science Journals. L1

Expected Outcomes

Students will observe and measure a pulse rate in an earthworm and compare this pulse rate with their own. The earthworm pulse rate is much slower.

Analyze and Conclude

1. Answers will vary. average earthworm rate = 23-30; average human rate = 60-80

2. The earthworm has a much slower pulse rate than the student by about half.

3. Pulse rate fluctuates with body activity. Reliable data require more than one or two readings. An average gives an accurate indication of the typical pulse.

4. Students will conclude that the more active the organism, the higher the pulse rate.

✔ Assessment

Process Have students work in small groups to collect data comparing the pulse rate of females and males. Make a table on the board and have students list their data in the appropriate columns. Students can then calculate an average pulse rate for the females and males in the class and compare the results. *In the general population, women tend to have a slightly higher pulse rate than men.* Have students make a bar graph demonstrating the comparison. Use the Performance Task Assessment List for Conducting a Survey and Graphing the Results in **PASC,** p. 35.

Time needed 10 minutes

Materials plastic squeeze bottle, stopper with plastic and rubber tubing, water

Thinking Processes observing and inferring, recognizing cause and effect

Purpose To observe how pressure moves a liquid.

Teaching the Activity

Discussion Ask students to predict what will happen when the bottle is squeezed. Then discuss their results. L1

LEP

Science Journal Students may describe that the sides of the bottle compress and water squirts from the tube as the bottle is squeezed.

Expected Outcome

Students will observe that the pressure caused by a pumping action can move a liquid.

Conclude and Apply

1. Students may describe water spurting from the bottle in pulses, not a steady stream.

2. by spurting when pressure is applied and by stopping when pressure was no longer applied

✔ Assessment

Oral Ask students to orally compare the movement of the water from the bottle to the movement of blood from the heart. Students should infer that bloods leaves the heart in spurts with each heartbeat.

When Fluids Are Under Pressure

Your pulse rate tells you the rate of your heartbeat. With every beat of your heart, blood is pushed through the blood vessels in your circulatory system. What happens to the blood and the vessels when this takes place?

What happens to a liquid under pressure?

Blood is a fluid that responds to pressure the way other fluids do.

What To Do

1. Fill a plastic squeeze bottle with water.
2. Fit a bendable plastic straw or rubber tubing into a stopper.
3. Put the stopper into the water-filled bottle.
4. Hold the bottle and direct the open end of the tube into a sink. Then squeeze the bottle rhythmically. Try different pressures.

Conclude and Apply

1. *In your Journal,* describe what happened to the bottle when you squeezed it.
2. How did the stream of water respond to differing pressures?

Connect to...

Physics

Many systems make use of Pascal's Principle. Find out what hydraulic lifts are and what they are used for. How is a hydraulic lift similar to your circulatory system?

Each time you squeezed the bottle, you applied a force that pushed the water out of the bottle into the tubing. Between squeezes, when you weren't exerting a force, the water didn't move out. Your hand acted as a pump to move water out of the bottle. How does this activity relate to the heart as it contracts and relaxes during a heartbeat?

Your heart is a pump that moves blood through the blood vessels of your body with pressure that rises and falls. Like the rhythmic squeezing of the water bottle, each time your heart beats, it exerts a force on your blood and pushes it along. What happens to blood vessels when the heart applies and releases pressure?

■ Pascal's Principle

If you have ever squeezed a water-filled balloon, you know that the walls push out in all directions. A water-filled balloon illustrates Pascal's Principle. Pascal's Principle states that any change in pressure applied to a fluid in a confined space is sent unchanged throughout that fluid. What does this mean for your blood and blood vessels?

94 Chapter 3 Circulation

Program Resources

Laboratory Manual, pp. 9–12, Blood Pressure L2
Science Discovery Activities, 3-2, 3-3

Blood: A Fluid Under Pressure

Blood exerts pressure against the walls of the blood vessels in which it is confined. The pressure blood exerts against the inner walls of blood vessels is called **blood pressure**.

■ Variations in Pressure

While there is pressure on blood in arteries, capillaries, and veins, it is usually measured in arteries where it is highest. When the left ventricle contracts, it rapidly forces blood under the highest pressure into your aorta. The aorta expands and swells, then contracts, forcing the blood along down the artery.

The graph in **Figure 3-5** shows how blood pressure changes as it travels from the aorta. By the time it enters the right atrium, its pressure is almost zero. How does the body handle these differences in blood pressure?

■ The Walls of Blood Vessels

The walls of arteries and veins have structural differences that tell you something about how they respond to changes in blood pressure. Arteries have thick, muscular walls that stretch and contract a little each time your heart pumps blood into them. Arteries exert and withstand great pressure. Veins, on the other hand, have thinner and more elastic walls. Blood in veins and capillaries does not exert such great pressure.

In the next Investigation, you can learn about your circulatory system.

Figure 3-5

The force of blood flowing through your circulatory system puts pressure on the walls of all your blood vessels—arteries, capillaries, and veins.

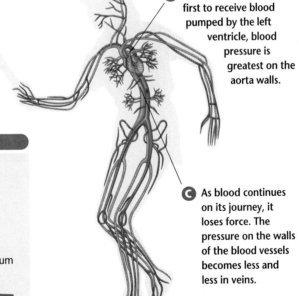

A This graph shows how the force of your blood puts less and less pressure on the walls of your blood vessels as it makes its round-trip journey from the heart and back to the heart.

Blood Pressure in the Circulatory System

High / Low — Blood pressure

a. Aorta
b. Large arteries
c. Small arteries
d. Capillaries
e. Small veins
f. Large veins
g. Vein to right atrium

a b c d e f g

B Because the aorta is the first to receive blood pumped by the left ventricle, blood pressure is greatest on the aorta walls.

C As blood continues on its journey, it loses force. The pressure on the walls of the blood vessels becomes less and less in veins.

A System Under Pressure **95**

INVESTIGATE!

3-2 Explore Your Circulatory System

Planning the Activity

Time Required 40 minutes

Purpose To observe the working of the human circulatory system.

Process Skills observing and inferring, measuring, comparing and contrasting, recording data

Materials See the reduced student text.

Preparation To be prepared for student difficulties and questions, try these activities before the class period.

Teaching the Activity

Students should work in pairs. Everyone should practice taking their pulse on their wrist before starting the activity. Most students' pulses will be from 60 to 80 beats per minutes. L1

COOP LEARN

Troubleshooting If students have trouble getting the knee pulse to respond, have them relax back against the chair. The movement of the leg and foot is subtle. They must observe closely.

Process Reinforcement To carry out these activities, students need to follow directions, observe, and record data carefully.

Science Journal Have students record their pulse rates and describe the results of the Harvey experiment in their journals. L1

Explore Your Circulatory System

You've learned that blood pulses throughout your body through arteries and returns to the heart through veins. In this investigate, you will observe some of the external features of your circulatory system.

Problem

Can you observe the working of your circulatory system?

Materials

clay or putty
coffee stirrer
small mirror

What To Do

1 Feel your pulse on the inside of your wrist. Where it is strongest, place a blob of clay or putty. Then push a coffee stirrer into the clay. When you lay your arm flat on a desk, you should see the stirrer move with your pulse. Measure and record the pulse *in your Science Journal.*

2 Now use your leg as a pulse meter. Sit down, and cross your legs so that the artery at the back of the knee is pressed against the other knee. Relax. Your top leg should pulse gently. Record your pulse.

3 Your heart muscle pumps blood around your body. Find out what other force affects your circulation. For one minute, hold a hand above your head and the other straight down. Then compare your hands. What happened? Describe your observations *in your Science Journal.*

4 In strong light, use a mirror to look under your tongue. Can you see veins? Do you see the pink arteries at the base of the tongue? It is possible to see capillaries on your body.

5 You are going to conduct an experiment similar to one William Harvey, an English physician, performed in the 17th century to show how blood returned to the heart. Have a person with prominent veins hold his/her hand straight down until veins stand out in the hand. Then, place the hand palm down on a table. Place a finger on an unbranched vein near the wrist. With another finger, *very gently* stroke the vein going up the arm in the direction the blood flows back to the heart. *In your Science Journal,* record what happens to the vein.

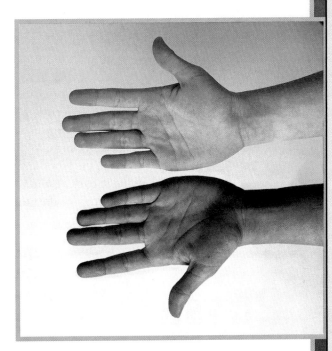

Analyzing

1. Why do you feel a pulse in an artery?

2. In step 3, what caused the hands to be different colors? Explain why this happened.

3. What force makes it difficult to return blood from your feet to your heart?

Concluding and Applying

4. Why do you have capillaries in your eyelids instead of arteries or veins?

5. Why is there no pulse in the veins on the back of your hand?

6. In step 5, when the vein collapsed, why did this happen? Look ahead on page 98 for help.

Program Resources

Activity Master, pp. 17–18, Investigate 3-2 **L3**

The veins in your legs contain the greatest number of valves. **How is this helpful?** *The veins in the legs have the greatest number of valves because the blood there must travel a long distance against the force of gravity. As the text notes, muscular action helps move the blood along.*

Uncovering Preconceptions

Students may be confused about why a valve does not work like a door, opening and closing and permitting people to go through it in either direction. Compare the valve to a revolving exit gate at a park. Those gates have bars blocking a person from coming back in after he or she has exited, just as the valve blocks blood from moving in the wrong direction.

Blood pressure is highest in arteries, because they are the first to receive the force of the blood pushed from the heart. If blood pressure is zero in an artery, blood is not being pumped from the heart effectively. This could indicate severe damage to the heart. It would also mean that body cells would not get the nutrients and oxygen usually delivered by an adequate blood supply. Zero pressure in a vein would not be cause for alarm because by the time blood has circulated through veins on its way back to the heart, much of the pressure has dissipated.

Observing and Inferring

A blood pressure of zero in a vein is not necessarily cause for alarm, but a blood pressure of zero in an artery would be. Explain. If you need help, refer to the **Skill Handbook** on page 642.

■ **Going Against Gravity**

Have you ever wondered how blood gets from your feet to your heart? It has to move against the force of gravity, doesn't it? In Chapter 1, you learned that there is a gravitational force that pulls things toward Earth. Because of gravity and because pressure is lower in veins, blood can collect in veins and enlarge them. Without help, the blood in veins wouldn't make it back to the heart. How does blood return to the heart?

■ **Valves in Veins**

Blood in veins receives some help from one-way valves, similar to the valves in your heart. As you can see in **Figure 3-6**, the one-way valves in your veins stop the blood from flowing backward. If there is a backward movement of blood, the pressure of the blood itself closes the valves. The veins in your legs contain the greatest number of valves. How is this helpful?

■ **Muscle Power**

Blood in veins also receives help getting back to the heart from the skeletal muscles in your body as they contract. When the muscles surround-

Physics CONNECTION

The Physics of Blood Pressure Measurement

A sphygmomanometer and a stethoscope are used to measure blood pressure.

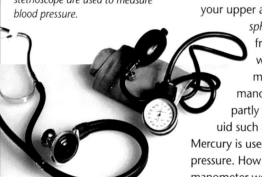

Have you ever had your blood pressure measured? If so, then you've had the cuff of an instrument called a *sphygmomanometer* (sfig moh muh NAHM uh ter) wrapped around your upper arm. The prefix *sphygmo-* comes from the Greek word *sphygmos*, meaning pulse. A manometer is a tube partly filled with a liquid such as mercury. Mercury is used to measure gas pressure. How does a sphygmomanometer work?

Taking Blood Pressure

To take blood pressure, the rubber bulb is squeezed until the air pressure inside the cuff is greater than the blood pressure inside the arm artery. As you would guess, when this happens the artery walls collapse, temporarily stopping the flow of blood.

Then the air pressure inside the cuff is slowly reduced by allowing the cuff to deflate. When air pressure becomes less than blood pressure inside the artery, blood surges through the artery again in a pulsating fashion. The pressure at which

Physics CONNECTION

Purpose

Physics Connection extends the discussion in Section 3-2 of blood pressure by describing how it is measured. The theme is stability and change, for a stable blood pressure is maintained within a range of values. This is essential to normal activity in all animals.

Content Background

Blood pressure indicates the condition of the overall circulation system, so the ability to measure it is very important. Many sphygmomanometers are pressure transducers and now display blood pressure digitally, instead of a mercury column. But the mercury-column type still is used, and blood pressure readings still are expressed in mm of mercury because many people know these numbers. As long as the tube remains sealed, the mercury—which is a poisonous substance—poses no hazard.

ing veins contract, they exert pressure on the veins and push blood along   toward the heart.

Figure 3-6

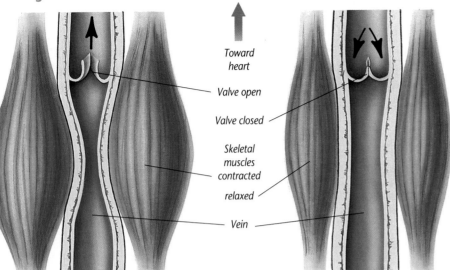

Toward heart

Valve open

Valve closed

Skeletal muscles
contracted

relaxed

Vein

When muscles surrounding veins contract, they squeeze the veins, forcing the blood within to move forward. One-way valves keep the blood moving to the heart.

blood flow resumes is called the systolic pressure. It is the pressure exerted when the left ventricle contracts.

As the pressure in the cuff is further reduced, blood moves through the artery freely. The pressure at which blood can move through the artery is the diastolic pressure.

As you can see from this procedure, blood pressure is not really being measured—air pressure inside the cuff is measured! Blood pressure is measured only indirectly.

Healthy Signs

In a healthy young person, systolic pressure is about 110-120 mm Hg, and diastolic pressure is about 70-80 mm Hg. What does this mean? Hg is the symbol for the element mercury. Millimeters of mercury refers to the pressure needed to raise a column of mercury up to a certain level. Therefore, normal systolic blood pressure raises a column of mercury 120 mm, and normal diastolic blood pressure raises a column of mercury 80 mm.

What Do You Think?

Find out why mercury is used and not water or some other fluid. How is mercury different from most other fluids?

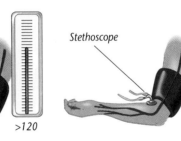

Cuff

Artery

>120

Stethoscope

120

80

Blood Pressure in Other Closed Systems

As you recall from Section 3-1, reptiles, fish, amphibians, and mammals all have closed circulatory systems where a heart pumps blood through a continuous loop of blood vessels. But not all closed circulatory systems are the same. There are differences in the structure of the heart and blood vessels among these animals. These differences tell you something about the lifestyle of the organism and its needs. These differences also affect blood pressure and the efficiency with which oxygen and nutrients are supplied to cells.

Figure 3-7

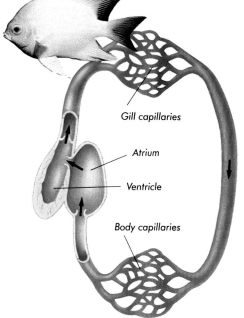

Gill capillaries

Atrium

Ventricle

Body capillaries

Fish hearts have only two chambers, an atrium and a ventricle. The atrium is the pumping chamber. Blood pumped by a fish's ventricle flows through a single loop with two capillary networks before it returns to the heart.

A Blood carrying carbon dioxide is pumped from the ventricle and travels through the network of capillaries at the gills. Here the blood releases carbon dioxide and picks up oxygen from the water.

B The blood, now rich with oxygen, continues to the second network of capillaries, which supplies the body cells. Here blood leaves oxygen and nutrients and picks up carbon dioxide and wastes from the body cells. The blood returns and a new cycle begins.

Figure 3-8

The circulatory system of frogs and most other amphibians is powered by a heart with three chambers—two atria that collect blood and a ventricle that is the main pumping organ.

A The left atrium receives blood rich with oxygen from the lungs, and the right atrium receives blood rich with carbon dioxide from the body cells.

B Blood from both atria empty into the ventricle. The ventricle contracts and sends blood in two directions. Even though the atria share the same ventricle, most of the oxygen-rich blood moves on to the body and most of the oxygen-poor blood flows to the lungs.

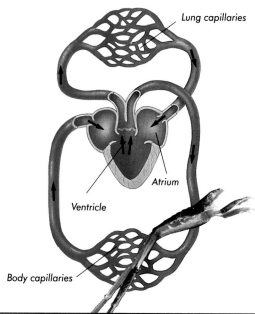

Lung capillaries

Atrium

Ventricle

Body capillaries

100 Chapter 3 Circulation

Figure 3-9

Birds have a four-chambered heart that works just as your four-chambered heart works. The two sides of the heart are entirely separate. There is no mixing of blood. The right side of the heart pumps blood only to the lungs, and the left side of the heart pumps blood only to the body.

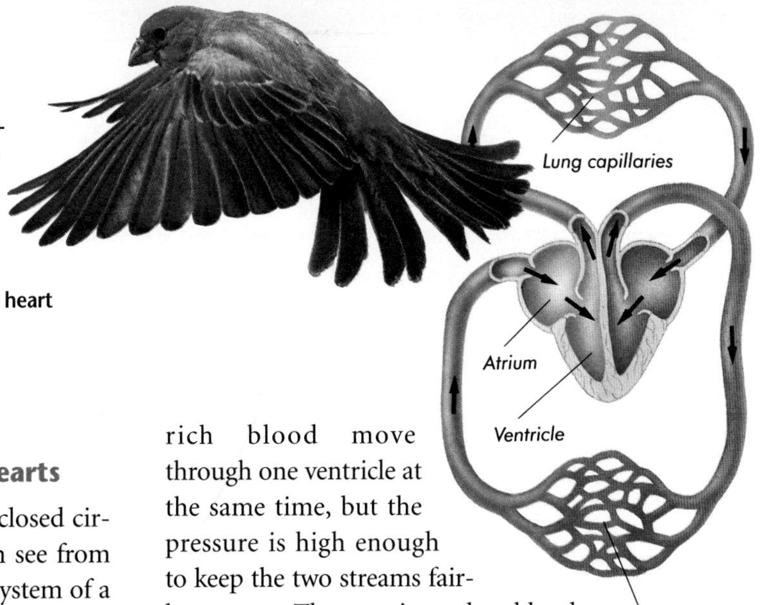

Lung capillaries

Atrium

Ventricle

Body capillaries

■ Two- and Three-Chambered Hearts

Fish have a very simple closed circulatory system. As you can see from **Figure 3-7**, the circulatory system of a fish consists of a two-chambered heart and a single loop of blood vessels. In a fish, pressure builds as blood moves into the atrium by suction, and then into the ventricle. Once the ventricle contracts, pressure drops and the body receives blood under low pressure. Blood pressure drops off as blood moves into the numerous narrow capillaries in the gills. Blood moves slowly as it passes to the body organs in these animals who have low energy requirements. **Figure 3-8** shows circulation in amphibians such as frogs. Oxygen-poor and oxygen-rich blood move through one ventricle at the same time, but the pressure is high enough to keep the two streams fairly separate. The one time when blood does mix in a frog is when the animal is submerged. Then, oxygen tends to move into the blood through its skin, not through its lungs.

■ Four-Chambered Hearts

In birds and in mammals like yourself, not only is blood kept separate in the heart because the ventricles are separated, but there are pressure differences as well. Blood moving to the lungs is at a very low pressure, while blood moving out to the body is under a much higher pressure.

check your UNDERSTANDING

1. Explain how the force of gravity and lower blood pressure are handled in veins.
2. If you receive a serious injury and cut a vein, the blood would flow out smoothly. How would you expect blood to flow out of a cut artery? Explain your answer.
3. Diagram the hearts of fish, amphibians, and birds. Label the following: atria, ventricles, direction of blood flow and pressure differences.
4. **Apply** Explain how the functions of lungs and gills are related.

check your UNDERSTANDING

1. One-way valves prevent blood from flowing backward. Muscle contractions help push blood upward, against the force of gravity.
2. Blood pressure in arteries is higher so you would expect the blood to flow out more rapidly. The blood in the arteries is also more responsive to the pumping action of the heart, so you would expect it to spurt, a reflection of the heartbeat.
3. Students' diagrams should resemble Figures 3-7 and 3-8.
4. Lungs and gills are both able to take oxygen from the environment.

PREPARATION

Planning the Lesson

Refer to the Chapter Organizer on pages 82A–D.

Concepts Developed

In this section, students learn of some of the problems that prevent the circulatory system from accomplishing its task, how fatty deposits contribute to heart disease, and how some lifestyles lead to high blood pressure.

1 MOTIVATE

Bellringer

Before presenting the lesson, display **Section Focus Transparency 9** on an overhead projector. Assign the accompanying **Focus Activity** worksheet. [L1]
LEP

Discussion Ask students to share any experience they have had with heart disease and what they know about the changes in lifestyle that such problems require. They also can share what they have heard about heart disease from articles or television.

2 TEACH

Tying to Previous Knowledge

In the previous sections, students learned the function of the circulatory system. Ask what would happen if something were to narrow or block an artery. Tell students that this section will explain several disorders that can affect the heart.

3-3 Disorders in Circulation

Section Objectives

In this section, you will

- Describe the role of fatty deposits in heart disease.
- Relate lifestyles to high blood pressure.

Key Terms

atherosclerosis
hypertension

Circulation to the Heart

The term *heart disease* is used to describe any of the health problems that affect the heart. Your heart functions all your life to keep your body supplied with nutrients and oxygen, but what keeps the heart muscle itself functioning? How do your heart cells get the oxygen and nutrients they need? What happens when these materials aren't supplied?

Figure 3-10A shows blood is supplied to the heart by several coronary arteries.

Heart disease can occur when problems arise in the coronary vessels. One leading cause of heart disease is **atherosclerosis** (a thuh roh skluh ROH suhs), a condition in which fatty deposits and calcium build up inside the coronary arteries. **Figure 3-10B-D** shows the progressive stages of atherosclerosis in a coronary artery.

As heart muscle tissue dies, scar tissue forms. The ability of the heart muscle to contract and relax is severely affected.

Figure 3-10

A Like the rest of the body, the heart receives the oxygen and nutrients it needs and rids itself of waste by way of blood flowing through blood vessels. On the diagram, you can see the coronary arteries, which nourish the heart.

B This cross section of a healthy coronary artery shows a clear, wide-open pathway through which blood easily flows.

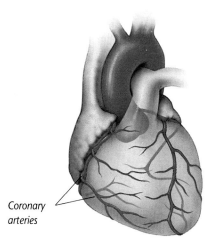

Coronary arteries

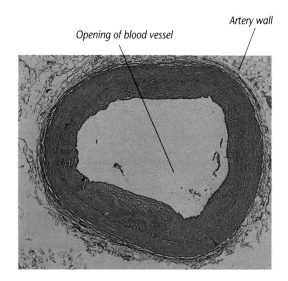

Opening of blood vessel

Artery wall

Program Resources

Study Guide, p. 15

Critical Thinking/Problem Solving, p. 11, Smoking and Heart Disease [L2]

Multicultural Connections, p. 9, A Pioneer in Cardiology [L1]

How It Works, p. 6, How a Pacemaker Works [L2]

Section Focus Transparency 9

ENRICHMENT

Research Have students find out about the Framingham Study, and summarize their findings for the class. The people of Framingham, Massachusetts, took part in a long-range study of risk factors and heart disease. Data was collected for more than 40 years. Ask students to infer the importance of large samples as a basis for a statistical study.

What happens to liquid flow in a clogged tube?

Model a blocked artery to show its effect on blood flow.

What To Do

1. Insert a dropper full of mineral oil into a piece of plastic tubing.

2. Squeeze the oil through the tube.

3. Observe how much oil comes out the tube.

4. Next, refill the dropper and squeeze oil through a piece of plastic tubing that has been clogged with cotton.

5. How much oil comes out of the clogged tube?

Conclude and Apply

1. *In your Journal*, explain how the addition of the cotton to the tube changed the way oil flowed through the tube.

2. How does this activity demonstrate what takes place when arteries become clogged?

3. What differences in pressure did you notice in the squeeze bulb?

As you observed in this activity, clogging the plastic tube with cotton severely restricted the flow of oil through it. In the same manner, fatty deposits clog arteries so that blood is restricted from flowing. When this occurs, the heart muscle cells begin to die.

C Here the blood-flow pathway has been narrowed by a buildup of fatty deposits. Blood flow is slowed. The heart muscle does not get enough oxygen and nutrients to do its work. The muscle begins to die.

D If the deposit continues to build, blood flow through the artery may stop. The person will suffer a heart attack.

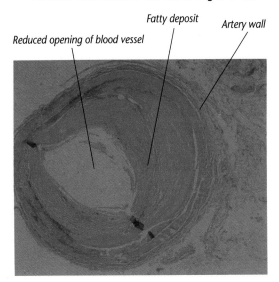

Reduced opening of blood vessel
Fatty deposit
Artery wall

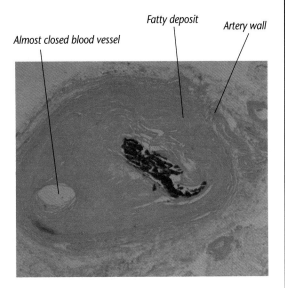

Almost closed blood vessel
Fatty deposit
Artery wall

3-3 Disorders in Circulation **103**

What happens to liquid flow in a clogged tube?

Time needed 10 minutes

Materials two 10-cm pieces of plastic tubing, dropper, mineral oil, cotton, tweezers

Thinking Processes observing and inferring, comparing and contrasting, recognizing cause and effect

Purpose To compare and contrast the flow of a liquid through an open tube and a tube that is partially blocked.

Preparation Be sure there is enough cotton to impede the flow of the oil, without blocking it entirely.

Teaching the Activity

Science Journal Students will describe the cotton slowing the flow of oil through the tube. L1 LEP

Expected Outcome

Students will observe slowing and infer that it is because of the blockage

Conclude and Apply

1. The addition of cotton slowed down the movement of oil in the tube.

2. Cotton represents deposits that may collect on the walls of blood vessels.

3. It may take more pressure in the squeeze bulb to move materials through a clogged tube.

✔ Assessment

Content Have students write a brief explanation of the effects of atherosclerosis and predict what lifestyle might contribute to the development of the disorder. *Atherosclerosis causes the arteries to become narrow and less flexible. As a result, body cells do not get enough blood supply and suffer from a lack of nutrients and oxygen. Students may infer that a diet high in fat and cholesterol might contribute to the disease.*

Multicultural Perspectives

Jokichi Takamine

The human heart is subject to many potential disorders. The hormone adrenaline or epinephrine occurs naturally in the human body, secreted by the body's adrenal glands. Have students research how it is used medicinally in case of cardiac arrest or to dilate the blood vessels of a patient in shock. Jokichi Takamine (1854–1922) was born in Japan but came to the United States in 1884. A chemist, he was the first person to extract adrenaline from the adrenal glands of animals in 1901. Eventually researchers learned to synthesize it, and it is now used by physicians to stimulate the heart when necessary. Takamine also founded the Japanese Association of New York and helped other Issei (first-generation Japanese immigrants) who wished to work in chemistry.

When Heart Rates Change

Heart rates and blood pressure change for many reasons. Occasional increases in blood pressure are normal. During strenuous activity, your heart beats faster and your breathing rate increases, thus increasing oxygen delivery to your body cells. This increased heart rate is absolutely necessary if your circulatory system is to be able to pick up and deliver the additional oxygen. These activities also increase your blood pressure for a short time. After resting awhile, heart rate and blood pressure in a healthy person return to normal.

■ Hypertension

In some people, however, blood pressure remains high after exercise, or even without exercise. A disorder of the circulatory system when blood pressure is higher than normal is known as **hypertension**, or high blood pressure.

Although the exact causes of most hypertension are not known, diets high in fat or sodium are linked to it. One known cause of hypertension is atherosclerosis. Clogged arteries cause the pressure within a blood vessel to increase by reducing the elasticity of the artery walls and narrowing the pathway of blood flow. This extra pressure puts a strain on the heart, which has to beat faster in an attempt to keep oxygen-rich blood flowing to the body tissues. Many people suffer from hypertension—as many as 50 million people in the United States. However, many of them are never aware of it. Why is high blood pressure dangerous? The extra pressure exerted on your blood vessels is transferred to the organs of your body, and they begin to suffer damage over time.

Figure 3-11

While skating, the blood pressure of these rollerbladers is higher than usual. Blood pressure varies in relationship to activity. If you bike, swim, or shovel heavy snow, your muscle cells use up oxygen at a faster rate than when you are resting.

Your muscle cells require an increased oxygen-rich blood supply. In response to this need for more oxygen, your heart beats faster. When your heart beats faster, your blood pressure temporarily goes up.

Be Good to Your Heart

One of the best things you can do for your circulatory system is to exercise regularly. But not all exercise is the same. To benefit your circulation, you need to regularly participate in walking or aerobic exercises, such as running, swimming, bicycling, or cross-country skiing. Aerobic exercises promote the efficient use of oxygen by your body. When you exercise, your lungs take in more oxygen, and your heart muscle becomes stronger. It then pumps more blood with each beat. This, in turn, allows your heart rate to decrease while still sending the same amount of oxygen-containing blood into your arteries.

B For exercise to be both safe and effective, it should raise and keep your heartbeat rate to a level that ranges between 70 percent and 85 percent of your maximum heartbeat rate. A rate lower than 70 percent will not help you develop a fit heart and lungs. A higher rate can be dangerous to your heart. Use the chart to calculate your target pulse range during exercise.

Figure 3-12

A Doing regular aerobic exercise such as cycling over a period of several weeks increases your heart's ability to pump blood. Your heart, even at rest, will send out more blood with each contraction.

The increased flow of blood to the heart provides more oxygen. The heart is able to beat at a slower rate and still send the body cells all the oxygen-rich blood they need.

How to Find Your Heart Rate Range

1. Subtract your age from 220 to find your maximum heart rate.
   For example, if you are 12 years old,
   220 − 12 = 208.
2. Multiply your maximum heart rate by 0.85 to find the top of your range.
   208 × 0.85 = 177
3. Multiply your maximum heart rate by 0.70 to find the bottom of your range.
   208 × 0.70 = 146

check your UNDERSTANDING

1. What are the risk factors of heart disease?
2. Explain how atherosclerosis can lead to hypertension.

3. **Apply** How can the risk factors of heart disease be turned around or prevented?

check your UNDERSTANDING

1. Diets high in fats and cholesterol can lead to buildup of fats on the inside of coronary arteries. This in turn can damage the heart tissue.
2. Clogging of the arteries increases blood pressure by reducing elasticity in the walls of the arteries and reducing the amount of space the blood has to flow through.

3. The risk factors of heart disease can be prevented by leading a healthy lifestyle. Eat less fat and more fruits, vegetables, and grains. Regular aerobic exercise helps prevent heart disease.

Visual Learning

Figure 3-12 Have students use the information from Figure 3-12 to find their own target heartbeat rate ranges. Ask students to use this information to design a safe exercise program.

3 ASSESS

Check for Understanding

Have students answer the Check Your Understanding questions. With regard to the Apply question, ask students to relate diet and exercise to the functioning of the circulatory system.

Reteach

Activity To help students practice classifying skills, have them divide a piece of paper into two columns. In the first column, ask them to list different disorders of the circulatory system. In the second column, opposite each item in the first column, ask them to list causes of those disorders. L1

Extension

Activity Ask students to brainstorm activities that people often engage in, such as smoking, watching hours of television, and bike riding. Students can determine a new method for advertising the effects of these habits on cardiovascular health. L2

4 CLOSE

Writing About Science

Students can use communication skills to write a short article describing the importance of the heart and circulatory system to their own health. Encourage them to use concrete examples from their own lives. L1

Science *and* Society

Purpose

Science and Society extends the discussion in Section 3-3 of the effect of lifestyle on the circulatory system by examining how people are reducing cholesterol intake and exercising more.

Content Background

Cholesterol ($C_{27}H_{46}O$) has several functions in animals. It does not occur in plants. Our bodies manufacture some, and it is in many foods as well. Scientists identify two cholesterols: "good" (HDL, High-Density Lipoprotein), which helps clean the blood, and "bad" (LDL, Low-Density Lipoprotein), which clogs arteries.

Teaching Strategies

To reinforce classifying skills, provide a list of cholesterol-rich foods available from health-food stores, doctors, and books on diet and heart disease. Have students examine their daily diets for cholesterol. Ask them to list everything they ate yesterday, and put a C to the left of every cholesterol-rich item. Then have them compare lists.

Using Computers

Students can develop lists of questions to ask, so that each person they survey answers the same questions. Students must survey a large enough number of people (50 or more) so that they can draw more accurate conclusions from their results.

Going Further ▮▮▮▮➡

Divide your class into an even number of groups. Have half of the groups plan healthful, balanced, low-cholesterol diets for a week, including recipes. Have the other half plan a reasonable exercise program for a week,

Science *and* Society

Changing a Nation's Lifestyle

In 1980, a United States Department of Health and Human Services report recommended that people in the United States cut back on the fat and fatty foods they consume in order to reduce their cholesterol levels and reduce the risk of heart attack.

A Country on the Move

Because aerobic exercise was found to lower cholesterol levels, some Americans have changed the way they exercise and how much they exercise. Many people participate in programs that involve brisk walking, running, bicycling, and swimming, as well as the familiar dance-type aerobics. The Bicycling Institute of America, a group that monitors and promotes the sport of bicycling in the United States, estimates that in 1990 there were 25 million adults bicycling regularly (at least once a week) compared to only 10 million in 1983.

It's Never Too Late

Ruth Anderson, pictured here, is a living example of the fact that it's never too late to begin new ways of moving. Ruth, a nuclear chemist, began running at age 44. She won the women's division for her age group of the first marathon she competed in. Physical and mental endurance have enabled her to participate in nearly 100 marathons and compete in numerous ultramarathons.

Using Computers

Using a survey and a computer spreadsheet, calculate the percentage of people you know who actually participate in regular exercise and graph the results. Find out why some people resist regular exercise.

CAREER connection

Dieticians plan menus and supervise food preparation using the principles of good nutrition. They deal with dietetics—the relationship between food and health. Dieticians work in hospitals, schools, restaurants, and the food services of businesses. They have a bachelor's degree and have studied nutrition, foods, institution management, biology, and chemistry.

including length of time for each exercise and where it could be done (home, gym, etc.). Then have each group create a bulletin board display describing the benefits of proper diet or exercise. Use the Performance Task Assessment List for Bulletin Board in **PASC,** p. 59. L1

HISTORY CONNECTION

"Sewed Up His Heart"

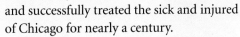

"Sewed Up His Heart" was the newspaper headline on the day in 1893 that Dr. Daniel Hale Williams performed the world's first open-heart surgery. Williams (1856-1931) believed he had no other choice but to operate when a man with a severe knife wound to the chest was brought into Chicago's Provident Hospital.

Surgical Breakthrough

Williams's surgery involved cutting and suturing the pericardium—the sac that surrounds the heart. This marked the first time a surgeon had ever entered the chest cavity. The risk of infection while the chest cavity was open during surgery was great. Williams's patient recovered completely.

It would be many years before open-heart surgery became common. Over the next fifty years, fewer than ten of these operations were recorded. The first woman to operate on the heart was Dr. Myra Logan (1908-1977), an African American physician at Harlem Hospital in New York.

In addition to their surgical achievements, Drs. Williams and Logan made important contributions toward advancing medical practices for African Americans. Dr. Williams had founded Provident Hospital in 1891. Provident started one of the first training programs in the country for minority nurses and successfully treated the sick and injured of Chicago for nearly a century.

Dr. Logan graduated from New York Medical College in 1933. In the course of her medical career, she researched the early detection of breast cancer and pioneered the idea of group practice so that patients could benefit from a variety of services in one place.

Science Journal

In your Science Journal, write and design a pamphlet directed toward increasing student awareness of heart health.

Science Journal

Ask students to start with any measures they currently take to ensure a healthy heart, such as low-fat, low-cholesterol diets and/or regular exercise. Students can take their own photographs or use photographs from magazines to illustrate their pamphlets.

Going Further ⫸

Have students research some aspect of heart surgery. Students may be interested in artificial hearts, such as the one received by Barney Clark in 1982. Other students may wish to research the artificial heart developed, in part, by Charles Lindbergh in the 1930s. Have students write newspaper articles detailing their findings. Students can imagine that they are reporting the event as it happens. Use the Performance Task Assessment List for Newspaper Article in **PASC**, p. 69. [L2] [P]

HISTORY CONNECTION

Purpose

History Connection extends Section 3-1. In this section, students saw the structure of the heart. In this article, they learn about the first surgery ever done on this organ.

Content Background

In the hundred years since Daniel Hale Williams performed emergency heart surgery, several surgical techniques have been performed on the heart. Many advances were made in the 1960s, including valve replacements and coronary bypass surgery. In 1967, Dr. Christiaan Barnard performed the first successful heart transplant. His patient lived only 18 days, but in 1968, the second heart transplant patient survived for more than two months.

Teaching Strategies

Have students list the risks involved in the surgery performed by Dr. Williams. Students can analyze why the patient did not develop any infections as a result of the surgery. Because Dr. Williams believed in germs as the cause of disease, he probably kept conditions as clean as possible, for the time. Although the germ theory was not fully accepted at that time, many doctors did believe in the use of disinfectants.

GLENCOE TECHNOLOGY

Videodisc

STVS: Human Biology
Disc 7, Side 1
Heart By-Pass Surgery (Ch. 7)

Computer-Aided Heart Surgery (Ch. 8)

HOW IT ⓌⓄⓇⓀⓈ

Purpose

How It Works compares the way nutrients move through plants to how blood moves through the circulatory system. The theme is stability, for a plant's transport system maintains a stable supply of nutrients and water to all of the structures of the plant.

Content Background

Plants lose water from their leaves in a process called transpiration. Water escapes through the stomata. Guard cells in the leaves control how fast water leaves the stomata. Two guard cells sit side by side at the opening of the stomata. They look like two kidney beans. The inner walls of the guard cells are more rigid than the outer walls. When the cells get moist, they swell and buckle outward, causing the stomata to open and release moisture from the plant.

Teaching Strategies Obtain prepared slides of stem cells from a laboratory supply store and allow students to observe the structure of the cells under a microscope. Ask students: **What about the structure allows the cells to form transport vessels?** Students should observe that the cells have rigid cell walls that are linked to form columns.

Answer to You Try It!

Students may say that water moves from fiber to fiber in the paper towel. When water reaches the end of the towel, it will continue to move and begin to drip off the end of the towel as long as the other end of the towel is still in water. The model demonstrates that water moves up in a plant from roots to leaves. As water begins to drip out of the top of the towel, this demonstrates the fact that water can be lost through leaves.

HOW IT ⓌⓄⓇⓀⓈ   How Does Water Move in a Plant?

Most of the water that moves about in a plant is supplied through the roots of the plant. Have you ever wondered how water gets to the top of a tree? You have learned that your body has certain adaptations that assist blood in its movement against gravity. What adaptations do plants have that enable water to move upward?

Vessels in Plants

Water and nutrients move about in plants for the same reasons that blood circulates in an animal body. Raw materials and food need to be supplied to the cells in the roots, stems, and leaves of a plant just as your arms, legs, and internal organs are supplied with the materials they need. However, if you were to dissect a plant stem, you wouldn't find a heart, nor would you find arteries, capillaries, and veins. You would find that plants do contain a system of transport vessels called xylem and phloem. Xylem is a type of tissue in plants that carries water and nutrients from roots to leaves. Phloem is a type of tissue that carries food made in leaves to other parts of the plant.

The Basic Plan

The basic plan for the movement of water in plants is that water moves up through xylem cells in the stem and out of the plant through openings in the leaves called stomata. The water molecules stick together in the xylem in a continuous threadlike stream such as in the diagram to the left. Water loss is controlled by the stomata, which close up as temperature rises and at night. Water movement up the stem is then slowed.

You Try It!

Make a model to demonstrate how water moves up in a plant. Roll a piece of paper toweling and dip it into about 2 cm of water. *In your Journal,* describe how water moves in the paper towel. What happens to the water when it reaches the far end of the paper towel? How is this model an example of how water moves in a plant?

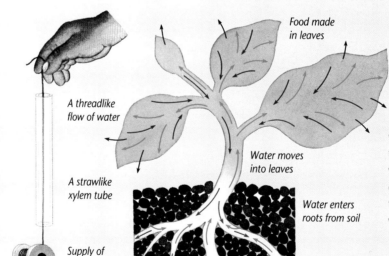

A threadlike flow of water

A strawlike xylem tube

Supply of

Food made in leaves

Water moves into leaves

Water enters roots from soil

Going Further ⫸

Ask students to draw a diagram of a green plant and label its transport vessels. Student diagrams should show that nutrients are stored in the roots, and that the xylem transports nutrients from the roots to the leaves. Have students do research to identify the structures that allow plants to release water vapor and exchange gases such as oxygen and carbon dioxide. *stomata* Students can create a display that explains what they have learned about plant structure. Use the Performance Task Assessment List for Display in **PASC,** p. 63. **P**

Chill Out— It's Good For You

Your teacher has placed the final exam face-down on your desk. You've studied for weeks and feel confident that you will do well. So why does your heart begin to beat a little faster as you turn the test over?

The Stress Factor

According to 16-year-old high school junior Tamika Walker, the answer is stress.

"Some people think that teenagers don't feel stress as deeply as adults. But that's just not true," Tamika said. To prove her point, Tamika conducted a survey to measure the level of stress reported by classmates at her high school in Currie, North Carolina. She wrote and distributed a questionnaire asking the teenagers to rate their stress levels under different types of situations, both positive and negative.

Results Based on Data

"The results of the survey showed that there is a lot of stress in most teenagers' lives. Of course, big things like a divorce or a death in the family scored the highest. But I was surprised to discover that even smaller events, like getting a bad grade on a test or losing a textbook, scored high."

Tamika also researched the physical effects of stress. "It's a chain of events," she explained. "Stress affects hormones, which in turn affect the pituitary gland, which eventually begins to wear down a person's immune system. That's why some people get sick when they are under a lot of stress. In fact, my survey showed that kids who had high stress scores also had the highest numbers of sick days. I think that it's time that someone started teaching kids how important it is to learn how to relax."

Sharing the Results

Tamika entered her report in a regional science fair. Tamika's report was the most widely read exhibit in the fair. "It makes me feel good to know that so many kids took the time to look over the report," Tamika said. "It's time for us to begin learning everything we can about our bodies. After all, the more we know, the better we are likely to feel."

You Try It!

Design a survey to test whether stressful factors are responsible for absenteeism on a certain day of the week.

Going Further ⅢⅢⅢ▶

Dr. Thomas Holmes, a professor of psychiatry, found that 80% of those whose lives dramatically change in a year can expect a major illness within the next two years. He devised a rating scale to assess a person's stress risk. On Holmes' scale, death of a family member is 63 points; the gain of a new family member, 39; outstanding personal achievement, 28; change in school, 20; vacation, 13; Christmas, 12.

Have students work in small groups to create their own rating scale, based on a top score of 100 points. Which experiences do teenagers find most stressful? The least stressful? Then have students administer the scale to their peers to assess its validity. Use the Performance Task Assessment List for Group Work in **PASC**, p. 97. **COOP LEARN**

Purpose Teens in Science extends Section 3-2, in which students learn about pulse and blood pressure.

Content Background
In the face of danger, tension, or conflict, adrenaline pours into the bloodstream. This elevates blood pressure, narrows blood vessels, and causes the heart to race. Doctors cite these changes as proof that emotional stress can harm the heart.

On the positive side, stress can help people cope with physical pain. When under stress, the brain releases chemicals called *endorphins* and *enkephalins*. These "stress-induced analgesias" (pain-reducers) produce their most dramatic effects only under conditions of extreme stress.

Teaching Strategies Invite pairs of students to read the selection together and then plan a campaign to help teenagers reduce stress. Students may wish to consider such means as sports and meditation.

Discussion
Students can read the selection and then discuss whether they have more or less stress than teenagers in other countries.

Answer to You Try It!
The surveys students design will vary, but should include making a list of appropriate survey questions, deciding how many people need to be surveyed, and choosing a method for presenting the result such as graphs, tables, or written reports.

The following activities will help students review the main ideas of the chapter.

Teaching Strategies

Copy each main idea statement in the student text onto a separate sheet of paper. For each statement, write two false statements, each on a separate sheet of paper. Fold all the papers and put them into a box. Divide the class into two teams and have team members choose one paper at a time. Ask the student to read the statement and declare whether it is true or false. You may also wish to require students to provide the correct statement if they choose a false statement and an additional fact if they choose a correct statement.

Answers to Questions

1. Answers will vary. Response might include that a healthy heart is necessary to maintain delivery of nutrients and oxygen to body cells and removal of wastes.

2. gills, skin, and lungs

3. Answers will vary. Animals having hearts with four chambers in which oxygen-rich and oxygen-poor blood is kept separate seem to support bodies that are larger and more active.

4. Clogged blood vessels increase blood pressure in vessels and decrease that amount of blood reaching the organs of body systems.

Science Journal

Review the statements below about the big ideas presented in this chapter, and answer the questions. Then, re-read your answers to the Did You Ever Wonder questions at the beginning of the chapter. *In your Science Journal*, write a paragraph about how your understanding of the big ideas in the chapter has changed.

1 Circulatory systems deliver nutrients and oxygen to body cells and remove wastes, such as carbon dioxide. *What does this statement tell you about the importance of a healthy heart?*

2 Both open and closed circulatory systems exist in animal groups. Circulation and body structures that take in oxygen are usually closely related. *Name three organs that take up oxygen in various organisms.*

3 Amphibians, reptiles, fish, birds, and mammals all have adaptations for circulation in the number of chambers of the heart. *How is the type of heart an animal has related to the lifestyle of the organism?*

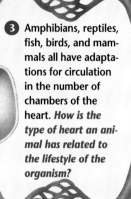

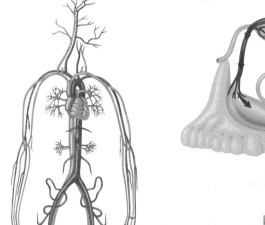

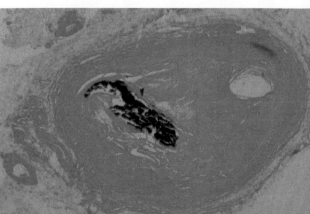

4 Diseases resulting in clogged arteries, such as the one shown below, also affect body organs. *Why would damaged blood vessels affect the rest of the body systems?*

Science at Home

Ask students to read *Fantastic Voyage* by Isaac Asimov or obtain the video. In this science fiction novel, a team of miniaturized scientists (in a miniaturized submarine) is injected into a patient to remove a blood clot. Students should keep a journal of the parts of the circulatory system through which the scientists travel and evaluate the accuracy of the descriptions, based on what they have learned in class.

Science Journal

Did you ever wonder...

• When the heart contracts, it creates pressure that forces the blood through a network of blood vessels. (p. 86)

• Different animals have either an open or closed circulatory system. (p. 85)

• Blood pressure causes blood vessels to expand. (p. 95)

Using Key Science Terms

artery
atherosclerosis
blood pressure
capillary
hypertension
pulse
vein

Answer the following questions using what you know about the science terms.

1. Explain the difference between arteries, veins, and capillaries.
2. How is hypertension related to the circulation of blood?
3. How does atherosclerosis affect arteries?
4. Describe what causes the pulse felt in various parts of your body.
5. How is blood pressure different in arteries and veins?

Understanding Ideas

Answer the following questions in your Journal using complete sentences.

1. Compare and contrast open and closed circulatory systems.
2. What is the relationship between oxygen and blood?
3. Why is hypertension a serious health problem?
4. How does availability of oxygen in the blood of an animal with a three-chambered heart differ from the availability of oxygen in the blood of an animal with a four-chambered heart?
5. Ventricles have thicker walls than atria. What does this adaptation tell you about the work each type of chamber does?

Developing Skills

Use your understanding of the concepts developed in each chapter to answer each of the following questions.

1. **Concept Map** Complete the following spider concept map on the circulatory system.

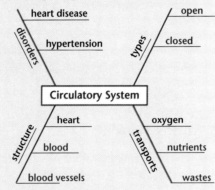

2. **Design an Experiment** Refer to the Investigate on page 92 to design an experiment to increase an earthworm's heart rate.
3. **Predicting** Take your at-rest pulse as in the Explore activity on page 91. If your health permits, run in place for two minutes. Check to see how long it takes the pulse to return to the at-rest rate. Then predict how long it would take to return to the at-rest rate if you ran for three minutes.
4. **Making and Using Graphs** Research the pulse rates of several different animals, including the earthworm from the Investigate on page 92. Make a bar graph to compare these different pulse rates.
5. **Making and Using Graphs** The graph on the next page shows the pulse rate of a boy before, during, and after bicycling. When did the pulse rate increase most rapidly?

Using Key Science Terms

1. arteries—carry blood away from the heart; veins—carry blood to the heart; capillaries—network of vessels and arteries between veins
2. Blood circulation may be slowed down and blood pressure increased.
3. Results in the buildup of fats and calcium in vessels, restricts blood flow.
4. Rhythmic expansion and contraction of arteries; heartbeat.
5. Higher in arteries.

Understanding Ideas

1. Answer can reflect captions for Figure 3-1.
2. Blood carries oxygen throughout the body in many organisms.
3. Hypertension leads to damaged blood vessels, and a decrease of oxygen to cells.
4. In amphibians some oxygen-rich blood moves through the heart. In birds and mammals oxygen-rich blood never mixes with oxygen-poor blood.
5. Students can infer that ventricle walls are more muscular, contract harder and perform more work than atrial walls.

Developing Skills

1. See reduced student page for red answer annotations.
2. Students' experimental designs should be approved; experiments might include lowering or raising temperatures from room temperature.
3. Most will predict that it will take longer for pulse to return to at-rest rates as exercise time increases.
4. The Merck Veterinary Manual in a library gives average rates. Students can refer to Making and Using Graphs in the Skill Handbook for information about bar graphs.
5. Rate increased most rapidly right after starting; 160 beats per minute; pulse rate slowed after he stopped.

Critical Thinking

1. Dehydration lowers blood pressure. The body reacts by increasing the heart rate.

2. Closed systems under pressure deliver a consistent flow of oxygen more efficiently.

3. Possible answers: It would need to pump regularly. It could not leak. It would need to pump with a constant force. Blood would need to flow smoothly through it.

4. Massaging simulates the action of working leg muscles, which squeezes blood in veins to help push it back to the heart.

Problem Solving

The pulse rate would be lowest during sedentary activities; highest after eating, during exercise, and when excited. The pulse rate changes with the body's need for oxygen and nutrients.

Connecting Ideas

1. Pressure is force divided by the area it acts on. If force remains the same in a blood vessel, pressure will increase—acting over a smaller area.

2. The weight of a column of blood in the veins of your legs is greater than it is in any other part of your body. Thus the tendency of blood to back up in your legs toward your feet is great. Thus a large number of valves are needed to prevent the backing up of the blood.

3. Student answers might include that an artificial valve is less likely to wear out or be rejected.

4. The cuff and the pump work together first to shut off the flow of blood in the arm and then let it flow again under conditions that can be measured. The gauge allows you to measure the blood pressure.

5. Stomata close down when temperatures drop and at night.

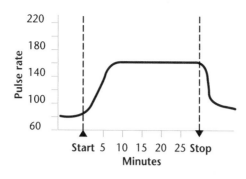

What was the boy's pulse rate after fifteen minutes of bicycling? How did his pulse rate change when he stopped biking?

Critical Thinking

Use your understanding of the concepts developed in the chapter to answer each of the following questions.

1. If a person becomes dehydrated, the lack of fluids decreases the volume of blood in the body. How would dehydration affect blood pressure and heart rate?

2. Larger, active organisms, like yourself, require oxygen to release energy for movement. Why is a closed circulatory system under pressure?

3. Imagine that you are designing an artificial heart. What three factors will your artificial heart have to have or do in order to work well in the body?

4. Doctors often recommend massage of the legs for people who cannot move about as a result of illness or injury. How is massage beneficial to circulation?

Problem Solving

Read the following problem and discuss your answers in a brief paragraph.

A special device that attaches to your wrist can be used to constantly measure your pulse rate and display the number like a digital clock. If you were to wear one of these devices for 24 hours, when do you think you would find your pulse rate the lowest? When would your pulse rate be the highest? Why would your pulse rate change during a 24-hour period?

CONNECTING IDEAS

Discuss each of the following in a brief paragraph.

1. Theme—Scale and Structure Explain how a narrowed blood vessel affects the pressure of blood within it.

2. Theme—Systems and Interactions Using your knowledge of the force of gravity, explain the importance of the valves in the circulation of blood in your legs.

3. A Closer Look Suggest some benefits of using an artificial valve instead of one from another organism.

4. Physics Connection Explain how a sphygmomanometer works.

5. How It Works Describe an adaptation that plants have for conserving water.

✔ Assessment

Portfolio Review the portfolio options that are provided throughout the chapter. Encourage students to select one product that demonstrates their best work for the chapter. Have students explain what they learned and why they chose this example for placement into their portfolios.

Additional portfolio options can be found in the following **Teacher Classroom Resources:**

Making Connections: Integrating Sciences, p. 9
Multicultural Connections, pp. 9, 10
Making Connections: Across the Curriculum, p. 9
Concept Mapping, p. 11
Critical Thinking/Problem Solving, p. 11
Take Home Activities, p. 8
Laboratory Manual, pp. 9–12
Performance Assessment

Forces in Action

In this unit, you investigated forces producing motions and changes of motion. You learned about Newton's Laws of Motion and how everyday events, such as picking up a bowling ball, are affected by the laws of motion.

You learned how these same kinds of forces within Earth produce earthquakes and volcanic activity. You also learned that forces within your body cause movement of blood and that water moves up through a plant under force.

Try the exercises and activity that follow. They will challenge you to use and apply some of the ideas you learned in this unit.

CONNECTING IDEAS

1. Sometimes when you are in the car, on an elevator, or on a ride at an amusement park your stomach "drops" as you go down a hill. Explain why that might happen.

2. Why do movements of only two meters along the San Andreas fault cause such great forces to be generated that the ground can vibrate for many miles around?

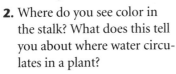

Exploring Further ACTIVITY

Going Up? How does water move up in a plant?

What To Do

1. Obtain a hand lens and a stalk of celery with leaves. Make sure that it has been in a glass of water with blue or red food coloring for several hours.

2. Where do you see color in the stalk? What does this tell you about where water circulates in a plant?

3. Do leaves have anything to do with how water moves up in plants? Design an experiment to show what happens to the movement of water through a plant (a) in light, (b) in the dark, (c) if the leaves are cut off, and (d) if you coat the leaves with something.

Forces in Action

THEME DEVELOPMENT

This unit supports the themes of systems and interactions and stability and change. Motion and change in motion are the result of forces derived from energy acting on matter, for example, pushing a grocery cart down the aisle. The same forces are the main factors behind the motion of blood inside the human body.

Connections to Other Units

The main concepts in unit 1, namely forces and pressure and their effects, form a foundation for the concepts of work and energy in unit 2, the effects of pressure on rocks in unit 3, and the activities of air in the atmosphere and in the process of breathing discussed in unit 4.

CONNECTING IDEAS
Answers

1. Blood moves to the base of your stomach causing pressure. The pressure makes it feel as if your stomach has dropped.

2. Force is equal to mass times acceleration. Although the acceleration is relatively small, the mass of Earth involved is great. Therefore, the vibrations can be felt for miles around.

Exploring Further

Going Up? How does water move up in a plant?

Purpose To observe the circulation of water in a plant.

Background Information The xylem transports nutrients from the roots to the leaves. The phloem transports nutrients from the leaves to the rest of the plant.

Materials food coloring, water, tall glasses, hand lens, celery with leaves

Answers to Questions

2. Students may say they see color in the rib-like vessels, showing that water circulates up through the vessels.

3. Possible results are (a) they should see no difference; (b) they may see some difference; (c) they should see a slowdown; (d) they should see a slowdown.

✔ Assessment

Process Have students design an experiment that demonstrates that fluids move through plants, using a white carnation and food coloring. Have them give oral presentations explaining how fluids move through plants. Use the Performance Assessment Task List for Oral Presentation in **PASC**, p. 71. **P**

Energy at Work

UNIT OVERVIEW

UNIT FOCUS

In Unit 2, students will learn about ways in which forces are used to do work. Machines and the human body both use forces to do work. Simple machines, such as levers and inclined planes, can be used to move objects. The muscles and bones of the body act as simple machines as well. Thermal energy, or the transfer of heat, also moves objects.

THEME DEVELOPMENT

Unit 2 supports the themes of energy and stability and change. The theme of energy is developed in the physical sciences to show how machines and the human body use forces to do work and to make things move. Chapters 7 and 8 focus on stability and change to reveal how the body's nervous system works with the muscle system in responding to forces outside the body.

Connections to Other Units

Information on the skeletal, muscle, and nervous systems presented in this unit is useful for better understanding the respiratory system, presented in Unit 4, Chapter 16. These chapters also provide background for chapters on cells and cellular reproduction in Unit 5. In addition, students will apply what they learned in this unit on thermal energy (Chapter 6) to exploring energy resources (Unit 3, Chapter 13).

Energy at Work

Energy springs you into action. Without the energy generated by her muscles, a gymnast couldn't do a handspring ... or much of anything! Without energy, nothing would happen at work or at play. As you might guess, energy is vital to everything that happens in the world around you. This unit will help you understand the energy connection.

114

GLENCOE TECHNOLOGY

 Videodisc

Use the *Science Interactions, Course 2* **Integrated Science Videodisc** lesson *Machines and Forces: The Great Bicycle Challenge,* after Chapter 5 of this unit.

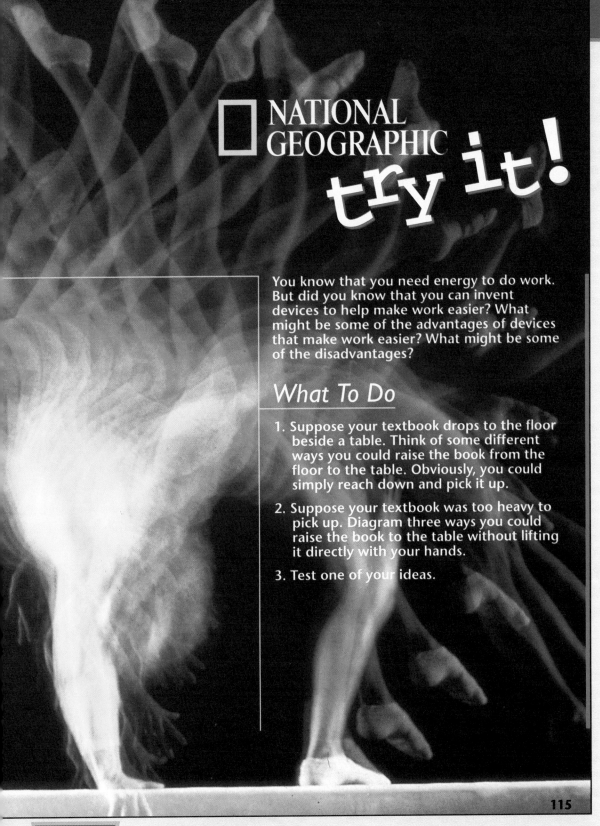

NATIONAL GEOGRAPHIC
try it!

You know that you need energy to do work. But did you know that you can invent devices to help make work easier? What might be some of the advantages of devices that make work easier? What might be some of the disadvantages?

What To Do

1. Suppose your textbook drops to the floor beside a table. Think of some different ways you could raise the book from the floor to the table. Obviously, you could simply reach down and pick it up.

2. Suppose your textbook was too heavy to pick up. Diagram three ways you could raise the book to the table without lifting it directly with your hands.

3. Test one of your ideas.

115

Try It!

Purpose Students will analyze the actions necessary to lift a heavy object and then they will make a model of a machine that could be used to do so.

Background Information Many machines are based on combinations of simple machines, such as the lever and the inclined plane. Bicycles, can openers, screwdrivers, tow-truck winches, flagpoles, and mountain roads, as well as the bones and muscles of the human body, all operate on the same basic principles.

Materials Materials needed will depend on ideas students choose to test but may include a board, a chair, a rope, a jack, and clothesline pulleys.

Troubleshooting If students have trouble getting started, have them act out lifting a heavy object while describing each step of the process.

✔ Assessment

Content Ask students to think of machines or devices (because students may not think of some simple machines as machines at this stage) that they have used that resemble some of their diagrams. Use the Performance Task Assessment List for Model in **PASC**, p. 51.

Chapter Organizer

SECTION	OBJECTIVES	ACTIVITIES & FEATURES
Chapter Opener		**Explore!**, p. 117
4-1 Work (2 sessions; 1 block)	1. **Define** work. 2. **Calculate** work done on an object. **National Content Standards: (5-8) UCP2, A1, B3**	**Explore!**, p. 118 **Skillbuilder:** p. 120 **How It Works**, p. 143 **History Connection**, p. 144
4-2 Forms of Energy (3 sessions; 1 1/2 blocks)	1. **Describe** how work can produce kinetic energy, potential energy, and thermal energy. 2. **Specify** that energy is transferred when work is done. **National Content Standards: (5-8) UCP2-4, A1, B3, D3**	**Explore!**, p. 123 **Design Your Own Investigation 4-1:** pp. 126–127 **Explore!**, p. 129 **Find Out!**, p. 132 **A Closer Look**, p. 128
4-3 Conservation of Energy (2 sessions; 1 block)	1. **Describe** how energy changes from one form to another. 2. **Understand and apply** the Law of Conservation of Energy. **National Content Standards: (5-8) UCP1-5, A1-2, B3, D3, E1, F5, G1, G3**	**Find Out!**, p. 134 **Investigate 4-2:** pp. 136–137 **Earth Science Connection**, pp. 138–139 **Technology Connection**, p. 142

ACTIVITY MATERIALS

EXPLORE!

p. 117* 2 books
p. 118* several books, watch with second hand, backpack
p. 123* softball, tent stake, bucket of dirt
p. 129* 20-N spring scale, metric ruler

INVESTIGATE!

pp. 136-137* ring stand, right-angle clamp, support rod and clamp, 2 metersticks, 2-hole medium rubber stopper, 100 cm of string, masking tape

DESIGN YOUR OWN INVESTIGATION

pp. 126-127* 20-N spring scale, meterstick, stopwatch, 1.0- to 2.0-m ramp, masking tape, 1.0-kg mass, cart, several books

FIND OUT!

p. 132 20-N spring scale, shallow box, several books, meterstick, rug or rough surface
p. 134 a marble, foam cup, grooved ruler, balloon, tape, wood splint, or tongue depressor

KEY TO TEACHING STRATEGIES

The following designations will help you decide which activities are appropriate for your students.

- **L1** Basic activities for all students
- **L2** Activities for average to above-average students
- **L3** Challenging activities for above-average students
- **LEP** Limited English Proficiency activities
- **COOP LEARN** Cooperative Learning activities for small group work
- **P** Student products that can be placed into a best-work portfolio
- 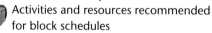 Activities and resources recommended for block schedules

Need Materials? Call Science Kit (1-800-828-7777).

`00:00` **OUT OF TIME?** We recommend that students do the activities with an asterisk.

Chapter 4 Work and Energy

Student Masters	Transparencies
Study Guide, p. 16 **Making Connections: Integrating Sciences**, p. 11 **Science Discovery Activities**, **4-1** **Multicultural Connections**, p. 12 **Critical Thinking/Problem Solving**, p. 12	**Teaching Transparency 7**, Work **Section Focus Transparency 10**
Study Guide, p. 17 **Concept Mapping**, p. 12 **Take Home Activities**, p. 10 **Critical Thinking/Problem Solving**, p. 5 **Activity Masters**, Design Your Own Investigation 4-1, pp. 19–20 **Multicultural Connections**, pp. 11–12 **Science Discovery Activities**, **4-2**	**Section Focus Transparency 11**
Study Guide, p. 18 **How It Works**, p. 7 **Making Connections: Across the Curriculum**, p. 11 **Activity Masters**, Investigate 4-2, pp. 21-22 **Making Connections: Technology and Science**, p. 11 **Science Discovery Activities**, **4-3** **Laboratory Manual**, pp. 13–16, The Energy of a Pendulum	**Teaching Transparency 8**, Pendulum **Section Focus Transparency 12**

ASSESSMENT RESOURCES	TEACHING & TECHNOLOGY
Review and Assessment, pp. 23–28 **Performance Assessment**, Ch. 4 **PASC*** **MindJogger Videoquiz** **Alternate Assessment in the Science Classroom** **Computer Test Bank**	**Spanish Resources** **Cooperative Learning Resource Guide** **Lab and Safety Skills** **Science Interactions, Course 2, CD-ROM** **Computer Competency Activities**

*Performance Assessment in the Science Classroom

NATIONAL GEOGRAPHIC TEACHER'S CORNER

Index to National Geographic Magazine

The following articles may be used for research relating to this chapter:

- "Searching for the Secrets of Gravity," by John Boslough, May 1989.

National Geographic Society Products Available From Glencoe

To order the following products for use with this chapter, contact your local Glencoe sales representative or call Glencoe at 1-800-334-7344:

- *Newton's Apple Physical Sciences* (Videodisc)
- *GTV: Planetary Manager*
- *STV: Atmosphere*

GLENCOE TECHNOLOGY

The following multimedia resources are available from Glencoe.

Science and Technology Videodisc Series (STVS)
Physics
    Seismic Simulator
    Images of Heat
Chemistry
    Solar Food Dryer
Ecology
    Greenhouse Effect

The Infinite Voyage Series
A Taste of Health
Living with Disaster

Physical Science CD-ROM

Teacher Classroom Resources

These are key components of the classroom resources package.

TEACHING AIDS

Section Focus Transparencies

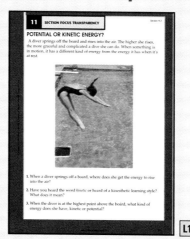

10 SECTION FOCUS TRANSPARENCY — Section 4-1

WORK IT OUT

Whew! Studying is hard work! But is it work by the scientific definition? In science, *work* has a definite meaning, somewhat different from its everyday meaning. How is the term *work* used differently in the three photos?

1. How do you think scientists define *work*?

2. Describe some examples that would show *work* in the scientific sense.

L1

11 SECTION FOCUS TRANSPARENCY — Section 4-2

POTENTIAL OR KINETIC ENERGY?

A diver springs off the board and rises into the air. The higher she rises, the more graceful and complicated a dive she can do. When something is in motion, it has a different kind of energy from the energy it has when it's at rest.

1. When a diver springs off a board, where does she get the energy to rise into the air?

2. Have you heard the word *kinetic* or heard of a kinesthetic learning style? What does it mean?

3. When the diver is at the highest point above the board, what kind of energy does she have, kinetic or potential?

L1

12 SECTION FOCUS TRANSPARENCY — Section 4-3

BATTERIES NOT INCLUDED

This toy doesn't need batteries to work. A few turns of the winder and the dinosaur roars to life.

1. What kind of energy does a wound-up toy have? Where did it get this energy?

2. If you release the wound toy, what kind of energy does it have? How do you know?

3. If a child forgets about the toy and leaves it wound up, what happens to the energy? Why?

L1

Teaching Transparencies

7. WORK

L1

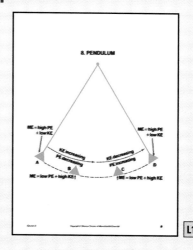

8. PENDULUM

L1

HANDS-ON LEARNING

Science Discovery Activity*

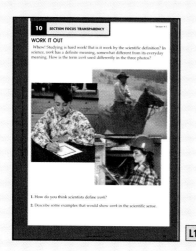

ACTIVITY 4-1 Work, Work, Work

It's a job! It's a task! It's *F × d*! It's all of these! It's work. But you know that only the third statement is the scientific meaning of work. What does work accomplish? Let's do some work and find out.

Getting Started

If you moved an object, you know you've done work on it. To know how much work, you need to know the size of the force you used to move the object and the distance the object moved in the direction of the force. Can you compare the work you do to perform different tasks?

Thinking Critically

Think of how work can be done on an object. How would you supply the force? Could you use your muscles? How would you measure this force? What other measurements would you have to make to find out how much work was done? What do you think would happen to the amounts of work you would perform in lifting objects with different masses? How could you compare the work? How could you do the same amount of work in lifting different masses?

Try It!

You may want to use the following:
• calibrated masses   • spring scale
• meterstick   • wire ties

1. *In your Journal*, write a procedure that you can use to determine the amount of work done in lifting each mass. What

20

L1

Laboratory Manual*

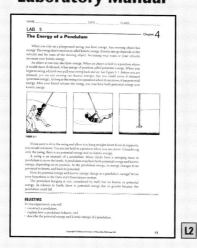

NAME _____ DATE _____ CLASS _____

LAB 5 — Chapter 4

The Energy of a Pendulum

When you ride on a playground swing, you have energy. Any moving object has energy. The energy due to motion is called kinetic energy. Kinetic energy depends on the velocity and the mass of the moving object. Increasing your mass or your velocity increases your kinetic energy.

An object at rest may also have energy. When an object is held in a position where it would move if released, it has energy of position called potential energy. When you begin to swing, a friend may pull your swing back and up. See Figure 5-1. Before you are released, you are not moving (no kinetic energy), but you could move if released (potential energy). As long as the swing is in a position where it can move, it has potential energy. After your friend releases the swing, you may have both potential energy and kinetic energy.

FIGURE 5-1

If you were to sit in the swing and allow it to hang straight down from its supports, you would not move. You are not held in a position where you can move. Considering only the swing, there is no potential energy and no kinetic energy.

A swing is an example of a pendulum. Many clocks have a swinging mass or pendulum to move the hands. A pendulum may have both potential energy and kinetic energy, depending on its position. As the pendulum swings, its energy changes from potential to kinetic to potential.

How do potential energy and kinetic energy change as a pendulum swings? Write your hypothesis in the Data and Observations section.

The pendulum hanging at rest, considered by itself, has no kinetic or potential energy. In relation to Earth, there is potential energy due to gravity because the pendulum could fall.

OBJECTIVES

In this experiment, you will
• construct a pendulum.
• explain how a pendulum behaves.
• describe the potential energy and kinetic energy of a pendulum.

13

L2

Take Home Activity

Chapter 4

AMAZING BUTTON WHIZZER

You need: one large, two-hole button; about 2 feet of string; scissors

Explain to your adult partner that this activity shows how energy can be transferred, stored, and released. When string is twisted, it can store potential energy. When the string unwinds, the energy is converted to kinetic energy that is capable of doing work. The button also helps store kinetic energy.

• Thread the string through one button hole and back through the other hole on the same side.
• Tie the ends of the string into a knot. Trim the excess string ends with the scissors.
• Slide the button to the middle of the string loop.
• Grasp the string so that you hold one end in each hand, with the button centered on the string.
• Holding your hands out in front of you, twist the button and string away from you. After several twirls, gently begin to tug outward on the string.
• As the spinning button twists the string and begins to put pressure on your fingers, release some of the tension on the string. The button begins to twist in the opposite direction.
• Each time the button reverses its twirling direction, alternate your pulling or relaxing on the string.
• Practice the procedure until you can keep the button spinning. See if you can hear a soft whizzing or whirring sound as the button spins.

• When you first twirled the string and "wound up" the button and string, you performed work and transferred energy. After that, you had only to tug gently on the string to keep the button twirling. Explain this. You have made an amazing button whizzer! Experiment with buttons of different sizes and strings of different thicknesses. How does button size and string thickness affect your button whizzer?

10

L2

Chapter 4 Work and Energy

Study Guide*

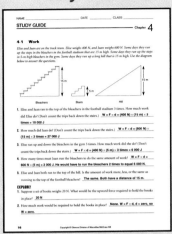

Concept Mapping

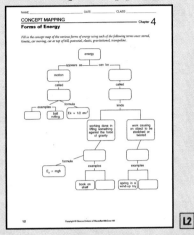

Critical Thinking/ Problem Solving

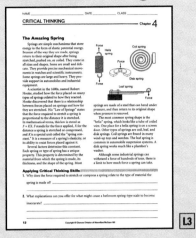

Integrating Sciences

Across the Curriculum

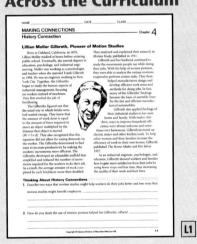

Technology and Society

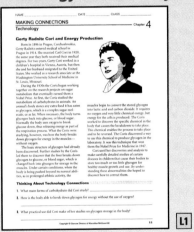

Multicultural Connection**

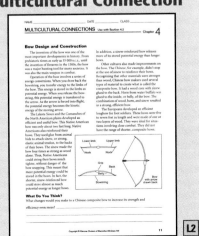

Performance Assessment

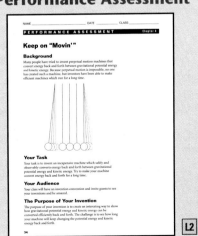

Review and Assessment

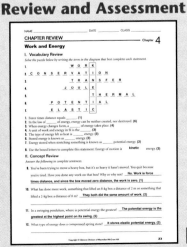

---

*One per section **Two per chapter

Work and Energy

THEME DEVELOPMENT

The themes of this chapter are systems and interactions and energy. Students learn that energy can be used to do work. They learn about the transfer of energy, and study the interactions between energy and work.

CHAPTER OVERVIEW

In this chapter, students will discover how the concepts of force and motion developed in Chapter 1 apply to work.

Potential and kinetic energy are described, and the concept of conversion from one form of energy to another is developed.

Tying to Previous Knowledge

Have students recall the definition of force from Chapter 1. Ask what happens when a force is applied to an object. *Its motion changes.*

INTRODUCING THE CHAPTER

Have students look at the photograph of the girl on page 117. Ask students if she would be doing more work if she carried hand weights.

Uncovering Preconceptions

Students may have developed the idea that any use of muscles constitutes work. However, work is done only if an object is moved in the direction of an applied force.

⏱ 00:00 OUT OF TIME?

If time does not permit teaching the entire chapter, use the Chapter Overview on this page, Reviewing Main Ideas at the end of the chapter, and the Chapter 4 audiocassette to point out the main ideas of the chapter.

Work and Energy

Did you ever wonder...

✓ If a football could store energy?

✓ Where energy goes after you use it?

✓ Why a ball bounces lower each time it hits the floor?

📝 Science Journal

Before you begin to study work and energy, think about these questions and answer them *in your Science Journal.* When you finish the chapter, compare your journal write-up with what you have learned.

Answers are on page 145.

When you hear the words *work* and *energy,* what do you think of? You might say that work is what you do at school or at a job. If you take books home with you, you call it homework. But when do you think that you're not working?

When you're enjoying some sport or activity, it's usually too much fun to be called work. For most of us, work consists of the things we have to do. The things that we do for our own pleasure we call recreation or play.

What about energy? What is energy? You might say that you're full of energy. Television advertisements tell you that a certain food will give you energy. It takes energy to do just about anything. What's the relationship between work and energy?

▶ *In the activity on the next page, begin to explore what makes work work!*

Learning Styles		
Kinesthetic	Explore, pp. 117, 118, 123; Activity, pp. 122, 133, 135, 141; A Closer Look, pp. 128-129; Discussion, p. 131; Find Out, p. 132	
Visual-Spatial	Visual Learning, pp. 119, 123, 130, 131, 132, 139, 140; Demonstration, p. 121; Activity, pp. 125, 128, 130; Explore, p. 129	
Interpersonal	Activity, p. 137; Project, p. 145	
Logical-Mathematical	Visual Learning, pp. 122, 125; Investigate, pp. 126-127, 136-137; Find Out, p. 134	
LS Linguistic	Across the Curriculum, pp. 119, 124, 128; Multicultural Perspectives, p. 119; Visual Learning, p. 124; Research, p. 125; Discussion, pp. 130, 134, 138; Earth Science Connection, pp. 138-139	

Explore! ACTIVITY

What is work?

Believe it or not, both weeding the garden and playing basketball are work! What makes some activities work? Are all activities work? Have you done work every time your muscles get tired? Begin to explore work with the activity below.

What To Do

1. Stand up and hold your arms out in front of you at waist level, with your hands together, palms up.

2. Have a classmate stack two books on your hands.

3. Raise the books to shoulder level, then lower them.

4. Now try raising them above your head. Is this more work than raising them to shoulder level?

5. Hold the books at shoulder level until you get tired. Are you exerting force? Do you think you're doing work on the books?

6. *In your Journal,* answer the questions and record your observations. Also record your ideas of why some activities are work and others aren't.

Explore!

What is work?

Time needed 5 minutes

Materials 2 books

Thinking Processes observing and inferring, forming operational definitions

Purpose To help students infer that work involves force and motion, and that force can be exerted and energy expended without any work being accomplished.

Teaching the Activity

In this activity, direct students to use two books of equal mass.

Science Journal Encourage students to compare and contrast the factors involved in the different trials and to record their observations in their journals. Have them tell when balanced and unbalanced forces are involved in this activity. L1

Expected Outcomes

Students are likely to infer that work is done whenever a force is exerted, until they understand that they are doing work only when they are moving the books.

Answers to Questions

4. Yes, raising the books higher is more work.

5. Yes. No, because the books are not moving.

✔ Assessment

Performance Have students exert forces against different objects and surfaces, such as desks, chairs, doors, and walls. Have them describe when work is accomplished and when it is not and explain their answers. Use the Performance Task Assessment List for Writing in Science in **PASC,** p. 87.

ASSESSMENT PLANNER

PORTFOLIO
Refer to page 147 for suggested items that students might select for their portfolios.

PERFORMANCE ASSESSMENT
Process, pp. 123, 129, 132
Skillbuilder, p. 120
Explore activities, pp. 117, 118, 123, 129
Find Out activities, pp. 132, 134
Investigate, pp. 126–127, 136–137

CONTENT ASSESSMENT
Oral, pp. 127, 137
Check Your Understanding, pp. 120, 133, 141
Reviewing Main Ideas, p. 145
Chapter Review, pp. 146–147

GROUP ASSESSMENT
Opportunities for group assessment occur with Cooperative Learning Strategies.

PREPARATION

Planning the Lesson

Refer to the Chapter Organizer on pages 116A-D.

Concepts Developed

Work is the transfer of energy through motion. Students learn to calculate the work done on an object. The section explores the relationship between energy and work.

1 MOTIVATE

Bellringer

Before presenting the lesson, display **Section Focus Transparency 10** on an overhead projector. Assign the accompanying **Focus Activity** worksheet. L1 LEP

Have two students pantomime sitting still, then playing baseball. Ask the class which activity showed work. Explain that you will examine further the scientific meaning of work in this section. L1

Explore!

Is it work?

Time needed 10 minutes

Materials backpack, books, watch with second hand

Thinking Processes classifying, observing and inferring

Purpose To observe and infer when an applied force accomplishes work and when it does not.

Teaching the Activity

Science Journal Have students compare and contrast the two trials and record their observations in their journals.

 Work

Section Objectives
- Define work.
- Calculate work done on an object.

Key Terms
work

Work, Force, and Motion

In the Explore at the beginning of the chapter, you lifted books and then held them still. Were both of these actions work? Let's explore again.

Explore! ACTIVITY

Is it work?

What To Do

1. Put several books into a backpack and place it on the floor.
2. Now, pick it up and put it on the table. Did that action take energy? Do you think you did work on the backpack?
3. Now, hold your hands out in front of you and have another student hand you the backpack. Hold it in place for 10 to 20 seconds. Did you use energy? Did you do work on the backpack?
4. Record your observations and answers *in your Journal.* Also write a paragraph using this activity to explain what you think work means.

What was the difference between the two actions in the Explore activity you just did? In the first, there was motion. You moved the backpack from the floor to the table. In the second, there was no motion. You held the backpack in one position.

What did the actions have in common? For both actions you used energy, and you exerted an upward force on the pack.

In the first case, there was both force and motion. In the second case, there was force but no motion. For work to be done on an object, both force and motion must be present. If there is force but no motion, there is no work. If there is motion but no force, there is no work. See **Figure 4-1** for an example of this.

Figure 4-1

When the girl releases the flying disk she stops doing work on the disk.

Expected Outcomes

Students are likely to respond that both lifting and holding the backpack are work, until they understand that work must involve both force and motion. L1 LEP

Answers to Questions

2. yes, yes

3. yes, no

✔ Assessment

Performance Have students remove half of the books from the backpack and repeat the activity. Have them compare the amount of work done with the work done in the activity. P

■ Defining Work

In the Explore activity, you transferred energy to the backpack as you picked it up. In scientific terms, **work** is energy transferred through both force and motion. Since there was no motion when you held the backpack steady, no work was done.

For work to be done, one more condition must be met. There must be force, there must be motion, and the motion must be in the direction of the force. **Figure 4-3** shows you when an action can be considered work. When you pick up an object, you exert an upward force, and the object moves in an upward direction. This is work. If you then carry the object across the room while holding it level, you are still exerting an upward force, but the movement is at right angles to that force. In scientific terms, no work is

Figure 4-2

A These people are rushing to their jobs. Scientifically speaking, they are already working.

B Their muscles exert an upward force to climb the stairs. Force and motion in the direction of the force are required for an action to be called work.

Figure 4-3

 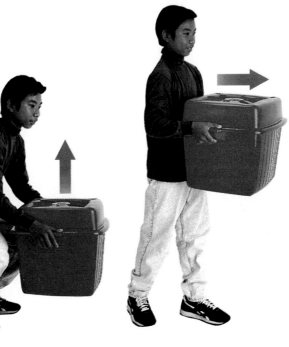

A Sothila pushed a box of old records forward. The box slid along the floor straight in front of him all the way across the room. As he pushed, did Sothila do work on the box? Why?

B Sothila bent down, grabbed hold of the box and lifted it from the floor. As he lifted, did he do work on this box? Why?

C Holding the box level, Sothila carried it across the room. Did he do work on the box? Why?

---

2 TEACH

Tying to Previous Knowledge

Review with students what they learned about balanced and unbalanced forces and the relationship between force and motion in Chapter 1.

Across the Curriculum

Daily Life

Have students volunteer daily experiences that they consider to be work. Then discuss each with the class to determine whether work, as defined in this chapter, was done.

Visual Learning

Figure 4-2 Ask students if people in the photograph are doing work. Then lead students to understand that the people are using their muscles to apply force to their bones and the "something" being moved is their bones.

Figure 4-3 In A and B, Sothila does work on the box because the box moves in the direction of the force he is applying. In C, Sothila exerts a force up on the box, but the box moves at right angles to the direction of the force. According to the scientific definition of work, no work is being done on the box.

Program Resources

Study Guide, p. 16
Critical Thinking/Problem Solving, p. 12, The Amazing Spring
Teaching Transparency 7
Science Discovery Activities, 4-1
Making Connections: Integrating Science, p. 11, Hungry as a Bear
Section Focus Transparency 10

The everyday meaning of *work* includes any time effort is expended. The scientific meaning requires that a force be exerted through a distance.

3 ASSESS

Check for Understanding

Give each student two index cards. On one, students draw a picture of a situation where it looks like someone is doing work but no work is being done. On the other, the students draw a picture of work being done. Have students mix all their cards, then sort the cards by whether or not they show work being done. Use the Performance Task Assessment List for Group Work in **PASC**, p. 97. L1
COOP LEARN

Reteach

Students may have less difficulty understanding the concept of work if familiar terms are used. Explain that if a one-pound weight is moved one foot, one foot-pound of work is done. L1

Extension

Have students calculate the amount of work done in some favorite personal activities by using their weight as one of the factors. L3

4 CLOSE

Ask students why they feel tired holding books if they are not doing any work. Even though they are not doing work, they are expending energy to keep their muscles contracted. This causes fatigue. L1

SKILLBUILDER

Comparing and Contrasting

Compare and contrast the everyday meaning of the word *work* and the scientific definition of that term. Give examples of work in the everyday sense that would not be considered work in the scientific sense. If you need help, refer to the **Skill Handbook** on page 642.

Figure 4-4

Amad lifted a stack of books that weigh 40 N from the floor to a shelf 1.75 m high. How much work did Amad do on the books? You can find the answer by using the equation Work = Force × distance.

---

done on the object as you walk.

Work is done when an object moves while there is a force acting in the direction of motion. The amount of work done is found by multiplying the force times the distance

that the object moves. The mathematical formula is:

$$W = F \times d$$

W means work done, *F* means force in the direction of motion, and *d* means distance moved.

When force is expressed in newtons (N), and distance in meters (m), the unit for work is the newton·meter (N·m). The N·m is also called a joule (J). One joule is the work done when a force of one newton acts through a distance of one meter.

Here's an example of how to calculate work. A student's backpack weighs 10 N. She lifts it from the floor to a table 0.75 m high. How much work is done on the backpack of books?

$$W = F \times d$$
$$W = 10 \text{ N} \times 0.75 \text{ m}$$
$$W = 7.5 \text{ N·m or } 7.5 \text{ J}$$

We've said that work is transferring energy through motion. If a machine uses fuel and does work, it loses energy. Where does that energy go? What forms does it take? As you can see, there's much more to learn about energy and its relationship to work and motion.

check your UNDERSTANDING

1. If you went outside and pushed against the school building as hard as you could for five minutes, how much work would you do on the building? Explain your answer.
2. While performing a chin-up, Carlos raises himself 0.8 m. If Carlos weighs 600 N, how much work does he do? If Carlos holds the chin-up for 10 seconds before letting himself down, how much more work does he do?
3. **Apply** Jill is doing chin-ups on the same bar as Carlos. Jill weighs only 400 N. Compare the amount of work Jill does in doing 10 chin-ups to the work Carlos does in doing 10 chin-ups.

check your UNDERSTANDING

1. None; there is no motion.
2. 600 N × 0.8 m = 480 J; none
3. Jill does two-thirds as much work as Carlos.

Energy of Motion: Kinetic Energy

Look at **Figure 4-5** that shows a sailing iceboat with steel runners. Imagine that a wind pushes on the sail in one direction with a constant force of 400 N. When a constant, unbalanced force pushes on an object, what happens?

You learned in Chapter 1 that a force causes an object to accelerate. The longer the force acts, the faster the iceboat will be going. Say that the iceboat travels 20 m before the wind stops blowing. The wind accelerated

the iceboat by doing (20 m × 400 N) = 8000 joules of work ($F \times d = W$).

What was the difference in the iceboat after the work was done? Before the work was done, the iceboat was at rest. Energy was transferred from the wind to the iceboat. Work resulted in the iceboat gaining speed.

Figure 4-5

Iceboats travel on energy from wind. Iceboats can skim across ice at speeds greater than 160 km per hour.

The wind pushes on the sail with a constant force of 400 N.

The wind accelerated the iceboat 20 m. You can calculate the work done on the iceboat by using the formula $F \times d = W$. 400 N × 20 m = 8000 J

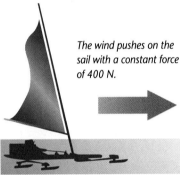

A A calm has left this iceboat at rest.

B A gust of wind blowing with a force of 400 N comes up from behind the iceboat. The sail catches the wind and the iceboat accelerates. The iceboat travels 20 m.

C The wind transferred energy to the iceboat by doing work on it. As the iceboat moved, it gained kinetic energy—the energy of motion.

PREPARATION

Planning the Lesson

Refer to the Chapter Organizer on pages 116A-D.

Concepts Developed

This section builds on the concepts of work and energy developed earlier in the chapter and includes demonstrations and discussion of kinetic energy, potential energy, and elastic potential energy. It introduces the concept of the transfer of energy from one form to another.

1 MOTIVATE

Bellringer

Before presenting the lesson, display **Section Focus Transparency 11** on an overhead projector. Assign the accompanying **Focus Activity** worksheet. [L1] [LEP]

Demonstration To illustrate potential and kinetic energy, use a marble and a cardboard ramp. Place the marble at the top of the ramp and have students observe its motion as it rolls down the ramp. Repeat several times, altering the slope of the ramp each time. Have students compare and contrast the results of each trial. Ask how altering the slope affected the marble's movement.

2 TEACH

Tying to Previous Knowledge

Review Newton's laws of motion, as presented in Chapter 1. Have students suggest how energy and motion are related.

Theme Connection Each form of energy can be changed into another form. However, during such changes, the total amount of energy remains the same. No energy is created or destroyed. In this section, the focus is on two forms of energy—potential energy and kinetic energy. The transformation of these forms supports the themes of energy and stability and change.

Visual Learning

Figure 4-6 To reinforce mathematical thinking skills have student volunteers lead the class through the calculations in the caption.

Figure 4-7 Help students recognize that most of the hammer's kinetic energy is transferred to the nail. Evidence of this is the fact that the nail moves into the wood. Some students may also know that some of the hammer's kinetic energy is changed to thermal energy.

GLENCOE TECHNOLOGY

CD-ROM

Science Interactions
Course 2, CD-ROM

Chapter 4

You could say that the iceboat is given energy of motion. This energy of motion is called **kinetic energy**. You can see common activities that involve kinetic energy in **Figures 4-6** and **4-7**.

Figure 4-6

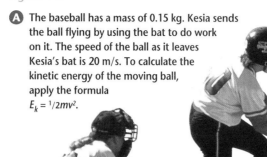

A The baseball has a mass of 0.15 kg. Kesia sends the ball flying by using the bat to do work on it. The speed of the ball as it leaves Kesia's bat is 20 m/s. To calculate the kinetic energy of the moving ball, apply the formula $E_k = 1/2mv^2$.

B Start by substituting numbers for the symbols. You know the mass of the ball is 0.15 kg. The formula also calls for the ball's velocity, which is a measure of speed in a given direction. Because direction is not important here, you can use the ball's speed: 20 m/s.

C Now work the formula. Start by calculating the square of the velocity. Remember that the little dot means multiply.

$E_k = 1/2mv^2 = 1/2 \cdot (0.15 \text{ kg}) \cdot (20 \text{ m/s})^2$
$E_k = 1/2 \cdot (0.15 \text{ kg}) \cdot (400 \text{ m}^2/\text{s}^2)$
$E_k = 1/2 \cdot 60 \text{ kg} \cdot \text{m}^2/\text{s}^2$
$E_k = 30 \text{ J}$

How Do We Know?

Kinetic Energy Equation

Where did the equation for kinetic energy come from? It comes from quantities that you are already familiar with. The work done on an object can appear as the object's kinetic energy. You'll remember that work equals force times distance ($W = Fd$). You also know that force equals mass times acceleration ($F = ma$) and distance equals the average velocity times time ($d = vt$). Finally, you know that acceleration is a change in velocity with time. You can combine all these relationships mathematically to get the equation for kinetic energy: $E_k = 1/2mv^2$.

The kinetic energy of a moving object may be calculated using a formula. It is

$$E_k = 1/2mv^2$$

where E_k is kinetic energy in joules, m is mass in kg, and v is velocity in m/s. Kinetic energy is measured in the same unit as work—joules. The How Do We Know explains how this equation is derived. The amount of kinetic energy an object possesses can't be greater than the amount of work that was done on the object. You'll see why later in the chapter.

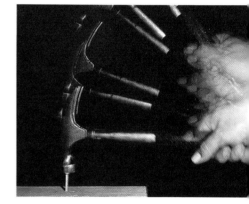

Figure 4-7

When it is moving down toward a nail, a hammerhead has a lot of kinetic energy. Once the head hits the nail, the hammer stops and no longer has any kinetic energy. Where does the energy go?

Program Resources

Critical Thinking/Problem Solving, p. 5, Flex Your Brain

ENRICHMENT

TECH PREP

Activity Have students drop a ball bearing or a golf ball into a large cube of clay from various heights and note the impression made in the clay. Students should infer from their observations that the kinetic energy of the ball increases with height. They should then examine the equation for kinetic energy to determine which variable increased the kinetic energy. *Since m is constant, the variable is v.* **L3**

Energy of Position: Potential Energy

It's fairly easy to tell that a moving object has energy. But how can you tell if an object has energy when it isn't moving?

Does a softball have energy?

How can you tell if an object at rest has energy? You can explore this question in the activity.

What To Do

1. Hold a softball in your hand. Does it have any kinetic energy? Does it have any other kind of energy that you can tell?

2. Press a tent stake about halfway into a bucket of dirt.

3. Hold the softball about 1 m above the stake. What kind of energy will the ball have when you drop it?

4. Drop the ball onto the stake. How did the ball transfer its energy? Did the ball's energy perform work?

5. Lift the ball back up to about 1 m. Does the ball have the ability to do work, to produce change now? Does the ball have energy? If so, where did the energy come from?

6. Record your observations and answers *in your Journal.*

When you held the ball above the stake, it looked the same as it did on the ground. There was no obvious change in the ball and there was nothing to indicate that there was any energy in the ball. But as you saw, when you released the ball, it began to move toward Earth. The ball gained kinetic energy. Where did the energy come from? The only change in the ball was a change in its location. Energy was stored in the ball because of its position above the ground. Stored energy is called **potential energy**. **Figure 4-8** shows one example of potential energy in nature.

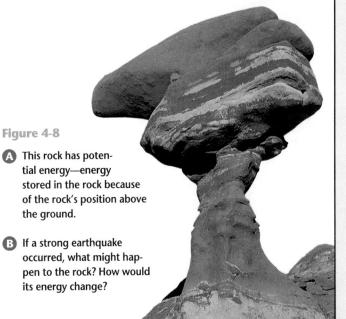

Figure 4-8

Ⓐ This rock has potential energy—energy stored in the rock because of the rock's position above the ground.

Ⓑ If a strong earthquake occurred, what might happen to the rock? How would its energy change?

To reinforce dictionary skills and increase student awareness of word origins have students use the dictionary to check the origins of the words *kinetic* and *potential*. They should discover that the former comes from the Greek root *kinein* (to move) and the latter comes from the Latin root *potere* (to be powerful). Have students explain how these origins relate to the meanings used to describe energy. L1

Visual Learning

Figure 4-9 Have students who have ridden on a roller coaster describe their experiences. Have them tell how the car is pulled up to top of the first hill.
Figure 4-10 Have students read the text material on page 125. To reinforce calculation skills ask a volunteer to lead the class through the calculation of the potential energy of the car at the top of the first hill.

Inquiry Questions What happens to the kinetic energy of the car when it reaches the bottom of one hill and starts to climb another hill? *It changes to potential energy, which increases as the height of the car increases.* Why is the first hill on a roller coaster the highest one? *The gravitational potential energy of the car at that point must provide the kinetic energy of the car for the rest of the ride.*

DID YOU KNOW?

The gravitational potential energy stored in a block of stone when it was lifted to the top of a pyramid 2500 years ago is still there, unchanged.

■ Gravitational Potential Energy

In the Explore activity you did, where did the potential energy come from? Originally, the ball was on the ground. Did the ball have any potential energy relative to the ground when it was there? When you lifted the ball against the force of gravity, you did work against gravity and transferred some of your energy to the ball. The energy you transferred was stored in the ball as potential energy because of the ball's position above the ground. This kind of stored energy is gravitational potential energy. The force of gravity was able to act on the ball when you released it, increasing the ball's speed until it reached the ground. As the ball fell, the kinetic energy increased and the potential energy decreased. Just before the ball hit the stake, almost all of the ball's energy was kinetic. That kinetic energy did the work of driving the stake into the ground.

As you may have guessed, the higher you lift an object, the more potential energy it gains. Also, the more an object weighs, the more potential energy it gains as it is lifted. In each case, the ball will be capable of doing more work when it falls. Another way to think of this is that you must do work to lift the object

Figure 4-9

At some time, work was done on the roller coaster car to get it to the top of the hill. The work done is stored in the coaster car as gravitational potential energy.

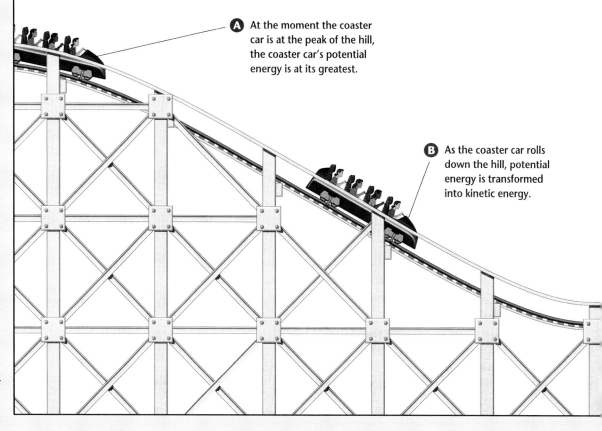

A At the moment the coaster car is at the peak of the hill, the coaster car's potential energy is at its greatest.

B As the coaster car rolls down the hill, potential energy is transformed into kinetic energy.

Meeting Individual Needs

Learning Disabled Many of the adjectives used to modify "energy" are readily understood by most students. However, "potential" may be misunderstood even though its scientific meaning coincides with common usage. Ask students to construct sentences that include the word *potential* or *potentially*. Have them relate the way they have used the word to the concept of potential energy. L1

Visually Impaired Have students hold a baseball or other relatively heavy ball at various heights over an empty hand. Students should drop the ball into the empty hand and, using the sense of touch, determine the relationship between height and potential energy. *The ball dropped a longer distance will have a greater impact and, thus, will have started with greater potential energy.*

against gravity. The object then possesses the energy you used to do that work and, in turn, can do work as it falls back to Earth.

A roller coaster car at the top of a hill, as shown in **Figure 4-9**, is a great example of gravitational potential energy.

■ Calculating Gravitational Potential Energy

Gravitational potential energy is the work done in lifting something against the force of gravity. Work equals $F \times d$. When you lift something, the force you use–F–must equal the weight of the object. Therefore, gravitational potential energy–E_p–is given by:

$$E_p = W \times h,$$

where W is the object's weight and h is the distance lifted.

What is the relationship between the potential energy stored in a car at the top of a hill and the kinetic energy that it has when it reaches the bottom of the hill? Let's investigate this relationship in the following activity.

C Even though it doesn't fall straight down toward Earth, the coaster car is being affected by gravity. Gravity is pulling the coaster car down, not forward.

Figure 4-11

Which water slide would you climb to the top of if you wanted to gain the most gravitational potential energy? If a friend weighing 40 newtons more than you climbed the same slide, would his potential energy at the top of the slide be the same as yours? Why?

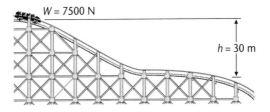

$W = 7500$ N

$h = 30$ m

Figure 4-10

A When using SI units, the gravitational potential energy of an object is expressed in joules. Gravitational potential energy is equal to the weight of the object multiplied by the height the object is lifted. The formula is stated $E_p = W \times h$.

B Use the formula to calculate the gravitational potential energy of the roller coaster car.

$E_p = W \times h = 7500$ N $\times 30$ m
$E_p = 225\ 000$ J

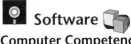

DESIGN YOUR OWN
INVESTIGATION

4-1 Potential and Kinetic Energy

Preparation

Purpose To design an experiment to calculate gravitational potential energy and the resultant kinetic energy of an object.

Process Skills designing an investigation, sequencing, observing and inferring, measuring in SI, making and using graphs and tables, calculating data

Time Required 40 minutes

Materials See reduced student text.

Possible Hypotheses Students should hypothesize that the greater the potential energy of an object (W × h), the greater will be the kinetic energy transferred to the object. They may also hypothesize that the distance traveled by an object will be directly related to its kinetic energy.

Plan the Experiment

Process Reinforcement When approving student plans, check to make sure that students understand the relationships involved in determining potential energy and kinetic energy. Check their graphs and/or tables for proper construction and inclusion of data.

Possible Procedures Observe and measure the motion of an object rolling down a ramp. The potential energy of the object will be varied by changing the height (slope) of the ramp, the length of the ramp, and the weight of the object. Use the formula $E_p = W \times h$ to calculate potential energy. To calculate weight, multiply mass by the acceleration due to gravity. (They can use the approximation of 10 m/s².) The average velocity of the moving object can be found by measuring the time it takes to travel the length of the ramp. Use this value in the formula $E_k = \frac{1}{2} mv^2$ to calculate the kinetic energy of the object.

Potential and Kinetic Energy

At the top of a hill, your bicycle may have considerable potential energy. As you roll down the hill, that energy is transferred into kinetic energy. Does a measurable relationship exist between these two types of energy? What is that relationship?

Preparation

Problem
What is the relationship between the potential energy of an object and the kinetic energy of the same object?

Form a Hypothesis
How does the potential energy of an object compare to its kinetic energy?

Objectives
• Calculate potential and kinetic energy from experimental observations.

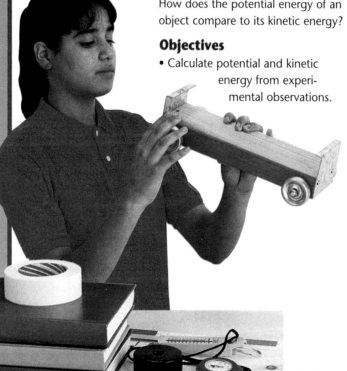

• Construct a graph to demonstrate the relationship found between potential energy and kinetic energy.

Materials
20-N spring scale
stopwatch
masking tape
cart
meterstick
1.0- to 2.0-m ramp
1.0-kg mass
several books

Safety
Make sure your cart has a clear path that is free from people before you let your cart go.

Program Resources

Activity Masters, pp. 19-20, Design Your Own Investigation 4-1 [L3]

Multicultural Connections, p. 11, Bow Design and Construction

INVESTIGATION

Plan the Experiment

1 How will you set up your trial? Will you use the same setup throughout the experiment? Will you run more than one trial?

2 How will you calculate or measure velocity? What is the formula for potential energy? For kinetic energy?

3 Are mass and weight the same? Will you need to convert mass into weight (kilograms to newtons)?

4 Write out your plan with a rough drawing of your setup and a method for recording data *in your Science Journal.*

Check the Plan

1 How are you organizing all of your data and calculations? Do you have everything marked so you don't confuse numbers and trials?

2 Present your plan to your teacher before you do it. Revise your plan as necessary and then carry out the experiment.

Analyze and Conclude

1. Calculate Calculate the potential energy and the kinetic energy of each object in each of your trials.

2. Graph Construct a bar graph that shows the relationship between the kinetic energy and the potential energy in each trial. Do the data support your hypothesis? Explain.

3. Interpret Describe the relationship between the two types of energy by interpreting your data.

Going Further

Predict how increasing the mass of the object you used would affect its velocity, its potential energy, and its kinetic energy. Test your prediction.

The Soap Box Derby got its name long ago when many of the race cars were built from wooden soap boxes. Derby race cars have no motors. Drivers rely solely on the conversion of gravitational potential energy to kinetic energy to bring their cars down the race-course hill.

Meeting Individual Needs

Behaviorally Disordered Students who are proficient in using a yo-yo may demonstrate some tricks they can do with it. For each trick, have them identify the kind of energy in the yo-yo in various positions.

Teaching Strategies

Troubleshooting Have the students check their units to make sure they agree in all calculations. Remind them that a quick check of units can catch an incorrect equation.

Science Journal Check the students' journals for drawings of their setups.

Expected Outcomes

Students will discover the mathematical relationships between the potential and kinetic energies of a vehicle of known mass and the variables that affect the relationship.

Analyze and Conclude

1. The calculations should show potential and kinetic energy about equal.

2. Check student graphs to see that potential energy is always greater than, or at least equal to, kinetic energy.

3. Student responses should indicate a direct relationship between potential and kinetic energies.

✔ Assessment

Oral Assign photos of high divers, skiers, ski jumpers, acrobats, or other athletes who use height and motion in individual sports. Have students work in small groups to prepare a presentation in which they describe how they would calculate the gravitational potential energy and the kinetic energy of each athlete in the photos given to their group. Use the Performance Task Assessment List for Oral Presentation in **PASC**, p. 71.

Going Further

Students should predict that increasing the weight will increase the potential energy and kinetic energy. The velocity will not change.

Have students check the meanings of *work* in the dictionary. Then ask them to list a series of tasks they perform each morning or evening, such as getting out of bed, brushing teeth, eating breakfast. Ask them to analyze these tasks according to whether they fit the scientific definition of work or another meaning.

Activity Have students cut open an old softball or baseball and an old golf ball. Have students examine what is inside each. How does the concept of elastic potential energy help explain why golf balls go farther when hit than softballs? L2

■ Potential to Kinetic

As you did the Investigation, you may have found that the kinetic energy you were able to calculate was a bit less than the potential energy. Is this what you would have predicted? Why might this be so?

The work you put into the cart to raise it to the top of the ramp is stored in the cart as gravitational potential energy. As the car rolls down the ramp, gravity accelerates it, turning the potential energy into kinetic energy—motion. When the cart leaves the ramp, does it have any potential energy left? Where is that energy now? If you had been able to do this Investigation on a friction-free surface, what do you think the relationship between the potential energy at the top of the ramp and the kinetic energy at the bottom would have been? Later in this chapter, you'll learn what happened to the energy the cart lost by going down the ramp.

■ Elastic Potential Energy

Simply by looking around, you can see many things in your environment that possess gravitational potential energy. Is gravitational potential energy the only way to store work? Can you think of any other ways in which energy transferred by work done on an object can be stored? Let's explore.

Energetic Toy

Your baby sister is playing with her favorite roll-back toy. If you watch closely, you'll notice that this toy demonstrates how energy can be converted from one form to another. Let's build such a toy so we can take a better look.

Materials

one-pound coffee can
2 plastic coffee can lids
heavy rubber band—folded length about 17 cm
lead fishing weight, about 2.5 cm long

What To Do

1 Remove both ends from the coffee can. Beware of sharp edges!

2 With a skewer or the tip of a pair of scissors, poke two holes in each lid, about 2.5 cm apart. Make the holes just large enough so that the rubber band can pass through them.

3 Cut the rubber band at one end. From the inside of one lid, thread the rubber band up through one hole and back down through the other. Even the ends.

Purpose

A Closer Look reinforces Section 4-2 by demonstrating how energy can be converted from one form to another.

Content Background

The concept underlying this excursion is the conversion of potential energy to kinetic energy. Potential energy is the result of work that has already been done. As the corresponding movement takes place, energy undergoes a conversion to a different form.

Teaching Strategy

CAUTION: Make certain that students do not share their toys with younger siblings. The lead weight and rubber band can be dangerous. Encourage students to create other simple toys that use the conversion of energy to create motion or do work. Supply additional materials to use or have students work in pairs to construct a toy as a homework project.

Explore! ACTIVITY

How is work stored?

Can you think of any other ways to store work? What happens when you wind the propeller on a toy plane driven by a rubber band? Is energy stored? Explore another way of storing energy.

What To Do

1. Hold a 20-N spring scale flat on the table. Grasp the hook and pull it until the scale reads about 10 N.

2. Have another student measure the distance you moved the hook in centimeters.

3. Now, release the hook. What happens?

4. Where was this work after you stopped pulling? Did kinetic energy increase? Potential energy? Record your answers and observations *in your Journal.*

---

4 Stretching the rubber band slightly, attach the fishing weight to both ends of the rubber band. You may run the rubber band ends through the loop of the weight or tie the weight to both pieces with string.

5 Place the lid and rubber band assembly on the can.

6 Feed each end of the rubber band through the holes on the other lid and tie the ends together. Attach the second lid.

You Try It!

Place the toy on a smooth surface and push. What happens? How would you explain this motion? Make a list of the energy changes, beginning with your push. Be sure to state whether the energy is kinetic or potential and which type of potential energy is involved. Why does the toy eventually stop?

129

Explore!

How is work stored?

Time needed 15 minutes

Materials 20-N spring scale; metric ruler

Thinking Processes measuring in SI, comparing and contrasting, inferring

Purpose To infer that kinetic energy of a moving spring is converted to and stored as elastic potential energy.

Teaching the Activity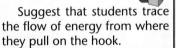

Suggest that students trace the flow of energy from where they pull on the hook.

Science Journal Suggest that students include labeled sketches of the spring before and after being stretched and after being released. They should include a written description of each stage of the activity. **L1**

Answers to Questions

3. The hook moves back to its original position.

4. The energy applied in doing the work was stored in the stretched spring. There was no increase in kinetic energy, but there was an increase in potential energy.

✔ Assessment

Process Have students conduct a different activity involving elastic potential energy, such as compressing an inflated balloon. Have them compare and contrast their observations with those of this activity. Use the Performance Task Assessment List for Carrying Out a Strategy/Collecting the Data in **PASC**, p. 25.

Answers to
You Try It!

The push is transferred into kinetic energy as the toy begins to roll. The rolling motion causes the rubber band to twist, storing elastic potential energy. As the toy changes direction, the energy in the rubber band is transformed into kinetic energy. Friction between the can and the surface causes the toy to stop rolling.

Going Further ⫸

Students can work in small groups to build a different version of the roll-back toy. After the roll-back toys have been built, provide an opportunity for each group to experiment with its toy. Different groups may want to display their roll-back toys to other groups and create some sort of competition for moving the toys.

Encourage students to make predictions in response to the following questions and then to test their predictions: **What would happen if the toy were used on a surface that is not smooth? How would changing the weights affect the motion of the toy?** *A rough surface or added weight would cause the toy to travel a shorter distance.* Use the Performance Task Assessment List for Group Work in **PASC**, p. 97. `COOP LEARN`

Discussion Many a tense and dramatic football game has been decided in the last seconds by a field goal made or missed. Have students discuss the variables that affect the success or failure of a field goal attempt. Students should, of course, focus on the property of elastic potential energy and consider such questions as: Was the football underinflated? Overinflated? Made of more or less elastic material? With regard to elastic potential energy, what would be the effect of striking the football other than dead center?

Figure 4-12

A The football player does work on the football by kicking it. When his foot connects with the ball, kinetic energy is transferred from his foot to the ball.

B The energy the ball receives is stored as elastic potential energy in the flattening of the ball.

C As the ball springs back to its original shape, the ball's elastic potential energy is converted to kinetic energy.

Figure 4-13

A Because the spring in this wind-up toy can go back to its original shape after it is tightened, the spring can store elastic potential energy.

B A loose spring has no elastic potential energy. A tightly wound spring has great elastic potential energy. Where did the energy in the tightly wound spring come from?

Because the scale didn't move after you did the work in the Explore activity, the work did not produce kinetic energy. However, you didn't lift the scale so it can't be gravitational potential energy. The energy is stored in the spring. When you released the hook, the scale popped back to its original setting. The spring had stored the energy you put in, and that energy showed itself as motion—kinetic energy—when you released it. This kind of stored energy is called elastic potential energy.

■ **Everyday Examples of Elastic Potential Energy**

In **Figure 4-12**, you can see an example of elastic potential energy in a football kick. Energy is stored as elastic potential energy when work causes an object to be stretched or twisted, or if its shape is changed, as in the objects in **Figures 4-13, 4-14, 4-15,** and **4-16**. The object must be capable of going back to its original shape in order to store energy in this way. Rubber bands, trampolines, the spring in a wind-up toy, and diving boards are other examples of objects that can store elastic potential energy.

Figure 4-14

The propeller of this model plane is attached to a long rubber band. Winding the propeller allows the rubber band to store elastic potential energy as it twists. What will happen when the propeller is let loose? Why?

130 Chapter 4 Work and Energy

Is Energy Used Up?

When you drop a tennis ball or golf ball from several feet above the floor, the ball bounces. Actually, it bounces several times, each time a little lower. The potential energy stored in the ball, because of its position above the ground, is converted to kinetic energy as it falls. When the ball hits the ground, it is compressed, storing the energy as elastic potential energy.

When the ball regains its shape, that potential energy does work on the ball, increasing its velocity in an upward direction. But why doesn't it bounce as high? Where did some of the energy go? Let's find out.

Figure 4-16

The spring on a jack-in-the-box gains elastic potential energy when you compress it. By closing the lid, you keep the spring compressed. What happens to Jack's elastic potential energy when you release the lid?

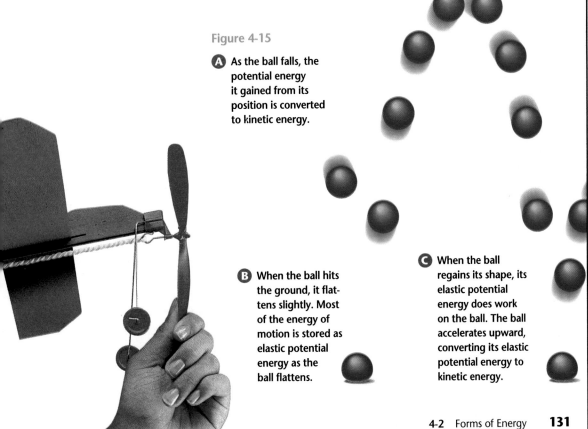

Figure 4-15

A As the ball falls, the potential energy it gained from its position is converted to kinetic energy.

B When the ball hits the ground, it flattens slightly. Most of the energy of motion is stored as elastic potential energy as the ball flattens.

C When the ball regains its shape, its elastic potential energy does work on the ball. The ball accelerates upward, converting its elastic potential energy to kinetic energy.

4-2 Forms of Energy **131**

Find Out! ACTIVITY

Where does the energy go?

If objects did not transfer energy, they would never come to a halt. But objects do stop, so where does the energy go?

What To Do

1. Attach a 20-N spring scale to the end of a shallow box.

2. Place several books in the box.

3. Pull the box with a steady force 1 m across the table or floor. Record the force needed.

4. Now pull the box with a steady force 1 m across a rug or other rough surface at the same speed you used on the table. Record your measurements and observations *in your Journal.*

Figure 4-17

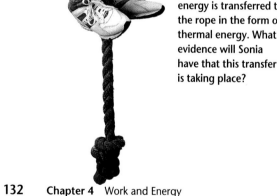

Ⓐ Sonia gained gravitational potential energy by climbing the rope. As she slides down the rope, her potential energy is converted into kinetic energy.

Ⓑ Some of the kinetic energy is transferred to the rope in the form of thermal energy. What evidence will Sonia have that this transfer is taking place?

132   Chapter 4   Work and Energy

Conclude and Apply

1. Calculate the amount of work done in each case. Compare these values. Why do you think you had to do more work on the box in one case than the other?

2. The work you did transferred energy from you to the box and books. Where is that energy now?

■ **Where the Energy Goes**

If you had put your hand on the bottom of the box immediately after you'd finished dragging it, you probably would have noticed that the box was a bit warmer than it was before. The surface over which you dragged it would also be warmer. This difference in temperature is so small that you might not be able to feel it, but here's an easy way to demonstrate what happens.

Rub your hands together rapidly. What happens to your hands? You're doing work, and the work shows itself in warming your hands. We'd say that the work done shows itself as thermal energy. For another example of this, see **Figure 4-17.**

infer that friction caused some energy to be converted to thermal energy.

✓ Assessment

When you drop a tennis ball, the ball strikes the ground and some of the energy is changed to thermal energy. The floor and the ball get a little warmer. That energy is no longer available to move the ball.

Think back to the Investigate you did with the cart. The kinetic energy of the rolling cart was less than the potential energy of the cart at the top of the ramp. Where did that energy go? The floor and the wheels of the cart warmed up. Some of the kinetic energy was converted to thermal energy.

■ **Energy from Work**

You've seen that doing work on an object can give different results. When you throw a softball, it gains kinetic energy. When you lift the ball, you increase its gravitational potential energy. When you hit the softball with a bat, it compresses and stores the work as elastic potential energy. When you roll the softball across the floor, it will eventually stop. The energy isn't used up or lost—it's present as thermal energy in the ball and floor.

In the next section, we'll track work and energy through several different situations.

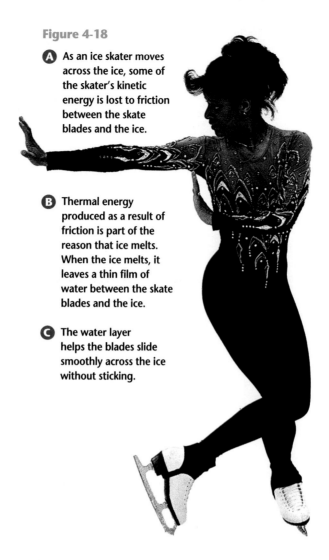

Figure 4-18

A As an ice skater moves across the ice, some of the skater's kinetic energy is lost to friction between the skate blades and the ice.

B Thermal energy produced as a result of friction is part of the reason that ice melts. When the ice melts, it leaves a thin film of water between the skate blades and the ice.

C The water layer helps the blades slide smoothly across the ice without sticking.

---

check your UNDERSTANDING

1. Give two examples where work done on an object produces kinetic energy.
2. A 15 kg model plane flies horizontally at 2.5 m/s. Calculate its kinetic energy.
3. Which of the following is an example of work producing potential energy? Explain.
   a. putting on your hat

   b. carrying a box across the room
   c. hitting a golf ball with a club
4. **Apply** You pick up a beanbag from the table and lift it over your head. Then you drop it to the floor. Discuss the work done and changes in energy that take place during these actions.

Check for Understanding

Have students create a list of situations where gravitational potential energy can be used constructively and when it is destructive. For example, it is used constructively in digging a garden but could be destructive if a brick falls off the side of a building. Use the Performance Task Assessment List for Group Work in **PASC**, p. 97. `COOP LEARN`

Reteach

Activity Have students use a Slinky to identify potential energy, kinetic energy, and elastic potential energy. Have students try dropping it directly to the floor, in a closed position, and then position it at the top of a staircase so that it "walks" down the stairs. Point out the different forms of energy it possesses at various heights, explaining the changes from potential to kinetic. `L1`

Extension

Activity Have students use a small model car and a rubber band to create a demonstration of kinetic and potential energy. `L3`

4 CLOSE

How does a bouncing ball demonstrate kinetic and potential energy? Bounce a small ball and have students describe the point at which it has the most kinetic energy and the point at which it has the most potential energy. `L1`

---

check your UNDERSTANDING

1. Possible responses include hitting a ball, pedaling a bike, pushing an eraser across a blackboard.
2. $E_k = \frac{1}{2}\,mv^2$; $E_k = \frac{1}{2} \times 15\ \text{kg} \times (2.5\ \text{m/s})^2 = \frac{1}{2} \times 15\ \text{kg} \times 6.25\ \text{m}^2/\text{s}^2 = 46.875$ J, rounded to two significant digits = 47 J.
3. a: The hat was raised to a higher level.

c: The club strikes the ball, which compresses it and lifts it into the air.
4. You do work on the beanbag, giving it potential energy. As it falls, the potential energy changes to kinetic energy. When it hits the ground, the kinetic energy changes to thermal energy and sound energy.

PREPARATION

Planning the Lesson

Refer to the Chapter Organizer on pages 116A-D.

Concepts Developed

In this section, students will learn how energy is transformed from one form to another. In addition, the Law of Conservation of Energy will be presented.

1 MOTIVATE

Bellringer

 Before presenting the lesson, display **Section Focus Transparency 12** on an overhead projector. Assign the accompanying **Focus Activity** worksheet. L1

LEP

Find Out!

How far will it go?

Time needed 20 minutes

Materials marble, foam cup, grooved ruler, balloon, tape, and wood splint

Thinking Process designing an experiment

Purpose To experiment to produce a device that will best take advantage of the transfer of potential energy to kinetic energy.

Teaching the Activity

Safety Caution students to direct the movement of the marble away from other students. L1

Science Journal Have students record their different trials and results in their journals. You might also suggest that they include sketches of their various setups.

Conclude and Apply

1. Answers may vary but students should be able to clearly explain their methods.
2. The kinetic energy of the

Section Objectives

- Describe how energy changes from one form to another.
- Understand and apply the law of conservation of energy.

Key Terms

law of conservation of energy

Energy on the Move

When you do work on an object, such as throwing, hitting, or rolling a ball, you are transferring energy from yourself to that object—the ball. You already know some ways in which energy can be transferred. Let's find out other ways in which you can transfer energy to move an object.

Find Out! ACTIVITY

How far will it go?

What's the best way to make something move? How can you give the most energy to an object? Experiment to find out!

What To Do

1. Obtain a set of materials, including a marble, foam cup, grooved ruler, balloon, tape, and wood splint or tongue depressor.

2. Use these objects to transfer energy in such a way that your marble travels the greatest distance when you release it. You may not throw the marble. You must simply let it go from wherever your starting point is.

3. Try to demonstrate as many different types of energy transfer as you can.

Conclude and Apply

1. What was the best way to make your marble move?

2. In terms of energy, why did your marble eventually stop moving?

3. Write a paragraph *in your Journal* to describe the kinetic and potential energy present in the system you constructed. Also describe any energy changes that took place.

134 Chapter 4 Work and Energy

---

marble was transformed into thermal energy by friction.

✔ Assessment

Performance Organize students into groups to experiment with the materials. Ask students to share methods they used to transfer energy. Have a "playoff" to see which method causes the marble to travel farthest. Use the Performance Task Assessment List for Group Work in **PASC**, p. 97. L1 COOP LEARN P

Program Resources

Making Connections: Across the Curriculum, p. 11, Lillian Moller Gilbreth, Pioneer of Motion Studies L1

Making Connections: Technology and Science, p. 11, Gerty Radnitz Cori and Energy Production

Section Focus Transparency 12

Energy Is Like Money

You can think about energy transfers as being similar to money exchanges. How? Examine the diagram on this page to start making your comparison.

Figure 4-19

The transfer of energy can be compared to the transfer of money among people.

Ⓐ You have twenty dollars in your pocket. The money has buying power, but for now you are not using that power. You could call the money in your pocket potential money. It has the potential to buy, just as potential energy has the potential to do work.

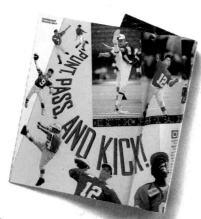

Ⓑ You spend two dollars for a magazine. By putting your money into action, you change its potential buying power to kinetic buying power. Those two dollars are now potential buying power for the person who sold you the magazine.

Ⓒ You transfer five of your potential dollars to a friend. He goes to a store and converts those potential dollars into kinetic dollars when he exchanges them for a gift for his sister.

Ⓓ You exchange a five dollar bill for 20 quarters. You spend eight quarters playing video games, and decide to save the rest.

Ⓔ Your money has changed form and moved from person to person, but the original twenty dollars still exist. No more. No less. Like the money, energy also neither increases nor decreases. Energy changes form and moves from object to object, but the original amount of energy stays the same.

You've seen that energy occurs in several forms and can be changed from one form to another. Try another activity to see if you can keep track of the energy of an object as it goes through a number of changes.

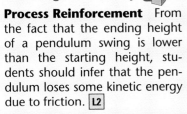

4-2 The Motion of a Pendulum

Planning the Activity

Time needed   40 minutes

Process Skills   making and using tables, observing and inferring, interpreting data

Materials   See activity in student text.

Purpose   To measure the motion of a pendulum and use the data to interpret how energy is transferred.

Teaching the Activity

Process Reinforcement   From the fact that the ending height of a pendulum swing is lower than the starting height, students should infer that the pendulum loses some kinetic energy due to friction. L2

Possible   Hypotheses   Students may hypothesize that friction is present at the point where the string is attached to the ring as well as between the pendulum and the air.

Troubleshooting   The cross arm should be moved out of the way for the first three trials.

Expected Outcomes

Students will recognize the relationship of initial pendulum height to the kinetic and potential energy conversions that occur during the swing of the pendulum.

INVESTIGATE!

The Motion of a Pendulum

What kinds of energy transfers take place in the motion of a pendulum?

Problem

What happens to the energy of a pendulum?

Materials

ring stand	cross arm
right-angle clamp	support rod and clamp
masking tape	2-hole rubber
2 metersticks	stopper, medium
	100 cm of string

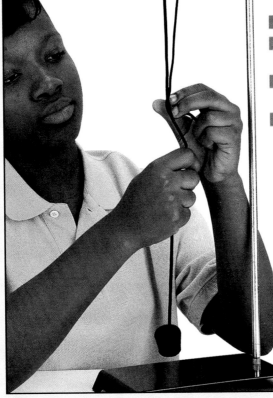

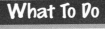

What To Do

1 Copy the data table *into your Journal.*

2 Set up the apparatus as shown, omitting the cross arm at this time.

3 Use the masking tape to mark the center of the stopper. Use this line to measure heights above the tabletop.

4 Pull the stopper to one side. Measure the height of the stopper above the table (see photo **A**). Record the measurement.

Sample data

Data and Observations		
Trial	Starting Height	Ending Height
1	0.4 m	0.35 m
2	0.3 m	0.27 m
3	0.2 m	0.15 m
4	The results for trials 4, 5,	
5	and 6 will depend on individual	
6	setups.	

Program Resources

Activity Masters,   pp. 21–22, Investigate 4-2 L2

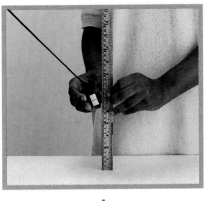

A   **B**

5 Release the stopper and let it swing. Observe carefully and measure the greatest height the stopper reaches just before it begins its return swing. Record.

6 Repeat Steps 4 and 5 twice, each time starting the stopper at a greater height.

7 Repeat Steps 4 through 6 with the cross arm in place (see photo **B**). Begin the first swing below the cross arm, the second level with the cross arm, and the third above it. Record all data and observations *in your Journal*.

Analyzing

1. For a single swing without the cross arm, is the ending height of the stopper exactly the same as its starting height? Explain.

2. What is the highest point that the stopper will reach when it hits the cross arm?

3. Write or draw the sequence of changes in kinetic and potential energy of the stopper at various points on its arc.

Concluding and Applying

 4. If you could calculate the maximum potential energy and kinetic energy of the stopper, what would you infer about their relationship?

5. What caused the string to wrap completely around the cross arm?

6. **Going Further** Would it make sense to build a roller coaster with its highest point near the end of the ride? Explain.

Answers to Analyzing/ Concluding and Applying

1. No, some of the kinetic energy is lost due to air friction and friction where the pendulum is tied.

2. As the string bends around the cross arm, the pendulum may continue upward to a vertical position above the arm.

3. Sequences should indicate maximum potential energy at the two high points and maximum kinetic energy at the lowest point. They should indicate kinetic energy increasing and potential energy decreasing on the downward swing and vice versa on the upward swing.

4. They should be nearly equal; potential energy would be slightly greater.

5. The stopper had more energy than could be converted into potential energy because the string was not long enough to let it rise that high. Therefore, the remaining kinetic energy caused the stopper to continue moving in the only direction left to it.

6. No; the energy used to move the car originates with the gravitational potential energy of the car at the top of the highest hill. Thus, it makes sense to put the highest hill near the beginning of the ride.

✔ **Assessment**

Oral Have students describe the energy conversions taking place at various points along the path of a pendulum's swing. Also have them explain the effects of friction on the system. Use the Performance Task Assessment List for Group Work in **PASC**, p. 97. COOP LEARN

ENRICHMENT

Activity Ask students to work in pairs to develop a poster or three-dimensional display of something that depicts the Law of Conservation of Energy. Have them use arrows and labels to indicate the conversion of energy from one form to another.

Life Science

Living things need energy not only to move, but also to maintain all life processes. Even when people are not active, their bodies continually supply energy to cells. Life-sustaining systems need energy to work. For example, the heart needs energy to pump blood throughout the body; the digestive system needs energy to digest food; the waste system needs energy to dispose of wastes; and the pulmonary system needs energy to maintain breathing.

Discussion Have students suggest ways in which engineers design machinery and vehicles to reduce friction. Examples include using lubricants and streamlining. Use this discussion to lead into an equally important discussion of situations where friction is useful or even essential. Examples include preventing slipping while walking or running, and the slowing down or stopping of vehicles. These would be impossible without friction. Finally, have students discuss environments in which there is virtually no friction and how scientists and engineers have learned to change the motion of vehicles in such environments. Space is essentially frictionless. The motion of space vehicles is controlled by using propellants to apply forces in various directions.

Connect to...

Life Science

All living things need a constant supply of energy to live. Use what you know about energy conversions to explain where the energy from food goes even when you aren't active.

The Law of Conservation of Energy

If there were no friction where the pendulum is tied and no loss of energy to the air, the pendulum would continue to swing forever. The potential energy you put in by lifting the stopper would change to kinetic energy and back to potential energy over and over again. When you include the thermal energy lost to friction, you find that when you add up all the energy at any point during the action, the total amount of energy remains the same. It is conserved as it changes from one form to another. This is called the **law of conservation of energy**. This law says that energy can't be created out of nothing, nor can it be destroyed. Energy may be changed to other forms or transferred to other objects, but the total energy remains unchanged. Energy is conserved.

Why did your pendulum wrap itself around the cross arm when you started it above the level of the cross arm? As the kinetic energy at the bottom of the swing was converted back into potential energy, the stopper had enough energy to reach nearly the same height as where it started.

Earth Science CONNECTION

To Jupiter and Beyond

In the recent past, we've sent robotic probes to Mercury, Venus, and Mars. After these missions had been accomplished, scientists wanted to send similar probes to investigate the far planets of the solar system. To achieve this goal, they designed two space probes.

Voyage of the Century

The twin robotic space probes *Voyager I* and *Voyager 2* traveled to Jupiter, Saturn, Uranus, and Neptune. Through the transmissions they beamed to Earth, the world had the incredible experience of riding along and seeing sights never before possible.

By June of 1989, *Voyager 2* had traveled more than 2.8 billion miles to reach Neptune. When the probe gave us our

A prototype Voyager space-craft is shown as it successfully passes vibration testing in 1977.

138

Earth Science CONNECTION

Purpose

The Earth Science Connection reinforces Section 4-3, by explaining how the initial energy imparted to a space vehicle can be augmented by directing the probe to fly close to planets.

Content Background

Space probes get their original energy boost from the conversion of chemical energy to heat energy when a fuel is burned. In some cases, they may pick up additional energy from the gravitational force of a large object in space, such as a planet or a satellite, as described in the selection.

Teaching Strategy

Have students read the selection in small groups and create a series of questions they would like to research further on energy in space exploration. 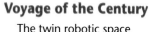 **COOP LEARN** **L1**

Demonstration

Encourage interested students to prepare a demonstration on rockets for the

Figure 4-20

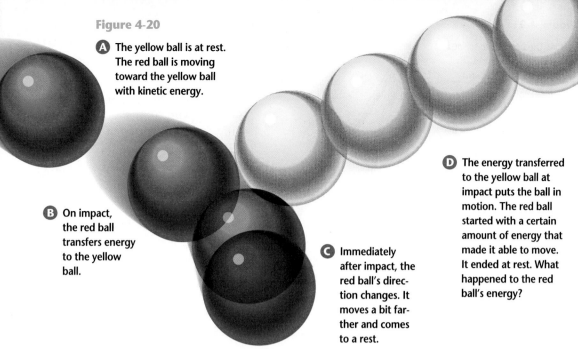

A The yellow ball is at rest. The red ball is moving toward the yellow ball with kinetic energy.

B On impact, the red ball transfers energy to the yellow ball.

C Immediately after impact, the red ball's direction changes. It moves a bit farther and comes to a rest.

D The energy transferred to the yellow ball at impact puts the ball in motion. The red ball started with a certain amount of energy that made it able to move. It ended at rest. What happened to the red ball's energy?

first look at this planet, *Voyager 2* was traveling at a speed greater than when it had left Earth nearly 12 years earlier. A photo of Neptune taken by *Voyager* is shown here.

Riding the Slingshot Effect

How do space probes such as *Voyager* get the energy they need for space travel? Space scientists use a slingshot effect to provide additional kinetic energy to space probes. This effect reduces the amount of fuel the probes must carry.

As a probe comes near a planet, it experiences the gravitational pull of the planet. As a result, potential energy is converted to kinetic energy.

The probe speeds up and reaches its greatest speed as it moves behind the planet.

Behind the planet, the probe's path then takes it away from the surface. The kinetic energy is again changed to potential energy. However, the probe's path leaving the planet is a different shape than the path of the probe approaching the planet. Not all of the kinetic energy it gains is needed to pull it away from the planet's surface. Therefore, the probe has more kinetic energy when it

leaves the planet than it had when it approached the planet.

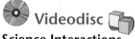

Visit NASA's Planets website and view the most recent images from space probes. How did the proposed missions of these probes affect their path around the planets?

class. Working rockets can be obtained from a variety of sources, such as scientific supply houses or specialty toy stores. Appropriate safety precautions must be taken when working with rockets. **COOP LEARN** **L3**

NASA's Planets website can be found at **http://pds.jpl.nasa.gov/planets**

Going Further ▐▐▐▐▐▐▶

This is a good opportunity for students to investigate the interaction between fuel and various forms of transportation, particularly those using new developments in technology, such as electric or solar cars.

Have groups of students work together to select a form of transportation, research their subject, and make a presentation to the class, using illustrations or models to explain how energy is used to move the

vehicle. Use the Performance Task Assessment List for Oral Presentation in **PASC,** p. 71. **L2**

Figure 4-21 Have students relate the diagram and the captions to their experience with pendulums.

Figure 4-22 Have students compare and contrast the motion and energy transfers of the roller coaster car to that of a pendulum. **At what point in its trip is the car's energy all kinetic?** *The car's energy is all kinetic at the lowermost point of the ride, just before it starts to climb.*

Figure 4-23 **In preparing to dive, why does a diver jump up in the air and land at the end of a diving board?** *By jumping, the diver increases the kinetic energy with which he or she lands on the board, causing the board to bend more and increasing its elastic potential energy. When released, the increased kinetic energy of the board sends the diver higher into the air.*

Figure 4-24 Ask students to explain how the flexibility of the pole helps the vaulter. This property makes it possible to build up elastic potential energy.

NATIONAL GEOGRAPHIC SOCIETY

Videodisc

Newton's Apple: Physical Sciences

Roller Coaster
Chapter 1, Side B
Changes in weight

5154-5556

What's centripetal force?

7107-7859

Where does the coaster get its energy?

9976-12216

Figure 4-21

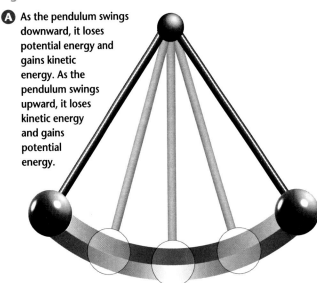

A As the pendulum swings downward, it loses potential energy and gains kinetic energy. As the pendulum swings upward, it loses kinetic energy and gains potential energy.

B At every point in the swing, the total mechanical energy of the pendulum (K.E. + P.E.) is the same. Energy is converted from one form to the other, but is never lost from the system.

Figure 4-22

A At the top of a hill, the roller coaster car has almost no kinetic energy. It has only potential energy.

B As the car rushes downhill, its potential energy is decreasing. At the same time, the car's kinetic energy is increasing.

C The instant the car begins climbing the next hill, the car's kinetic energy begins to decrease and its potential energy begins to increase.

D At the top of the hill, the car's kinetic energy has been converted to potential energy. Again, it has almost no kinetic energy at this point. At what point in its trip is the car's kinetic energy largest?

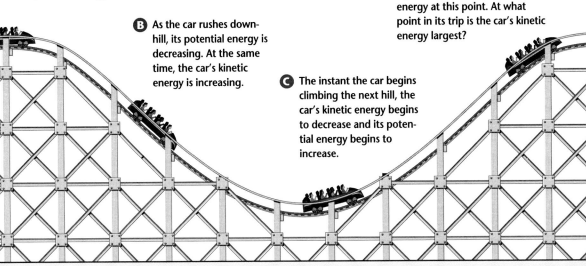

But the cross arm blocked it. Because the stopper hadn't completely converted its kinetic energy into potential energy, it still had kinetic energy left. Kinetic energy is the energy of motion. The stopper moved in the only direction left open to it by the string. It went around the cross arm. Eventually, motion ended when all of the stopper's energy had been changed into thermal energy. Other examples of energy conversions are present in **Figures 4-21, 4-22, 4-23,** and **4-24.**

■ Conservation and Forms of Energy

There are no cases in which energy appears or disappears without coming from or ending up somewhere else. Whenever energy seems to disappear, scientists have found another form of energy. Some other forms are electrical, electromagnetic, and

Figure 4-23

A When a diver's weight bends a diving board downward, the board gains elastic potential energy.

B When the board springs back to its original shape, the elastic potential energy is converted to kinetic energy. The board's kinetic energy is transferred to the diver. The diver moves upward.

Figure 4-24

A The kinetic energy of a pole vaulter is stored as elastic potential in the bending of the pole.

B As the pole straightens, the elastic potential energy is converted to kinetic energy, and the vaulter is lifted.

nuclear energy. Although we still use these terms, scientists now believe that there are only two basic kinds of energy. Regardless of where the energy comes from, it can be described as kinetic or potential. For example, what we have been calling thermal energy is another combination of kinetic and potential energy. You will learn more about this kind of energy in Chapter 6.

check your UNDERSTANDING

1. Give one example for each of the following energy changes:
   a. gravitational potential energy to kinetic energy
   b. kinetic energy to elastic potential energy
2. You are riding your bike on a level surface. Why do you have to keep pedaling to maintain the same speed?
3. Trace the energy changes as a heavy rock rolls down a hill. Is any energy lost?
4. **Apply** When a machine burns fuel to do a job, only part of the chemical potential energy in the fuel comes out as work on the other end. What can you do to reduce loss of energy to thermal energy?

4-3 Conservation of Energy **141**

check your UNDERSTANDING

1. Possible responses: **a.** dropping a ball; **b.** stretching a spring
2. You have to overcome friction as kinetic energy is transformed into thermal energy.
3. At the top of the hill, before the rock moves, it has potential energy but no kinetic energy. As it rolls, potential energy is converted to kinetic energy. As the rock hits the hillside, some kinetic energy is converted to thermal energy. Energy is not lost.
4. Cut down on friction by lubricating the parts.

4-3 Conservation of Energy **141**

3 ASSESS

Check for Understanding

Display pictures of different energy uses and have students work in small groups to identify the conversions illustrated. For example, a picture of a boulder rolling down a hill illustrates the conversion of potential energy to kinetic and sound energy. Use the Performance Task Assessment List for Group Work in **PASC**, p. 97. **COOP LEARN**

Reteach

Activity Roll a marble through a cardboard tube that is raised 1 cm at one end. Measure the distance the marble rolls after it leaves the tube. Place different types of materials at the bottom end of the tube. Ask students to predict the distance the marble will roll. Test the predictions and discuss the results. L1

Extension

Activity Ask students to design their own experiments using materials at hand to demonstrate in another way how energy is conserved as it changes from one form to another. L3

4 CLOSE

Ask students to summarize how kinetic energy, potential energy, and thermal energy interact to demonstrate the Law of Conservation of Energy. L1

Technology Connection

Purpose
Technology Connection reinforces Section 4-3 by explaining how people have tried to use the Law of Conservation of Energy to build a perpetual motion machine.

Content Background
Perpetual motion is impossible because some of the energy used to operate a machine is transferred to objects outside of the machine. Thus more energy must be put into the machine to keep it operating. For example, a car will not keep moving unless its supply of fuel is replenished. Even if the fuel seems inexhaustible, such as in solar powered cars, solar energy will be converted to thermal energy via friction and the car's parts will eventually wear out.

Teaching Strategy
Have students work in small groups to read and discuss the information on this page. Several group members may analyze for others why the perpetual motion machine illustrated on the page won't function forever.

Answers to
What Do You Think?

The wheel will eventually slow to a stop because of energy lost to the friction of the balls rolling within the spokes.

Going Further ⫸
Have small groups of students discuss any experiences they may have had using a machine or a toy that ran out of energy because of friction. Then have students brainstorm how to redesign the machine or the toy to decrease the amount of energy lost through friction. Use the Performance Task Assessment List for Group Work in **PASC**, p. 97. **COOP LEARN**

Technology Connection

The Search for Perpetual Motion

One idea that has fascinated inventors is the idea of a perpetual motion machine. This is a machine that, once set in motion, continues with no additional energy required. With the energy shortages in many parts of the world, this would be a wonderful accomplishment. But scientists know that this is impossible.

What's Wrong with Moving Forever?

The law of conservation of energy states that energy can neither be created nor destroyed—it can only be transformed into another form of energy. In other words, you can't get something from nothing. But what if energy could somehow be recaptured and used over and over? One example of such a machine is a battery that powers a motor that runs a generator that recharges the battery. It sounds good, but why would such a machine not continue forever?

There are moving parts in the device mentioned above. As parts of machines move against one another, the friction produces thermal energy that ends up being distributed through the particles in the air.

Some would say that people who try to invent perpetual motion machines are foolish and are ignoring the laws of science. Do you think that this is necessarily true? What, if anything, might these people accomplish?

What Do You Think?

Here's another idea for a perpetual motion machine. Can you find the problem with it? Remember that this is only an idea for a perpetual motion machine. This machine was not necessarily built!

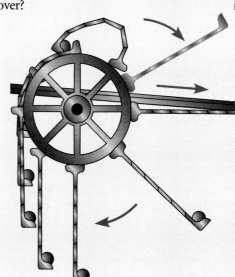

The wheel shown in the diagram is supposed to turn forever because it is always heavier on one side than the other. Steel balls are held in curved spokes that unwind as they reach the top of the wheel. The ball rolls toward the rim, forcing the right side of the wheel down. As the arm moves past horizontal, the ball rolls back to the end. Will the wheel keep turning, once started?

HOW IT WORKS

Vacuum Cleaner

Sometimes "home energy conservation" refers to the use of appliances to conserve human energy. Try telling that to the users of the first vacuum cleaners. One early model, invented around 1908, was made of steel and weighed a hefty 60 pounds. Today's cleaners usually weigh between 6 and 30 pounds, but they work on the same principle as the 1908 model.

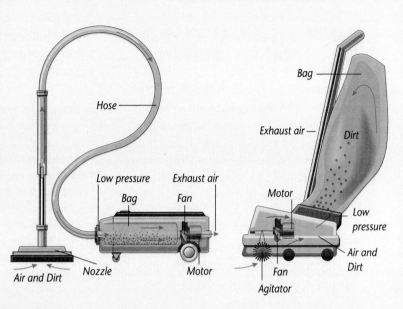

Low-Pressure Cleaning

A fan driven by an electric motor blows air through the unit. The moving air creates an area of low pressure. This reduces pressure inside the bag and hose. Since air pressure is now greater than the pressure at the nozzle, dirt and dust are gathered in the air that rushes in to even out the pressure. The fan forces this dirt-filled air into a bag, where the dust is trapped, and the air is blown out.

Tanks and Uprights

There are two main types of vacuum cleaners—canisters (or tanks) and uprights. A canister vacuum cleaner has a long, flexible hose that ends in a detachable nozzle. Usually a variety of nozzles come with the vacuum

cleaner as attachments. The body of this type of cleaner contains a bag and a powerful fan. Dirt is forced into the bag through the hose.

An upright vacuum cleaner has a small fan in its base. The base of the machine also contains an agitator—a rotating cylinder covered with bristles that loosen dirt. Dirt is forced upward into a bag attached to the vacuum cleaner's handle. In addition, there are vacuum cleaners that have both a canister unit for strong suction and an agitator in the nozzle.

Science Journal

In your Science Journal, make a list of some "labor-saving" devices. If labor is work, and work requires energy, are these devices really "savers"? Explain.

What's in a Name?

HISTORY CONNECTION

Purpose

History Connection supports Section 4-1, by reinforcing the relationship between work and energy.

Content Background

A wide variety of scientists have been honored by having their names used to describe a unit of measurement. For example, Marie Curie and Wilhelm Roentgen's names are used to describe units in the field of radiation. Volta, Ohm, and Ampere's names were given to the volt, ohm, and ampere, all units of electricity.

Teaching Strategies

This section is appropriate for students to read by themselves or with a partner. Students might work together in groups to come up with a list of scientists described in the first part of What Do You Think.

Science Journal

In addition to Joule, some scientists honored by having their names turned into words in reference to their discoveries are Marie Curie, Anders Angstrom, Andre Ampere, Alessandro Volta, James Watt, Georg Ohm, William Kelvin, Wilhelm Roentgen, Michael Faraday, Louis Pasteur, and Isaac Newton. In addition, a number of chemical elements derive their names from those of scientists. Among those elements are: curium, einsteinium, fermium, lawrencium, mendelevium, and nobelium.

Students' descriptions of their imaginary discoveries will vary.

HISTORY CONNECTION

Can you imagine how a scientist would answer this question? You've learned that in science, names are very important. However, sometimes for convenience, scientists substitute a shorter name for one that is longer. One example is the use of the term joule when referring to the newton meter. Who was Joule, and how did he come to have a unit of work named after him?

James Prescott Joule

James Prescott Joule was an English brewer who lived from 1818 to 1889. His hobby was physics. During the 1840s, he put the law of conservation of energy to a thorough test. He believed that if the law applied to all work and all forms of energy, then it had to be shown that one form of energy could be converted into another, quantitatively.

Energy Conversions

In other words, in energy conversions all energy must be accounted for—no energy should be lost in the process, and no energy created. Joule also measured thermal energy produced by an electric current, the friction of water against glass, and so on. He found that a fixed amount of one kind of energy was converted into a fixed amount of another kind. In fact, energy was neither lost nor created. It is in his honor that we call a unit of work a joule. His ideas were so fundamental to today's understanding of work and energy that most countries use the SI unit joule as the unit of energy.

Science Journal

It is common to name scientific quantities after the men and women who have made great contributions to science. In addition to Joule, make a list of other scientists honored in this way.

In your Science Journal imagine that your name is commonly used to describe an important quantity. Describe the quantity you discovered.

Science Journal

Review the statements below about the big ideas presented in this chapter, and answer the questions. Then, re-read your answers to the Did You Ever Wonder questions at the beginning of the chapter. *In your Science Journal*, write a paragraph about how your understanding of the big ideas in the chapter has changed.

1 Work is energy transferred through force and motion. When a force acts in the direction of motion, work is done. *You pick up a box, transfer it to a cart, push the cart across the room, pick up the box, carry it to a ramp, and push the box down a ramp. Decide when work is being done and when work is not being done. Explain.*

2 Kinetic energy is the energy of motion. Potential energy is the stored energy of position. Although there are many types of energy, all energy exists in these two forms. *List three common examples of both kinetic and potential energy.*

3 Energy can change from one form to another, but it cannot be created or destroyed. *Explain what happens to the energy of a soapbox derby car as it starts at the top of a hill, rolls down the hill, and eventually comes to a stop.*

Have students work in groups to illustrate the main ideas of the chapter.

1. Work is done both times you pick up the box, when you push the cart, and when you push the box down the ramp. Work is not done when you carry the box. Work is done only when a force is applied to an object and the object moves in the direction of the force.

2. Examples of kinetic energy may include a thrown ball, a person skating, and a ball rolling down a hill. Examples of potential energy may include a stretched rubber band, a wound-up spring, and books on the edge of a table.

3. The car has maximum potential energy at the top of the hill. Once the car starts to roll down the hill, potential energy is converted to kinetic energy. As the car continues rolling down the hill, its kinetic energy increases and its potential energy decreases. The car comes to a stop when frictional forces change the remaining kinetic energy of the car to thermal energy.

GLENCOE TECHNOLOGY

MindJogger Videoquiz

Chapter 4 Have students work in groups as they play the Video-quiz game to review key chapter concepts.

Science Journal

Did you ever wonder...

• A football stores energy when a force causes its shape to change. (p. 130)

• The energy you use is transferred to an object that you work on. (pp. 126–127)

• Some of the kinetic energy is changed to thermal energy, which is then no longer available to move the ball. (p. 133)

Project

Have students work in cooperative groups to build a device that transfers energy from one form to another. They might begin with a rolling marble, for example, and add components, such as levers or turnstiles, that are moved by the marble. Challenge students to see which device can do the most work or perform the greatest number of transfers of energy. **COOP LEARN** L2

Using Key Science Terms

1. In everyday speech, work refers to physical labor, a task to be completed, or a job. When scientists use the term work, they are referring to the motion of an object produced by a force acting on the object in the direction of its motion.

2. Explanations should include the idea that energy cannot be created or destroyed, but energy conversions can take place.

3. Kinetic energy is energy of motion. Potential energy is stored energy.

Understanding Ideas

1. Work is being done on the worm. The amount depends on the weight of the worm and how far the bird moves it. Work does not include the horizontal distance the bird carries the worm to her chicks. Work is measured in joules.

2. Work = force × distance = 18 N × 1.5 m = 27 joules

3. The formula $E_k = ½ \, mv^2$ shows that kinetic energy is directly related to mass and velocity. Thus, if either the mass or velocity of an object changes, kinetic energy changes.

4. The potential energy of water behind a dam can be increased by allowing the water behind the dam to rise much higher than the water below the dam.

5. Gravitational potential energy = weight × height = 6 N × 2.5 m = 15 J

6. The tennis ball has kinetic energy due to its motion and gravitational potential energy due to its height above the ground.

7. Some of the potential energy (20 J) is converted to thermal energy as the object rolls.

Developing Skills

1. See answers on student page.

2. Because the ball falls from a greater height, it has more kinetic energy when it strikes the

---

U sing Key Science Terms

kinetic energy   potential energy
law of conservation   work
  of energy

1. How is the scientific definition for work different from other work definitions?

2. Briefly explain the law of conservation of energy.

3. What is the difference between kinetic energy and potential energy?

U nderstanding Ideas

Answer the following questions in your Journal *using complete sentences.*

1. A robin picks up a worm and carries it to her chicks. On what is work being done? How much? What unit is used to measure work?

2. How much work is done when you lift an 18-N object 1.5 m off the ground?

3. How do velocity and mass affect an object's kinetic energy?

4. How might the potential energy of water behind a dam be increased?

5. What is the gravitational potential energy of an object weighing 6 N, lifted 2.5 m?

6. When a tennis ball sails over a net, what kind of energy does it have? Explain.

7. The gravitational potential energy of an object is 300 J. The maximum measured kinetic energy of the object rolling down a ramp is 280 J. How do you explain the difference between the two energies?

D eveloping Skills

Use your understanding of the concepts developed in this chapter to answer each of the following questions.

1. Concept Mapping Complete the concept map of energy.

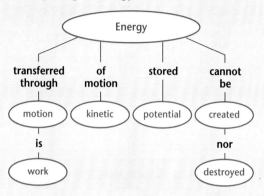

2. Observing and Inferring Repeat the Explore activity on page 123 holding the ball 2 meters above the stake. How did the results of this trial differ from the first trial? What caused the difference?

3. Forming a Hypothesis If you repeated the Find Out activity on page 132 on a surface of smooth ice, how would the results change?

4. Making Models Remake your model from the Find Out activity on page 134 to be the same as the group's whose marble traveled the greatest distance. Will the models work in the same way, so all the marbles travel the same distance? Can your group find a way to increase the distance traveled even more?

---

stake, and thus strikes it with greater force, driving the stake deeper.

3. Friction would be less of a factor. Therefore, less force would be needed to move the box and less work would be done on the box of books.

4. Changes to the model will vary with the differences in the models. Distances will probably vary slightly due to differences in workmanship and operation.

Program Resources

Review and Assessment, pp. 23–28 [L1]
Performance Assessment, Ch. 4 [L2]
PASC
Alternate Assessment in the Science Classroom
Computer Test Bank [L1]

Critical Thinking

In your Journal, *answer each of the following questions.*

1. List the examples of kinetic energy, gravitational potential energy, elastic potential energy, and thermal energy found in the picture below.

2. On Monday morning, you see a small weed growing through a crack in the sidewalk. On Thursday, you notice that the weed has grown bigger. How does this prove that the weed has used energy?

3. Which required more work: lifting one 100-N weight 3 m off the ground, or lifting each of three 50-N weights 2 m off the ground?

Problem Solving

Read the following problems and discuss your answers in a brief paragraph.

Jake loves swimming and wants to be a coach. He doesn't see how science can help swimmers.

1. How could the definition of work help him advise a swimmer how to hold his or her hands as he or she strokes through the water?

2. How could an understanding of potential energy and kinetic energy help a diver know when to flex his or her knees and when to leave the board?

CONNECTING IDEAS

Discuss each of the following in a brief paragraph.

1. **Theme—Energy** A dancer lifts a 400-N ballerina 1.4 m off the ground and holds her there for 5 seconds. How much work did he do?

2. **Theme—Stability and Change** In Chapter 1, you learned that opposing forces can cancel each other. Does this mean that their energies also cancel each other? Explain.

3. **A Closer Look** Why did the roll-back toy eventually come to a stop?

4. **Earth Science Connection** What are the advantages of using the slingshot effect in planning and executing the path of space vehicles such as *Voyager*?

5. **Technology Connection** Describe the obstacles facing the development of a perpetual motion machine.

✔ Assessment

Portfolio Review the portfolio options that are provided throughout the chapter. Encourage students to select one product that demonstrates their best work for the chapter. Have students explain what they learned and why they chose this example for placement into their portfolios.

Additional portfolio options can be found in the following **Teacher Classroom Resources:**
Making Connections: Integrating Sciences, p. 11

Multicultural Connection, p. 12
Making Connections: Across the Curriculum, p. 11
Concept Mapping, p. 12
Critical Thinking/Problem Solving, p. 12
Take Home Activities, p. 10
Laboratory Manual, pp. 13–16
Performance Assessment [P]

Critical Thinking

1. Kinetic energy—motion of the gymnast; gravitational potential energy—height of the gymnast above the ground; elastic potential energy—bending of the bar; thermal energy—heat produced by friction of gymnast's hands on the bars.

2. All work requires energy. The weed used energy to increase in size and to grow upward against the force of gravity.

3. An equal amount of work is done in both cases—300 joules.

Problem Solving

1. Jake must develop training techniques that will help his swimmers increase the amount of force they apply to the water, or decrease friction between them and the water.

2. The greater height a diver can reach, the better his or her technique can be. Therefore, a diver must know how to land on the end of the diving board to achieve maximum potential energy of body and board.

Connecting Ideas

1. $W = F \times d = 400 \text{ N} \times 1.4 \text{ m} = 560 \text{ J}$, rounded to 1 significant digit = 600 J.

2. No. Energy can't be created or destroyed. However, effects of energies can be canceled out as energies change forms.

3. All the energy stored in the rubber band was eventually changed to thermal energy due to friction.

4. Fuel can be conserved as gravitational forces provide additional energy to the spaceship.

5. A vacuum cleaner motor reduces air pressure inside the appliance. Air at higher pressure outside the appliance rushes in to equalize the pressure, carrying dirt and dust with it. Sucking on a straw reduces air pressure inside the straw. Air outside the straw pushes a liquid up the straw.

Chapter Organizer

SECTION	OBJECTIVES	ACTIVITIES & FEATURES
Chapter Opener		**Explore!**, p. 149
5-1 Simple Machines (2 sessions, 1 block)	1. **Identify** the six types of simple machines. 2. **Describe** how each simple machine makes completing a job easier. 3. **Explain** why machines do not reduce the amount of work that must be done. **National Content Standards: (5-8) UCP2, UCP5, A1, B2**	**Explore!**, p. 151 **Find Out!**, p. 153 **Find Out!**, p. 155 **Skillbuilder:** p. 156 **Leisure Connection**, p. 173
5-2 Mechanical Advantage (3 sessions; 1.5 blocks)	1. **Operationally define** mechanical advantage. 2. **Calculate** the mechanical advantage of several machines. **National Content Standards: (5-8) UCP2-3, A1, B2**	**Find Out!**, p. 158 **Investigate 5-1:** pp. 162–163 **A Closer Look**, p. 160
5-3 Using Machines (4 sessions; 2 blocks)	1. **Recognize** the simple machines that make up a compound machine. 2. **Describe** the relationship between work, power, and time. **National Content Standards: (5-8) UCP2, UCP5, A1, B2, E1-2, F1, F5**	**Explore!**, p. 166 **Skillbuilder:** p. 168 **Find Out!**, p. 169 **Design Your Own Investigation 5-2:** pp. 170–171 **Life Science Connection**, p. 168 **Science and Society**, p. 174 **Teens in Science**, p. 176

ACTIVITY MATERIALS

EXPLORE!

p. 149* 2 books, ruler or pencil
p. 151* spoon with handle, can with lid
p. 166 can opener

INVESTIGATE!

pp. 162-163* sheet of stiff paper, 20 cm × 28 cm (8 1/2 × 11 in); quarter, dime, nickel, balance, metric ruler

DESIGN YOUR OWN INVESTIGATION

pp. 170-171* meterstick, bathroom scale, watch with second hand or digital chronometer, flight of stairs

FIND OUT!

p. 153* brick, ring and ring stand, 2 pieces of string, pulley, spring scale
p. 155* board with a hole drilled partially through it, screwdriver, wood screw
p. 158* 2 or 3 books, meterstick
p. 169 stiff board, stack of books, wind-up toy car, meterstick, stopwatch, pan balance

KEY TO TEACHING STRATEGIES

The following designations will help you decide which activities are appropriate for your students.

L1 Basic activities for all students
L2 Activities for average to above-average students
L3 Challenging activities for above-average students
LEP Limited English Proficiency activities
COOP LEARN Cooperative Learning activities for small group work
P Student products that can be placed into a best-work portfolio
Activities and resources recommended for block schedules

Need Materials? Call Science Kit (1-800-828-7777).

OUT OF TIME? We recommend that students do the activities with an asterisk.

Chapter 5 Machines

TEACHER CLASSROOM RESOURCES

Student Masters	Transparencies
Study Guide, p. 19 **Critical Thinking/Problem Solving**, p. 13 **Science Discovery Activities**, 5-1 **Concept Mapping**, p. 13 **Laboratory Manual**, pp. 21–26, Pulleys	**Teaching Transparency 9**, Simple Machine **Section Focus Transparency 13**
Study Guide, p. 20 **Multicultural Connections**, p. 13 **Take Home Activities**, p. 11 **Making Connections: Across the Curriculum**, p. 13 **Activity Masters**, Investigate 5-1, pp. 23–24 **Science Discovery Activities**, 5-2	**Section Focus Transparency 14**
Study Guide, p. 21 **Critical Thinking/Problem Solving**, p. 5 **Making Connections: Integrating Sciences**, p. 13 **Multicultural Connections**, p. 14 **Making Connections: Technology & Society**, p. 13 **Science Discovery Activities**, 5-3 **How It Works**, p. 8 **Laboratory Manual**, pp. 17–20, Work and Power **Activity Masters**, Design Your Own Investigation 5-2, pp. 25–26	**Teaching Transparency 10**, Compound Machine **Section Focus Transparency 15**

ASSESSMENT RESOURCES	TEACHING & TECHNOLOGY
Review and Assessment, pp. 29–34 **Performance Assessment**, Ch. 5 **PASC*** **MindJogger Videoquiz** **Alternate Assessment in the Science Classroom** **Computer Test Bank**	**Spanish Resources** **Cooperative Learning Resource Guide** **Lab and Safety Skills** **Science Interactions, Course 2, CD-ROM** **Computer Competency Activities**

*Performance Assessment in the Science Classroom

NATIONAL GEOGRAPHIC TEACHER'S CORNER

National Geographic Society Products Available From Glencoe

To order the following products for use with this chapter, contact your local Glencoe sales representative or call Glencoe at 1-800-334-7344:

• *STV: Electricity and Simple Machines* (Videodisc)

Additional National Geographic Society Products

To order the following products for use with this chapter, call the National Geographic Society at 1-800-368-2728:

• *Everyday Science Explained* (Book)
• *Simple Machines* (Video)

GLENCOE TECHNOLOGY

The following multimedia resources are available from Glencoe.

Science and Technology Videodisc Series (STVS)
Physics
  Bipedal Robot
Human Biology
  Tricycle for the
  Handicapped

National Geographic Society Series
STV: Human Body Volume 2

Glencoe Physical Science Interactive Videodisc
Machines and Forces

Physical Science CD-ROM

Teacher Classroom Resources

These are key components of the classroom resources package.

TEACHING AIDS

Section Focus Transparencies

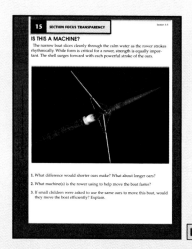

Teaching Transparencies

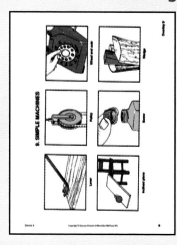

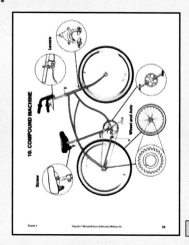

HANDS-ON LEARNING

Science Discovery Activity*

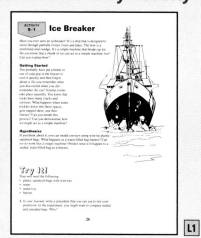

Laboratory Manual*

Take Home Activity

Chapter 5 Machines

REVIEW AND REINFORCEMENT

Study Guide*

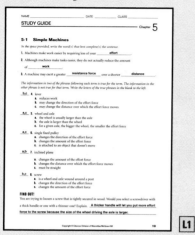

Concept Mapping

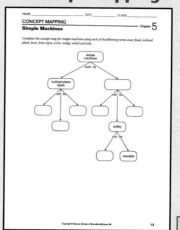

Critical Thinking/Problem Solving

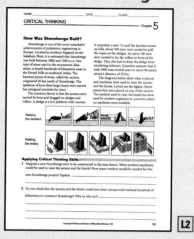

ENRICHMENT AND APPLICATION

Integrating Sciences

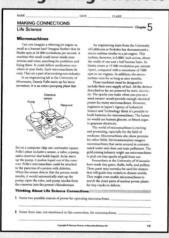

Across the Curriculum

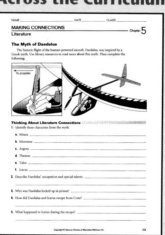

Technology and Society

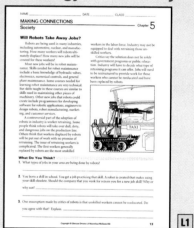

Multicultural Connection**

ASSESSMENT

Performance Assessment

Review and Assessment

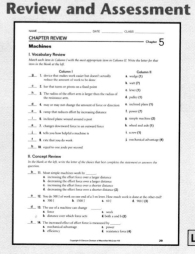

CHAPTER 5

Machines

THEME DEVELOPMENT

The themes that this chapter supports are energy and systems and interactions. Students learn how simple machines are used to perform work through the transfer of mechanical energy. Students learn to calculate the mechanical advantage of different systems of machines.

CHAPTER OVERVIEW

Simple machines make many everyday jobs easier. Calculating mechanical advantage is one way to decide how much a machine helps in completing a task. The power of a machine describes the machine's ability to accomplish work in a period of time. Compound machines are machines that combine two or more simple machines.

Tying to Previous Knowledge

In Chapter 4, students learned that both force and motion are involved when work is done. In this chapter, students will see how machines can be used to transmit force and help us to do work.

INTRODUCING THE CHAPTER

Ask the class to imagine a camping trip. Have them tell what they would take with them. Write the items on the board and circle the machines. Write "simple machine" next to each item that will be covered in the first section of this chapter.

00:00 OUT OF TIME?

If time does not permit teaching the entire chapter, use the Chapter Overview on this page, Reviewing Main Ideas at the end of the chapter, and the Chapter 5 audiocassette to point out the main ideas of the chapter.

MACHINES

Did you ever wonder...

✓ How using a crowbar helps you move a large rock?

✓ Why a screwdriver makes it easier to drive a screw into wood?

✓ Why mountain roads twist and turn?

Science Journal

Before you begin to study machines, think about these questions and answer them *in your Science Journal*. When you finish the chapter, compare your journal write-up with what you have learned.

Answers are on page 177.

148

The steep road to the campground curves back and forth as it twists up the mountain. Everyone helps set up camp. You pound tent stakes into the ground with a hammer, while a friend fills water jugs at the outdoor faucet. Two older campers use a crowbar to move a large rock that's in the way of a tent. A counselor is splitting a log with an ax.

It may not seem obvious, but these campers are using several machines. How many can you identify? We use machines to make jobs easier. Can you imagine trying to pound in a nail without a hammer or chop wood without an ax? What if you had to climb the flagpole to get the flag to the top?

▶ *In this chapter, you'll explore some simple machines, and find out how much easier they can make the jobs you do.*

Learning Styles		
Kinesthetic	Explore, p. 149; Find Out, pp. 153, 155, 158; Activity, p. 164	
Visual-Spatial	Explore, p. 151; Demonstration, pp. 152, 156, 159; Visual Learning, pp. 152, 154, 156, 159, 172; Activity, p. 154; Multicultural Perspectives, p. 163	
Interpersonal	Activity, p. 167	
Logical-Mathematical	Activity, p. 160; Visual Learning, pp. 161, 164, 167; Investigate, pp. 162-163, 170-171; Across the Curriculum, p. 164; Discussion, pp. 165, 167; Find Out, p. 169; Demonstration, p. 171	
LS **Linguistic**	Discussion, pp. 150, 152, 157, 161; Activity, p. 157; A Closer Look, pp. 160-161; Explore, p. 166; Life Science Connection, pp. 168-169	

How can a machine make it easier to move an object?

Books may not weigh very much, but how can you make it easier to lift them? Can a machine help?

What To Do

1. Stack two books on a flat desk or table.

2. Place your fingertips under the bottom book and lift the books with your fingers.

3. Using a ruler, pencil, or other materials, devise an easier way to lift the books. Experiment to find the easiest way you can.

4. *In your Journal,* describe the methods you used. Do you think your device is a type of machine?

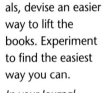

Explore!

How can a machine make it easier to move an object?

Time needed Ten minutes

Materials 2 books, ruler

Thinking Process recognizing cause and effect

Purpose To show how a simple machine can make work easier.

Teaching the Activity

Discussion Have each student perform the activity and then ask the students to describe the use of the ruler. Do they consider the ruler to be a machine? What type? **L1**

Science Journal Have students note how the ruler helped them move the books.

Expected Outcome

Students will find that less effort was required to lift the books with the ruler.

Answer to Question yes

✔ Assessment

Content Have students re-read the first paragraph of the text on page 148. Have them tell which of the activities is most like using a ruler to lift the books. *using a crowbar to move a large rock* Then have students suggest other similar activities that they have performed and justify their answers. Use the Performance Task Assessment List for Group Work in **PASC,** p. 97. **COOP LEARN**

ASSESSMENT PLANNER

PORTFOLIO
Refer to page 179 for suggested items that students might select for their portfolios.

PERFORMANCE ASSESSMENT
Process, pp. 163, 166, 169, 171
Skillbuilder, pp. 156, 168
Explore! Activities, pp. 149, 151, 166
Find Out! Activities pp. 153, 155, 158, 169
Investigate!, pp. 162–163, 170–171

CONTENT ASSESSMENT
Check Your Understanding, pp. 157, 165, 172
Reviewing Main Ideas, p. 177
Chapter Review, pp. 178–179

GROUP ASSESSMENT
Opportunities for group assessment occur with Cooperative Learning Strategies.

PREPARATION

Planning the Lesson

Refer to the Chapter Organizer on pages 148A–D.

Concepts Developed

This section is about simple machines and how they are used to perform work. Six types are introduced: lever, wheel and axle, pulley, inclined plane, screw, and wedge.

1 MOTIVATE

Bellringer

 Before presenting the lesson, display **Section Focus Transparency 13** on an overhead projector. Assign the accompanying **Focus Activity** worksheet. L1

LEP

Discussion Have students brainstorm simple jobs that are made easier by using a tool. Focus on identifying the task, the tool or device used, and how it functions. L1

2 TEACH

Tying to Previous Knowledge

Have students recall from Chapter 4 the scientific meaning of work. Point out that in this section they will learn how machines make work easier. Ask students if they have ever used a screwdriver to pry open a paint can. If so, they have used a simple machine (lever).

Student Text Questions

When campers use a crowbar to move a rock, what is the effort force? *The force the campers are exerting on the crowbar.* **What is the resistance force?** *The force exerted by the crowbar on the rock.*

 Simple Machines

Section Objectives

- Identify the six types of simple machines.
- Describe how each simple machine makes completing a job easier.
- Explain why machines do not reduce the amount of work that must be done.

Key Terms

effort force, resistance force, lever, wheel and axle, pulley, inclined plane, screw, wedge

Making Jobs Easier

Machines make work easier. Think back to the Explore activity at the beginning of the chapter. Although it wasn't very hard to lift the books with your fingertips, the device you invented made the job seem easier. The device you used to move the books was a machine. How does that machine compare to what you usually think of as a machine?

When a machine is used to do work, two kinds of force are involved. The force applied *to* the machine (the force you exert) is called the **effort force**. The force applied *by* the machine is called the **resistance force**. When campers use a crowbar to move a rock, what is the effort force? What is the resistance force?

Although machines make work easier, they do not actually reduce the amount of work that has to be done.

The weight of a rock doesn't change just because you use a crowbar to move it—the same amount of weight has to be moved. How can a machine make work easier if it doesn't reduce the amount of work that has to be done? Let's take a look at the six types of simple machines and find out.

■ Lever

Sitting around the campfire after dark seems like a perfect time for hot chocolate and toasted marshmallows. The hot chocolate can has one of those metal lids that fits tightly into the top. You find a spoon and slip the end of the handle under the edge of the lid. One push on the spoon, and the lid comes right off.

Could you have pried off the lid using just your fingers? Let's explore why it is easier with a spoon.

Figure 5-1

Another example of a common lever is a boat oar. Like the spoon used to open the hot chocolate can, the oar doesn't reduce the amount of work that has to be done, but it does increase the distance over which the force is exerted.

A When you use an oar, you exert an effort force on the oar.

Effort force

Fulcrum

Resistance force

150 Chapter 5 Machines

Content Background In this chapter, we have applied the terms *effort force* and *resistance force* to describe the dynamics of machines. The effort force is the force generally applied by a person to the machine. The resistance force is the force applied by the machine to something else. Commonly, these forces are called input and output forces, respectively. Whichever term is used, it is important that students are able to name both the source of the force and the object to which it is applied.

How does a lever work?

What To Do

1. Watch carefully as your teacher or classmate uses a spoon handle to pry the lid off a can. *In your Journal*, record how far he or she had to push down on the spoon. How far up does the tip of the spoon handle go to lift the lid? Are these two distances the same or different?

2. Again, watch as your teacher or a classmate pries the lid off the can. Does the spoon handle rest on a part of the can while the lifting is being done?

3. Look at the length of spoon between his or her hand and the can's edge. How does that length compare to the length of spoon between the can's edge and the lid? What direction is the force he or she exerts on the spoon? In which direction does the lid move?

A **lever** is a bar that turns or pivots on a fixed point called a fulcrum. What was the fulcrum for the spoon in the Explore activity? When you use a lever, such as the spoon handle, you exert a small force over a long distance. At the same time, the lever exerts a large resistance force over a short distance. Recall the definition of work you learned in Chapter 4:

force (N) × distance (m) = work (J)

Think about the Explore activity. Suppose you pushed down on the spoon with a force of 20 N, exerted over a distance of 0.1 m:

20 N × 0.1 m = 2 J

At the same time, the lever pushed up on the lid with a force of 200 N, over a distance of 0.01 m:

200 N × 0.01 m = 2 J

In each case, the work done was the same—2 J.

Connect to...
Life Science

Not only do humans *use* levers, but many parts of our bodies *are* levers. Think about how your bones and muscles move in your arm and identify the fulcrum, effort force and resistance force.

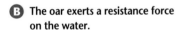
B The oar exerts a resistance force on the water.

C The resistance force pushes the boat through the water.

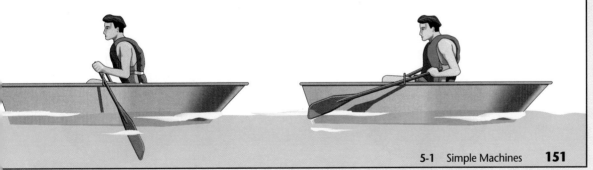

How does a lever work?

Time needed ten minutes

Materials A can with a pry-off lid (such as a cocoa can or tea tin), a spoon

Thinking Processes recognizing cause and effect

Purpose To observe how a force is applied to a lever.

Preparation Make sure the lid is tight on the can so that force must be exerted.

Teaching the Activity

Science Journal Have students note their observations of the forces involved in prying a can lid off. L1

Expected Outcomes

Students observe the pivoting action of the lever against the fulcrum and recognize the op-

---

Connect to . . .
Life Science

To help students relate levers to their body parts, use a model or skeleton to demonstrate examples of skeletal levers.

Answer: *The bones in our forearms, the radius and ulna, act together as a lever. The fulcrum is the elbow joint. The hand exerts the resistance force, while the effort force acts at the muscle's point of attachment.*

Uncovering Preconceptions

Some students may think the machines make work easier by reducing the amount of work that has to be done. Point out that machines do not change the amount of work that is done. Machines allow us to multiply our effort or speed but not our energy.

posing directions of the force on the lever and the work done on the lid.

Answers to Questions

1. The spoon has to be pushed down about 1 inch. The spoon handle goes up about 1/8 inch. The spoon has to be pushed down farther than the handle goes up.

2. The spoon rests on the can while the lifting is being done.

3. The length between the hand and the edge is greater than the length between the edge and the lid. The person exerts force downward on the spoon. The lid of the can moves upward.

✔ Assessment

Performance Provide students with many opportunities to manipulate levers. Examples would be sports equipment, such as a tennis racket, bat, or hockey stick; tools, such as a wrench, or a crowbar.

Allow ample time for students to locate the fulcrum and identify the effort arm and resistance arm in each lever. Use the Performance Task Assessment List for Group Work in **PASC**, p. 97.

GLENCOE TECHNOLOGY

CD-ROM

Science Interactions
Course 2, CD-ROM

Chapter 5

■ Wheel and Axle

There are two water faucets at the campground. It's your turn to fill the water jugs. When you get to the faucets, you find that the handle is broken off of one of them. Which would you use? Why?

Effort force

Resistance force

Figure 5-2

The handle of the water faucet is a wheel. The shaft attached to the center of the faucet wheel is an axle.

A When you use a wheel and axle, you exert a smaller effort force over a longer distance. The machine exerts a greater resistance force over a shorter distance.

The handle of the water faucet, like the knob of a doorknob, is a wheel. Each wheel rotates around its center. The shaft attached to the center of the wheel is an axle. You can think of a **wheel and axle** as a small wheel attached to the center of a larger wheel. The wheel and the axle always rotate together.

We have developed many uses for the wheel and axle. You already know this simple machine makes it easier to move cars, carts, wheelchairs, and wagons. But there are also other, less obvious uses for the wheel and axle. Can you think of any that you use or have seen?

B The bigger a wheel is, the longer the distance you must turn it to move the axle. But as turning distance increases, the amount of effort force needed to turn the wheel decreases. You need less effort force to turn a large wheel than you do to turn a small wheel.

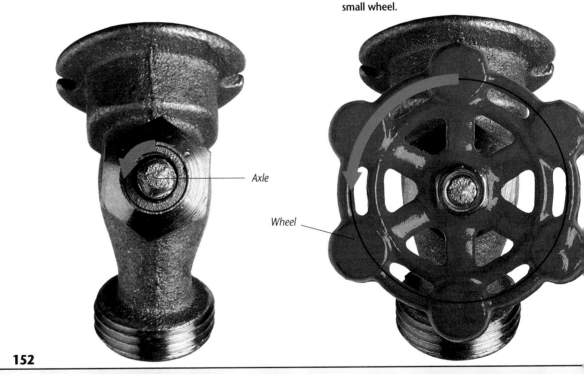

Axle

Wheel

152

■ Pulley

During a morning walk through camp, you find the ranger is preparing to raise the flag. You help unfold it and the ranger lets you hook the flag to the rope and hoist it up the flagpole. You pull down on one strand of rope while the flag goes up with the other strand. Is this easier than climbing up the flagpole? Have you done any work?

A **pulley** is a wheel that has a rope or chain passing over it. The pulley on the flagpole is the simplest kind of pulley. It's called a single fixed pulley because it's attached to something that doesn't move. Fixed pulleys change the direction of the force that's applied to the object, but not the amount of force. Can the force exerted to get a job done be reduced when a pulley is used? Let's find out.

Figure 5-3

Single fixed pulley

Find Out! ACTIVITY

Can a pulley reduce the force you have to exert to get a job done?

What To Do

1. Obtain a brick, spring scale, ring and ring stand, two pieces of string, and a pulley.

2. Tie one piece of string around the brick and lift it using the spring scale. Record the force required to lift the brick.

3. Tie one end of the other piece of string to the ring stand.

4. Place the pulley in the center of that piece of string as shown in the figure.

5. Attach the spring scale to the other end of the string, and hook the brick to the pulley.

6. Pull up on the spring scale. Record *in your Journal* the force required to lift

the brick using the pulley. Which requires more effort from you?

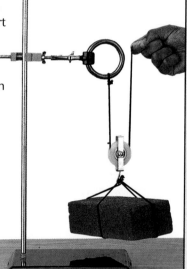

Conclude and Apply

1. How far did you pull up on the free end of the rope?

2. How far did the brick move?

3. Did the pulley reduce the amount of force you had to exert to move the brick?

4. What direction did you have to pull to lift the brick? How does this compare to the flagpole example?

You've just demonstrated how a single movable pulley works. A single movable pulley increases the effect of the effort force and reduces the

amount of effort you must exert to get a job done. But a single movable pulley does not change the direction of the force.

5-1 Simple Machines **153**

Find Out!

Can a pulley reduce the force you have to exert to get a job done?

TECH PREP

Time needed 20 minutes

Materials A brick, a ring and ring stand, two pieces of string, pulley, spring scale

Thinking Processes comparing and contrasting, recognizing cause and effect

Purpose To observe how a simple machine affects the force needed to do work.

Preparation Set up equipment at several stations and have students work in teams.

Teaching the Activity

Troubleshooting Make sure the knot will not come loose and the string will not break when the brick is lifted.

Demonstration After observing the movable pulley, attach a fixed pulley to the ring stand. Use the fixed pulley to raise the brick. Have students compare the direction of force with each

type of pulley. **L1** **COOP LEARN**

Science Journal Have students note their observations of the ways the two different pulleys affect the force needed to move the brick in their student journals.

Expected Outcome

Students will see that a single movable pulley does not change the direction of the force, but does reduce the effort required to get a job done, while a fixed pulley does the opposite. Thus, a fixed pulley is used for moving light objects, but a movable pulley is needed to move heavier objects.

Conclude and Apply

1. Several inches
2. Only about half as far
3. Yes
4. Up; to move the flag you would pull down.

✔ Assessment

Content Pulleys are generally used to either change direction of force or reduce the amount of effort force needed. A fixed pulley is attached to something that doesn't move, such as a flagpole or ceiling. Ask students which effect is obtained using a fixed pulley. *A fixed pulley changes the direction of an effort force.* Because the resistance arm and effort arm of a pulley are of equal length, a fixed pulley does not reduce the amount of effort force you use. A movable pulley is attached to the object being moved. The force of moving the object is divided between the two ends of the rope. Ask students what it does. *It reduces the amount of effort force you use.*

5-1 Simple Machines **153**

■ Inclined Plane

Suppose you need to lift a 100-pound box of camping gear a distance of 3 feet, from the ground into the back of a truck. Lifting the box straight up and into the truck would be difficult. But if you use a board to make a ramp, you could probably push the box up, even though you would have to exert force against the friction between the box and the board.

An **inclined plane** is a ramp or slope that reduces the force you need to exert to lift something. The inclined plane decreases the effort force. Does it change the direction of force?

Look at the diagrams in **Figure 5-4**. Describe the relationship between the length of the ramp and how much easier the job would be.

What are some of the inclined planes you see around you every day? Driveways? Ramps in parking garages? Most buildings with entrances that are above or below sidewalk level have a ramp in addition to steps. The ramp makes the building accessible to people in wheelchairs, and makes it easier for people to use carts to move heavy objects into and out of the building.

Figure 5-4

A Olivia decided to use an inclined plane to help her load boxes of camping gear into the truck. The inclined plane decreases the effort force Olivia needs by increasing the distance through which her effort force is applied. How could Olivia make her job even easier?

B After loading the first box, Olivia replaced the short ramp with a longer one. By making the slope more gradual, Olivia increased the distance she had to push the boxes, but she decreased the effort force she had to exert.

C To get to the mountaintop, the campers drove up a winding mountain road. A mountain road is a type of inclined plane that is used to raise cars up to the top of a mountain, much as Olivia's ramp was used to raise boxes up into the truck. Why do mountain roads take a zigzag course up the mountainside instead of going straight up?

Multicultural Perspectives

Building Pyramids

In ancient Egypt the pyramid builders were faced with the problem of how to lift the heavy stone blocks to enormous heights. Scientists have demonstrated that the method most likely used to build the pyramids would have used ramps. These ramps could have been built of mud brick and rubble, and sledges could have been used to drag the blocks up (wheeled transport was not used in the pyramid age). As the pyramid grew higher, the length of the ramp and the width of its base could have been increased to prevent it from collapsing. Several ramps approaching the pyramid from different sides were probably used.

■ Screw

You just learned that a winding mountain road is an inclined plane. Imagine a road that wound around a mountain like a spiral staircase, slowly making its way to the top. Now think of a screw. How could the winding road be compared to a screw? Is the screw a machine? What about the screwdriver used to turn the screw? Let's find out in this next activity.

Are screwdrivers and screws simple machines?

What To Do

1. Gather a board with a hole drilled partially through it, a screwdriver, and a wood screw.

2. Use the screwdriver to drive the screw part of the way into the hole.

3. Now try using your fingers to turn the screw into the hole. *In your Journal*, record your observations. Can you do it? Do you think you could have finished this task if the screwdriver did not have a handle?

Conclude and Apply

1. Is the screwdriver a simple machine? To decide, you need to compare what it does to what a simple machine can do. Did the screwdriver make the job easier?

2. Did the screwdriver reduce the amount of work that was accomplished?

3. What type of simple machine is the screwdriver? Figure 5-5 may help you decide.

4. What about the screw itself? What type of simple machine is a screw?

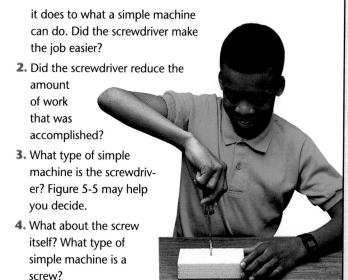

Figure 5-5

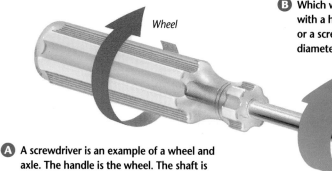

B Which would be easier to turn, a screwdriver with a handle larger in diameter than its shaft, or a screwdriver with a handle the same diameter as its shaft? Why?

Wheel

Axle

A A screwdriver is an example of a wheel and axle. The handle is the wheel. The shaft is the axle.

observations about the difference between turning a screw with a screwdriver and turning a screw with their fingers.

Expected Outcome

Students will see how the screwdriver is a type of simple machine classified as a wheel and axle.

Conclude and Apply

1. Yes; yes, using the screwdriver is easier.
2. No
3. Wheel and axle
4. A screw is an inclined plane wrapped around a post

✔ Assessment

Content Point out that a screw is an inclined plane that moves. Ask students how the motions of the object and the machine differ in a ramp and a screw. *Some inclined planes, such as ramps, stay in one place while objects move along their surfaces. With a screw, the object remains in one place while the inclined plane moves.*

Find Out!

Are screwdrivers and screws simple machines?

Time needed Fifteen minutes

Materials Board with hole partially drilled; a screwdriver; wood screw

Thinking Processes recognizing cause and effect, classifying

Purpose To identify characteristics of a simple machine.

Preparation Partial holes must be drilled into the boards; the hole should be only large enough to give the threads of the screw a bite into the wood. Make sure that the screwdrivers match the heads of the screws. Use a lightweight wood like pine. **L1**

Teaching the Activity

Safety Make sure that the wood is smooth to avoid splinters.

Science Journal Have students note their

Simple Machines	Changes Direction	Changes Force
Lever	Yes	Yes
Wheel and Axle	No	Yes
Pulley	Yes	Yes
Inclined plane	Yes	Yes
Screw	Yes	Yes
Wedge	Yes	Yes

1. All machines except wheel and axle

2. All machines except fixed pulley

3. All except fixed pulley and wheel and axle L1

Demonstration So students may observe that a jar lid is a type of screw show the students the mouth of a mayonnaise jar and its lid. Ask them which direction the lid must be turned to tighten.

Visual Learning

Figure 5-6 If the piece of string that is wound around the threads of this screw is unwound, which will be longer, the length of the string or the length of the screw? Why? *The length of the string because a screw increases the length of the incline to decrease the amount of effort needed to turn the screw.* **Are tops of bottles and jars screws? What about a light bulb? Why?** *Tops of bottles and jars and light bulbs are examples of screws because, like nuts and bolts, they have matching threads that fit together.*

GLENCOE TECHNOLOGY

Videodisc

The Infinite Voyage: The Search for Ancient Americans

Chapter 3
Ancient Weapons for Hunters

What about the screw itself? The ridges spiraling around a screw are called threads. As you drove the screw into the board, the threads seemed to pull the screw into the wood. Perhaps this leads you to describe a **screw** as an inclined plane wound around a post. In a way, the spiraling inclined plane helped to lift the wood up around the screw.

SKILLBUILDER

Making and Using Tables

Make three columns on a sheet of paper. Label them Simple Machines, Changes Direction, and Changes Force. In the first column, list the six simple machines. In the next two columns, write Yes or No. Then answer these questions. If you need more help, refer to the **Skill Handbook** on page 639.

1. Which machines change the direction of force?

2. Which machines multiply the effort force?

3. Which machines both change the direction of the effort force and increase the force?

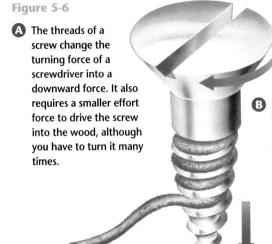

Figure 5-6

A The threads of a screw change the turning force of a screwdriver into a downward force. It also requires a smaller effort force to drive the screw into the wood, although you have to turn it many times.

Effort force

B If the piece of string that is wound around the threads of this screw is unwound, which will be longer, the length of the string or the length of the screw? Why?

Resistance force

Head

C This diagram shows how the screw's threads are like an inclined plane wrapped around the post of the screw.

Post

D Wood screws, sheet metal screws, and nuts and bolts are all examples of screws. Instead of sharp threads to cut into wood, bolts have rounded threads to match the threads on the inside of the nut. Are tops of bottles and jars screws? What about a lightbulb? Why?

Wood screw   *Sheet metal screw*   *Bolt and nut*

156   Chapter 5   Machines

Meeting Individual Needs

Learning Disabled So that students may more easily classify screws and wedges as inclined planes, cut two pieces of plain white paper in half diagonally to form a total of four right triangles. Starting with the narrow edge of one triangle, wrap it around a pencil to form a screw. When the pencil is removed, the paper will unwind slightly, but hold its spiral shape so that the students will observe the inclined plane of the spiral. Take two more triangles and tape together their bases to form a wedge. Have students compare the inclined plane, represented by the remaining triangle, the screw, and the wedge. Ask students to state the relationships between the inclined plane and the other machines. *The screw is a spiral inclined plane and the wedge is a double inclined plane.*

■ Wedge

Once more, think of the campers. In particular, think of the counselor who was splitting logs. Would it be easier to split a log with a baseball bat or with a sharp-edged axe? Why?

Figure 5-7 takes a closer look at an axe. Does it remind you of another simple machine? You can see that each side of the axe blade looks like an inclined plane. The blade of an axe is a wedge. A **wedge** is an inclined plane that uses the sharp, narrow edge to cut through materials.

Chisels, knives, the teeth of saw blades, and many other sharp-bladed tools use the wedge. Can you think of other types of wedges?

Even though the campers were on vacation, there was much work to do. Recall that work is done when a force is exerted to move an object. Luckily, the campers had some simple machines to help them—levers, wheels and axles, pulleys, inclined planes, screws, and wedges. The campers were probably too busy to think about effort or resistance forces. The next time you use a simple machine, try to identify these forces.

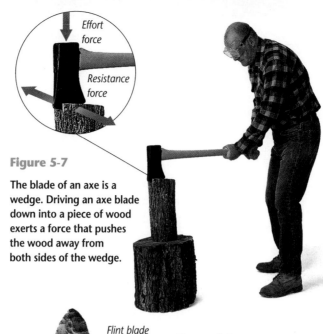

Effort force

Resistance force

Figure 5-7

The blade of an axe is a wedge. Driving an axe blade down into a piece of wood exerts a force that pushes the wood away from both sides of the wedge.

Flint blade dating from 20 000 B.C.E.

Figure 5-8

Wedge-shaped stone blades of different sizes were used by prehistoric people for a variety of cutting, scraping, and digging jobs. Some Middle Eastern cultures used a small wedge-shaped tool to press marks into clay tablets in a form of writing called cuneiform.

Sumarian cuneiform tablet from 2500 B.C.E.

check your UNDERSTANDING

1. Give one example not given in the text of each kind of simple machine.
2. How do simple machines make it possible for you to exert less force to get a job done?
3. Do any of the simple machines reduce the amount of work that must be done? Explain your answer.

4. **Apply** One counselor built a machine. A camper applied a downward effort force to one end of the machine. The force exerted by the other end of the machine was greater, and the direction of the force was up rather than down. What kind of machine did the counselor build? Explain.

check your UNDERSTANDING

1. Answers will vary, but may include nail clipper, window blinds, ramp, corkscrew, scissors, razor.
2. They can change a small effort force exerted over a longer distance into a larger resistance force exerted over a shorter distance.
3. No. The amount of work done does not change. What changes is the amount of force applied to do the work.
4. A lever or a combination of fixed and moveable pulleys. These two machines can both increase the force and change the direction of the force.

3 ASSESS

Check for Understanding

Have students answer the questions on their own and then discuss various answers in class. To help students answer the Apply question, suggest that they use arrows to diagram the forces involved. Use the Performance Task Assessment List for Group Work in **PASC**, p. 97. [L1]

Reteach

Activity Have students record in their journals every simple machine they use or observe someone else using in the course of a day or two. Have them note the name of the machine, how it was used, and how it reduces the amount of effort required to accomplish a purpose. [L1]

Extension

Discussion Help students compare the relative directions of effort and resistance force in various machines. In levers, pulleys, wheels and axles, and inclined planes, the effort and resistance forces are in the same or opposite directions. In screws, the resistance forces are perpendicular. For example, students should be able to say that the effort force applied to a wood screw is in the same place as the head of the screw. The resistance force is along the shaft of the screw. [L2]

4 CLOSE

Activity

Divide the class into small groups. Ask them to decide whether a zipper is a simple machine and to identify which type of machine it is. They should agree that the zipper is a simple machine, classify it as a wedge, and explain why it is a wedge.
[L1] **COOP LEARN**

5-2 Mechanical Advantage

PREPARATION

Planning the Lesson

Refer to the Chapter Organizer on pages 148A-D.

Concepts Developed

In the previous section, students learned about six simple types of machines and how they can be conceptualized into two broad categories: levers and inclined planes. In this section, students will develop an understanding of mechanical advantage and how to calculate the MA of different types of machines.

Section Objectives
- Operationally define mechanical advantage.
- Calculate the mechanical advantage of several machines.

Key Terms
mechanical advantage

How Much Does a Machine Help?

One morning, the counselors returned from gathering firewood. They were covered in mud! They told the story of how the truck got stuck, and after a few unsuccessful attempts to push it out, they decided to jack the truck up and put some logs under the stuck wheel. Unfortunately, they discovered they had left the jack back at camp. They decided to fashion a lever out of a sturdy branch. They found a large log and rolled it near the truck to use as a fulcrum. One of them found two possible branches to use as the lever. One was long while the other was short. Which one should they have used to get the most help from the lever? Try this next activity to find out.

Find Out!

Can the length of a lever affect the amount of force needed to do work?

Time needed 15 minutes

Materials two or three books and a meterstick

Thinking Processes observing and inferring

Purpose To analyze how the distance between effort force and fulcrum affect the mechanical advantage of a machine.

Teaching the Activity

Demonstration If spring scales are available, have students measure and compare the effort force needed to lift the books at the 30-cm and 90-cm marks. L1

Science Journal Have students note observations regarding the effects of the length of a lever in their journals.

Expected Outcome

Students will observe that less effort force is needed at the 90-cm mark.

Conclude and Apply
1. at the 30-cm mark
2. The closer the effort force is

Find Out! ACTIVITY

Can the length of a lever affect the amount of force needed to do work?

What To Do

1. Place a stack of two or three books near the edge of a table.
2. Slide a meterstick under the books so the entire width of the bottom book is resting on the stick, as shown in the figure.

3. Grasp the meterstick at the 30-cm mark. Push up on the meterstick to lift the books. Try again, this time holding the meterstick at the 90-cm mark. Record your observations *in your Journal.*

Conclude and Apply

1. Compare the force you used to lift the books each time. Which time did you need more force to lift the books?
2. How is the distance between the effort force and the fulcrum related to the amount of effort force needed?

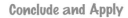

158   Chapter 5   Machines

to the fulcrum, the more effort force is needed.

✔ Assessment

Performance Ask students to reread the information at the top of the page about the stuck van. Then ask whether they think the short branch or the long branch will be the more effective in getting the wheel out of the mud. *The long branch will probably be better.*

Program Resources

Study Guide, p. 20
Multicultural Connections, page 13, Early Native American Hunters L1
Take Home Activities, p. 11, Brains vs. Brawn L1
Making Connections: Across the Curriculum, p. 13, *The Myth of Daedalus* LEP
Section Focus Transparency 14

Now that you've done the activity, which branch would you have chosen? The counselors decided the long one would give them the most help. After they had lifted the truck and driven it out of the mud, the counselors decided to figure out exactly how much the lever helped them.

They wanted to calculate the mechanical advantage of the machine they used. The **mechanical advantage (MA)** of a machine tells you how much the effort force is multiplied.

Figure 5-9

The branch-lever allows the counselors to lift the truck with less effort force than they would need to exert using just their bodies. The lever provides help with the lifting. But how much help? You can measure the amount of help a machine provides by calculating its mechanical advantage.

You can calculate this by dividing the resistance force by the effort force.

$$MA = \frac{\text{Resistance force}}{\text{Effort force}}$$

The larger the mechanical advantage, the more help the machine provides.

Don't forget, a machine does not change the amount of work that is done. It just increases the force on the object.

Calculating the mechanical advantage of a machine can help us make decisions. By comparing the MA of two different machines, we can decide which machine we would need to use to get the job done.

Effort force
500 N

Lever

Fulcrum

Resistance force
2500 N

A The mechanical advantage of a machine is the number of times the machine multiplies the effort force. The formula for finding mechanical advantage is:
mechanical advantage
(MA) = resistance force (F_r) divided by effort force (F_e).

$$MA = F_r/F_e$$

B The counselors applied an effort force of 500 N to the branch. The branch-lever applied a resistance force of 2500 N to the truck. Calculate the mechanical advantage of the branch-lever.

$$MA = F_r/F_e = MA = 2500 \text{ N}/500 \text{ N}$$
$$MA = 5$$

C The branch-lever multiplied the force exerted by the counselors 5 times. This means the branch-lever made the job of lifting the truck 5 times easier for the counselors.

5-2 Mechanical Advantage **159**

1 MOTIVATE

Bellringer

 Before presenting the lesson, display **Section Focus Transparency 14** on an overhead projector. Assign the accompanying **Focus Activity** worksheet. [L1]
LEP

Remind students that machines do not decrease the amount of work to be accomplished. They do, however, decrease the amount of force required to do the work.

2 TEACH

Tying to Previous Knowledge

Have students recall the concepts of effort force and resistance. Remind them of removing the lid from the can with a spoon. The effort force was applied to the bowl of the spoon. The resistance force was the force needed to lift the lid.

Uncovering Preconceptions

Students may equate multiplying force with multiplying the amount of work done. However, machines cannot multiply work. When a machine multiplies effort force, the distance that force must travel is greater than the distance traveled by the resistance force.

Visual Learning
Figure 5-9 Ask students if there might be an easier way for the people in the picture to exert an effort force on the lever. *They could stand on the lever if it were flat enough.*

ENRICHMENT

TECH PREP

Demonstration Help students conclude that each strand of a single, movable pulley system supports half the resistance force.

Materials needed are a single pulley, two spring scales, small weight, and a meterstick.

Weigh the pulley and a small attached weight. Attach a spring scale to each end of the cord that passes through the pulley and sus- pend the pulley and weight from the scales. Remove one spring scale and attach that end of the string to a fixed support. Have a volunteer read the spring scale supporting the pulley.

Stand a meterstick behind the pulley system and have a volunteer measure the distance the weight moves as you pull the spring scale upward 20 cm. From the values of distance effort and distance resistance, help students calculate that the MA of a single, movable pulley is 2. [L3]

Theme Connection Each type of simple machine is a system with a specific equation for calculating its mechanical advantage. Each equation is related to the distance the effort force moves, divided by the distance the resistance force moves. As students learn how to calculate the mechanical advantage of different machines, have them identify the forces and distances involved as each is used in the calculation. Point out that knowing how to calculate the mechanical advantage of different simple machines allows you to combine them to invent more complex machine systems and interactions.

Activity Have the students work in groups to solve the following problem: A worker uses a board that is 4.0 m long to pry up a large rock. A smaller rock is used for a fulcrum and is placed 0.5 m from the resistance end of the lever. **What is the MA of the lever? If an effort force of 250 N is needed, what is the weight of the large rock?** *The effort arm of the lever is 4.0 m - 0.5 m = 3.5 m. Therefore, the MA = effort arm/ resistance arm = 3.5 m/0.5 m, or MA = 7. For any machine, MA = resistance force/effort force. So, 7 = resistance force/250 N. Resistance force = 250 N × 7 = 1750 N.* COOP LEARN L2

Figure 5-10

Suppose you did not know the amount of resistance force or effort force. Could you still calculate the MA of a lever? Yes, MA can also be calculated by dividing the length of the effort arm (l_e) by the length of the resistance arm (l_r). The formula for this calculation is $MA = l_e/l_r$.

A Apply the formula to determine the mechanical advantage of the branch-lever shown here and the branch-lever on page 161. Which provides the most help with lifting the truck?

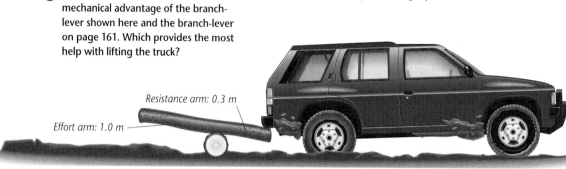

Resistance arm: 0.3 m

Effort arm: 1.0 m

Engineers use mechanical advantage as they design machines. They need to know how long to make a lever, or how steep an inclined plane should be to do the job they need done.

But you don't have to be an engineer to use mechanical advantage. Have you ever played on a seesaw with

Efficiency

When you squeeze the brake handles of the bicycle in the picture, the rubber brake pads clamp down against the rim of the tires, using friction to exert a force on the wheels and slow their speed. Friction occurs whenever two substances rub against each other.

Lost Effort

Some of the effort force put into any machine with moving parts must overcome friction. For example, some of the effort force you exert when you pedal a bike must overcome the friction of the pedal gear rubbing against the bicycle chain. The energy

used by a machine to work against the force of friction is lost rather than being able to do useful work. This reduces the efficiency of the machine. Efficiency is a comparison of how much work a machine can do and how much work must be put into the machine. You can figure out the efficiency of a machine by dividing the work output by the work input, then multiplying this by 100.

$$\text{efficiency} = \frac{W_{out}}{W_{in}} \times 100\% = \frac{F_r \times d_r}{F_e \times d_e} \times 100\%$$

Efficiency is usually expressed as a percentage. Low-efficiency machines lose much of the

Purpose

This excursion will reinforce students' understanding of the concepts of mechanical advantage and the ability of machines to do work, which is discussed in Section 5-2.

Content Background

A brake is a device that slows or stops the movement of a wheel, an engine, or an entire vehicle. Most brakes have a fixed part, called a brake shoe or block, that

presses against a turning wheel to create friction. The friction causes the wheel to stop or slow down.

There are three kinds of brakes that are used in most vehicles, mechanical, hydraulic, and air brakes. Bicycles use mechanical brakes which operate by using a lever that exerts force against the wheel, causing it to slow down or stop.

Teaching Strategy

Have students observe the operation of

someone who weighed a lot more or less than you? What did you have to do to balance your weights? Most likely, you had to move the seesaw so the heavier person was on the shorter side. You changed the mechanical advantage of the seesaw (which is a lever) by making one of its arms longer than the other. What other situations can you think of when knowing about mechanical advantage would help?

In the Investigate activity that follows, you will discover another use for mechanical advantage.

B What science principle about the relationship between lever length and mechanical advantage do these two examples demonstrate?

Effort arm: 3.0 m

Resistance arm: 0.3 m

work put into them to work against friction—high-efficiency machines do not. A machine that does no work against friction would have an efficiency of 100 percent. That is, 100 percent of the work put into the machine is used to do useful work. No machine operates at 100 percent efficiency because all machines must work against friction of some kind.

Boosting Efficiency

You can increase the efficiency of a machine by adding a lubricant, such as oil or grease, to the surfaces that rub together. If a bicycle's chain, gears, and other moving parts are cleaned and lubricated periodically, the bicycle will operate more efficiently. Also keeping the tires properly inflated will reduce friction between the road and tires.

📝 **Science Journal**
   The efficiency of an automobile affects its gas mileage. How can changing the engine oil increase the gas mileage of the automobile?

5-1 Measuring Mass With Levers

Planning the Activity

Time needed 20 minutes

Purpose To use mechanical advantage to find the mass of an object.

Process Skills observing and inferring, comparing and contrasting, recognizing cause and effect, measuring in SI

Materials 1 sheet of stiff paper, 20 cm × 28 cm (8½ in × 11 in), 3 coins (quarter, dime, and nickel), balance, metric ruler

Preparation Fold papers and tape the edges so the paper doesn't pop up and make the coins slide off when you get to step 5. Have students work in groups and ask one or two students in each group to do the calculations while the others take turns doing the activity with the different coins.

Teaching the Activity

Process Reinforcement Ask students to explain mass and what it measures. *the amount of material contained in an object, causing it to have weight in a gravitational field* L1

Possible Procedures Because of the multiple steps in the procedure, it is advisable to demonstrate the activity before students divide into groups.

Possible Hypotheses Students may hypothesize that the length of the effort arm increases when the resistance force increases.

INVESTIGATE!

Measuring Mass with Levers

You have seen that mechanical advantage can be calculated using either the forces or the length of the arms. Let's investigate another way that mechanical advantage can be used.

Problem
Can you measure mass with a lever?

Materials
1 sheet of stiff paper,
    20 cm × 28 cm (8 1/2 × 11 in)
3 coins (quarter, dime, and nickel)
balance
metric ruler

What To Do

1 Fold the paper in half lengthwise, then fold it lengthwise again, to make a lever 5 cm wide by 28 cm long.

2 Mark a line 2 cm away from one end of the lever. Label this line resistance.

3 Slide the other end of the lever over the edge of a table until the lever begins to teeter but doesn't fall off. Mark a line across the lever at the point where it crosses the table edge. Mark this line effort.

4 Measure the mass of the lever to the nearest 0.1 g. Write this mass on the effort line.

5 Center a dime on the resistance line.

Program Resources

Activity Masters, pp. 23–24, Investigate 5-1

Science Discovery Activities, 5-2

Integrated Science Videodisc, The Great Bicycle Challenge

ENRICHMENT

Have students construct a model of a seesaw using wood or other materials and work out a formula for balancing objects of different weights on the seesaw.

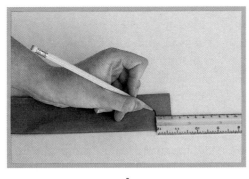

A

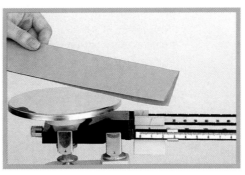

B

6 Once again, slide the lever over the edge of the desk until it teeters. Mark a line where the lever crosses the table edge and label it fulcrum #1.

7 *Measure* the length of the resistance arm from the center of the dime to the new fulcrum, and the effort arm from the new fulcrum to the effort line. Measure to the nearest 0.1 cm.

8 *Calculate* the MA of the lever. Multiply the MA by the mass of the lever to find the mass of the coin.

9 Repeat Steps 5 through 8 with the nickel and then with the quarter. Mark the fulcrum line #2 for the nickel and #3 for the quarter.

Analyzing

1. Is the total length of the lever a constant or a variable?

2. Describe the length of the effort arm and resistance arm. Identify whether the lengths are constant or variable.

3. *Infer* what provides the effort force.

4. What does it mean if the MA is less than 1?

Concluding and Applying

5. Is it necessary to have the resistance line 2 cm from the end of the lever?

6. *Going Further* Why can mass units be used in place of force units in this kind of problem?

Expected Outcomes

Students will understand how to use mechanical advantage to calculate the resistance force of a lever.

Answers to Analyzing/ Concluding and Applying

1. A constant

2. Lengths are variable.

3. Gravity (the weight of the lever hanging off the edge of the table)

4. The resistance force is less than the effort force.

5. No

6. Because both the resistance and effort forces are exerted by gravity and thus, the proportions of the masses are the same as the proportions of the weights.

✔ Assessment

Process Ask students to identify the variables in the Investigate experiment. *the lengths of the effort arm and the resistance arm and the mass of the coins* What is the constant? *the length of the lever* Then ask students to hypothesize how the results would differ if a different lever, such as a plastic ruler, were used instead of the folded paper. Use the Performance Task Assessment List for Analyzing the Data in **PASC**, p. 27. **COOP LEARN**

Multicultural Perspectives

Lubrication

Machines must be lubricated to reduce friction between moving parts. Elijah McCoy, an African American engineer, invented a device that automatically lubricates machines. He was working for a railroad company, when he noticed that machines could be lubricated only while they were not working. McCoy devised an oil-filled cup that would drip oil onto the machinery while it was moving. To analyze how McCoy's invention worked have interested students research the device, then draw and label a picture of the device. Use the Performance Task Assessment List for Scientific Drawing in **PASC**, p. 55. **L2** **P**

Activity

Activity Supply the students with silhouettes of several common wheels and axles, such as screwdrivers, house keys, and beaters from an electric mixer; or ask them to trace similar items at home and bring the tracings to class. Have them cut out the silhouettes or tracings, fold each along the center of the wheel and axle, unfold the paper, and measure the approximate radius of the wheel and the radius of the axle for each item. They should then discuss the significance of their findings. L2

Across the Curriculum

Mathematics

Have students use calculators to determine the MA of each item they measured for the activity above and relate this value to the task that the item is used for.

Inquiry Question An automobile steering wheel having a radius of 24 cm is used to turn the steering column, which has a radius of 4 cm. **What is the MA of this wheel and axle?**

$$MA = \frac{r_w}{r_a} = \frac{24\ cm}{4\ cm} = 6$$

Content Background

The mechanical advantages discussed in this section are *ideal* mechanical advantages of simple machines. The two terms are synonymous if we assume that no mechanical energy is converted to thermal energy.

Visual Learning

Figure 5-12 **What is the MA of the multiple pulley system?** *MA = 5* **From these examples, what can you infer about the relationship between number of supporting ropes and mechanical advantage?** *MA of a pulley system increases as number of supporting ropes used increases.*

If you had put two dimes on the resistance point, what would have happened to the fulcrum? What about four dimes? If you had done this Investigate activity with a number of different masses, you could have plotted a graph of the relationship between the length of the effort arm and the mass. This is a direct relationship. As one increases, the other increases, assuming the length of the lever stays constant. Using this relationship, you could find the length of a lever needed to lift different masses. What would happen to the effort force if you moved the fulcrum even closer to the resistance force?

Figure 5-11

What kind of simple machine is the ice cream maker? Do you know a formula to calculate its mechanical advantage?

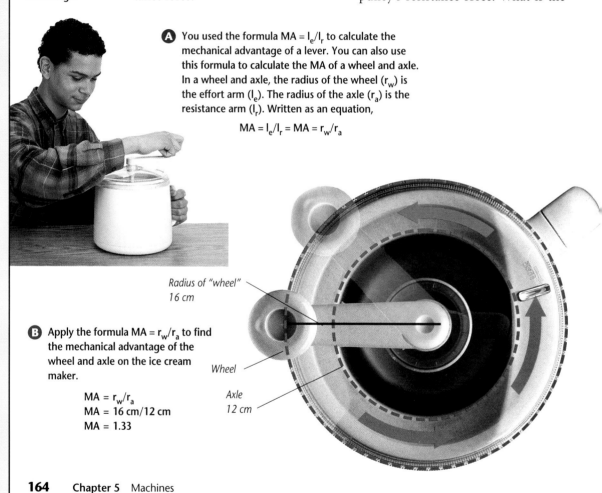

A You used the formula MA = I_e/I_r to calculate the mechanical advantage of a lever. You can also use this formula to calculate the MA of a wheel and axle. In a wheel and axle, the radius of the wheel (r_w) is the effort arm (I_e). The radius of the axle (r_a) is the resistance arm (I_r). Written as an equation,

$$MA = I_e/I_r = MA = r_w/r_a$$

Radius of "wheel"
16 cm

Wheel

Axle
12 cm

B Apply the formula MA = r_w/r_a to find the mechanical advantage of the wheel and axle on the ice cream maker.

MA = r_w/r_a
MA = 16 cm/12 cm
MA = 1.33

Use another method to calculate the MA of a wheel and axle. The radius of the wheel is the effort arm, and the radius of the axle is the resistance arm. Look at **Figure 5-11** for an example.

Does every machine offer a mechanical advantage? What about a machine that does not increase the force, but only changes the direction of force? Assume you're raising a flag a distance of 10 m to the top of a flagpole. You pull down on 10 m of rope in order to raise the flag 10 m. The effort arm is 10 m long, and the resistance arm is 10 m long. Your effort force moves the same distance as the pulley's resistance force. What is the

---

Meeting Individual Needs

Learning Disabled Have students give an example of a simple machine they have used recently and answer the following questions. **How did you apply effort force? How did the machine apply resistance force?** *Accept all reasonable answers. Be sure students identify a simple machine and indicate an understanding of the effort force and how the machine changes that force.*

Figure 5-12

A The ropes in pulley systems can be strung in a variety of ways. Some pulleys change direction of the force, but offer no mechanical advantage. Other pulleys multiply the effort force. You can estimate the mechanical advantage of a pulley by counting the number of supporting ropes. A supporting rope is one that leads upward from the load.

MA of the pulley? The fixed pulley has a mechanical advantage of one. It just makes raising the flag more convenient.

Mechanical advantage tells you the ratio between the resistance force and the effort force. The larger the mechanical advantage, the more help the machine is providing. You calculate MA by dividing resistance force by effort force.

You could use this knowledge to explain to the other campers why they wouldn't necessarily need the biggest,

B The single fixed pulley has an MA of 1. The single movable pulley has an MA of 2. What is the MA of the multiple pulley system? From these examples, what can you infer about the relationship between number of supporting ropes and mechanical advantage?

strongest campers to get the truck out of the mud. All they would need is a simple machine with a high mechanical advantage.

---

check your UNDERSTANDING

1. Define mechanical advantage.
2. An automobile steering wheel with a radius of 40 cm is used to turn the steering column, which has a radius of 4 cm. What is the mechanical advantage of the wheel and axle system?

3. **Apply** Each screwdriver in a set has a handle with a radius that is different from every other screwdriver in the set. The radius of each axle is also different. Which screwdriver would you use to get the greatest mechanical advantage?

---

check your UNDERSTANDING

1. the ratio of the resistance force to the effort force, or the effort arm to the resistance arm
2. MA = 40 cm/4 cm = 10
3. the one that has the largest ratio of handle radius to axle radius

---

Check for Understanding

Assign questions under Check Your Understanding. Have students support their answers to the Apply question by drawing a diagram of the wheel and axle. Use the Performance Task Assessment List for Group Work in **PASC**, p. 97.

Reteach

Group Work Do the following problem as a group activity. Formula: $MA = F_r/F_e$ Problem: A carpenter uses a claw hammer to pull a nail from a board. The nail has a resistance of 2500 N. The carpenter applies an effort force of 125 N. **What is the mechanical advantage of the hammer?** $\frac{2500\ N}{125\ N} =$ *20* **L1**

Extension

Calculation Machines with an MA of less than one are used to increase the distance an object moves or the speed at which it moves. Have students consider a tennis racket with a MA of ½. If a player swings the racket with a force of 60 N, the racket will deliver a force of only 30 N to the ball. **If the effort force delivered by the player to the handle is delivered at a speed of 5 m/s, what is the speed at which the ball will strike the racket?** *10 m/s* **L3**

4 CLOSE

Discussion

The mechanical advantage of any ideal pulley or pulley system is equal to the number of ropes that support the resistance weight. Have students solve the following problem: A person uses a block and tackle to lift an automobile engine that weighs 1800 N. The person must exert a force of 300 N to lift the engine. **How many ropes support the engine?**

Formula: $\frac{\text{resistance force}}{\text{effort force}} = MA$

$\frac{1800\ N}{300\ N} = 6$ **L1**

PREPARATION

Planning the Lesson

Refer to the Chapter Organizer on pages 148A–D.

Concept Development

In this section, students will apply what they have learned about simple machines to develop an understanding of compound machines. Compound machines are systems containing two or more simple machines. The relationship between work, power, and time is also explored. The rate at which a machine transfers energy is measured by its power.

Explore!

What kind of machine is a can opener?

Time needed 10 minutes

Materials traditional can openers

Thinking Processes observing and inferring, recognizing cause and effect

Purpose To formulate a model of a compound machine being composed of two or more simple machines.

Preparation To save time and increase interest, collect several different models of hand-crank can openers.

Teaching the Activity

Discussion Ask students to identify the simple machines in a can opener. *lever, wheel and axle, and wedge* L2

Science Journal Have students try to describe the way a can opener works in their journals.

Expected Outcomes

Students will recognize simple machines and see how they combine to make a complex tool.

 Using Machines

Compound Machines

Section Objectives
- Recognize the simple machines that make up a compound machine.
- Describe the relationship between work, power, and time.

Key Terms
compound machine
power

It's getting close to dinnertime. You hum a favorite song as you put paper plates, mustard, and potato chips on the picnic table. This is the best possible night to pull cooking duty. The menu says chili dogs, so all you have to do to prepare dinner is put a few things on the table and heat up some canned chili. Everyone will roast his or her own hot dog over the campfire. You pull out the can opener and start working on the chili.

Explore! ACTIVITY

What kind of machine is a can opener?

What To Do

1. Using what you know about simple machines, examine a can opener.

2. *In your Journal*, try to describe how it works. What kind of simple machines can you find? Is it a combination of simple machines?

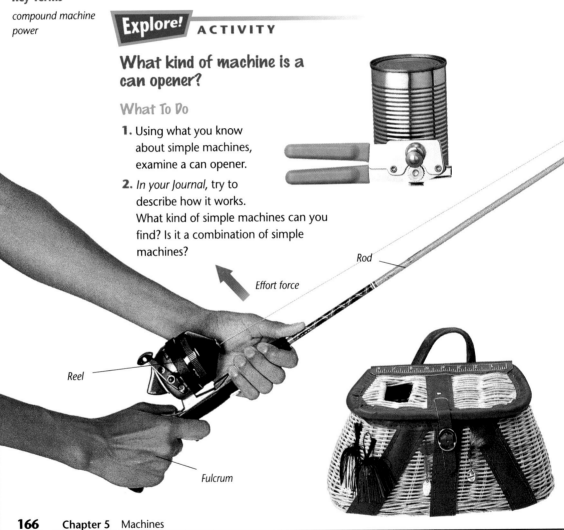

Rod

Effort force

Reel

Fulcrum

Answers to Questions

2. Lever, wheel and axle, wedge; yes

✔ Assessment

Process Have students draw a diagram of the can opener they observe, labeling each of the simple machines that is part of the can opener. Use the Performance Task Assessment List for Scientific Drawing in **PASC**, p. 55.

COOP LEARN P

Figure 5-13

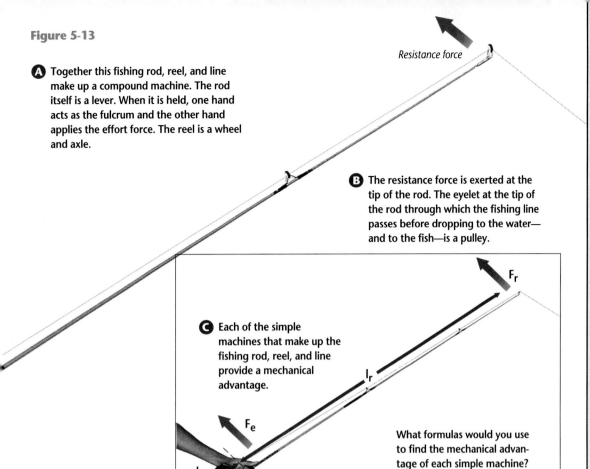

Resistance force

A Together this fishing rod, reel, and line make up a compound machine. The rod itself is a lever. When it is held, one hand acts as the fulcrum and the other hand applies the effort force. The reel is a wheel and axle.

B The resistance force is exerted at the tip of the rod. The eyelet at the tip of the rod through which the fishing line passes before dropping to the water—and to the fish—is a pulley.

C Each of the simple machines that make up the fishing rod, reel, and line provide a mechanical advantage.

F_r

I_r

F_e

I_e

What formulas would you use to find the mechanical advantage of each simple machine?

As you can see, a can opener is actually composed of several simple machines. The handles are two levers that make it easier to fasten the opener onto the edge of a can. The crank is a wheel and axle that turns a toothed wheel. The toothed wheel is called a gear. The first gear turns another gear that moves a circular wedge along the top of the can. The can opener combines the lever, the wheel and axle, and the wedge into a compound machine that makes it easier to open a can.

A **compound machine** is a combination of simple machines that makes it possible to do something one simple machine alone can't do. Can you think of some common compound machines? Look at **Figure 5-13** to see another common compound machine. A fishing rod and reel may look simple, but as far as machines go, it's compound! Before reading on, challenge yourself to identify the simple machines in the fishing rod and reel. If that's too easy, try identifying the machines in a bicycle!

DID YOU KNOW?

The tin can was invented early in the 1800s, but the can opener wasn't invented until more than a hundred years later. At first, people had to use a hammer and chisel to open tin cans!

ENRICHMENT

TECH PREP

Activity Have students brainstorm a list of 20 compound machines that can easily be identified as containing simple machines. Assign each student to research the invention of one of the machines listed. Then have students work as a group to create a timeline that depicts the machines and the date each was invented. L2

Program Resources

Study Guide, p. 21
Critical Thinking/Problem Solving, p. 5, Flex Your Brain
Teaching Transparency 10 L1
Making Connections: Integrating Sciences, p. 13, Micromachines L2
Section Focus Transparency 15

---

1 MOTIVATE

Bellringer

 Before presenting the lesson, display **Section Focus Transparency 15** on an overhead projector. Assign the accompanying **Focus Activity** worksheet. L1

LEP

Discussion Have students identify the simple machines that make up some common devices, such as a pencil sharpener, a rotary beater, a food mill, and a bicycle. L1

2 TEACH

Tying to Previous Knowledge

Have students recall what a simple machine does. In this section, they will see how combining simple machines also makes work easier.

Visual Learning

Figure 5-13 What formulas would you use to find the mechanical advantage of each simple machine? *Lever: MA = I_e/I_r; Pulley: MA = number of supporting lines; Wheel and axle: MA = r_w/r_a.*

Discussion To introduce the concept of power, pose the following questions. **Suppose two students are asked to unpack identical cartons of books and place them on identical sets of shelves. One student completes the job in 10 minutes, whereas the other takes 20 minutes. Which student worked "harder"? Which student did more work?** Students are likely to say the first student worked "harder," although they may admit that both students did the same amount of work. Develop the idea of power, as the rate of work, by asking students what they mean when they say someone worked "harder" at a task.

1. Feet apply force to pedals.

2. Pedals move in a circle around axle.

3. Force is transferred from pedals to wheels through gears and chain.

4. Wheels exert resistance force on road surface. L1

Find Out!

Can you measure the power of a toy car?

Time needed 20 minutes

TECH PREP

Materials three or four books, a stiff board, stopwatch, meterstick, balance, and wind-up toy car

Thinking Processes observing, forming a hypothesis, designing an experiment to test a hypothesis, isolating and controlling variables, interpreting data.

Purpose To determine the power of a compound machine.

Preparation Any wind-up toy that will travel up an inclined plane can be used.

Teaching the Activity

In order to complete the Conclude and Apply section of the activity, students must understand that power is the rate at which energy is converted.

The weight of the car is calculated by multiplying the mass, in kilograms, by 9.8 N. Measure the height of the inclined plane in meters. One joule = N · m. L3

Power

You and a friend borrow bikes from the park ranger for the afternoon. You and your friend weigh the same, and your bikes are identical. After pedaling several miles, you begin getting tired. You round a curve and find you're facing a half-mile of steep hill. You manage to pedal all the way up, while your friend has to get off and push the bike. You get to the top of the hill before your friend.

Did you both do the same amount of work? How do you know? Work = force × distance. Since force and distance (the height of the hill) were equal for you and your friend, your work was equal. The only variable was time. You reached the top of the hill first. You did the work faster.

If you divided the amount of work you did by the amount of time to do the work, you would know how much power was used. **Power** is the work done divided by the time interval. The formula for calculating power is

$$power = work/time$$

SKILLBUILDER

Sequencing

Make an events chain to show the sequence of how some of the simple machines in a bicycle work together to move the bicycle. Start with the feet applying force to the pedals. If you need help, refer to the **Skill Handbook** on page 636.

Life Science CONNECTION

The Machine/Body Interface

Have you ever tried to ride a bike that was the wrong size for you? It probably didn't go very fast, and it may have given you a backache. Or maybe you find that most bikes give you a backache. If so, you need to study ergonomics.

The art and science of making machines suitable for the people who use them is called ergonomics. From computers to office chairs, ergonomics makes many machines more comfortable to use. Applying ergonomic principles to bicycles results in new designs.

Look at the recumbent bicycle shown here. In what ways is it more ergonomically designed than an upright bike?

A Speedy, Comfortable Ride

The first thing people notice about recumbent bikes is the rider's posture. The seat is like a recliner. It gives firm support to

Life Science CONNECTION

Purpose

The Life Science Connection extends the discussion of machines by looking at machines that are designed with attention to user comfort as well as efficiency.

Content Background

The designs of many commonly used machines, such as bicycles and automobiles, have evolved from historical designs. With cars, for example, the design evolved from horse-drawn carriages and not from a study of the most efficient ways to design a car.

The science of ergonomics is growing increasingly important. Many types of workers, including assembly line workers, word processors and data-entry clerks, and musicians, suffer from repetitive strain injuries—injuries that result from performing a particular motion over and over. Many people also suffer from back pain and injuries partly due to poorly designed seating

Can you measure the power of a toy car?

What To Do

1. Place one end of a stiff board on a stack of books to create an inclined plane. Experiment with a wind-up toy car and the steepness of the plane until you find the angle at which the car will travel up the incline at the slowest possible speed.

2. Wind up the car, place it at the bottom of the plane, and then record *in your Journal* the number of seconds it takes to get to the top.

3. Measure the height of the inclined plane in meters. The height is the distance straight up from the floor or table to the top of the incline.

4. Measure the weight of the car in newtons (N). (A kilogram weighs about 9.8 N on Earth's surface.)

5. Multiply the weight of the car by the height of the incline to calculate the work done, in joules (J). Now divide J by the number of seconds it took for the car to climb the incline. Your result is the power of the car. It's expressed in a unit called the watt (W).

Conclude and Apply

What was the power of your toy car?

Find Out!

Materials stiffboard, stack of books, wind-up toy car, meterstick, stopwatch, pan balance

Science Journal Have students develop a hypothesis about the steepness of the plane their cars can travel and note it in their journal. Then have them make notes on their efforts to prove or disprove the hypothesis.

Conclude and Apply
Students should compare the power calculated for their toy car with the power of other students' cars or with the manufacturer's advertised power (if available).

✔ **Assessment**

Process Ask students to design and then carry out an experiment that would allow them to measure the power of two other machines. Use the Performance Task Assessment List for Designing an Experiment in **PASC**, p. 23.
COOP LEARN

the rider's back and lets him or her make strong forward thrusts to the pedals—so strong that every world land speed and endurance record for human-powered vehicles is held by recumbents.

The biggest advantage of recumbents is their aerodynamic efficiency. In comparison, the rider of an upright bike wastes energy fighting wind drag. Since most of the force a bicycle rider exerts is used to overcome wind resistance, a recumbent can save a lot of energy.

The recumbent rider's nearly reclining position has another benefit, too. It's easier to take deep breaths when you're leaning back than it is when you're crouching over underslung handlebars. Many recumbent bikes have their handlebars below the seat; this helps efficiency by leaving the rider's arms in a natural resting position.

Recumbents can easily be modified for physically challenged riders. For instance, foot pedals can become hand cranks. Paraplegics, amputees, some quadriplegics, and those adapting to the effects of multiple sclerosis or polio can ride modified recumbents and experience speed and maneuverability under their own muscle power.

Science Journal
Examine the recumbent bicycle shown here. Compare it to an upright bike in terms of air resistance. Discuss *in your Science Journal* the kind of bike that would create more drag and thus lose energy to friction. Think of a way to make the bike with less drag even more streamlined.

GLENCOE TECHNOLOGY

Videodisc

Science Interactions, Course 2 Integrated Science Videodisc

Lesson 2

---

and work space. Manufacturers and workplace designers are giving greater emphasis to ergonomics as they redesign products and work spaces.

Teaching Strategy
Show students pictures of bicycles from different eras, from the earliest large-wheel bicycles to modern designs. Have students hypothesize why the designs evolved as they have.

Science Journal
The upright bike would have greater air resistance and thus create more drag and lose energy to friction. Students may suggest to add an aerodynamic shell or alter the components or rider position so that there is less frontal area.

Going Further ▐▌▌▐▌▐▌
Have small groups of students discuss any experiences they have had using a machine that was well designed in contrast to one that was poorly designed. Encourage students to redesign a frequently used machine to make it more ergonomic. Use the Performance Task Assessment List for Group Work in **PASC**, p. 97. COOP LEARN

5-2 Calculating Power

Preparation

Purpose To calculate the amount of power used to do a job.

TECH PREP

Process Skills sequencing, making and using tables, recognizing cause and effect

Time Required 30-40 minutes

Materials See reduced student text

Safety Precautions Stress to students that this is not a race. Students should walk, not run, up the stairs. If you have students with disabilities that prevent them from climbing stairs, allow them to participate fully by recording and keeping time or alter the activity to have students go up a ramp or lift an object a certain height.

Possible Hypotheses Students may hypothesize that the power needed to climb the stairs will increase according to the weight of the person climbing the stairs and the speed with which the person climbs.

Plan the Experiment

Process Reinforcement Have students review the scientific definitions of *force, work,* and *power.*

Possible Procedures Measure the weight of a volunteer, multiplying by 4.45 to get newtons. The height of the stairs can be calculated using the height of one step and the number of steps in the flight. Time the volunteer as he or she walks up the steps. Calculate work and power.

Teaching Strategies

Troubleshooting You may need to talk about the difference between pounds and newtons and conversion factors as the students make their calculations.

Science Journal Have the students record all of their data and observations in their journals.

Calculating Power

After climbing a long flight of stairs, your heart beats a little faster, your legs may burn, and you probably breathe harder. You don't have to tell your body it just used some power! Is the power used the same for people of different weight and for climbing different numbers of steps? Is the power used the same if you walk or if you run?

Preparation

Problem
What factors determine the power used in going up a flight of stairs?

Form a Hypothesis
Is the amount of power required to climb a flight of stairs affected by weight, distance, and time? How?

Objectives
• Determine the factors that affect power.
• Calculate power using experimental data.

Materials
meterstick
stopwatch or watch with second hand
flight of stairs
bathroom scale

Program Resources

Laboratory Manual, pp. 17–20, Work and Power L3

Activity Masters, pp. 25–26, Design Your Own Investigation 5-2

Multicultural Connections, p. 14 L2

How It Works, p. 8, Going Up? L3

Making Connections: Technology & Society, p. 13 LEP

Science Discovery Activities, 5-3

DESIGN YOUR OWN
INVESTIGATION

Plan the Experiment

1 What variables will you measure for this test? What will remain constant? What will change?

2 How will you calculate the work done? How will you calculate power? What units must your measurements be in?

3 Who will walk or run the stairs? Will you test more than one individual?

4 Write out your plan *in your Science Journal.* Prepare a data table to record your data for each trial.

Check the Plan

1 Does your plan adequately test your hypothesis? What else do you need to include?

2 Check your plan with your teacher and make any suggested revisions. Carry out your experiment.

Analyze and Conclude

1. Compare and Contrast Compare and contrast the force exerted by different people. Did time play a factor in the force exerted?

2. Infer Who used the most power? Infer what factors affected power used. Did your experiment support or refute your hypothesis?

3. Using Math Determine whether the following variables are directly or

inversely dependent upon each other: power and weight, power and time, work and weight, time and work.

4. Predict Suppose the heaviest climber walked up the stairs much more quickly than the lightest climber. Predict the difference in power used by each person. Explain.

Going Further

Two people with different weights each decrease their time to climb the stairs from 10 seconds to 8 seconds. Which person has the greatest increase in power?

5-3 Using Machines **171**

Expected Outcomes

Students will see that power depends on the amount of resistance and the time needed to move it.

Analyze and Conclude

1. The greater the weight, the greater the force exerted. Time did not make a difference.

2. The heaviest and/or quickest person used the most power. Weight and time were the determining factors. Answers will vary.

3. Directly, inversely, directly, not dependent

4. The heavier person would use much more power because of more weight and less time.

Going Further

The one with the heaviest weight. The increase in speed creates a greater increase in power for the heavier person than for the lighter one.

✔ Assessment

Process Ask students to identify the variables and constants in this activity. Then have them formulate a different experiment in which the constant in this experiment (the distance) becomes the variable and a variable in this experiment (the force) becomes a constant. Use the Performance Task Assessment List for Designing an Experiment in **PASC**, p. 23.

ENRICHMENT

Demonstration Have students analyze the elements of a machine and identify their functions by observing this demonstration. Set up a meterstick as a lever with a fulcrum located at the 0 cm mark. Hang a heavy weight at the 33-cm mark and connect a single, movable pulley to the meterstick with a wire tied at the 99-cm mark. Raise and lower the lever using the pulley string. The resistance force of the pulley is actually driving the lever. Have students calculate its

MA. $MA = l_e/l_r = 99 \text{ cm}/33 \text{ cm} = 3$ Have students recall that the MA of a movable pulley is 2. Attach a spring scale to the pulley string and record the force needed to raise and lower the lever. Have a volunteer measure the weight of the heavy weight at the 33-cm mark. From these values, have students calculate the MA of the compound machine. **L3**

5-3 Using Machines **171**

What simple machines are examples of a wheel and axle? *The water faucet, the wheelbarrow, fishing reel* **Does the pulley used to raise the flag offer a mechanical advantage?** *No.* **What simple machine is an example of an inclined plane?** *The axe blade and the ramp for taking the boat out of the pond.* **How many levers are being used?** *4* **Where?** *The oars, the axe handle, the fishing pole, and the wheelbarrow*

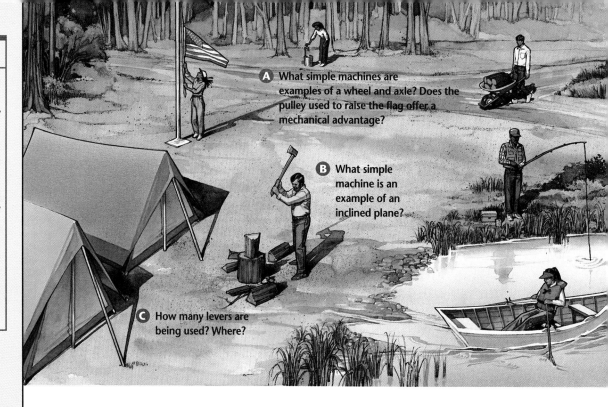

Ⓐ What simple machines are examples of a wheel and axle? Does the pulley used to raise the flag offer a mechanical advantage?

Ⓑ What simple machine is an example of an inclined plane?

Ⓒ How many levers are being used? Where?

3 ASSESS

Check Your Understanding

Assign the Check Your Understanding questions. Have student volunteers work out the answer to question 3 in front of the class. Use the Performance Task Assessment List for Group Work in **PASC,** p. 97.

Reteach

Power is the rate at which work is done. Ask students to list the three factors that affect the amount of power produced. *effort force, distance, time* Refer them back to Investigate 5-2. Ask how power would be changed if the person walked more slowly? *decrease* If the person ran? *increase* Ask what would happen if the person carried a stack of books? *power increased* Why? *More effort force was needed.* L1

Extension

Have students solve the following problem: A 500-N passenger is inside a 25 000-N elevator that rises 30 m in exactly one minute. **How much power is needed for the elevator's trip?** *Power = F × D/Time 25 500 N × 30 m/60 seconds = 12 750 watts* L3

Figure 5-14

You've calculated human power, but the same method could be used to calculate the power of simple and compound machines.

The original campsite scene contained many simple machines. Now, after completing this chapter, you can identify several simple and compound machines. You have determined how much easier these machines can make a task by computing the mechanical advantage. The MA is the ratio of resistance force to effort force.

Finally, you found the power of these machines by dividing the work done by the amount of time it took to do the work. Power = work/time. Now you can compute how much power you use to complete a 3-hour, 5-mile hike up the mountain. What a surprise to find all that work and power at a vacation campsite!

check your UNDERSTANDING

1. Give an example of a compound machine. What are the simple machines that make it up?
2. How are work, power, and time related?
3. **Apply** How much power does a person weighing 500 N need to climb a 3-m ladder in 5 seconds? How could the same person climb the ladder using less power?

4 CLOSE

Power can be defined as the rate at which work is done (work/time). Ask students to name other quantities that represent rates. Answers may include speed (distance/time) and wages (dollars/hour). L1

check your UNDERSTANDING

1. Answers will vary. Several examples are given in the chapter.
2. Power = Work/Time
3. 300 watts; by taking longer (increase the time) to climb the ladder

Leisure Connection

Spills, Chills, Waves, and Dunks

You're zooming down a 60-foot-high water slide at speeds of up to 40 miles per hour. You reach the bottom in as little as 15 seconds, but an engineer may have worked on the design of the slide's inclined plane for years. Every bend and twist has been carefully constructed for both speed and safety.

The latest computer technology is used to calculate such factors as the angle of the curves and the lubricating effect of the flowing water. Even the amount of resistance that will be created by your bathing suit is considered.

If your water park has a pool for body surfing and boogie boarding, you can thank a machine for those perfect waves. In the ocean, the wind causes the waves to rise and fall. In a water park, waves are usually created by a computer-operated wave machine. By intermittently pumping air into the pool, the computer generates waves of many sizes and shapes.

You've learned how machines can help to make work easier. The next time you go to a water park, be sure to notice how machines can also work to help you have fun.

You Try It!

Draw the water slide of your dreams. How tall would your slide stand? How long would it take a person to ride it from top to bottom?

Going Further ▐▐▐▶

Have students explore occupations related to the construction and maintenance of recreational facilities. Have students find out how the equipment is maintained and what the newest inventions are. Students should report their findings by writing a newspaper article that they can illustrate with photographs or drawings. Use the Performance Task Assessment List for Newspaper Article in **PASC**, p. 69. **P**

Leisure Connection

Purpose

This excursion gives students the opportunity to explore how machines are used in recreational settings. It uses an example of an inclined plane which students learned about in the first section of the chapter. The concept of how inclined planes use curves to reduce effort force is reinforced in this material.

Content Background

Computer-aided design and computer-aided manufacturing systems are widely used in manufacturing industries for both design and manufacturing tasks. Special computer systems are used by engineers and architects. And other designers use the computer to assist in designing products by generating drawings on a graphic display screen and then manipulating that drawing until a final design is obtained. Changing a design, which once took hundreds of hours, can be done in seconds.

Teaching Strategies

Have students discuss a variety of amusement park rides. Students should identify the kinds of machines that are involved in each ride and the methods that are employed to make them more efficient and safe.

Answers to You Try It!

Answers will vary. All things being equal, the ride from top to bottom would depend on the length and slope of the slide.

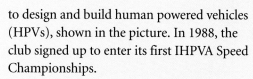

Science and Society

Purpose

This excursion reinforces the concept of power and the mechanical advantage of compound machines, which are discussed in Section 5-3.

Content Background

Bicycle racing, also called cycling, is one of the most popular sports in the world, especially in Europe where millions of fans follow the feats of cyclists the way Americans follow their favorite spectator sports. Cycling has been an event in the Olympic games since the modern games began in 1896.

The most popular form of bicycle racing is the original form—road racing. Thousands of cyclists may participate in a race that may cover extremely long distances. The best known cycling road race of this kind is the Tour de France, which takes place annually. The race course is 4,000 kilometers (2500 miles) and takes place over a 24-day period.

Other types of bicycle racing include track races, held on oval tracks called velodromes, mountain bike races held on trails and dirt roads, and motocross races, which are held on dirt tracks that have many bumps and sharp turns. Each type of race requires a different style of bicycle to produce the greatest speed on the various courses.

Science and Society

Pedal Power!

You've spent months fine-tuning your vehicle. You're all set. At the sound of the starting gun, you jump on your pedals.

Jump on your pedals? That's right. How else do you compete in the International Human Powered Vehicle Association's (IHPVA) Annual Speed Championships?

Most people agree that the bicycle is the most efficient means of transportation on Earth. With its annual races, the IHPVA sets out to prove that bicycles can also be fast. A streamlined bicycle called *Cheetah* set the IHPVA speed record at 68.7 mph. However, a group of high school students in Saginaw, Michigan, just might give the *Cheetah* a run for its money.

On Your Marks ...

The Arthur Hill High School Technology Club was founded by drafting teacher and club advisor, Bruce Isotalo. Members of the club use the principles of compound machines

to design and build human powered vehicles (HPVs), shown in the picture. In 1988, the club signed up to enter its first IHPVA Speed Championships.

Mr. Isotalo recalls how the club created that first HPV, the *da Vinci*. "First, we had to decide what we wanted the vehicle to accomplish. We chose speed as our main goal." To meet their goal, the club members paid

Engineers apply the discoveries of scientists to design, develop, and produce products and systems. Mechanical engineers are unique in the engineering field because they create many of the tools and machines required by other engineers. Mechanical engineers work in industry, business, government service, and universities.

ROAD CLOSED TO THRU TRAFFIC

CAREER connection

Have volunteers contact a local manufacturing company or university. Have them find out about the education and experience requirements for engineering employees and the varieties of jobs available for various degrees or backgrounds. Have students share findings with the class.

special attention to the *da Vinci's* body design. As you know, when energy is used to overcome friction, efficiency is reduced. With this in mind, the club chose a streamlined body design to allow air to flow smoothly around the *da Vinci*. To keep the weight of the *da Vinci* low, they selected lightweight aluminum as the building material. Once the design had been finalized, it was time to begin the task of transforming the *da Vinci* from an idea into a roadworthy machine. The students worked for months to perfect the wheel-and-axle system in the vehicle's powerful pedaling mechanism.

Get Set ...

The IHPVA Speed Championships were scheduled for August and would be held in Michigan. During summer vacation, many students worked on the *da Vinci* eight hours a day. "We sent out for a lot of pizza that summer," laughs Mr. Isotalo. "But we stuck with it. I think that was one of the greatest things about this project. We learned that there is nothing that can't be accomplished when you put your mind to it."

Go!

On a hot August day in 1989, the *da Vinci* took its place at the starting line of the International Human Powered Vehicle Speed

Championships. The club members watched with pride as the *da Vinci* placed among the top 20 in the sprint event.

In 1990, the Technology Club radically altered the design and the *da Vinci 2* placed third in a grueling 24-hour marathon event. The club continues to explore the potential of human power with a new model every year.

Designing the Future

As engineers, like the students in the Technology Club, push the limits of human power, there may be a day when human-powered vehicles are a common sight on roadways. The dwindling supply of fossil fuels is pushing us toward clean, alternative sources of power. As the IHPVA Annual Speed Championships have shown, many of the shortcomings of alternative power sources can be overcome through innovative design.

Science Journal

Imagine what it may be like getting on the school bus and having to pedal your way to school! Think about the possibility of human-powered cars, trains, and school buses. Discuss *in your Science Journal* what problems there might be with human-powered transportation. What else could human power be used for besides transportation?

Purpose

This excursion provides students with insights into the satisfaction that can be derived from seeing a project through to its completion. The chapter emphasized the use of machines to make work easier. Here students can combine this concept with the idea that work can also be fun.

Content Background

The scope of the mechanical engineer's work includes the generation of power and its transmission, the design and production of all kinds of goods, the movement of people and goods by means of all types of transportation, and the material handling of equipment, such as automotive, aeronautic, and conveying. Without mechanical engineers, much of the equipment and machinery that we take for granted would not be available. When an inventor or industrial designer comes up with a new idea, it usually takes a mechanical engineer to make it operational. Mechanical engineers are involved in research, development, invention, design, construction, operation, and maintenance of machinery, transportation equipment, generators, fuel engines, steam engines, diesel engines, power plants, airplanes, jets, and rockets.

Teaching Strategy

Have students agree on a group project in which they will build a new mechanical device. Have them develop a plan for completing the project, including assigning roles in teams, listing materials needed, setting time frames, and making a list of steps from beginning to the end of the project. **COOP LEARN**

Answers to
You Try It!

Answers will vary, but may

Hard Work—the Easy Way to Have Fun

Have you ever been surprised by how much fun you were having while doing a difficult task? For Stephanie Ostler, an original member of the Technology Club, helping to design and build a human powered vehicle was both a challenge and a thrill.

"For me, working on the *da Vinci* was like putting together a puzzle. Only it was harder, because we had to invent each piece of the puzzle first."

With each stage of the *da Vinci's* production, the club members discovered that a new set of problems developed.

"We all worked together to solve problems. Every decision was made by vote."

"By the time summer came, the *da Vinci* was the only thing any of us thought about."

Stephanie did, however, devote some thought to seeing the world around her in a different way.

"I had just gotten my driver's license when I joined the club. At that time, I thought cars were about the greatest thing on Earth. I never really thought about how they affect the environment. After I drove the *da Vinci*, I was really sold on HPVs as an alternative to cars. Not only do they save energy, but they are good exercise, too!"

As the summer drew to an end, Stephanie realized

that the experience of working on the *da Vinci* had an important effect on her. "This may sound weird, but I think one of the best things about the project was completing it. We started out with nothing more than an idea, but we just kept working at it until we had something we could be proud of. I think everybody should explore things they don't know anything about. It really feels good when you figure it out."

The greatest satisfaction, however, was to come on the day of the race.

"I don't think I'll ever forget the race. There we were with inventors and inventions from all over the world. It was awesome."

You Try It!

Even though her HPV did not win the IHPVA Speed Championships, Stephanie is very proud of her work on the *da Vinci*. Which experiences in your life have affected you most? Stephanie Ostler encourages people to try new things. Make a list of some new things you would like to try.

include trying new leisure activities, such as learning a new sport or learning to play a new musical instrument.

Going Further ⟩⟩⟩⟩⟩⟩→

Divide the class into groups and have them brainstorm inventions that would improve life in our society. Encourage the students to think about a broad range of things from small conveniences in everyday life to inventions that would improve the world we live in. After students have come up with all the ideas they can think of, have them narrow their lists to items they think the class could work on as a project. Have each group share their final list with the class and let the entire class vote on selecting one or two class projects. Refer students' attention to the "Career Connection" box on page 174, which points out the different roles people play in designing, developing, and producing a new product. Use the Performance Task Assessment List for Group Work in **PASC**, p. 97. **COOP LEARN**

Science Journal

Review the statements below about the big ideas presented in this chapter, and answer the questions. Then, re-read your answers to the Did You Ever Wonder questions at the beginning of the chapter. *In your Science Journal*, write a paragraph about how your understanding of the big ideas in the chapter has changed.

1 Machines make work easier. *What are some machines that make work you need to do easier?*

2 There are six simple machines—lever, wheel and axle, pulley, inclined plane, screw, and wedge. They can increase force, change the direction of force, or both. *What kind of simple machine is a nutcracker?*

3 Machines do not reduce the amount of work that has to be done. *If a machine doesn't change the amount of work that needs to be done, how does it make the work seem easier?*

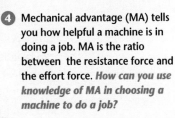

4 Mechanical advantage (MA) tells you how helpful a machine is in doing a job. MA is the ratio between the resistance force and the effort force. *How can you use knowledge of MA in choosing a machine to do a job?*

5 Power is the work done divided by the time it took to do the work. *Does it require more power to lift a book quickly or to lift the same book the same distance slowly?*

Science Journal

Did you ever wonder...

• The crowbar is used as a lever. When you push down on the lever, you apply an effort force, which increases the resistance force acting on the rock. (p. 150)

• A screwdriver is a wheel and axle. A small effort force applied to the wheel (handle) is multiplied at the axle (the point). (p. 155)

• It takes more effort to go up a steep slope than a less steep one. Curving roads lengthen the slope, making it less steep. (p. 154)

Project

Have students select a machine with which they are familiar, perhaps one they use at home, such as a lawnmower, a bicycle, or a kitchen appliance. Ask students to prepare a presentation tracing the history of the machine. They should show pictures of succeeding versions of the machine. They should also include information on the use of the machine and how people performed the same tasks before the machine was invented. Use the Performance Task Assessment List for Oral Presentation in **PASC,** p. 71.

Have students study the four illustrations and accompanying captions to review the main ideas of this chapter.

Teaching Strategies

Using the lists of simple and compound machines that students compiled while they were studying the chapter, ask them to identify the type of work for which each machine is used. Then ask volunteers to use machines from their lists as examples that support the main ideas listed on the page.

Answers to Questions

1. What are some machines that make work you need to do easier? *Answers will vary; hammers, crowbars, nuts and bolts, ramps, and scissors are possible answers.*

2. What kind of simple machine is a nutcracker? *It is a lever.*

3. If a machine doesn't change the amount of work that needs to be done, how does it make the work seem easier? *By requiring a smaller effort force exerted over a greater distance, or by changing the direction of the force, the work seems easier.*

4. How can you use knowledge of MA in choosing a machine to do a job? *You can use the machine with the appropriate MA required to do the job.*

5. Does it require more power to lift a book quickly or to lift the same book the same distance slowly? *Because power= work/time, the shorter the time, the more power would be required to do the work.*

GLENCOE TECHNOLOGY

MindJogger Videoquiz

Chapter 5 Have students work in groups as they play the Videoquiz game to review key chapter concepts.

Using Key Science Terms

1. Power; others are simple machines.

2. Watt; others are concepts, watt is a unit measurement.

3. Pulley; others are types of inclined planes.

4. Power; others are units of measurement.

5. Lever; others are compound machines.

Understanding Ideas

1. A screw is an inclined plane with grooves that spiral downward and move into wood. Like an inclined plane, it reduces the effort by increasing the distance.

2. MA = 17

3. Machines change a small effort force exerted over a long distance into a larger resistance force exerted over a shorter distance, therefore the force is increased and the direction or distance is changed.

4. No, machines make jobs easier by changing the size or direction of the force and never increasing the amount of work.

5. Because, the pulley changes the direction of the effort force, but does not provide a mechanical advantage.

6. The resistance force divided by the effort force.

7. Because this increases the force of the lighter person by having a longer effort arm than the resistance arm.

Developing Skills

1. See concept map on student page.

2. The pulley in Find Out activity is a movable pulley where it increases the force but does not change the direction. A pulley on a fishing rod is fixed and the direction of the force is changed.

3. Students should indicate that the claw hammer is a lever because it makes the job of pulling nails out easier by producing a greater resistance force by moving a smaller effort force over a

greater distance.

4. Students should graph data of height and the calculated power. They should infer by the graph that the greater the height, the more power required to move the car up the ramp.

5. By interpreting data students should infer that the more force required, the more work required per second; therefore the greater the weight, the more work is done to climb the stairs.

U sing Key Science Terms

compound machine	pulley
effort force	resistance force
inclined plane	screw
lever	wedge
mechanical advantage	wheel and axle
power	

For each set of terms below, choose the one term that does not belong and explain why it does not belong.

1. wheel and axle, power, lever
2. watt, power, mechanical advantage
3. wedge, pulley, inclined plane, screw
4. joule, watt, power, newton
5. bicycle, can opener, lever

U nderstanding Ideas

Answer the following questions in complete sentences.

1. How is a screw an inclined plane?
2. The radius of a bicycle wheel is 35 cm. The gear that is the axle has a radius of 2.0 cm. Find the mechanical advantage.
3. Explain why machines make things easier.
4. Do machines ever increase the amount of work done? Explain your answer.
5. Why doesn't using a pulley on a flagpole increase the force?
6. How is mechanical advantage calculated?
7. Explain why a seesaw works best when the heavier person sits closer to the middle than the lighter person.

D eveloping Skills

Use your understanding of the concepts developed in this chapter to answer each of the following questions.

1. **Concept Map** Complete the following concept map of simple machines using the following terms: *compound machines, mechanical advantage, resistance force, work.*

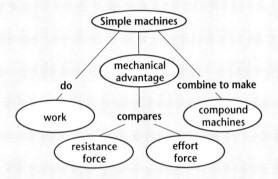

2. **Comparing and Contrasting** Compare and contrast the pulley made in the Find Out activity on page 153 and a pulley on a fishing rod.

3. **Forming a Hypothesis** Form a hypothesis stating why a claw hammer used for removing nails is or is not a simple machine.

4. **Making and Using Graphs** Refer to the Find Out activity on page 169. Graph the measurements of height versus power. First change the height of the ramp several times, and calculate the power each time. What relationship is there between the height and power?

5. **Interpret Data** From the data gathered in the Investigate on page 170, interpret the relationship between force and work.

Program Resources

Review and Assessment, pp. 29–34 **L1**
Performance Assessment, Ch. 5 **L2**
PASC
Alternative Assessment in the Science Classroom
Computer Test Bank **L1**

Critical Thinking

Use your understanding of the concepts developed in the chapter to answer each of the following questions.

50N

1. What is the MA of the axe the woodcutter is using to split the log in the figure?

350N

2. To screw a 5-cm screw into a piece of wood requires turning a screwdriver 15 times. For each turn of the screwdriver, your hand moves 15 cm. What is the MA of the screw?

3. A bricklayer is carrying boxes weighing 400 N up a flight of stairs 10 m high. It takes him 90 seconds. A carpenter is hammering in one nail every 30 seconds. He exerts a force of 500 N over a distance of .05 m. How many watts of power does the bricklayer produce in the 90 seconds it takes him to climb the stairs? How much power does the carpenter produce in 90 seconds?

4. A claw hammer has a mechanical advantage equal to 8. It is used to pull a nail that exerts a force of 2500 N. What is the effort force needed to pull the nail?

Problem Solving

Read the following paragraph and discuss your answers in a brief paragraph.

Suppose you and a friend are pushing boxes of camping gear up a ramp to load them into a van. All the boxes weigh the same, but you push faster than your friend. You can move a box up the ramp from the ground to the van in 30 seconds. It takes your friend 45 seconds.

1. Are you both doing the same amount of work? Explain your answer.

2. Assume the boxes weigh 100 N and the ramp is 1 m high. How many watts of power are you each producing?

CONNECTING IDEAS

Discuss each of the following in a brief paragraph.

1. **Theme—Systems and Interactions** Explain why scissors are a compound machine.

2. **Theme—Systems and Interactions** Explain how a shovel works like a lever.

3. **A Closer Look** Using a ramp 4 meters long, workers apply an effort force of 1250 N to move a 2000 N crate onto a platform 2 meters high. What is the efficiency of the ramp?

4. **Science and Society** What design features did the *da Vinci* bicycle have that allowed it to conserve more energy than a conventional bicycle?

5. **Life Science Connection** Why do you think it is helpful to design some machines to work with the human body?

Critical Thinking

1. The woodcutter applies an effort force of 50 N to the axe. The resistance of the log is 350 N. MA = 7

2. 15 × 15 cm = distance of the work put into the system. MA = 45 = 225 cm/5 cm

3. The bricklayer, W = 4000 J P = 44.4 watts

The carpenter, W = 75 J P = 0.8 watts

4. F_c = 312.5 N

Problem Solving

1. Yes, you are both doing the same amount work because the boxes weigh the same and are being moved the same distance. (work = force x distance)

2. You produce 3.3 watts (100 N × 1 m / 30 sec = power), and your friend produces 2.2 watts (100 N × 1 m / 45 sec = power)

Connecting Ideas

1. Both halves are levers that turn on the screw that holds them together. The screw and the blades are inclined planes.

2. A shovel changes the distance over which the force moved, just as a lever does. The hand in the middle of the shovel is the fulcrum. As you exert an effort force, the dirt you are shoveling moves (the resistance force). You can change the mechanical advantage of the shovel by adjusting your hand placement.

3. efficiency = 2000 N × 2 m / 1250 N × 4 m × 100% efficiency = 80%

4. Answers will vary. Students should indicate that the design was streamlined to allow the bicycle to operate more efficiently. The frame was made of a lightweight material.

5. Answers will vary. Student should include information concerning efficiency.

Chapter Organizer

SECTION	OBJECTIVES	ACTIVITIES & FEATURES
Chapter Opener		**Explore!**, p. 181
6-1 Thermal Energy (3 sessions; 1.5 blocks)	1. **Describe** how two objects achieve thermal equilibrium. 2. **Determine** when two objects have the same temperature. 3. **Use** the Celsius temperature scale. **National Content Standards: (5-8) UCP1-4, A1-2, B3, G3**	**Explore!**, p. 182 **Design Your Own Investigation 6-1:** pp. 184–185 **Find Out!**, p. 187 **A Closer Look**, pp. 188–189
6-2 Heat and Temperature (3 sessions; 1.5 blocks)	1. **Describe** heat as the transfer of energy. 2. **Distinguish** temperature from heat. 3. **Give examples** of heat transfer by conduction, convection, and radiation. 4. **Identify** materials that reduce heat transfer most effectively. **National Content Standards: (5-8) UCP2, A1, B3, D1, D3**	**Explore!**, p. 190 **Investigate 6-2:** pp. 194–195 **Earth Science Connection**, pp. 192–193 **Consumer Connection**, p. 206
6-3 Making Heat Work (2 sessions; 1 block)	1. **Describe** how heat can produce work. 2. **Explain** why a heat engine can't be 100% efficient. 3. **Describe** how an air conditioner or refrigerator works. **National Content Standards: (5-8) UCP2, UCP5, A1-2, B3, C1, D3, E1, F2, F4, G3**	**Explore!**, p. 200 **Explore!**, p. 202 **Science and Society**, pp. 204–205

ACTIVITY MATERIALS

EXPLORE!

p. 181* coin, carpet
p. 182* 3 pans, lukewarm water, pan of cold water and crushed ice, pan of hot water, watch or clock with second hand
p. 190* wooden spoon, plastic spoon, metal spoon, 3 plastic beads, butter, beakers, water, hot plate or electric immersion heater
p. 200 car and driver
p. 202* eyedropper, rubbing alcohol, inner tube, air pump

INVESTIGATE!

pp. 194-195* colored ice cubes, tongs, thermometers, 250-mL beaker, salt, stirring rod, water

DESIGN YOUR OWN INVESTIGATION

pp. 184-185* 3 thermometers, 400 mL beakers (3), 300 mL water, hot plate or heating element, stopwatch or clock, 3 stirring rods

FIND OUT!

p. 187 beaker, ice water, crushed ice, unmarked alcohol thermometer, hot plate, stirrer, fine-point marker, metric ruler

KEY TO TEACHING STRATEGIES

The following designations will help you decide which activities are appropriate for your students.

L1	Basic activities for all students
L2	Activities for average to above-average students
L3	Challenging activities for above-average students
LEP	Limited English Proficiency activities
COOP LEARN	Cooperative Learning activities for small group work
P	Student products that can be placed into a best-work portfolio

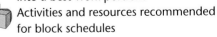 Activities and resources recommended for block schedules

Need Materials? Call Science Kit (1-800-828-7777).

[00:00] **OUT OF TIME?** We recommend that students do the activities with an asterisk.

Chapter 6 Thermal Energy

TEACHER CLASSROOM RESOURCES

Student Masters	Transparencies
Study Guide, p. 22 **Critical Thinking/Problem Solving,** p. 5 **Activity Masters,** Investigate 6-1, pp. 27–28 **Science Discovery Activities, 6-1**	**Teaching Transparency 12,** Fahrenheit and Celsius Temperature Scales **Section Focus Transparency 16**
Study Guide, p. 23 **Concept Mapping,** p. 14 **Take Home Activities,** p. 12 **Activity Masters,** Investigate 6-2, pp. 29–30 **How It Works,** p. 9 **Making Connections: Integrating Sciences,** p. 15 **Science Discovery Activities, 6-2** **Laboratory Manual,** pp. 27–32, Heat Transfer **Laboratory Manual,** pp. 33–34, Conduction of Heat **Laboratory Manual,** pp. 35–38, Specific Heats of Metals	**Teaching Transparency 11,** Conduction, Convection, and Radiation **Section Focus Transparency 17**
Study Guide, p. 24 **Critical Thinking/Problem Solving,** p. 14 **Multicultural Connections,** pp. 15–16 **Making Connections: Across the Curriculum,** p. 15 **Making Connections: Technology & Society,** p. 15 **Science Discovery Activities, 6-3**	**Section Focus Transparency 18**

ASSESSMENT RESOURCES	TEACHING & TECHNOLOGY
Review and Assessment, pp. 35–40 **Performance Assessment,** Ch. 6 **PASC*** **MindJogger Videoquiz** **Alternate Assessment in the Science Classroom** **Computer Test Bank**	**Spanish Resources** **Cooperative Learning Resource Guide** **Lab and Safety Skills** **Science Interactions, Course 2, CD-ROM** **Computer Competency Activities**

*Performance Assessment in the Science Classroom

NATIONAL GEOGRAPHIC TEACHER'S CORNER

Index to National Geographic Magazine

The following articles may be used for research relating to this chapter:

- *Energy: Facing Up to the Problem, Getting Down to Solutions,* A Special Edition, February 1981.
- "The Promise and Peril of Nuclear Energy," by Kenneth F. Weaver, April 1979.

Additional National Geographic Society Products

To order the following products for use with this chapter, call the National Geographic Society at 1-800-368-2728:

- *Everyday Science Explained* (Book)
- *Solar Energy* (Kids Network Curriculum Unit)

GLENCOE TECHNOLOGY

The following multimedia resources are available from Glencoe.

Science and Technology Videodisc Series (STVS)
Physics
  Images of Heat
  Solor Food Dryer

Glencoe Physical Science Interactive Videodisc
Behavior of Gases

Physical Science CD-ROM

Teacher Classroom Resources

These are key components of the classroom resources package.

TEACHING AIDS

Section Focus Transparencies

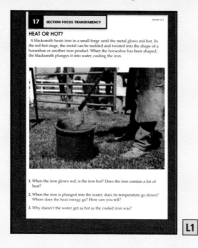

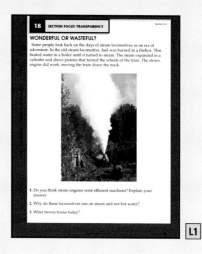

Teaching Transparencies

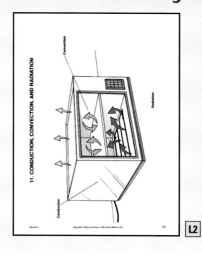

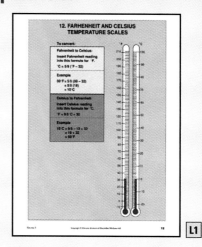

HANDS-ON LEARNING

Science Discovery Activity*

Laboratory Manual*

Take Home Activity

*There may be more than one activity for this chapter.

Chapter 6 Thermal Energy

REVIEW AND REINFORCEMENT

Study Guide*

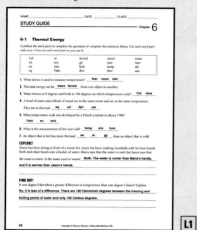

L1

Concept Mapping

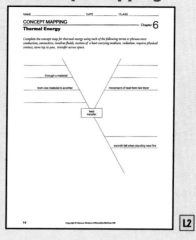

L2

Critical Thinking/Problem Solving

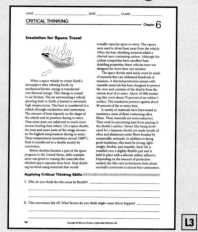

L3

ENRICHMENT AND APPLICATION

Integrating Sciences

L2

Across the Curriculum

L2

Technology and Society

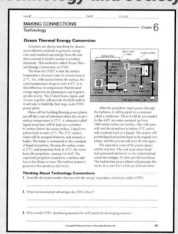

L2

Multicultural Connection**

L1

ASSESSMENT

Performance Assessment

L2

Review and Assessment

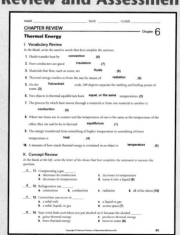

L1

*One per section **Two per chapter

Chapter 6   Thermal Energy   **180D**

Thermal Energy

THEME DEVELOPMENT

Energy is the predominant theme in this chapter. The transfer of thermal energy causes work to be done and provides warmth. When the transfer of energy takes place in a system such as a thermometer or heat engine, the theme of systems and interactions is evident.

CHAPTER OVERVIEW

In the first section of this chapter, students will learn that thermal equilibrium exists when two materials in contact with each other are at the same temperature. In the second section, students will learn about the three methods by which heat is transferred: conduction, convection, and radiation. In the final section, heat engines are introduced and used to demonstrate how thermal energy can be used to do work.

Tying to Previous Knowledge

Have groups of students classify ways energy of any kind does work in machines, then identify machines (if any) in which thermal energy does work.

INTRODUCING THE CHAPTER

Have students identify other extremely hot and cold objects that appear in nature, and then have them identify hot and cold manufactured items.

Did you ever wonder...

✓ **How temperature scales are created?**

✓ **Why insulation keeps things hot or cold?**

✓ **How an air conditioner or refrigerator works?**

Science Journal

Before you begin to study thermal energy, think about these questions and answer them *in your Science Journal.* When you finish the chapter, compare your journal write-up with what you have learned.

Answers are on page 207.

W e live our lives in climates that range from hot to cold. For many people on Earth, the seasons bring regular changes. Temperatures will vary from hot to mild to cold and then back again.

Hot and cold are things you feel, things your body reacts to. You sweat under a scorching sun and shiver in freezing wind. Before you even go outside, you'll react to the temperature when you read a thermometer. How you dress might depend on what the thermometer tells you about the outside temperature. A cold day means more clothes and a hot day fewer. But what actually is heat? What is temperature? Are they the same thing?

▶ ***In the activity on the next page, explore your body's built-in thermometer.***

180

Learning Styles	Kinesthetic	Explore, pp. 181, 182, 200, 202; A Closer Look, pp. 188-189; Activity, p. 191; Investigate, pp. 194-195
	Visual-Spatial	Visual Learning, pp. 183, 186, 191, 193, 197, 198; Activity, pp. 189, 193, 196, 203; Explore, p. 190; Demonstration, pp. 192, 196
	Interpersonal	Activity, p. 199
	Logical-Mathematical	Design Your Own Investigation, pp. 184-185; Activity, p. 186; Find Out, p. 187; Visual Learning, p. 188
LS	**Linguistic**	Discussion, pp. 183, 198, 201, 203; Earth Science Connection, pp. 192-193; Across the Curriculum, p. 196; Multicultural Perspectives, pp. 197, 198; Activity, p. 201

Explore! ACTIVITY

How do we feel "heat"?

Have you ever put your hand in your bath or shower water to make sure it's at a comfortable temperature, or felt your forehead to see if you have a fever? If you have, you were relying on your sense of touch to estimate temperature. But how accurate is your sense of touch?

What To Do

1. Collect a group of familiar objects made of various materials, and place them on a table together, leaving them for several minutes.

2. Use the underside of your wrist to feel which objects feel warmest and which ones feel coldest. This area of your skin is very sensitive to hot and cold objects.

3. *In your Journal,* estimate the temperature of each object.

ASSESSMENT PLANNER

PORTFOLIO
Refer to page 209 for suggested items that students might select for their portfolios.

PERFORMANCE ASSESSMENT
Process, pp. 182, 185, 190, 195, 200
Explore! Activities, pp. 181, 182, 190, 200, 202
Find Out! Activities, pp. 187
Investigate!, pp. 184–185, 194–195

CONTENT ASSESSMENT
Oral, pp. 187, 202
Check Your Understanding, pp. 189, 199, 203
Reviewing Main Ideas, p. 207
Chapter Review, pp. 208–209

GROUP ASSESSMENT
Opportunities for group assessment occur with Cooperative Learning Strategies.

Uncovering Preconceptions

Some students may not know the difference between temperature and thermal energy. Explain that a large pot of boiling water has more thermal energy than a small pot of boiling water even though they both have the same temperature. Temperature is a measure of the average energy of molecules within an object.

Explore!

Time needed 5 minutes

Materials coin, carpet, other objects; if carpeting is not available at your school, a small scatter rug can be used.

Thinking Processes observing and inferring, comparing and contrasting, forming operational definitions

Purpose To experience that different materials at the same temperature may feel like they are at different temperatures.

Teaching the Activity

Science Journal Students should record their temperature estimates. They should also record the perceived differences in the temperature of the two materials. **L1**

Expected Outcome

Students should notice that the coin and other metallic materials feel cooler than the carpet. They should realize that their sense of touch is not a reliable way to determine the temperature of an object.

Answer to Questions

If all the materials are at room temperature, they should all have the same temperature.

✓ Assessment

Content Have students explain why the words *warm* and *cool* are relative terms. Use the Performance Task Assessment List for Making Observations and Inferences in **PASC**, p. 17.
COOP LEARN

PREPARATION

Planning the Lesson

Refer to the Chapter Organizer on pages 180A–D.

Concepts Developed

In this section, students will differentiate between hot and cold. They will use thermometers and the concept of temperature to make a more precise differentiation. They will also investigate the concept of thermal equilibrium.

1 MOTIVATE

Bellringer

 Before presenting the lesson, display **Section Focus Transparency 16** on an overhead projector. Assign the accompanying **Focus Activity** worksheet. L1

LEP

Explore!

Is your sense of touch accurate for judging temperature?

Time needed 5 minutes

Materials 3 deep pans, lukewarm water, ice, cold water, hot water, watch with second hand

Thinking Processes observing, comparing and contrasting

Purpose To observe that relying on sense of touch to judge temperature can be inaccurate.

Teaching the Activity

Safety Do not use water hotter than 60°C.

Section Objectives
- Describe how two objects achieve thermal equilibrium.
- Determine when two objects have the same temperature.
- Use the Celsius temperature scale.

Key Terms
thermal equilibrium

How Do We Know If Something Is Hot or Cold?

You're on the telephone with a friend in another city on a hot summer afternoon. You're both trying to describe how hot it is outside, but neither of you has a thermometer. How could you accurately describe how hot it is?

Like most ideas in science, heat and temperature are connected to our everyday lives. You use your sense of touch to tell you whether something is hot or cold. It can even tell you when one object is hotter or colder than another object. Or can your sense of touch be fooled? In the next activity, you will explore how accurate your sense of touch is for judging the temperature of three containers of water.

Explore! ACTIVITY

Is your sense of touch accurate for judging temperature?

What To Do

1. Fill a pan with lukewarm water. Fill a second pan with cold water and crushed ice. Fill a third pan with very warm water.

2. Now, put one hand into the cold water and the other hand into the warm water. Hold them in the water for 15 seconds.

3. Quickly remove both hands from the water and then put both into the pan of lukewarm water. How do they feel now? Do both hands feel the same way? Can you explain why your hands feel as they do? Record your observations *in your Journal.*

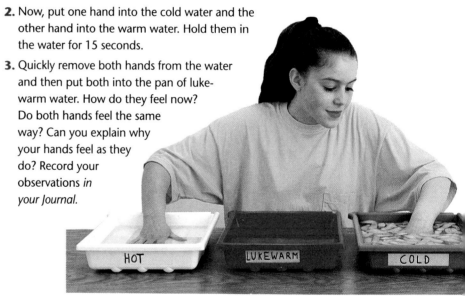

Science Journal Have students describe how each hand feels as it is moved from one pan to another. L1

Expected Outcome

To the hand removed from the cold water, the lukewarm water felt hot; to the hand removed from the hot water, the lukewarm water felt cold.

Answers to Questions

3. No; The cold hand is warmed by the lukewarm water, while the warm hand is cooled by it.

✔ Assessment

Process Have students predict how their hands would feel if both hands were placed in water of the same temperature, and then moved to the lukewarm water. Then have them test their predictions.

As you've just seen, your sense of touch can be fooled. What does that tell you about your earlier Explore activity with the various objects? If two items feel different, does that really mean that they're at different temperatures? Relying on our sense of touch doesn't help us tell someone else exactly how hot something is. We need a more precise way to indicate temperature.

Around the year 1600, Galileo observed that air takes up more space when it's warm than when it's cold. Galileo used that information in making an instrument to indicate a change in temperature, shown in **Figure 6-1**. About 100 years later, German scientist G. D. Fahrenheit improved Galileo's design by using alcohol or mercury instead of air. Like air, these liquids also take up more space when hot than when cold. But they don't expand as much as air does. Fahrenheit let the liquids expand in a very narrow tube. Why do you suppose he had to do this? Fahrenheit's invention, in **Figure 6-2**, is the thermometer you've used to find the temperature outside or to check your temperature when you're sick.

Figure 6-1

Galileo's thermoscope consisted of a bent column of glass connected to a glass bubble. The column contained liquid. The bubble contained air.

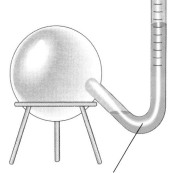

As the room temperature rose, the air in the bubble warmed. The warmed air, needing more room, pushed the liquid up the column.

Figure 6-2

A *Liquid-In-Glass Thermometers* consist of a fluid contained in a glass bulb connected to a narrow tube. The liquid—usually mercury or alcohol—expands as it warms, causing it to flow upward into the tube as temperature increases.

B *Thermistor Thermometers* indicate temperature by measuring the resistance of a ceramic bead, called a thermistor, to an electric current. The thermistor's resistance decreases as temperature increases. A microchip converts the resistance data to a temperature reading, which is displayed in digital form.

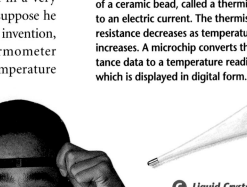

C *Liquid Crystal Thermometers* are made of individual containers, each holding a different type of crystal, which melts and changes color at a specific temperature. The containers are fastened together. As the crystals in each container react to the warmth and cold, color change indicates the temperature.

6-1 Thermal Energy **183**

6-1 How Cold Is It?

Preparation

Purpose To observe changes in temperature as thermal energy is transferred to and from samples of water.

Process Skills making graphs, making and using tables, observing, comparing and contrasting, measuring in SI, predicting, forming operational definitions

Time Required 30 minutes

Materials See reduced student text.

Possible Hypotheses Students should predict that the temperature of the three samples of water will change to closer to room temperature as they sit at room temperature.

The Experiment

Troubleshooting On their graphs, have students use a red pencil for hot water, a blue for cold water, and a black for room temperature.

Process Reinforcement At the beginning of the experiment, make certain that students take the temperature of the cold and hot water. Then as students construct their graphs, make sure they mark off enough increments to include the highest and lowest water temperatures.

Possible Procedures Fill one beaker with water and ice. In a second beaker, add room temperature water. In the third beaker, heat water to about 70°C. Take the temperature of the three beakers and record the data. Then every minute, take temperatures again in the same order as the first time. Record the data.

Teaching Strategies

Troubleshooting In the cold beaker, after water gets to the cold temperature desired, make certain that students remove the extra ice or the water temperature will continue to remain at the same temperature.

How Cold Is It?

Do you remember the last time you had to wait for soup or hot chocolate to cool enough so you could sip it? How long did it take? Was it several minutes or did the liquid cool instantly? In this activity, investigate how the temperature of water changes over time.

Preparation

Problem
What happens to the temperature of hot, cold, and room temperature water left at room temperature?

Form a Hypothesis
Predict what will happen to hot, cold, and room-temperature water left at room temperature for a certain period of time.

Objectives
• Predict how cold and warm water will change over time in a room-temperature environment.
• Design an experiment that tests the hypothesis and collects data that can be graphed.
• Interpret the data.

Materials
thermometers (3)
stirring rods (3)
hot plate
ice
400-mL beakers (3)
water, 300-mL
stopwatch or watch with second hand

Safety

Do not use mercury thermometers. Use caution when heating water with a hot plate.

184

Program Resources

Activity Masters, pp. 27–28, Design Your Own Investigation 6–1 L2
Teaching Transparency 12 L1

DESIGN YOUR OWN
INVESTIGATION

Plan the Experiment

1 With the materials provided, design an experiment to test your hypothesis. This is a group activity, so make sure that everyone gets to contribute to the discussion.

2 What is the variable? Have you provided a control?

3 How hot should the water be at the start of the experiment? How cold?

4 How long are you going to take measurements? How often will you take them?

5 Design a data table *in your Science Journal* to record the observations you make.

6 It may be more efficient for one person to take the measurements and another to record them.

Check the Plan

1 To see the pattern of how the temperature of water changes, you will need to graph your data. What kind of graph will you use? Make certain you have taken enough measurements during the experiment.

2 The time intervals between measurements should be the same. Be sure to keep track of time as the experiment goes along.

3 Before you proceed with the investigation, be certain your teacher approved your plan.

4 Carry out your investigation and record your observations.

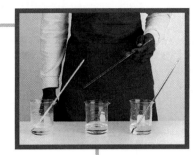

Analyze and Conclude

1. **Graph** Graph your data. Use one graph for all three beakers of water.

2. **Analyze** Did the water temperatures change as you had predicted? Was your hypothesis supported? Use your data and graph to explain your answers.

3. **Infer** Why did the temperature of the hot water change? Where did the energy go?

4. **Infer** Why did the temperature of the cold water change? Where did the energy come from?

5. **Conclude** Do the lines on the graph curve the same way? Explain why or why not?

Going Further

If you doubled the amount of hot, cold and room temperature water you used in your experiment, how would the cooling curves be affected?

Discussion Have students formulate explanations for the predictions of their hypotheses. Where does the heat energy go and where does it come from?

Science Journal Have students design data tables in their journal and record their data in them.

Expected Outcomes

Students will observe that hot water cools and cold water warms over time. They will find that the water temperature of each beaker stabilizes at room temperature. Their graphs should show this.

Analyze and Conclude

1. The curves on the graph tend to come together at room temperature.

2. Responses may vary, but most students probably made accurate predictions, which they can support with the information on their graphs.

3. The temperature of the hot water changed because the water lost heat energy to the cooler surrounding air.

4. The cold water grew warmer as heat energy flowed into the water from the surrounding warmer air.

5. The hot water line on the graph curves opposite of the cold water line. The room temperature graph is a straight line. The hot water line and the cold water line curve the opposite way because they are doing opposite things. The cold water is getting warmer while the hot water is getting cooler.

Going Further ⅲⅲⅲ➤

The time needed to cool or warm the water would be increased because more energy would have to flow into or out of the room temperature air.

✔ Assessment

Process Have students carry out the experiment in the Going Further to check their predictions. Use the Performance Task Assessment List for Designing an Experiment in **PASC**, p. 23.

Figure 6-3 Ask students to tell in which direction the transfer of energy took place. Students should realize that the question cannot be answered with the information available.

Figure 6-4 Ask students to identify which glass of iced tea is freshly poured and which has been left for three hours. Have students explain their thinking. **Which is in thermal equilibrium with the air in the room?** *If either reached thermal equilibrium, it would be the one that has been left for 3 hours.* **Will they warm up or cool down? Why?** *As long as they stay in the water, they will cool down. Thermal energy will transfer from their bodies to the water, making them feel cooler.*

SKILLBUILDER

Liquid-in-glass thermometers rely on the expansion and contraction of a liquid in response to thermal energy. Thermistor thermometers use the changing resistance of a ceramic bead at different temperatures. Liquid crystal thermometers use various crystals to indicate temperature.

Thermal Equilibrium

Figure 6-3

The glass of water has been in the room long enough for the water temperature to become equal to the temperature of the air in the room. The water and air are in thermal equilibrium.

In the Investigate you just completed, the temperatures of the warm and cold water were changing at every data point. When you extended your graph, however, you could predict when the temperature change would stop. It stopped when it reached the temperature of the air surrounding the water. When two objects are in contact, as the water and air were, and the temperature of one is the same as the temperature of the other, as in **Figure 6-3**, they are said to be in **thermal equilibrium**. We call the temperature of the two objects the equilibrium temperature.

■ Thermal Equilibrium and Energy

You observed that a warm beaker of water will gradually cool off until it reaches thermal equilibrium with the air around it. You might wonder where the warmth goes. Does it just disappear, or does it go somewhere else?

For a clue to help you answer this question, think back to when you were studying work and energy. You pulled a heavy box across a surface, doing work on the box. When you stopped pulling, the box stopped moving so there was no observable kinetic energy. According to the law of conservation of energy, energy cannot be lost, so you looked for where the energy you put into the box went. It hadn't gained potential energy. However, when you felt the bottom of the box, it was warm. You were feeling evidence of another kind of energy, called thermal energy.

This means that when the beaker of warm water cooled, thermal energy was transferred to the surrounding air. A similar thing happened to the cool water. As it warmed, it gained thermal energy from its surroundings.

Figure 6-4

A The photographs show two glasses of iced tea. One is just poured, the other has been left for three hours. Which is in thermal equilibrium with the air in the room?

B These people are playing in cool water on a very hot day. Will they warm up or cool down? Why?

186 Chapter 6 Thermal Energy

Activity Students can use the following formulas to convert from the Fahrenheit scale to the Celsius scale, and vice versa.

$$°F = \frac{9}{5}°C + 32 \qquad °C = \frac{5}{9}(°F - 32)$$

Example at 25°C:

$$°F = \frac{9}{5} \times 25 + 32$$

$$°F = 77$$

Example at 41°F:

$$°C = \frac{5}{9}(41 - 32)$$

$$°C = 5$$

Behaviorally Disordered Use wrapping paper or newsprint, yardstick, felt-tip pen, and tape to have students practice making divisions on a scale.

Draw a line about 2 feet long on the wrapping paper or newsprint taped to the floor. Have students use the yardstick and the diagram to divide the line into three equal parts. Have them verify with the yardstick that each part is about 8 in. long. **LEP** **COOP LEARN**

Temperature Scales

As you wake up one morning, you turn on the radio and the weather reporter says it's 30 degrees outside. Should you put on a sweater or a T-shirt? In order to answer this, you need to know what scale the weather reporter was using to measure the temperature. In the next activity, you will find out how to make a temperature scale.

 Find Out! ACTIVITY

How do you make a temperature scale?

What To Do

1. Prepare a beaker of ice water. Place an unmarked alcohol thermometer in the ice water and put the beaker on a hot plate.

2. Using a stirring rod, stir the mixture as the ice melts and watch the thermometer. What happens to the level of the thermometer while the ice is melting? Mark that level on the thermometer.

3. As the water warms and eventually boils, keep watching the thermometer level. What happens while the water boils? Mark that level, too.

4. Now divide the space between the two points into 20 equal divisions. Mark each division on the thermometer with a fine-point marker and a metric ruler, as shown in the figure. Number each division starting at the melting point mark. Each division is one degree on your scale.

5. Name your scale and use it to measure the temperature in various places. *In your Journal,* record these measurements, making sure you identify the scale.

Conclude and Apply

1. How would your measurements be different if there were twice as many divisions on your thermometer?

2. Would the thermometer be as helpful if the divisions were not equal? Why?

 **SKILLBUILDER**

Comparing and Contrasting

Compare and contrast the method used to measure temperature in the following thermometers: liquid-in-glass, thermistor, and liquid crystal. If you need help, refer to the **Skill Handbook** on page 642.

Would it be easy to tell whose city were warmer? Why? *No, because if it were 10° on your thermometer, it would be 20° on your friend's. You could convert the temperature by multiplying your degrees by 2.*

Visual Learning

Figure 6-5 Does the temperature at which water melts or boils actually change from scale to scale? *No.*

Purpose

A Closer Look relates to the concepts of Section 6-1, in which students observed temperature changes as hot or cold water reached its equilibrium temperature.

Teaching Strategies

A burn caused by steam at 100°C is much more severe than one caused by water at 100°C. Ask students to discuss this difference and to hypothesize why this is true. The difference is in the latent heat of steam. When steam condenses on the skin, the latent heat released in the process is absorbed by the skin.

Explore! ACTIVITY

How do we feel "heat"?

Have you ever put your hand in your bath or shower water to make sure it's at a comfortable temperature, or felt your forehead to see if you have a fever? If you have, you were relying on your sense of touch to estimate temperature. But how accurate is your sense of touch?

What To Do

1. Collect a group of familiar objects made of various materials, and place them on a table together, leaving them for several minutes.

2. Use the underside of your wrist to feel which objects feel warmest and which ones feel coldest. This area of your skin is very sensitive to hot and cold objects.

3. *In your Journal,* estimate the temperature of each object.

Answers to
You Try It!

The beaker with water only will reach 20°C before the beaker with ice and water because some of the heat added to the ice and water was used to melt the ice.

Going Further ⫸

Have small groups of students discuss how latent heat helps explain how perspiring cools a body. Students might experiment by putting water on the back of their hands and observing what happens to the temperature as the water evaporates. Use the Performance Task Assessment List for Group Work in **PASC,** p. 97.

COOP LEARN

begin with? How did the heat outdoors get into your homes? How do you think you could make your homes cooler in the hot weather? Changing things from hot to cold or from cold to hot requires us to look at the concept of transferring thermal energy. You'll explore that in the next section.

check your UNDERSTANDING

1. Suppose you remove a spoon from a cup of hot cocoa. The cocoa has a temperature of 80° C and the room is at 23° C. Sketch a graph of the temperature of the spoon versus time after the spoon is removed from the cocoa.
2. Describe the flow of thermal energy between the spoon and the air when the spoon was the same temperature as the room.

3. Using the Celsius scale, what temperature would be midway between the freezing point and boiling point of water? What are the temperatures of the following: melting ice, boiling water, and room temperature?
4. **Apply** How would the graph you sketched in Question 1 be different if you had measured temperature using the Fahrenheit scale instead of the Celsius scale?

steam has more internal energy than 100° C water.

Latent heats are the same for freezing as for melting. The same is true for boiling and condensation. You need to add 334 J to 1 gram of ice at 0° C to make it melt, or remove 334 J from 1 gram of water to make it freeze. You must add 2260 J to 1 gram of water at 100° C to make it boil, or remove 2260 J from 1 gram of steam at 100° C to turn it back to water.

You Try It!

In a large beaker, add 200 mL of water to a beaker full of crushed ice. Stir with a stirring rod until the temperature is 0° C. Pour 100 mL of the water only into a second beaker. Pour 100 mL of the water and crushed ice mixture into a third beaker. With a thermometer in each, place both of these beakers on a hot plate and turn it on. Which beaker will reach 20° C first? Record *in your Journal* the temperature of each beaker every 10 seconds. Record how long it takes for each beaker to reach 20° C. How can latent heat explain what you observed?

6-1 Thermal Energy **189**

check your UNDERSTANDING

1. The temperature of the spoon drops to room temperature.
2. Thermal energy would be flowing to and from the spoon at the same rate. There would be no net loss or gain in thermal energy.

3. 50°C; melting or just-formed ice is 0°C; boiling water is 100°C; room temperature is about 20°C to 23°C.
4. The graph would have a similar shape, but it would have a different slope.

3 ASSESS

Check for Understanding

Assign questions 1 to 4 to small groups of students. These questions are all related and you may want students to answer them as an end-of-section project. Use the Performance Task Assessment List for Group Work and the Individual in **PASC**, p. 97.

Reteach

Discussion Students can discuss how they can tell whether something is hot or cold. Have them list items they consider to be hot and those they consider to be cold. Then have them describe how each item would reach thermal equilibrium in a room and whether that would require a loss or gain of thermal energy. L1

Extension

Activity Students can prepare glasses of iced tea and hot tea and leave the liquids to sit undisturbed for 30 minutes. At the end of that time, ask students to explain what happened to the liquid in each glass. *Both tend toward thermal equilibrium.* L3

4 CLOSE

Be sure that students understand that 100°C and 212°F mean the same thing—the temperature at which water boils. Ask students to identify other situations in which a measurement for the same thing is described in different ways. *For example, 8:45 is the same as a quarter to nine.* L1

PREPARATION

Planning the Lesson

Refer to the Chapter Organizer on pages 180A–D.

Concepts Developed

In this section, students will learn about heat as a transfer of energy and will differentiate between heat and temperature. They will explore heat transfer by conduction, convection, and radiation, and will identify materials that reduce heat transfer.

Explore!

How does thermal energy transfer from one object to another?

TECH PREP

Time needed 15 minutes

Materials wooden spoon, plastic spoon, metal spoon, 3 plastic beads, butter, beaker, water, hot plate or electric immersion heater

Thinking Processes comparing and contrasting, observing

Purpose To observe the rates at which different materials transfer thermal energy.

Teaching the Activity

Safety Caution students to work carefully around heat and heated materials. If an immersion coil is used, be sure students unplug it before removing it from the water. The coils should be placed in a heat-resistant container. **L1**

Troubleshooting Spoons should be leaning over the edge of the beaker to prevent convection from the hot water from melting the butter.

Science Journal Have students record their predictions in their science journal. Then have them record the results of the activity and tell if their predictions were accurate.

Section Objectives

- Describe heat as the transfer of energy.
- Distinguish temperature from heat.
- Give examples of heat transfer by conduction, convection, and radiation.
- Identify materials that reduce heat transfer most effectively.

Key Terms

heat, conduction, convection, radiation

What Is Heat?

You may have heard the expression, "If you can't stand the heat, stay out of the kitchen." In fact, you've probably used the word "heat" many times. But what exactly is heat? Is it the same as temperature?

Heat is related to temperature, but it is not the same as temperature. Temperature measures how much thermal energy there is in an object. **Heat** is the energy transferred from something of higher temperature to something of lower temperature.

When you feel a beaker of warm water, thermal energy is transferred from the water to your skin.

Suppose you place warm water in a cold container. The heat moves from the warm water to the cold container. The transfer of thermal energy will continue until the two objects are in thermal equilibrium.

If heat is energy in transit, how does it move? What is it that takes thermal energy from one place to another? Let's explore.

Explore! ACTIVITY

How does thermal energy transfer from one object to another?

What To Do

1. Gather a wooden spoon, a metal spoon, and a plastic spoon, all about the same length. Stick a bead to the handle of each spoon the same distance from the other end, using a dab of butter.

2. Stand the spoons in a short beaker so that the beads are hanging over the edge of the beaker. If the butter melts, the beads will fall. *In your Journal*, predict the order in which the beads will fall if hot water is added to the beaker.

3. Boil water in another beaker and carefully pour about 5 cm of the hot water into the beaker holding the spoons. Observe the order in which the beads fall. Did other groups get the same results?

Expected Outcome

Students will observe that metal transfers thermal energy best of the three materials tested. The bead will drop first from the metal spoon, then the plastic spoon, then the wooden spoon.

✔ Assessment

Process Tell students that they have three coins that have been painted to look alike. However, one is made primarily of silver, another is made primarily of aluminum, and the third is made primarily of brass. Have students devise an experiment based on what they learned about heat conduction to tell what each coin is made of. Use the Performance Task Assessment List for Designing an Experiment in **PASC**, p. 23. **COOP LEARN**

From the Explore activity, you could infer that thermal energy traveled from the hot water to the spoons. Did you see any evidence of energy moving from the water to the spoons? If you held a spoon about 10 centimeters above the water, do you think it would get warm?

■ Conduction

In the Explore activity you just completed, you saw that heat was transferred along the spoons. This process—heat moving through a material or from one material to another—is known as **conduction**. That's what happened with the spoons. The heat flowed up the spoon to melt the butter.

In order for thermal energy to transfer by conduction from one object to another, objects must be in physical contact. Conduction transfers heat from a beaker of hot water to the butter via the spoons.

Figure 6-7

Ⓐ The properties of various materials, including their ability to transfer heat are taken into account when designing cooking utensils. Because metals, such as copper and aluminum, are good conductors of heat, they are natural choices for pots and pans.

Ⓑ The metal bottom of the pan transfers thermal energy from the burner to the food. Why are the handles of many pots and pans made of wood?

Figure 6-6

Ⓐ To measure how well various substances transfer heat, you need a standard to measure them against. Silver is a good standard because it conducts heat very well. Silver is given a value of 1.

SILVER
1.00

WOOD
0.01

Ⓑ The ability of other substances to transfer heat is given as a number that compares that substance's ability to conduct heat with the conducting ability of silver.

Ⓒ The ability of wood to conduct heat is only about 0.01 of the conducting ability of silver.

WATER
0.0014

Ⓓ Which is a better conductor of heat: air or water?

AIR
0.00005

Visual Learning

Figure 6-6 **Which is a better conductor of heat: air or water?** *water*
Figure 6-7 **Why are the handles of many pots and pans made of wood?** *Wood is a poor conductor of heat. This allows a person to handle a hot metal pot without getting burned.*

Program Resources

Study Guide, p. 23
Concept Mapping, p. 14, Thermal Energy L2
Teaching Transparency 11 L2
Take Home Activities, p. 12, How Heat Travels in Air L1
Section Focus Transparency 17

1 MOTIVATE

Bellringer

 Before presenting the lesson, display **Section Focus Transparency 17** on an overhead projector. Assign the accompanying **Focus Activity** worksheet. L1
LEP

Activity This activity allows students to see how they can affect the movement of thermal energy.

Materials needed are lamp with bare bulb, paper plate, foam plate, china plate, metal plate or pie pan.

Have students turn on the lamp and hold a hand near it. CAUTION Be sure students do not touch the bulb. Then have students put each plate, one at a time, between the bulb and their hands and compare the warmth they feel. Ask if some materials block heat better than others. L1

2 TEACH

Tying to Previous Knowledge

In the previous section, students learned that heat moves from warmer to cooler objects. In this section, they will learn about three methods of heat transfer.

Student Text Questions
Did you see any evidence of energy moving from the water to the spoons? *You couldn't see the energy moving, but you could infer that it had because the butter melted.* **If you held a spoon about 10 centimeters above the water, do you think it would get warm?** *no*

Connect to...
Life Science

The amount of wind affects the rate at which heat is lost. It results in cooling a person and can lead to hypothermia. Prepare a short talk to inform your classmates of the symptoms and dangers associated with hypothermia.

■ Convection

Liquids and gases are called fluids because they flow. Fluids conduct thermal energy poorly. However, you've seen many cases where thermal energy is transferred by a fluid. When you feel heat from a hair dryer, the heat is contained in the moving air. Warm winds carry heat from one place to another on Earth. Warm ocean currents carry heat from the tropics to cooler northern climates. Warm air is blown through ducts to heat a home. In all of the above examples, something carries heat as it moves from one place to another. A material that carries something else is known as a medium. A fluid such as air or water can be a medium for carrying heat. **Convection** is heat transfer by motion of a heat-carrying medium. Can you think of any other examples where you have seen heat transferred by convection?

You know that the air near the ceiling of a room is generally warmer than the air near the floor. This can be an important thing to remember if you are in a burning building. The warmer smoke will rise to the ceiling, and the cooler air will be found near the floor. That means to avoid the smoke, you can crawl along the floor.

In the next Investigate, you will explore convection using a colored ice cube.

Convection Beneath Your Feet

In Chapter 2 you learned about plate tectonics, the theory that explains how mountains form and why earthquakes and volcanic eruptions occur in some places but not in others. But what makes plate tectonics work? If you look back to pages 68-69, you'll learn that Earth's outer crust is made of many rigid pieces, called plates, that move around because they float on the denser molten rock forming the mantle beneath. This denser molten rock is called Earth's mantle. It surrounds the extremely dense core.

The Energy Comes from the Core

The pressure at Earth's center is calculated to be about 3.7 million atmospheres—that is, about 3 700 000 times the surface pressure at sea level. This tremendous pressure makes the core very hot. Heat radiates outward from the core and warms the mantle, which is a putty-like solid.

Since the mantle surrounds the core, the energy heating

Purpose

Earth Science Connection illustrates how the transfer of thermal energy affects the solid planet on which we live. In this connection, students learn that energy transferred from the very hot center of Earth to regions near Earth's crust contributes to the movement of Earth's crustal plates.

Content Background

Although scientists are not in total agreement about the actual mechanism, most agree that convection cells within Earth's interior are responsible for the movement of tectonic plates. The uppermost portions of the convection cells in the mantle are in contact with the bottom of the lithosphere. As the mantle material moves, friction causes the material of the lithosphere to move along with it.

Teaching Strategy

Use the analogy of an object floating in water to represent the plates "floating" in

Figure 6-8

Heat transfer by convection is possible because warm air is less dense than cold air. When a radiator is hot, it warms the air touching it by conduction.

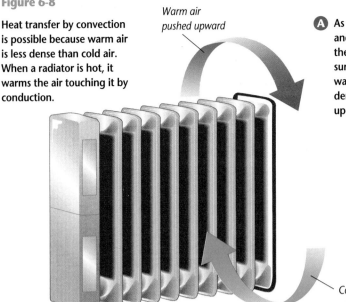

Warm air pushed upward

Ⓐ As the air warms, it expands and becomes less dense than the surrounding air. The cool surrounding air slides under the warm air. Because it is less dense, the warm air is pushed up above the cold air.

Ⓑ In time, the cold air is also warmed and is pushed upward by more cold air. The stream of rising warm air is called a convection current.

Cool air slides underneath the warm air

the mantle comes from underneath. You already know what happens when a fluid is heated from the bottom: convection occurs. As the mantle material warms, it expands and becomes less dense. It is pushed upward by cooler, denser material flowing in underneath. In the circulation that results, heat is carried upward toward Earth's surface.

The convection currents move slowly, compared to those you can see in a pot of boiling water. Earth's convection currents are estimated to move at less than a few centimeters a year. But slow or not, Earth's convection currents are an important part of the theory of plate tectonics.

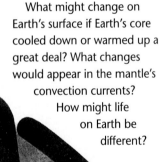

What Do You Think?

What might change on Earth's surface if Earth's core cooled down or warmed up a great deal? What changes would appear in the mantle's convection currents? How might life on Earth be different?

the mantle. As water around the floating object moves, it carries the object with it. Basically, this is what happens to cause the plates of the lithosphere to move.

6-2 Watching Ice Melt

Planning the Activity

Time needed 30 minutes

Purpose To observe heat transfer by convection and to devise a method of reducing heat loss from a fluid by convection.

Process Skills making and using tables, observing and inferring, comparing and contrasting, measuring in SI, interpreting data, predicting

Materials See student activity.

Preparation Use a few drops of vegetable food coloring to dye the ice cubes. Prepare these dyed cubes several hours before students start the experiment.

Teaching the Activity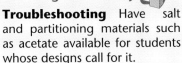

Troubleshooting Have salt and partitioning materials such as acetate available for students whose designs call for it.

Process Reinforcement When students are measuring temperature at different depths, make sure they hold the thermometer at each level until the liquid in the thermometer has stopped moving. L1

Possible Hypotheses Students may hypothesize that increasing the density of the water or insulating the beaker will slow down the convection.

Possible Procedures Students may suggest adding salt to the warm water to increase its density or partitioning the beaker to reduce convection currents.

Science Journal Have students record their hypotheses and results in their journals.

Expected Outcomes

Students should observe that the temperature of the water and ice mixture decreases as they measure lower in the beaker. This is because the warmer water rises while the colder water sinks in the beaker. Results of the student-devised activity will vary with the designs.

INVESTIGATE!

Watching Ice Melt

By now you can guess that if you put an ice cube into a glass of water, it will melt. And you can explain why—thermal energy moves from the warm water to the colder ice cube. But you may not be able to explain how the thermal energy moves from the water to the ice cube. Try this next activity to discover how convection transfers heat from one place to another.

Problem

How does convection help an ice cube melt?

Materials

You may want to choose from the following list of materials.

colored ice cubes	tongs
thermometer	250-mL beaker
water	salt
stirring rod	

Safety Precautions

Use caution if using a mercury thermometer.

What To Do

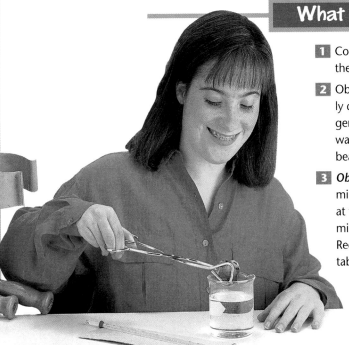

1 Copy the data table *into your Journal.* Fill the beaker with warm water.

2 Obtain an ice cube that has been strongly dyed with food coloring. Using tongs, gently place the ice cube into the warm water. Do not stir or mix. Keep the beaker as still as possible.

3 *Observe* the ice-water mixture for several minutes. Then, *measure* the temperature at the surface and at the bottom of the mixture, and at three levels in between. Record the temperatures in your data table.

Program Resources

Activity Masters, pp. 29–30, Investigate 6-2 L2

How It Works, p. 9, How Does a Vacuum Bottle Work? L2

Making Connections: Integrating Sciences, p. 15, Your Body's Thermostat L2

Science Discovery Activities, 6-2

As glaciers flow into the sea, huge masses of ice break off. These chunks of broken off glaciers are called icebergs. Convection in the water causes the portion of the iceberg below water to melt much faster than the part above water.

4 Empty the beaker and refill it with warm water. Work with your group to plan a way to reduce the heat transfer through convection. Refer to the definition of convection to help you devise a way.

5 Show your design to your teacher. If you are advised to revise your plan, be sure to check with your teacher again before you begin. Look at all of the safety precautions before you begin. Carry out your plan.

Sample data

Data and Observations	
Depth	Temp (°C)
Surface	22
2 cm	21
4 cm	19
6 cm	18.5
Bottom	15.5

Analyzing

1. As the ice melts, what happens to the meltwater? What happened to it when you tried to prevent convection?

2. Does the meltwater blend readily with the warm water, or does it tend to remain separate from it?

3. Describe the changes in temperature within the meltwater-warm water mixtures.

Concluding and Applying

4. Infer why convection occurs between the water and meltwater. How did your design interfere with this process?

5. ~~Going Further~~ How could you assist the process of convection and make the ice cube melt faster? Refer to the definition of convection to help you think of a way.

Answers to Analyzing/ Conclude and Apply

1. The meltwater sinks to the bottom of the beaker. Increasing the density of the warm water slows down the rate of energy transfer by convection.

2. The meltwater and warm water do not mix readily.

3. Temperature changes sharply at the boundary of the colored layer.

4. Convection occurs when a dense fluid overlies a layer of less dense fluid. Student responses will vary according to their designs.

5. Students may suggest heating (adding thermal energy to) the water in the bottom of the beaker.

✔ Assessment

Process Have students predict the effect on the rate of heat transfer by convection within the water if they place a beaker of water and ice on a block of ice. Use the Performance Task Assessment List for Group Work in **PASC**, p. 97. `COOP LEARN`

GLENCOE TECHNOLOGY

 Software

Computer Competency Activities
Chapter 6

Meeting Individual Needs

Visually Impaired Allow these students to put a finger gently in the beaker of warm fresh water with ice cubes added to see if they can feel a temperature difference between the top and bottom layers of water. Have them do the same for the beaker of salt water with ice added. Have them describe what they feel. `LEP`

Program Resources

Laboratory Manual, pp. 27–32
Laboratory Manual, pp. 33–34
Laboratory Manual, pp. 35–38

Demonstration This activity will show the presence of a convection current in water. Fill a heat-proof glass beaker about three-quarters full with water. Add two or three spoonfuls of sawdust. Heat the beaker and have students describe the motion of the sawdust.

Across the Curriculum

Language Arts

Students can research how arctic animals can survive in extremely low temperatures. Students can use what they learn to write stories from the animal's point of view about living in a cold environment.

Content Background

In order for conduction or convection to take place, matter must be present. Radiation, the third type of heat transfer, does not require matter. Radiant energy is transferred in invisible waves.

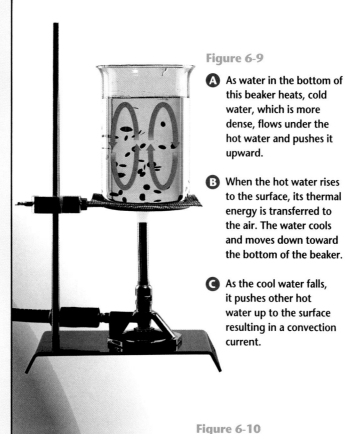

GLENCOE TECHNOLOGY

Videodisc

STVS: Physics

Disc 1, Side 1

Images of Heat (Ch. 8)

Solar Food Dryer (Ch. 12)

Figure 6-9

A As water in the bottom of this beaker heats, cold water, which is more dense, flows under the hot water and pushes it upward.

B When the hot water rises to the surface, its thermal energy is transferred to the air. The water cools and moves down toward the bottom of the beaker.

C As the cool water falls, it pushes other hot water up to the surface resulting in a convection current.

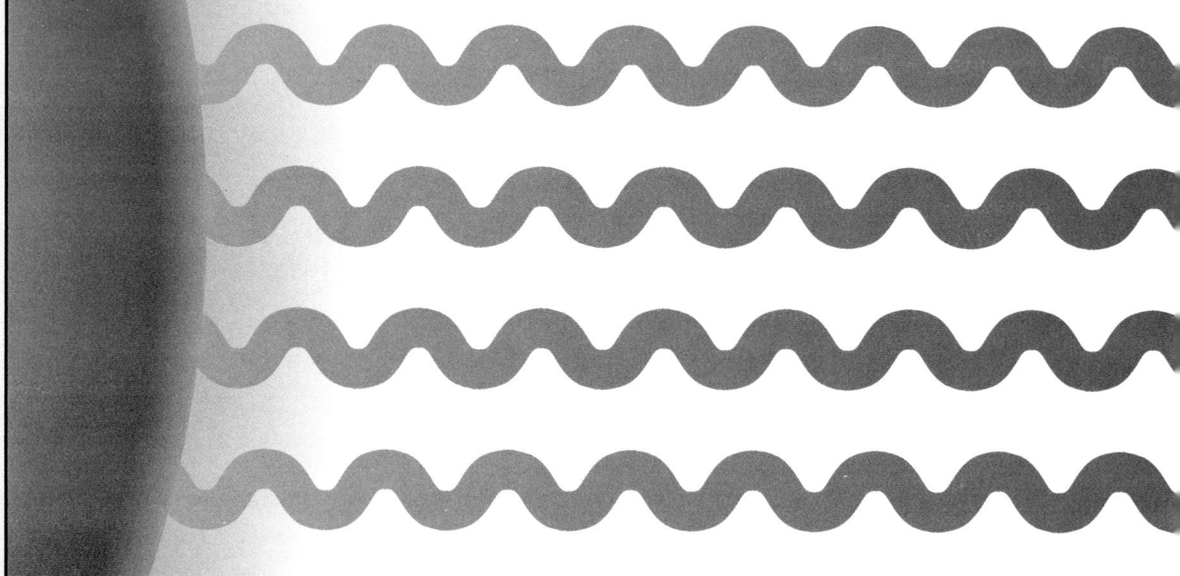

You can see the same process you saw in the Investigate in a pan of boiling water. **Figure 6-9** shows how convection occurs in a boiling pot. Convection is also a major cause of winds. Sailplanes are gliders that soar on the convection currents in the atmosphere. Many currents in the ocean result from the process of convection. As you can see, convection affects us in many ways.

Radiation

Imagine walking outside on a bright, sunny day. As you look up at the sky, you can feel the sun's warmth against your face. But how did the thermal energy from the sun reach your face? How could the heat you feel get across the millions of kilometers of space? In outer space, there is not enough matter to transfer heat through conduction or convection. So

Figure 6-10

A Radiant energy from the sun travels 150 million kilometers through mostly empty space to reach Earth.

Meeting Individual Needs

Visually Impaired Have visually impaired students feel samples of insulating materials such as plastic, foam, quilts, and down jackets. Emphasize materials in which students can feel air pockets. Have students suggest where the use of each material is appropriate.

ENRICHMENT

TECH PREP

Activity Students can visit building supply stores and investigate the types of insulating materials available. They can make a chart naming the material, its R-value, and its uses. L1

Conduction Conduction is the transfer of thermal energy by two objects in contact.

Examples:
• A pan of food on an electric stove gets warm.
• A spoon in boiling water gets hot.
• A glass of ice water is warmed by the hand holding it.

Conduction can occur in a solid, liquid, or gas.

Convection Convection is the transfer of thermal energy by the movement of matter.

Examples:
• Water boiling in a pot
• Warm air rising above a fire
• Moving air currents causing wind

Convection can occur in a liquid or a gas.

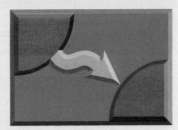

Radiation Radiation is the transfer of thermal energy across a space.

Examples:
• The heat from the sun
• The heat from candles
• The heat that browns marshmallows held near a fire.

Radiation does not require intermediate matter to transfer thermal energy.

Teacher F.Y.I.

Radiant energy includes light, radio waves, and other forms of radiation besides heat. Of the radiant energy Earth receives from the sun, about 19 percent is absorbed by the atmosphere, 34 percent is reflected by the atmosphere and Earth's surface, and about 47 percent is absorbed by Earth's surface.

Visual Learning

Figure 6-9 Help students trace the transfer of energy in the beaker through conduction and convection.
Figure 6-10 Help students compare and contrast radiant energy and thermal energy. Emphasize that sunlight must be absorbed before it can be changed to thermal energy.

B When this energy reaches Earth, some of it is reflected toward space and some is absorbed. Only radiant energy that is absorbed is changed to thermal energy.

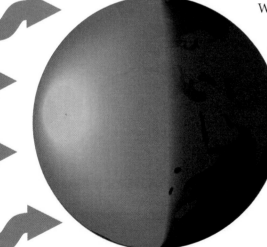

that means the heat gets here another way. That way is called radiation.

Radiation is the transfer of thermal energy across space. No matter is needed for radiant energy to flow as it is for conduction and convection. When you're playing outside on a sunny day, the heat you feel on your face radiated across the vast space between you and the sun.

You've felt also the effects of radiant heat when you put your hand underneath a lighted bulb, or sat next to a campfire. What other examples can you think of that are evidence of radiation?

GLENCOE TECHNOLOGY

 Videodisc

STVS: Chemistry
Disc 2, Side 2
Solar House (Ch. 10)

Solar Tower (Ch. 13)

Geothermal Wells (Ch. 17)

Multicultural Perspectives

Homes Around the World

In different parts of the world, homes are insulated from heat or cold in different ways. Display photographs of different types of homes such as pueblos built of adobe and igloos built of ice blocks. Have students identify the insulating material and classify whether it blocks heat from entering or leaving the dwelling. Ask students to share their knowledge of insulating materials in their countries of origin.

Discussion Have students list times when they do not want heat to be transferred and how they might accomplish their goal. Some suggestions:

Item	Slows Down Heat Transfer
hot cocoa	foam cup, thermos
cold drink	foam cup, thermos

For each example, have students discuss in which direction the heat transfer is blocked. *The foam reduces the rate at which heat leaves the cocoa. The foam slows down the rate at which room heat warms the cold drink.*

Visual Learning

Figure 6-11 Students can compare and contrast the vacuum bottle with large coolers: although some coolers are constructed like vacuum bottles, others are made of foam insulation.

Figure 6-12 Students can relate a penguin's insulation to clothing. In cold climates, people often wear several layers of clothes; the air trapped between layers acts as an insulator.

Minimizing Heat Transfer

Sometimes, you'd rather stop thermal energy from being transferred as it normally would. When it's hot outside, it would be nice to keep the heat out of your room. When you're packing a cold or hot drink to take on an outing, you want to keep it from becoming lukewarm. When you heat your home in winter, you want to keep the energy inside. To do all these things, you need to reduce all three types of thermal energy transfer as much as possible.

You can reduce conduction loss from an object by surrounding it with insulating materials. Because insulators conduct heat very poorly, heat is conducted in or out of an insulated object or room very slowly. Many homes use fiberglass insulation, which contains many pockets of air. Air, as you saw, is a very poor conductor. Containers that keep food hot or cold also use air as a barrier to heat transfer, as shown in **Figure 6-11**.

Heat can't escape from a home by convection unless the air carrying the heat leaves the house. To keep heat inside in winter and outside in

Figure 6-11

A vacuum bottle keeps hot liquids hot or cold liquids cold for several hours. Vacuum bottles are constructed to reduce the rate of all three forms of heat transfer.

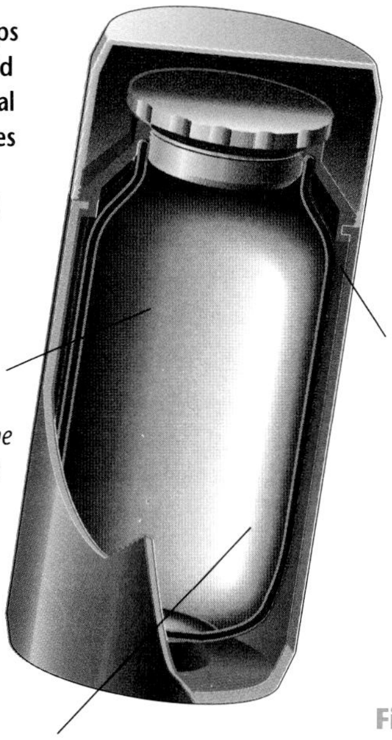

Vacuum bottles are made of two glass bottles, one within the other. Glass does not transfer heat well, so heat transfer by conduction is slow.

Most of the air in the space between the two bottles is pumped out and then the bottles are sealed together at the top. There is so little air in the partial vacuum between the bottles that heat transfer by convection or conduction cannot take place.

The sides of the bottles that face each other are coated with a reflecting material, which reflects most of the heat coming from the liquid or from outside air back to its source. The reflecting coating cuts down on heat transfer by radiation.

Figure 6-12

Feathers and layers of fat enable penguins to withstand the cold. An outer layer of oily feathers keeps them waterproof. An inner layer of fluffy down feathers and thick layers of fat act as insulators and keep penguins from losing body heat through conduction.

Multicultural Perspectives

Universal Taps

Markings on hot and cold water faucets are not the same in all countries. For example, in France, C stands for *chaud* (French for "hot"). If you have students from different cultures in your class, ask them to say and write their words for *hot* and *cold.* Ask students to devise universal markings for hot and cold water faucets.

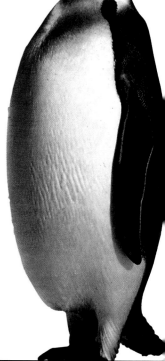

Figure 6-13

A The film used in a thermogram records the radiation from thermal energy instead of light. In the photograph, warm temperatures look orange or red. Cooler temperatures are blue or green.

B Many home insulation materials have a shiny foil backing that reflects radiation. Escaping heat is reflected back into the home. Heat from the sun that strikes the building is reflected away. Thus, insulation helps homes stay warmer in winter and cooler in summer.

summer, doors and windows need to be well sealed.

This brings us back to the discussion about heat that you and your friend were having on the phone. You should now know where all the thermal energy has come from. But start to think about this: If all this heat you feel is a form of energy, and energy can do work, how much work could the thermal energy in your room do? You'll find out about that in the next section.

check your UNDERSTANDING

1. Describe the transfer of energy as a pan of water is heated to boiling on a gas stove.
2. Explain how a marshmallow can cook even when you hold it next to, but not in a fire.
3. Suppose you needed to keep water hot inside a container for as long as possible. How would you construct the container to meet your goal? Explain your design.
4. Water has a very low thermal conductivity. Yet, when you heat water in a pan, the surface gets hot very fast—even though you're applying the heat to the bottom of the water. Why?
5. **Apply** Which of these materials would best reduce the rate of loss of thermal energy by radiation: iron, wood, or silver? Why?

6-2   Heat and Temperature   **199**

check your UNDERSTANDING

1. Energy transfers by conduction from the flame to the pan bottom when the flame is touching the pan. Conduction transfers heat from the pan bottom to the water. Energy transfers throughout the water by convection.
2. It will cook by heat transferred by radiation.
3. Responses may vary, but students may suggest ways to limit one or all three methods of heat transfer such as constructing the container of a material that does not conduct thermal energy well.
4. The hot water moves to the top by convection.
5. Silver; although it is a good conductor, silver is very shiny, so it reflects radiation.

3 ASSESS

Check for Understanding

Make a chart on the chalkboard with columns labeled conduction, convection, radiation, insulation. Students can give examples for each column. Use the Performance Task Assessment List for Group Work in **PASC**, p. 97. **COOP LEARN**

Reteach

Activity Have students do the following activities to differentiate among types of heat transfer. Seat six students in a row across the front of the classroom to represent a solid. Have them pass a ball along the row to represent heat being conducted through a solid. Now have the six students represent a fluid by moving their chairs apart so that their fingertips can barely touch those of their neighbors. Have them pass the ball to represent conduction of heat through a fluid. Have students walk in a circle around the room, carrying the balls, to represent convection. Have three of the students stand on one side of the room and the other three on the opposite side. Have them toss the ball back and forth to represent thermal energy being transferred by radiation. **L1** **LEP** **COOP LEARN**

Extension

Activity Form teams to design an energy-efficient home. Each member of the team draws a house plan, and the team selects the best one. Then each member researches one aspect of the new house for the best energy conservation ideas. **L3**

4 CLOSE

Have each student list at least one item he or she uses at home or in school that transfers heat by conduction, convection, and radiation. Make a three-part master list of all correct suggestions. **L1**

6-2   Heat and Temperature   **199**

PREPARATION

Planning the Lesson

Refer to the Chapter Organizer on pages 180A–D.

Concepts Developed

In this section, students will describe some of the work heat can do in heat engines and learn how cooling machines such as refrigerators and air conditioners transfer heat. Students will also discuss the efficiency of these machines.

1 MOTIVATE

Bellringer

Before presenting the lesson, display **Section Focus Transparency 18** on an overhead projector. Assign the accompanying **Focus Activity** worksheet. L1

LEP

Explore!

Why is a heat engine inefficient?

Time needed 10 minutes

Materials car and driver

Thinking Processes observing and inferring, recognizing cause and effect

Purpose To observe the increase in temperature of a car engine resulting from operating the engine and to infer that much of the thermal energy of the engine is not converted to useful work.

Teaching the Activity

Science Journal Students should infer what happens to much of the thermal energy produced by a heat engine.

Expected Outcome

Students should observe that thermal energy produced by operating the car engine is transferred to the hood of the car. They should infer that this heat is "lost" energy—it does not accomplish useful work.

Answer to Question

The hood became warmer.

✔ Assessment

Process Have students investigate other heat engines used at home to determine if heat is a by-product. Students can make a poster with a picture of the device in which the engine is found, a description of how it is used, and the results of their investigation. Use the Performance Task Assessment List for Posters in **PASC**, p. 73. **COOP LEARN**

Making Heat Work

Section Objectives
- Describe how heat can produce work.
- Explain why a heat engine can't be 100 percent efficient.
- Describe how an air conditioner or refrigerator works.

Key Terms
heat engine

Work from Heat

You've already seen heat do work throughout this chapter. Every time you used a thermometer, the transfer of heat did some work on the column of liquid. The energy made the liquid expand as it warmed.

Explore! ACTIVITY

Why is a heat engine inefficient?

What To Do

1. You'll need some help from a teacher or parent for this exercise. Feel the hood of a car.

2. Have the car's driver start the engine and let it run for 5 minutes. Then feel the hood again. What's different?

Figure 6-14

Automobiles run as a result of the transfer of thermal energy in their gasoline engines. Most automobile engines have pistons that move either up and down or back and forth. A part called a crankshaft changes this up-down or back-forth motion to rotary motion, which turns the automobile's wheels. The power to move the pistons comes from the energy released by burning gasoline.

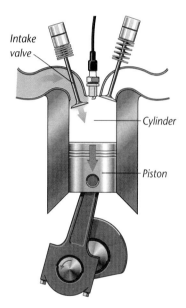

Intake valve

Cylinder

Piston

A *Intake Stroke* The piston moves down the cylinder and draws in the fuel-air mixture—fine droplets of gasoline mixed with air.

What makes heat engines inefficient?

In 1824, the French scientist/engineer Sadi Carnot investigated the amount of work a heat engine produces. He found that even a perfect engine with no friction at all could only use a certain fraction of thermal energy from the fuel. The exact fraction depends upon how much heat the fuel creates and how much the hot exhaust gases can cool in the engine while it

runs. But even if the exhaust gases cooled until they were in thermal equilibrium with the surrounding air, there would still be thermal energy left in them. That thermal energy cannot do work because, to make it do so, heat would have to move from a place of lower temperature to a place of higher temperature.

Running on Thermal Energy

One way thermal energy is used to do work is in a car engine. A car engine is an

example of a **heat engine.** That's an engine that uses fuel to make thermal energy do work. Heat engines are very inefficient because most of the heat they produce never does work. The heat produced by an engine expands gas, doing work, but it also gets transferred to various engine parts and the air surrounding them. Look at Figure 6-14 to see how a heat engine works. What parts do you think will heat up, wasting thermal energy?

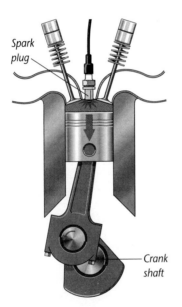

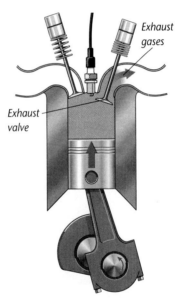

B *Compression Stroke* The piston moves up. The fuel-air mixture is compressed into a smaller space.

C *Power Stroke* When the piston is almost at the top, a spark ignites the mixture. Hot gases expand, forcing the piston down. Energy is transferred from the piston to the wheels of the automobile.

D *Exhaust Stroke* The piston moves up again, compressing and pushing out the waste products left over from burning the fuel-air mixture.

6-3 Making Heat Work **201**

Discussion Challenge students to name a kind of heat engine that makes things cold. Remind students that the transfer of heat is an example of work and that heat engines do work.

2 TEACH

Tying to Previous Knowledge

In the previous section, students learned that thermal energy does work in a thermometer. In this section, students will explore other systems that use thermal energy to do work. They will also explore the low efficiency of these systems.

Theme Connection In this section, the themes of energy and systems and interactions are closely integrated in the descriptions of heat engine systems that use thermal energy to produce work or to transfer heat.

How Do We Know?

What parts do you think will heat up, wasting thermal energy? *Students should identify those parts that have contact with the exploding gas. Also, moving parts in contact with each other will heat up due to friction.*

How can doing work cool something?

Time needed 20 minutes

Materials eyedropper, rubbing alcohol, inner tube, air pump

Thinking Processes observing and inferring, comparing and contrasting, recognizing cause and effect

Purpose To observe changes in temperature due to the transfer of thermal energy that takes place during the process of evaporation and during the expansion of a gas.

Teaching the Activity

Students will note that the air pump becomes warm as the air in it is compressed. This compression causes the air to gain thermal energy and become warmer. This warm air then moves into the inner tube. **L1**

Expected Outcome

Students should observe that their wrists cooled as the alcohol evaporated, that the inner tube became warmer as air inside it was compressed, and that the valve stem cooled as air was released from the tube.

Answers to Questions

1. The valve feels warm. Work is done by moving air from outside to inside the tube.

2. The valve is cool. The cooling produced by the transfer of thermal energy as a gas expands can be used in devices to produce a cooling effect.

✔ Assessment

Oral Have students explain the differences in the transfer of thermal energy that occur when an inner tube expands and when the air expands as it escapes into the surroundings. Use the Performance Task Assessment List for Group Work in **PASC**, p. 97.

Keeping Cool with Thermal Energy

Another kind of heat engine is one that makes air cold, such as a refrigerator or air conditioner. Instead of using fuel to produce heat to do work, a refrigerator uses fuel to do work to transfer energy as heat from a lower temperature to a higher temperature. You can examine the basic principles involved in refrigeration by doing the next activity.

Explore! ACTIVITY

How can doing work cool something?

What To Do

1. Fill an eyedropper with alcohol and put a few drops on the inside of your wrist. Record your observations *in your Journal.*

2. Next, inflate an inner tube with an air pump. After the tube is filled, feel the valve. Is it warm or cool? Have you done any work? Let it stand for about 10 minutes. Open the valve and let the air out. Feel the valve again. Is it warm or cool? How might this be helpful in cooling something?

DID YOU KNOW?

Laboratory measurements have shown that the maximum efficiency of gasoline engines is less than 30 percent.

You learned two important properties of fluids that the inventors of refrigerators noticed. The first property is liquids evaporate more as they become warmer. They gain thermal energy from their surroundings. Your wrist cooled when the alcohol gained thermal energy as it evaporated.

The second property has to do with the compression and expansion of air. When air is compressed, it gains thermal energy. When it expands, it loses thermal energy and gets cooler. The work you did while pumping, compressed and warmed the air. When you opened the valve, the expanding air cooled.

Figure 6-15

The icebox was the first home appliance to keep foods cool. By the 1920s, mechanical refrigeration began to replace iceboxes on a large scale.

202

Meeting Individual Needs

Behaviorally Disordered Most students will enjoy doing the Explore activity since the results are easy to achieve and the time span for the first part is short. While waiting the 10 to 15 minutes for the tube to cool you may want to let these students try to evaporate other liquids such as water or salt water on their wrists.

Program Resources

Study Guide, p. 24

Critical Thinking/Problem Solving, p. 14, Insulation for Space Travel **L3**

Multicultural Connections, p. 15, Stove Builders and People's Architects **L1**

Multicultural Connections, p. 16, The Real McCoy **L1**

Figure 6-16

Liquids absorb heat when they vaporize. Refrigerators are designed to take advantage of this property of liquids. The cooling part of a refrigerator is a closed system of tubes in which a cooling liquid, usually Freon 12, is caused to vaporize, to condense back into liquid, and then again to vaporize. This is a continuous cycle.

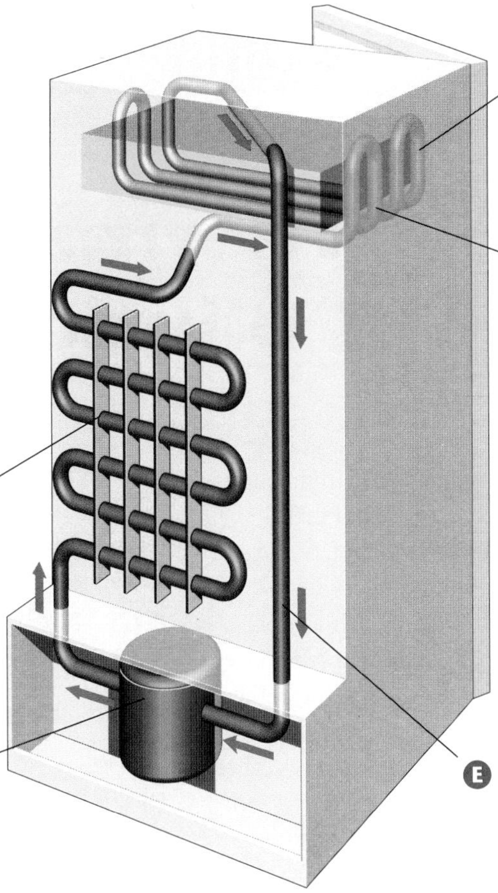

C The liquid is pumped through coils inside the refrigerator.

D The liquid absorbs heat from the food stored in the refrigerator and evaporates, turning back into vapor. This is where the refrigeration takes place.

B The hot vapor moves to the condenser. In the condenser, heat from the vapor is transferred to the surrounding air through the coils along the outside back of the refrigerator. The vapor condenses into a liquid.

A Vapor is compressed by the compressor. Being compressed makes the vapor hot.

E The vapor flows back to the compressor. The cycle begins again.

By the time you and your friend have finished your discussion of heat, the sun has begun to set. You open a window to let cooler air in and the warmer air out. Thermal energy is leaving your room by convection through the open window and—very slowly—by conduction and radiation through the walls and glass. Tomorrow that energy might be almost anywhere. But the sun will return to provide a new supply.

check your UNDERSTANDING

1. Why can heat do work on a piston in a car engine?
2. What prevents heat engines from being perfectly efficient?
3. In an air conditioner or refrigerator, what is the process that actually removes heat from inside the room or refrigerator?
4. **Apply** Could an air conditioner also be used as a heater? Explain what you would have to do and where the heat would come from.

6-3 Making Heat Work **203**

check your UNDERSTANDING

1. Heat causes materials to expand. An expanding fluid or gas can press on a piston, making it move.
2. Some energy is always "lost" in the form of thermal energy transferred to the outside. This energy is not converted to useful work.
3. Heat is transferred from the refrigerator or room to a liquid, where it is used to evaporate the liquid.
4. Yes; when vapor condenses in the coils of the air conditioner, heat is released to the surrounding air. If the air conditioner is placed backward in a window, it would heat indoors, cooling the outside.

6-3 Making Heat Work **203**

3 ASSESS

Check for Understanding

Have students answer questions 1 to 3 individually and then work in groups to answer the Apply question.

Reteach

Activity Have students draw diagrams to show how a refrigerator works. **L1**

Extension

Discussion Have students answer this question. **Before refrigerators were invented, food was kept cold in an icebox, which consisted of an insulated box containing shelves and a door. Why was ice placed on the top shelf of the icebox rather than on the bottom shelf?** *Convection currents formed in the air within the icebox. The rising warm air from the food was cooled when it came into contact with the ice. The cool air dropped and cooled the food.* **L2**

4 CLOSE

Discussion

Pose the following problem: **On a hot day, you leave the refrigerator door open to cool your house. Is this a good idea? Explain.** *No. Any thermal energy the refrigerator removes from the food compartment is transferred to the back of the refrigerator and back into the room. The inside of the refrigerator is getting colder, but the back of the refrigerator is getting hotter.* **L1**

Science and Society

Purpose

Science and Society relates to the concepts of Section 6-3, in which students learned how thermal energy is used in heat engines. Power plants use steam generators, which are a kind of heat engine.

Content Background

In most power plants, water is boiled to make steam. Expanding steam turns a turbine, which turns the generators that make electricity. The energy needed to boil the water comes from the burning of fossil fuels, such as coal or oil, or from a nuclear reactor. After the steam has driven the turbine, it still contains thermal energy. However, there is not enough energy to turn another turbine, so this energy is regarded as waste heat. To efficiently return the steam to the boiler for reheating, it must be condensed back to the liquid state. This process removes the waste heat from the steam and transfers it to the coolant water, which is usually a nearby lake or river.

Teaching Strategy

Waste heat from power plants can be used to heat the power plant or nearby buildings. Students can compare and contrast the positive side, that thermal pollution is eliminated, and the negative side, that there are usually no other buildings close to a power plant. Ask students to compare and contrast the advantages and disadvantages of their own ideas.

Science and Society

Thermal Pollution

Electric power plants use tremendous amounts of water to cool their machines and equipment. Some plants take water from natural bodies of water and return the water once it is used. When used or excess heated water is returned to natural rivers, lakes, or oceans, the higher temperatures may damage the ecology of that water. This dumping of heated water is called thermal pollution.

Power plants are not the only facilities responsible for thermal pollution. Manufacturing plants often use water from a nearby river or lake to cool equipment during the manufacturing process. The warm water is often returned to the river or lake. Researchers have several ways to tell if

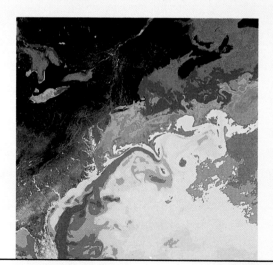

thermal pollution is occurring. One way relies on satellite photographs taken with temperature-sensitive film. In the photo shown here, red areas are warm. This photograph shows warm water from the Gulf Stream current, not from thermal pollution.

Ecological Impact

Hot, or even warm, water drained into a lake or river, or a section of ocean, may raise the temperature of that body of water by as much as 5° to 11° C. This increased temperature may make it difficult for fish to breathe or to incubate their eggs. Also, over time, this increased temperature can disrupt the total ecosystem of a body of water.

An ecosystem is a community of organisms interacting with one another and with the environment. A self-contained area shared by various animal and plant species is an ecosystem. Microorganisms, plants, and animals found in any body of water occur in groups. Some of these groups are dependent on each other. For example, certain kinds of fish may need to be present to preserve certain species of microorganisms that feed in the water. Heated water can allow the introduction of new organisms, such as species of fish and snails normally associated with areas of warmer water.

In essence, the community changes, the balance of life of various species is disrupted, and the original ecosystem is changed.

Thermal pollution also contributes to eutrophication in a body of water. Eutrophication occurs when an excessive amount of nutrients, such as nitrates and phosphates, enters the water and causes an overgrowth of weeds, algae, and other plant life. The nutrient-rich water overfeeds the fast-growing weeds. If too many weeds and algae develop, they use a large percentage of the oxygen in the water.

Eventually fish and other marine life die from suffocation. The body of water becomes overgrown with unwanted plants, and soon the lake or river changes to a dark murky color. The hydrogen in the water combines with other elements and forms foul-smelling gases. When there is no more oxygen in the water, the remaining algae and unwanted weeds die and decay.

Solutions

Today, most nuclear and fossil fuel power plants have cooling towers like those in the photograph. Heated water is stored in these towers until its temperature is similar to the body of water it is to be returned to. Some facilities send the heated water through sectioned-off rivers or down slow-moving winding creeks, whose angles slow the water and give it time to cool. Other facilities have built their own cooling lakes so that they don't have to rely on natural sources of water for cooling.

interNET CONNECTION

Use the World Wide Web to contact the U.S. Environmental Protection Agency, Office of Water, and answer these questions. How does the EPA test for thermal pollution of water? In what other aspects of water management is the EPA involved? Describe how these EPA activities affect water quality in your community.

CAREER connection

Water quality engineers, such as the one shown below, are trained in biology, chemistry, and engineering. They study the composition of lakes, streams, rivers, and oceans. They observe effects of pollution and seek ways to protect the available water on the planet.

interNET CONNECTION

The home page for the U.S. EPA Office of Water can be found at **http://www.epa.gov/OW/**

Going Further ⟩⟩⟩⟩⟩▶

Thermal pollution has caused highway accidents. When the hot water contacts cool air, fog can form. If this happens near a highway, serious accidents can result. Students can write newspaper articles discussing this possibility. Some students can contact state or local police to ask if any such accidents have occurred in your state or community. Other students can do library research to find out about such accidents in other states. One such accident occurred in Tennessee in 1991. Local people blamed it on the steam released by a local paper mill. Use the Performance Task Assessment List for Newspaper Article in **PASC**, p. 69.

CAREER connection

Have volunteers contact the local water company. Have them find out about the education and experience requirements for new employees and the varieties of jobs available for various degrees or backgrounds. Have students share findings with the class.

Consumer Connection

Consumer Connection

Solar Energy for Solar Homes

Purpose

The Consumer Connection expands the ideas of radiation and heat transfer presented in Section 6-2 with an introduction to the types of solar heating systems used in homes.

Content Background

One important factor in planning a solar house is the position of the house. In the northern hemisphere, the solar panels must face south. Also, the angle at which the panels are tilted is important. The angle of the sun's rays, relative to Earth, varies with latitude. The sun's rays are at a steep angle near the equator and at a greater slant as you move toward the poles.

Teaching Strategy

Ask students which side of the school building the sun shines on at noon. They should be aware that the sun is not directly overhead, but in the southern sky. Therefore, it shines on the southern wall of any building for much of the day. Ask how this is important to a solar design. Students should see that the collectors must have a southern exposure to collect the most energy.

Demonstration

Obtain a solar-powered music box, calculator, or other device. Show students the device, pointing out that the solar cell on a calculator is different from the solar collector on a building. The calculator contains a device that converts the sun's radiant energy into electricity. The solar panels on a home collect thermal energy, but they do not change it to another form of energy.

In our search for thermal energy, the sun has been one of the most obvious sources throughout human history. As the ultimate source of much of the energy found on Earth, the sun's heat can be felt radiating from a dark rock long after dusk.

For thousands of years, people have tried to take advantage of this abundant energy source in everything from heating water for baths, to generating electricity and heating their homes.

Today, solar energy is considered a valuable alternative to dwindling fossil fuel supplies. The challenge in using solar energy, though, is in harnessing the radiation that only strikes Earth during the daylight hours.

Collecting Thermal Energy

The most common kind of solar collector is a flat plate, usually installed on rooftops. It works much like a parked car that sits in the open on a sunny day. The sun's thermal energy radiates through the windows and is absorbed by the car's seat. The thermal energy is then trapped inside and cannot easily pass through the glass window. The solar collector works in a similar way. A shallow, black box with a clear lid sits facing south on the roof of a house. The sunlight passes through the collector and hits a plate. The plate absorbs the radiant energy, which then is temporarily trapped in the collector.

There are basically two types of solar heating systems: active and passive. In an active system, the trapped thermal energy is transferred through convection by either water or air. The water or air is circulated throughout the house with the help of pumps or fans. Excess energy is stored in a water reservoir and used when needed.

In a passive solar system, the sun's thermal energy is collected and absorbed by energy-absorbing masses like walls or water-filled drums. At night, these masses radiate the thermal energy. Through convection, this radiated energy can heat a particular space.

You Try It!

Using a shoe box, some plastic wrap, and other materials you can think of, design your own mini solar collector. How can you collect the most thermal energy with your box? How could you use that energy?

Answers to You Try It!

Student responses will vary depending on their designs. They should indicate that collectors should be located to receive maximum hours of direct radiant energy. The energy collected might be used to heat water or heat a room.

Science Journal

Review the statements below about the big ideas presented in this chapter, and answer the questions. Then, re-read your answers to the Did You Ever Wonder questions at the beginning of the chapter. *In your Science Journal*, write a paragraph about how your understanding of the big ideas in the chapter has changed.

1 All objects contain some amount of thermal energy. Energy transferred as the result of temperature difference is heat. *Explain in terms of temperature why objects feel warm and cold to our touch.*

2 Heat always flows from the area of higher temperature to areas of lower temperature. *What happens to a glass of iced tea left in the sun on a warm day?*

3 Thermal energy transfers in three ways. Heat flowing through an object is conduction. Heat carried by a moving fluid is convection. Heat traveling across space is radiation. *Why does blowing on hot food cool it faster?*

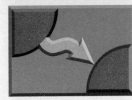

4 Thermal energy may be used to do work, as in an engine. *How could you use thermal energy to turn the blades of a propeller?*

Have students work in groups to illustrate the main ideas of the chapter.

Answers to Questions

1. If an object has a higher temperature than our hands, it will feel warm as heat is transferred from the object to our hands. If an object has a lower temperature than our hands, heat will transfer from our hands to the object, making our hands feel cooler.

2. Heat is transferred from the warm air to the ice-water mixture until all the ice has melted and the liquid is at the same temperature as the surrounding air.

3. Heat is transferred from the food to the moving air, which carries the heat away from the food.

4. Responses may vary. If the propeller is held above a heat source and facing the source, rising convection currents would turn the propeller.

Science Journal

Did you ever wonder...
- Temperature scales are created by finding the freezing and boiling points of water and dividing the space between them into units. (p. 187)
- Insulation reduces the rate of transfer of thermal energy from a hotter object to a cooler object. (pp. 198–199)
- Air conditioners and refrigerators are heat transfer engines. They use evaporation and expansion principles to cool. (pp. 202–203)

Science at Home

Have each student list as many uses as possible for thermal energy at home. After individuals have made lists, have students form groups and consolidate their lists. You may ask students to find photographs from magazines or to make drawings of the ways thermal energy is used in their homes. Encourage creative suggestions that may be made into posters for display. **L1**

GLENCOE TECHNOLOGY

MindJogger Videoquiz

Chapter 6 Have students work in groups as they play the Videoquiz game to review key chapter concepts.

Using Key Science Terms

1. They are all methods of heat transfer.

2. Thermal energy is a form of energy related to the energy of moving particles that make up a material. Heat is a transfer of thermal energy from a region of higher to lower energy.

3. The potential energy in fuel is converted to thermal energy, which is used to do work.

Understanding Ideas

1. Two objects are in thermal equilibrium when they are in contact and are at the same temperature.

2. Metals are good conductors of heat and transfer energy from the heat source to all parts of the pot. Handles should be made of a nonconducting material so that heat from the pot is not transferred to a person's hand.

3. Conduction is the transfer of heat through materials by direct contact; convection is the transfer of heat by moving particles of a material; radiation is the transfer of heat across space with no medium involved.

4. Radiant energy must be absorbed to be changed to thermal energy. Aluminum is a shiny metal that reflects most of the radiant energy that reaches it, and therefore remains cool and keeps the house cooler in warm weather.

5. Insulations that have pockets of air are used to reduce heat loss due to conduction. Insulations that slow down the movement of fluids are used to reduce heat loss due to convection. Insulations that reflect energy are used to reduce heat loss due to radiation.

Developing Skills

1. See student page for concept map.

2. The lower volume of water will take less time to reach thermal equilibrium. The lower

Using Key Science Terms

conduction   convection

heat   heat engine

radiation   thermal equilibrium

Use the terms from the list to answer the following questions.

1. What do convection, conduction, and radiation all have in common?

2. Explain the relationship between heat and thermal energy.

3. Explain the relationship of thermal energy to the work of a heat engine.

Understanding Ideas

Complete the following exercise in your Journal using complete sentences.

1. What do we mean when we say that two objects are in thermal equilibrium?

2. Explain why metal cooking pots generally have handles made of nonmetal materials.

3. What are the differences between conduction, convection, and radiation?

4. At one time, many older homes in the southern part of the United States had aluminum roofs. Explain the benefit of this type of roofing using the principle of radiation.

5. Even the best insulation can't completely prevent the transfer of thermal energy. Describe ways insulations are designed to reduce the rate at which heat is lost by conduction, convection, and radiation.

Developing Skills

Use your understanding of the concepts developed in this chapter to answer each of the following questions.

1. Concept Mapping Complete the concept map of thermal energy.

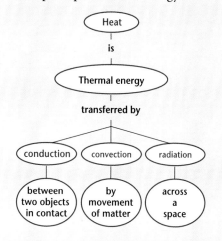

2. Making and Using Graphs Repeat the Investigation on pages 184–185 by adding half the volume of water to the crushed ice. Chart and graph your data and compare with the original results. Did the addition of the new volume of water decrease or increase the time required to change the temperature of the container of water? Why?

3. Predicting Predict what will happen if, on a hot day when the outside temperature is 90° F, you leave all the windows open while the air conditioner is set at 70° F.

volume contains less thermal energy, even though both beakers are the same temperature. If the rate of thermal energy transfer is the same, then less time is required for the new volume of water to reach thermal equilibrium.

3. Due to convection, hot air will move into the house continuously, making it virtually impossible to lower the temperature inside the house to 70°F.

Program Resources

Review and Assessment, pp. 35–40 L1
Performance Assessment, Ch. 6 L2
PASC
Alternate Assessment in the Science Classroom
Computer Test Bank L1

Critical Thinking

In your Journal, answer each of the following questions.

1. Examine the photo below. Name at least one place where heat is traveling by conduction, convection, and radiation.
2. Explain this statement: Unless a thermometer is already in thermal equilibrium with the fluid being tested, the act of measuring the temperature either adds or removes thermal energy from the fluid.
3. Before refrigerators were common, many kitchens had an icebox where food was kept cool. The box had one or two shelves

in it and a large block of ice was placed at the top of the icebox. The ice block usually weighed 20-30 pounds, and it was often a struggle to place it on top of the box. Why didn't people take it easy and put the ice at the bottom of the icebox?

Problem Solving

Read the following problem and discuss your answers in a brief paragraph.

Raphael and his mother are planning to add a new room onto their home. They plan to use a forced-air furnace to heat the new room. Raphael thinks the room will look nicer if the heating vents are near the ceiling, out of sight. His mother says the heating bills will be lower if the vents are in the floor.

1. How does Raphael's mother know the heating bill will be less if the hot-air vents are near the floor?
2. How could Raphael use a ceiling fan to make his plan, with vents at the ceiling, more energy efficient?

CONNECTING IDEAS

Discuss each of the following in a brief paragraph.

1. **Theme—Energy**
   Anyone who has used ropes much, for any purpose, has heard of rope burn. Explain rope burn in terms of thermal energy.
2. **Theme—Systems and**

Interactions How does heat do work on a hot-air balloon?

3. **A Closer Look** Could you use a plastic container that melts at 102° C to boil water? Explain.
4. **Science and Society** What is thermal pollution

doing to plant and animal life in rivers and lakes? Are there any ways to keep this from happening?

5. **Consumer Connection** Explain how solar energy can be a valuable alternative to some of our present energy sources.

✔ Assessment

Portfolio Review the portfolio options that are provided throughout the chapter. Encourage students to select one product that demonstrates their best work for the chapter. Have students explain what they learned and why they chose this example for placement into their portfolios.

Additional portfolio options can be found in the following Teacher Classroom Resources:
Making Connections: Integrating Sciences, p. 15

Multicultural Connections, p. 15–16
Making Connections: Across the Curriculum, p. 15
Concept Mapping, p. 14
Critical Thinking/Problem Solving, p. 14
Take Home Activities, p. 12
Laboratory Manual, pp. 27–32, 33–34, 35–38
Performance Assessment P

Critical Thinking

1. The icebox walls transfer a small amount of heat by conduction. Sunlight shining through the window is transferring heat through radiation. Convection can take place within the air in the room, or inside the icebox.
2. If the thermometer is cooler than the fluid, heat will transfer from the fluid to the thermometer. If the thermometer is warmer than the fluid, heat will transfer from the thermometer to the fluid.
3. Ice placed near the top of an icebox cools the air around it. The cool air sinks allowing it to cool food throughout the icebox.

Problem Solving

1. Raphael's mom knows that hot air rises. The resulting convection current will warm the entire room.
2. A ceiling fan can force warm air downward.

Connecting Ideas

1. As a rope slides along your hand, friction produces thermal energy in your hand.
2. Heat warms air, causing it to push against the sides of the balloon. This expands the balloon. Any exertion of force through a distance is work.
3. Yes; as long as the temperature of the container is raised to 100 degrees and kept at that temperature, the energy will be transferred to the water until the water and the container reach thermal equilibrium.
4. Thermal pollution is depriving some aquatic plants and animals of oxygen. Higher temperatures may also encourage the growth of weeds and algae, further depleting the oxygen. Cooling water before it is released can help.
5. Using solar energy can reduce our dependence on fossil fuels and reduce pollution.

Chapter Organizer

SECTION	OBJECTIVES	ACTIVITIES & FEATURES
Chapter Opener		**Explore!**, p. 211
7-1 Living Bones (3 sessions; 1.5 blocks)	**1. Identify** the major functions of bone. **2. Describe** bone and its features. **National Content Standards: (5-8) UCP1-2, UCP4-5, A1, B1, C1**	**Explore!**, p. 212 **Find Out!**, p. 214 **Design Your Own Investigation 7-1:** pp. 216–217 **Physics Connection,** pp. 218–219
7-2 Your Body in Motion (2 sessions; 1 block)	**1. Compare and contrast** types of joints and their movements. **2. Describe** the functions of ligaments and cartilage in joints. **National Content Standards: (5-8) UCP1-2, UCP5, A1, C1, E2**	**Explore!**, p. 221 **A Closer Look,** pp. 222–223 **How It Works,** p. 233
7-3 Muscles (4 sessions; 2 blocks)	**1. Describe** the three kinds of muscle. **2. Explain** muscle action and how it results in movement of body parts. **National Content Standards: (5-8) UCP1-2, UCP5, A1, C1, E1, F1**	**Find Out!**, p. 225 **Investigate! 7-2:** pp. 228–229 **Find Out!**, p. 230 **Skillbuilder:** p. 231 **Science and Society,** p. 232 **Teens in Science,** p. 234

ACTIVITY MATERIALS

EXPLORE!

p. 211 measuring tape
p. 212* clay, small wooden sticks, tissue paper or fabric
p. 221* wide masking tape, pencil

INVESTIGATE!

pp. 228–229 apron, paper towels, dissecting pan, human bone and muscle chart, forceps, scalpel, scissors, boiled chicken leg and thigh

DESIGN YOUR OWN INVESTIGATION

pp. 216–217 index cards, tape, books

FIND OUT!

p. 214* 2 clean chicken bones, 2 pint jars with lids, labels, vinegar, water
p. 225* heavy book
p. 230 index card, scissors, hole punch, metal fastener, string, metric ruler, tape

KEY TO TEACHING STRATEGIES

The following designations will help you decide which activities are appropriate for your students.

- **L1** Basic activities for all students
- **L2** Activities for average to above-average students
- **L3** Challenging activities for above-average students
- **LEP** Limited English Proficiency activities
- **COOP LEARN** Cooperative Learning activities for small group work
- **P** Student products that can be placed into a best-work portfolio
- 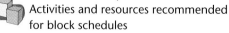 Activities and resources recommended for block schedules

Need Materials? Call Science Kit (1-800-828-7777).

00:00 **OUT OF TIME?** We recommend that students do the activities with an asterisk.

Chapter 7 Moving the Body

TEACHER CLASSROOM RESOURCES

Student Masters	Transparencies
Study Guide, p. 25 **Activity Masters,** Design Your Own Investigation 7-1, pp. 31–32 **Critical Thinking/Problem Solving,** p. 5 **Multicultural Connections,** pp. 17–18 **How It Works,** p. 10 **Science Discovery Activities, 7-1, 7-2** **Laboratory Manual,** pp. 39–42, Analyzing Bones	**Teaching Transparency 13,** Human Skeletal System **Section Focus Transparency 19**
Study Guide, p. 26 **Critical Thinking/Problem Solving,** p. 15 **Making Connections: Integrating Sciences,** p. 17 **Concept Mapping,** p. 15	**Section Focus Transparency 20**
Study Guide, p. 27 **Take Home Activities,** p. 13 **Making Connections: Across the Curriculum,** p. 17 **Making Connections: Technology & Society,** p. 17 **Activity Masters,** Investigate 7-2, pp. 33–34 **Science Discovery Activities, 7-3** **Laboratory Manual,** pp. 43–44, Testing for Proteins	**Teaching Transparency 14,** Human Muscle System **Section Focus Transparency 21**

ASSESSMENT RESOURCES	TEACHING & TECHNOLOGY
Review and Assessment, pp. 41–46 **Performance Assessment,** Ch. 7 **PASC*** **MindJogger Videoquiz** **Alternate Assessment in the Science Classroom** **Computer Test Bank**	**Spanish Resources** **Cooperative Learning Resource Guide** **Lab and Safety Skills** **Science Interactions, Course 2, CD-ROM** **Computer Competency Activities**

*Performance Assessment in the Science Classroom

NATIONAL GEOGRAPHIC TEACHER'S CORNER

National Geographic Society Products Available From Glencoe

To order the following products for use with this chapter, contact your local Glencoe sales representative or call Glencoe at 1-800-334-7344:

- *Newton's Apple Life Sciences* (Videodisc)
- *STV: Human Body Series,* "Nervous, Muscular, Skeletal Systems."(Videodisc)

Additional National Geographic Society Products

To order the following products for use with this chapter, call the National Geographic Society at 1-800-368-2728:

- *Everyday Science Explained* (Book)
- *The Incredible Machine* (Book)
- *The Incredible Human Machine* (Video)
- *Man: The Incredible Machine* (Video)

- *The Robotic Revolution* (Video)
- *Your Body Series,* "Muscular and Skeletal Systems." (Video)

Teacher Classroom Resources

These are key components of the classroom resources package.

TEACHING AIDS

Section Focus Transparencies

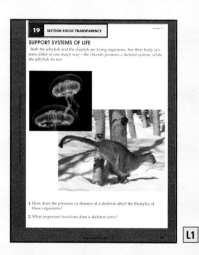

19 SECTION FOCUS TRANSPARENCY

SUPPORT SYSTEMS OF LIFE

Both the jellyfish and the cheetah are living organisms, but their body systems differ in one major way—the cheetah possesses a skeletal system, while the jellyfish do not.

1. How does the presence or absence of a skeleton affect the lifestyles of these organisms?
2. What important functions does a skeleton serve?

L1

20 SECTION FOCUS TRANSPARENCY

READY . . . SET . . . ACTION!

The pool area becomes silent as the stands poised, focusing on the dive ahead of her. Then whoosh . . . off she goes, spinning and twisting on her descent. Another beautiful dive! Platform diving is a difficult art that requires a healthy body and mind in order to be competitive.

1. What body parts are used in the sport of platform diving?
2. What physical skills are needed for such a career?

L1

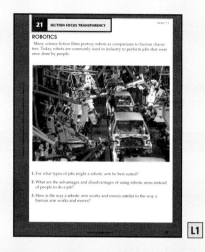

21 SECTION FOCUS TRANSPARENCY

ROBOTICS

Many science fiction films portray robots as companions to human characters. Today, robots are commonly used in industry to perform jobs that were once done by people.

1. For what types of jobs might a robotic arm be best suited?
2. What are the advantages and disadvantages of using robotic arms instead of people to do a job?
3. How is the way a robotic arm works and moves similar to the way a human arm works and moves?

L1

Teaching Transparencies

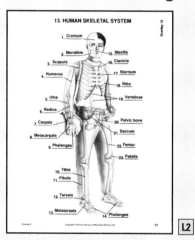

13. HUMAN SKELETAL SYSTEM

L2

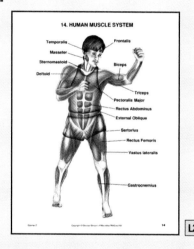

14. HUMAN MUSCLE SYSTEM

L2

HANDS-ON LEARNING

Science Discovery Activity*

ACTIVITY 7-1

Built-In Shock Absorbers

L1

Laboratory Manual*

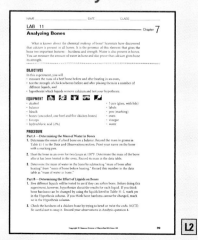

LAB 11 Chapter 7

Analyzing Bones

L2

Take Home Activity

Chapter 7

FOOLING YOUR MUSCLES

L1

*There may be more than one activity for this chapter.

Chapter 7 Moving the Body

Study Guide*

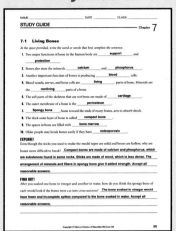

Concept Mapping

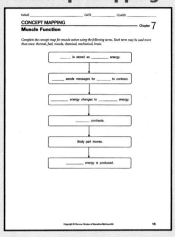

Critical Thinking/Problem Solving

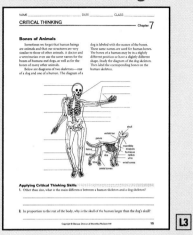

Integrating Sciences

Across the Curriculum

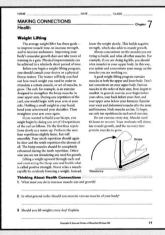

Technology and Society

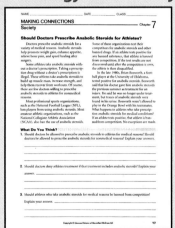

Multicultural Connection**

Performance Assessment

Review and Assessment

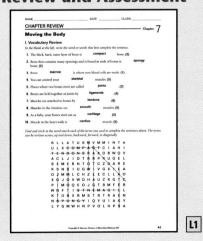

*One per section **Two per chapter

Moving the Body

THEME DEVELOPMENT

Several systems have to work together and interact to provide motion to the body. While the skeletal and muscular systems work in an integrated manner to produce movement, the respiratory and digestive systems are also required to provide the energy necessary to accomplish that work, thus supporting the theme of energy.

CHAPTER OVERVIEW

In this chapter, students will study the bones and muscles that give the body shape and structure and allow it to move. The bones of the skeleton support the body, protect internal organs, make blood, and provide places of attachment for muscles. The power needed to move the skeleton comes from muscles, which contract to pull your bones.

Tying to Previous Knowledge

Arrange students in small groups and have them compare and contrast the machines described in Chapter 5 to parts of the human skeletal and muscular systems. **COOP LEARN**

INTRODUCING THE CHAPTER

Ask students if they have ever participated or have seen a motocross. Have them describe the event. Tell them that as the cycle has many moving parts, so does the human body.

00:00 OUT OF TIME?

If time does not permit teaching the entire chapter, use the Chapter Overview on this page, Reviewing Main Ideas at the end of the chapter, and the Chapter 7 audiocassette to point out the main ideas of the chapter.

MOVING THE BODY

Did you ever wonder...

✓ **Whether bone is alive?**
✓ **Why your knee can't bend backwards?**
✓ **If athletes have more muscles than couch potatoes have?**

Science Journal

Before you begin to study bones and muscles, think about these questions and answer them *in your Science Journal.* When you finish the chapter, compare your journal write-up with what you have learned.

Answers are on page 235.

This is Felípe's first bicycle motocross. His gloved hands clasp the upturned handlebars. His body jets forward, eyes straight ahead. Then, with legs pumping like pistons, Felípe speeds into the turn, leaving his fellow racers in a dusty dirt cloud.

You may not race, or even cycle, but even if listening to music is your favorite hobby, you move. You walk and run, sit and stand, twist and turn. You may have even changed positions while reading this paragraph.

That's what motion is—the process of changing place or position. The kind of movements your body makes depends on the structure of your body, which has more than 200 bones and more than 600 muscles. Muscles move bones. In this chapter, you'll explore the structures that support your body, move it, and give it shape—your body's shapers and movers—bones and muscles.

▶ **In the activity on the next page, explore what happens to your muscles as you bend and straighten your arm.**

210

Learning Styles **LS**	Kinesthetic	Explore, pp. 212, 221; Investigate, pp. 216-217, 228-229; Activity, pp. 221, 225, 227; Find Out, pp. 225, 230
	Visual-Spatial	Visual Learning, pp. 213, 215, 219, 220, 222, 224, 226, 227; Find Out, p. 214
	Interpersonal	Explore, p. 211; Enrichment, pp. 217, 227; A Closer Look, pp. 222-223; Project, p. 235
	Logical-Mathematical	Across the Curriculum, p. 219; Physics Connection, pp. 218-219
	Linguistic	Across the Curriculum, pp. 215, 223; Multicultural Perspectives, p. 229
	Intrapersonal	Multicultural Perspectives, p. 230

Explore! ACTIVITY

How big is your muscle?

You use your muscles to move your bones, which in turn move you to where you want to go. But did you ever stop to think about how your muscles work—how they get those bones in motion?

What To Do

1. Straighten your arm into a relaxed position. Use a measuring tape to determine the size (circumference) of your upper arm when it is relaxed.

2. Flex your arm by bending it and making a fist. Measure your upper arm again.

3. How did the size of your upper arm change when you flexed your arm?

4. What happened to the muscle to cause this change? Did the muscle actually become larger? Or did it shorten and bunch up?

5. *In your Journal*, write a paragraph about what you think is happening to the muscles in your arm as you straighten and flex your arm.

Explore!

How big is your muscle?

Time needed 10–15 minutes

Materials measuring tape

Thinking Processes observing, comparing and contrasting, measuring, analyzing data

Purpose To measure the change in a muscle when part of the body moves.

Teaching the Activity

Have students work in pairs so they can measure one another's muscles in both a relaxed and a flexed state.
COOP LEARN

Science Journal Suggest that students include a sketch in their journals of the way they think their muscles change when they move their arms. **L1**

Expected Outcome

Students should find that the diameter of the arm grows thicker when the arm's muscles are flexed.

Answers to Questions

3. It became larger.

4. The muscle shortened and bunched up; it did not become larger.

5. The upper arm thickens when the muscle is flexed because the muscle contracts, or shortens. When the muscle shortens, its mass is compressed and the muscle becomes thicker.

✔ Assessment

Process Ask students to observe and decide where the largest muscles in their bodies are located. *The largest muscles are located where they can move the limbs and other large parts of the body, such as the back.* Use the Performance Task Assessment List for Making Observations in **PASC**, p. 17.

ASSESSMENT PLANNER

PORTFOLIO
Refer to page 237 for suggested items that students might select for their portfolios.

PERFORMANCE ASSESSMENT
Process, pp. 211, 225, 229
Skillbuilder, p. 231
Explore! Activities, pp. 211, 212, 221
Find Out! Activities, pp. 214, 225, 230
Investigate, pp. 216–217, 228–229

CONTENT ASSESSMENT
Check Your Understanding, pp. 220, 224, 231
Reviewing Main Ideas, p. 235
Chapter Review, pp. 236–237

GROUP ASSESSMENT
Opportunities for group assessment occur with Cooperative Learning Strategies.

PREPARATION

Planning the Lesson

Refer to the Chapter Organizer on pages 210A–D.

Concepts Developed

In this section, students will compare the three functions of the skeletal system and observe bone structure.

1 MOTIVATE

Bellringer

 Before presenting the lesson, display **Section Focus Transparency 19** on an overhead projector. Assign the accompanying **Focus Activity** worksheet. L1

LEP

Explore!

How does your skeletal system support your body?

Time needed 15–20 minutes

Materials clay, small wooden craft sticks, tissue paper or fabric

Thinking Processes observing and inferring, recognizing cause and effect

Purpose To observe the role of an internal framework in providing support to a structure and infer how the human skeleton does the same.

Preparation Straws may be used in place of wooden sticks.

Teaching the Activity

Troubleshooting If the tepee falls apart, point out how important a strong structure is.

Expected Outcome

Students will infer that shape and stability depend on

7-1 Living Bones

Section Objectives
- Identify the major functions of bone.
- Describe bone and its features.

Key Terms
spongy bone, compact bone, bone marrow, cartilage

Your Body's Framework

In the Explore activity on page 211, you observed the changes that took place in your muscle as you straightened and bent your arm. How do these changes in the muscle affect the bones? Before you can answer this question, you need to understand how the bones themselves form a framework for your body. Your skeleton is made up of many different shapes of bones that together support the body and help it move in a coordinated manner.

Even as you read this page, the bones in your neck are being moved so that you can turn your head from side to side. The bones in your neck also perform another important job. Can you guess what it is? They are holding up your head.

Explore! ACTIVITY

How does your skeletal system support your body?

What To Do

1. Using clay and small wooden sticks, build a small model of a tepee.

2. Cover your tepee with tissue paper or fabric. How much of the frame of the model can you see?

3. What is the job of the wooden sticks in your model? What would happen to your model tepee if you removed the sticks?

4. *In your Journal,* write a paragraph about how your body is like the model tepee you built.

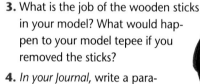

212 Chapter 7 Moving the Body

---

a well-structured frame.

Answers to Questions

2. The sticks cannot be seen.

3. The sticks determine shape and provide support. The tepee would fall apart.

4. Bones are like the sticks; skin is like the tissue paper or fabric.

Science Journal Have students list in their journals other structures that depend on an in-

ternal framework for support. L1

✔ Assessment

Performance Have students experiment with the materials and build different frameworks. Ask them to prepare oral reports that illustrate how changing the internal framework affects the outside structure. Use the Performance Task Assessment List for Oral Reports in **PASC**, p. 71.

Jobs of Bones

A skeletal system is an amazing structure with many functions. Blood is formed in bone, and the body's supply of calcium and phosphorus is warehoused throughout the skeleton. But the most obvious job of your skeleton is to enable you to have a particular shape that doesn't change, by supporting your body and giving it shape. Bones protect internal organs, such as the brain, heart, and lungs. These organs are surrounded by bone. In Section 7-3, you'll learn that bones also function as the place of attachment for many of the body's muscles.

Figure 7-1

Human and animal skeletal systems can be compared to the internal framework systems that support large structures, such as the Statue of Liberty. For example, if the framework of the statue and a person's skeletal system were removed, the result for each would be the same: both outer coverings would collapse.

Cutaway drawing of the Statue of Liberty

A The Statue of Liberty, pictured at left, is made of more than 300 copper shells, which are bolted to a framework of iron beams. To which body parts can each of these parts of the statue be best compared?

B Bones not only add shape and support, they surround and protect organs such as the brain, heart, and lungs. Without this protection, these organs could be easily injured.

7-1 Living Bones **213**

Discussion Discuss how the skeleton gives general shape to the body. Have students list any parts on their skeleton that they can actually feel. *Students may indicate ribs, wrist, knuckle, elbows, ankles, and vertebrae.* L1

2 TEACH

Tying to Previous Knowledge

In Chapter 1, students observed that a force is a push or a pull. When you stand on the ground, you and the ground exert forces on each other. Bones must be strong enough to withstand these forces, but also have a spongy nature which serves as a shock absorber by spreading the forces all over a greater area.

Theme Connection The theme that this section supports is energy. The skeletal and muscular systems work in an integrated manner to produce movement. Energy is used, and work is done.

NATIONAL GEOGRAPHIC SOCIETY

 Videodisc

STV: Human Body Vol. 2
Muscular and Skeletal Systems
Unit 1, Side 1
Axial Skeleton

6729-10400

Appendicular Skeleton

10419-16673

Skeletal system, front and back view (art)

48673

Program Resources

Study Guide, p. 25 L1
How It Works, p. 10
Teaching Transparency 13
Science Discovery Activities, 7-1, 7-2
Section Focus Transparency 19

Visual Learning

Figure 7-1 Ask students to compare and contrast the human skeleton shown in Figure 7-1B with the framework of the Statue of Liberty shown in Figure 7-1A. L1

To which body parts can each of these parts of the statue be best compared? *Copper shells can be compared to skin; iron beams can be compared to bones.*

What makes bones hard?

Time needed 10–15 minutes

Materials 2 chicken bones, 2 jars with lids, labels, vinegar, water

Thinking Processes thinking critically, observing and inferring, practicing scientific methods, forming a hypothesis

Purpose To use scientific method in order to infer what makes bones hard.

Preparation Rinse bones in a mild bleach solution beforehand to remove all meat and prevent rotting.

Teaching the Activity

Troubleshooting To reduce the number of variables, have students choose two chicken bones of similar size and shape. Make sure bones are completely covered by vinegar or water.

Science Journal Have students write their predictions in their journals. After four days have passed students can record their observations next to their predictions. L1

Expected Outcome

Students will observe that the bone in the jar of vinegar becomes flexible. The vinegar dissolves the mineral salts from the bone, leaving only the elastic tissue.

Conclude and Apply

1. Yes; the bone in the water changed very little, while the bone in the vinegar became flexible.

2. The vinegar dissolved something that makes bone hard.

✔ Assessment

Content Ask students to list the ways that strong bones are essential to health. *Answers may include: Strong bones support the body and give it shape; they protect internal organs, such as the brain, heart, and lungs.*

Parts of a Bone

You've seen bones at one time or another. Maybe they were in last night's fried chicken, or you may have watched (and heard) your dog as she chewed on one. You probably know that bones are hard, but did you ever wonder what bones are made of that gives them that hardness?

Find *Out!* ACTIVITY

What makes bones hard?

To support the weight of your body, bones must be hard. If bones were soft, you would not be able to run, to stand, or even to sit.

What To Do

1. Get two pint jars that have lids. Fill one with vinegar, and the other with water. Label each jar.

2. Put one chicken bone in the jar of vinegar, and another chicken bone in the jar of water. Cover the jars and store them for four days.

3. Predict what will happen to the bones at the end of four days.

4. After four days, remove the bones

from their containers and wash them thoroughly. How do the bones feel now? Try bending both bones.

Conclude and Apply

1. Is the bone that was in vinegar different from the bone that was in water?

2. Can you suggest a reason for any changes that took place? Write an explanation for your observations *in your Journal.*

■ Living and Nonliving Parts of Bone

Your bones, like the chicken bones you examined, are unusual because they are made up of both living and nonliving materials. Look at **Figure 7-2** to see the inside and outside parts of a bone. On the outside, a bone is covered by a thin, living membrane called the periosteum. This membrane has many blood vessels in it. The blood vessels are living and carry food and oxygen to the other living parts of bone, which include nerves and bone cells.

You learned that soaking a chicken bone in vinegar makes it flexible. You dissolved something that had made the bone very hard. The nonliving parts of the bone—the minerals calcium and phosphorus—dissolved in the vinegar. What would happen if you were to lose the minerals from your bones? Rickets and osteoporosis

---

Meeting Individual Needs

Gifted Have students make clay or papier-mâché models of the human skull using diagrams as references. Students can search for diagrams of different animals and make similar models. Challenge students to make the lower jaw articulate. L3

Multicultural Perspectives

Animal Bones
Animal bones have been used in many cultures. The Chinese of the Shang dynasty used bones called "oracle bones" to record historical events. Over 8000 years ago, Africans from the region now known as Zaire carved a bone called the Ishango in a way that indicates a knowledge of doubling and primes.

are two diseases that result from lack of minerals in bone. Rickets occurs in children and prevents normal bone growth. Bones become bent during development. Osteoporosis occurs mainly in older people. Bone material in people with this disease becomes fragile and breaks easily because the bones lack minerals.

■ Spongy Bone

Do you know what a sponge looks like? Sponges have many little openings in them that hold liquids. One part of a bone looks like a sponge. It's called **spongy bone**. As you see in **Figure 7-2**, spongy bone contains many openings and is found toward the ends of many bones. Unlike a sponge, however, spongy bone is not soft. The fine network of bone you see in the spongy bone is made of minerals and is hard.

■ Compact Bone

You know that calcium and phosphorus are the nonliving parts of bone that make it hard. Compounds of calcium and phosphorus are concentrated in the thick outer layer called **compact bone**. Besides minerals, compact bone also contains living bone material—blood vessels, bone cells, and nerves—and elastic fibers. Elastic fibers keep bone from being too rigid.

■ Bone Marrow

Besides providing support for your body, bones also make blood cells. Have you ever broken a chicken bone? If you have, you may have been

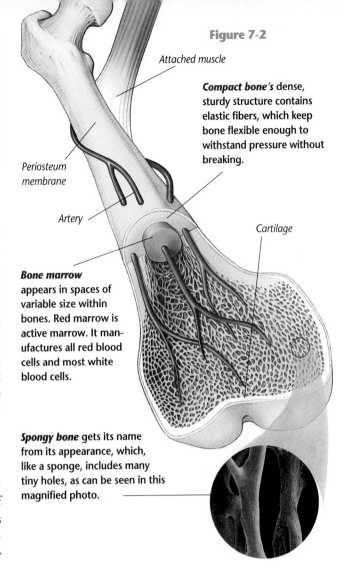

Figure 7-2

Attached muscle

Compact bone's dense, sturdy structure contains elastic fibers, which keep bone flexible enough to withstand pressure without breaking.

Periosteum membrane

Artery

Cartilage

Bone marrow appears in spaces of variable size within bones. Red marrow is active marrow. It manufactures all red blood cells and most white blood cells.

Spongy bone gets its name from its appearance, which, like a sponge, includes many tiny holes, as can be seen in this magnified photo.

surprised to find that the bone was not solid. Many bones have a hollow area or cavity. This space, as well as the spaces in spongy bone, are filled with a gel-like substance called **bone marrow**. Bone marrow, another living part of bone, is red or yellow in color. Yellow marrow is found in the long part of bone and is made mostly of fat. Red marrow is found in spongy bone. New blood cells are made in red marrow.

Program Resources

Multicultural Connections, p. 17, Living and Nonliving Bone in Different People; p. 18, Growing Bones Need the "Sunshine" Vitamin L1
Critical Thinking/Problem Solving, p. 5, Flex Your Brain

Inquiry Question Why are our bones not completely solid? *The extra material in the bones would increase their weight and make it much harder to move them.*

Across the Curriculum

Health

Although bones are strong and flexible, they can break. A break in bone is a **fracture.** A simple fracture occurs when a bone has broken but the broken ends do not break through the skin. In a compound fracture, the broken ends of the bone break through the skin.

Have students recall any experiences with fractures. Ask students to research how broken bones heal. *To promote healing, the broken ends of the bone are brought into contact with each other. The periosteum starts to make new bone-forming cells, called osteoblasts (AHS tee oh blastz). A thick band of new bone, called a callus, forms around the break, acting like a built-in splint. Over time, the thickened band disappears as bone is reshaped with the help of osteoclasts. Osteoclasts (AHS tee oh klastz) are bone cells that break down bone tissue. Breaks are usually held immobile with a cast.*

Flex Your Brain Use the Flex Your Brain activity to have students explore the SKELETON. L1

7-1 Strong Bones

Preparation

Purpose To formulate models to determine how bones are able to support weight.

Process Skills observing and inferring, making models, making tables, practicing scientific method, forming hypotheses

Time Required 30 minutes

Materials See reduced student text.

Possible Hypotheses Students may hypothesize that the models most shaped like bones (tubes) will be the strongest.

The Experiment

Process Reinforcement Ask students what shapes they would use to support a heavy structure. For example, how could they use plastic building blocks to create a platform strong enough to hold a big dictionary? Invite students to draw some of their ideas on the board.

Possible Procedures Make at least three different models that can be tested. Take turns actually sketching a model to build, building the model, and recording data about the strength of the models.

Teaching Strategies

Troubleshooting Make certain that models are labeled.

Discussion Make certain that students relate making these models to the design of bones. Bones can be hollow in parts and spongy like the example mentioned in question 6.

Science Journal Have students draw their final design of models they intend to test in their journals. In their journals, students should also prepare tables and add model numbers to the table.

Expected Outcome

Students may roll note cards into tubes or fold them into squares, or tent (triangular) shapes. The tube-shaped model should be the strongest.

Strong Bones

Bones are hard, hollow structures that support your body. Inside many bones are hollow areas or cavities that contain marrow. How can bones that are not completely solid be so strong? This activity will give you some insight into bone strength.

Preparation

Problem
How are bones able to support weight?

Form a Hypothesis
Make a hypothesis about how it is possible for our bones to support so much weight. Build a model to test the hypothesis.

Objectives
- Make a model that illustrates why bone design is strong.
- Determine which bone design is the strongest.

Materials
index cards
tape
books

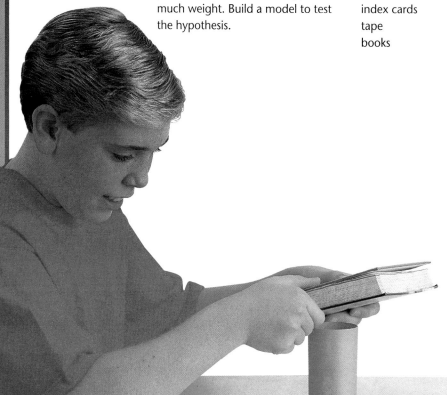

ENRICHMENT

Activity Invite an architect, carpenter, or structural engineer to the class to talk about what kinds of shapes are used to support skyscrapers, bridges, and other large structures. Prior to the visit have students make a list of questions about structure that they would like to ask. [L1]

DESIGN YOUR OWN
INVESTIGATION

Plan the Experiment

1 Examine materials and decide how your group will test your hypothesis by building models.

2 Write out a procedure showing possible designs and how you will test these models to see which design is the strongest.

3 Assign a letter to each of your models.

4 *In your Science Journal*, prepare a space to draw your models, then make a data table to record your tests.

Check the Plan

1 How many models will you make?

2 How will you test each model under exactly the same conditions?

3 When you have completed your plan, and it has been approved by your teacher, build your models. Draw your designs and record your observations.

Analyze and Conclude

1. **Conclude** Explain why you developed the models as you did.

2. **Measure** How did you measure the strength of the models?

3. **Observe** Which model was stronger than the rest?

4. **Infer** Why was this model stronger than the others?

5. **Compare and Contrast** Compare your models to the structure of bone. Which model was closest to bone in structure?

6. **Analyze** Which do you think would be stronger: a note card tube; straws clustered together and stuck in clay; or straws clustered together inside a note card tube? Explain your answer.

7. **Analyze** Was the model in question 5 the strongest model? Explain.

Going Further

Test to see what is stronger, a hollow tube or a solid rod. Both must be of the same material and weight.

Program Resources

Activity Masters, pp. 31–32, Design Your Own Investigation 7-1

Laboratory Manual, pp. 39–42, Analyzing Bones **L2**

Analyze and Conclude

1. Rectangular models because a house or building is rectangular in shape; tent-shaped models because the triangular supports seem to share the load at the top; tube-shaped models because columns are made like that

2. Placing the same books on top of each model

3. A tube-shaped model should be the strongest; then, a rectangular model; then, tent-shaped. Other models are possible.

4. It is all one piece of material and it is not folded.

5. If students did not make a tube, answers may vary. Rolling the card into a tube most closely resembles a bone.

6. The straws clustered together inside a note card would be the strongest because it has the added strength of many tubes within a tube.

7. The tube was the strongest model. For its weight, it can support the most weight. It has no seams, it is not folded or bent, causing loss of strength, and in some places the tube will be double strength.

✔ Assessment

Performance Have students work in pairs to create the strongest and most economical chair, applying what they have learned about the structure and shape of bones. Students may use only paper as a building material. When all the models are finished, have the class work together to devise a test of the chairs' strength. Then assess the strength of each chair and have the students explain how they adapted what they know about bones into their design. Use the Performance Task Assessment List for Model in **PASC**, p. 51.
L1 **COOP LEARN** **P**

Going Further

If the tube is made of the same material and is the same weight as the rod, the material of the tube will be more dense. The denser material can support more weight.

Content Background

Bone tissue is composed chiefly of collagen and the minerals hydroxylapatite $[Ca_3(PO_4)_2]_3$ and calcium hydroxide $Ca(OH)_2$.

In an embryo, the skeleton begins as a cartilaginous-membrane system. In the last months of pregnancy, the cartilage and membranes are replaced by bone. Bone tissue is constantly being reabsorbed and reformed. Calcium and vitamin D are necessary for the deposition of new bone material.

Connect to . . .
Physics

Answer: *Students should infer from the core text and the Physics Connection that bone density and strength are related. For example, bird wings are lightweight and very porous (less dense) and, therefore, have less strength to withstand outside pressures than do bones from steer or pigs.*

NATIONAL GEOGRAPHIC SOCIETY

Videodisc

STV: Human Body Vol. 2
Muscular and Skeletal Systems
Unit 1, Side 1
Bone Structures

16674-21750

Connect to...
Physics

Compact bone has a strength of about 19 000-30 000 lb/in². It would take you and about 200 of your classmates all standing on a single one-inch cube of compact bone to crush it. How does this strength relate to a bone's density?

■ Bone Strength

Healthy bones are hard and flexible, and the arrangement of minerals and fibers in bones gives them great strength. Your bones must be strong to withstand the forces that act on them. Forces that are directed toward an object, such as your body, cause materials in the object to be pressed together. Think of how much force is exerted on your bones when you jump up and down. Forces acting on the body are concentrated at contacts between bones. This action places a lot of pressure on these contact points.

In the Investigate, you made models of bones and discovered one of the

Figure 7-3

To jump upward, you must push against the ground with enough force to overcome the pull of gravity. When you push against the ground, it pushes back with equal force, placing pressure on your bones.

Physics CONNECTION

Bone Density

The bones of various animals have many similarities, but as you might expect, they also have differences. A cow's bones are strong enough to support the cow as it lumbers across pastures. On the other hand, a finch's bones are light, enabling the finch to flit from tree to tree in search of food. This activity will show you one way in which animals' bones are different.

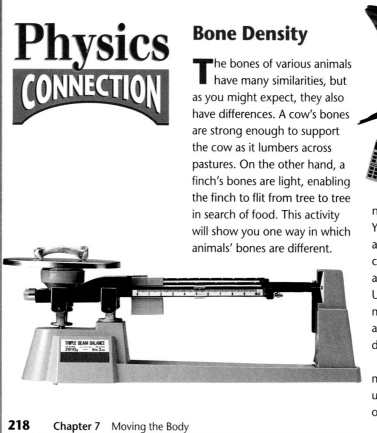

USING MATH

For this exercise, you will need to copy the data table. You'll also need water, a balance, a 100-mL graduated cylinder, and bones from a pig, a steer, a turkey, and a chicken. Use the balance to find the mass in grams of a steer bone and record this amount in the data table.

Use the displacement method to determine the volume of the bone. Pour 50 mL of water into the cylinder and

---

Physics CONNECTION

Purpose

The Physics Connection reinforces the description of bones and their functions, presented in Section 7-1, by describing how the density of bones varies by species and how this is related to variations among species in movement or size.

■ Teaching Strategy

Have students bring in animal bones from home, or ask a local butcher to save them for the class. Have students measure the density of pig, steer, turkey, and chicken bones. Make sure students label the bones with the name of each animal to ensure that their conclusions are valid. Densities should appear as follows:

steer 1.6 g/mL	chicken 1.1 g/mL
pig 1.3 g/mL	turkey 1.1 g/mL

Steer bones will be the most dense; chicken and turkey bones the least dense.

features of bones that give them strength. What was it? Although bones must be strong, they must also be lightweight enough for you to move. The arrangement of materials in spongy bone helps keep your skeleton lightweight because it has many spaces. The spongy nature of the ends of bone also helps you in another way. The ends of bone function as shock absorbers. A shock absorber is anything that absorbs a force and spreads it out over a large area. You may know that many running shoes have cushioned soles and some have air or a soft gel in their soles. These soles are designed to absorb shock. They work like spongy bone works.

Figure 7-4

Landing from a jump, you hit the ground with considerable force. The ground pushes back with equal force. What characteristics of your bones and your skeletal system make them able to absorb the forces of jumping and landing without breaking?

Visual Learning

Figures 7-3 and 7-4 To help students recognize that force is exerted on bones during exercise, have students walk a few steps and then try jumping and landing as illustrated in Figures 7-3 and 7-4. Ask students to describe the pressure they feel being exerted on the bones in their legs when they walk and when they jump up and down. **What characteristics of your bones and your skeletal system make them able to absorb the forces of jumping and landing without breaking?** *Movable joints, such as knees, bend to absorb the force of pressure when you land. Spongy bone, which is concentrated at the ends of long bones, acts as a shock absorber. Also, long bones are wider at the ends where the pressure on these bones is the greatest.* L1

then add the bone to it. Record the volume of the water plus the bone. Find the volume of the bone by subtracting the volume of water (50 mL) from the water-plus-bone volume reading. Record this amount in your data table. Calculate the density of the bone by dividing its mass by its volume.

Repeat the measurements using the bones of a pig, chicken, and turkey. Record the measurements and find the density of each bone sample. Which kind of bone had the highest density? The lowest?

Why Don't Cows Have Wings?

What makes one bone more or less dense than another? The

Sample data

Data and Observations

	Steer	Pig	Chicken	Turkey
Mass				
Water and bone volume				
Bone volume				
Density	1.6 g/mL	1.3 g/mL	1.1 g/mL	1.1 g/mL

answer lies in the structure of the bone. The bones in a bird's skeleton have numerous air pockets, much like the spongy portion of your own bones. The air pockets make the bone lighter and more buoyant because air is less dense than bone. This lightness helps the bird remain airborne. A cow's skeleton, on the other hand, is made up of solid bones that

lack air spaces. Thus, a cow's bones lack the buoyancy of a bird's bones. Even if a cow had wings, it couldn't fly!

What Do You Think?

Bones from several types of dinosaurs have been found in the United States. How can bone density help researchers determine how large a dinosaur was?

Across the Curriculum

Math

Tell students that an adult male is 3.84 times as tall as the length of his thighbone. Ask them to determine the height of a man with a thighbone that is 45 centimeters long. *About 173 centimeters, or roughly 5 feet 8 inches.* L1

7-1 Living Bones **219**

Troubleshooting Demonstrate how to determine the mass and volume of a bone, and how to calculate the density of one sample so that students can observe the process step by step.

Answer to
What Do You Think?
The more dense a dinosaur's bone, the larger the creature would have been.

Going Further ▌▌▌▌▌
Suggest that students work in small groups to continue the experiment by using fish bones. Before students begin, have them predict where they think the fish bones would fit on the chart they already constructed. **Would the fish bones be denser than the pig bones? The turkey bones? Why?** Then have students figure the density of the fish bones, using the same method they learned in the excursion. How close were their predictions? **COOP LEARN**

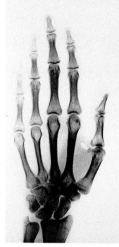

Visual Learning

Figure 7-5 Ask students to compare the amount of cartilage in each X ray and identify the X ray showing the least amount of cartilage. *X ray of adult hand* Tell students that the replacement of cartilage with bone is not complete until a person is about twenty years old.

Student Text Question

Which hands have more cartilage? *the hands of the two-year-old child*

3 ASSESS

Check for Understanding

Have students answer the questions in Check Your Understanding using Figure 7-5 to answer the Apply question.

Reteach

Activity Use Figure 7-2 to review with students the parts of a bone. Then have students cover the captions with self-stick notes or masking tape. Tell students to identify and label compact bone, spongy bone, bone marrow, cartilage, and the periosteum. Ask students to describe the features of bones. **L1** **LEP**

Extension

Model Have students who have mastered the concepts in this section make a life-sized model of the human skeleton by drawing the major bones in the skeletal system on tagboard or lightweight cardboard. Then have them cut out and connect the bones with brads to form the skeleton. **L3**

4 CLOSE

Activity

Have an athletic trainer, sports medicine physician, or physical therapist come to the class to discuss athletic injuries common to adolescents and adults.

220   Chapter 7   Moving the Body

Your Bone Development

Not all parts of your skeletal system are hard. Wiggle the end of your nose with your fingers and feel the external part of your ear. They feel soft and flexible, don't they? These parts of your body are part of your skeleton, but they are not bone. They are made of **cartilage**, a soft, flexible material. In your early development, before you were born, your entire skeleton was made up of cartilage. During development, most of your cartilage was replaced by hard bone.

Look at the X rays of child and adult hands in **Figure 7-5**. Cartilage doesn't show up in the X rays the way hard bone does. Which hands have more cartilage?

Without your skeleton, you would have no body support. Your shape might constantly change, and your internal organs wouldn't be protected against injury. Moving would be a challenge. In the next section, you will explore how your body is able to move.

Figure 7-5

All bone begins as a tough, rubbery substance called cartilage. As a person ages, most of his or her cartilage is gradually replaced by bone. These X-ray pictures of the hands of a two-year-old, an older child, a young adult, and an adult show the gradual replacement of cartilage with bone as people age.

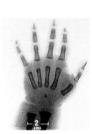 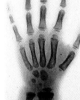 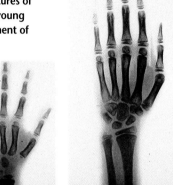

| Two-year-old | Older child | Young adult | Adult |

check your UNDERSTANDING

1. What are three functions of the skeletal system?
2. Why is bone important to your circulatory system?
3. How do compact and spongy bone differ?

4. How is cartilage like compact bone?
5. **Apply** Why do you think certain bones break more easily in an older adult than in a small child?

220   Chapter 7   Moving the Body

check your UNDERSTANDING

1. Answers may include support, body movement, make blood cells, protect vital organs.
2. Bones make blood cells; new blood cells are produced in red marrow, a part of bone.
3. Compact bone is harder than spongy bone and is found in the long shaft of a bone, whereas spongy bone is usually found in the ends.
4. Both give the body support.
5. The older adult's bones are less elastic. A child has more cartilage in the skeleton and cartilage is more flexible than bone.

The Body in Action

Your body is an amazing collection of bones and muscles, all working together to keep you going. Most of the time you don't think about all the moves you make. But, think of the movements your skeletal system goes through when you turn a page in a book, walk, or throw a ball. When you walk, your leg moves up and down at your hip. Your legs bend at the knee. Your foot bends at the ankle, and your toes bend, too. When you throw a ball, your shoulder twists around. Your arm bends and then straightens as the ball is released. Your wrist also bends, and your fingers straighten. Would any of these actions be possible if your skeleton couldn't move?

Section Objectives
■ Compare and contrast types of joints and their movements.
■ Describe the functions of ligaments and cartilage in joints.

Key Terms
joints
ligaments

Explore! ACTIVITY

Do you need joints?

What To Do

1. On the hand you write with, use masking tape to tape your thumb and fingers together so that your fingers can't bend and your thumb can't move.

2. Pick up a pencil. How easy was it?

3. Try some other activities. Button your shirt. Tie your shoelaces. Try to write your name. Why is it difficult to do these simple, everyday activities with your fingers taped?

4. *In your Journal*, write a paragraph about how important it is for you to be able to bend your fingers.

221

Explore!

Do you need joints?

Time needed 10 minutes

Materials wide masking tape, pencil

Thinking Processes thinking critically, recognizing cause and effect

Purpose To recognize the importance of skeletal motion in doing everyday tasks.

Preparation Wrap the tape around the second joints of the fingers. Wrap toward the wrist, taping the thumb against the fingers.

Teaching the Activity

Demonstration Use a model of a skeletal hand to show where the bones and joints are and the variety of motion possible. **LEP**

Science Journal Students can record their activities and observations in a chart. **L1**

Expected Outcome

Students should find fingers useless when

PREPARATION  **7-2**

Planning the Lesson
Refer to the Chapter Organizer on pages 210A–D.

Concepts Developed
Bones are connected to one another by joints and ligaments. The different types of joints allow for a variety of movement.

1 MOTIVATE

Bellringer
Before presenting the lesson, display **Section Focus Transparency 20** on an overhead projector. Assign the accompanying **Focus Activity** worksheet. **L1**
LEP

Activity To compare and contrast types of joints and their movements, have each student stand up and demonstrate a simple movement. Ask the class to describe the way the bones move in relation to one another and what mechanical model moves in the same way. **L1**

joints cannot move.

Answers to Questions
2. not easy
3. When taped, your finger cannot bend around objects.
4. The bones of the fingers need to be able to move to pinch, grasp, and hold.

✔ Assessment

Content Have students observe the number of joints in their hand and contrast that with the number of joints in their arm. Ask students to explain the difference. Use the Performance Task Assessment List for Making Observations in **PASC**, p. 17.

Joints

Stand up and move your arm in a circle. Now, bend your elbow and your fingers. You can make a complete circle with your shoulder, but you can't do that with your elbow or fingers. The same is true for your legs. You can move your hip joint in a circle, but you can't do the same with your knee. Body movements are allowed by movable joints, shown in **Figure 7-6**. **Joints** are places in your skeleton where two or more bones meet or are joined together.

Figure 7-6

Match the joint types shown on these two pages with the examples labeled on the baseball player.

Elbow: hinge

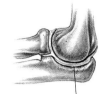

A *Hinge joints* move back and forth like the movement of a door hinge. These joints are located at your fingers, toes, elbows, and knees.

Cartilage

B *Pivot joints* located at your elbows enable you to twist your lower arms. Pivot joints allow the movement of rotation, or turning on an axis.

Arm: pivot

One Step at a Time

Have you ever heard anyone refer to the human body as a machine? If so, you may have wondered what happens when a part of the body no longer works correctly. You've probably seen automobile mechanics replacing old or worn-out parts. Can human parts be replaced?

Artificial Joints

Yes, thanks to bionics—the science of designing artificial replacements for parts of the human body. Today, the knee is one of the most common

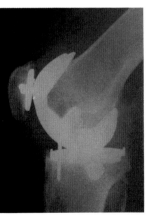

parts of the human body to be replaced by an artificial device.

To understand how an artificial knee joint works, you must first understand how a healthy knee joint functions.

How a Real Knee Works

The knee joint, as shown in the X ray, connects the bones in your upper and lower leg. With every step you take, these bones rotate, roll, and glide on each other. Connective tissue, called cartilage, forms a smooth weight-bearing surface between the bones that allows for painless movement. However, with injury or aging, the cartilage begins to wear

---

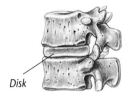

Skull: immovable

C *Immovable joints* are the point at which the bones of the skull come together. Immovable joints allow no range of movement at all.

Vertebrae: gliding

D Vertebrae are small bones that make up your backbone. Vertebrae are separated from one another by disks of cartilage, which allow them to glide over one another at *gliding joints*.

Disk

Hip: ball-and-socket

E *Ball-and-socket joints* allow for the greatest range of movement. At these joints, the large round end of one long bone fits into the circular shaped hollow of another bone.

---

out. Eventually the bones begin to rub together, creating friction and chronic pain.

Technology Takes Us One Step Further

In the late 1960s, bioengineers developed an artificial joint with a wide enough range of movement to replace the knee. Since that time, many designs and materials have been tested. Today, the most common artificial knee joints are made of chrome-cobalt, a hard yet lightweight metal. A plastic material is used to create a durable weight-bearing surface.

Replacing a Joint

Replacing a knee joint requires surgery, in which the artificial joint is fastened to the ends of the bones. One part of the artificial joint has a shaft that is inserted into the upper leg bone. The other half is attached to the lower leg bone and cemented into place. Each half of the joint is coated with a plastic that helps to reduce friction between the two parts of the joint. Joint replacement in other joints, such as the hip, is similar. A shaft is inserted into

the leg bone, and a plastic cup is cemented to the hip socket. Once in place, an artificial hip or knee joint works just like the human joint it replaces.

What Do You Think?

Make a list of body parts in each of these categories:

organs
limbs
senses

Some people think that we are tampering with nature when we use artificial body parts to replace hearts (organs), legs (limbs), or eyes (sense organs). Other people believe that, whenever possible, technology should be used to improve health. What do you think?

---

place? Which would be harder? Encourage students to work together in small groups to design bionic hands, feet, or other body parts. **L2**

Answers to
What Do You Think?

Possible answers: Organs—heart, kidneys, lungs; Limbs—hands, arms, legs; Senses—stapes bone in ear, hearing aid; Sight—contact lenses. Lead the class through a discussion of the controversy

posed, and cover reasonable opinions on both sides.

Going Further ▪▪▪▪▪▪▶

Have students further explore the concept of the human body as a machine. Have them list parts of the body that function as "machines." Examples are the heart, valves in veins, muscles, the lungs, and various joints. Students should identify the mechanical principles of such biological "machines" and discuss their efficiency. **P**

Theme Connection Have students explore the scale and structure of the human skeleton through questions like these. Is the skeleton of the body really similar to the framework of a house? After all, the joints provide for movement, while steel girders appear solid. Flexibility is needed in the structure of buildings as well as bodies. High-rise buildings that do not bend in high winds or earthquakes would be in great danger of collapse. San Francisco's Transamerica Building is expected to sway only 24 inches in a major quake; buildings of similar size may sway as much as 36 inches. Buildings cannot be completely rigid or they would break in an earthquake. They are equipped with joints. Our bodies cannot be rigid or they would be incapable of movement.

Inquiry Question Why do you think movable joints have a fluid-filled cavity between the two cartilage surfaces? *The fluid lubricates the cartilage and also acts as a shock absorber.*

Across the Curriculum

Health

Ask students if they have heard of someone who has had arthroscopic surgery. Have them find out about this procedure and how this surgery is different from other types of surgery. **L3**

NATIONAL GEOGRAPHIC SOCIETY

 Videodisc

STV: Human Body Vol. 2

Ankle and foot (art)

48680

Hand

48886

3 ASSESS

Check for Understanding

As students review their answers to the Apply question, discuss friction and the effect of excess friction on the ends of bones if it were not for the presence of cartilage.

Reteach

Activity Have students perform various sports or dance movements. Then have them describe specific joints that were involved. LEP COOP LEARN

Extension

Model Challenge students who have mastered the concepts in this section to construct models of the four different kinds of movable joints using wood, cardboard, small pieces of hardware, etc. L3

4 CLOSE

Demonstration

Wrap a rubber band tightly between two pencils to hold them in the shape of an X. Then move the pencils as much as possible. The pencils represent bones; their intersection, joints; and the rubber band, ligaments. Show students how the ligaments allow restricted movement of the joint. But actually to move the bones, force is required. In the next section, students will learn how muscles provide this force. L1

How Bones Are Held Together

Figure 7-7

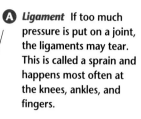

Ⓐ **Ligament** If too much pressure is put on a joint, the ligaments may tear. This is called a sprain and happens most often at the knees, ankles, and fingers.

Ⓑ **Fluid** Located under the ligaments are small pouches which contain tiny amounts of fluid. This fluid nourishes and lubricates the surfaces of the joint.

Ⓒ **Cartilage** helps reduce bone-on-bone friction and cushions bones against pressure.

Bone

If bones meet one another at joints, what keeps them from separating or bumping into one another? Bones are held together with very strong bands of tissue called **ligaments**. A fluid, also found in the joint, keeps the joint lubricated. This fluid reduces friction, the force that slows down the motion of surfaces that touch. A thin layer of cartilage over the ends of the bones further reduces friction. Cartilage in healthy joints is like firm, soft plastic, so the bones move smoothly against each other. Because cartilage is flexible, it also acts as a built-in shock absorber in a joint. That helps reduce the effect of forces on the joint by spreading the forces out.

Think about what would happen if you couldn't move your bones. You wouldn't be able to eat, run, pick up things, bend, throw, or do any number of things you probably don't even think about. In this section, you learned how the structure of joints allows for the different movements of your skeleton. In the next section, you will investigate what moves the joints in your body.

check your UNDERSTANDING

1. What makes it possible for you to play the piano? To do sit-ups? Identify the different joints in your body that make these motions possible and the types of joints they are.
2. How can a torn ligament affect movement?
3. Compare ligaments and cartilage in terms of structure and function.
4. **Apply** If a football player tears cartilage in the knee, how might this affect the bones at the knee joint?

check your UNDERSTANDING

1. Playing the piano uses hinge joints in your fingers, gliding joints in the wrists, and hinge joints in your elbow. Doing situps uses the gliding joints in the back.
2. Bones would not stay in position for smooth motion if a ligament were torn.
3. Ligaments are tough bands of tissue that join bones together at the joints; cartilage is a flexible material that covers the ends of the bones in the joints.
4. There may be excessive wear and tear at the joint; since one job of cartilage is to reduce friction, pain may occur.

Muscles

Skeletal Motion

Your bones and joints are designed and assembled to fit together like the parts of a motion machine. Your bones and joints have no power to move by themselves. Where does their power come from? Your muscles move your bones and joints and that enables you to slam-dunk a ball, back flip off a diving board, or even run a marathon, if you want to. Together, all your bone-muscle systems are like machines.

Section Objectives

- Describe the three kinds of muscle.
- Explain muscle action and how it results in movement of body parts.

Key Terms

skeletal muscles, tendons, cardiac muscle, smooth muscle

Find Out! ACTIVITY

What are the levers in your body?

In your body, your bones and muscles are arranged to form levers. Recall that a lever is a rigid structure that transmits forces by turning at a point called the fulcrum.

What To Do

1. Rest your elbow, forearm, and hand, palm up, flat on your desk. Lay a heavy book on your hand. Raise your hand, keeping your elbow on the table.

2. You are now using a lever in your body. The book exerts a resistance force that acts downward. Your elbow is the fulcrum. The muscles in your forearm provide the effort force that acts against the resistance force.

3. Now, rest your hand on the table and try to lift the book with your fingers.

Conclude and Apply

1. Why is it more difficult to lift the book with your fingers?

2. *In your Journal,* compare and contrast the leverage provided by your arm and your fingers.

Resistance force

Effort force

Fulcrum

Like all machines, the levers formed by your bones and muscles have mechanical advantage. Mechanical advantage is defined as the number of times a machine multiplies the effort force. The mechanical advantage of your forearm is greater than that of your fingers because your arm is longer and, therefore, creates a greater effort force.

7-3 Muscles **225**

Find Out!

What are the levers in your body?
Time needed 5–10 minutes

Materials heavy book

Thinking Processes thinking critically, observing and inferring

Purpose To observe how certain body parts function as levers.

Teaching the Activity

Science Journal Have students draw diagrams to illustrate their comparison of the leverage provided by the arm and fingers. L1

Expected Outcomes

Students will observe that fingers are not strong enough to lift the book. Lifting should take a minimum of force when using upper and lower arm bones and biceps.

7-3

PREPARATION

Planning the Lesson
Refer to the Chapter Organizer on pages 210A–D.

Concepts Developed
Now students are ready to learn how muscles act as the motors that move body parts.

1 MOTIVATE

Bellringer
Before presenting the lesson, display **Section Focus Transparency 21** on an overhead projector. Assign the accompanying **Focus Activity** worksheet. L1
LEP

Activity Tell students to measure their pulse rates by finding their pulse on their wrists. Have them begin counting while you start timing. Have them stop counting after one minute and write their pulse rates. Repeat the activity running in place. The increase in pulse rates is due to the increase in the work of the cardiac muscle trying to circulate more blood. L1

Conclude and Apply
1. The muscles of the fingers do not exert as much force as the muscles of the forearm.
2. The arm is longer than the fingers so it provides more leverage.

✔ Assessment

Process Have students work with a partner to locate other levers in the body and formulate an experiment to test their effectiveness in lifting weights. Use the Performance Task Assessment List for Designing an Experiment in **PASC**, p. 23.
COOP LEARN

7-3 Muscles **225**

Tying to Previous Knowledge

Everyone has had sore muscles when he or she overworked them. Have students recall that the soreness often appears in muscles that are not properly used or are infrequently exercised.

Theme Connection The theme that this section supports is energy. Muscles use energy to move the body by providing effort force, making the human "machine" do work. The bones are the levers and the joint is the fulcrum.

Student Text Questions
How much control do you think you have over cardiac and smooth muscle? *none*

Content Background

As a nerve impulse travels to a muscle, chemicals called neurotransmitters transmit the impulse from one neuron to the next and finally to an adjacent muscle. Muscle cells respond to any stimulus by contracting to their maximum; there is no partial contraction. If a muscle partially contracts, it is because only some of its cells are contracting.

Visual Learning

Figure 7-8 To compare and contrast the three kinds of muscles, have students complete the following list for each type of muscle as they study Figure 7-8: (1) involuntary or voluntary control, (2) striped or smooth appearance, and (3) location in the body. **Which is the only body organ made of cardiac muscle?** *The heart is cardiac muscle.* **Do you control these actions [of smooth muscle]?** *No, not directly.* L1

Types of Muscle

Different types of muscle in your body perform different functions. The muscles that move bones are **skeletal muscles**. At movable joints, skeletal muscles are attached to bones by tendons. **Tendons** are strong elastic bands of tissue.

You're most aware of skeletal muscles, probably because you can control the movement of these muscles. But you have two other kinds of muscles, too. The walls of your heart are made of cardiac muscle. **Cardiac muscle** pumps blood through the heart and forces blood through the rest of the body. **Smooth muscle** is found in many places inside your body, such as your stomach and intestines. Food moves through your digestive system by smooth muscle. How much control do you think you have over cardiac and smooth muscle? Compare the three kinds of muscles in **Figure 7-8**.

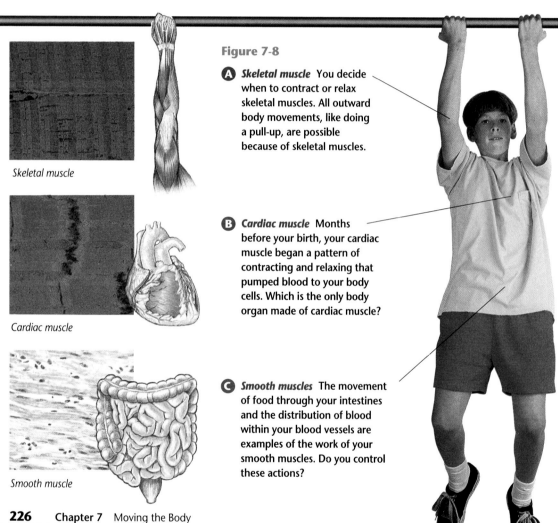

Figure 7-8

A *Skeletal muscle* You decide when to contract or relax skeletal muscles. All outward body movements, like doing a pull-up, are possible because of skeletal muscles.

B *Cardiac muscle* Months before your birth, your cardiac muscle began a pattern of contracting and relaxing that pumped blood to your body cells. Which is the only body organ made of cardiac muscle?

C *Smooth muscles* The movement of food through your intestines and the distribution of blood within your blood vessels are examples of the work of your smooth muscles. Do you control these actions?

Skeletal muscle

Cardiac muscle

Smooth muscle

226   Chapter 7   Moving the Body

Program Resources

Study Guide,   p. 26
Laboratory Manual,   pp. 43–44, Testing for Proteins L1
Teaching Transparency 14 L2
Take Home Activities,   p. 13, Fooling Your Muscles LEP
Making Connections: Across the Curriculum,   p. 17, Weight Lifting LEP

Making Connections: Technology & Society,   p. 17, Should Doctors Prescribe Anabolic Steroids for Athletes? L2
Science Discovery Activities, 7-3
Section Focus Transparency 21

Muscle Action

Body movement depends on the action of your muscles. But how do muscles work? Muscles are made up of bundles of long, stringlike structures called fiber that can contract. When a muscle contracts, it gets shorter. In doing so, it pulls on the attached bone. The force of a muscle pulling on a bone moves a body part. Since work is done when a force moves an object a distance, the muscle works when it contracts.

Like all things that do work, your muscles need energy. Fuel of some kind is always needed to obtain the energy to do work.

Glucose is your muscles' main fuel. In your muscles, chemical energy stored in glucose changes to mechanical energy, and your muscles contract.

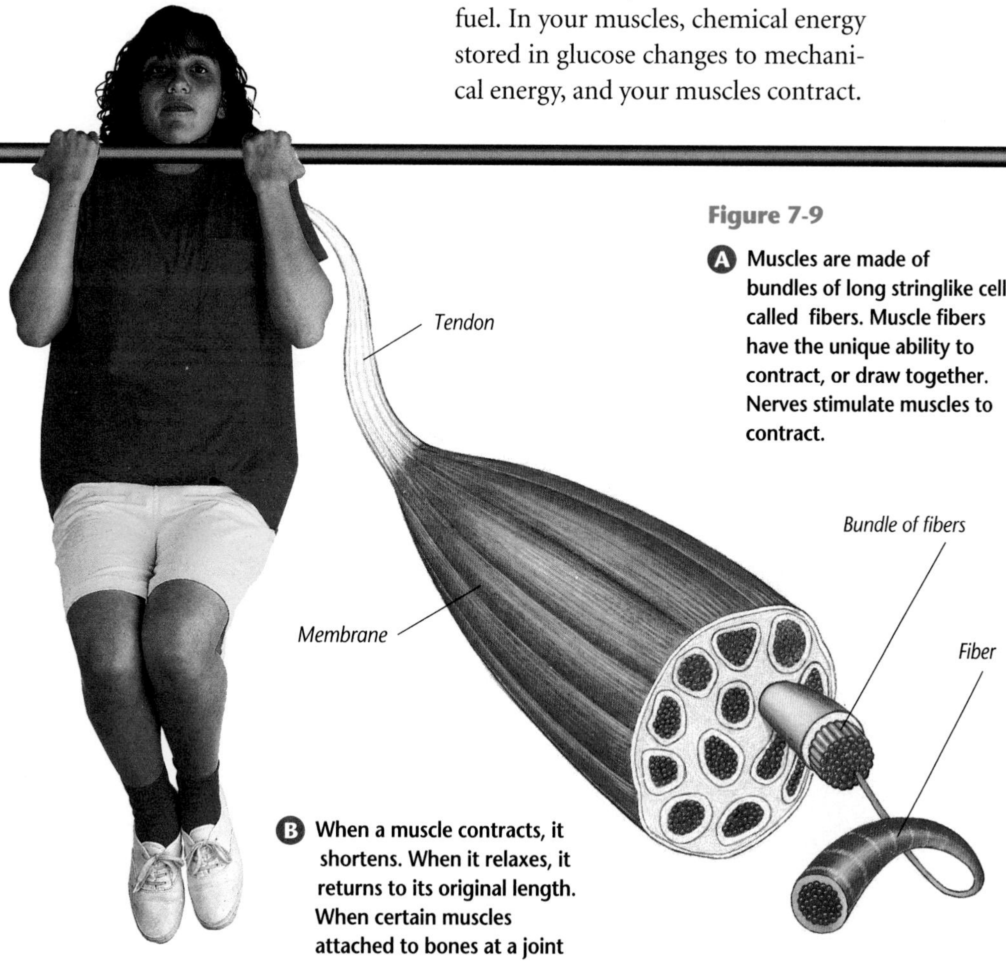

Figure 7-9

A Muscles are made of bundles of long stringlike cells called fibers. Muscle fibers have the unique ability to contract, or draw together. Nerves stimulate muscles to contract.

Tendon

Membrane

Bundle of fibers

Fiber

B When a muscle contracts, it shortens. When it relaxes, it returns to its original length. When certain muscles attached to bones at a joint contract, the bones move. Do muscles do work? Why?

7-3 Muscles **227**

Planning the Activity

Time needed 35–45 minutes

Materials See student text activity.

Process Skills thinking critically, observing and inferring, representing and applying data, interpreting scientific illustrations

Purpose To observe the way muscles, bones, tendons, joints, ligaments, and cartilage work together.

Preparation Divide students into groups of four. Have one member organize materials used for the dissection and put on the apron, while another makes a copy of the table. Have one student do the dissection while another draws and records the data. **COOP LEARN**

Teaching the Activity

Safety Art knives or sharp kitchen knives may be used in place of the scalpel, and tweezers can replace the forceps.

Troubleshooting In step 2, students should use the scalpel carefully if they need help in removing the skin with the forceps. This will prevent cutting through any ligaments. To prevent overeager students from cutting into the chicken leg before observing all the parts, do not distribute scalpels or other cutting instruments before all students have completed the first four steps of this activity.

Process Reinforcement Ask students to list the body parts that allow movement. *muscles, bones, tendons, joints, cartilage, ligaments*

Science Journal Have students record their data and observations in their journals. L1

Muscles and Bones

Skeletal muscles pull bones at joints. That is how body parts are moved. In this activity, you will observe the relationships among bones, muscles, tendons, and joints.

Problem
How do muscles move bones?

Materials

apron	paper towels
dissecting pan	forceps
scalpel	scissors
human bone and	boiled chicken
muscle chart	leg and thigh

Safety Precautions

Use care with the scalpel. It can slice your skin as readily as it can slice chicken muscle.

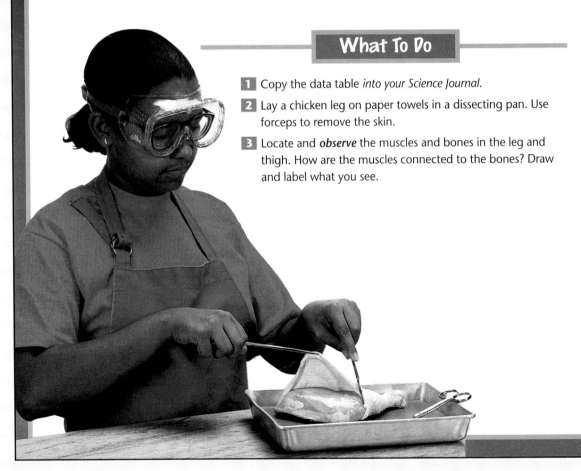

What To Do

1. Copy the data table *into your Science Journal.*

2. Lay a chicken leg on paper towels in a dissecting pan. Use forceps to remove the skin.

3. Locate and *observe* the muscles and bones in the leg and thigh. How are the muscles connected to the bones? Draw and label what you see.

Program Resources

Activity Masters, pp. 33–34, Investigate 7-2

Meeting Individual Needs

Visually Impaired Visually impaired students may require special assistance during the Investigate on pages 228–229. Have a fully sighted partner do the dissection of the chicken leg, providing detailed oral descriptions of the muscles, bones, joints, ligaments, and cartilage.

A

Sample data

Data and Observations	
Body Systems	Body Parts
Muscular	muscles, tendons
Skeletal	legbone, thighbone, joint, cartilage, ligaments

4 Bend and straighten the leg and thigh at the joint. *Observe* what happens to the muscles.

5 Use a scalpel and scissors to remove the muscles from the bones (see photo **A**). Locate and *observe* the bones, joints, ligaments, and cartilage. Draw what you see and label these parts.

6 Record the parts of the skeleton and muscle systems that you located.

7 Give all sharp instruments and dissected specimens to your teacher after completing your investigation.

8 Wash your hands.

Analyzing

1. *Identify* which muscles remain the same length and which shorten when a chicken picks up its leg to walk. *Compare and contrast* their movements.

2. What connects muscles to bones?

3. Did you locate any cartilage or ligaments? Where and how did they appear?

Concluding and Applying

4. What kind of joint is between the thigh bone and the lower leg bone of the chicken?

5. ~~Going Further~~ How is your arm similar to a chicken wing in the way it moves?

Expected Outcome

Students should see the way the muscles of the leg and thigh lengthen and shorten when the joint is bent and clearly locate the bones, joints, ligaments, and cartilage.

Answers to Analyzing/ Concluding and Applying

1. When the chicken flexes its leg the muscle on the front side of the thigh shortens and the muscle on the back side of the thigh remains the same length.

2. tendons

3. Cartilage is found on the ends of bones; ligaments connect the two bones at the joint.

4. hinge joint

5. The bones in both are moved by muscle contraction that pulls on bones at joints.

✔ Assessment

Process Have students, working in pairs, take turns observing their arm muscles as they lift a heavy book or an object weighing between 2 and 5 pounds. As one partner slowly moves his or her arm, the other notes which muscles remain the same length, and which shorten during the movement. Then have students make a poster that illustrates the relationship among bones, muscles, tendons, and joints in humans. Use the Performance Task Assessment List for Posters in **PASC**, p. 73.

COOP LEARN **P**

Multicultural Perspectives

The First Aspirin?
Ask students what muscle diseases they have heard of. Some may mention muscular dystrophy or rheumatism. Tell students that people around the world have tried to treat muscular diseases. The Bantu-speaking people of Africa were among the first to use *Salix capensis* (willow) to treat the pains of rheumatism. This plant provides salicylic acid, the active ingredient in aspirin.

Muscles Work Together

In the Investigate, you observed two different muscles bending and straightening a chicken leg. Muscles in your body work in a similar way. Your biceps muscle bends your arm. But how does your arm straighten out?

Find Out! ACTIVITY

How do muscles work together?

To find out how muscles work together, make a model with an index card, a metal fastener, string, and tape.

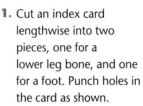

What To Do

1. Cut an index card lengthwise into two pieces, one for a lower leg bone, and one for a foot. Punch holes in the card as shown.

2. Attach the leg to the foot using a metal fastener as an ankle joint.

3. Cut two 25-centimeter pieces of string and tape one in front and one behind the ankle as shown. Thread one string through each hole at the upper end of the leg.

4. Pull up on the string behind the ankle and observe the action of the foot.

5. Now let go of this string and pull up on the other string.

Conclude and Apply

1. What happens to the foot when you pull up on the string behind the ankle? What happens to the string in front of the ankle?

2. What happens to the foot when you pull up on the string in front of the ankle? What happens to the string behind the ankle?

3. Can both strings be pulled at the same time? Why or why not?

4. What do the strings in your model represent?

5. *In your Journal,* explain what you have demonstrated about the action of muscles.

DID YOU KNOW?

Skeletal muscles are so powerful that if both muscles of a pair that work together were to contract at the same time, they could easily break a bone.

In this chapter, you have explored how muscles work together. When one muscle of a pair contracts, the other muscle relaxes, or returns to its original length. Muscles always work in pairs this way. Because skeletal muscles move bones only when muscles contract, muscles only *pull* bones. Muscles never push on bones.

Bend and straighten your arm a few times to simulate the muscle action shown in **Figure 7-10.** What happens when your biceps contracts? When it relaxes?

You can help your muscles be better coordinated and stronger by using them. Your muscles will become larger or smaller depending on how much you

Multicultural Perspectives

Storing Blood

Charles Drew (1904–1950), an African-American scientist, developed important new techniques to separate and store blood. Thanks to his research during the 1940s many lives have been saved. While doing advanced training at Columbia University, Dr. Drew began research into the properties of blood plasma. He devised techniques for preserving and banking blood. During World War II, Drew organized the British Blood Bank from which wounded soldiers received lifesaving infusions of blood plasma. Later, he led the American Red Cross project to collect and bank the blood of 100 000 donors. Have students find out what the needs of blood banks in your community are. `L3`

Figure 7-10

Skeletal muscles work in pairs. When one contracts, its partner relaxes. Your biceps and triceps are classic examples of this teamwork.

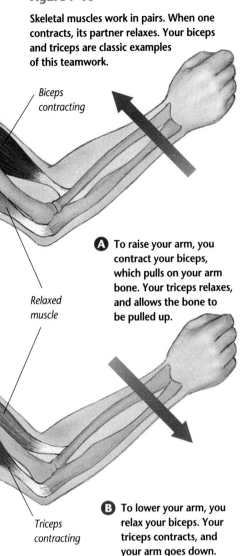

Biceps contracting

A To raise your arm, you contract your biceps, which pulls on your arm bone. Your triceps relaxes, and allows the bone to be pulled up.

Relaxed muscle

B To lower your arm, you relax your biceps. Your triceps contracts, and your arm goes down.

Triceps contracting

use them. Regular exercise increases the size of your muscles as well as their strength. When you exercise, the size of your muscle fibers increases, and this increases your muscles' strength. Exercise also increases muscle tone. *Tone* refers to the state of readiness of your muscles to contract. Muscles are usually slightly contracted so they're ready to go when you need them. Even when you think you aren't moving, some of your muscle fibers are contracting and relaxing to maintain tone. A well-toned body can help you have a better shape and increase your ability to do physical activity. Muscles, along with bones, support and shape your body.

Simple movements you probably take for granted, such as walking, aren't so simple after all. To move the bones and joints required to take just one step can involve the teamwork of up to 200 different muscles from your shoulders to your toes. That's one major reason why it takes a toddler so long to master walking. But once learned, it can be expanded into a variety of movements and activities that can enrich your life.

SKILLBUILDER

Concept Mapping

How do your thigh muscles function to bend your leg at the knee? To straighten your leg? Draw a concept map of the sequence of muscle movement. Refer to the diagram of body muscles. If you need help, refer to the **Skill Handbook** on page 637.

check your UNDERSTANDING

1. What do your heart, your stomach, and your thigh muscles have in common? How do they differ?
2. How do your arm muscles function to bend your elbow? To straighten your elbow?
3. How do the muscles in your body attach to the bones?
4. **Apply** Suppose the biceps muscle was removed. What arm movement would not be possible?

7-3 Muscles **231**

SKILLBUILDER

Muscle in back of thigh contracts while → muscle in front of thigh relaxes → muscle in front contracts while → muscle in back relaxes.

3 ASSESS

Check for Understanding

Have students answer the questions in Check Your Understanding.

Reteach

Activity Have students make flash cards with the names of body organs or parts that require muscle action. On the reverse side of each card, identify the type of muscle (skeletal, smooth, or cardiac) involved. Partners can then test each other using the flash cards. **LEP**
COOP LEARN

Extension

Research Investigate what happens when muscles become fatigued. **L3**

4 CLOSE

Activity

The human hand contains more than 50 bones and 28 muscles. Ask students to attempt to identify as many of these bones and muscles as they can by exploring the various movements they can make with their fingers. **L1**

check your UNDERSTANDING

1. All cause movement when they contract, but your heart is made of cardiac muscle, your stomach of smooth muscle, and your thigh muscles are skeletal muscles. You can control skeletal muscles, but not cardiac or smooth muscles.
2. The skeletal muscles on the front contract to bend your elbow, while the muscles on the back of your arm relax. The back muscles contract to straighten your elbow, while the muscles on the front of your arm relax.
3. with tendons
4. flexing the arm

Science and Society

Purpose Science and Society reinforces the discussion of muscle strength and tone presented in Section 7-3 by explaining what happens to a person's muscles when they are not exercised regularly.

Content Background

Muscles lose strength through inactivity; bones do, as well. Prolonged bed rest during convalescence from an illness makes bones weaken. After a few weeks of being idle, bones lose a substantial proportion of their calcium content. These studies were confirmed in an unlikely place—space!

In the early days of the space programs, astronauts who endured prolonged periods of inactivity in outer space were found to have lost significant bone strength. The lack of exercise contributed to the bone loss. Now, astronauts follow an exercise program while in space, and their bones, as well as muscles and cardiovascular systems, fare much better. All this seems to confirm that lack of exercise is bad for bones as well as muscles.

Teaching Strategies

Have students list the physical and social advantages of regular exercise. Then construct a class list with at least one contribution from each of the students' lists. **L1**

Discussion

Have students discuss whether they think they are more or less physically fit than their parents were as teenagers.

Answer to
What Do You Think?

While answers will vary, most students should paint a bleak picture of isolated, flabby individuals.

EXPAND
your view

Science and Society

Spud Dud!

This afternoon after school, keep track of how much time you spend watching television or playing video games. Do you think this is more or less time than you spent a year ago? Five years ago?

The terms *couch potato* and *spud dud* describe people who spend a lot of their leisure time just sitting.

The couch potato trend is taking its toll on children and teens in the United States. Medical researchers are concerned that, while there is a trend toward physical fitness in adults, young people are less physically fit today than ever before.

Get a Move On

If you sit for a long time, your pulse rate slows, much the way it does when you sleep. When you do get up and move—to get a snack—your system is moving slowly, and so you feel listless.

Muscles that don't get enough exercise lack flexibility. Infrequent or sudden strenuous activity may cause injury. Muscles kept in shape through regular exercise give you the capacity for taking on unexpected activities. Well-developed muscles also use more energy than poorly developed ones, even when you are resting. The result is that you feel more energetic, and you have less chance for fat cell buildup.

Physical activity builds cardiovascular endurance. Cardiovascular endurance is the ability of your heart, lungs, and circulatory system to deliver oxygen to your cells and to take away wastes. To build this endurance, you have to condition your heart and lungs to work more efficiently. Good activities are those that require continuous movement, such as running, swimming, or cycling.

What Do You Think?

Write a description of what you think the world might be like in the year 2020 if everyone became a couch potato today and remained one.

Going Further ⅢⅢ➤

Have students work in small groups to create a comprehensive physical fitness routine for people their age. Caution students to make their routines realistic as well as effective. What daily exercises are teenagers likely to do? Students may wish to speak with their physical education teachers, look at exercise videos, or write to:

President's Council on Physical Fitness and Sports
701 Pennsylvania Ave., N. W. Suite 250
Washington, DC 20004

Then have each group explain its program to the class, demonstrating the exercises and the advantages of each one. Use the Performance Task Assessment List for Group Work in **PASC,** p. 97.
COOP LEARN

Robot Arms

Most robot arms used in industry are jointed like the human arm. A typical robot arm may have five or more different movements. Each movement is controlled by a separate power source, usually an electric motor. Motors for these individual movements are controlled by computers. An interface connects the power source, the motors, and the computer. With an intricate electrical circuit, the interface directs power, switching motors on and off, as instructed by the computer.

Nuts and Bolts

A robot arm has a waist (or base), a shoulder, an elbow, and a wrist. You may expect that each of these joints might hold its own motor. However, in most robot arms, especially the smaller ones, the weight of these motors would make efficient movement of the arm difficult. Instead, many robot arms have their motors located near the base. Much like tendons, cords link the motors to the joints they operate. Pulleys are used to hold the cords and ease movement within the sections of the arm.

Electronic Wizardry

Frequently, the motors used to control joint movements are "stepping motors" or "steppers." These motors can turn on and off very rapidly, for quick changes in movement. Extremely fast changes in motion and function happen with step-by-step instructions from the computer. The robot arm can grip, lift, turn, and twist in seconds, much like your own arm.

Science Journal
Imagine that you are designing a robot to assist you with your physical tasks at home. Think about the movements that the robot's arm could make. Then describe *in your Science Journal* which chores the robot could help you accomplish.

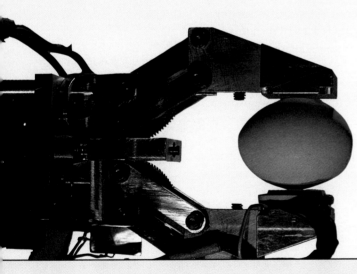

Science Journal
Discuss robots students might have read about in other sources or seen in films and television shows. Ask them how similar these robots were to the devices described in this selection.

Going Further ▪▪▪▪▪▶
Students can find out in which industries robot arms are used and for what tasks. Students might consider how robot arms might be used in various aspects of automobile production, such as welding and painting. After students collate their findings, they can create a display showing one use of robot arms in an industry of their choice. Use the Performance Task Assessment List for Display in **PASC**, p. 63. **P**

EXPAND your view

HOW IT WORKS

Purpose

This excursion adds to the discussion of joints in Section 7-2, by describing how robot arms function to re-create the motions of living arms.

Content Background

The creation of highly sophisticated robot arms owes a great deal to silver-gray flecks of silicon called integrated circuits, or chips. Barely the size of a newborn's thumbnail, the chip is capable of holding a million electronic components. Thanks to the microprocessor, a "computer on a chip," robot arms have moved from industry to the body. Medical prosthetic designers are making bold advances with circuits often as small as nerves and neurons. Researchers at the University of Utah's Center of Biomedical Design, for example, have built electronic arms that are activated by motion sensors and electrical signals from the skin.

Teaching Strategies Have student pairs create a chart showing how robot arms, as described in this excursion, are similar to human arms, as described in Section 7-2. **COOP LEARN**

GLENCOE TECHNOLOGY

Videodisc
STVS: Physics
Disc 1, Side 2
Bipedal Robot (Ch. 2)

Teens in SCIENCE

Purpose

Teens in Science reinforces the discussion of muscle tone in Section 7-3 by explaining how sixteen-year-old Jennifer November helps patients in a convalescent home maintain their muscle tone. These experiences helped convince Jennifer to pursue a career in the medical field.

Content Background

Systematically manipulating body tissues with the hands to improve muscle tone is the earliest form of physical therapy, used more than 3000 years ago by the Chinese. In the 1800s, Henrick Ling, a Swedish doctor, developed a system of massage to treat problems with muscles and joints. Today, there are three methods of massage for helping people keep their muscle tone. *Effleurage,* light or hard stroking, relaxes muscles and improves circulation of the small surface blood vessels. *Petrissage,* kneading and squeezing, helps stretch muscles to improve the range of movement. *Tapotement* uses the sides of the hands to strike the surface of the skin rapidly to improve circulation.

Teaching Strategies

Hold a class discussion of ways that students might volunteer in their community to work with people of any age group.

Activity

Invite a representative from your high school's vocational education department or a technical school in your area to discuss his or her programs, especially in health care.

Teens in SCIENCE

Lending a Helping Hand

Have you ever spent a few days in bed because you were ill? If so, you may have been surprised at how quickly muscles in your body began to lose their strength. Imagine how weak you might become if you were bedridden for months!

As a volunteer in a convalescent home, 16-year-old

Jennifer November helps her patients keep their muscle tone. "I start at the hands, moving each finger five times. I'll move on until the entire body has been worked. This motion keeps the muscles from getting contractures or shrinking."

Changing Goals

Jennifer became involved in this work as part of a special vocational education program at her high school in Hacienda Heights, California.

"I had no plans to sign up for the Nursing Assistant Program. In fact, I did it because all the other elective courses were filled. I'd always thought I wanted to be a journalist. But as soon as I got involved in medicine, I knew that I'd found my career. I'm going to nursing school first and then on to become a doctor."

Although Jennifer is very comfortable on the job now, the first few days were tough. "I was frightened. Like most people, I hadn't spent a lot of time with disabled or elderly people. I didn't want to be afraid of them, but I just was. It was confusing. But as soon as I got to know some of my patients, I found out that we had a lot in common. We're people."

In addition to making new friends, Jennifer has found a good deal of satisfaction in her work. "At first, it seemed like I was the one doing all the helping. But after a while, I realized how much I was being helped. I've learned a lot about life and courage from my patients."

You Try It!

Jennifer describes stumbling into nursing by accident. Write about an unexpected event that had a big impact on your life.

Do you know any disabled or elderly people? Write a brief description of your relationship with them. If you don't have any friends or relatives who fit this description, write about how you might meet such a person.

Going Further ▌▌▌▌▶

Invite students to become involved in programs for the elderly, lending their expertise or acquiring new skills and interests. Students may wish to volunteer in their community as Jennifer did or investigate regional or national avenues. For further exploration, students may wish to contact their state Commission on Aging or these organizations:

American Society on Aging
833 Market St., Suite 516
San Francisco, CA 94103

Rehabilitation Research and Training Center on Aging
University of Southern California
c/o Rancho Los Amigos Hospital
7600 Consuelo St.
Downey, CA 90242

Science Journal

Review the statements below about the big ideas presented in this chapter, and answer the questions. Then, re-read your answers to the Did You Ever Wonder questions at the beginning of the chapter. *In your Science Journal*, write a paragraph about how your understanding of the big ideas in the chapter has changed.

1 Bone is a vital, living tissue with numerous functions. It gives the body shape and supplies it with minerals. *What are the parts of bone?*

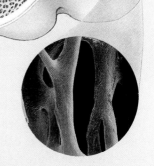
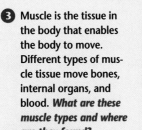

2 The skeletal system is able to move at certain points throughout the body because of joints. *What type of movement does each joint allow?*

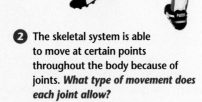

3 Muscle is the tissue in the body that enables the body to move. Different types of muscle tissue move bones, internal organs, and blood. *What are these muscle types and where are they found?*

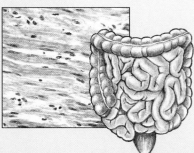
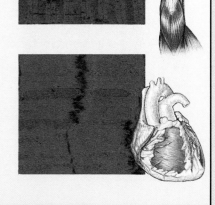

chapter 7
REVIEWING MAIN IDEAS

Students should examine the illustrations and text on this page to draw the shape, bones, and muscles that make up parts of the human body.

Teaching Strategies

Divide the students up into four-member teams. Assign each team one part of the body to diagram. One member of the team should draw the external shape of the body part on poster paper. The second member should use tracing paper to draw the underlying skeletal structure. The third member of the team should make a diagram of the muscles on tissue paper. Stack the drawings so they form a complete picture. The fourth member can make an illustration showing a cross section of the bones. **COOP LEARN**

Answers to Questions

1. Bone consists of compact bone, spongy bony, marrow, and the periosteum membrane.

2. Hinge joints allow back-and-forth movement. Pivot joints allow rotation on an axis, like the neck. Vertebrae glide over one another on disks of cartilage. Ball-and-socket joints allow the fullest range of vertical and circular motion.

3. Skeletal muscles are attached to bones. Smooth muscle is found in the body's internal organs like the stomach. Cardiac muscle is in the heart.

Science Journal

Did you ever wonder...
• Bones are made up of living and nonliving materials. (pp. 214-215)
• Your knee contains a hinge joint that gives it a limited range of motion in one plane. (p. 222)
• Everyone has the same number of muscles. Muscles become larger or smaller depending on how they are used. (pp. 230-231)

Project

Divide the class into three-member teams to prepare reports on one of the following: the hand, the back, the arms, the legs, or the neck. One member will draw and identify the major bones; another member will draw and identify the major muscles; the third member will research and demonstrate exercises to strengthen and tone the muscles. Use the Performance Task Assessment List for Group Work in **PASC**, p. 97.

COOP LEARN **P**

GLENCOE TECHNOLOGY

 **MindJogger Videoquiz**

Chapter 7 Have students work in groups as they play the Video-quiz game to review key chapter concepts.

Using Key Science Terms

1. Ligament is not part of bone but holds bones together.

2. Smooth muscles are not attached to bone by tendons as skeletal muscles are.

3. Cardiac muscle is not related; it is not found at a joint.

4. Stomach muscles are smooth; the others are skeletal muscles.

5. Bone marrow produces blood cells; it does not support the body.

Understanding Ideas

1. Tendons and ligaments are both connecting tissues. Tendons connect skeletal muscles to bones, ligaments connect bones at joints.

2. cardiac muscles–heart; smooth muscles–internal organs; skeletal muscle–attached to bone

3. support and shape body; protect organs; produce blood cells; store minerals

4. Muscles can only contract. One muscle contracts while another relaxes.

5. ball-and-socket joints–hip and shoulder; pivot joints–neck; immovable joints–skull; hinge joints–knee and fingers; gliding joints–vertebrae.

6. living–periosteum, a thin membrane that surrounds bone; blood vessels that carry oxygen; bone cells; nerves; bone marrow, a gel-like substance that makes blood cells; non-living–calcium and phosphorus minerals in compact bone

7. Bone has minerals–calcium and phosphorus–that provide strength. The structure of bone makes it strong–the lattice network of spongy bone and the tubular construction of long bones.

Developing Skills

1. See reduced student page.

2. a hinge joint

3. Skeleton of model can't move, isn't living; similarities between

U sing Key Science Terms

bone marrow	ligament
cardiac muscle	skeletal muscle
cartilage	smooth muscle
compact bone	spongy bone
joint	tendon

For each set of terms below, choose the one term that does not belong and explain why it does not belong.

1. compact bone, spongy bone, bone marrow, ligament

2. tendon, skeletal muscle, smooth muscle, bone

3. joint, ligament, cardiac muscle, cartilage

4. jaw muscles, hand muscles, skeletal muscles, stomach muscles

5. support, compact bone, bone marrow, skeletal muscles

U nderstanding Ideas

Answer the following questions in your Journal using complete sentences.

1. How are tendons and ligaments similar? How are they different?

2. List the three types of muscles and an example of where each is found.

3. Describe three functions of bones.

4. Why is it important for muscles to work together in a team?

5. List the types of joints and an example of each.

6. List and describe the living and nonliving parts of bone.

7. What features of bone make it possible for bone to be so strong?

D eveloping Skills

Use your understanding of the concepts developed in this chapter to answer each of the following questions.

1. Concept Mapping Using the following terms, complete the concept map of body movement: *cardiac, compact, joints, muscles, skeletal, smooth, spongy.*

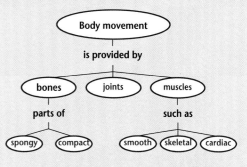

2. Making Models Determine what kind of joint the foot model from the Find Out activity on page 230 represents.

3. Comparing and Contrasting Thinking about the model you used in the Explore activity on page 212, what are some differences between the model and your body? Now think about your house or apartment. In what ways can the systems or features of that structure be compared to your systems or features?

4. Predicting Predict the results of the Find Out activity on page 214 using beef bones rather than chicken bones. Repeat the activity using beef bones and compare your prediction with your results. How do these results compare with those from the original activity?

house and body systems include: plumbing system for moving water, hinged doors (joints), need for repair.

4. The vinegar will take a longer time to dissolve the minerals in the bone.

5. fulcrum–knee; effort force–muscles in upper leg; resistance force–weight of lower leg

Program Resources

Review and Assessment, pp. 41–46 L1
Performance Assessment, Ch. 7 L2
PASC
Alternate Assessment in the Science Classroom
Computer Test Bank L1

5. Observing and Inferring After doing the Find Out activity on page 225, stand up and bend one of your legs at the knee, keeping your upper leg straight. Where is the fulcrum? Which part of your body provides the effort force? What provides the resistance force?

Critical Thinking

In your Journal, answer each of the following questions.

1. Pick up this book, and then put it down again. What enables you to do both with the same bones and joints in your arm?

2. What is the advantage of having a dome-shaped skull with immovable joints rather than a square skull with gliding joints?

3. Why is there always some movement in your body?

4. How do the foods you eat affect your bones?

5. A bone marrow transplant involves killing the cells in a person's bone marrow and replacing it with bone marrow from a donor. How does a bone marrow transplant help a person with a blood disorder?

Problem Solving

Read the following problem and discuss your answers in a brief paragraph.

On a fossil hunting expedition, Dr. Susman uncovers a remarkable find—an almost complete skeleton of an ape ancestor! Dr. Susman examines the fossil and discovers something important to her research. She finds that the fossil's leg joints are perfectly intact and undamaged. She also finds that the leg joints are similar to those of certain living species of apes. Dr. Susman wants to investigate how the ape ancestor used its legs to move around.

Use your knowledge of the skeletal system to describe what Dr. Susman might do to find out how the leg joints of the fossil worked.

CONNECTING IDEAS

Discuss each of the following in a brief paragraph.

1. Energy What forms of energy are involved in a muscle contraction?

2. Stability and Change Why does your face turn red when you exercise?

3. Systems and Interactions How are your bones and muscles like machines?

4. Science and Society Describe some of the benefits of exercise for your muscular system.

5. How It Works Compare a robot arm with a human arm. What supplies the energy to move the joints of a robot arm? How is this source of energy similar to that for a human arm?

Assessment

PORTFOLIO Review the portfolio options that are provided throughout the chapter. Encourage students to select one product that demonstrates their best work for the chapter. Have students explain what they learned and why they chose this example for placement into their portfolios.

Additional portfolio options can be found in the following **Teacher Classroom Resources:**

Making Connections: Integrating Sciences, p. 17
Multicultural Connections, pp. 17–18
Making Connections: Across the Curriculum, p. 17
Concept Mapping, p. 15
Critical Thinking/Problem Solving, p. 15
Take Home Activities, p. 13
Laboratory Manual, pp. 39–44
Performance Assessment P

Critical Thinking

1. The paired action of the biceps and triceps; the biceps contracts to bend the arm, while the triceps relaxes, and vice versa.

2. The dome-shaped, immovable skull is the strongest shape to afford the most protection.

3. The heart is always beating; smooth muscle is contracting in internal organs; some fibers in skeletal muscles are always in a state of contraction.

4. You need adequate amounts of the minerals calcium and phosphorus in the foods you eat for strong bones. If deficient, your bones may break easily.

5. Bone marrow is responsible for making blood cells. Replacing unhealthy bone marrow with healthy bone marrow may help restore the healthy condition of the blood.

Problem Solving

Some students may suggest that Dr. Susman may build a working model based on the structures in the fossil. Her model would be exact since the fossil is in good condition.

Connecting Ideas

1. The chemical energy stored in glucose is changed to mechanical energy, which is used to do work and release thermal energy, which is used to warm the body and lost as heat to surroundings.

2. Muscle contractions produce thermal energy. To prevent the body from becoming overheated, an increased amount of blood flows to the capillaries in the skin, carrying away heat.

3. They form levers. Bones are the levers; muscles exert the effort force.

4. Exercise keeps muscles in shape, which promotes cardiovascular endurance.

5. Electricity; the motor moves only as much as needed, just as you control the contraction of skeletal muscles.

Chapter Organizer

SECTION	OBJECTIVES	ACTIVITIES & FEATURES
Chapter Opener		**Explore!**, p. 239
8-1 The Nervous System: Master Control (3 sessions; 1.5 blocks)	1. **Demonstrate** the relationship between a stimulus and a response. 2. **Describe** the function of the nervous system. 3. **Diagram** the basic structure of a neuron. 4. **Explain** how impulses travel along nerves. **National Content Standards: (5-8) UCP1-2, A1, C1, C3**	**Explore!**, p. 240 **Find Out!**, p. 242
8-2 The Parts of Your Nervous System (3 sessions; 1 block)	1. **Explain** how the cerebrum and cerebellum work together during complex activities. 2. **Compare and contrast** the roles of the central and peripheral nervous systems. 3. **Trace** the pathway of a reflex. **National Content Standards: (5-8) UCP1-2, UCP5, A1-2, C1, C3, E2, F5**	**Explore!**, p. 244 **Design Your Own Investigation 8-1:** pp. 248–249 **Skillbuilder:** p. 251 **Investigate 8-2:** pp. 252–253 **Find Out!**, p. 254 **A Closer Look**, pp. 246–247 **Science and Society**, pp. 260–261
8-3 Your Endocrine System (2 sessions; 1 block)	1. **Explain** the functions of hormones. 2. **List** three endocrine glands and explain the effects of their hormones. 3. **Explain** how the endocrine system is involved in human growth. **National Content Standards: (5-8) UCP1, A1-2, B1, C1, F1, F4-5**	**Life Science Connection**, pp. 256–257 **SciFacts**, p. 259

ACTIVITY MATERIALS

EXPLORE!

p. 239 notebook, animal to observe
p. 240* balance beam, low curb, or secured 2-by-4 board to walk across
p. 244 slips of paper, pencils, reference books (optional)

FIND OUT!

p. 242 relay race baton or stick
p. 254 blindfold, flashlight (preferably a penlight)

INVESTIGATE!

pp. 252–253* large paper clip, metric ruler

DESIGN YOUR OWN INVESTIGATION

pp. 248–249* coin, eraser, other small object, meterstick

KEY TO TEACHING STRATEGIES

The following designations will help you decide which activities are appropriate for your students.

L1	Basic activities for all students
L2	Activities for average to above-average students
L3	Challenging activities for above-average students
LEP	Limited English Proficiency activities
COOP LEARN	Cooperative Learning activities for small group work
P	Student products that can be placed into a best-work portfolio
	Activities and resources recommended for block schedules

Need Materials? Call Science Kit (1-800-828-7777).

[00:00] OUT OF TIME? We recommend that students do the activities with an asterisk.

Chapter 8 Controlling the Body Machine

TEACHER CLASSROOM RESOURCES

Student Masters	Transparencies
Study Guide, p. 28 **Multicultural Connections**, p. 19 **Making Connections: Technology & Society**, p. 19 **Science Discovery Activities**, 8-3	**Teaching Transparency 15**, Neuron **Section Focus Transparency 22**
Study Guide, p. 29 **Concept Mapping**, p. 16 **Critical Thinking/Problem Solving**, p. 16 **Making Connections: Integrating Sciences**, p. 19 **Activity Masters**, Design Your Own Investigation 8-1, pp. 35–36 **Activity Masters**, Investigate 8-2, pp. 37–38 **Science Discovery Activities**, 8-1 **Laboratory Manual**, pp. 45–46, Demonstration of Reflexes	**Section Focus Transparency 23**
Study Guide, p. 30 **How It Works**, p. 11 **Take Home Activities**, p. 14 **Making Connections: Across the Curriculum**, p. 19 **Multicultural Connections**, p. 20 **Science Discovery Activities**, 8-2, 8-3	**Teaching Transparency 16**, Plant Movement **Section Focus Transparency 24**

ASSESSMENT RESOURCES	TEACHING & TECHNOLOGY
Review and Assessment, pp. 47–52 **Performance Assessment**, Ch. 8 **PASC*** **MindJogger Videoquiz** **Alternate Assessment in the Science Classroom** **Computer Test Bank**	**Spanish Resources** **Cooperative Learning Resource Guide** **Lab and Safety Skills** **Science Interactions, Course 2, CD-ROM** **Computer Competency Activities**

*Performance Assessment in the Science Classroom

NATIONAL GEOGRAPHIC TEACHER'S CORNER

Index to National Geographic Magazine	National Geographic Society Products Available From Glencoe	Additional National Geographic Society Products
The following articles may be used for research relating to this chapter: • "Quiet Miracles of the Brain," by Joel L. Swerdlow, June 1995.	To order the following products for use with this chapter, contact your local Glencoe sales representative or call Glencoe at 1-800-334-7344: • *Newton's Apple Life Sciences* (Videodisc) • *STV: Human Body Series*, "Nervous, Muscular, Skeletal Systems" (Videodisc)	To order the following products for use with this chapter, call the National Geographic Society at 1-800-368-2728: • *The Incredible Machine* (Book) • *The Incredible Human Machine* (Video) • *Marvels of the Mind* (Video) • *Your Body Series*, "Nervous System" (Video)

Teacher Classroom Resources

These are key components of the classroom resources package.

TEACHING AIDS

Section Focus Transparencies

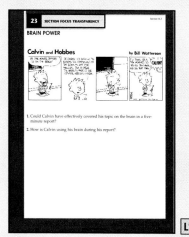

Teaching Transparencies

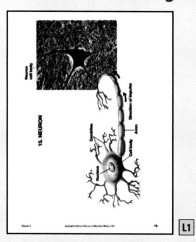

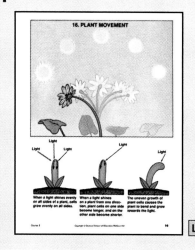

HANDS-ON LEARNING

Science Discovery Activity*

Laboratory Manual*

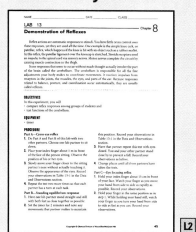

Take Home Activity

*There may be more than one activity for this chapter.

Chapter 8 Controlling the Body Machine

Study Guide*

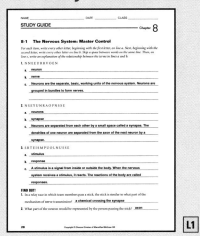

L1

Concept Mapping

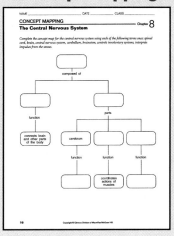

L1

Critical Thinking/ Problem Solving

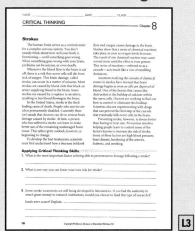

L3

Integrating Sciences

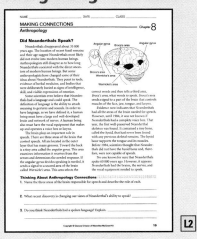

L2

Across the Curriculum

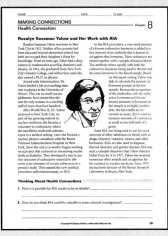

L1

Technology and Society

L1

Multicultural Connection**

L1

Performance Assessment

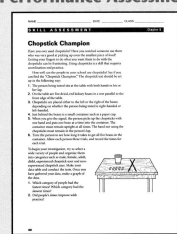

L2

Review and Assessment

L1

*One per section **Two per chapter

Controlling the Body Machine

THEME DEVELOPMENT

The themes this chapter supports are systems and interactions, and stability and change. The nervous system interacts with other body systems to produce body movement. The nervous system coordinates all responses to the environment, providing stability, and responding quickly to any changes.

CHAPTER OVERVIEW

In this chapter, students will discover that the nervous and endocrine systems are the body's control and communication systems.

Tying to Previous Knowledge

Ask students to describe how their bodies respond to various stimuli in the environment such as a hot stove, a pin prick, or a loud noise. Ask them whether their response is deliberately thought out or automatic.

INTRODUCING THE CHAPTER

Ask students to suggest the different movements people riding bicycles need to make to keep from falling.

00:00 OUT OF TIME?

If time does not permit teaching the entire chapter, use the Chapter Overview on this page, Reviewing Main Ideas at the end of the chapter, and the Chapter 8 audiocassette to point out the main ideas of the chapter.

CHAPTER 8

CONTROLLING THE BODY MACHINE

Did you ever wonder...

✓ How you can walk on a beam without losing your balance?

✓ Why your stomach sometimes growls when you smell food?

✓ Why, if you touch something hot, you pull your hand away before you even think about it?

Science Journal

Before you begin to study about controlling the body machine, think about these questions and answer them *in your Science Journal*. When you finish the chapter, compare your journal write-up with what you have learned.

Answers are on page 235.

238

"*T*he first time I tried to ride a bike, I wondered how anyone could do it. There was so much to remember all at once. Look straight ahead. Pedal. Don't lean sideways. Keep the bike straight. Time after time, the bike and I crash-landed. Then one morning, I got on, pushed away from the curb, and pedaled down to the corner without stopping or falling!"

Every activity you do requires coordination of bones, joints, and muscles in your body. In this chapter, you'll explore the body systems that control and coordinate most of your behaviors and body processes. You'll examine how these systems work to help you interact with the world around you.

▶ *In the activity on the next page, explore some behaviors of an animal that lives near you or with you.*

Learning Styles	Kinesthetic	Explore, p. 240; Investigate, pp. 248-249, 252-253; Find Out, p. 254
	Visual-Spatial	Explore p. 239; Demonstration, pp. 241, 243, 251; Visual Learning, pp. 241, 242, 245, 247, 250, 257, 258; Activity, pp. 245, 247
	Interpersonal	Find Out, p. 242
	Logical-Mathematical	Life Science Connection, pp. 256-257
	Linguistic	Discussion, p. 256; Explore, p. 244; Across the Curriculum, pp. 247, 250; Multicultural Perspectives, p. 251
LS	Intrapersonal	Science at Home, p. 262

Explore! ACTIVITY

Why does your dog bark?

Why do animals do the things they do? Why do some pets run to the kitchen when they hear a can opener? Why do some fish come to the top of the tank when you're about to feed them? Such activities are called behaviors.

What To Do

1. Choose an animal that lives near you or one that you see every day. It can be a pet or an outside animal, such as a bird or squirrel.

2. Observe and record the behaviors of the animal for 30 minutes. Be sure to watch carefully what goes on around the animal.

3. *In your Journal*, record whether the animal did things at random, or whether there seemed to be a pattern to its behaviors.

PREPARATION

Planning the Lesson

Refer to the Chapter Organizer on pages 238A–D.

Concepts Developed

In this section, students examine the structure of a neuron and learn how messages are transmitted from one neuron to the next.

1 MOTIVATE

Bellringer

 Before presenting the lesson, display **Section Focus Transparency 22** on an overhead projector. Assign the accompanying **Focus Activity** worksheet. L1

LEP

Explore!

How do you keep your balance?

Time needed 20 minutes

Materials a balance beam, low curb, or two-by-four board

Thinking Processes observing and inferring, recognizing cause and effect

Purpose To observe how the body responds to stimuli.

Teaching the Activity

Safety Be sure that the board is secured. If a balance beam is used, adjust it to be close to floor level. L1

Science Journal Students may write that the arms are important in maintaining balance and that their upper tor-

---

Section Objectives

- Demonstrate the relationship between a stimulus and a response.
- Describe the function of the nervous system.
- Diagram the basic structure of a neuron.
- Explain how impulses travel along nerves.

Key Terms

neuron
synapse

The Nervous System: Master Control

The Role of Your Nervous System

While learning to ride a bike, you probably found yourself a little confused by all the things you had to remember to do all at the same time. As you ride a bike, your body makes adjustments for each of these things. All living organisms are constantly faced with situations in which they have to make adjustments to the changing world around them.

Explore! ACTIVITY

How do you keep your balance?

What To Do

1. Find a long, secured two-by-four board that you can walk across. Make sure the part you are walking on is no wider than your foot. This will be your balance beam.

2. Try walking along your balance beam, slowly at first and then try to walk more quickly.

3. *In your Journal,* record some of the things you have to watch out for on a balance beam. What kinds of body adjustments do you make to keep yourself in balance?

In order to make adjustments to changes around them, living organisms have systems that control all the parts of their bodies. On the balance beam, you probably had to constantly adjust the way you held your body so you wouldn't tumble off. Did you start to tip over or did you lean too far to one side? Maybe you stuck your arms out or bent your knees to keep yourself in balance. Your body has a system that permits you to make these rapid adjustments. The body system that makes adjustments by controlling parts of the body is called your nervous system.

■ Stimuli and Responses

Each behavior you perform during a typical day is the result of the work of your nervous system. A nervous system receives signals, or

---

sos often swayed or tipped from side to side.

Expected Outcome

Students will observe how they use their bodies to maintain balance.

Answer to Question

3. Watch out for tipping over or leaning too far to one side. Adjustments include changing position, especially of the arms, to regain balance.

✔ Assessment

Oral Have students choose a sport such as bicycling or ice skating and describe the importance of maintaining balance when performing the sport. Use the Performance Task Assessment List for Oral Presentation in **PASC**, p. 71.

stimuli, from inside and outside your body. The smell of lunch is a stimulus. The ring of a telephone is also a stimulus. Once your nervous system receives a stimulus, it reacts. The reactions made by your body are called responses. Your stomach growling is a response to the smell of lunch, just as reaching for the telephone is a response to its ring. When you ride a bike or walk on a balance beam, you are faced with stimuli related to balance. You respond by adjusting how your body is held. What stimuli and responses did you observe in the animal you watched?

■ How the Nervous System Works

In many ways, your nervous system can be compared to an emergency response system. An example of such a response system is shown in **Figure 8-1** in which a sprinkler system is often used to monitor the safety of buildings.

Like an emergency response system, the job of your nervous system is to receive stimuli, process them, and give directions to various parts of your body so that there is a coordinated response. On an average day, you are bombarded by thousands of stimuli. Your body handles all of these stimuli at the same time, and your body systems respond in a coordinated way. Your nervous system and your endocrine systems are the main systems which control these activities.

Figure 8-1

Your nervous system can be compared to a type of fire department response system. Both receive stimuli, process information, give directions, and respond with action. What part of the fire response system might be compared to your brain?

A For the fire department, the heat of a fire, a "stimulus", triggers ...

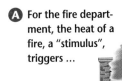

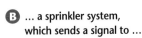

B ... a sprinkler system, which sends a signal to ...

C ... the fire department dispatcher. The dispatcher processes the information and makes an appropriate response. She then gives directions to the appropriate fire vehicles.

D The crews of the alerted fire vehicles spring into action to provide a coordinated response to the original stimulus of the fire.

8-1 The Nervous System: Master Control **241**

Demonstration Observing a sequence of dominoes fall can introduce students to the idea of stimulus/response. Arrange a set of dominoes on their edges so that each domino is just a short distance from the next. Give the first domino a push and observe the results. Discuss with students how the stimulus results in a response. Ask students to compare the movement through the domino pattern to the transmission of a nerve impulse. LEP L1

2 TEACH

Tying to Previous Knowledge
In Chapter 7, students learned that body movement is a result of muscles contracting to move the bones. Review this information, then ask whether anyone knows how the muscles get the information that tells them when to move.

Theme Connection As students learn about the nervous system, they will see that it is responsible for the way an individual interacts with the environment. Students will develop a model of the workings of this system as they do the Find Out activity in this section.

Visual Learning

Figure 8-1 Have students make a flow chart similar to Figure 8-1 illustrating the stimuli and responses involved in walking a balance beam. **What part of the fire response system might be compared to your brain?** *the dispatcher*

GLENCOE TECHNOLOGY

CD-ROM
Science Interactions, Course 2, CD-ROM
Chapter 8

Visual Learning

Figure 8-2 Have students study Figure 8-2. Draw a similar figure, without labels, on the chalkboard and ask students to identify each part: cell body, nucleus, axon, dendrites, and the direction of the impulse. Have students compare the branching of the dendrites with the branching of a tree.

NATIONAL GEOGRAPHIC SOCIETY

Videodisc

STV: Human Body Vol. 2

Axons and dendrites 1

48718

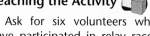

How does information move through your nervous system?

Time needed 30 minutes

Materials relay race baton or stick

Thinking Processes observing and inferring, comparing and contrasting, recognizing cause and effect

Purpose To compare the transmission of a nerve impulse to a relay race and infer how a nerve impulse travels through the nervous system.

Preparation Arrange to use the track or schoolyard and to move your class outdoors.

Teaching the Activity

Ask for six volunteers who have participated in relay races or like to run. **COOP LEARN** **L1** **LEP**

Science Journal Encourage students to compare factors that allow successful completion of a relay race to factors that allow the successful transmission of nerve impulses.

Neurons: Your Body's Relay Team

The sprinkler sends its signal to the fire department using telephone lines. To understand how messages called impulses travel throughout the nervous system, we have to look at the working unit of the nervous system, the **neuron**. Neurons have three main parts—a cell body and branches called dendrites and axons. Each neuron is a separate cell, but neurons usually are grouped together in a bundle called a nerve. **Figure 8-2** demonstrates how impulses are transmitted from one neuron to the next.

Figure 8-2

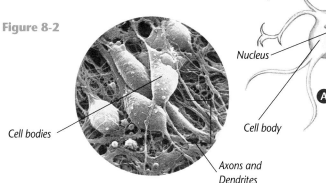

Dendrites

Axon

Nucleus

Cell body

Cell bodies

Axons and Dendrites

Direction of impulse

A Neurons are made up of dendrites, cell bodies, and axons. Dendrites carry messages, in the form of impulses, toward the cell body. The cell body, which is the largest part, directs the action of the cell. The axon then carries impulses away from the cell body.

Find Out! ACTIVITY

How does information move through your nervous system?

What To Do

1. Out on the track field, or in the school yard, organize a relay team of six students spaced around the track.

2. At the word *Go*, the student at the starting line runs and passes a stick to the second team member.

3. The second team member runs to the next member and passes the stick. The race continues in this fashion until the last team member crosses the finish line.

Conclude and Apply

1. Record *in your Journal* whether the second runner can begin before the first has come along.

2. What enables the second and third to run their parts of the race?

Expected Outcome

Students will infer that a nerve impulse is passed from one neuron to the next, just as the stick was passed from one runner to the next.

Conclude and Apply

1. no

2. They can run when they receive the stimulus (the stick).

✔ Assessment

Oral Ask students to describe how a nerve impulse travels from one neuron to the next. Students can then act out this process in a skit. Use the Performance Task Assessment List for Skit in **PASC**, p. 75.

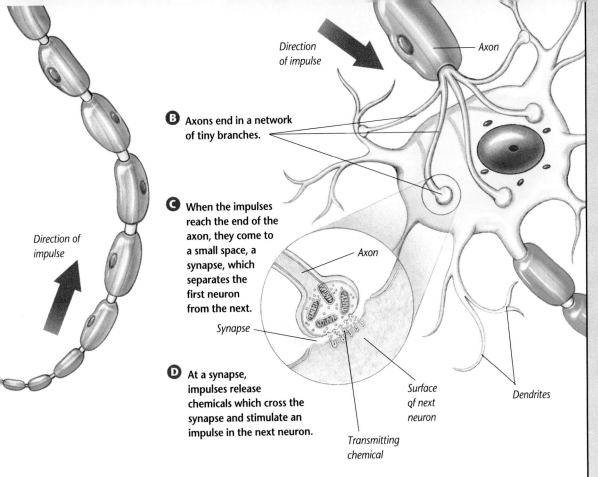

Direction of impulse

B Axons end in a network of tiny branches.

C When the impulses reach the end of the axon, they come to a small space, a synapse, which separates the first neuron from the next.

D At a synapse, impulses release chemicals which cross the synapse and stimulate an impulse in the next neuron.

Direction of impulse

Axon

Axon

Synapse

Surface of next neuron

Transmitting chemical

Dendrites

In your relay race, you never actually touched the next team member. The stick was simply passed to that person. Neurons work in a similar way. As you can see in **Figure 8-2C**, there is a small space called a **synapse** found between the neurons. The impulse from the axon of one neuron causes a chemical to cross the synapse, and an impulse starts in the next neuron.

check your UNDERSTANDING

1. Give an example of a stimulus you encounter every day and your response. What role does your nervous system play in this situation?
2. Diagram two neurons. Include the following: axon, cell body, dendrites, synapse. Show the direction an impulse travels.

3. Neurons never touch one another, so how does an impulse move from one neuron to the next?
4. **Apply** Certain drugs prevent axons from releasing chemicals into a synapse. What effect would drugs like this have on the transmission of an impulse?

check your UNDERSTANDING

1. Example: The stimulus of seeing a Don't Walk sign results in a person ceasing to walk. The nervous system detects the stimulus and then transmits messages to your muscles.
2. Diagrams should resemble Figure 8-2.
3. The impulse enters through a dendrite, then travels through the cell body to the axon, which releases a chemical that crosses the synapse to a dendrite of the next neuron.
4. The impulse couldn't be transmitted. Muscles controlled by the neuron wouldn't contract.

3 ASSESS

Check for Understanding

Ask students to write the answers to Check Your Understanding. Discuss the Apply question. Ask how this is like arriving at a drawbridge and finding it has been pulled up, and they cannot cross it.

Reteach

Activity To represent a stimulus/response reaction, have four students demonstrate this activity. Have the students pass a note from one to another. Each student creates a stimulus (a signal) to get the next student's attention, such as tapping the shoulder. Have the class identify the note as a nerve impulse, the signal as a stimulus, and taking the note as a response. L1

Extension

Model Reinforce students' ability to formulate models by having students make three-dimensional models of neurons from clay and pipe cleaners. (Use Figure 8-2 as a guide.) Ask students to show how a nerve impulse would travel through the models. L3

4 CLOSE

Demonstration

The classic children's game of Mousetrap is an excellent example of the stimulus/response relationship. Each of the objects on the board moves only when activated by the previous object. Set up the game and set the sequence in action. Then ask students to describe the stimulus and response that set each piece in motion. Discuss with students how this sequence resembles the transmission of a nerve impulse from one neuron to the next. L1

PREPARATION

Planning the Lesson

Refer to the Chapter Organizer on pages 238A–D.

Concepts Developed

In this section, students will see how the brain interprets incoming stimuli and decides what message to send. Students will study the role played by the cerebrum and cerebellum in coordinating complex activities.

1 MOTIVATE

Bellringer

  Before presenting the lesson, display **Section Focus Transparency 23** on an overhead projector. Assign the accompanying **Focus Activity** worksheet. L1
LEP

Explore!

What does a central information computer do for you?

Time needed 15 minutes

Materials slips of paper, pencils, reference books

Thinking Processes sequencing, observing and inferring, recognizing cause and effect

Purpose To observe how a central organizing agent responds to a variety of stimuli and infer how the nervous system responds to stimuli.

Teaching the Activity

Give science reference books to students so the operator can direct specific questions to an individual holding a book on that subject. After five questions, switch opera-

---

8-2 **The Parts of Your Nervous System**

Section Objectives

- Explain how the cerebrum and cerebellum work together during complex activities.
- Compare and contrast the roles of the central and peripheral nervous systems.
- Trace the pathway of a reflex.

Key Terms

cerebrum,
cerebellum,
brain stem,
spinal cord, reflex

The Central Nervous System

Whether you're sitting alone quietly or playing football, your nervous system receives and acts on many different stimuli all at the same time. Imagine playing football—the clock is running down and you're being chased by opposing team members.

You need to throw the ball! There are so many stimuli. They all seem to happen at once, yet your body responds to all of them. What parts of the nervous system take care of the incoming stimuli? How does your body know what responses to make?

Explore! ACTIVITY

What does a central information computer do for you?

Try the following game to help you understand how your nervous system responds to a stimulus. In this model, you and your classmates will simulate the working of a telephone information service. Students will take turns playing a central computer operator. The operator wins points by either answering science questions or redirecting the caller to where he or she can find the answer.

What To Do

1. Choose a classmate to be the operator and others to be callers.

2. Callers will each write one question and pass it to the

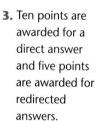

operator, who will respond to each of the calls.

3. Ten points are awarded for a direct answer and five points are awarded for redirected answers.

4. *In your Journal,* describe how well the operator handled the questions. Were many questions redirected?

---

tors. COOP LEARN L1 LEP

Science Journal Students may want to record in their journals each question and answer asked of the operator, and note how the operator handles each one.

Expected Outcome

Students should infer that the brain receives incoming messages and decides what responses to send and where to send them.

✔ Assessment

Oral Have students explain how the brain acts like a central computer operator and give two examples of how the nervous system responds to stimuli. Use the Performance Task Assessment List for Oral Presentation in **PASC**, p. 71.

The Brain: Operator of the Nervous System

Like the central computer operator who receives and processes hundreds of incoming calls a day, your nervous system contains structures that receive and process dozens of stimuli every second. This processing division of your nervous system is called the central nervous system. As shown in **Figure 8-3**, the central nervous system is made up of the brain and the spinal cord.

Your brain, protected by your skull, is divided into three parts, as shown in **Figure 8-4**. The parts are the cerebrum, the cerebellum, and the brain stem, and each has a different function.

■ The Cerebrum

The **cerebrum** is the largest part of the brain. You may have noticed that the surface of the cerebrum is wrinkled. Different areas, or centers, in the wrinkled surface of the cerebrum interpret impulses that nerves carry to it from different parts of the body. There is a center for each of your senses, in addition to centers for speech, memory, and for motor activities, which involve muscle movement.

Figure 8-3

Brain

Spinal cord

Figure 8-4

One of the three parts of the brain, the cerebrum, carries out complex functions such as memory, thought, and speech. It receives nerve impulses from your skin, eyes, ears, tongue, and nose and changes them into the sensations of touch, sight, sound, taste, and smell. The cerebrum also sends out signals that help control many muscles.

Skull

Cerebrum

Brain stem

Cerebellum

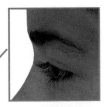

A When playing basketball, nerve impulses from a player's eyes travel to the vision center of the cerebrum. The information is processed and the player recognizes a teammate.

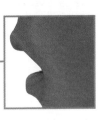

B When the player decides to call out to a teammate, the cerebrum forms the idea and sends nerve impulses to the vocal cords and to the muscles of the tongue and lips to make the player speak.

C When a player wants to pass the ball to a teammate, the motor center of the cerebrum sends out signals to the muscles needed to carry out that action.

8-2 The Parts of Your Nervous System **245**

〔 Program Resources 〕

Study Guide, p. 29
Concept Mapping, p. 16, The Central Nervous System L1
Critical Thinking/Problem Solving, p. 16, Strokes L3
Science Discovery Activity, 8-1
Section Focus Transparency 23

Meeting Individual Needs

Learning Disabled You may want to make the function of the central computer operator in this Explore activity more closely resemble the brain's function. Ask callers to write down a stimulus, such as a *loud noise*. The operator reads the message and writes a response, such as *jump*. The operator then gives the paper to another student who carries out the instruction on the paper. L1

Activity Help students refine their observation skills by choosing a few volunteers to observe and record behavior during the first five minutes of class. Have the observers share their observations. Ask students how they think their bodies are capable of keeping track of everything that goes on around them. L1

2 TEACH

Tying to Previous Knowledge

In Chapter 3, students learned that the heart is the center of the circulatory system. Blood is either heading for the heart or leaving it. Ask students to compare the heart to the brain and describe why the brain is the center of the nervous system. *Nerve impulses are either heading for the brain or leaving it.*

Theme Connection In this section, students will learn that the brain, spinal cord, and nerves function together as a system and that it is the interaction among these parts that enables that system to work. Some nerves accept stimuli, the brain interprets these stimuli, and other nerves carry impulses to parts of the body to cause a response to the initial stimulus.

Visual Learning

Figure 8-4 Have students identify the three parts of the brain shown in the figure. *the cerebrum, the cerebellum, and the brain stem* Discuss reasons why students think the cerebrum is the largest part of the brain. *Answers might include that it is the seat of intelligence.* Point out that the cerebrum is divided into two equal halves called the right and left cerebral hemispheres. Each hemisphere controls different functions.

NATIONAL GEOGRAPHIC SOCIETY

Videodisc

STV: Human Body Vol. 2

Cerebellum, cross section

48692

*Muscular and Skeletal Systems
Unit 2, Side 2
The Brain and Its Parts*

5757-9356

GLENCOE TECHNOLOGY

Videodisc

STVS: Human Biology

Disc 7, Side 1
PETT Scanner (Ch. 19)

The Infinite Voyage: Fires of the Mind

Chapter 4
Positron Emission Tomography (PET)

■ The Cerebellum

When you ride a bike, the motor center in your cerebrum sends impulses to the muscles. Think for a moment about how many muscles you use when you ride a bike. You can control a bike with ease, even though it requires the coordination and control of many muscles at the same time. How do you know how hard to pedal, where to steer, and how far you should tilt your body to balance?

During complex activities like riding a bike, your **cerebellum** coordinates the actions of all your muscles and maintains balance. The cerebellum is much smaller than the cerebrum. It is located toward the back and bottom of the brain, as shown in **Figure 8-5**. When you're riding a bike, the cerebellum sends messages to the cerebrum that direct and coordinate the activity. The cerebrum then sends out impulses to your muscles.

Your cerebrum and cerebellum work together when you need to move muscles in response to stimuli. For example, if somebody throws a ball toward you in gym class, you may move the muscles in your arms and hands to catch that ball. This type of response must be quick and well timed. In the next Investigate activity, you will explore how quickly your brain causes you to respond to such stimuli.

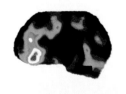

Mysteries of the Brain

The brain is the last and greatest biological frontier," says James Watson, the codiscoverer of DNA. He calls the brain "the most complex thing we have yet discovered in our universe."

How Can We Study a Working Brain?

It would help if hair, scalp, skull, and gray matter were all transparent. But they aren't.

Before 1972, what we knew of the brain was learned by studying brain-damaged people. But in 1972 a new technology was introduced. It's called positron emission tomography scanning, or PET scanning.

In a PET scan, the patient's head is surrounded by scanner cameras while he or she sits in something like a dentist's chair. The patient is injected with water containing sugar molecules that have radioactive tracers attached. The patient is given a thought problem, and the cameras record gamma rays being emitted by the radioactive tracers. Where the tracers are found inside the brain shows what part of the brain is working (and using the sugar molecules for fuel) to solve that problem. The computer in the scanner then generates color-coded images of brain activity. Some thought problems researchers

246 Chapter 8 Controlling the Body Machine

Purpose

A Closer Look shows how the PET scanner helps doctors understand brain function and diagnose disorders in the brain and nervous system. A PET scanner can help doctors pinpoint problems within the brain's millions of cells without using exploratory surgery.

Content Background

The *electroencephalograph* (EEG) is a forerunner of the PET scanner. While an EEG detects electrical activity of the brain, PET detects the brain's chemical activity. Both produce printed records or interpretable images from the data. An early use of the EEG was in the detection of brain tumors, which were detected through the lack of electrical activity in their tissues and by the distortions they produced in nearby normal tissues.

Teaching Strategies

Ask students to describe what they see

Figure 8-5

When you're riding a bike, your cerebellum sends messages to your cerebrum, which directs the activity. The messages tell your cerebrum in what order and with how much force your arm, hand, and leg muscles need to move for you to ride successfully. Your cerebrum processes the messages and sends out impulses to the appropriate muscles to carry out the needed actions.

Skull

Cerebrum

Brain stem

Cerebellum

have used include finding words that rhyme, playing a guitar, reading nonsense words, and remembering a fact.

Look at the brain scans accompanying this article. First, picture each of the four images as the left profile of someone's head as shown in the diagram on page 246. Now, compare the colored areas, which indicate brain activity. From the key you can judge whether the brain activity was weak (at the minimum) or strong (maximum).

PET scans also have medical uses. Since tumors grow more rapidly (and metabolize sugar faster) than normal tissue, PET scans can find tumors. Other uses are predicted for the near future.

PET scans are expensive, though. Each scan costs about $1700. The equipment is a big investment for a hospital, too: about $6 to $7 million to build and $1 million a year to run it. A PET setup includes the scanner and its cameras and computer support along with a cyclotron and its radiation shielding. Why a cyclotron? The radioactive tracers injected into the patient decay in just a few minutes so they have to be manufactured on site.

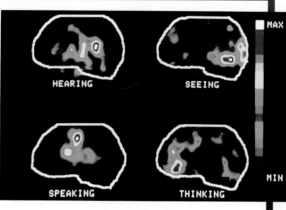

HEARING   SEEING

SPEAKING   THINKING

MAX

MIN

inter NET CONNECTION

Information on PET scanner operation is available on the World Wide Web. Locate operator training information and find out which isotopes are used for various PET procedures.

8-1 Reaction Time

Preparation

Purpose To design an experiment to test reaction time, which is the time it takes for the brain to respond to a stimulus.

Process Skills observing and inferring, comparing and contrasting, making and using tables, interpreting data

Time Required 30 minutes

Materials See reduced student page. Let the students choose their own small object to use in their tests.

Possible Hypotheses Students may hypothesize that a person's reaction time improves with practice. They may also hypothesize that a person's writing hand reacts faster than that person's non-writing hand.

The Experiment

Process Reinforcement Suggest that students set up a data table before beginning to measure their reaction times. If students have trouble setting up their data table, have them list the questions they need answered and then go over the mechanics of creating a table with them. Instruct students to take turns—one student performs the activity while another records the results. **L1**

COOP LEARN

Possible Procedures Possible methods of testing reaction time may include: (1) Hold out an arm, palm down, and place an object on the back of the hand. Turn the hand so that the penny slides off, and then try to catch the penny in the same hand before the penny lands. Vary the distance between the penny and the hand and repeat the test. (2) One partner holds the penny above the other experimenter's hand. When the penny drops, try to move the hand before the penny hits it. Vary the distance between the penny and the hand and repeat the test. (3) Hold a meterstick between the thumb and index finger. The partner observes the stick as it is released. The experimenter tries to catch the meterstick before it hits the ground. Record the distance the meterstick fell.

Repeat these experiments several times to test whether practice affects reaction time. Repeat these experiments using the non-writing hand.

Reaction Time

The time it takes the brain to react to a stimulus is called reaction time. Observing reactions doesn't always require using a timer. In a race, you can tell who is faster by seeing who reaches the finish line first. In this experiment, you will design methods of measuring reaction time.

Preparation

Problem
How fast do you react?

Form a Hypothesis
What affects reaction time? Will your reaction time be the same as your classmates'? Do you react faster with practice? Does your writing hand respond faster than your non-writing hand?

Objectives
- Predict the differences in reaction time under different circumstances.
- Compare reaction times by measuring distances.

Materials
coin
eraser
other small object
meterstick

248

Program Resources

Activity Masters, pp. 35–36, Design Your Own Investigation 8-1

Laboratory Manual, pp. 45–46, Demonstration of Reflexes **LEP**

Critical Thinking/Problem Solving, p. 5, Flex Your Brain

Making Connections: Integrating Sciences, p. 19, Did Neanderthals Speak? **L2**

DESIGN YOUR OWN
INVESTIGATION

Plan the Experiment

1 Choose materials to use in your tests and devise two different methods of measuring individual reaction time.

2 Will you be able to answer the following questions after carrying out your plan? Is the response of one hand quicker than the other? Can you affect reaction time by practicing?

3 How will you keep track of your data? Record all observations *in your Science Journal.*

Check the Plan

1 Will you measure each person's reaction time the same way? Why is that important?

2 Show your testing plan to your teacher. Make any changes that are necessary and carry out your plan.

Analyze and Conclude

1. Organize Information *In your Science Journal,* list the stimulus used in each of your activities.

2. Compare and Contrast Compare and contrast your reaction times for each hand. Is there a connection between your response and your writing hand?

3. Analyze Did your reaction time improve after a few trials?

4. Form a Hypothesis Compare your results with those of your classmates. Were your tests the same? Hypothesize why some people had faster reaction times.

8-2 The Parts of Your Nervous System **249**

Meeting Individual Needs

Visually Impaired For visually impaired students, place a bell (the type that rings when you tap the button at the top) under the student's hand. Make a loud noise, such as clapping the hands, and time how long it takes the student to react to the noise by ringing the bell. See if there is a difference between the right and left hands, and try other stimuli, such as tapping the shoulder.

Going Further ⫸

Answers may include requiring certain keystrokes to be made within a time limit, or maneuvering the mouse to certain places according to instructions. Check to be sure students have something measurable in their design.

Teaching Strategies

Troubleshooting Have the students think carefully about finding something measurable in their tests. Even though they are not measuring time in seconds, they need to have some evidence to support their hypothesis.

Science Journal Have students record their hypotheses and results (data tables) in their journals.

Expected Outcome

Students will design experiments that allow them to measure and compare reaction times, and to observe how factors such as experience and practice affect reaction time.

Analyze and Conclude

1. Possible answers could include: a penny sliding off the hand, dropping the penny, and dropping the meterstick.

2. Most likely the student's reaction time will be quicker with the writing hand.

3. Reaction time probably improved.

4. The ability to perform any task varies from individual to individual. Practice can improve anyone's ability, but some people are born with predispositions toward certain skills.

✔ Assessment

Process Have students design an experiment that shows how a stimulus from the environment, such as a loud or distracting noise, might negatively affect reaction time. Have students make a graph and record these new data as well as the data from the Design Your Own Investigation activity. Ask students what the graph shows about how external stimuli can affect reaction time. Use the Performance Task Assessment List for Graph from Data in **PASC**, p. 39. **P**

Going Further

How might you use a computer in designing a test of reaction times?

■ The Brain Stem and the Spinal Cord

In the Investigate, you had direct control over the muscles of your hand. Activities that are under your direct control, such as moving arm or leg muscles, are called voluntary activities. But there are many body activities that you do not have control over. Digestion, heartbeat, and breathing, for example, occur without you having to think about them and are called involuntary activities.

Even though involuntary body activities occur without you thinking about them, they are still controlled by a part of your brain, called the **brain stem**. The brain stem is the part of the brain that connects with the spinal cord. The **spinal cord** is a long cord that extends from the brain stem down in the back. It acts like the connection between the brain and parts of the body.

Figure 8-6

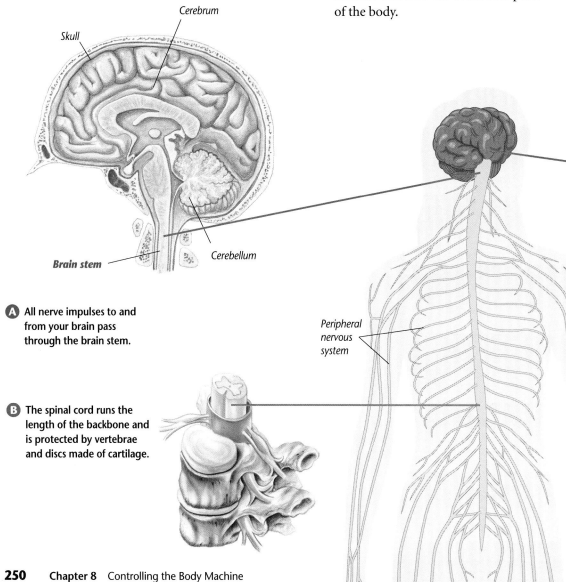

Skull

Cerebrum

Brain stem

Cerebellum

A All nerve impulses to and from your brain pass through the brain stem.

B The spinal cord runs the length of the backbone and is protected by vertebrae and discs made of cartilage.

Peripheral nervous system

250 Chapter 8 Controlling the Body Machine

The Peripheral Nervous System

Your brain and spinal cord form the central nervous system (CNS). All of the nerves outside the CNS make up the peripheral nervous system. *Peripheral* means to the side and away from. The nerves of the peripheral nervous system, as shown in the **Figure 8-6** body outline, extend to and away from the central nervous system.

Different stimuli are processed in different centers of the cerebrum. The cerebrum sends out impulses along motor nerves to activate muscles. During complex activities, the cerebellum coordinates the speed and timing of muscle action, making the activity run smoothly.

Impulses to and from the central nervous system are carried by the nerves of the peripheral nervous system. Sensory nerves carry impulses from a stimulus to the central nervous system for processing. Motor nerves carry impulses from the central nervous system to activate the muscles of the body. In the following Investigate, you will explore the structures that gather stimuli.

Making and Using Tables

Make a table of the divisions and functions of the parts of the nervous system. Include the following: central nervous system, cerebrum, cerebellum, brain stem, spinal cord, and peripheral nervous system. If you need help, refer to the **Skill Handbook** on page 639.

Figure 8-7

Ⓐ Sensory neurons carry the impulses from a stimulus to the central nervous system. The sensory neurons of the children send impulses that their cerebrums process to give the children the sensation of smelling the fragrance of the flowers.

Ⓑ The sight and fragrance of the flowers inspire the thought to pick them. The girl's cerebrum stimulates motor neurons to carry impulses from her central nervous system to the muscles of her arm and hand. She reaches out, grasps the flower, and pulls. The flower is picked. Her thought has been put into action.

251

8-2 Testing for Skin Sensitivity

Planning the Activity

Time needed 30 minutes

Purpose To observe how the sensory nerves in the skin respond to pressure.

Process Skills observing and inferring, comparing and contrasting, recognizing cause and effect, making and using tables, interpreting data

Materials large paper clip, metric ruler

Preparation Prepare data tables ahead of time. Have paper clips bent before class.

Teaching the Activity

Assign each student a partner. Demonstrate the procedure outlined in the student text and then ask students to conduct their own experiments. Discuss the text questions when everyone has completed the experiment. **L1** **COOP LEARN**

Safety Warn students not to apply heavy pressure with the paper clips.

Process Reinforcement Be sure that each partner carefully records the data and that the student being tested has his or her eyes closed.

Possible Hypotheses Students may hypothesize that the fingertips are more sensitive than other parts of the body.

Science Journal Have students record their predictions in their journals. You may want to ask students to compare their predictions with one another, having them give reasons why they ranked areas differently. Where there is a discrepancy between the prediction and the result, ask students to suggest reasons for their result.

Testing for Skin Sensitivity

Pressure receptors, scattered throughout your body, permit you to feel the objects you come into contact with. This activity will help you determine the relationship between the sensitivity of your skin and the location, number, and spacing of pressure receptors.

Problem
How sensitive is your skin?

Materials
large paper clip   metric ruler

Safety Precautions

Do not apply heavy pressure when using the paper clip.

What To Do

1. Copy the data table *into your Journal.*

2. Look at the test areas listed in the data table. *Predict* which parts of your arm will be the most sensitive to touch. Record your predictions.

3. Open a large paper clip and bend it into a U shape. Push the two tips of the paper clip together until they are 1 cm apart (see photo *A*).

4. **CAUTION:** *Lightly touch your partner's fingertip with both points of the paper clip.* Make sure your partner does not see what you are doing (see photo *B*).

5. Ask your partner whether one or two points were felt. Record this response in your data table.

Program Resources

Activity Masters, pp. 37–38, Investigate 8-2

A

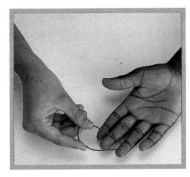

B

6 Fix the paper clip so that the points are farther apart and repeat Steps 2 and 3. Do this for 3 cm, 5 cm, and 7 cm. Record all responses in the data table.

7 Repeat Steps 4 through 6 for each location listed in the data table.

Sample data

Data and Observations

Distance	Predictions	1 cm	3 cm	5 cm	7 cm
Fingertip	5	2	2	2	–
Palm	4	1	2	2	2
Back of Hand	2	1	1	2	2
Forearm	1	1	1	1	2
Back of Neck	3	1	1	1	2

Analyzing

1. *Interpret* the data in your table to determine which area of your arm was the most sensitive. Which was the least sensitive?

2. *In your Journal,* **compare** the sensitivity of different parts of your arm.

3. How well did your predictions match your results?

Concluding and Applying

4. Can you suggest a reason why it might be beneficial for your fingertips to have many receptors?

5. Can you suggest a reason why your upper arm and neck have fewer receptors than your fingertips?

6. **Going Further** What parts of your body, besides your arm, would you *predict* to be the least sensitive? Explain your predictions.

Expected Outcomes

Students will observe their own sensitivity to touch and compare the sensitivity of different parts of the skin. They will find that certain areas of the skin are more sensitive than others.

Answers to Analyzing/ Concluding and Applying

1. The fingertip is the most sensitive area and the back of the hand or forearm is the least sensitive.

2. Students should rank the areas. One possible ranking is: fingertip, palm, back of hand, and forearm.

3. Responses will vary, depending on students' predictions.

4. Most tactile exploring is done with the fingertips. Thus it is helpful if the fingertips have many receptors to gather information.

5. The upper arm and neck are rarely used to gather new information about an object.

6. Going Further Students may mention the back and the legs because they are unlikely to be used to gather new information about an object.

✔ Assessment

Process Have students predict which body parts will be most sensitive to temperature and design an experiment to test the skin's sensitivity to temperature. Ask students to write a brief report detailing the experiment and the results. Use the Performance Task Assessment List for Designing an Experiment in **PASC,** p. 23. **P**

Reflexes

Sometimes you encounter stimuli in your environment that are so strong they may be harmful to you. How does your nervous system protect you?

Find Out! ACTIVITY

How are your eyes protected from light?

What To Do

1. After your teacher dims the lights in the classroom, lightly tie a blindfold around your partner's head so that the eyes are completely covered. Wait several minutes.

2. Remove the blindfold and quickly shine a flashlight into your partner's right eye for about one second. **CAUTION:** *Do not shine the flashlight any longer than required.*

3. Carefully observe the changes in the pupil of your partner's eye.

4. Repeat the procedure on the left eye. What happens to the pupil?

Conclude and Apply

1. How does the eye respond to the light?

2. Is this a protective response?

Connect to...

Physics

Nerve impulses can travel at speeds of up to 120 kilometers per second in the body. By comparison the speed of electricity is about 300 000 kilometers per second. How many times faster is an electric impulse than a nerve impulse?

Your eyes respond to brightness by controlling the size of the pupil. When you walk around in a dimly lit room, your pupils become larger. As a result, more light enters the eye to strike the retina. If there is a lot of light, such as on a sunny day, your pupils get smaller. The changes in your pupil size happen automatically and involuntarily. You have no control over them. How is this response helpful? Can you identify another response that protects your eyes?

Other parts of your body also respond automatically to stimuli.

Step on a sharp object and you jump instantly! Do you think about your reaction before it happens? An automatic body response to a potentially harmful stimulus, such as a bright light or a hot object, is called a **reflex**. Can you think of some other reflexes that protect you? How are reflexes important for the survival of an organism?

Reflexes must occur very quickly and in the same way each time. These impulses follow the shortest possible pathways through the nervous system and do not involve the brain. These pathways are called "reflex arcs".

254 Chapter 8   Controlling the Body Machine

Figure 8-8

If you picked up a piece of burning hot pizza, which would happen first: would you yell because of the pain or would you drop the pizza? A reflex arc would be triggered.

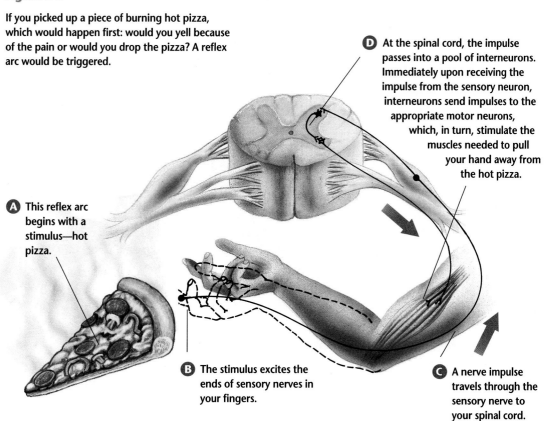

D At the spinal cord, the impulse passes into a pool of interneurons. Immediately upon receiving the impulse from the sensory neuron, interneurons send impulses to the appropriate motor neurons, which, in turn, stimulate the muscles needed to pull your hand away from the hot pizza.

A This reflex arc begins with a stimulus—hot pizza.

B The stimulus excites the ends of sensory nerves in your fingers.

C A nerve impulse travels through the sensory nerve to your spinal cord.

This whole reflex arc can take less than a second to complete. It is completed even before you feel the pain. To be aware of the sensation of pain, nerve impulses must travel up the spinal cord to your brain. This takes a few milliseconds. By the time you're ready to yell about the pain, your hand is safely pulled away.

check your UNDERSTANDING

1. Describe the roles of your cerebrum and cerebellum during a swimming exercise.
2. Describe the path of an impulse associated with lifting your arm. What part of your brain starts this impulse?
3. Describe the role the nerves of your peripheral nervous system play in responding to a stimulus.
4. Discuss some reasons why reflexes are important for the survival of organisms.
5. **Apply** After a severe accident, a person can talk and write, but has to learn to walk all over again. What parts of the nervous system were probably affected by the accident? What parts of the nervous system were not affected?

check your UNDERSTANDING

1. The cerebellum coordinates the speed and timing of the muscle contractions that produce coordinated arm strokes and leg kicks. It sends this information to the cerebrum's motor center, which sends out impulses to activate muscles.
2. The impulse originates in the motor center of the cerebrum. It travels along the spinal cord and to the arm muscles.
3. Stimuli are carried along sensory nerves of the peripheral nervous system to the central nervous system for processing.

Then, impulses are carried along motor nerves of the peripheral nervous system to the muscles.
4. Organisms face many different stimuli which may be harmful. Reflexes, because they happen automatically, protect the organism.
5. The cerebellum, spinal cord, and some motor nerves may have been affected. The cerebrum and brain stem were not affected.

Check for Understanding

As you discuss the Apply question, review the function of each part of the nervous system, identifying whether it belongs to the central or peripheral nervous system.

Reteach

Activity Have students refer to Figure 8-4 to make papier-mâché or clay models of the brain and its major parts. Use the models to review the parts and their functions. L1 LEP

Extension

Activity Ask students to rub their stomachs and pat their heads at the same time. The cerebellum has difficulty coordinating both movements simultaneously. Have students note how much easier the task becomes with practice. L2

4 CLOSE

Discussion

Ask students to imagine that they have been on their feet all day. Have them picture themselves sinking into a comfortable chair, taking off their shoes, and soaking their feet in warm water. Discuss with students the roles of the cerebrum and cerebellum in these actions.

8-3 ◆ Your Endocrine System

PREPARATION

Planning the Lesson

Refer to the Chapter Organizer on pages 238A–D.

Concepts Developed

In this section, students will study the endocrine system. Students will learn about the glands and hormones that make up the endocrine system and learn how hormones affect body functions. Students also will study the role of calcium in regulating body functions.

1 MOTIVATE

Bellringer

Before presenting the lesson, display **Section Focus Transparency 24** on an overhead projector. Assign the accompanying **Focus Activity** worksheet. L1

LEP

Discussion Ask students if they have grown taller in the past year, and if so, what might have stimulated this growth. Explain that endocrine glands and hormones are involved in many of the physical changes that happen as a person gets older.

2 TEACH

Tying to Previous Knowledge

Many students will have knowledge of the fact that some human behaviors are influenced

Section Objectives

■ Explain the function of hormones.
■ List three endocrine glands and explain the effects of their hormones.
■ Explain how the endocrine system is involved in human growth.

Key Terms

hormone
target tissue

Body Control

"The tallest man in the world!" and "the shortest woman in the land!" were commonly seen entertainers in circuses of the past. These people were ordinary persons except for their extraordinary height or lack of height. In most cases, their sizes were the result of a malfunction in their endocrine systems.

The endocrine system is another system for sending messages through the body, but it does not use neurons. This system is made up of tissues throughout the body called ductless glands. A **hormone** is a chemical made by a ductless gland in one part of the body that brings about a change in another part of the body. Hormones are needed in very small quantities and move directly from the

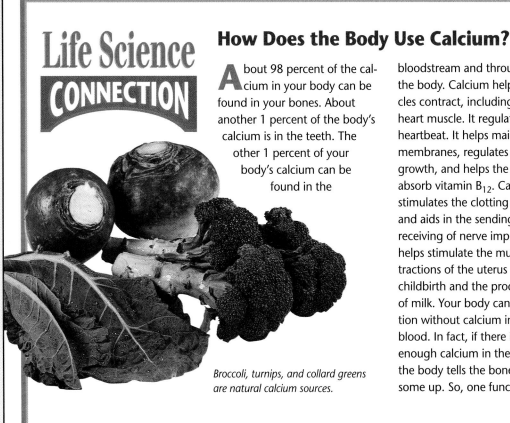

How Does the Body Use Calcium?

About 98 percent of the calcium in your body can be found in your bones. About another 1 percent of the body's calcium is in the teeth. The other 1 percent of your body's calcium can be found in the

bloodstream and throughout the body. Calcium helps muscles contract, including the heart muscle. It regulates the heartbeat. It helps maintain cell membranes, regulates cell growth, and helps the body absorb vitamin B_{12}. Calcium stimulates the clotting of blood and aids in the sending and receiving of nerve impulses. It helps stimulate the muscle contractions of the uterus during childbirth and the production of milk. Your body can't function without calcium in the blood. In fact, if there isn't enough calcium in the blood, the body tells the bones to give some up. So, one function of

Broccoli, turnips, and collard greens are natural calcium sources.

Purpose

Calcium is crucial for the generation of nerve impulses described in Section 8-1. The Life Science Connection details the role of calcium in maintaining body functions and describes how the endocrine system helps maintain blood calcium levels.

Content Background

Explain that vitamin D helps the body absorb calcium from foods. Most people get the vitamin D they need from exposure

to sunlight and from the foods they eat. The skin contains an inactive form of vitamin D which is activated by exposure to sunlight. Foods such as dairy products, sardines, salmon, liver, and some cereals contain vitamin D.

Teaching Strategies

To emphasize classifying skills, show students a copy of the school lunch menu for the week and have them classify foods that are rich in calcium. Then have students

Figure 8-9

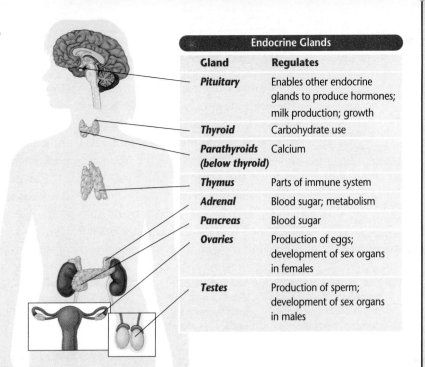

Endocrine Glands	
Gland	**Regulates**
Pituitary	Enables other endocrine glands to produce hormones; milk production; growth
Thyroid	Carbohydrate use
Parathyroids (below thyroid)	Calcium
Thymus	Parts of immune system
Adrenal	Blood sugar; metabolism
Pancreas	Blood sugar
Ovaries	Production of eggs; development of sex organs in females
Testes	Production of sperm; development of sex organs in males

cells of the glands into your bloodstream. The specific tissue affected by a hormone is its **target tissue**. A target tissue may be located far from the gland that makes the hormone. **Figure 8-9** shows where eight endocrine glands are located and what each gland regulates in the body.

by hormones. They may also know about the role of hormones in human growth, especially as it relates to athletes. Encourage students to share their information.

Theme Connection In this section, students will learn that the endocrine glands and hormones work together as a system to help the body function. The pituitary gland releases hormones that affect other glands in the system. These glands release hormones which affect target tissues. This interaction between the pituitary, other endocrine glands, and target tissues helps regulate growth, sexual development, metabolism, and other body functions.

bone tissue is storing calcium for the blood to use.

What Controls Calcium?

Two endocrine glands, the thyroid and the parathyroid, work together to keep levels of calcium in the blood at equilibrium. How does it work?

Eating calcium-rich foods causes a high level of blood calcium. This cues the thyroid to release a hormone that causes calcium to be deposited in the bones and to be excreted in urine from the kidneys.

On the other hand, a low level of blood calcium stimulates the parathyroid gland to secrete a hormone that causes bone to partially dissolve and causes the kidneys to conserve

calcium, not excrete it.

If you regularly avoid eating calcium-rich foods, the result may be weakened bones. Research has found that calcium deficiency may also be a possible cause of high blood pressure and of colon cancer.

USING MATH

Discover which foods are rich in calcium. Then find out the amount of calcium

Products such as cheese, milk, and yogurt also provide your body with calcium.

recommended for your age group. Analyze how much calcium you have ingested in the past three days. Should you make an adjustment to your present diet?

8.3 Your Endocrine System **257**

make up their own lunch menus, including foods from the five food groups and at least one serving of a calcium-rich food. Ask students to list some of the benefits of including calcium-rich foods in their diets.

USING MATH

Calcium-rich foods include dairy products, tofu, leafy green vegetables, and canned salmon and sardines eaten with the

bones. The recommended amount of calcium for children over the age of 4 is 1000 to 1500 milligrams per day.

Going Further ▥▥▥▶

Have students form teams to research one of the functions of calcium described in the Life Science Connection. Have each team create a scientific poster and use it to present their findings to the class. Use the Performance Task Assessment List for Poster in **PASC**, p. 73. COOP LEARN P

GLENCOE TECHNOLOGY

 Software

Computer Competency Activities

Chapter 8

3 ASSESS

Check for Understanding

Have students work with a partner to answer the Check Your Understanding questions. Then have student pairs create a public service announcement promoting healthy habits that can help ensure normal growth and development. L1 COOP LEARN

Reteach

Diagram Sketch a human outline on the chalkboard. Draw, but do not label, the endocrine glands. Ask students to identify each gland and the hormone(s) it releases. Have volunteers label the glands. Ask students to relate the regulatory role of each endocrine gland and hormone. L1

Extension

Activity Have student teams design flow charts to show the interaction between the pituitary gland and other endocrine glands. L2 COOP LEARN

4 CLOSE

Discussion

Have students describe the changes humans undergo during puberty. Discuss the endocrine glands and hormones that cause or influence these changes. L1

Figure 8-10

In addition to sex organs, which are primary sex characteristics, adult males are distinguished from adult females by specific physical traits called secondary sex characteristics.

During a lifetime, a person passes through stages. A newborn baby grows through infancy, childhood, adolescence, adulthood, and old age followed eventually by death. You are now in the stage of adolescence, which follows puberty.

Puberty is brought about by an increased output of hormones from the pituitary gland. These hormones stimulate the sex organs in both males and females to produce sex hormones. Such development is also influenced by general health, nutrition, and heredity of the individual.

The nervous and endocrine systems control the activities of your body. The effects of hormones often take longer and last longer than those of a nerve impulse.

A One secondary sex characteristic of females, which occurs during puberty, is developed breasts and widened hips.

B Increased muscle development and facial hair are secondary sex characteristics of males.

C Soon after puberty, the female reproductive system becomes active. Menstrual flow, known as a woman's period, repeats about every month.

D Enlargement of the larynx, which at first causes a cracking voice, is a secondary sex characteristic of males.

check your UNDERSTANDING

1. What is the function of a hormone?
2. List three endocrine glands and explain what each gland regulates in the body.
3. What are secondary sex characteristics? Give two examples of such characteristics.
4. **Apply** Why do some adolescents exhibit secondary sex characteristics earlier than others?

check your UNDERSTANDING

1. Hormones help control body functions such as growth and repair, metabolism, sexual maturity, response to stress, and the use of nutrients and minerals by the body.

2. Answers should concur with the information given in the chart on page 257.

3. Secondary sex characteristics are physical traits that distinguish females from males. In males, they include increased muscle development, facial hair, and an enlarged larynx resulting in a deeper voice. Breasts and widened hips are examples in females.

4. How fast a person develops depends on heredity, nutrition, and general health.

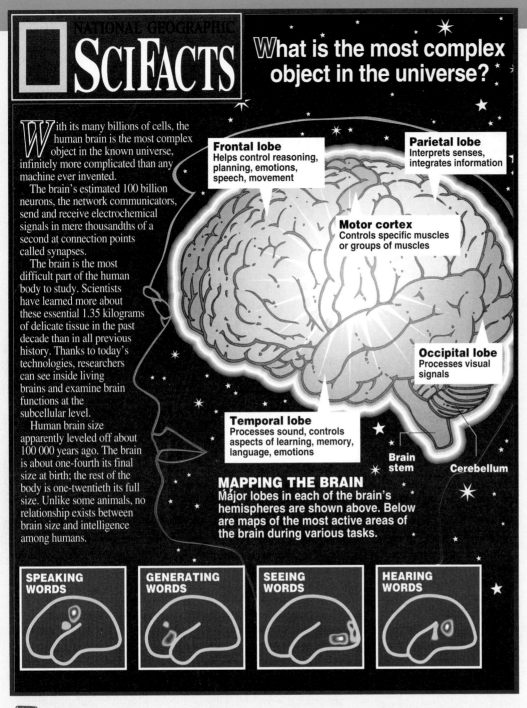

NATIONAL GEOGRAPHIC SCIFACTS

What is the most complex object in the universe?

With its many billions of cells, the human brain is the most complex object in the known universe, infinitely more complicated than any machine ever invented.

The brain's estimated 100 billion neurons, the network communicators, send and receive electrochemical signals in mere thousandths of a second at connection points called synapses.

The brain is the most difficult part of the human body to study. Scientists have learned more about these essential 1.35 kilograms of delicate tissue in the past decade than in all previous history. Thanks to today's technologies, researchers can see inside living brains and examine brain functions at the subcellular level.

Human brain size apparently leveled off about 100 000 years ago. The brain is about one-fourth its final size at birth; the rest of the body is one-twentieth its full size. Unlike some animals, no relationship exists between brain size and intelligence among humans.

Frontal lobe
Helps control reasoning, planning, emotions, speech, movement

Parietal lobe
Interprets senses, integrates information

Motor cortex
Controls specific muscles or groups of muscles

Occipital lobe
Processes visual signals

Temporal lobe
Processes sound, controls aspects of learning, memory, language, emotions

Brain stem

Cerebellum

MAPPING THE BRAIN
Major lobes in each of the brain's hemispheres are shown above. Below are maps of the most active areas of the brain during various tasks.

SPEAKING WORDS

GENERATING WORDS

SEEING WORDS

HEARING WORDS

Science Journal
Research methods scientists use to study the brain and describe these methods with diagrams *in your Science Journal*.

NATIONAL GEOGRAPHIC SCIFACTS

What is the most complex object in the universe?

Purpose
These SciFacts are designed to help students grasp the vast complexity of the human brain. They expand upon information provided in Section 8-3 by discussing how the brain processes information, and new technologies aimed at mapping the brain.

Content Background
The brain is divided into three main parts: the cerebrum, the cerebellum, and the brain stem. Generally, the lower portions of the brain control basic functions while the upper portions are involved in higher processes. For instance, the brain stem relays information between the brain and the body and regulates basic processes such as breathing. The cerebellum, positioned slightly higher than the brain stem, helps control motor activities. The topmost portion of the brain, the cerebrum, governs reason and language. The cerebrum is divided into the right and left cerebral hemispheres. In most people, the left hemisphere governs development and language, and the right hemisphere controls vision and artistic endeavors. Each hemisphere is divided into four sections; the frontal lobe, which controls movement and reason; the parietal lobe, which interprets sensory information; the temporal lobe, which processes sound and memory; and the occipital lobe, which processes visual stimuli.

Science Journal
Answers will vary, but most students will probably mention positron emission tomography scanning or PET scanning, which records which part of the brain performs various tasks; and electroencephalographs or EEGs, which analyze brain waves. Some students may mention relatively new research, such as studies on the re-growth of nerve fibers, that examines the brain on a microbiological or biochemical level.

Activity
Show students a "map" of the brain, such as the one illustrated in SciFacts. On the chalkboard or on a poster board, write a list of various functions, such as regulating heartbeat, playing basketball, giving an oral report, studying for a test, or watching a sunset. Have student volunteers point to the sections of the brain that are involved in the various functions. In some cases, more than one section of the brain will play a role in processing the information. Take this opportunity to stress to students that the brain functions as a whole.

Science and Society

Purpose

Science and Society relates some of the known facts about Alzheimer's Disease, its possible causes, and the attempts made so far to cure it. While the disease may begin in the brain, it eventually affects all the parts of the nervous system described in Section 8-2. This explains the loss of physical function that can come to victims of the disease. Finally, the article outlines hopes and problems concerning the drug THA, which has had mixed success against Alzheimer's Disease.

Content Background

The use of drugs like THA to combat a disease like Alzheimer's is an example of the relatively new science of *psychobiochemistry*. Just as nerve cells in the body depend on the presence of certain chemicals to transmit pulses, brain cells' interconnected functions depend on complex electrochemical activities in the synapses between them. The discovery and helpful manipulation of these chemical balances began in part with the work of 20th century researchers like Dr. Paul E. Bleuler of Switzerland. His work with schizophrenic patients, including those with forms of "senile dementia," led others to look for biochemical causes for mental disorders. Pellagra, a widespread form of mental dementia in the Mediterranean region, was found to be caused by a dietary deficiency. When it was cured by adding milk to a person's diet, this began a period of tremendous progress in the biochemical treatment of mental illnesses.

Science and Society

Alzheimer's Disease

Mrs. Greeley was surprised and confused the other day when she realized she was outside in her nightgown. She couldn't remember coming downstairs or unlocking the door.

Early Symptoms

Every day, scenes like this one take place across the country. These people feel frightened, confused, and ashamed. After accumulating a lifetime of memories, memory has begun to slip away like sand along a beach. These people are experiencing Alzheimer's disease.

Cause of Disease

At the present time, researchers have only a few concrete answers about the cause of Alzheimer's disease. The nervous system uses a chemical called acetylcholine to move impulses from one neuron to the next. In Alzheimer's patients, the brain cells fail to produce this vital chemical. The neurons become inactive, then begin to die. It may be the inactivity that causes neurons to die, but researchers are not sure.

Role of Heredity

Many researchers believe that heredity plays a role in causing the disease. They point to a specific type of chromosomal abnormality in Alzheimer's patients as a probable cause. Statistics also show that children of Alzheimer's patients are more likely to develop the disease themselves.

Who Can Get It?

People with Alzheimer's are generally over 65. However, some people in their 40s have been affected by the disease. It doesn't occur more often in certain ethnic or economic groups, nor has it been linked to the occurrence of any other disease.

Diagnosis

Because some brain cells die, the Alzheimer's patient suffers a loss of mental powers. Although 60 to 70 percent of the patients who suffer a loss of mental powers due to a physical cause have Alzheimer's disease, there is no specific test for it. When

other possible causes have been eliminated, Alzheimer's disease is usually diagnosed.

Symptoms

While some patients have occasional periods of full awareness, others have difficulty feeding and dressing themselves. Many Alzheimer's patients suffer severe personality changes, leaving them unable to recognize family members or to control violent outbursts toward these people and other caregivers. The ultimate outcome for all, however, is the eventual loss of physical function, and then death. In the United States each year, Alzheimer's disease takes the lives of more than 100 000 people.

Will There Be a Cure Soon?

Although intensive research has been done only since the early 1970s, much knowledge of the disease has been gained. Still, no real progress toward a cure has yet been made. Drugs that produce the essential acetylcholine have been tested. However, these drugs have been found to cause serious side effects in most patients.

Treatment Drugs

In 1986, test results were published that showed the drug THA produced a significant reduction in symptoms in Alzheimer's patients. The drug was studied by the Food and Drug Administration, the government agency that determines whether drugs work and whether they are safe and can be marketed in the United States.

Some patients were helped greatly by THA. Other patients showed no improvement, and others even suffered a harmful side effect that could cause liver damage. The FDA lowered the legal dosage of the drug to an amount that would reduce the risk of liver damage. At the lower dosage, however, the drug helped so few people that the FDA decided THA is not an effective treatment for Alzheimer's disease and did not approve the use. Other drugs are being developed and tested.

Science Journal

The FDA's decision not to approve the use of THA was very controversial. *In your Science Journal* discuss how you think Alzheimer's patients and their families would evaluate the possible risk of liver damage against the possible benefit of THA.

Teaching Strategies

If students know of someone with Alzheimer's disease or a similar disease of the mind, ask if they will describe how that person meets the daily challenges of the disease. What help and special treatments does the person receive? Has the person shown any positive responses to treatment since his/her diagnosis? What do students feel are the best ways in which they can help such people?

Student Writing

Ask students to write a paragraph on the question, "What is memory?" Do they believe it to be a part of the human "spirit," a mechanical function of brain cells, or something else? Ask them also to sum up what they think memory's most important function is in human life and culture.

Science Journal

1. Encourage students to imagine themselves personally connected to an Alzheimer's patient, so that the person's overall welfare is the primary consideration.

2. Some students may say that the relief from Alzheimer's symptoms is more important than the risk of liver damage. Others may say that one should never take potentially harmful medication. Opinions on the FDA decision will vary, but should be supported with specific information.

Going Further ⫸

Groups of students can form research teams and then pool information in a conference on the mechanics of memory. Each group should prepare to illustrate or explain how one aspect of human memory contributes to the brain's overall ability. Possibilities include the function of RNA (ribonucleic acid); how brain synapses are involved in memory; how one part of the brain can compensate for another damaged part. Use the Performance Task Assessment List for Writing in Science in **PASC**, p. 87. COOP LEARN P

Have students do the following activity to be sure they can identify the parts of the nervous and endocrine systems.

Teaching Strategies

Write the key terms for this chapter on slips of paper. Fold each paper and place it in a container. Divide the class into two teams. Have a member of one team draw a paper from the container. Give that person one minute to describe the function of the term while his or her team members try to guess the term. Keep track of terms which were not guessed for review later.

Answers to Questions

1. The nervous system enables you to make organized responses to stimuli. Throughout this system, stimulus and response impulses are carried by neurons.

2. Your sensory nerves pick up the message that your shoe needs to be tied, which is transmitted to the brain. The cerebrum interprets the message and decides what to do. The cerebellum coordinates arm muscles and sends information to the cerebrum's motor center, which sends out impulses to activate muscles.

3. Reflexes do not involve your brain. They occur very quickly and always in the same way. They enable your body to very quickly remove itself from dangerous situations.

4. The pituitary gland increases its output of hormones and stimulates the production of sex hormones. This signals the beginning of puberty.

GLENCOE TECHNOLOGY

 MindJogger Videoquiz

Chapter 8 Have students work in groups as they play the Videoquiz game to review key chapter concepts.

Science Journal

Review the statements below about the big ideas presented in this chapter, and answer the questions. Then, re-read your answers to the Did You Ever Wonder questions at the beginning of the chapter. *In your Science Journal*, write a paragraph about how your understanding of the big ideas in the chapter has changed.

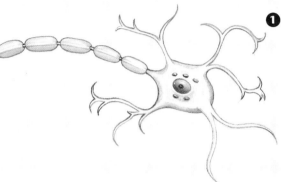

1 The neuron is the basic unit of the nervous system. Messages called impulses travel along neurons from dendrites to cell body to axons. *How do neurons help you to respond to changes in your environment?*

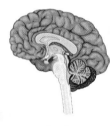

2 The central nervous system contains the brain and spinal cord, highly specialized organs containing billions of neurons that control and coordinate body activities. The spinal cord acts as a connection between the brain and the nerves of the body, which make up the peripheral nervous system. *When you tie your shoe, how are the different parts of your nervous system involved?*

3 Reflexes are automatic body responses to potentially harmful stimuli. The pathway of a reflex bypasses the brain. *How are reflexes important for your survival?*

4 Endocrine glands control body activities by secreting hormones directly into the bloodstream. These hormones affect specific tissues throughout the body. *How is the pituitary gland involved in the beginning of puberty?*

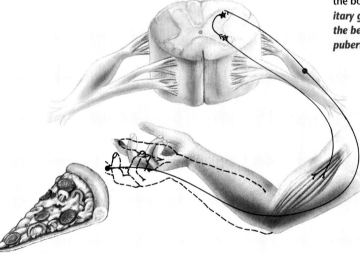

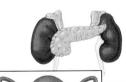

Science at Home

Ask students to keep a personal journal for the next two weeks. Each day they should record one or two incidents where their bodies had to respond to a complex situation. They should record the stimuli they encountered and what their body did to respond to that stimuli. They should identify any reflex actions. Use the Performance Task Assessment List for Science Journal in **PASC,** p. 103.

Science Journal

Did you ever wonder...
• The cerebrum decides what needs to be done and the cerebellum coordinates the action of the muscles. (p. 246)
• The brain recognizes a stimulus that usually precedes eating. The brain sends a signal to the stomach. (pp. 240–241)
• With stimuli that could injure, the body responds automatically. This is called a reflex action. (p. 254)

Using Key Science Terms

brain stem	reflex
cerebellum	spinal cord
cerebrum	synapse
hormone	target tissue
neuron	

Answer the following questions using what you know about the science terms.

1. Explain the relationship between a neuron and a synapse.
2. Differentiate by function among the three parts of the brain: brain stem, cerebellum, and cerebrum.
3. What is the function of the spinal cord?
4. Describe and give an example of a reflex.
5. How do hormones and target tissues interact?

Understanding Ideas

Answer the following questions in your Journal using complete sentences.

1. What role do hormones play in the body?
2. List the part of the nervous system responsible for the following activities.
   walking on a balance beam
   reading a book
   breathing
   playing soccer
   circulating blood through the circulatory system
3. If neurons don't touch each other, how does the impulse continue on its course?
4. Compare your central nervous system to a fire emergency situation. Which part of your CNS acts like the fire dispatcher?

5. Discuss the role of the cerebellum for performing activities involving many skeletal muscles all working at the same time.

Developing Skills

Use your understanding of the concepts developed in this chapter to answer each of the following questions.

1. **Concept Mapping** Using the following events, create an events chain concept map of the nervous system: *central nervous system processes stimulus, response, stimulus, impulse reaches central nervous system, sensory neurons carry impulse, central nervous system sends impulses to appropriate muscles or tissues.*

Initiating event

Stimulus

Event 1

Sensory neurons carry impulse

Event 2

Impulse reaches central nervous system

Event 3

Central nervous system processes impulse

Event 4

Central nervous system sends impulses to appropriate muscles or other tissues

Final outcome

Response

2. **Predicting** Refer to the Explore activity on page 240 and predict ways to improve your balance. Test your prediction.
3. **Comparing and Contrasting** Compare and contrast the two body control systems, the endocrine system and the nervous system, using a table.

Using Key Science Terms

1. Nerve cells are neurons and a synapse is the small space between neurons by which impulses travel.
2. The brain stem controls involuntary activities. The cerebellum coordinates muscle activity. The cerebrum is the largest part where thinking, sensory perception, and memory take place.
3. The function of the spinal cord is to carry impulses from all parts of the body to the brain and back.
4. An automatic body response to a potentially harmful stimulus is a reflex. An example is when you blink your eyes when an object is in front of them.
5. Target tissues are tissues affected by hormones.

Understanding Ideas

1. Hormones assist and regulate body processes. Hormones maintain levels of carbohydrate use, calcium, blood sugar, and the production of eggs and sperm.
2. cerebellum or cerebrum; cerebrum; brain stem; cerebellum or cerebrum; brain stem
3. The axon of the one neuron releases chemicals into the synapse, stimulating an impulse in the next neuron.
4. The spinal cord acts like the sprinkler and the fire truck, receiving and sending all messages. The brain acts like the fire dispatcher, responding to all incoming signals.
5. The cerebellum coordinates the timing of muscle action. The cerebellum sends these instructions to the cerebrum's motor center.

Developing Skills

1. See reduced student page for completed concept map.
2. Accept any reasonable response. A sample response might be to practice by carrying books on my head.

3.

Body Control Systems	
Endocrine System	Nervous System
Produces hormones carried in body fluids	Produces impulses carried by neurons
Affects target tissues far from glands	Affects far muscles and glands
Effects may be fast or slow	Effects are rapid
Effects may last hours, days, years	Effects pass quickly

Critical Thinking

1. When hormones are produced by the pituitary, other glands are stimulated to produce other hormones.

2. The speech center in the cerebrum may have been damaged.

3. The bloodstream transports hormones that regulate growth from the pituitary to target tissues in the feet.

4. The leg muscles are most likely to be affected, because the nerves in the legs are attached to that marked location on the spinal cord.

Problem Solving

Accept any reasonable answer. A sample response might be that the skeletal muscles in the dog's tail may have been damaged. There could be damage to the reflex arc, either in a motor nerve, an interneuron, or a sensory nerve. The sense organs that are sensitive to the stimulus may have been damaged.

Connecting Ideas

1. Dendrites are short, twiglike projections out from a neuron's cell body that receive messages and send them to the cell body. The cell body, the largest neuron part, receives messages, or impulses, and carries them away by the axons to the next neuron. Axons are long, stringlike projections that connect neurons.

2. The whole body is coordinated by the brain, which is the nervous system's control center. The nervous and muscular systems interact in the reflex arc to react to sensed dangers. The nervous system makes it possible for food to be digested, cells to get energy, breathing to continue, and heart rate to be regulated.

3. An X ray shows the structure of body tissues. A PET scan can display the brain's energy levels. By charting the energy level changes in the brain, a researcher can tell which parts of the brain are being used.

Critical Thinking

In your Journal, *answer each of the following questions.*

1. Why is the pituitary gland known as the "master gland"?

2. Why would a person have to learn how to speak again after a serious brain injury?

3. How do feet grow when the gland that controls growth is the pituitary, located in the head?

4. Look at the illustration of the nervous system. Which body parts might be affected if the spinal cord were injured at the area marked "X"—the arm muscles or the leg muscles?

Problem Solving

Read the following problem and discuss your answers in a brief paragraph.

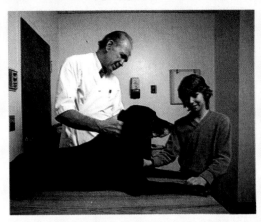

During your dog's visit to the veterinarian's office, the doctor checks your pet's nervous system by testing a reflex that makes the dog's tail wag. During the test, the doctor finds that your dog's tail doesn't wag after the proper stimulus is given. Using your knowledge of the nervous system and skeletal muscles, suggest two reasons your dog's tail-wagging reflex may not have worked.

CONNECTING IDEAS

Discuss each of the following in a brief paragraph.

1. **Theme—Scale and Structure** Describe the parts of a neuron and their function: cell body, axon, and dendrite.

2. **Theme—Systems and Interactions** Describe the functions of the nervous system and how it interacts with other systems.

3. **A Closer Look** Compare a PET (positron emission tomograph) scan to a normal X ray. How does PET technology allow researchers to learn more about the brain?

✔ Assessment

Portfolio Review the portfolio options that are provided throughout the chapter. Encourage students to select one product that demonstrates their best work for the chapter. Have students explain what they learned and why they chose this example for placement into their portfolios.

Additional portfolio options can be found in the following **Teacher Classroom Resources:**
Multicultural Connections, pp. 19, 20

Making Connections: Technology and Society, p. 19
Concept Mapping, p. 16
Critical Thinking/Problem Solving, p. 16
Take Home Activities, p. 14
Making Connections: Across the Curriculum, p. 19
Laboratory Manual, pp. 45–46
Performance Assessment, Ch. 8 **P**

ENERGY AT WORK

In this unit, you learned that work is done on an object when force is applied and the object is moved in the direction of the force. You also learned that the Law of Conservation of Energy means that work can only be done if energy is put into a system. Systems such as machines or the bones and muscles of your body convert energy into work. Energy for machines may come from various sources, while your body relies on energy from food.

Try the exercises and activity that follow. They will challenge you to use and apply some of the ideas you learned in this unit.

CONNECTING IDEAS

1. As you ride your bike, an animal darts in front of you. You apply the brakes and come to a stop. Explain how your nerves and muscles acted, how machines in your body and bike stopped the bike, and where the bike's energy went.

2. An inventor claims that she can make a car more efficient by using the waste heat from the exhaust pipe. In fact, she says the modified car will be twice as efficient as the original car. Is this possible? Explain.

Exploring Further ACTIVITY

How can you model your forearm?

What To Do

1. Build a model of your forearm using wood for the bones, a metal hinge for the joint, and cords for muscles.

2. What is the mechanical advantage of your model? How does it compare to that of your arm?

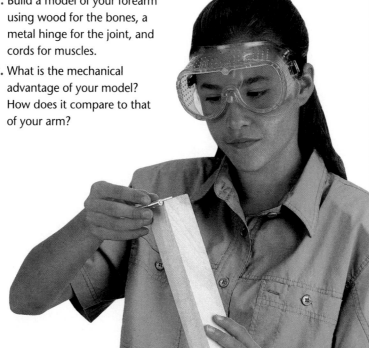

Energy at Work

THEME DEVELOPMENT

This unit supported the themes of energy and stability and change. The unit explored energy as an essential component of force that allows people and machines to work. Students learned how the skeletal, nervous, and muscular systems change to control the body's responses to outside forces, thereby maintaining a stable system.

Connections to Other Units

The definition of force, presented in Unit 1, is the basis for students' understanding of work and energy (Chapter 4) and how machines and the body work (Chapters 5, 7, 8).

Connecting Ideas
Answers

1. Nerves in your eye detected movement and sent a signal to your brain. Nerves carried your reaction to your muscles. They squeezed on the brakes. The rider's bones and muscles are levers; the bike's pedals and wheels are axles and wheels. The energy of motion was converted into thermal energy.

2. There is much less thermal energy in the exhaust gases than there is in exploding gasoline. Therefore you could not double the amount of mechanical energy necessary to double the efficiency of the engine.

Exploring Further
How can you model your forearm?

Purpose Students will observe how parts of the human body act as simple machines.

Materials wood, metal hinge, cords, or rubber bands

Preparation Provide pieces of balsa wood if possible, cut to appropriate lengths.

Safety Caution students to wear goggles while working on and testing their models.

Answers to Questions

The mechanical advantage is the ratio of the distance from the hinge to where the "muscle" is connected to the distance from the hinge to the end of the arm.

✔ Assessment

Performance Have students use their knowledge of simple machines to design catapults. Students can test their catapults outdoors by seeing how far they can propel a small object. (Remind students to be careful not to propel the object toward another person.) Use the Performance Task Assessment List for Invention in **PASC**, p. 45. **P**

Earth Materials and Resources

UNIT FOCUS

In Unit 3, students will learn to compare and contrast metallic and nonmetallic materials. They will learn about the special properties of some minerals and how Earth's surface changes over time.

THEME DEVELOPMENT

The two themes in this unit are systems and interactions and energy. Students will learn that minerals are specific components of a system of minerals that make up rocks. They will also learn that energy from the sun becomes stored in plants. Under certain conditions, decayed, compacted plants become coal over time. The energy is released later when coal is burned.

Connections to Other Units

In Unit 4, students will study how atoms relate to the classification of Earth material they read about in this unit.

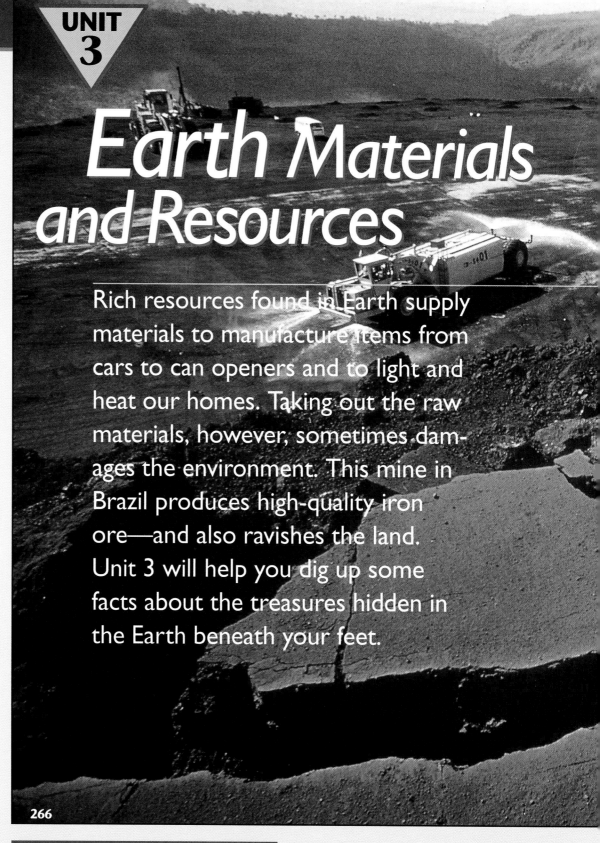

UNIT 3

Earth Materials and Resources

Rich resources found in Earth supply materials to manufacture items from cars to can openers and to light and heat our homes. Taking out the raw materials, however, sometimes damages the environment. This mine in Brazil produces high-quality iron ore—and also ravishes the land. Unit 3 will help you dig up some facts about the treasures hidden in the Earth beneath your feet.

266

GLENCOE TECHNOLOGY

Videodisc

Use the *Science Interactions, Course 2* **Integrated Science Videodisc** lesson, *Petroleum Formation: A Long Slow Process,* in conjunction with Chapter 11 and Chapter 13.

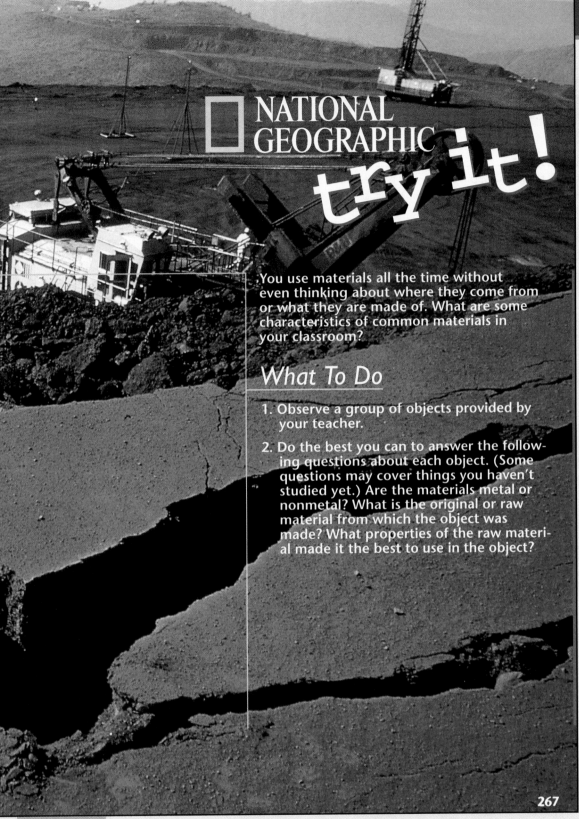

NATIONAL GEOGRAPHIC try It!

You use materials all the time without even thinking about where they come from or what they are made of. What are some characteristics of common materials in your classroom?

What To Do

1. Observe a group of objects provided by your teacher.

2. Do the best you can to answer the following questions about each object. (Some questions may cover things you haven't studied yet.) Are the materials metal or nonmetal? What is the original or raw material from which the object was made? What properties of the raw material made it the best to use in the object?

267

GETTING STARTED

Discussion Some questions you may want to ask your students are

1. What is an element? Most students will know that an element is a basic substance but will probably not be able to describe it as made up of only one kind of atom.

2. Why don't all rocks look alike? Students will probably suggest obvious physical properties without yet being able to describe how a rock's formation affects its looks.

3. How is electricity made from coal? Many students will know that coal is burned to make electricity and that coal is a nonrenewable energy source.

Responses to these questions will help you determine misconceptions that students have.

Answers to Questions

Students will probably realize that through careful observation, they can classify many of the materials. They may be surprised to discover that some materials, such as rocks, contain metallic and nonmetallic components.

✔ Assessment

Content Have students describe how they classified the sample materials. Then have students specify characteristics of a particular material (for example, metallic or nonmetallic).

Try It!

Purpose Students will compare and contrast common objects to form hypotheses about their origins. Students rely on their prior experiences and their sense of sight as they attempt to classify materials.

Background Information You may wish to relate this activity to students' knowledge of food sources. Some foods, such as fruits and vegetables, are similar to raw materials that are used without processing. Other foods, such as flour, are made from a source material such as wheat. This can be related to extracting a metal such as silver from its ore.

Materials Items may include common objects such as pencils, coins, chalk, lightbulbs, a cooking pot with a copper bottom, a lump of coal, a can of engine oil, and jewelry made of gold (or gold plating) and silver and precious or semi-precious gems. You may also wish to include rocks and minerals.

Chapter Organizer

SECTION	OBJECTIVES	ACTIVITIES & FEATURES
Chapter Opener		**Explore!,** p. 269
9-1 Discovering Metals (4 sessions; 2 blocks)	1. **Describe** the physical properties of a typical metal. 2. **Compare and contrast** the terms *malleable* and *ductile*. 3. **Explain** how the properties of metals determine their uses. **National Content Standards: (5-8) UCP1-2, UCP5, A1, B1, F2**	**Explore!,** p. 271 **Explore!,** p. 271 **Design Your Own Investigation 9-1:** pp. 274–275 **Skillbuilder:** p. 281 **A Closer Look,** pp. 276–277 **Science and Society,** p. 293 **Consumer Connection,** p. 294 **How It Works,** p. 295
9-2 Discovering Nonmetals (4 sessions; 2 blocks)	1. **Describe** the physical properties of a nonmetal. 2. **Compare and contrast** metals and nonmetals. 3. **Relate** the properties of nonmetals to their uses. **National Content Standards: (5-8) UCP2, UCP5, A1, B1, D1, D3, F2**	**Explore!,** p. 282 **Explore!,** p. 283 **Investigate 9-2:** pp. 286–287 **Earth Science Connection,** pp. 284–285 **Technology Connection,** p. 296
9-3 Understanding Metalloids (1 session; 5 blocks)	1. **Distinguish** among metals, nonmetals, and metalloids. 2. **Relate** the unique properties of metalloids to their uses. **National Content Standards: (5-8) UCP1-2, UCP5, A1, B1, F2, F5**	**Explore!,** p. 292

ACTIVITY MATERIALS

EXPLORE!

p. 271* element samples
p. 283 small piece of sanded magnesium ribbon, test tube, white vinegar, cork stopper, wooden splints, matches, safety goggles
p. 292* one or more bottles of mineral supplements

INVESTIGATE!

pp. 286–287 test tubes, test-tube holders, wooden splints, matches, 0.5 g cobalt chloride, balance, cork stopper, graduated cylinder, liquid laundry bleach (5% sodium hypochlorite), spatula, 5-cm × 5-cm square of notebook paper

DESIGN YOUR OWN INVESTIGATION

pp. 274–275* 7 test tubes and rack, 7 metal salt solutions, candle or laboratory burner, wooden splints, matches, marking pen, tweezers

KEY TO TEACHING STRATEGIES

The following designations will help you decide which activities are appropriate for your students.

L1 Basic activities for all students
L2 Activities for average to above-average students
L3 Challenging activities for above-average students
LEP Limited English Proficiency activities
COOP LEARN Cooperative Learning activities for small group work
P Student products that can be placed into a best-work portfolio
Activities and resources recommended for block schedules

Need Materials? Call Science Kit (1-800-828-7777).

00:00 **OUT OF TIME?** We recommend that students do the activities with an asterisk.

Chapter 9 Discovering Elements

TEACHER CLASSROOM RESOURCES

Student Masters	Transparencies
Study Guide, p. 31 **Take Home Activities,** p. 16 **Concept Mapping,** p. 17 **Activity Masters,** Design Your Own Investigation 9-1, pp. 39–40 **Multicultural Connections,** p. 21 **Making Connections: Across the Curriculum,** p. 21 **Science Discovery Activities,** 9-1, 9-2 **Laboratory Manual,** pp. 47–48, Reclamation of Copper from Mine Wastes	**Teaching Transparency 17,** Composition of the Body **Section Focus Transparency 25**
Study Guide, p. 32 **Making Connections: Technology & Society,** p. 21 **Activity Masters,** Investigate 9-2, pp. 41–42 **Making Connections: Integrating Sciences,** p. 21 **Science Discovery Activities,** 9-3	**Teaching Transparency 18,** CFCs and Ozone **Section Focus Transparency 26**
Study Guide, p. 33 **Critical Thinking/Problem Solving,** p. 17 **Multicultural Connections,** p. 22	**Section Focus Transparency 27**

ASSESSMENT RESOURCES	TEACHING & TECHNOLOGY
Review and Assessment, pp. 53–58 **Performance Assessment,** Ch. 9 **PASC*** **MindJogger Videoquiz** **Alternate Assessment in the Science Classroom** **Computer Test Bank**	**Spanish Resources** **Cooperative Learning Resource Guide** **Lab and Safety Skills** **Science Interactions, Course 2, CD-ROM** **Computer Competency Activities**

*Performance Assessment in the Science Classroom

NATIONAL GEOGRAPHIC TEACHER'S CORNER

Index to National Geographic Magazine

The following articles may be used for research relating to this chapter:

- "Advanced Materials—Reshaping Our Lives," by Thomas Y. Canby, December 1989.
- "The Miracle Metal—Platinum," by Gordon Young, November 1983.
- "Silver: A Mineral of Excellent Nature," by Fred Ward, September 1981.
- "Aluminum: The Magic Metal," by Thomas Y. Canby, August 1978.

GLENCOE TECHNOLOGY

The following multimedia resources are available from Glencoe.

Science and Technology Videodisc Series (STVS)
Chemistry
  Images of Atoms
Human Biology
  Measuring Calcium Deficiency

The Infinite Voyage Series
Miracles by Design

Glencoe Physical Science Interactive Videodisc
Periodicity

Physical Science CD-ROM

National Geographic Society Series
STV: Atmosphere

Teacher Classroom Resources

These are key components of the classroom resources package.

Section Focus Transparencies

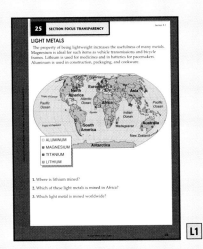

25 SECTION FOCUS TRANSPARENCY

LIGHT METALS

The property of being lightweight increases the usefulness of many metals. Magnesium is ideal for such items as vehicle transmissions and bicycle frames. Lithium is used for medicines and in batteries for pacemakers. Aluminum is used in construction, packaging, and cookware.

1. Where is lithium mined?
2. Which of these light metals is mined in Africa?
3. Which light metal is mined worldwide?

L1

26 SECTION FOCUS TRANSPARENCY

SWIMMING POOLS

The water in a public swimming pool has a distinctive odor. The element responsible for this odor is chlorine. The people who maintain the pool test the water and add chlorine compounds as needed.

1. Why are chlorine compounds added to water in pools?
2. Given that chlorine gas is poisonous to humans, how is it possible to use chlorine compounds safely in pools?

L1

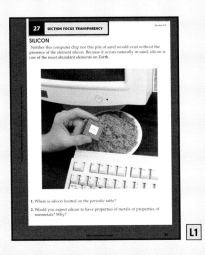

27 SECTION FOCUS TRANSPARENCY

SILICON

Neither this computer chip nor this pile of sand would exist without the presence of the element silicon. Because it occurs naturally in sand, silicon is one of the most abundant elements on Earth.

1. Where is silicon located on the periodic table?
2. Would you expect silicon to have properties of metals or properties of nonmetals? Why?

L1

Teaching Transparencies

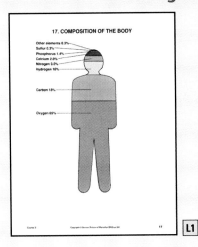

17. COMPOSITION OF THE BODY

L1

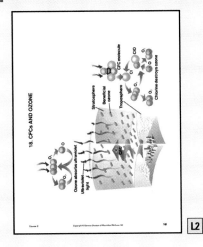

18. CFCs AND OZONE

L2

Science Discovery Activity*

ACTIVITY 9-1

Twang or Snap?

L1

Laboratory Manual*

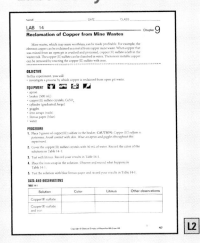

LAB 14 — Chapter 9

Reclamation of Copper from Mine Wastes

L2

Take Home Activity

Chapter 9

MONEY METALS

L1

*There may be more than one activity for this chapter.

Chapter 9 Discovering Elements

REVIEW AND REINFORCEMENT

Study Guide*

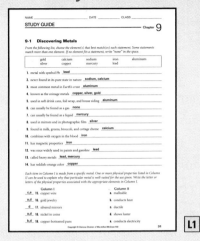

Concept Mapping

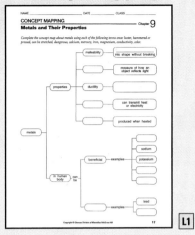

Critical Thinking/ Problem Solving

ENRICHMENT AND APPLICATION

Integrating Sciences

Across the Curriculum

Technology and Society

Multicultural Connection**

ASSESSMENT

Performance Assessment

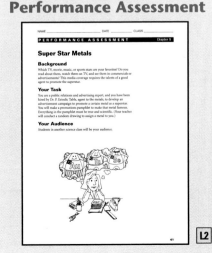

Review and Assessment

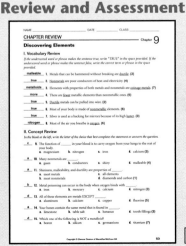

Discovering Elements

THEME DEVELOPMENT

The themes that are supported by this chapter are scale and structure and systems and interactions. Elements make up the structure of all matter. The properties of an element do not depend on the size or scale of the sample. Elements interact to form compounds. Some elements form compounds readily. Others rarely form compounds.

CHAPTER OVERVIEW

In this chapter, students find out about metals, nonmetals and metalloids. Their properties are explored and the importance of specific examples is discussed.

Tying to Previous Knowledge

In this activity students will observe that metals conduct heat.

Fill a beaker with very hot tap water. Put a metal spoon and a wooden spoon into the beaker of water. After a minute, have volunteers carefully touch the handle of each spoon simultaneously and describe what they feel. The metal should feel hot, but not the wood.

INTRODUCING THE CHAPTER

Have students study the collections of items on these two pages and list properties of the materials from which they are made. Look for terms such as shiny, hard, brittle.

00:00 OUT OF TIME?

If time does not permit teaching the entire chapter, use the Chapter Overview on this page, Reviewing Main Ideas at the end of the chapter, and the Chapter 9 audiocassette to point out the main ideas of the chapter.

DISCOVERING ELEMENTS

Did you ever wonder...

✓ **Why pennies contain copper?**

✓ **What happens to plants and animals after they die?**

✓ **How video games can be so small?**

Science Journal

Before you begin to study about elements, think about these questions and answer them *in your Science Journal*. When you finish the chapter, compare your journal write-up with what you have learned.

Answers are on page 297.

O mar collects stamps, Yolanda collects shells, and Luis collects model trains. What do you collect? Whether buttons or baseball cards, the objects in your collection are made up of elements or a combination of elements. Coins, for example, may be made up of gold and silver but more likely contain a combination of elements, such as copper, zinc, and nickel. Omar's stamps, Luis' trains, and Yolanda's shells also are made up of combinations of elements.

Everything around you, in fact, everything in the known universe, is made of elements. What are these elements and what combinations do they form? After completing this chapter, you'll think the answer to that question is elementary!

▶ **In the activity on the next page, explore some elements that may be found in your environment.**

Learning Styles	Kinesthetic	Investigate, pp. 274-275; Demonstration, p. 276; Activity, p. 285
	Visual-Spatial	Demonstration, pp. 270, 272, 282, 289, 290; Visual Learning, pp. 270, 277, 285; Explore, pp. 271, 283; Activity, pp. 273, 280; Earth Science Connection, pp. 284-285; Investigate, pp. 286-287; Across the Curriculum, p. 288; Project, p. 297
	Interpersonal	Across the Curriculum, p. 277; Research, p. 278
	Logical-Mathematical	Across the Curriculum, p. 278; Activity, p. 280
LS	Linguistic	Explore, pp. 269, 271, 282, 292; Multicultural Perspectives, p. 272; A Closer Look, pp. 276-277; Discussion, pp. 277, 285, 289; Across the Curriculum, pp. 279, 280, 291; Research, p. 279; Debate, p. 284; Activity, p. 288

Explore! ACTIVITY

What elements are in you and in your environment?

The oceans and the forests, automobiles and clouds, stars and planets, even your body and the air you breathe—all are composed of fewer than 100 different elements. What are these elements?

What To Do

Think about what elements are in your body and your environment.

1. *In your Journal*, make a list of as many elements as you can think of that are found in your body—in your muscles, bones, teeth, and blood.

2. Where in your environment are these elements found?

3. What other elements exist in your environment? Where are they found?

4. *In your Journal*, use your list of elements to suggest why you think each item is an element.

269

Uncovering Preconceptions
Many students don't realize that some elements exhibit properties of both metals and nonmetals.

Explore!

What elements are in you and in your environment?

TECH PREP

Time needed 20 minutes

Thinking Processes comparing and contrasting

Purpose To compare and contrast elements in the human body and in different environments.

Teaching the Activity

Before they begin this activity, have students brainstorm the names of as many elements as they can. **L1**

Science Journal Suggest that, when possible, students include where the element is found. For example, iron is in blood, calcium in bones, etc. **L1**

Expected Outcome

Students should be able to list several elements found in and around them.

Answers to Questions

Answers will vary, depending upon students' knowledge of elements. Students may list oxygen in the air, aluminum kitchen utensils. Guesses may be incorrect, but can lead to a discussion of how to identify an element.

✔ Assessment

Oral Have students form small groups and discuss their answers to the first three questions of the Explore! Then have each group make posters showing elements in the environment. Use the Performance Task Assessment List for Poster in **PASC**, p. 73. **COOP LEARN**

 Discovering Metals

PREPARATION

Planning the Lesson

Refer to the Chapter Organizer on pages 268A–D.

Concepts Developed

An element is the simplest form of matter. Chemical symbols are abbreviations of the names of elements. Metals are shiny and good conductors of heat and electricity. Most metals are extremely ductile and malleable. Metals are found in many objects as well as within living organisms. Some are toxic.

1 MOTIVATE

Bellringer

 Before presenting the lesson, display **Section Focus Transparency 25** on an overhead projector. Assign the accompanying **Focus Activity** worksheet. L1
LEP

Demonstration To have students predict and observe properties of the metal aluminum, put on leather gloves and show students a clean, empty soda can. Ask students to predict if you will be able to tear the can. Demonstrate that you are able to twist the can but not tear it. Explain that although aluminum is a lightweight metal, it is "strong" enough to resist tearing.

Visual Learning

Figure 9-1 Use the illustration to point out that a chemical symbol is most often one or two letters. Have students consult the periodic table on pp. 630-631 to find other examples of symbols. L1

Section Objectives

■ Describe the physical properties of a typical metal.
■ Compare and contrast the terms *malleable* and *ductile.*
■ Explain how the properties of metals determine their uses.

Key Terms

metals, malleable, ductile, coinage metal

Elements

As you have learned, substances are materials that are made up of one kind of matter. Elements, such as sulfur, are the simplest type of substance. An element can't be broken down into anything simpler by ordinary physical or chemical means. Every other form of matter is either a compound, another type of substance that is produced when elements combine, or a mixture.

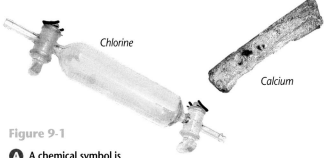

Chlorine

Calcium

Figure 9-1

A A chemical symbol is one, two, or three letters, taken from the name of the element it represents. The letter H, for example, stands for the element hydrogen.

Hydrogen ➡ H
Calcium ➡ Ca
Chlorine ➡ Cl
Sodium ➡ Na
Iron ➡ Fe
Lead ➡ Pb

Lead

■ Symbols

Sometimes, when we want to discuss these elements we use a kind of shorthand. You're already familiar with some shorthand notations. For instance, we use St. to stand for the word street, h to stand for the word hour, L to stand for the word liter, or lb to stand for the word pound.

In a similar type of shorthand, chemical symbols are the abbreviations that stand for the names of different elements. These symbols are needed so that scientists from different countries with different languages can communicate. For example, the element gold will be called different names in different languages, but the chemical symbol for gold is the same all over the world.

B The first letter in a symbol is always capitalized. Any other letter is always lowercase. Ca stands for calcium and Cl stands for chlorine. As you can see, two-letter symbols are not always the first two letters of the element.

For some elements, the letters of the symbol come from the Latin, not the English name that is used for the element. Lead, for example is Pb. Iron is Fe.

Program Resources

Study Guide, p. 31
Concept Mapping, p. 17, Metals and Their Properties L1
Laboratory Manual, pp. 47–48, Reclamation of Copper from Mine Wastes L2
Take Home Activities, p. 16, Money Metals L1
Science Discovery Activity, 9-1, 9-2
Section Focus Transparency 25

Meeting Individual Needs

For the first Explore!, students with weak reading skills may find it best to copy the elements' names into one vertical column in their journals and the symbols into an adjacent column. Students should use pencils to draw lines to match the symbols with the elements' names. Using pencils will allow students to change any answers, if necessary. L1 LEP

ACTIVITY

What element is it?

What To Do

1. Look at the following list of elements and symbols.

Oxygen	Copper	Hg	Zn
Zinc	Aluminum	Al	Au
Nitrogen	Carbon	Ca	N
Gold	Mercury	C	Cu
Magnesium	Calcium	O	Mg

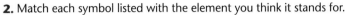

2. Match each symbol listed with the element you think it stands for.

3. Can you think of a reason why symbols aren't always just the first letter of an element name? Record your answer *in your Journal.*

4. Which symbols were the easiest to match to the name of an element? Which were the most difficult? Explain your answer *in your Journal.*

Aluminum

Some of the elements listed in the Explore activity may be unfamiliar to you, and you do not know how they look. Others are easy to picture in your mind. As you picture the familiar elements that were named in the Explore activity, you may notice that each element has its own set of properties. You may notice that some elements differ in color, but some elements have similar colors. You may notice that some are solids, some are gases, and one, mercury, is a liquid at room temperature. Could you use any other properties to place elements into groups?

 ACTIVITY

How can you identify a metal?

What To Do

Examine the element samples provided by your teacher.

1. Separate the elements into two or more groups by using their properties.

2. What properties did you choose? Did other students choose the same properties? How did the properties chosen affect the grouping?

3. In which group would you place gold? Would you place nitrogen in the same group? Explain your answer *in your Journal.*

Explore!

What element is it?

Time needed 10 minutes

Thinking Process observing and inferring

Purpose To infer which symbols correspond to elements.

Teaching the Activity

Pair students with strong reading skills with those who need help reading element names. COOP LEARN LEP

Science Journal To reinforce their classification skills, suggest that students first eliminate the obvious answers. L1

Expected Outcomes

Oxygen–O; zinc–Zn; nitrogen–N; gold–Au; magnesium–Mg; copper–Cu; aluminum–Al; carbon–C; mercury–Hg, calcium–Ca

Answers to Questions

There may be several names that start with the same letter.

✔ Assessment

Content Have students define an element. Then have small groups work to write a newspaper article describing it and the origin of its symbol. Use the Performance Task Assessment List for Newspaper Article in **PASC,** p. 69. COOP LEARN L2

Explore!

How can you identify a metal?

Time needed 20 minutes

Materials metallic and nonmetallic elements; vial of helium or neon

Thinking Processes classifying, comparing and contrasting

Purpose To compare and contrast metallic and nonmetallic elements.

Teaching the Activity

If nonmetals other than carbon are unavailable, explain that air is mainly nitrogen and oxygen and use their properties.

Science Journal Students' choices might be based on luster, color, state. L1

Expected Outcomes

Metals generally are shiny and flexible. Solid nonmetals are dull and brittle.

Answers to Questions

Properties may include state, luster, or hardness. Have students discuss choices.

✔ Assessment

Content Small groups can write and perform a skit describing the properties of a metal. Use the Performance Task Assessment List for Skit in **PASC,** p. 75. COOP LEARN L1

2 TEACH

Tying to Previous Knowledge

Ask students why they wear metallic jewelry. Most will say because it is shiny. Explain that jewelry is made of metal because of two properties of metals—luster and malleability.

Theme Connection The general structure of all atoms is similar, but the atoms of different elements vary in the details of their scale and structure. Remind students that elements are classified according to their atomic structure. Sketch an atom of sodium and point out the nucleus, protons, neutrons, and electrons. Explain that metals are elements with three or fewer electrons in their outermost energy levels.

Demonstration Using BBs, a clear plastic box lid, and an overhead projector, demonstrate why a metal is malleable.

Explain that malleability is due to atoms sliding past one another. Fill the lid with BBs. Put the lid on the overhead projector and have students observe that the shot will form rows that you can easily move with a short piece of wood.

Properties of Metals

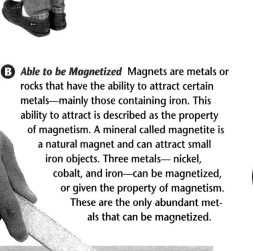

Figure 9-2

A *Luster* When light strikes an object, the object either absorbs or reflects the light. Ability to reflect light is described as the property of luster. Most metals reflect a large amount of the light that strikes them and appear shiny. Silver reflects almost all light that strikes it, and is, for this reason, used as backing for mirrors.

B *Able to be Magnetized* Magnets are metals or rocks that have the ability to attract certain metals—mainly those containing iron. This ability to attract is described as the property of magnetism. A mineral called magnetite is a natural magnet and can attract small iron objects. Three metals— nickel, cobalt, and iron—can be magnetized, or given the property of magnetism. These are the only abundant metals that can be magnetized.

There are several ways you might have grouped your elements in the Explore activity. Although all groups of elements are important, one of the most important groups is **metals**. Picture a metal. What does it look like? How do you know that it's a metal? Is it shiny and hard?

There are several properties that most metals have in common. Many metals are shiny solids. The shine or sheen of an object is called luster. Many metals are **malleable**, that is, they can be hammered or pressed into shape without breaking. Have you ever seen a roll of copper wire? Many metals are also **ductile**, which means that they can be pulled into a wire without breaking. **Figure 9-2 A-E** shows examples of these, and other properties of metals.

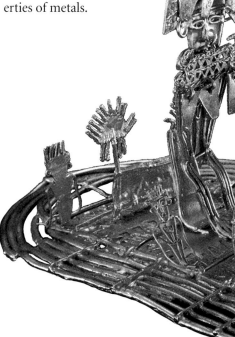

Multicultural Perspectives

Metallurgy

Discuss with students the contributions of African, Native American, and Oriental cultures to the development of metallurgy. Africa is a continent rich in mineral deposits. Iron ore is Liberia's most important export. About one-third of the world's supply of gold comes from gold mining in South Africa. Explain to students that Africa's terrain has precluded a viable transportation system, so many of Africa's rich mineral deposits have yet to be mined. Asians found gold and silver and learned to mold the metals into different kinds of ornaments. Native Americans found large copper deposits near Lake Superior and discovered ways to work copper into weapons and tools.

■ How Do Metals Differ?

Not all metals are equally malleable or ductile. There are metals, such as sodium and potassium, soft enough to be cut with a knife!

Metals do not all conduct electricity or heat as well as copper, iron, or aluminum. Metals differ in other properties as well.

Metals such as gold, silver, and copper react less readily and can be found in their natural state in Earth, while other metals, such as sodium and calcium, immediately form compounds when they come in contact with air or water. Sodium and calcium combine so easily with other substances that they are never found as elements in nature.

When metals form compounds, they no longer have properties of metals. Let's see if we can still identify those elements.

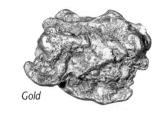

Gold

Iron

Copper

E *Color* The colors of many metals are so different from one another that color is one of the properties used to identify specific metals. For example, gold is yellow, iron may be grayish-silver, and copper is reddish-orange.

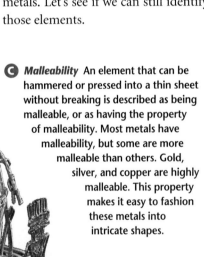

C *Malleability* An element that can be hammered or pressed into a thin sheet without breaking is described as being malleable, or as having the property of malleability. Most metals have malleability, but some are more malleable than others. Gold, silver, and copper are highly malleable. This property makes it easy to fashion these metals into intricate shapes.

Gold Muisca raft with figures from Colombia

D *Ductility and Conductivity* A material that can be stretched into a wire without breaking is described as ductile, or having the property of ductility. Most metals that are malleable are also ductile. An ounce of gold can be pulled into a wire more than 46 miles long! Many metals are also good conductors of electricity. Metals that are both ductile and able to conduct electricity are used to carry electrical signals. Copper, which has both properties, is a common material for electrical wire.

Copper wire

9-1 Discovering Metals **273**

Uncovering Preconceptions
Make sure students realize that most metals are shiny on newly exposed surfaces. Chemical reactions, however, cause many metals to tarnish. Have students list examples of metals becoming dull due to chemical reactions. *Answers might include the formation of rust when iron combines with oxygen, the tarnishing of silver jewelry, or the dulling of copper-bottomed kettles when exposed to high heat.*

Inquiry Question Explain that silver conducts electricity better than gold and is less expensive than gold. Have students communicate why, then, gold is used to plate electrical contacts in high-quality switches and in computers. *Silver is more reactive than gold and tarnishes by combining with pollutants in the air.*

GLENCOE TECHNOLOGY

 CD-ROM

Science Interactions, Course 2, CD-ROM
Chapter 9

 Videodisc

The Infinite Voyage: Miracles by Design

Chapter 1
Auto Racing: Advancements in the Field of Metallurgy

Chapter 2
Improving Steel: Examining Chemical Makeup

Chapter 9
"Smart Materials" and Their Construction

ENRICHMENT

Activity Have students experiment to see how two metal cans will react with oxygen in the air. Have them compare and contrast a clean, empty aluminum can with a clean, empty steel can. *Both are 'silver' in color. The aluminum is more lightweight than the steel.* Have them then predict which can, if either, will change if left outside for a few days. Have students record their predictions in their Student Journals. Have students place the cans outside where they won't be disturbed. When oxidation has occurred, i.e., when rust has formed on the steel can, have students compare and contrast their observations with their predictions. Have students hypothesize why the steel can rusted. *Steel contains iron and iron oxidizes quickly.* L2

9-1 Identifying Metals

Preparation

Purpose To identify metals by observing flame tests.

Process Skills observing and inferring; using variables, constants, and controls; predicting; forming operational definitions

Time Required 40-45 minutes

Materials Chloride salts of lithium, calcium, potassium, copper, strontium, sodium, and barium should be available from scientific supply houses. Table salt may be used for the test for sodium. To make solutions, dissolve about 5 g of solid in 100 mL of distilled water. Prepare the solutions beforehand. Depending on time constraints, you may want to have the testtube rack already prepared with measured solutions.

Safety Precautions Have students wear goggles and protective gloves.

Possible Hypotheses The students may realize that different metals cause different flame colors and may have some guesses as to what colors go with different metals.

Plan the Experiment

Process Reinforcement To reinforce the principles involved in a scientific experiment, have students identify the constants and variables in this experiment. Constants include the amount of solution, the length of the splint, and the heat source. The variable being tested is the metal.

Possible Procedures Soak short pieces of splint thoroughly in each solution. Using the tweezers, hold each splint in the flame and observe.

Identifying Metals

Centuries ago, the Chinese discovered what happens when a compound containing a metal is exposed to an open flame. The Chinese put this discovery to a use in the fireworks that have been enjoyed for centuries. Use this same concept to identify what metal is present in a solution by soaking a wooden splint in the solution and testing it in a flame.

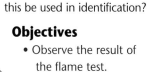

Preparation

Problem
How can we identify the presence of certain metals in a solution?

Form a Hypothesis
What do you think will happen when solutions containing metal are exposed to a flame? How might this be used in identification?

Objectives
- Observe the result of the flame test.
- Identify the presence of metals in solution.

Materials
test tubes, numbered (7)
metal salt solutions, numbered (7)
test-tube rack
candle or laboratory burner
wooden splints
matches
tweezers
goggles
laboratory apron
oven mitt

Safety
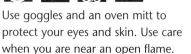

Use goggles and an oven mitt to protect your eyes and skin. Use care when you are near an open flame.

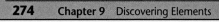

274 Chapter 9 Discovering Elements

Program Resources

Activity Masters, pp. 39–40, Design Your Own Investigation 9-1
Teaching Transparency 17

Plan the Experiment

1 How do you plan to expose the solutions to the flame? How will you treat each solution and each wooden splint?

2 Determine how you will keep track of your results. Plan for this *in your Science Journal.*

3 Will you use a candle or a laboratory burner?

4 With your teacher's guidance, explain in your plan how you will dispose of the solutions.

Check the Plan

1 Did you explain the safety precautions you will take?

2 Will you use clean glassware and tweezers for each trial?

3 Check the plan with your teacher and make any changes that are suggested. Proceed with the experiment.

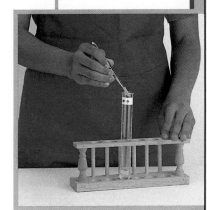

Analyze and Conclude

1. Observe What did you observe when you heated each of the solutions? How was this different than your hypothesis?

2. Compare Compare the colors of the flames with the colors of the solutions.

3. Infer Infer how the Chinese used this property of metals in fireworks.

4. Describe If you were given an unknown metal compound solution, describe how you would identify the metal present.

Going Further

Predict what would happen if your test solution were a mixture of the compounds of two different metals. Could each metal be identified?

Meeting Individual Needs

Learning Disabled Have learning-disabled students work in small groups to discover that metallic elements are present in the school's cafeteria lunches. Each pair of students should choose one lunch selection and use reference books to find out the essential dietary metals present in each food. For example, if a meal consists of mashed potatoes, meat loaf, a salad, and an orange, students should determine that the potatoes and orange are good sources of potassium, the meat is a good source of iron, and the salad a good source of calcium. Class results can be tallied and discussed. **COOP LEARN** **L1**

Teaching Strategies

Troubleshooting Remind students to handle the splints only with the tweezers, as there is salt (sodium chloride) in perspiration. Touching the splints or the tweezers with the fingers may contaminate them. The distinct yellow sodium flame will mask most other colors.

Science Journal Have students record their observations and the answers to the questions in their journals.

Expected Outcomes

Lithium should produce a crimson flame; calcium a yellow-red flame; potassium a violet flame; copper a blue-green flame; strontium a red flame; sodium a yellow flame; and barium a green-yellow flame.

Analyze and Conclude

1. Different colors were produced. Answers will vary.

2. All the solutions are colorless except for the solution containing copper, which is blue-green. The other solutions are not the same color as their corresponding flames.

3. Their fireworks contain different metals.

4. by doing a flame test of the unknown solution and comparing it with known flame tests

✔ Assessment

Process Have students observe the flame produced by several unmarked solutions used in this activity. Have students analyze the results from this experiment in order to determine the composition of their solutions. Use the Performance Task Assessment List for Making Observations and Inferences in **PASC,** p. 17.

Going Further

No, one metal could mask the other or the two colors might combine to produce a color not characteristic of either metal.

Demonstration Demonstrate two uses of the metal magnesium with magnesium wire, tongs, burner, and an empty coffee can. Allow students to pull a piece of magnesium ribbon. Caution students to be careful as they handle the ribbon so that they do not cut their hands. They should discover that it is very light, yet strong. Then, use the tongs and burner to burn a piece of the ribbon in the darkened classroom. Do not allow students to burn the magnesium. Hold the glowing ribbon inside the can. The reflected light will not injure students' eyes. Now ask the question below. L2

Inquiry Question Based on your observations during this demonstration, what are some potential uses of magnesium? *Magnesium is used in flares to light large areas at night and is used in alloys to make some airplane frames.*

Figure 9-3

Ⓐ When certain metal salts are heated to a high temperature, they produce light of specific colors. For example, strontium salts produce a red flame; sodium salts, yellow; and copper salts, blue-green.

Ⓑ Pyrotechnics, people who make fireworks, use their knowledge of these properties of metal salts to design and make fireworks of various colors. How might scientists use these properties of metals to produce colors to identify unknown metals?

■ Uses of Metals

In the Investigate, you saw that when certain metal compounds are heated, they produce different, characteristic colors. The colors can then be used to identify the metals. Then, you can see how many different roles metals play in your life.

Think of the properties of metals described in this section. What common objects are lustrous, malleable, or ductile? You know from the Investigation that not all substances containing metals look like metals. Your body contains metals. Can you guess where these metals might be found? Where else are metals found?

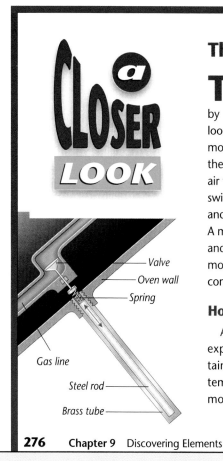

Valve
Oven wall
Spring
Gas line
Steel rod
Brass tube

Thermostats

The temperature in your home is probably regulated by a thermostat. Have you ever looked closely at one? A thermostat is a combination of a thermometer that measures the air temperature in a room and a switch that turns the heating and cooling systems on and off. A modern automated heating and cooling system uses a thermostat to provide safety and comfort and to conserve fuel.

How It Works

A thermostat uses the expansion of metal to maintain the house at a constant temperature. The most common thermostat is a bimetal thermostat. This type uses two layers of metal held together. An increase in the temperature of the room causes each layer to expand at a different rate. The strip (metal layers) bends, closing a switch that controls the heating and cooling systems.

A clock thermostat allows you to change the temperature in your home automatically at certain times of the day. This feature allows you to keep your home at a comfortable temperature when you are there and conserve fuel at other times.

Gas ovens and heaters use a rod thermostat to control the temperature. The control

Purpose

To allow students to **explore** how the property of expansion of a metal when heated is used in their daily lives. This feature extends the discussion of properties of metals.

Teaching Strategy

Have students apply what they have learned about expansion by describing the following situations and having students suggest ways to solve the problems. L2

How might you use the property of expansion to remove a metal lid that is "stuck" on a glass jar? *Warming the lid with hot water will make the lid expand, facilitating its removal.* **Suppose a drinking glass is "stuck" inside another. How can you get them apart without breaking them?** *Pouring hot water over the outer glass and filling the in-*

Metals in Your Body

It may seem strange to think that you have metal in your body. Well, maybe in a tooth filling, but in your bones? Or blood?

■ Calcium

Most of the calcium in your body is in the compound calcium phosphate, which strengthens your bones and teeth. Some calcium is also in your muscles and in the fluid between your body cells. At work in your body, calcium contracts muscles and regulates your heartbeat.

Calcium, along with sodium and

Figure 9-4

Milk and milk products, such as cottage cheese, are the most outstanding source of calcium. Other sources include greens and broccoli. Fish in which the bones are eaten, such as sardines, are also good sources of calcium.

Spinach

Broccoli

Cottage cheese

potassium, is essential for the proper working of the nervous system. The concentration of these elements as they move in and out of your nerve cells determines what signals are

mechanism in this kind of thermostat is attached to a steel rod that sits inside a brass tube. When the temperature increases, the brass tube expands more than the steel rod expands. As the right temperature is reached, a spring closes the valve that controls the gas supply. When the temperature decreases, the brass tube contracts, pushing the rod back so that it opens the valve, allowing gas to flow.

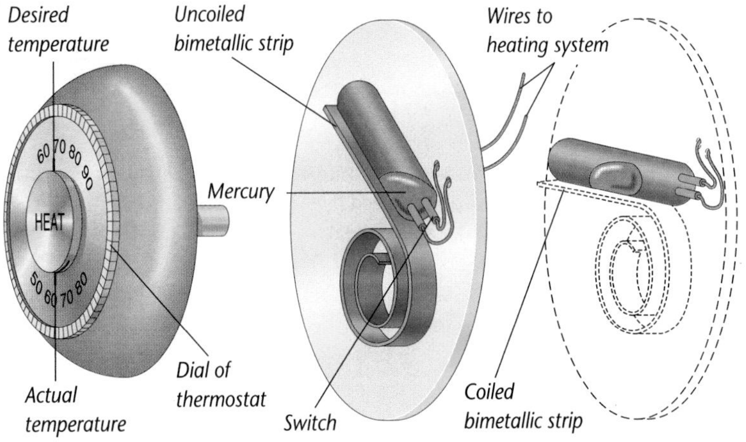

Desired temperature

Uncoiled bimetallic strip

Wires to heating system

Mercury

Actual temperature

Dial of thermostat

Switch

Coiled bimetallic strip

A Part of the Whole

The thermostat is part of an automated system. An automatic furnace heats a home, but the temperature is kept in

check by the thermostat. The thermostat measures temperature in the air and adjusts the heat by switching the furnace on when room temperature is below the desired temperature and off when the room temperature is too warm.

You Try It!

Can you identify all the thermostats in your home? Remember, thermostats are switches that control heating or cooling. Make a list of the appliances and machines in the home that use thermostats.

9-1 Discovering Metals **277**

ner glass with cold water will allow easy separation of the glasses.

Answers to
You Try It!

Answers may include air conditioners, heaters, electric blankets, clothes dryers, ovens, refrigerators, car radiators, clothes irons, and fish tank heaters, among others.

Going Further ⑊⑊⑊▶

Reinforce observing by bringing a thermo-

stat used to control the temperature of a house to class and allowing students to disassemble it. If the thermostat contains a mercury switch, caution students to carefully handle the mechanism so as to avoid possible breakage. Have students identify the various parts of the thermostat including the switch and the bimetal coil. Use the Performance Task Assessment List for Making Observations and Inferences in **PASC**, p. 17. `COOP LEARN`

Discussion Have students hypothesize why calcium, potassium, and sodium are essential to the proper functioning of the body's nervous system. Lead a discussion that helps students to conclude that their cells are not able to conduct electricity, so they rely on these metals to transfer signals within the nervous system.

Student Text Questions

What common objects are lustrous, malleable, or ductile? *Answers might include jewelry, wiring, paper clips, cutlery, cooking utensils, and so on.* **Can you guess where these metals might be found?** *Metals are found in the blood, tissues, bones, and so on of the human body.* **Where else are metals found?** *Metals are found in many items around homes, schools, and other places.*

Across the Curriculum

Fine Arts

Have someone with goldsmithing experience talk to students about his or her craft. Consider arranging a field trip to a local jewelry store.

Visual Learning

Figure 9-3 To reinforce their observational skills, have students use their results from the activity on pages 274–275 to determine the metals used in the compounds in the fireworks in this photograph. *Lithium produces a crimson flame; calcium a yellowish-red flame; potassium a violet or purple flame; copper a bluish-green flame; strontium a red flame; sodium a yellow flame; and barium a yellowish-green flame.* **How might scientists use these properties of metals to produce colors to identify unknown materials?** *They could burn the metals and match the colors to those of known metals.*

Some students may think that any amount of table salt, NaCl, in the diet is harmful. Explain that too much sodium can cause health problems, but so can too little. The sodium ion from salt is needed for proper nerve function and for maintaining the body's fluid balance. Have students examine food labels to determine the amount of sodium per serving. Inform them that the recommended dietary allowance (RDA) of sodium chloride has a range of 80–150 mg per day.

Inquiry Questions When physicians prescribe a low-salt (-sodium) diet, they often suggest **potassium chloride as a salt substitute. Can you explain why?** *The metals are similar and react in similar ways to help keep the nervous system working properly.* **You have just learned that iron is contained in hemoglobin. What is the function of the iron?** *The iron combines with the oxygen. The blood then provides a source of oxygen for cell respiration.*

Across the Curriculum

Mathematics

Have students estimate their daily intake of iron, magnesium, calcium, and sodium using the information provided on food labels. Then, have them discuss how to take into account the metals ingested in drinking water, homemade baked goods, foods eaten in restaurants, and so on. From nutritional information on food labels students can generalize about the content of similar foods that are not labeled. Note, most food labels give percent of recommended daily allowances for metals, not mg., sodium being the exception. L2

Figure 9-5

Magnesium is part of the green pigment compound known as chlorophyll. Without magnesium, plants could not produce food for their survival or oxygen for our survival. Magnesium also stimulates the body processes that release energy from food. Whether you are doing something active, such as swimming, or are just resting, your need for energy never stops.

Figure 9-6

When your muscles are working hard, such as when you're doing sit-ups, they require a lot of oxygen. Iron in your blood carries oxygen to your muscles. You can get the iron you need by eating a variety of foods including meats, especially liver; peas; beans; and prunes.

Prunes

Kidney beans

Peas

transmitted from one nerve cell to another. We get sodium from foods that contain sodium chloride, or table salt. Bananas, oranges, and potatoes are good sources of potassium.

■ Iron—The Blood Element

Iron is another metal important to the proper functioning of your body. A small amount of iron is contained in hemoglobin, a substance in red blood cells. When the iron in hemoglobin combines with oxygen from the lungs, blood turns from dark red to very bright red. It is this ability of iron to combine with oxygen that makes it important. The iron in your blood picks up oxygen in your lungs and carries it to the rest of your body.

278 Chapter 9 Discovering Elements

ENRICHMENT

Research Have students use their communication skills to conduct a telephone interview with a radiologist or an X-ray technician to find out about the barium compounds used to make the intestinal tract easier to see on X rays. Students will find out that even though barium compounds are quite toxic, the barium sulfate used in labs does not dissolve well, so very little enters the tissues of the body. L3

Metals Dangerous to Your Health

Not all metals are beneficial for your body. Lead and mercury are both metals that can be poisonous to living things. Lead and mercury, called heavy metals, can take the place of iron in your red blood cells. Because they don't have the same ability to carry oxygen through your system as iron does, they can produce some of the same symptoms as anemia.

In ancient Rome, the water pipes were made of lead. Some historians believe that many Romans died from lead poisoning, because the lead contaminated the water as it flowed through the pipes. In fact, the name *plumber* comes from the Latin word for lead, *plumbum*. The symbol for lead is Pb.

At one time, workers who made felt hats treated them with mercury and mercury-containing materials to preserve the felt. Many of these people later went insane because of mercury poisoning. That's where the expression "mad as a hatter" came from.

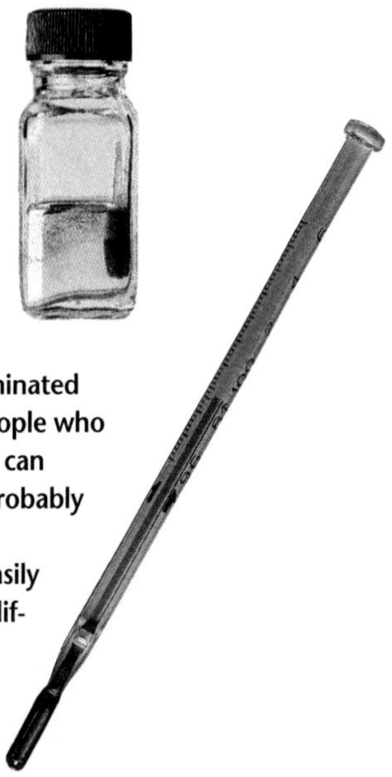

Figure 9-7

Mercury, which can be inhaled or absorbed through the skin and collect in the body, is poisonous. Because mercury compounds have long been used in agriculture and industry, they have contaminated the environment in some places. People who eat fish contaminated with mercury can become ill and may die. You have probably seen mercury in some types of thermometers and barometers. What easily seen property of mercury makes it different from all other metals?

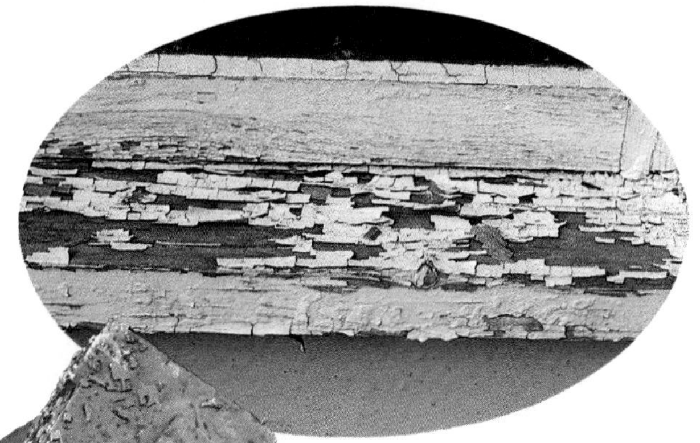

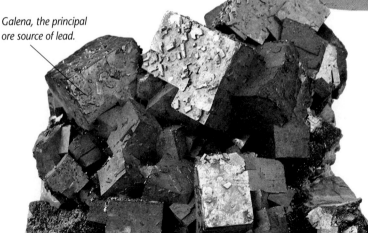

Galena, the principal ore source of lead.

Figure 9-8

Lead particles can enter the body by being breathed in, swallowed, or absorbed through the skin. Lead interferes with the formation of red blood cells and if it builds up in the body may damage the brain and other body organs. Lead is no longer used to make paint, but old paint containing lead still injures thousands of children each year.

279

Uncovering Preconceptions

Ask students to identify the soft gray material at the tips of their pencils. Most will likely respond "lead." Explain that this material is not lead but rather a mixture of graphite, which is carbon, and clay. Lead students to conclude that contrary to popular belief, a person does not get "lead" poisoning when the skin is poked or punctured with a pencil tip.

Across the Curriculum

Language Arts

To help students begin to analyze words and word meanings, have them use dictionaries to find out that the root *therm-* or *thermo-* means "heat." Point out that although it is poisonous, mercury is used in thermostats and laboratory thermometers because of its high boiling point (357°C). L1 LEP

Student Text Question **What easily seen property of mercury makes it different from all other metals?** *It is a liquid at room temperature.*

ENRICHMENT

Research Have interested students use reference books and current journals to communicate to the rest of the class the problems encountered by children who suffer from high levels of lead. Encourage discussion of proper preventative measures. L3

Coinage Metals

For centuries, three metals have been widely used as coins. Study **Figure 9-9**. Can you name these metals? If you said copper, silver, and gold, you're right. Together, these are called **coinage metals**.

Figure 9-9

1897
$5 Gold half-eagle

1922
Peace dollar

1981
Lincoln penny

A The current price of metals has a lot to do with how they are used. For many years gold, silver, and copper were the metals most widely used as coins. As the price of the gold and silver went up, this use of the metals was discontinued in the United States. Gold has not been minted as legal currency since 1933. "Silver" coins, once 90 percent silver, have been minted with no silver at all since 1970. Even the penny, once pure copper, now contains zinc.

B Copper is still used as a base for many coins. Sometimes it is mixed with other metals and the coin is minted from the mix. Another way copper is used is as a base coin that is covered with a thin layer of a more expensive metal. The Susan B. Anthony dollar—a copper base covered by nickel—is an example of such a coin.

Copper, silver, and gold are often uncombined in nature and can be mined. They are also malleable and ductile. These metals are easily shaped and stamped into coins. These coins have a denomination imprinted on them, but the actual value may be different. By international agreement, the value of gold and silver is determined by the amount of these metals available in the marketplace.

In addition, silver is used as a backing for mirrors and in photographic film. Silver chloride or silver bromide is used in photographic film and turns dark when exposed to light. That's why film turns dark when it is exposed.

Silver, gold, and copper are not the only elements used in coins. Although there is no silver in new silver money, there is nickel inside the nickel coin you may have in your pocket. Nickels are actually 25 percent nickel and 75 percent copper. Zinc is also used in some pennies.

Copper
Nickel

280 Chapter 9 Discovering Elements

Aluminum—Jack of All Trades

The most common metal in Earth's crust is aluminum. Aluminum is strong, light, and not easily affected by oxygen or other substances that can destroy many metals. Aluminum is in soft drink cans, in the foil wrap in your kitchen, and may be in the siding you have on your house. Aluminum compounds may be used in medicines, deodorants, pigments, and dyes.

Aluminum alloys, which are homogeneous mixtures of aluminum with another element, are used in many products. Aluminum-lithium alloys provide a strong, lightweight material used in aircraft.

■ Recycling Aluminum

When an aluminum can is recycled, some aluminum ore is conserved. But of equal importance, much less energy is needed to process the recycled can. To produce one aluminum can from ore takes much more electricity than is needed to produce the same can from recycled aluminum.

Look around you. How many of the objects that you can see right now contain metals? They are in the things that you eat, touch, ride in, and wear. Think about how different your life would be without metals.

Making and Using Tables

Make a table listing six of the metals described in this section. Include column headings for four properties and a heading for uses of each metal. If you need help, refer to the **Skill Handbook** on page 639.

Every month there are about 31 aluminum cans used per person in the United States. 21 of those cans are recycled and the other 10 are just thrown away.

Figure 9-10

When recycling metal is mentioned, what do you think of? Most people think of aluminum beverage cans, and for good reason. In 1992, 6.8 billion pounds of aluminum were recycled. Of that amount, 2.14 billion pounds came from used beverage cans. That comes to about 253 cans per person, per year.

21 cans   *10 cans*

---

check your UNDERSTANDING

1. Given a substance, how would you test it to see if it is a metal?
2. Name some elements that are both malleable and ductile.
3. Why would silver be a good coinage metal, while calcium would not?
4. **Apply** Early civilizations used salt, glass, and seashells as coins or money. Why do you think metals have replaced these materials as coins?

Tables will vary. Properties should include four of the following: hardness, luster, malleability, ductility, and conductivity. If possible, ask students to rank the extent to which each metal shows these properties. L1

Extension

Activity Have students determine the properties and uses of a metal of their choice. Have students record their findings on index cards. Provide a periodic table for reference and have students use tape to arrange their metals onto the chalkboard or wall as they appear on the table. Have students compare and contrast properties of the metals. COOP LEARN L2

4 CLOSE

Have students study the periodic table on pp. 630–631 to determine how the known elements are grouped into three categories—metals, nonmetals, and an area that has properties of both metals and nonmetals. L1

---

check your UNDERSTANDING

1. Check properties such as luster, malleability, ductility, and conductivity.
2. Answers might include iron, copper, aluminum, gold, and silver.
3. Silver is harder, more durable and found as a free element; calcium is not.
4. Metals have properties more suited for coins—they don't break or dissolve and can be made consistent in size and shape.

PREPARATION

Planning the Lesson

Refer to the Chapter Organizer on pages 268A–D.

Concepts Developed

Most nonmetals are gases at room temperature. Solid nonmetals are dull, brittle, and are insulators rather than conductors. The most reactive nonmetals include fluorine, chlorine, and oxygen.

1 MOTIVATE

Bellringer

Before presenting the lesson, display **Section Focus Transparency 26** on an overhead projector. Assign the accompanying **Focus Activity** worksheet. **L1** **LEP**

Demonstration Help students distinguish a nonmetal from a metal by using sulfur, a hammer, and a towel. Students should note the element's dull surface. Wrap the sulfur in the towel and, using the hammer, break off a piece. Students should observe that sulfur is brittle and powdery, not malleable.

Section Objectives

- Describe the physical properties of a nonmetal.
- Compare and contrast metals and nonmetals.
- Relate the properties of nonmetals to their uses.

Key Terms

nonmetal

DID YOU KNOW?

At today's prices, if the human body were reduced to its elements, it would be worth about $3.88.

Properties of Nonmetals

In the last section, you grouped several elements together. These elements, metals, usually share properties such as luster, malleability, ductility, and conductivity of heat and electricity. But what about the elements that do not have those properties? Can we observe the properties that these other elements have? How can they be grouped together?

Explore! ACTIVITY

How can you identify nonmetals?

What To Do

1. Think back to the second Explore in Section 9-1 on page 271. If we remove all the metals from the group of elements we examined, we're left with the elements shown in this photo.

2. What properties might you use to describe the elements that are not metals? Write your answer *in your Journal*. Did other students choose the same properties?

3. What other elements can you think of that are not metals? What, if anything, do all these elements have in common?

The elements examined in the Explore activity are called **nonmetals**. Solid nonmetals, such as sulfur and carbon, are dull rather than lustrous. They are brittle and break into pieces instead of being malleable and ductile. Most are poor conductors of heat and electricity. Many nonmetals, such as oxygen and nitrogen, are gases. You might say that most of the properties of nonmetals are just the opposite of most of the properties of metals.

Although there are many more metals than nonmetals, most living material is composed of nonmetallic compounds. Most of your body is made up of compounds of nonmetallic elements.

Explore!

How can you identify nonmetals?

Time needed 10 minutes

Materials No special materials or preparation are required.

Thinking Processes observing, comparing and contrasting, forming operational definitions

Purpose To differentiate metals and nonmetals.

Teaching the Activity

Troubleshooting If necessary, lead students to conclude that being invisible is a property.

Science Journal Have students write their answers in their journals. **L1**

Answers to Questions

Answers should mention properties common to nonmetals, such as nonmetallic solids are brittle and dull; many nonmetals are gases.

✔ Assessment

Portfolio Have students work in small groups to compile properties of nonmetals. Students then can draw cartoons to describe nonmetals and their properties. Use the Performance Task Assessment List for Cartoons in **PASC**, p. 61.

COOP LEARN **P**

Common Nonmetals

Many common and important elements are nonmetals. Oxygen, for example, is the most common element in Earth's crust. Most common nonmetals are gases.

■ Hydrogen

Just as metals have properties you could test for, nonmetallic gases, such as hydrogen, have properties you can test for.

Explore! ACTIVITY

What are some properties of hydrogen?

What To Do

1. Drop a small piece of sanded magnesium ribbon into a test tube containing about an inch of white vinegar.

2. *In your Journal*, record what you observe.

3. When the reaction is going well, put a cork stopper loosely over, not into, the top of the tube.

4. After a few moments, hold the tube at a slant, making sure you aren't pointing it at anyone.

5. Just as you remove the cork, bring a lighted splint to the mouth of the test tube. Record your observations *in your Journal*.

Based on your observations, what are three properties of hydrogen? You probably noticed that it is a colorless, odorless gas that explodes when lit.

Hydrogen is also much less dense than air. Most of the hydrogen on Earth is found in the compound water.

How Do We Know?

What's the universe made of?

How do researchers know that the universe is 99 percent hydrogen? The process of spectroscopy is the chemist's most important tool in identifying what kind of material—and how much of a material—is contained in a sample. You may recall that using spectroscopy is a little like looking at a rainbow. In a rainbow, the colors appear because the light is affected as it passes through the raindrops. In spectroscopy, a researcher can look at light given off by a sample such as stars. The colored bands of light the substance produces can be matched to the colored bands given off by identified substances. The brightness of the colored bands helps researchers know how much of a material is present.

9-2 Discovering Nonmetals **283**

Program Resources

Study Guide, p. 32

Teaching Transparency 18 L2

Making Connections: Technology & Society, p. 21, Hydrogen—Inexhaustible and Clean L2

Section Focus Transparency 26

✔ Assessment

Content Have students find out how hydrogen gas is being investigated as an alternative, nonpolluting fuel for cars, buses, and airplanes. Students can then give Oral Presentations to the rest of the class presenting the advantages and disadvantages of using the fuel. Use the Performance Task Assessment List for Oral Presentation in **PASC,** p. 71.

Inquiry Question Plant rotation is a farming technique in which a different crop is planted in a field each year. Why do you think farmers include in the rotations plants whose roots contain nitrogen-fixing bacteria? *The plants are included in the rotation so that they will replace nitrogen that other plants remove from the soil.*

Debate Trinitrotoluene (TNT) is a nitrogen compound used in bombs and rockets. Still other nitrogen compounds are used in poisonous gases. Have students debate whether scientists should work on the development of explosives and poisons. Students who are against development will probably focus on the harmful aspects of these products. Those in favor can focus on the possibility of useful materials that are developed as by-products of the research. L3

Connect to...

Life Science

Living organisms on Earth contain carbon. Carbon-12 makes up 99 percent of this carbon. Carbon-13 and carbon-14 make up the other 1 percent. What is meant by carbon-12, carbon-13 and carbon-14? Make a poster that explains this and shows the difference between these carbons.

■ Nitrogen

Each breath you take is about 80 percent nitrogen. However, your body can't use nitrogen directly, but uses compounds of nitrogen. Bacteria and plants play a key role in producing nitrogen your body can use. First, bacteria change the nitrogen in the soil into nitrogen compounds. Then plants take in these compounds and change them to proteins that your body can use. This process is called the nitrogen cycle, which you have studied previously.

Although it supports life, nitrogen can produce problems in your environment. Nitrogen compounds from car and truck exhausts react with water in the air to form nitric acid. This contributes to acid rain, which corrodes metal and poisons soil and water in which living organisms exist.

Although, like hydrogen, nitrogen itself is colorless and odorless, nitrogen and hydrogen combine to form a strong-smelling compound, ammonia. Ammonia is found in many household cleaners. Nitrogen makes up a major portion of fertilizers because it promotes plant growth.

■ Carbon

Carbon is found in all things that are living or that lived in the past.

Earth Science CONNECTION

Diamonds and Pencils

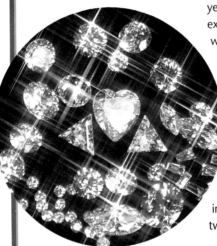

Did you know that a diamond and the graphite in your pencil have a lot in common? Diamond is one of the hardest substances known. And yet, if exposed to temperatures exceeding 1830° F, a diamond will turn into graphite—the soft, black material found in a pencil. The pure natural form of carbon we know as a diamond is different from graphite only in the way its crystals formed. Although diamond and graphite are both forms of carbon, their value and use in society are as different as two materials can be.

How Diamonds Formed

Diamonds formed billions of years ago in dying volcanoes. When molten lava in the volcanoes became solid, heat and pressure changed the carbon present into diamond crystals. As Earth's crust and upper mantle moved, volcanic eruptions pushed the diamond deposits closer to the surface. Not all volcanoes contained carbon. For this reason, diamonds formed in only certain regions of the world.

Where are Diamonds Found?

The earliest diamond mine is believed to have been in central

284 Chapter 9 Discovering Elements

Earth Science CONNECTION

Purpose

The Earth Science Connection extends the discussion of the properties of nonmetals by allowing students to explore how the same element, in this case carbon, can have such different uses.

Content Background

On Mohs' scale of relative hardness, (which ranges from 1 to 10 with 1 being the softest and 10 being the hardest), diamonds have a hardness of 10 and graphite has a hardness of 1–2.

Diamonds belong to the isometric crystal system while graphite crystals are classified as hexagonal. Diamond crystals are found in altered igneous rocks called kimberlites.

Teaching Strategy

Ask students to brainstorm to compile a list of other uses for graphite. Responses might include as a lubricant, in batteries, bearings, rubber stoppers, and in the man-

When plants and animals die, they decompose. As layers of soil are deposited on top of them, the pressure produces carbon-containing products, such as coal and petroleum. These fuels are the basic form of energy used in our society.

■ Oxygen

Oxygen is another nonmetal that is necessary for you to survive. Let's investigate some of the properties of this element.

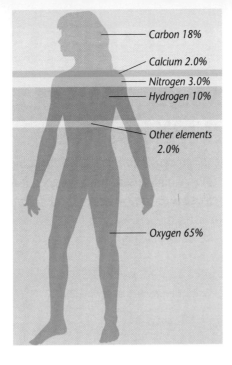

Carbon 18%
Calcium 2.0%
Nitrogen 3.0%
Hydrogen 10%
Other elements 2.0%
Oxygen 65%

Figure 9-11

A Although your body contains several metals, most of it is composed of elements that are nonmetals. This diagram shows the major elements that make up your body. The *Other elements* category contains both metals and nonmetals.

B Calculate the percentage of your body that is made of nonmetals by identifying the metals discussed in the previous section. The remaining elements are nonmetals. How much do the nonmetals add up to?

Activity Have students compare three mixtures that contain carbon using Nos. 1, 2, and 3 pencils. Graphite is soft and leaves a mark when it is rubbed on a surface. Have students apply the same amount of pressure to draw a line with each of the three pencils. Help students recognize that the softest pencil—the No. 1—contains more graphite than the other two pencils contain. L1

Discussion Ask students if they have heard of black lung disease and what causes it. This disease is caused by inhaling coal dust and is common among coal miners. Ask students to infer how mining practices could cause such a disease. Ask students to suggest ways to reduce the amount of dust inhaled. L2

> **Visual Learning**
>
> **Figure 9-11** Calcium is the only metal shown on the diagram and along with the category "Other elements" makes up 4 percent of the human body. **How much do the nonmetals add up to?** *About 96 percent of the human body is nonmetals.*

India. By the early 1700s, mines had been dug in Brazil. Over a century later, young children playing along the Orange River in South Africa discovered a South African diamond. Today, most of the world's diamond production comes from Canada, Australia, Congo, Botswana, and South Africa.

Uses of Diamonds

Of the five tons of diamonds mined yearly, only one ton can be used for jewelry. The most valued diamonds are colorless or pale blue stones and are relatively free of impurities.

Diamonds are useful in industry. Their hardness makes them resistant to chipping or cracking. For this reason, better quali-

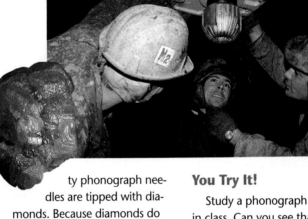

ty phonograph needles are tipped with diamonds. Because diamonds do not corrode and are resistant to temperature changes, they are excellent as mechanical parts. For example, hospital and science laboratories use diamond bearings in their machines.

You Try It!

Study a phonograph needle in class. Can you see that the needle is tipped with a diamond crystal? Test the hardness of the diamond by scratching it across an old glass bottle. Can you see why diamonds are very useful in industry?

ufacture of certain protective paints, among others.

Answers to You Try It!

On the scale of relative hardness described in the content background, glass has a hardness of about 5.5 and will easily be scratched by the diamond.

Going Further ▰▰▰▰➧

Have students use chemistry books or mineralogy books to compare the structures of the two forms of carbon. Students should discover each of the atoms in a diamond's structure is covalently bonded to four carbon neighbors. The structure of graphite consists of six-membered rings in which each carbon atom has three near neighbors. Have students use toothpicks and small foam balls to model these atomic arrangements. Then discuss how the arrangements determine properties of each substance. Use the Performance Task Assessment List for Models in **PASC**, p. 51.

9-2 Preparing and Observing Oxygen

Planning the Activity

Time needed 45 minutes

Purpose To prepare oxygen and observe some of its properties.

Process Skills observing and inferring, comparing and contrasting

Materials See student text.

Preparation Crystals of copper(II) sulfate ($CuSO_4$) or iron(II) sulfate ($FeSO_4$) can serve as substitutes for cobalt chloride.

Before you have the class do this activity, try the reaction using the materials students will be using. Make sure you have test tubes that are large enough to contain the reaction. Test tubes should have a capacity of 50 mL.

Teaching the Activity

Safety With a volunteer, demonstrate the proper technique for removing the cork from the test tube and lighting the splint.

Process Reinforcement
Check students' journal entries to see that they observed a change in color as well as bubbles in the test tube and were able to infer that a gas was formed when the cobalt chloride was added. They should also have observed that the splint relit when it reacted with the oxygen in the test tube.

Science Journal Have students record their observations and answers in their journals. L1

INVESTIGATE!

Preparing and Observing Oxygen

Oxygen makes up about 20 percent of air. It is the most abundant element in Earth's crust. Animals and humans need it for respiration and plants release it as they produce food during photosynthesis. In this Investigate activity, you'll observe some of the important properties of this element.

Problem
What are some of the properties of oxygen?

Materials

test tube	balance
test-tube holder	cork stopper
wooden splint	graduated cylinder
matches	0.5 g cobalt chloride
spatula	

liquid laundry bleach (5% sodium hypochlorite)
5-cm × 5-cm square of notebook paper

Safety Precautions

Wear protective clothing during this investigation. Dispose of substances as directed by your teacher.

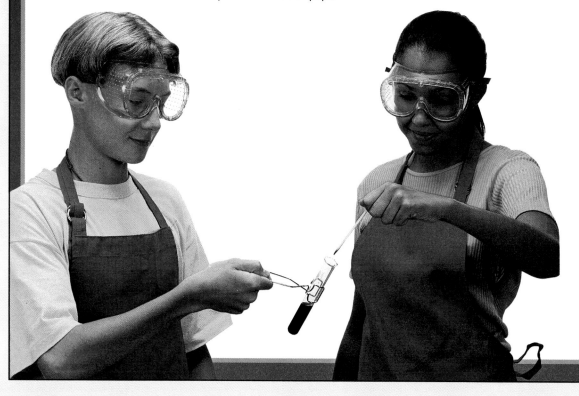

Meeting Individual Needs

Visually Impaired Have visually impaired students observe Procedure 5 more than one time. After students see the glowing splint relight, remove it from the test tube. Blow it out, and lower it into the test tube again to relight. If students still cannot see the splint relight, have sighted students describe what happens.

Program Resources

Activity Masters, pp. 41–42, Investigate 9-2

Making Connections: Integrating Science, p. 21, War Against Auto Rust L3

Science Discovery Activities, 9-3

A

B

What To Do

1 Measure 20 mL of bleach with the graduated cylinder and pour it into the test tube. **CAUTION:** *Wash your hands thoroughly after working with bleach.*

2 Place the paper on the balance pan. Using the spatula to transfer the cobalt chloride to the paper, measure 0.5 g of cobalt chloride (see photo **A**).

3 Add the cobalt chloride to the test tube and *observe*. Record your observations *in your Journal.*

4 Place the cork stopper loosely over the opening of the tube (see photo **B**).

5 Tilt the tube at a 45° angle. **CAUTION:** *Do not point at anyone.* Be sure you are holding the tube with a test-tube holder as shown in the picture. Remove the stopper. Have your partner light a splint, then blow it out. Then your partner should carefully lower the glowing splint into the mouth of the tube. *Observe,* and record your observations.

Analyzing

1. How did you infer that a gas was being formed in the test tube?

2. What happened when the glowing splint was placed in the tube?

3. What properties of oxygen did you *observe*?

Concluding and Applying

4. *Compare and contrast* the properties of oxygen with those of hydrogen.

5. Why do you think you used a glowing splint rather than a lit one? What property of oxygen does this demonstrate?

6. **Going Further** Sunlight can also release oxygen from bleach. How are bleach containers designed to stop this?

Theme Connection The
difference between oxygen and
ozone is an example of the
theme of scale and structure. As
students read about ozone, they
learn that it is oxygen in a different form. The difference in
structure (oxygen is O_2; ozone is
O_3) causes different properties.
Have students contrast the properties of these two gases. L1

Across the Curriculum

Social Studies

On a world map (pp. 628-
629), show students where
Antarctica is. Tell them that
there is a hole in the ozone layer over Antarctica. An unusual-
ly large number of cases of
blindness have been reported
in herders in southern Chile.
Ask students if these two facts
could be related.

Content Background

Research suggests that the
ozone layer has been thinning at
a rate of 2.5 percent over the last
decade. The chlorofluorocarbons, or CFCs, in Earth's atmosphere that are responsible for
the damage will persist for many
years. These chemicals were used
for many years as propellants in
aerosol cans.

Activity Have students find
out about statewide ozone levels.
Have them relate the values to
the population density of an
area as well as to the numbers of
industries and motor vehicles in
the area, which are contributing
to the levels. L2

Oxygen is another colorless, odorless gas. In the Investigate activity, you
used a glowing splint rather than a lit
one to test the gas. The splint relit,
didn't it? How do you make the fire
burn more brightly? You fan it, don't
you? That circulates more oxygen
around the coals. Putting the glowing
splint in the oxygen serves the same
purpose. Burning takes place in the
presence of oxygen, so the more oxygen present, the better the burn.

Figure 9-12

A Ozone is a form of oxygen that develops
near Earth's surface from the action of sunlight on pollutants, which result from burning gasoline. Ozone damages the leaves of
plants, and irritates the
human respiratory
system.

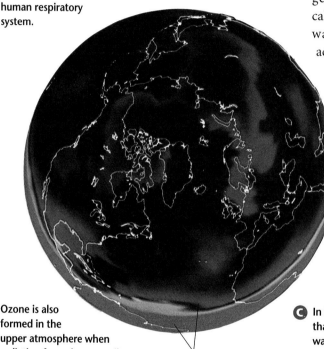

B Ozone is also
formed in the
upper atmosphere when
radiation from the sun strikes
regular oxygen. The ozone high in
the atmosphere is helpful. It forms
a layer that protects Earth from
harmful rays of the sun that can
cause sunburn and skin cancers.

*Areas where ozone has
decreased are colored
blue-purple. Increased
ozone areas are red.*

The oxygen that you inhale in
every breath is carried throughout
your body by your blood. Unlike
nitrogen, which must be changed into
useful compounds, your body can use
oxygen from the air. This oxygen is
used by your body to burn the foods
you eat and the fats and carbohydrates you have stored. It does this by
slowly combining these foods with
oxygen from the blood. In this way,
digestion provides you with the energy you need to live.

Oxygen easily combines with many
other elements to form compounds
called oxides. When the other element
is a nonmetal, such as sulfur or nitrogen, the oxide forms a substance
called an acid when it is dissolved in
water. As you have learned, these
acids cause problems in the environment when they fall as acid rain.

Oxygen is the other element
present in water. Isn't it interesting that one element that
burns explosively (hydrogen)
and another that allows things
to burn (oxygen) can com-
bine into something that puts
out fires—water?

Another form of oxygen is
ozone, which is described in
Figure 9-12.

C In the 1970s, scientists noticed
that the protective ozone layer
was thinning. Since then they have
carefully watched and measured
the ozone layer for changes. This
image compares ozone amount
differences between March 1993
and the average amount for 1979
through 1990.

288 Chapter 9 Discovering Elements

Multicultural Perspectives

Rain Forests

Ask students how the
clearing of tropical rain forests contributes to the loss of Earth's ozone layer. Discuss concerns such as: the corporate world
clears vast areas of rain forest land for purposes
of ranching; traditional uses of the forest include local inhabitants cutting down the
forests for firewood for fuel. Some local farmers grow crops without clearing large areas of
forest. Ask students to compare these uses
of the rain forests. Students may be aware of
some of the rain forest products, such as nuts
and nut candies that are produced without destroying the rain forests.

Other Nonmetals

You have learned from **Figure 9-12** that ozone can harm your lungs. Two other nonmetals that shouldn't be inhaled are fluorine and chlorine.

Chlorine is described in **Figure 9-13A**. Chlorine is a greenish-yellow, poisonous gas. You smell chlorine and observe a yellow color when you open a bottle of laundry bleach. Because chlorine is poisonous, you should never use another cleaner with a chlorine bleach unless you know it is safe to use, such as a laundry detergent. Some cleaners may contain substances that cause bleach to release enough chlorine gas to harm your health.

Figure 9-13B shows a common use of another gas, fluorine. Fluorine also has a strong odor. Its compounds are commonly used to help your teeth stay healthy. Most toothpastes and many water supplies use fluorine compounds because they help prevent tooth decay.

Life on Earth has developed around the elements that are present

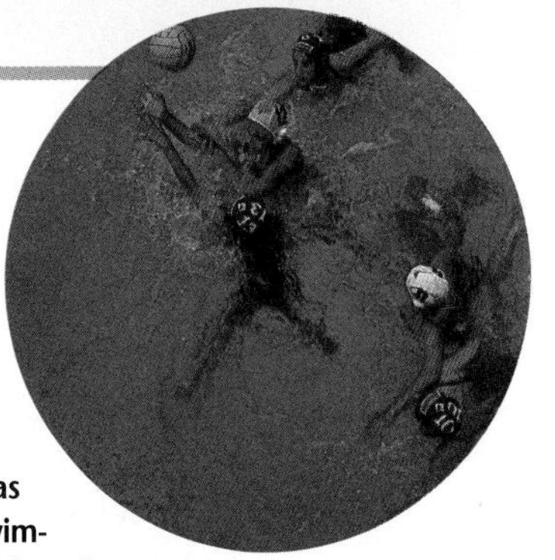

Figure 9-13

Ⓐ Because it reacts easily with other substances, chlorine is found in nature in compounds. Chlorine is used as a laundry bleach and a swimming pool disinfectant. Although chlorine as an element is poisonous, there is one chlorine compound that you probably eat every day—sodium chloride, common table salt.

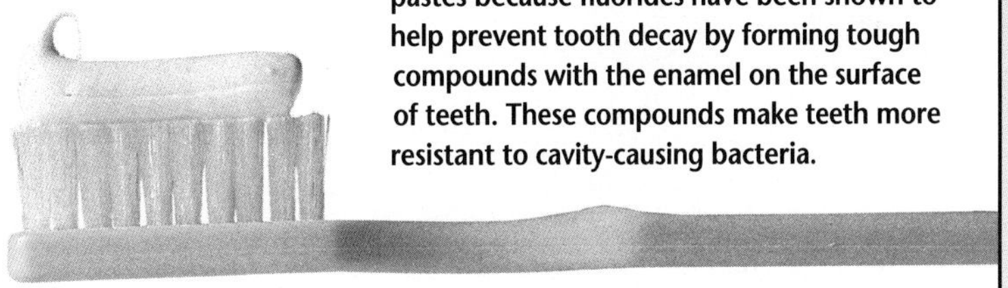

Ⓑ Compounds that contain fluorine are called fluorides. Fluorides are added to some toothpastes because fluorides have been shown to help prevent tooth decay by forming tough compounds with the enamel on the surface of teeth. These compounds make teeth more resistant to cavity-causing bacteria.

in the land, sea, and air—our environment. You've explored the properties of metals and nonmetals.

In the next section, you'll explore the properties of elements that are neither metals nor nonmetals.

check your UNDERSTANDING

1. If you were given a piece of the element sulfur, how would you test it to see if it is a metal or a nonmetal?
2. Compare and contrast the properties and uses of metals and nonmetals.
3. Many of the nonmetals are gases. How does this observation relate to the most common uses of nonmetals?
4. **Apply** If you wanted to protect a historical document from damage, you might put it in an airtight case. Which of the nonmetals would you choose to replace the air in the case? Explain why you would choose this particular nonmetal.

check your UNDERSTANDING

1. Determine its luster and test for conductivity, ductility, and malleability.
2. Metals are lustrous, malleable, ductile, and conduct heat and electricity. Most are solids at room temperature. They are used for jewelry and electrical wiring. Nonmetals have properties that are opposite those of metals.

Nonmetals are used in fertilizers and signs.
3. Possible answers include oxygen being breathed and chlorine being added to water to kill bacteria.
4. The best choice is nitrogen or a noble gas because they are the least reactive of the gases.

Demonstration Have students observe that some nonmetallic elements called halogens have strong odors. You will need a piece of cloth, chlorine bleach, and a pail. Shortly before class begins, pour a few tablespoons of liquid chlorine bleach on the cloth. Then ask students if they smell anything familiar. Help students to conclude that chlorine and the other halogens have very strong odors. This property can be used to identify the presence of chlorine in a solution. Caution students not to inhale the chlorine gas. [L1]

3 ASSESS

Check for Understanding

Assign questions 1–3 and the Apply question from Check Your Understanding. Have students discuss their answers to the Apply question. If they need a hint, ask what gases would not react with the paper.

Reteach

Activity Have pairs of students work to compare and contrast properties and uses of metals and nonmetals. Results can be put into a table. **COOP LEARN** [L1]

Extension

Acquiring Information Fluorine and chlorine both belong to a group of elements called halogens. Find them on the periodic table (pp. 630-631) and research the common properties of halogens. [L2]

4 CLOSE

Discussion

Have the class list three nonmetallic elements—carbon, oxygen, and nitrogen—required for life on Earth. Based on what they know about stars and the universe, have them discuss whether or not they think it is possible for other life-forms to exist in the universe. [L1]

PREPARATION

Planning the Lesson

Refer to the Chapter Organizer on pages 268A–D.

Concepts Developed

Metalloids are elements that have properties of both metals and nonmetals. Metalloids behave like metals in the presence of some elements or substances and like nonmetals in the presence of other matter.

1 MOTIVATE

Bellringer

 Before presenting the lesson, display **Section Focus Transparency 27** on an overhead projector. Assign the accompanying **Focus Activity** worksheet. L1

LEP

Demonstration Have students observe one of the common uses of the metalloid silicon. Bring to class a broken calculator. Have students remove the circuit board with the computer chip on it. Silicon chips are used in most electrical devices because they conduct electricity under some conditions, but not under others. Thus, they can serve as electronic switches. L1

2 TEACH

Tying to Previous Knowledge

Ask volunteers to list properties of metals and nonmetals. Then have students describe an element that is neither a metal nor a nonmetal but has properties of both. Accept any reasonable answer.

Section Objectives

- Distinguish among metals, nonmetals, and metalloids.
- Relate the unique properties of metalloids to their uses.

Key Terms

metalloids

Properties and Examples of Metalloids

What would you call something that was part human and part alien? You'd probably refer to this being as a humanoid, because it has some human characteristics but is not human. In the same way, what would you call an element that had some characteristics of metals and some of nonmetals? Right! A metalloid. As a group, **metalloids** are elements that have properties of both metals and nonmetals. Although the properties of metalloids vary from one element to another, all metalloids show metallic luster. It is the one property all metalloids have in common. Although there are eight metalloids, we'll talk about the two most common ones.

■ Boron

Do you hate scrubbing away the dreaded bathtub ring? Borax, a compound of boron, can be added to laundry products to soften the water so that the minerals in the water don't form soap scum. If you add a little borax to your bathwater, it helps prevent bathtub ring!

Figure 9-14

The major sources of the metalloid boron and boron compounds are mineral deposits from the beds of evaporated lakes and other bodies of water. Boron and its compounds are used in making varied products including cleaning agents, antiseptics, and rocket fuels.

Program Resources

Study Guide, p. 33
Critical Thinking/Problem Solving, p. 17, Taming Oil Fires in Kuwait Using Nitrogen L2
Multicultural Connections, p. 22, James Harris-The Discovery of Elements 104 and 105 L1
Section Focus Transparency 27

Boron is also found in boric acid, a mild disinfectant that is sometimes used to treat infections. Boron is also used in rocket fuels.

■ Silicon

Although the name may not be familiar, almost everyone has been in contact with silicon. Combined with oxygen, it may be called sand. When sand is melted and allowed to cool, it forms glass. Ordinary glass is melted silicon dioxide with at least one metal, such as sodium, calcium, or aluminum, added.

The element silicon is a perfect example of a metalloid. It comes in shiny gray chunks that look somewhat like a metal. But the chunks are rough and full of little holes—not smooth and dense like many metals. Silicon breaks apart easily; it is not malleable or ductile, but it does conduct an electric current.

Silicon

Figure 9-15

Ⓐ Silicon makes up about one-fourth of Earth's crust. Silicon and its compounds have many important uses.

Ⓑ Pure silicon is a semiconductor and is used to make integrated circuits and similar electronic devices.

Ⓒ The most common compound of silicon is silica, the main ingredient of most sand. Glass, mainly formed from sand, has many important household and industrial uses including optical fibers. Optical fibers are taking the place of copper telephone cables because one tiny optical fiber of glass can transmit more information than a much larger and more expensive copper cable.

Ⓓ Ceramic tiles made from silica are fixed to the outside of spacecraft to protect the vehicle from heat the spacecraft encounters during its re-entry into the atmosphere.

Ⓔ The properties of silicon make it possible for scientists to make complex, yet very small electronic devices, such as the experimental robot shown here. Scientists have developed a process to carve entire motors, called micromotors, into tiny flecks of silicon.

291

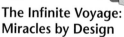

If you have any kind of electronic or video game, a calculator, or even a watch that runs on batteries, you are probably carrying around some silicon right now. Because silicon doesn't conduct electricity as well as a metal, it is called a semiconductor. This property makes it important in the computer and electronics industry.

The tiny electronic parts contained in the device in **Figure 9-15B** are based on silicon and other metalloids, such as germanium. Metalloids are used to manufacture items we rely on every day. But are metalloids, like metals and nonmetals, important to your health? The next Explore will help you answer that question.

Explore! ACTIVITY

What elements are in your mineral supplements?

What To Do

1. Examine the label on one or more bottles of mineral tablets.
2. *In your Journal*, write the names of all the elements listed on the label.
3. Classify each element as a metal, a nonmetal, or a metalloid.

From the observations you made in the Explore, you saw that some metalloids, as well as metals and nonmetals, are important to your health and well-being. Silicon, in particular, is needed in very small amounts for proper use of calcium in bones. But no matter what was listed on the label, the ingredients were all elements or contained elements.

Everything within you and outside of you is composed of metals, nonmetals, and metalloids. No matter how complex a substance seems, it is a combination of these approximately 100 elements. From the gold, copper, and iron objects at the beginning of the chapter to the video games and experimental robots at the end, it's an elemental world.

check your UNDERSTANDING

1. Classify each of the following elements as 1) a metal, 2) a nonmetal, or 3) a metalloid: chlorine, carbon, calcium, boron, neon, sodium, aluminum.
2. In addition to metals, which group of elements is likely to contain elements that can conduct electricity?
3. **Apply** The basic component of most small video games is the silicon chip. Why is this component called a chip rather than a wire?

Science and Society

Recycling Aluminum

Did you know that the energy saved from recycling one aluminum can could keep a television running for three hours? It requires only five percent of the energy needed to mine aluminum to recycle it.

Aluminum is found in the form of bauxite. Although bauxite is still plentiful, easily mined deposits will disappear one day if we do not recycle the aluminum we use.

Uses of Aluminum

World production of aluminum in 1990 exceeded 17 million metric tons. This lightweight and malleable metal is an excellent conductor and is useful when substituted for heavier, more costly materials.

Aluminum is used in making many cooking utensils. Heat spreads more evenly in aluminum pans. Aluminum foil is used in roofing and insulating homes.

The packaging industry uses about 30 percent of the total United States production of aluminum. Most of that is used for soft drink cans. More than 85 billion cans are produced yearly.

Aluminum as Waste

Aluminum products don't rust and break down. Aluminum combines with the oxygen in air and a thin coat forms over the metal, protecting it from further change.

When aluminum is used and thrown away, it becomes solid waste. Solid waste includes all discarded products that biodegrade slowly or not at all. All materials in aluminum products can be used and reused. Like some other kinds of industrial waste materials, aluminum can be melted down and reused in the manufacturing of new products.

Science Journal

Find out how recycling programs in your community recycle aluminum and other materials. What other materials does your community recycle? Summarize your findings *in your Science Journal.*

Science and Society

Purpose
To discuss uses of aluminum, one of the metals mentioned in Section 9-1.

Content Background
Aluminum is not found in its pure form in nature. It wasn't until 1886 that an inexpensive process was discovered to extract aluminum from its ore, bauxite.

Teaching Strategies Lead a discussion that enables students to determine the properties that make aluminum such a popular metal. Students should list that the metal is very lightweight, strong, resistant to atmospheric corrosion, and conducts electricity.

Science Journal

Have students obtain information on recycling from trash removal companies. Metals that are commonly recycled include aluminum, copper, and in some states, steel. Other commonly recycled items include plastics and newspapers. **L1**

Have students find out about the numerous uses for aluminum by writing to The Aluminum Association at 900 19th Street, NW, Washington, DC 20006. **L1**

Going Further ▮▮▮▮▶

Have students compare and contrast the chemical reactivity of iron and aluminum. Have each pair of students obtain an empty steel can and an empty aluminum can. Instruct them to carefully remove the tops of each with a can opener. Put the two cans outside for a few weeks. Have students make daily observations. Then organize and summarize their data. They will observe that the steel will react with oxygen in the air to form rust. Aluminum forms a protective coating on its surface when it combines with oxygen. Use the Performance Task Assessment List for Making Observations and Inferences in **PASC,** p. 17.

COOP LEARN **L1**

Consumer Connection

Purpose

The purpose of this feature is to allow students to learn about a select group of metals—the precious metals—and why these elements are so valuable. The feature expands the information presented in Section 9-1.

Content Background

Gold, silver, and platinum are all considered precious metals, even though about 80 percent of the platinum used in this country is for industrial purposes. The foremost use of gold is as a monetary unit, much of which is kept in reserve for notes issued.

Teaching Strategies

Have students simulate the mining of placer gold deposits with a shallow metal pie pan, sand, and BBs. Have them put about ½ cup of sand and 15 BBs into the pan. Then have them add about 1½ cups of water. Instruct them to swirl the pan while carefully tipping it above a sink or plastic tub in order to remove the sand from the "placer." [L1]

Answers to You Try It!

Prices increased in 1983, 1986, 1987, and 1990. Most fluctuations in precious metals are due to demand as well as the political/economical situations in major countries.

Going Further ⫸

Have students use the business section of the newspaper to make bar graphs of the prices of gold, silver, copper, and aluminum over the past year. Discuss factors affecting fluctuations. Use the Performance Task Assessment List for Graph from Data in **PASC**, p. 39. [L2]

Consumer Connection   The Prices of Precious Metals

The prices of precious metals, such as gold, silver, and platinum, are listed daily in *The Wall Street Journal* and other major urban newspapers around the country. Under the heading Commodities Futures Prices, readers all over the world can see how these prices change from day to day.

New York Gold Prices, 1980–1990

Why Do These Prices Change?

Each day, individuals or groups of people buy and sell these metals. The prices of gold and silver increase when more people wish to buy than sell. Prices go down if more people want to sell than buy. Everyone is affected by these prices.

As Good as Gold

Most of us have seen a gold ring or necklace. Gold has been valued as jewelry for thousands of years. It is also accepted all over the world as currency.

Recently, gold has become an important material in the industrial community. The metal's durability, luster, malleability, and electrical conductivity make it useful in several ways. For example, gold is used to coat windows to reduce heating and cooling costs. It has been used in space equipment to reflect light. And, of course, dentists use gold to repair teeth.

You Try It!

Study the graph on this page. Can you tell which years gold prices increased? Why do you think the price of gold went up during these years? Ask a parent or teacher what might have happened during those years when gold prices increased.

A **metallurgist** is trained in chemistry and engineering. Metallurgists study the way metals are extracted from ore and how they might be used to make different products. An extractive metallurgist explores the ways to remove and refine metal. A production metallurgist would work to develop the metal into some real product.

For more information on careers in metallurgy, students can write to:
The Metallurgical Society
420 Commonwealth Drive
Warrendale, PA 15086

Metalworking in Jewelry

Jewelry is made from a variety of rare and not-so-rare metals and gems. Expensive jewelry is usually made from gold or platinum. Most gold contains some copper, zinc, or silver. Less expensive jewelry is made from bronze or tin. Because consumers like gold, costume jewelry is sometimes given a gold wash or covered with a thin sheet of gold to make it look like pure gold.

The photos show one of the most popular metals used in jewelry today—silver. Most silver jewelry will have other metals mixed in. According to United States law, sterling silver must consist of at least 92.5 percent silver.

Historically, metalworkers made jewelry by hand. Today there are several methods used to make jewelry.

Casting

One frequently used method is called casting. Once a design is developed, a master model, usually made of metal, is formed. A rubber mold is then made of the metal model. Molten wax is poured into the rubber mold, creating a wax model. After completing this process, a second mold is made by dipping the wax model in thinned clay, letting the clay dry, and then baking it in an oven until hard. The melted wax drips through a hole left in the clay. Liquid metal is then poured through the hole, filling up the cavity left by the wax. Once the metal has cooled and hardened, the ceramic mold is cracked open to reveal the cast piece. While the rubber mold can be used a number of times, each wax model can only be used once.

Stamping

In another method called stamping, metal is squeezed between two steel pieces called dies. Stamping and other methods of mass production are used to make huge quantities of inexpensive costume jewelry.

You Try It!

Study some of your family's jewelry at home. Can you tell what kinds of metals were used to make the jewelry?

Answers to You Try It!

Answers will vary but most of the jewelry will probably contain some gold and/or silver and perhaps nickel, copper, tin, and zinc.

Going Further ▦▦▶

Arrange a field trip to a local jewelry store so that students can talk to the goldsmith and perhaps view a demonstration of his or her craft. Have students prepare a list of questions before the visit. If a field trip cannot be arranged, invite a jeweler to visit the class to discuss the composition of various types of jewelry. Use the Performance Task Assessment List for Asking Questions in **PASC**, p. 19.

HOW IT WORKS

Purpose

This excursion demonstrates how the properties of metals, discussed in Section 9-1, are applied in the making of jewelry.

Content Background

Pure gold melts at a little over 1064°C and boils at 2807°C. Gold has a very high specific gravity of 19.3 and a relative hardness of 2.5–3.0 on Mohs' scale of relative hardness. This scale is used by geologists, mineralogists, gemologists and jewelers to identify unknown minerals. The scale ranges from 1 to 10 with 1 being the softest and 10 being the hardest. Talc is one mineral with a hardness of 1. Diamond is the only mineral with a hardness of 10.

Gold alloys, which are used in jewelry, are measured in karats. One karat equals one twenty-fourth part gold. Pure gold is 24K. Eighteen- and fourteen-karat gold are 18 parts pure gold and six parts other metals, and 14 parts pure gold and ten parts other metals, respectively. White gold is a combination of gold, nickel, zinc, and copper. Yellow gold is an alloy of silver, gold, and copper.

Teaching Strategies
Have students compare gold-plated, 14K, 18K, brass, and bronze jewelry to observe how the color of jewelry differs with the amount of gold present. Have students contrast the color of the pieces. Have them note that the 18K piece is much more yellow than the other pieces. The yellow color of the brass piece is much more "golden" than the actual gold pieces. Explain the use of the karats to students. ▣ L2

Technology Connection

Purpose

This feature expands the discussion of nonmetals in Section 9-2 by describing the commercial use of noble gases, which were once thought inert and useless.

Content Background

The noble gases—helium, neon, argon, krypton, xenon, and radon—occur as single atoms in nature and seldom react to form compounds. Compounds that do form with these gases break down very easily.

Neon is used chiefly for filling lamps and luminous tubes of signs. Very little neon is needed to produce the "neon" signs—1 liter of neon will fill about 65 to 97 meters of tubing. Tubes are filled after the air has been removed. When electricity is applied to the tube of gas, electrical discharge occurs and the tube glows with the vivid colors with which we are familiar. The tube contains two electrodes between which the gases form luminous bands.

Teaching Strategy
Ask students how neon lights might be used in foggy cities. Students should be able to conclude, perhaps with some direct questioning from you, that neon lights are used in airplane beacons because the light waves of neon can penetrate the fog. In fact, the neon beacons can be seen at a distance of about 32 kilometers. L2

Answers to
You Try It!

Many businesses use neon lights to advertise. Bright colors, such as red and yellow, are most commonly used because they attract attention.

Technology Connection   Neon Lights

Before fluorescent lights became part of our culture, neon and lamps of other gases were used. The gas tubes weren't practical inside homes and offices because the color of the light is so unlike sunlight. These special gases were best used in creating glowing, lettered signs. Although they are called neon lights, after neon gas, the colored signs contain a number of different gases including helium, argon, krypton, and xenon.

By 1923, neon lights were used in Los Angeles to advertise automobiles. And in the 1970s, artists created a new medium using glass sculptures filled with different noble gases.

How Do These Signs Work?

Electricity passes through a tube that contains one of these gases, causing the gas to glow. The color of the glow depends on the type of gas in the tube. For example, argon gives off a

purple glow, neon gives off a red glow, and krypton gives off a pale violet glow.

Discovery

These gases were discovered late in the 19th century.

One reason that they went undiscovered for so long is that they make up a small fraction of Earth's atmosphere.

Noble gases are also difficult to observe. They are colorless, tasteless, odorless, and do not easily combine chemically to make compounds with other elements.

Sources

The chief source of the noble gases is air. To extract noble gases from air, the air is first chilled to a very low temperature, then liquefied. The liquid air is gradually heated. As the air boils, each noble gas separates out.

You Try It!

Look at the neon signs in your community or in a photo. Identify what colors are most common. Why might these colors be used for advertising?

Going Further ▸

Have students determine whether the use of neon lights in your city or town is a form of light pollution. Light pollution can be defined as the excessive use of any artificial lights. Students can do their research by reading newspapers or by polling neighbors. Groups can debate this issue if opinions on the subject vary considerably.

Interested students may wish to find out about the noble gas radon, which may cause cancer. High concentrations of the gas have been found in many homes and buildings. Have students find out how the radon can be detected and what should be done in case the levels exceed certain limits. Have students prepare a pamphlet describing the dangers of radon and preventative measures for their community. Use the Performance Task Assessment List for Booklet or Pamphlet in **PASC**, p. 57.

Science Journal

Review the statements below about the big ideas presented in this chapter, and answer the questions. Then, re-read your answers to the Did You Ever Wonder questions at the beginning of the chapter. *In your Science Journal,* write a paragraph about how your understanding of the big ideas in the chapter has changed.

1 All matter is made of elements. Elements can be grouped into metals, nonmetals, or metalloids. *Why must we conserve these resources?*

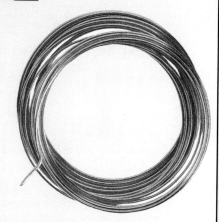

2 Metallic elements are prized for their luster, hardness, ability to be easily shaped, and ability to conduct heat and electricity. *What properties of a metal make it useful as a material for electrical wiring?*

3 Nonmetals are important ingredients in many common products, such as toothpaste. In addition to these uses, nonmetals form the basis of life cycles and play important roles in the processes of living organisms. *What is the most common nonmetal?*

4 Metalloids have properties of both metals and nonmetals. *How are metalloids used in our daily lives?*

Science Journal

Did you ever wonder...
- Copper is inexpensive, malleable, and resistant to change. (pp. 273, 280)
- Plants and animals decompose and are covered by soil. The pressure from layers of soil and decaying organisms produces coal and petroleum. (pp. 284–285)
- Properties of semiconductors enable manufacturers to make the devices small. (pp. 291–292)

Project

Bring in collections of buttons, marbles, and/or beans and have students group the objects in each collection based on similarities and differences among the items. At various points in the chapter, have students look at their classification schemes and make any changes they feel necessary. Relate this to the classification of elements into three major groups. Use the Performance Task Assessment List for Making and Using a Classification System in **PASC**, p. 49. L1

---

chapter 9
REVIEWING MAIN IDEAS

Students will perform skits that will reteach one of the main topics given on this page.

Teaching Strategies

Group students into fours or fives. Assign one of the topics and have students in each group prepare a skit demonstrating the information in the topics. Insist that each student have a speaking role. For example, for the nonmetals topic, students may opt to demonstrate the nitrogen cycle by pretending to be plants and animals that contribute to the cycle. For the metalloids topic, students can role-play various electronic devices that use semiconductors. Evaluate students on their abilities to work together, as well as the originality shown.

Answers to Questions

1. The supply of each is limited.

2. ductility and the ability to conduct electricity

3. oxygen (on Earth) or hydrogen (in the universe)

4. Metalloids are used in some fuels and in the semiconductors found in computers and transistorized electronics. They are needed in small amounts in our bodies.

GLENCOE TECHNOLOGY

MindJogger
Videoquiz

Chapter 9 Have students work in groups as they play the Videoquiz game to review key chapter concepts.

Using Key Science Terms

1. Ductility; the other terms listed are categories of elements.

2. Metal; the properties listed are those of nonmetals.

3. Magnetic; the other terms are properties of almost all metals.

4. Nonmetal; the properties listed are those of metals.

Understanding Ideas

1. Copper, silver, and gold are malleable and easily shaped and stamped. They are quite expensive.

2. Because hydrogen is less dense than air, it rises. Hydrogen is flammable.

3. There are more elements than there are letters of the alphabet.

4. Nitrogen oxides from exhausts react with water in the air. It corrodes metals and pollutes soil and water.

5. Blowing air circulates more oxygen around the fuel.

6. It is a liquid at room temperature.

7. Hydrogen, nitrogen, and oxygen because they, too, are nonmetals.

8. Silicon is shiny and a relatively good conductor.

Developing Skills

1. Concept maps should show that elements consist of metals, which have luster and are malleable and good conductors; nonmetals, which are dull, brittle, and poor conductors; and metalloids, which have properties of both metals and nonmetals.

2. Look for students' experiments to be based on identifying the properties of the element.

3. All are elements. Metals are generally solids that are malleable, ductile, lustrous, and good conductors. Nonmetals are generally gases or dull, brittle, nonconducting solids. Metalloids have some properties of both nonmetals and metals.

4. Neither; oxygen would have made the match burn brighter. Hydrogen would have burned.

Using Key Science Terms

coinage metal	metalloid
ductile	metal
malleable	nonmetal

For each set of terms below, choose the one term that does not belong and explain why it does not belong.

1. metalloid, ductility, nonmetal, metal

2. brittle, metal, nonmetal, gas

3. conductor, magnetic, malleable, ductile

4. metal, ductile, malleable, nonmetal

Understanding Ideas

Answer the following questions in your Journal using complete sentences.

1. Why were copper, silver, and gold used for coins years ago? Why aren't coins used today made out of silver and gold?

2. Why was hydrogen once used to fill aircrafts, such as blimps? Why isn't hydrogen used anymore?

3. Why do some symbols for elements have more than one letter?

4. How is nitric acid formed from pollutants in the air? What effect does it have on the environment?

5. Why does a fire burn more brightly when air is blown on it?

6. How does mercury differ from other metals?

7. Which of these elements are most likely to have properties similar to those of chlorine? Why?
   a. boron
   b. hydrogen
   c. nitrogen
   d. oxygen
   e. silicon

8. In what ways is silicon like a metal?

Developing Skills

Use your understanding of the concepts developed in this chapter to answer each of the following questions.

1. Concept Mapping Using the following terms, create a concept map of the three major groups of elements and their properties: *brittle, dull, elements, good conductors, luster, malleable, metalloids, metals, nonmetals, poor conductors.* Some of the terms may be used more than once.

2. Design an Experiment Suppose you just discovered an unknown element. How would you find out if it were a metal or a nonmetal? Refer to the Investigate on page 274 and the Explore activity on page 282.

3. Comparing and Contrasting Compare and contrast metals, nonmetals, and metalloids.

4. Observing and Inferring Water is added to baking powder and a gas is given off. A lighted match is held in the gas and the flame goes out completely. Is the gas either hydrogen or oxygen? Explain your answer.

5. Making and Using Graphs Use the data in Figure 9-11 to make a pie graph.

Program Resources

Review and Assessment, pp. 53–58 L1
Performance Assessment, Ch. 9 L2
PASC
Alternate Assessment in the Science Classroom
Computer Test Bank L1

Critical Thinking

In your Journal, *answer each of the following questions.*

1. Explain the relationship between oxygen and iron in your body.
2. Compare and contrast the properties of the elements sodium and chlorine with those of salt.
3. From the properties shown in the picture, is this element a metal, nonmetal, or metalloid? Explain your answer.

4. What is the most abundant element in your body? In the universe? In Earth's crust? Are they metals or nonmetals?
5. Why is it more useful to divide elements into metals, nonmetals, and metalloids than into solids, liquids, and gases?
6. Describe the relationship between ordinary glass and semiconductors in computers.

Problem Solving

Read the following problem and discuss your answers in a brief paragraph.

You've been hired to design a new fireworks display for the 4th of July. The display must be a flag in red, white, and blue.

1. Which of the flag's colors would be easily made using the metals from the Investigation on page 274? Why?
2. A sodium compound was mixed into your American flag display. What color will be added to the flag?

CONNECTING IDEAS

Discuss each of the following in a brief paragraph.

1. **Theme—Systems and Interactions** How can metals be dangerous to your health?
2. **Theme—Scale and Structure** Why are silver and copper good materials for making electric wires?
3. **How It Works** Explain why more inexpensive costume jewelry is made by stamping than by molding.
4. **Technology Connection** What physical property of matter is used to help collect noble gases from air? Explain the process that is most commonly used.
5. **Earth Science Connection** Describe how diamonds in nature are formed.

✔ Assessment

Portfolio Review the portfolio options that are provided throughout the chapter. Encourage students to select one product that demonstrates their best work for the chapter. Have students explain what they learned and why they chose this example for placement into their portfolios.

Additional portfolio options can be found in the following **Teacher Classroom Resources: Making Connections: Integrating Sciences,** p. 21

Multicultural Connections, pp. 21, 22
Making Connections: Across the Curriculum, p. 21
Concept Mapping, p. 17
Critical Thinking/Problem Solving, p. 17
Take Home Activities, p. 16
Laboratory Manual, pp. 47-48
Performance Assessment P

5. The graph will show: oxygen = 234°, carbon = 65°, hydrogen = 36°, calcium = 7°, other elements = 7°, nitrogen = 11°.

Critical Thinking

1. Iron in hemoglobin in the red blood cells carries oxygen throughout the body.
2. Sodium is a soft metal; chlorine is a yellow-green, poisonous gas. Both are dangerous to human health. Yet, salt, a compound of sodium and chlorine, is an edible compound that is necessary in the human body.
3. It is a nonmetal because it does not conduct electricity and it is brittle and dull.
4. oxygen; hydrogen; oxygen; nonmetals
5. The state of matter can vary depending on the temperature.
6. Both contain silicon.

Problem Solving

1. Red can be obtained by using either lithium or strontium compounds.
2. yellow

Connecting Ideas

1. Lead and mercury can take the place of iron in red blood cells but they don't have the ability to carry oxygen to your systems.
2. Silver and copper are ductile and both good conductors of electricity.
3. Molding requires more patience and skill than mass production of jewelry by stamping. Thus, jewelry made by stamping is less expensive.
4. Boiling point; air is cooled until it is a liquid. Then it is heated until each noble gas boils off.
5. Carbon-containing magma cools and solidifies deep within Earth. Heat and pressure compact the carbon to form diamonds.

Chapter Organizer

SECTION	OBJECTIVES	ACTIVITIES & FEATURES
Chapter Opener		**Explore!**, p. 301
10-1 Minerals and Their Value (2 sessions, 1 block)	1. **Name** four conditions that define minerals. 2. **Explain** how rarity and beauty can affect the value of a particular mineral. **National Content Standards: (5-8) UCP2, UCP5, A1, D1**	**Find Out!**, p. 303 **Life Science Connection**, pp. 304-305 **History Connection**, p. 324
10-2 Identifying Minerals (3 sessions, 1.5 blocks)	1. **Describe** how physical characteristics are used to identify minerals. 2. **Differentiate** between mineral characteristics such as hardness, color, streak, cleavage, and fracture. **National Content Standards: (5-8) UCP2-3, A1-2, B1, D1**	**Find Out!**, p. 308 **Find Out!**, p. 310 **Design Your Own Investigation 10-1:** pp. 312-31 **Explore!**, p. 314 **Science and Society**, pp. 322-323
10-3 Mineral Formation (2 sessions, 1 block)	1. **Examine** mineral formation from solution. 2. **Describe** mineral formation by the cooling of magma. **National Content Standards: (5-8) UCP2, UCP4, A1, D1, E2, F2, F4-5**	**Investigate 10-2:** pp. 316-317 **A Closer Look**, pp. 320–321

ACTIVITY MATERIALS

EXPLORE!

p. 314* samples of clear minerals, including a clear piece of calcite; light

INVESTIGATE!

pp. 316-317 salt solution, sugar solution, large test tube, toothpick, cotton thread, hand lens, shallow pan, thermal mitt, test-tube rack, cardboard, table salt, granulated sugar, hot plate, beaker

DESIGN YOUR OWN INVESTIGATION

pp. 312-313 mineral samples, hand lens, steel file, goggles, apron, streak plate, 5% hydrochloric acid, dropper, Mohs scale of hardness

FIND OUT!

p. 303* table salt, microscope, quartz sample, sugar, microscope slide
p. 308* streak plate; pyrite, hematite, graphite
p. 310* samples of calcite, quartz, and talc

KEY TO TEACHING STRATEGIES

The following designations will help you decide which activities are appropriate for your students.

- **L1** Basic activities for all students
- **L2** Activities for average to above-average students
- **L3** Challenging activities for above-average students
- **LEP** Limited English Proficiency activities
- **COOP LEARN** Cooperative Learning activities for small group work
- **P** Student products that can be placed into a best-work portfolio
- 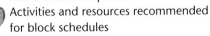 Activities and resources recommended for block schedules

Need Materials? Call Science Kit (1-800-828-7777).

00:00 **OUT OF TIME?** We recommend that students do the activities with an asterisk.

Chapter 10 Minerals and Their Uses

TEACHER CLASSROOM RESOURCES

Student Masters	Transparencies
Study Guide, p. 34 **Multicultural Connections,** p. 23 **Making Connections: Across the Curriculum,** p. 23 **Making Connections: Technology & Society,** p. 23 **Science Discovery Activities, 10-1** **Laboratory Manual,** pp. 49-50, Mineral Resources	**Section Focus Transparency 28**
Study Guide, p. 35 **Concept Mapping,** p. 18 **Multicultural Connections,** p. 24 **Take Home Activities,** p. 17 **Activity Masters,** Design Your Own Investigation 10-1, pp. 43–44 **Science Discovery Activities, 10-2**	**Teaching Transparency 19,** Mohs Scale of Hardness **Teaching Transparency 20,** Nonsilicate Mineral Groups **Section Focus Transparency 29**
Study Guide, p. 36 **Critical Thinking/Problem Solving,** p. 18 **How It Works,** p. 13 **Making Connections: Integrating Sciences,** p. 23 **Activity Masters,** Investigate 10-2, pp. 45–46 **Science Discovery Activities, 10-3** **Laboratory Manual,** pp. 51-52, Removal of Waste Rock **Laboratory Manual,** pp. 53-56, Crystal Formation	**Section Focus Transparency 30**

ASSESSMENT RESOURCES	TEACHING & TECHNOLOGY
Review and Assessment, pp. 59–64 **Performance Assessment,** Ch. 10 **PASC*** **MindJogger Videoquiz** **Alternate Assessment in the Science Classroom** **Computer Test Bank**	**Spanish Resources** **Cooperative Learning Resource Guide** **Lab and Safety Skills** **Science Interactions, Course 2, CD-ROM** **Computer Competency Activities**

*Performance Assessment in the Science Classroom

NATIONAL GEOGRAPHIC TEACHER'S CORNER

Index to National Geographic Magazine

The following articles may be used for research relating to this chapter:

- "Rubies and Sapphires," by Fred Ward, October 1991.
- "Emeralds," by Fred Ward, July 1990.
- "Air: An Atmosphere of Uncertainty," by Noel Grove, April 1987.
- "The Incredible Crystal: Diamonds," by Fred Ward, January 1979.

National Geographic Society Products Available From Glencoe

To order the following products for use with this chapter, contact your local Glencoe sales representative or call Glencoe at 1-800-334-7344:

- *NGS PictureShow Geology* (CD-ROM)

Additional National Geographic Society Products

To order the following products for use with this chapter, call the National Geographic Society at 1-800-368-2728:

- *Every Stone Has a Story* (Video)
- *Splendid Stones* (Video)

Teacher Classroom Resources

These are key components of the classroom resources package.

TEACHING AIDS

Section Focus Transparencies

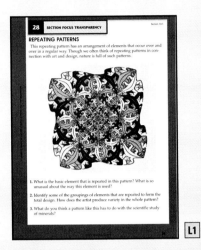

28 SECTION FOCUS TRANSPARENCY Section 10-1

REPEATING PATTERNS
This repeating pattern has an arrangement of elements that occur over and over in a regular way. Though we often think of repeating patterns in connection with art and design, nature is full of such patterns.

1. What is the basic element that is repeated in this pattern? What is so unusual about the way this element is used?
2. Identify some of the groupings of elements that are repeated to form the total design. How does the artist produce variety in the whole pattern?
3. What do you think a pattern like this has to do with the scientific study of minerals?

L1

29 SECTION FOCUS TRANSPARENCY Section 10-2

WHO'S TOUGHER?
Which of the two minerals do you think might win this fight—the lovely diamond or the tough-looking quartz? If you said the diamond, you're right. Diamonds are the hardest minerals around.

1. Name some other uses for diamonds besides jewelry.
2. Describe a way to sort minerals other than by hardness.

L1

30 SECTION FOCUS TRANSPARENCY Section 10-1

MODERN ART?
Some minerals look dull on the outside. But when you slice them open and magnify them many times, inside lies a kaleidoscope of color.

1. What do you think is responsible for the different colors in this mineral?
2. Why would scientists study minerals at such a close range?
3. How do you think minerals form?

L1

Teaching Transparencies

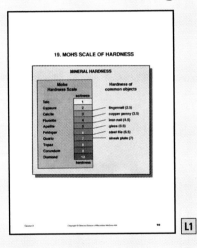

19. MOHS SCALE OF HARDNESS

L1

20. NONSILICATE MINERAL GROUPS

L2

HANDS-ON LEARNING

Science Discovery Activity*

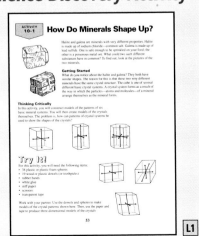

ACTIVITY 10-1

How Do Minerals Shape Up?

L1

Laboratory Manual*

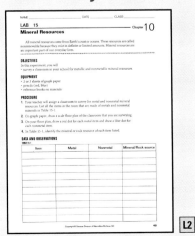

LAB 15 Chapter 10
Mineral Resources

L2

Take Home Activity

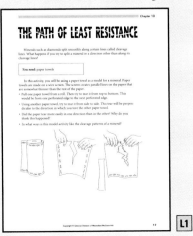

Chapter 10

THE PATH OF LEAST RESISTANCE

L1

*There may be more than one activity for this chapter.

Chapter 10 Minerals and Their Uses

REVIEW AND REINFORCEMENT

Study Guide*

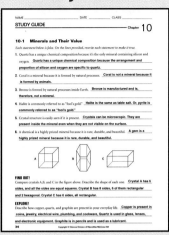

Concept Mapping

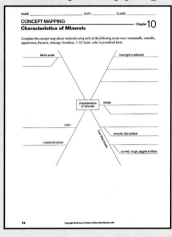

Critical Thinking/Problem Solving

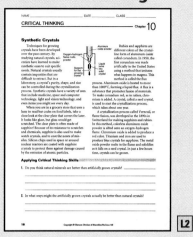

ENRICHMENT AND APPLICATION

Integrating Sciences

Across the Curriculum

Technology and Society

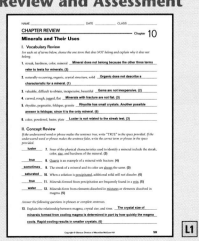

ASSESSMENT

Multicultural Connection**

Performance Assessment

Review and Assessment

Minerals and Their Uses

THEME DEVELOPMENT

One important theme of this chapter is systems and interactions. Students learn the system of classification of minerals and interactions involved in mineral formation.

CHAPTER OVERVIEW

Minerals are naturally occurring, inorganic crystalline solids with a definite composition. Minerals are formed when magma cools, when water evaporates, or by precipitation from a solution.

Tying to Previous Knowledge

Set up a display of minerals. Next to each put a representative sample of a product made from each mineral. Do not label the display. Let groups of students infer the connections and match the product to the mineral.

L1 COOP LEARN

INTRODUCING THE CHAPTER

Have students look at the photograph on page 300 and identify some of the materials used in these objects. Ask students why they think the Egyptians might have used these materials.

MINERALs
AND THEIR USES

Did you ever wonder...

✓ **Where gold comes from and why it's so valuable?**

✓ **Why some stones are called gems and others are called rocks?**

✓ **How to tell the difference between real gold and "fool's gold"?**

Science Journal

Before you begin to study minerals, think about these questions and answer them *in your Science Journal.* When you finish the chapter, compare your journal write-up with what you've learned.

Answers are on page 325.

Here you are, looking at the museum's display of relics from the tomb of King Tutankhamen—an ancient Egyptian pharaoh. The first thing you notice is the gleaming gold.

You catch your breath a little as you move to the next section of the exhibit. Blue sapphires, red rubies, and green emeralds glisten at you from necklaces and bracelets. Light seems to stream from every point on their surfaces.

Virtually every item on display here is made of minerals. What makes the minerals of King Tut's treasure different from the minerals around you?

▶ **In the activity on the next page, explore what characteristics make some items more valuable than others.**

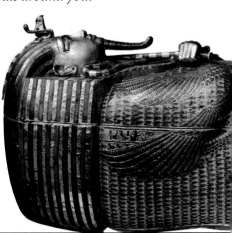

300

00:00 **OUT OF TIME?**

If time does not permit teaching the entire chapter, use the Chapter Overview on this page, Reviewing Main Ideas at the end of the chapter, and the Chapter 10 audiocassette to point out the main ideas of the chapter.

Learning Styles		
Kinesthetic	Find Out, pp. 308, 310; Activity, p. 311; Design Your Own Investigation, pp. 312-313; Investigate, pp. 316-317	
Visual-Spatial	Find Out, p. 303; Activity, pp. 304, 309, 314, 315, 319, 321; Visual Learning, pp. 305, 306, 307, 309, 319; Explore, p. 314; Demonstration, p. 318	
Interpersonal	Activity, p. 306	
Linguistic	Science Journal, p. 318; Explore, p. 301; Activity, pp. 302, 304, 307; Life Science Connection, pp. 304-305; Research, pp. 306, 321; Across the Curriculum, p. 319; A Closer Look, pp. 320-321	

LS

Explore! ACTIVITY

What makes jewels valuable?

Just as a museum director gathers similar items together to make an exhibit, many of us have collections of our own.

Perhaps you have a collection of leaves, shells, or baseball cards. No matter what you collect, you probably have one item in your collection that is special.

What To Do

1. Place the items of your collection around you. *In your Journal*, describe the common characteristics of all the items.

2. Pick out the one special item you have—the "gem" of your collection. Describe what makes it different.

3. Why is your collection important to you? Would it be valuable to others?

Explore!

What makes jewels valuable?

Time needed 15 minutes

Thinking Processes classifying, comparing and contrasting

Purpose To compare the relative value of items in a collection.

Teaching the Activity

Science Journal Have students note what they have chosen to collect, why, and the criteria they use to add an object to their collection. **L1**

Troubleshooting Some students may not have a collection. Have them think of any collection with which they are familiar.

Expected Outcomes

Students will choose an item to describe that has special attributes that make it more "prized" than other similar items.

Answers to Questions
Answers will vary but should include that every person has different aesthetic values, so what one person may see as beautiful may not be beautiful to another.

✓ Assessment

Oral Have some students who have collections bring a part of their collection to school. Students can explain to the class why they included certain items in their collection and what they value in the objects they collect. Use the Performance Task Assessment List for Oral Presentation in **PASC,** p. 71.

ASSESSMENT PLANNER

PORTFOLIO
Refer to page 327 for suggested items that students might select for their portfolios.

PERFORMANCE ASSESSMENT
Process, pp. 308, 310
Explore! Activities, pp. 301, 314
Find Out! Activities pp. 303, 308, 310
Investigate!, pp. 312–313, 316–317

CONTENT ASSESSMENT
Check Your Understanding, pp. 306, 314, 321
Reviewing Main Ideas, p. 325
Chapter Review, pp. 326–327

GROUP ASSESSMENT
Opportunities for group assessment occur with Cooperative Learning Strategies.

PREPARATION

Planning the Lesson

Refer to the Chapter Organizer on pages 300A-D.

Concepts Developed

This chapter is the foundation for Chapter 11, The Rock Cycle, in which students learn that rocks are comprised of minerals.

In this section, students recognize four conditions that define minerals.

1 MOTIVATE

Bellringer

 Before presenting the lesson, display **Section Focus Transparency 28** on an overhead projector. Assign the accompanying **Focus Activity** worksheet. L1

LEP

Activity Obtain a piece of granite from a scientific supply company or a local tombstone maker. Have students examine the granite and then describe two or three components (minerals) of which the rock is made. L1

2 TEACH

Tying to Previous Knowledge

Have students recall the meanings of *element* and *compound*. Ask students to name a metal as discussed in Chapter 9. Tell students that platinum, copper, silver, and gold occur as metallic minerals.

GLENCOE TECHNOLOGY

 CD-ROM

Science Interactions, Course 2 CD-ROM

Chapter 10

10-1 Minerals and Their Value

Section Objectives

- Name four conditions that define minerals.
- Explain how rarity and beauty can affect the value of a particular mineral.

Key Terms

gem
mineral

Minerals Can Be Familiar

Without asking anyone, you know that the gold and jewels you see in the museum are worth a good deal of money. You may decide that they are valuable because they are so beautiful. You may also take into account that gold and jewels are rare and hard to find. Even the most beautiful, rare, and costly jewels, however, fall into a broad category of substances. These substances are called minerals.

Minerals are part of your life. Write your name with a pencil, and you use the mineral graphite, which is used in pencil lead. Toss a penny, and you toss the mineral copper, one of several substances used in coin currency.

There are more than 4000 different minerals on Earth. As you can imagine, we have found many uses for

Quartz

Figure 10-1

When stimulated with an electric charge from a battery, microthin quartz crystal slices vibrate more than 30 000 times each second. Because the vibrations are so regular, they can be used as a measure of time.

Titanium

Figure 10-2

Because of its light weight and great strength, titanium is a favored material for making aircraft frames and engines.

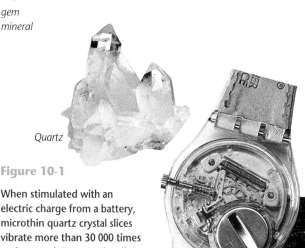

302 Chapter 10 Minerals and Their Uses

Program Resources

Study Guide, p. 34
Laboratory Manual, pp. 49–50, Mineral Resources L2
Multicultural Connections, p. 23, Early Iron Smelting in Africa L1
Making Connections: Across the Curriculum p. 23, Gold! L1

Making Connections: Technology and Society, p. 23, Exotic Gems—The Rarest of the Rare L2
Science Discovery Activities, 10-1
Section Focus Transparency 28

---

them. Some of these uses are shown on this and the following pages. Can you think of other common uses of minerals?

How can you know what substances are minerals? Even though there are thousands of minerals, they all meet several basic conditions. A **mineral** is a naturally occurring, inorganic solid with a definite crystalline structure, and a definite composition. In the following Find Out activity, you'll learn more about the crystalline nature of minerals.

Halite

Figure 10-3

When you munch on a salted pretzel, you eat the mineral halite. Halite is also known as rock salt and is the source of common table salt.

 Find Out! ACTIVITY

How do minerals look up close?

What To Do

1. Slide a tiny amount of table salt—the mineral halite—under your microscope and separate the granules.

2. Focus on just one granule as you examine the salt under your microscope.

3. Now, examine a quartz sample with evidence of its crystal structure visible on the outside. How does the shape of the quartz crystal compare with the shape of the halite crystal?

4. Count the number of sides each crystal has. Make a sketch of the quartz crystal and of the halite crystal.

5. The diagram below shows the six major crystal systems of minerals. Use it to answer the following questions.

Conclude and Apply

1. Which type of crystal is quartz?

2. Which type of crystal is halite?

3. Examine sugar grains with your microscope. Determine which type of crystal shape they are. Sugar is formed by plants. Is it a mineral?

 Cubic   Tetragonal   Hexagonal   Orthorhombic   Monoclinic   Triclinic

---

Meeting Individual Needs

Visually Impaired Give students good representative samples of minerals that show the characteristic crystal shapes.

Have them try to count the number of sides each crystal has.

How do minerals look up close?

Time needed 20–30 minutes

Materials quartz specimens, table salt, microscope, microscope slide, sugar

Thinking Processes observing and inferring, comparing and contrasting

Purpose To compare and contrast differences in two different minerals' crystalline shapes.

Teaching the Activity

Ask students to list those physical properties of the minerals that they can see without the aid of the microscope. Have them compare those properties with properties they observe in the minerals under magnification. **L1** **LEP**

Troubleshooting Be sure that the students know how to operate the microscopes. Remind students to be careful to avoid breakage when lowering the microscope lens toward the stage.

Science Journal Have students note their observations of halite and quartz crystals.

Expected Outcomes

Students will directly observe small crystals firsthand and can describe their appearance. They will be able to compare a cube-shaped crystal to a hexagonal crystal.

Conclude and Apply

1. Quartz is a hexagonal crystal.

2. Halite is cubic.

3. Sugar crystals have an orthorhombic shape. Sugar is not a mineral because it is an organic substance.

✔ Assessment

Performance Have students use toothpicks, pipe cleaners, or construction paper to construct the six major types of crystals illustrated on this page. Have them label each crystal shape they construct. Use the Performance Task Assessment List for Model in **PASC**, p. 51.

How Do We Define Minerals?

Let's look at the conditions that define minerals more closely.

1. Minerals Occur Naturally When salt deposits form by natural processes, the mineral halite is formed. However, salt manufactured by humans is not considered a mineral.

2. Minerals Are Inorganic Solids Minerals are not alive, nor are they formed from anything that ever was alive. For example, coral is formed from skeletons of tiny sea animals, therefore, it is organic. On the other hand, diamonds form deep underground from inorganic matter. A diamond is a mineral, but coral is not.

3. Minerals Have Unique Chemical Compositions A mineral can be an element or a compound. Each type of mineral has a chemical composition that is unique to that mineral. The mineral quartz, for example, is a combination of two elements, silicon and oxygen. Although other minerals may also contain silicon and oxygen, the arrangement and proportion of the elements in quartz are unique to quartz.

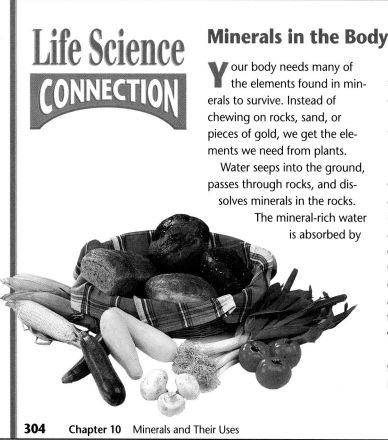

Life Science CONNECTION

Minerals in the Body

Your body needs many of the elements found in minerals to survive. Instead of chewing on rocks, sand, or pieces of gold, we get the elements we need from plants.

Water seeps into the ground, passes through rocks, and dissolves minerals in the rocks. The mineral-rich water is absorbed by plant roots and carried throughout the plant. The elements in minerals help the plant grow by assisting in photosynthesis and other processes.

The human body needs more than 20 different kinds of elements, but only in very small amounts. Iron, which is found in such minerals as magnetite and pyrite, helps blood carry oxygen throughout our bodies. Calcium, found in calcite and dolomite, helps make bones and teeth strong. Sodium, found in halite, helps regulate water in the body's cells.

The elements needed to form minerals can be recycled. Plants absorb minerals from the

Life Science CONNECTION

Purpose
Life Science Connection extends Section 10-1 by describing how minerals form.

Teaching Strategy
Explain that water and air break down minerals into simpler substances, compounds, and elements. These elements and compounds enter the soil and are absorbed by plants. Plants contain "minerals" which are actually the elemental remains of what used to be minerals. Have students identify the mineral contents listed on the labels of empty vitamin and mineral supplement bottles. Then make a master list of the minerals found in these supplements, the recommended daily allowance, and the daily percentage.

Answers to
You Try It!
A slice of whole wheat bread has about 18 mg of calcium, and white bread has 30

4. Minerals Have Crystalline Structure

The atoms in a mineral are arranged in a regular geometric pattern repeated over and over again. Substances with this kind of inner structure are crystalline.

■ Classifying Minerals

Most minerals are nonmetallic. Graphite, quartz, and halite are all examples of nonmetallic minerals.

However, other substances such as gold, silver, and copper also meet the definition of a mineral. They are also defined as metals. Metals are materials that are used in many different ways. Can you think of a few ways we use metals?

Figure 10-4

A Topaz is a naturally occurring inorganic solid. It has a unique chemical composition and a crystal structure. Having these four characteristics qualifies topaz to be classified as a mineral.

Topaz

Coal

B Coal is a solid, naturally occurring substance composed of the remains of plants and animals. Is coal a mineral? Why?

soil, then animals eat the plants. When the animals—including humans—die and decompose, they return the elements to the soil. But most of the plants humans eat come from large production farms, where there isn't enough time for this gradual cycle.

Farmers use different methods of returning elements to the soil. Often, organic or inorganic fertilizers are used. Organic fertilizers, often referred to as compost, are fertilizers made from plant and animal remains and waste. Inorganic fertilizers are made from essential elements and minerals that have been extracted from rock.

Crop rotation is another method by which minerals in soil can be preserved or replaced. Farmers rotate the types of crops they plant each year to ensure that the minerals used by one type of crop are replaced the following year by a different crop.

You Try It!

Calcium is the most abundant element in the body, and is found in teeth and bones. Calcium is found in many foods, including whole grains. Read and compare the labels on the foods you eat.

3 ASSESS

Check for Understanding

Direct the students to complete Check Your Understanding. Hold up several items in front of the class. By reviewing the characteristics of minerals, help students to determine whether or not each item is a mineral. Use the Classroom Assessment List for Making Observations in **PASC**, p. 17.

Reteach

Activity Have students reread Section 10-1 and write two facts for each page. Have paired partners compare facts and quiz each other. L2 **COOP LEARN**

Extension

Research Have students who have mastered the concepts use chemistry books, geology books, or encyclopedias to find out what elements are in the minerals named in Section 10-1. L3

4 CLOSE

Have students list the names of as many minerals as they can in three minutes. The students should then write one fact about each mineral. Have students share their lists. L1

Figure 10-5

People have been combining tin and copper to make bronze objects since 3500 B.C. Is bronze a metal? Is it a mineral? Why?

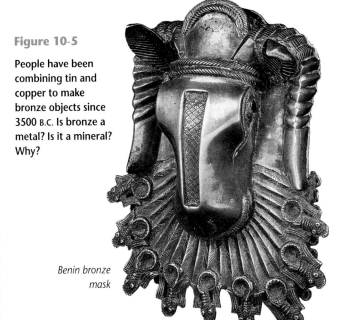

Benin bronze mask

Figure 10-6

A It takes a skilled stonecutter to bring out the sparkle in a gemstone. A stonecutter begins by cutting the mineral crystal in half, and rounding its edges.

B The next step is faceting the gemstone on various grinding wheels. Faceting is a process of grinding tiny, polished sides, called facets, into the gemstone.

C The number of facets varies according to cut style. The standard brilliant cut, shown here, has 58 facets.

Are all metals minerals? To find the answer to that question, read the caption of **Figure 10-5**.

■ Some Minerals Are Gems

Think back to your visit to the museum at the beginning of the chapter. Why were the minerals on display considered "gems"? A **gem** is a valuable, highly priced mineral that is rare, or is difficult to obtain.

The treasures of the pharaohs are easily identified as gems because they've been cut and polished. But suppose you were seeking gems in their natural form. Would you know what to look for? How could you tell a valuable find from a nonvaluable one? In the next section, you'll learn how we can distinguish one mineral from another.

check your UNDERSTANDING

1. What four conditions do minerals satisfy?
2. What factors help determine the value of a mineral?
3. **Apply** Opal is a substance that occurs naturally, is an inorganic solid, and has a unique composition. But opal does not have a definite crystalline structure. Is opal a mineral? Explain your answer.

check your UNDERSTANDING

1. Minerals are inorganic solids with unique chemical compositions and crystal structures.
2. rarity, beauty, durability, usefulness
3. **Apply** Opal is not a mineral because it does not have a definite crystal structure.

10-2 Identifying Minerals

Properties of Minerals

Section Objectives
- Describe how physical characteristics are used to identify minerals.
- Differentiate between mineral characteristics such as hardness, color, streak, cleavage, and fracture.

Key Terms
streak, hardness, cleavage, fracture

Picture a time in history when people left their jobs and families, took all they could carry, and ran off in search of their fortunes. That's what happened during the California Gold Rush.

A gold rush is a mass, rapid migration of people to an area where gold has been discovered. The largest and most famous gold rush in U.S. history began with the discovery of gold in Sutter's Mill in California in 1848.

Suppose you had been living during the 1840s. Would you have been part of the rush for gold? If so, would you have known what to look for? Many people of that day thought they, too, would know gold when they saw it, but they were fooled. The mineral pyrite, or "fool's gold," looks a lot like true gold. There are several simple tests you could have performed to know whether or not you had "struck it rich." You'll learn about one such test in the following Find Out activity.

Figure 10-7

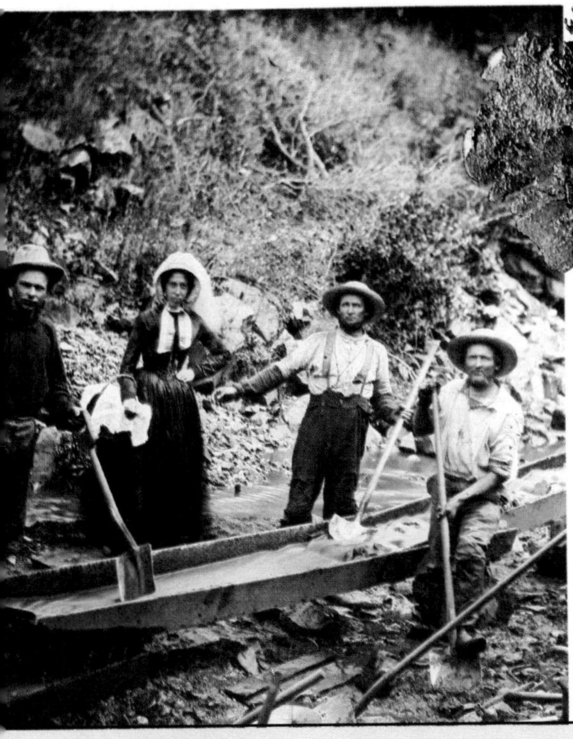

A The first California gold fields were discovered near present day Sacramento in 1848. Word of the find spread. The gold rush was on! In the following year thousands of people, hoping to get rich, rushed into California.

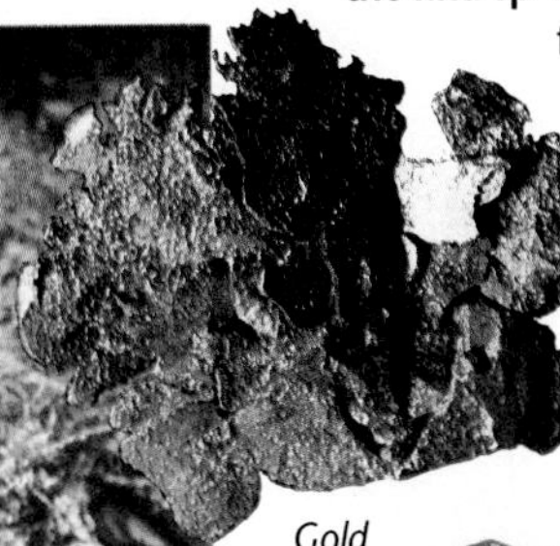

Gold

B The gold rush made a few people very rich. But many were disappointed. Some found too little gold to make a profit. Some found only pyrite, a mineral that looks so similar to what inexperienced prospectors think gold looks like that it is called "fool's gold."

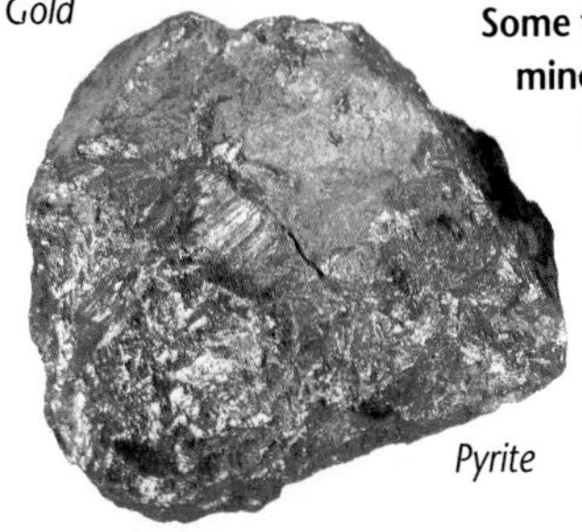

Pyrite

C The gold fever faded in a few years, but the social and economic development that it started continued to thrive.

Program Resources

Study Guide, p. 35.
Concept Mapping, p. 18, Characteristics of Minerals L1
Multicultural Connections, p. 24, Mound Builders and Minerals
Teaching Transparency 19 L1
Teaching Transparency 20
Section Focus Transparency 29

10-2 PREPARATION

Planning the Lesson

Refer to the Chapter Organizer on pages 300A-D.

Concepts Developed

In Section 1 students learned that there are four conditions that all minerals meet. In this section, students will apply this knowledge to comparing some physical properties that distinguish one mineral from another.

1 MOTIVATE

Bellringer

Before presenting the lesson, display **Section Focus Transparency 29** on an overhead projector. Assign the accompanying **Focus Activity** worksheet. L1 LEP

Activity Ask students to imagine that they are prospectors who have discovered the two minerals shown in the photos. How would they compare which was valuable and which was worthless? L1

Visual Learning

Figure 10-7 Have students look at Figure 10-7 and read the captions. Show students an example of gold, such as a piece of jewelry or other gold object, and a sample of pyrite. Choose a pyrite sample which shows the characteristic greenish tinge. Have students compare and contrast the samples. Point out that many samples of pyrite are greenish gold. Also tell the students that pyrite is harder than gold.

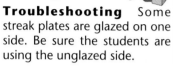
Find Out! ACTIVITY

What is a streak test?

Your teacher will give you a piece of unglazed porcelain called a streak plate and several samples of minerals, including pyrite, hematite, and graphite.

What To Do

1. First, run the graphite across the streak plate. This is called streaking the graphite.

2. Look at the color of graphite's streak on the streak plate. How does it compare with the color of the graphite?

3. Before you streak the pyrite and hematite, hypothesize what color they will streak on the streak plate.

4. Now streak the pyrite and hematite. What color are the streaks? How do the streak colors compare with the actual colors?

Conclude and Apply

1. Were you correct in the hypothesis you made before you streaked the pyrite?

2. When streaked, gold leaves a yellow mark. How could you use the streak test to make sure you weren't fooled by "fool's gold"?

■ Streak

When a mineral is rubbed across a streak plate, as you did in the Find Out activity, a streak is left behind. The color of a mineral when it is broken up and powdered is called its **streak**. Was the red-brown streak of hematite what you expected to see?

■ Color

A mineral's color is another clue to its identity. As you can see from the photos on this page, sulfur and azurite are easy to distinguish based on color alone. But is this always the case? Think back to what you know about gold and pyrite. Are they easy to identify based on color?

Color may give you a clue as to a mineral's identity, but you'll need more information before you can be sure.

Figure 10-8

A Color is one of the first characteristics you notice about a mineral, and it can often give a valuable clue to its identity. Sulfur, for example, has a very distinctive yellow color.

Sulfur

B You could never confuse sulfur with azurite, which is always a shade of blue. The blue of azurite, seen at the right, is the source of the color term azure blue.

Azurite

■ Luster

Look at **Figure 10-9**. How do these two minerals differ in appearance? Which of the two would you label as having a "metallic" luster? Which has a nonmetallic luster?

Luster refers to the way in which a mineral reflects light. If you refer to Appendices K and L of your book, you'll see that minerals are divided into two groups based on their luster.

Do you think all minerals with a metallic luster are metals? You may be surprised to learn that they are not. For example, graphite is metallic in luster. Yet, you know that the graphite in your pencil is not a true metal.

■ Hardness

So far, you've learned that a mineral's color, luster, or streak can help you identify it. Unfortunately, these characteristics sometimes vary among samples of the same mineral. A more useful property you can use to identify a mineral is its hardness.

Figure 10-9

A Minerals with a metallic luster shine like metal. Galena, which is an ore of lead, has a bright metallic luster.

Galena

B The appearance of minerals with a non-metallic luster vary. Feldspar, used in making glass and pottery, has a glossy to pearly appearance. Talc, another mineral with a nonmetallic luster, has a pearly surface.

Feldspar

Yellow sapphire   *Blue sapphire*

C A single color alone may not be enough to distinguish one mineral from another. Some minerals come in various colors. Sapphires, shown above, are forms of the mineral corundum and come in blue, purple, green, pink, and yellow.

Meeting Individual Needs

Learning Disabled Have students collect rocks and bring them to class. Have them pick representative samples and classify them according to color. Help students identify minerals present in the rocks. Obtain samples of pink and gray granite. Show students that mineral colors in rocks can vary. Show them samples of the minerals present in each of the rock samples (pink granite is colored with pink feldspar, gray with gray feldspar). **LEP**

---

2 TEACH

Tying to Previous Knowledge

Have students recall what physical properties are. Remind students that physical properties can be observed without changing a substance into a new substance. Show students several different mineral samples and have them list several different properties of each.

Theme Connection The physical properties of the different minerals are directly related to their chemical properties, and thus relate to the theme of interactions and systems.

Activity Explain that minerals have unique physical properties. Have students compare and contrast physical properties of minerals. Give some sulfur to students and explain that its bright yellow color is unique. Give students some lodestone and a paper clip to show them the magnetic properties of the mineral. Give students pieces of mica. Have them try to break the mica with their fingers parallel to its flat faces to show cleavage planes. **L2**

Visual Learning

Figure 10-9 Ask students to compare and contrast the minerals in the illustrations. Help students name and describe other minerals they have encountered.

GLENCOE TECHNOLOGY

Software
Computer Competency Activities
Chapter 10

How does a scratch test help identify minerals?

Time needed 15 minutes

Materials samples of calcite, quartz, and talc

Thinking Processes comparing and contrasting, sequencing, recognizing cause and effect, forming operational definitions

Purpose To determine the hardness of three minerals by using the scratch test.

Teaching the Activity

Troubleshooting Have students use two minerals to demonstrate the proper technique. Remind the students not to gouge the mineral samples.

Science Journal Have students describe the results of their scratch tests of calcite, quartz, and talc in their journals. L1

Expected Outcomes

Students will observe that different minerals have different hardnesses. They will sequence the minerals used in the activity from softest to hardest: talc, calcite, quartz.

Conclude and Apply

1. Talc can be scratched with a fingernail.

2. Gold is softer than pyrite, so a scratch test could help distinguish between them. Gold will be scratched by pyrite, but pyrite will not be scratched by gold.

3. Minerals that are harder than a piece of copper will scratch the copper. Minerals that are softer than the copper will be scratched by the copper.

✔ Assessment

Process Have students perform scratch tests on other minerals. Then have them make a data table comparing and contrasting the hardness of the minerals they have tested. Use the Performance Task Assessment List for Analyzing the Data in **PASC**, p. 27.

How does a scratch test help identify minerals?

Your teacher will give you samples of the three minerals listed in the table.

What To Do

1. Copy the table *in your Journal.*

2. Choose one of the mineral samples and try to scratch the other minerals with it. For example, does calcite scratch quartz? If so, make a check mark in the calcite row where it intersects with the quartz column.

3. Repeat the scratch test for each mineral in the chart.

4. Analyze the completed data table, and sequence the minerals from softest to hardest: 1 = softest, 3 = hardest.

Conclude and Apply

1. Can you scratch any of the minerals with your fingernail?

2. Hypothesize how a scratch test might help you identify pyrite or gold.

3. Describe how you could use a scratch test to determine which minerals are harder than a copper penny.

Sample data

Data and Observations			
	Calcite	Quartz	Talc
Calcite	✕		✔
Quartz	✔	✕	✔
Talc			✕

Figure 10-10

The Mohs scale of mineral hardness lists 10 minerals in order of hardness. The softest mineral is 1; the hardest is 10.

The Find Out activity shows one characteristic that helps distinguish one mineral from another. **Hardness** is a measure of how easily a mineral can be scratched. Talc is one of the softest known minerals and can be scratched with your fingernail. If you have ever rubbed talcum powder on your skin, you know how smooth and soft talc can be. In contrast, diamonds are so hard that they are used in sharpening and cutting tools. A diamond

1	2	3	4	5
Talc	Gypsum	Calcite	Fluorite	Apatite
Easily scratched by fingernail.	*Can be scratched by fingernail.*	*Barely can be scratched by copper penny.*	*Easily scratched with steel knife blade.*	*Can be scratched by steel knife blade.*

can be scratched only by another diamond.

Remember how you sequenced the minerals in the Find Out activity according to hardness? You were working with minerals that had three distinctive hardnesses, so you used a scale of 1 to 3. In a similar way, you can use a scale of 1 to 10 to compare the hardnesses of all minerals. This scale is shown in **Figure 10-10**. The scale is named for the German scientist who devised it, Friedrich Mohs.

■ Cleavage and Fracture

Minerals also differ in the way they break. You might perform this test by tapping a mineral sample against a hard surface. Minerals that break along smooth, flat surfaces have **cleavage**. Mica is a mineral that has perfect cleavage.

Minerals that have curved, rough, or jagged surfaces when they break apart have **fracture**. Quartz is a good example of a mineral with fracture.

Mica

Figure 10-11

Ⓐ Cleavage is the result of the orderly arrangement of the atoms that form the crystals of that mineral. Mica is a mineral that has perfect cleavage.

Ⓑ Barite is another mineral with perfect cleavage. If you examine a barite crystal, you can see the cleavage planes, along which the crystal would split.

Barite

6	7	8	9	10
Feldspar	Quartz	Topaz	Corundum	Diamond
Barely scratches glass.	Easily scratches both glass and steel.	Scratches quartz.	Scratches topaz.	Scratches all common materials.

Multicultural Perspectives

Mineral Art

Several Native American peoples of Latin America excelled at making artwork from precious stones and metals. The ancient Aztecs of Central Mexico panned for gold and collected nuggets which skilled artisans shaped into beautiful ornaments. Turquoise was used to fashion mosaic masks, knives, and other items. They also made items from obsidian, the black natural glass product of volcanic eruptions.

The Incas of the South American Andes Mountains also mined and worked precious metals and gems, including turquoise. Items made of copper appeared in Peruvian graves by about 2000 B.C.E. Incan artisans fashioned life-sized statues made entirely of gold or silver.

Content Background

The presence of trace elements can sometimes change the appearance of a mineral. The mineral corundum, which is often used as an abrasive, can also be found as rubies and sapphires. Rubies contain trace amounts of chromium, and sapphires contain trace amounts of cobalt or titanium. Corundum has a dull luster, while its other forms have a brilliant luster.

Some gemstones, like diamond, are not as beautiful in their native forms as they are when they are cut. Cutting the gemstone produces flat faces from which light can reflect. Quartz, in its natural crystalline form, is more lustrous than diamond. It has been said that inexperienced diamond hunters have thrown away diamonds and kept quartz!

Activity Have students estimate relative specific gravities of minerals by using the heft technique. The heft technique determines whether a mineral is heavier or lighter than another mineral or object of the same size. Give students two different minerals of approximately the same size. Choose one with a high specific gravity, such as galena, copper, or hematite, and one with a low specific gravity, such as calcite, dolomite, graphite or talc. Tell students to hold a mineral with a high specific gravity in one hand and a mineral with a low specific gravity in the other. Be sure students know the names of the minerals. Have students compare the two and write the name of the mineral with the greatest heft. Give students more minerals with specific gravities ranging from one to five and have them heft them. Tell students to rank the minerals from highest to lowest heft and write a list. Have them compare their list with the actual specific gravities as given in Appendices K and L. [L2]

10-1 Mineral Identification

Preparation

Purpose To design and carry out an investigation to identify unknown mineral samples.

Process Skills communicating, classifying, making and using tables, comparing and contrasting, forming a hypothesis, interpreting data, using numbers, observing and inferring, separating and controlling variables, designing an experiment

Time Required 60–80 minutes

Materials See student activity. Obtain several sets of mineral samples. Label the minerals and use a key so you can distinguish them. For a 5% HCl solution, add 50 g concentrated HCl to 950 mL of distilled water.

Possible Hypotheses Students may hypothesize that either hardness, streak, or luster would be the most useful in identifying minerals. A few may state that a combination of properties works best to identify unknown minerals. Studying the minerals' general appearance will direct you to other properties.

Plan the Experiment

Process Reinforcement Ask students to name the mineral properties that can be used for identification. List their answers on the board. Then ask them how each property can be observed or tested.

Possible Procedures Begin by scratching each mineral using the Mohs scale of hardness, or obtain a streak color for each mineral, or separate the minerals according to luster. Alternately, first lay out the minerals, study their appearance and heft (specific gravity), and decide which property would be most helpful in identifying each mineral.

Mineral Identification

Some minerals are easy to identify because they have one distinctive physical property. Other minerals require the testing of several properties before it's possible to identify them. Suppose it was your job to identify mineral samples. Where would you start?

Preparation

Problem
How can minerals be identified from tests?

Form a Hypothesis
As a group, recall what you have already learned about the physical properties of minerals. Then form a hypothesis about which property you think will be most useful in identifying minerals.

Objective
• Design an experiment that uses specific properties to help in identifying minerals.

Materials
mineral samples
steel file
streak plate
hand lens
goggles
apron
Mohs scale of hardness
(pp. 310-311)
5% hydrochloric acid (HCl)
with dropper
Appendices K and L

Safety

Review the safe use of acids. Wear goggles and an apron. **CAUTION:** *HCl may cause burns. If spillage occurs, rinse with water.*

Program Resources

Activity Masters, pp. 43–44, Design Your Own Investigation 10-1
Take Home Activities, p. 17, Path of Least Resistance [L1]
Science Discovery Activities, 10-2

DESIGN YOUR OWN
INVESTIGATION

Plan the Experiment

1 As a group, agree upon a way to test your hypothesis.

2 Examine the materials you will use to identify your mineral samples. Determine how each item will be used. Write down what you will do at each step of your test.

3 Make a list of any special properties you expect to observe or test.

4 How will you record the characteristics of each mineral sample as you perform various tests on it?

3 Make sure your teacher approves your experiment before you proceed.

4 Carry out your experiment. Record your observations.

Check the Plan

1 As a group, decide whether you will test any of the properties more than once for any of the minerals.

2 How will you determine whether certain properties indicate a certain mineral?

Analyze and Conclude

1. **Observe and Infer** What mineral samples were you able to identify? What samples were you unable to identify? Explain why. For instance, were there tests you needed to perform but were unable to?

2. **Compare and Contrast** Which property was most useful in identify-ing your samples? Which property was least useful? Did this support your hypothesis?

3. **Draw a Conclusion** Discuss reasons why one property is useful for identification purposes and others are not.

Going Further

If all you had were a piece of paper, a steel knife, and a glass bottle, could you distinguish between calcite and quartz? Explain. Be sure to mention what other test could help you identify calcite.

Meeting Individual Needs

Physically Challenged Divide the class into groups of two. Assign a mineral to each student. Have them write the properties of their minerals on file cards, one property per card. Have students write the name of the mineral on the back of the last card. Each student should trade cards with his or her partner who will then go through the cards and try to infer the name of the mineral from its properties.

COOP LEARN

Teaching Strategies

Troubleshooting Tell students that the side of the file is to be used in the hardness test.

Science Journal Students should record their data table, observations, and graph or chart of their results in their journal.

Expected Outcome

Students will discover that whichever property they choose, it will probably not be enough to identify all the mineral samples. Usually, they will discover that using a combination of properties is best when identifying unknown minerals.

Analyze and Conclude

1. Answers will depend on the mineral samples used and how well students succeeded in identifying them.

2. Although answers will vary, hardness and streak will generally be the properties commonly cited as most useful in identifying minerals. Color is usually the least helpful property.

3. Discussion among groups will vary, but hardness and streak will likely be cited as most useful because these properties tend to be the same for all samples of the same mineral. Color, by contrast, can be misleading because it can vary so much between and among mineral samples.

✔ Assessment

Portfolio Have students select two or three pairs of minerals that often look similar. Ask them to write a letter to a friend, describing how to distinguish samples of each mineral from the others. Use the Performance Task Assessment List for Data Table in **PASC**, p. 67. **P**

Going Further ⅢⅢ➡

Students should realize that calcite and quartz can be easily distinguished based on hardness. Quartz will scratch all of the items listed; calcite will scratch none of them. Adding HCl to the samples would cause calcite to fizz, but not quartz.

3 ASSESS

Check for Understanding

To answer the Apply question, have students write a paragraph describing the tests they would use to identify an unknown mineral.

Reteach

Activity Review each property used to identify minerals. Obtain a sample of magnetite. Have students observe that magnetite has a metallic luster, a hardness greater than glass, a black color, a black streak, and is magnetic. `L1` `LEP`

4 CLOSE

Have the students list the minerals they have used today and how they've used them. Create a master list on the chalkboard. `L1`

Other Properties of Minerals

Some minerals have unique traits that set them apart from other minerals. They have unique and interesting properties. In the following Explore activity, you'll be able to discover one of them.

Explore! ACTIVITY

How do clear minerals compare?

What To Do

1. Your teacher will give you samples of some clear minerals. Place each sample over the print on this page.
2. Describe the appearance of print through each mineral.
3. Which mineral can be identified by the way it changes light passing through it?

Figure 10-13

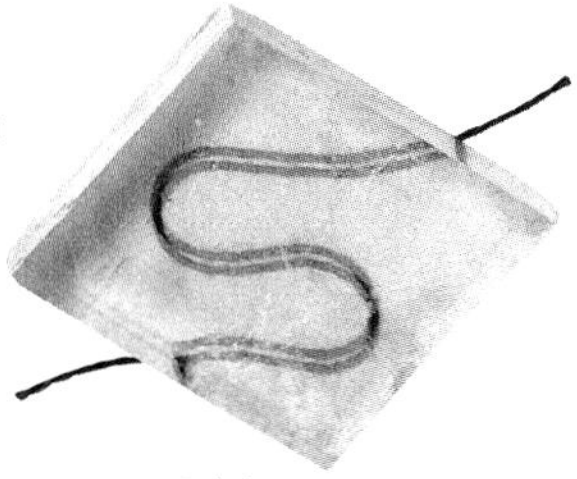

Ⓐ Light passing through calcite is split into two rays. Each ray produces its own image. When you look at something through calcite you see both images. This unusual property of calcite is called double refraction.

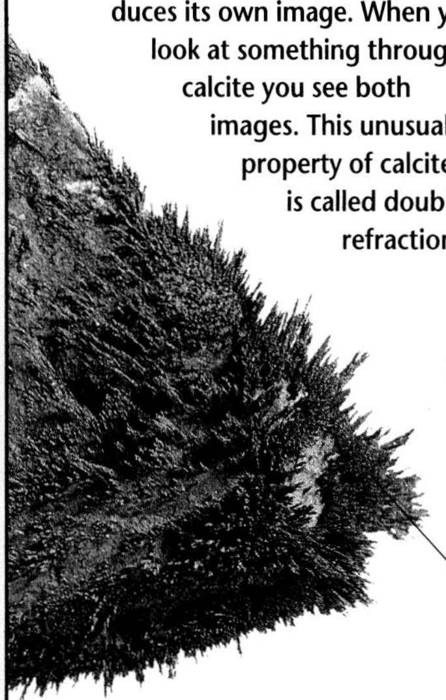

Calcite

Ⓑ The mineral magnetite is a natural magnet and has the ability to attract small iron and steel objects.

Iron filings

You discovered in the Explore activity that calcite has a unique property. Magnetite is another mineral with its own unique trait. What property of magnetite can help you identify it?

■ **Looking Ahead**

Since you first started thinking about minerals you've learned that four features set minerals apart from all other substances on Earth. You've probably been wondering where minerals come from and how we collect them. You'll find the answers to your questions in the next section.

check your UNDERSTANDING

1. Name four characteristics that can help distinguish one mineral from another.
2. **Apply** Pretend you're a prospector looking for silver. You find a nugget that you think may be silver. What tests can you use to determine whether you've "struck it rich"?

10-3 Mineral Formation

How Minerals Form

As you know, a substance can be classified as a mineral only if it occurs naturally. Yet the photos on this page show human-made "minerals." People were able to produce these synthetic "minerals" by studying the way true minerals formed.

There are two main ways that minerals form. One is from the cooling of magma. As the magma cools, elements in the magma may form minerals. Minerals can also form from elements dissolved in a liquid. When the liquid evaporates, the elements stay behind and may form minerals. When a solution becomes saturated, dissolved material may also precipitate, or start to crystallize, out of solution.

In the next activity, you'll watch crystals form. You will be able to see a process of mineral formation in action.

Section Objectives

- Examine mineral formation from solution.
- Describe mineral formation by the cooling of magma.

Figure 10-15

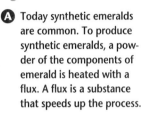

Ⓐ Today synthetic emeralds are common. To produce synthetic emeralds, a powder of the components of emerald is heated with a flux. A flux is a substance that speeds up the process.

Ⓑ The flux melts and the powdered components dissolve. As the mixture cools, crystals form and begin to grow. It takes several months for emerald crystals to grow. Is a synthetic emerald a mineral?

Figure 10-14

Ⓐ In 1988, U.S. astronaut mission specialists George D. Nelson (left) and David C. Hilmers conducted an experiment that involved growing crystals in the zero gravity of space.

Ⓑ The experiment, carried out aboard the space shuttle *Discovery,* was proposed by then high school student Richard Cavoli of Marlboro, N.Y. as part of the Shuttle Student Involvement project.

Program Resources

Study Guide, p. 36

Laboratory Manual, pp. 51–52, Removal of Waste Rock L1 , pp. 53–56, Crystal Formation L1

Critical Thinking/Problem Solving, p. 18, Synthetic Crystals L2

Section Focus Transparency 30

Planning the Lesson

Refer to the Chapter Organizer on pages 300A-D.

Concepts Developed

In Sections 1 and 2, students discovered what a mineral is and what properties distinguish one mineral from another. In Section 3, students will compare and contrast these processes: mineral formation from a solution and from the cooling of magma.

1 MOTIVATE

Bellringer

Before presenting the lesson, display **Section Focus Transparency 30** on an overhead projector. Assign the accompanying **Focus Activity** worksheet. L1 LEP

Activity   Obtain several geodes, the larger the better. Give groups of students a geode and a hand lens. Have students observe the individual crystals in the hollow part of the geode and compare and contrast them in shape and size. Have students write their observations and infer whether the crystals are all made of the same mineral.
COOP LEARN L1

2 TEACH

Tying to Previous Knowledge

Have students recall the appearance of several of the minerals with which they have worked so far in this chapter. Have them describe the shapes of some of the mineral crystals. Tell students that the chemical makeup of minerals determines the shapes of their crystals and that the environment in which they grow determines their size.

Planning the Activity

Time needed 20 minutes on two different days, one week apart

Purpose To demonstrate two methods of growing crystals.

Process Skills observing and inferring, interpreting data, comparing and contrasting

Materials See student text activity. For a saturated salt solution, mix about 400 g of salt with 1000 mL of distilled water. Not all of the salt will dissolve. Pour off the clear, saturated solution. For a saturated sugar solution, mix about 2100 g of sugar with 1000 mL of distilled water. Not all of the sugar will dissolve. Pour off the clear, saturated solution.

Preparation Make saturated sugar and salt solutions in advance. Be sure the hot plates are in good working order.

Teaching the Activity

Process Reinforcement Ask students to review the two ways that mineral crystals can form.

Safety Remind students to observe the Caution notice in the student text.

Possible Hypotheses Students may hypothesize that the rate of evaporation affects the size of the crystals formed.

Troubleshooting The test tubes and pans need to be left undisturbed for at least a week in order to have large crystals grow. When solutions are cooled suddenly, only small crystals will grow.

Science Journal Students should record their observations in their journals. Encourage students to make drawings of their procedures and the crystals they grow. L1

INVESTIGATE!

Growing Crystals

So far, you've read about "minerals in solution." Still, you may be unsure of what that means. You know something about both minerals and solutions. This activity will help you put the two together and then take them apart again!

Problem
What are two ways that crystals can form from solutions?

Materials

salt solution	sugar solution
large test tube	toothpick
cotton thread	hand lens
1 shallow pan	thermal mitt
test-tube rack	cardboard
table salt	granulated sugar
hot plate	beaker

Safety Precautions

During this Investigate you will be handling materials that will be hot.

What To Do

1 Pour the sugar solution into a beaker. Use the hot plate to gently heat the solution.

2 Place the test tube in the test-tube rack. Using a thermal mitt to protect your hand, pour some of the hot sugar solution into the test tube. **CAUTION:** *The liquid is hot. Do not touch the test tube without protecting your hands.*

3 Tie the thread to the middle of the toothpick. Place the thread in the test tube (see photo **A**). Be sure that it does not touch the sides or bottom of the tube. Let the toothpick rest across the top of the test tube.

A

B

C

4 Cover the test tube with a piece of cardboard (see photo **B**). Place the rack containing the test tube in a location where it will not be disturbed.

5 Pour a thin layer of the salt solution into the second shallow pan (see photo **C**).

6 Place the pan in a warm area of the room.

7 Leave both the covered test tube and the shallow pan undisturbed for at least one week.

8 Examine sample grains of table salt and sugar with the hand lens. *Observe* any similarities or differences.

9 At the end of one week, *observe* each solution and see if crystals have formed. Use a hand lens to examine any crystals.

Analyzing

1. *Compare and contrast* the crystals that formed from the salt and sugar solutions. Make a sketch of each type of crystal.

2. *Infer* what happened to the salt water in the shallow pan.

3. Did this same process occur in the test tube? Explain.

Concluding and Applying

4. What caused the formation of crystals in the test tube and in the shallow pan?

5. Are salt and sugar both minerals? Explain your answer.

6. Going Further *Hypothesize* the results of your experiments if you had switched the salt solution with the sugar solution. Explain your hypotheses.

Expected Outcomes

Students will be able to describe the growth process of salt and sugar crystals from solution.

Answers to Analyzing/ Concluding and Applying

1. Salt crystals are cubic and sugar crystals are orthorhombic.

2. The water evaporated. The salt stayed behind as a precipitate.

3. No. The liquid in the test tube cooled but didn't evaporate. The sugar precipitated from a saturated solution.

4. As the sugar solution cooled, the water was not able to hold all the sugar, so crystals formed. As water evaporated from the salt solution, the water became saturated and salt crystals formed.

5. No, the sugar is an organic compound and not a mineral. Salt is a mineral.

6. Going Further The sugar solution would have formed a thin glasslike sheet in the pan and the salt in the test tube would have formed larger crystals. The more slowly a solution evaporates, the larger the crystals will be.

✔ Assessment

Performance Have students form small groups and discuss their observations. Then have each group decide how they might produce and sell rock candy (sugar crystals). Students could then write their own newspaper articles about their new business. Use the Performance Task Assessment List for Newspaper Article in **PASC**, p. 69. **COOP LEARN**

ENRICHMENT

Activity Materials needed are a burner, a test tube, a test-tube holder, a watch glass or a glass petri dish, an apron, goggles, and sugar.

Have students conduct an experiment on making sugar crystal glass from a melt. Have the students place about 1 tablespoon of sugar in a test tube. Direct them to light the burner and gently heat the sugar until it melts. When the sugar is thoroughly melted, tell them to pour the liquid into the watch glass or petri dish. Have them describe the appearance of the cooled sugar. Have them speculate what sugar crystal glass might be used for. [L2]

■ **Minerals Form from Water Solution**

You may remember from what you've learned about solutions that a given volume of water can dissolve only a certain amount of solid before the water becomes saturated. When the water is saturated, the next teaspoonful of solid will not dissolve.

If the water is hot, you can dissolve even more solid. As you saw in the last activity, as the solution cools, crystals form from the solution onto the string.

Salt crystals formed from solution because as the water evaporated, the solution left couldn't hold the same amount of salt, and the salt crystals formed.

You read in the first section that many minerals are composed of more than one element. Before they become minerals, however, these elements are in solution. In some cases, this solution forms when elements are dissolved in water. When the water evaporates, the elements may combine in a mineral's characteristic crystal structure. If the mineral forms in an open space, crystals will form that show their crystal shape on the outside. If there is no open space in which the mineral can form, the crystals will overlap. When this happens, the elements have still combined in an orderly pattern, but we just cannot see it on the outside of the crystal.

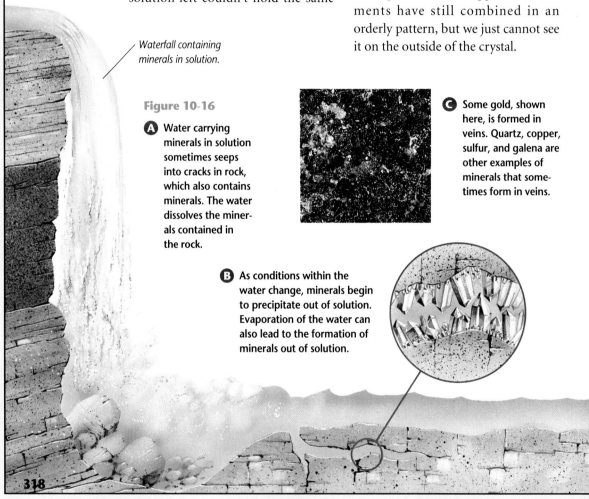

Waterfall containing minerals in solution.

Figure 10-16

Ⓐ Water carrying minerals in solution sometimes seeps into cracks in rock, which also contains minerals. The water dissolves the minerals contained in the rock.

Ⓒ Some gold, shown here, is formed in veins. Quartz, copper, sulfur, and galena are other examples of minerals that sometimes form in veins.

Ⓑ As conditions within the water change, minerals begin to precipitate out of solution. Evaporation of the water can also lead to the formation of minerals out of solution.

318

Figure 10-17

A Both rhyolite and granite are formed from the same type of magma, but rhyolite, which forms from rapidly cooling lava, is finer grained.

Rhyolite

B If magma cools slowly, as it often does in Earth's interior, large mineral crystals form. Pegmatites are examples of materials that form with large crystals.

Pegmatite

■ Minerals Form from Cooling Magma

The elements and compounds that make up minerals are also in solution deep below Earth's surface. The solution is magma—the hot, molten material found beneath Earth's crust. When magma is forced upward into cooler layers of the planet's interior, it cools. As the magma cools, the elements in magma may combine chemically and form minerals. Again, if there is enough open space, the minerals form large, visible crystal shapes.

The cooling process can be slow or quick. This cooling rate determines crystal size. If magma cools slowly, as it often does in Earth's warm to hot interior, large crystals form. Minerals such as mica, feldspar, and quartz are examples of minerals that often form large crystals. If the magma cools quickly, minerals form as small crystals. Lava, which is magma that reaches Earth's surface, is exposed to air and sometimes water. Lava cools quickly in air, so small mineral crystals form. If lava runs into water, it may cool so quickly that no crystals form at all.

Sometimes flowing magma or lava will fill cracks in surrounding rock.

10-3 Mineral Formation **319**

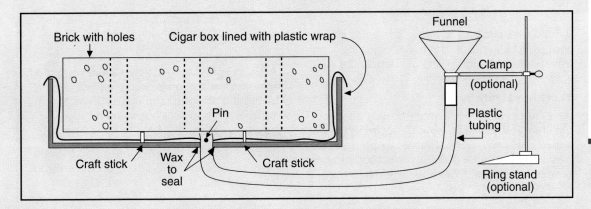

Brick with holes

Cigar box lined with plastic wrap

Funnel

Clamp (optional)

Pin

Plastic tubing

Craft stick

Wax to seal

Craft stick

Ring stand (optional)

Figure 10-18

A As hot magma flows, it sometimes fills cracks in surrounding rock. The rock keeps the heat from escaping, and the molten material takes a very long time to cool. The slow cooling time allows deposits of large mineral crystals to form in the cracks. These mineral-filled cracks are called veins. Veins of some gemstones such as ruby, topaz, sapphire, and beryl form this way.

B Veins of beryl crystals are found most often in granite rocks. Pure beryl is colorless, but impurities in the crystals add color. The most common colors of beryl are bluish-green and yellow-green. Emeralds, seen in the photo at the right, are dark green beryl crystals.

1.36 carat cut emerald (not to scale)

Emerald crystal

Retrieving Minerals

You know a lot about minerals and how to identify them. But before you can classify a mineral sample, you have to find one! Where would you look for minerals in their natural state?

Very few of Earth's minerals are found just lying on the ground, "waiting" to be picked up and used. Instead, they occur deep underground, trapped in rocks or soil. To get to these mineral deposits, the overlying rocks and soil must be removed.

There are many methods of removing the unwanted materials and collecting the minerals we desire. Strip mining, open pit mining, and room and pillar mining are just a few of the methods that we use.

Let's take a closer look at some mines. At an open pit mine, the mine is made deeper as the minerals and the rocks

320

the side resulting in a large, open pit. Room and pillar mining is done underground and leaves only walls and pillars to support the rock above.

The rock traps the heat from the molten material, causing it to cool very slowly and form large mineral crystals. Veins of some gems and rare minerals form in this way. Some examples are rubies and sapphires which are types of the mineral corundum; and emeralds and aquamarines which are types of the mineral beryl.

Minerals spend much of their existence in rocks. Humans go to great length and expense to extract them from the rocks. In fact, rocks are made of minerals. Rocks are the next group of materials you will investigate as you go on to Chapter 11.

check your UNDERSTANDING

1. What is the connection between evaporation and mineral formation?
2. Describe two ways in which a vein of a mineral may form in a rock.
3. **Apply** A volcano erupts in the middle of the ocean. Lava slips down the sides of its cone and eventually enters the ocean water. Will the mineral crystals formed from the lava be large or small? Explain your answer.

that contain them are hauled away. Usually, terraces are left on the slope of the deep pit.

You may think of a mine as a sort of underground "cave." People dig rooms in the sides of mountains or under flat ground. They then remove the minerals from the walls of the mine.

Not long ago, people thought that Earth's mineral supplies would last forever. Today, we know mineral supplies are limited.

Our demand for minerals increases as we find more and more uses for them. The answer to our mineral needs probably isn't simply to open more mines.

Mines can destroy land, water, and the plants and animals that live nearby.

Instead, we must find ways to make mining less destructive. We also need to use minerals more efficiently.

*inter*NET CONNECTION

Find a mining information site on the World Wide Web and learn what the job requirements are for a mining environmental engineer. What about the job appeals to you? Why?

check your UNDERSTANDING

1. Elements and compounds in solution may precipitate as minerals when water evaporates.
2. Magma flows into a crack in a rock. As the magma cools, the elements dissolved in it combine chemically to form a mineral vein. Veins can also form when a solution evaporates and the minerals that were dissolved in the solution are left behind.
3. Small. They are cooled too rapidly to grow large.

*inter*NET CONNECTION

Find a mining information site on the World Wide Web and learn what the job requirements are for a mining environmental engineer. What about the job appeals to you? Why?

3 ASSESS

Check for Understanding

To help students with the Apply question, have them write a paragraph describing the relationship between magma's cooling rate and the size of resultant mineral crystals.

Reteach

Activity To reinforce classification skills give students mineral samples, some containing large crystals and some containing small crystals. Have students sort the crystals according to relative cooling or evaporation rates. Students should understand that the longer it takes for a crystal to form in either situation, the larger the crystal will be. L1

Extension

Research Have students research the formation of black Hawaiian beach sand. *This sand forms when lava cools rapidly and forms very fine-grained, dark-colored rock. Ocean waves weather this dark rock into dark-colored sand. Interested students could also research other colors of beach sand and their origins.* L2

4 CLOSE

Students can draw diagrams or pictures to explain the processes by which minerals may crystallize. P

Science and Society

Purpose
Science and Society reinforces Section 10-2 on characteristics of minerals by describing how the characteristics of asbestos make it both a very useful and potentially harmful mineral.

Content Background
Asbestos can be found as the mineral chrysotile, a member of the serpentine family, and crocidolite, a member of the amphibole family. Chrysotile fibers are curly and do not penetrate lung tissue. The mineral is also water soluble, making it easy for the body to assimilate as ionic substances. On the other hand, crocidolite fibers are long and straight and can invade lung tissue. It is relatively insoluble, which means that it persists in human lung tissue for a long time. Only about five percent of the asbestos used in building is crocidolite. Most of the asbestos used in building is the non-dangerous chrysotile. Chrysotile mines in Quebec have been in existence since before 1900, and over forty million tons of chrysotile have been removed from them. Miners' wives were exposed to massive amounts of chrysotile dust over long periods of time, but four statistical studies have shown that there has been little increase in disease rates among them.

Teaching Strategy Conduct a class debate on asbestos removal. Should all asbestos be removed? What about the chrysotile variety?

Science and Society — Asbestos Debate

Asbestos has been the subject of debate for nearly 20 years. What is asbestos, and why is it so controversial?

Asbestos is a mineral sometimes found in metamorphic rock. It is a lightweight, fibrous mineral that is white to green in color. The use of asbestos dates back to ancient times, when Egyptians and Romans realized that asbestos was resistant to fire and heat. They wove asbestos into clothing and pressed it into paper.

Asbestos

More recently, asbestos has been used to produce insulation for buildings and water pipes. It is well suited for those uses because asbestos doesn't conduct heat or electricity very well. Since it doesn't burn easily and is resistant to acids, asbestos is used in products such as automobile brakes and fireproof clothing.

As far back as the early 1900s, scientists were concerned with the possible harmful effects of inhaling asbestos fibers. Airborne asbestos fibers are so small they can't be seen without a microscope. Even when asbestos is mixed with other materials, such as cement, the products can break down over time and release the fibers into the air.

Scientists first realized the dangers of asbestos by studying workers in plants where asbestos products were manufactured. Breathing some types of asbestos fibers can cause asbestosis, a disease that stiffens the lungs and makes breathing difficult. It can also cause lung cancer or cancer of the stomach lining. Sometimes these diseases don't show up until 20 or 30 years after a person has been exposed to asbestos.

The Environmental Protection Agency (EPA) has researched the effects of asbestos for the United States government since the early 1970s. Beginning with a ban on the use of asbestos in public schools in 1973, the EPA

CAREER connection

The **economic geologist** uses a broad knowledge of all areas of geology and applies it to the exploration and development of mineral deposits. Working with other specialists, such as engineers and financial analysts, the economic geologist determines whether and how a mineral or fuel can be developed.

CAREER connection

Have volunteers contact the local utility company. Have them find out about the education and experience requirements for new employees and the varieties of jobs available for various degrees or backgrounds. Have students share findings with the class.

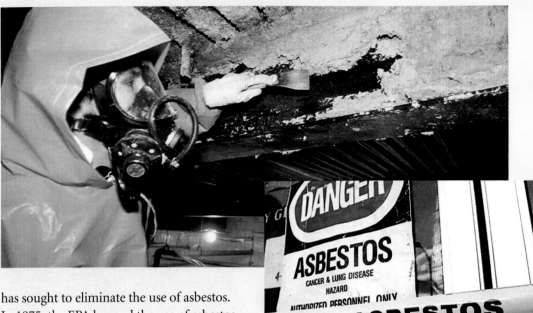

has sought to eliminate the use of asbestos. In 1975, the EPA banned the use of asbestos in new public buildings. However, this did not solve the problem of asbestos already in the buildings.

In 1986, Congress passed the Asbestos Hazardous Emergency Response Act, which directs local school districts to control or remove asbestos from public school buildings. The schools could cover the asbestos with sealants to prevent the fibers from becoming airborne, or could remove the asbestos altogether.

The debate over the government's anti-asbestos activities focuses on two issues. First, some studies show that the removal process may release more fibers into the air than if the asbestos were left in place.

Second, the asbestos industry and product manufacturers are concerned with the costs of eliminating asbestos. They say that substitute materials in the manufacture of products would cost far more in dollars than society would gain in health benefits.

The asbestos industry won a major court victory in the fall of 1991, when a federal appeals court overturned the EPA ban on the use and manufacture of asbestos. The court said that the EPA failed to prove that the dangers of asbestos outweighed the costs of the ban and the possible health effects of removal. The EPA plans to do further research on asbestos, which may set the stage for yet another court battle in the years to come.

Science Journal

Research has shown that although asbestos is dangerous to people who mine or manufacture products from it, the risk of dying from exposure is very small. *In your Science Journal* discuss whether you agree with the ruling that the EPA went too far in trying to ban asbestos.

Science Journal

Accept all reasonable answers that the students can support.

Going Further ⚞⚞⚞➧

Have students write a newspaper article about any asbestos removal in your school system or community. For any one particular project, have students find out where the asbestos is located, what is involved in its removal, and the costs involved. Students may also write to:

Occupational Safety and Health Administration
U.S. Department of Labor
200 Constitution Ave.
Washington, DC 20210
(202) 219-8148

Environmental Protection Agency
401 M Street SW
Washington, DC 20460
(202) 260-2090

Use the Performance Task Assessment List for Newspaper Article in **PASC**, p. 69.

HISTORY CONNECTION

Purpose
Section 10-1 discusses some minerals that are highly valued. The History Connection discusses the historical and economic importance of some of these minerals.

Content Background
As an uncombined element, tin is not too common in Earth's crust. The tin oxide cassiterite, however, is relatively common. This oxide is white, but usually contains impurities such as iron that make it dark in color. Sulfur and arsenic are also common impurities. To extract the metal from the ore, the ore is enriched and roasted at a temperature above 1100°C to remove certain impurities. It is then reduced with carbon in a kiln to obtain elemental tin.

Teaching Strategies
Have students examine samples of copper, tin, and bronze. Have them compare and contrast physical properties of the metals, such as ease of bending, color, and hardness. Emphasize that an alloy frequently has more desirable characteristics than any of the elements of which it is composed.

Answers to
You Try It!
Reporting of an event reflects the reporter's point of view. To obtain an objective picture, varied sources should be used.

Going Further ⫸
All bronze is not alike. Have interested students investigate how bronze samples differ, depending on their compositions. Students can create posters that include composition, any specific names, and uses. Examples are bell metal, which is 75% copper and 25% tin, and gunmetal, which is 90% copper and 10%

HISTORY CONNECTION

Rewriting Prehistory

From 5000 B.C.E. to 2500 B.C.E., human culture in Europe, Africa, and Asia depended on bronze, a combination of tin and copper, to make everything from tools and weapons to hairpins.

Tablets written by the Assyrians 4000 years ago claim that they were very shrewd traders with their trade partners to the north in Anatolia—present day Turkey. The Anatolians never got to tell their side of the story because they left no written record. However, a Turkish-American archaeologist, Kutlu Aslihan Yener, is writing another chapter in this story based on newly found archaeological evidence.

Her original research involved identifying the sources of lead used in bronze objects. While searching for trade patterns for some of these sources, she made a far greater find in the Taurus Mountains

of Turkey. Dr. Yener first found a cluster of 850 silver mines in a six-square-mile area. In 1989, further research in this area yielded what she felt was a Bronze Age tin mine.

When the tin mine was found, little tin was left. Dr. Yener knew she would have to find tin oxide to prove that the Taurus mine was actually originally mined for tin. Dr. Yener and her colleagues painstakingly searched for proof for six years before finding Turkish tin. The tin turned out to be burgundy colored rather than the usual black—one reason it may have been overlooked. As a final piece of evidence, in 1990 Dr. Yener discovered a vast underground city and tin processing center.

Dr. Yener's work has shown that the Bronze Age must have had some complex economic relationships. With competing sources of tin, there were probably trade wars even then.

You Try It!
Many scholars relied upon the Assyrian tablets to give them an accurate picture of Bronze Age trade. Can it be misleading to learn about an event from only one source?

tin. Some bronzes may have trace elements present also, such as phosphor bronze, which has a small amount of phosphorus in it and is harder than regular bronze. Use the Performance Task Assessment List for Poster in **PASC**, p. 73.

Science Journal

Review the statements below about the big ideas presented in this chapter, and answer the questions. Then, re-read your answers to the Did You Ever Wonder questions at the beginning of the chapter. *In your Science Journal,* write a paragraph about how your understanding of the big ideas in the chapter has changed.

1 All minerals occur naturally, are inorganic solids, have a unique chemical composition and have a crystalline structure. *Give examples of items that are minerals.*

2 All minerals have characteristics that we can use to help identify them. *What are some of these characteristics? Give an example of each.*

3 Minerals form in several ways. *Describe how minerals form from magma and compare a mineral that cooled quickly to one that formed more slowly.*

Science Journal

Did you ever wonder...

• Gold is a metallic mineral formed from natural processes in Earth. It is valued for its beauty and rarity. (p. 305)

• A rock is a combination of minerals. Gems are minerals that are prized for their beauty, durability, and rarity. (p. 306)

• Both are gold in color but pyrite is harder than gold and leaves a greenish-black streak. Gold is much rarer than pyrite. (p. 307)

Project

Have students make a chart on a large poster board and display it on the bulletin board. Students should design a chart to keep track of the minerals they learn about each day. On the chart list each mineral's name, hardness, luster, color, streak, whether it shows cleavage or fracture, and other special properties. Use the Performance Task Assessment List for Poster in **PASC,** p. 73.
L1

Have students look at the three pictures on this page. Direct them to read the statements to review the main ideas of this chapter.

Teaching Strategies

Arrange the students in three groups. Have one group write a list of the terms which define a mineral, have another group list the physical properties of minerals, and have the third group list the ways a mineral can form. Direct the groups to write two to three examples of each item on their list. Then have each group create a poster to illustrate their list and examples. When the posters are completed, display them in the classroom.
COOP LEARN

Answers to Questions

1. examples include: table salt, diamonds, quartz used in watches and clocks, pencil lead

2. Answers will vary, but may include: hardness, calcite is 3; color, sulfur is yellow; streak, pyrite is dark green; luster, galena is metallic.

3. When magma cools, minerals are formed. Those that cool quickly have small crystals; those that cool slowly have large crystals.

GLENCOE TECHNOLOGY

 MindJogger Videoquiz

Chapter 10 Have students work in groups as they play the Videoquiz game to review key chapter concepts.

Using Key Science Terms

1. A mineral is any naturally occurring, inorganic solid with a definite structure and composition. A gem is a rare mineral.

2. Cleavage and fracture refer to how minerals break apart. Minerals that break along smooth, flat surfaces have cleavage. Minerals with curved, rough, or jagged surfaces when they break apart have fracture.

3. Hardness is a measure of a mineral's hardness. Streak is the powdered mineral's color.

Understanding Ideas

1. Diamond jewelry would be less expensive. Diamonds could be more easily used in computers, tools, and furniture.

2. No. Metals that are human-made, such as bronze or pewter, are not considered minerals because they don't occur naturally. Some minerals are metals, but many, such as feldspar and quartz, are not.

3. Hardness is measured by a scratch test and rated on the Mohs hardness scale; luster is evaluated by visual comparisons; color is evaluated by visual observations; streak is evaluated by rubbing the mineral on a streak plate and observing the color of the powder; cleavage or fracture is evaluated by observing how the mineral breaks apart.

4. Minerals are formed by the cooling of magma or from water solutions.

5. The sample with larger crystals formed more slowly than the sample with smaller crystals.

6. Minerals form from evaporation when the water evaporates and the solution can no longer contain all the solid. Minerals that form from precipitation do not require loss of water. The solids come out of a supersaturated solution and form solids.

7. its rarity and beauty

Using Key Science Terms

cleavage   hardness
fracture   mineral
gem   streak

Explain the difference between the terms in each of the following sets:
1. mineral, gem
2. cleavage, fracture
3. hardness, streak

Understanding Ideas

Answer the following questions in your Journal using complete sentences.

1. If an enormous diamond mine were discovered, resulting in diamonds becoming common rather than rare, how might diamond usage change?
2. Are all metals minerals? Why or why not? Are all minerals metals? Why or why not?
3. List four characteristics of minerals and how each characteristic is measured or evaluated.
4. What are two ways minerals are formed?
5. If two samples of the same mineral have crystals of different sizes, what does that tell you about the samples?
6. What is the difference between evaporation and precipitation in the formation of minerals from water solutions?
7. What characteristics make a mineral a gem?

Developing Skills

Use your understanding of the concepts developed in this chapter to answer each of the following questions.

1. Concept Mapping Complete the concept map of minerals.

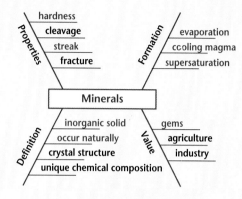

2. Observing and Inferring Repeat the crystal system identification activity on page 303 using the mineral calcite. Identify and sketch the crystal system of calcite.

3. Comparing and Contrasting Add the mineral fluorite to your data table from the Find Out activity on page 310. Perform the scratch test and make a new list sequencing the minerals. Refer to the Mohs scale of hardness on pages 310 and 311 to see if your results agree.

4. Predicting Predict what would happen if you used a streak plate to test a diamond. What would you expect to see on the streak plate? Why did you make the prediction you did?

Developing Skills

1. See student page for answers.
2. Calcite is hexagonal.
3. talc, calcite, fluorite, quartz
4. Diamond would not leave a streak on the plate because it is harder than the plate.

Program Resources

Review and Assessment, pp. 59–64 L1
Performance Assessment, Ch. 10 L2
PASC
Alternate Assessment in the Science Classroom
Computer Test Bank L1

Critical Thinking

In your Journal, *answer each of the following questions.*

1. Suppose you decide to enter the mineral mining business. You discover that there are five relatively abundant types of minerals near where you live. In the table, these five are ranked according to their abundance—5 represents the most abundant and 1 represents the least abundant mineral. The table also gives the ease with which they may be extracted—5 represents the most difficult and 1 represents the least difficult.

Mineral	Abundance Ranking	Ease of Extraction
Bauxite	5	2
Hematite	4	1
Halite	3	3
Quartz	2	5
Sphalerite	1	4

With two of your classmates, decide which mineral you would mine and explain why. Keep in mind the fact that the harder a mineral is to extract, the more it will cost to pay for the equipment and labor needed.

2. List at least five ways in which minerals are important to your daily life.

3. Hypothesize what might happen to the value of gold jewelry if someone found an inexpensive way to manufacture gold. What would happen if we found more sources of gold?

4. Hypothesize what might happen to the value of automobiles if someone found an inexpensive way to manufacture or mine the aluminum and iron used in automobiles.

Problem Solving

Read the following problem and discuss your answers in a brief paragraph.

While hiking in the mountains, Helen found a pink rock and a gold rock. She took them home and tested them. The pink one had no metallic luster and broke along a flat, smooth plane. The streak test was colorless. The other rock had a metallic luster and left a black streak. Which minerals did Helen decide her rocks contained? Why?

CONNECTING IDEAS

Discuss each of the following in a brief paragraph.
1. **Theme—Energy**
   Explain how mineral formation relates to forces inside Earth and thermal energy.
2. **Theme—Scale and Structure** Explain why observing physical properties is important in identifying minerals.

Critical Thinking

1. The most logical decision would be to mine hematite. It has a high abundance and is the easiest and cheapest to extract.

2. They give us such things as medicines, cleaning agents, salt, paints, materials for glass and metals, etc.

3. Its monetary value would decrease but people would still enjoy it for its beauty.

4. Since these metals are used in automobiles, cheaper sources might lower the cost of new cars.

Problem Solving

Based on color alone, students may decide that the pink rock could be dolomite, topaz, feldspar, or corundum. Because dolomite leaves a white streak and corundum fractures rather than shows cleavage, the sample is probably topaz or feldspar. The gold rock could be chalcopyrite or pyrite. Both are gold to yellow and have a blackish streak.

Connecting Ideas

1. Forces inside Earth produce thermal energy which causes rock material to become molten. When this molten material cools, mineral deposits can form.

2. Physical properties are different for different minerals. When the distinguishing differences are known, minerals can be compared, contrasted, and thus identified.

✔ Assessment

Portfolio Review the portfolio options that are provided throughout the chapter. Encourage students to select one product that demonstrates their best work for the chapter. Have students explain what they learned and why they chose this example for placement into their portfolios.

Additional portfolio options can be found in the following **Teacher Classroom Resources:**

Making Connections: Integrating Sciences, p. 23

Multicultural Connections, pp. 23-24

Making Connections: Across the Curriculum, p. 23

Concept Mapping, p. 18

Critical Thinking/Problem Solving, p. 18

Take Home Activities, p. 17

Laboratory Manual, pp. 49–56

Performance Assessment P

Chapter Organizer

SECTION	OBJECTIVES	ACTIVITIES & FEATURES
Chapter Opener		**Explore!**, p. 329
11-1 Igneous Rocks (3 sessions, 1.5 blocks)	1. **Distinguish** between a rock and a mineral. 2. **Explain** how igneous rock is formed. 3. **Identify and classify** igneous rocks. **National Content Standards: (5-8) UCP2, UCP4-5, A1, D1-2, F3**	**Explore!**, p. 330 **Find Out!**, p. 331 **Skillbuilder:** p. 335 **Design Your Own Investigation 11-1:** pp. 336-337 **A Closer Look,** pp. 334-335
11-2 Metamorphic Rocks (2 sessions, 1 block)	1. **Explain** how metamorphic rock is formed. 2. **Identify and classify** metamorphic rocks. **National Content Standards: (5-8) UCP2, UCP4-5, A1, B2, D1**	**Find Out!**, p. 339 **Explore!**, p. 341 **Life Science Connection,** pp. 350-351 **Science and Society,** pp. 354-355
11-3 Sedimentary Rocks (4 sessions, 2 blocks)	1. **Explain** how sedimentary rock is formed. 2. **Identify and classify** sedimentary rocks. 3. **Use** a diagram of the rock cycle to explain how rocks form and change. **National Content Standards: (5-8) UCP2, UCP4-5, A1, B1, C5, D1-2, F2, F4**	**Find Out!**, p. 344 **Find Out!**, p. 346 **Investigate 11-2:** pp. 348-349 **Leisure Connection,** p. 353 **SciFacts,** p. 356

ACTIVITY MATERIALS

EXPLORE!

p. 329* 8 or 9 rocks, magnifying glass
p. 330* several samples of pink granite, gray granite, quartz, feldspar, hornblende, and biotite, magnifying lens, nongranite rock
p. 341 4 metamorphic rocks, 4 non-metamorphic rocks

INVESTIGATE!

pp. 348-349 sedimentary rock samples, 5% hydrochloric acid (HCl), dropper, hand lens, paper towels, water, goggles, apron

DESIGN YOUR OWN INVESTIGATION

pp. 336-337* igneous rock samples: obsidian, basalt, granite, pumice, rhyolite, gabbro; hand lens

FIND OUT!

p. 331* salol (phenyl salicylate), microscope slide, thermal glove, hot plate, microscope or magnifying lens
p. 339* red, green, blue, and yellow crayons; pencil sharpener; aluminum foil; vise or 2 C-clamps; 2 boards
p. 344* 2 paper cups, sand, white glue, water, pin, bowl, ring stand
p. 346* 2 g alum, water, shallow pan, teaspoon, tablespoon

KEY TO TEACHING STRATEGIES

The following designations will help you decide which activities are appropriate for your students.

L1	Basic activities for all students
L2	Activities for average to above-average students
L3	Challenging activities for above-average students
LEP	Limited English Proficiency activities
COOP LEARN	Cooperative Learning activities for small group work
P	Student products that can be placed into a best-work portfolio
	Activities and resources recommended for block schedules

Need Materials? Call Science Kit (1-800-828-7777).

⏰ OUT OF TIME? We recommend that students do the activities with an asterisk.

Chapter 11 The Rock Cycle

TEACHER CLASSROOM RESOURCES

Student Masters	Transparencies
Study Guide, p. 37 **How It Works,** p. 14 **Making Connections: Integrating Sciences,** p. 25 **Making Connections: Across the Curriculum,** p. 25 **Activity Masters,** Design Your Own Investigation 11-1, pp. 47–48 **Concept Mapping,** p. 19 **Science Discovery Activities, 11-1**	**Section Focus Transparency 31**
Study Guide, p. 38 **Multicultural Connections,** p. 25 **Science Discovery Activities, 11-2** **Laboratory Manual,** pp. 57–58, Metamorphic Processes	**Section Focus Transparency 32**
Study Guide, p. 39 **Critical Thinking/Problem Solving,** pp. 5,19 **Take Home Activities,** p. 18 **Making Connections: Technology and Society,** p. 25 **Activity Masters,** Investigate 11-2, pp. 49–50 **Multicultural Connections,** p. 26 **Science Discovery Activities, 11–3** **Laboratory Manual,** pp. 59–60, Concretions	**Teaching Transparency 21,** The Rock Cycle **Teaching Transparency 22,** Underground Mining **Section Focus Transparency 33**

ASSESSMENT RESOURCES	TEACHING & TECHNOLOGY
Review and Assessment, pp. 65–70 **Performance Assessment,** Ch. 11 **PASC*** **MindJogger Videoquiz** **Alternate Assessment in the Science Classroom** **Computer Test Bank**	**Spanish Resources** **Cooperative Learning Resource Guide** **Lab and Safety Skills** **Science Interactions, Course 2, CD-ROM** **Computer Competency Activities**

*Performance Assessment in the Science Classroom

NATIONAL GEOGRAPHIC TEACHER'S CORNER

National Geographic Society Products Available From Glencoe

To order the following products for use with this chapter, contact your local Glencoe sales representative or call Glencoe at 1-800-334-7344:

• *NGS Picture Show Geology* (CD-ROM)

Additional National Geographic Society Products

To order the following products for use with this chapter, call the National Geographic Society at 1-800-368-2728:

• *Every Stone Has a Story* (Video)

GLENCOE TECHNOLOGY

The following multimedia resources are available from Glencoe.

Science and Technology Videodisc Series (STVS)
Earth and Space
  Fibers from Rocks

National Geographic Society Series
STV: Restless Earth
STV: Water
GTV: Planetary Manager

Glencoe Earth Science Interactive Videodisc
The Rock Cycle

Earth Science CD-ROM

Teacher Classroom Resources

These are key components of the classroom resources package.

TEACHING AIDS

Section Focus Transparencies

31 SECTION FOCUS TRANSPARENCY Section 11-1

FLOWING LAVA

The Hawaiian Islands are really volcanic mountains that rise far above the ocean. Lava from the volcanoes built up layer upon layer to form the islands.

1. What do you think happens to the lava when it reaches the ocean?
2. What kind of rocks might be found in Hawaii?
3. Why is the sand on this Hawaiian beach black?

L1

32 SECTION FOCUS TRANSPARENCY Section 11-2

MOUNTAIN VIEW

Deep within Earth, powerful forces are at work. These forces are so strong, they can create mountains.

1. What forces are acting on the layers of rock located under Earth's surface?
2. How do you think the materials under Earth's crust might change in response to these forces?
3. What happens if you step on a sponge? How is that like what is happening in this picture?

L1

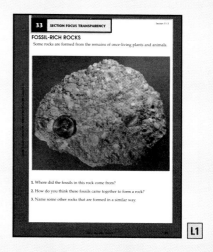

33 SECTION FOCUS TRANSPARENCY Section 11-3

FOSSIL-RICH ROCKS

Some rocks are formed from the remains of once-living plants and animals.

1. Where did the fossils in this rock come from?
2. How do you think these fossils came together to form a rock?
3. Name some other rocks that are formed in a similar way.

L1

Teaching Transparencies

21. THE ROCK CYCLE

L1

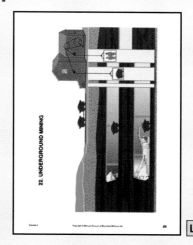

22. UNDERGROUND MINING

L1

HANDS-ON LEARNING

Science Discovery Activity*

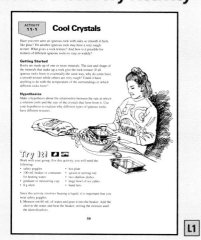

ACTIVITY 11-1 Cool Crystals

Have you ever seen an igneous rock with sides so smooth it feels like glass? Yet another igneous rock may have a very rough texture. What gives a rock texture? And how is it possible for textures of different igneous rocks to vary so widely?

Getting Started

Rocks are made up of one or more minerals. The size and shape of the minerals that make up a rock give the rock texture. If all igneous rocks form in essentially the same way, why do some have a smooth texture while others are very rough? Could it have anything to do with the temperature of the surroundings in which different rocks form?

Hypothesize

Make a hypothesis about the relationship between the rate at which a solution cools and the size of the crystals that form in it. Use your hypothesis to explain why different types of igneous rocks have different textures.

Try It!

Work with your group. For this activity, you will need the following:

- safety goggles
- 100-mL beaker or container for heating water
- graduate or measuring cup
- 8 g alum
- hot plate
- spoon or stirring rod
- two shallow dishes
- large bowl of ice cubes
- hand lens

Since this activity involves heating a liquid, it is important that you wear safety goggles.

1. Measure out 60 mL of water and pour it into the beaker. Add the alum to the water and heat the beaker, stirring the mixture until the alum dissolves.

59

L1

Laboratory Manual*

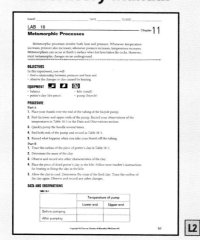

NAME _____ DATE _____ CLASS _____

LAB 18 Chapter 11

Metamorphic Processes

Metamorphic processes involve both heat and pressure. Whenever temperature increases, pressure also increases; whenever pressure increases, temperature increases. Metamorphism can occur at Earth's surface when hot lava bakes the rocks. However, most metamorphic changes occur underground.

OBJECTIVES

In this experiment, you will
- find a relationship between pressure and heat and
- observe the changes in clay caused by heating.

EQUIPMENT
- balance
- potter's clay (dry piece)
- kiln (small)
- pump (bicycle)

PROCEDURE

Part A

1. Place your thumb over the end of the tubing of the bicycle pump.
2. Feel the lower and upper ends of the pump. Record your observations of the temperatures in Table 18-1 in the Data and Observations section.
3. Quickly pump the handle several times.
4. Feel both ends of the pump and record in Table 18-1.
5. Record what happens when you take your thumb off the tubing.

Part B

1. Trace the outline of the piece of potter's clay in Table 18-2.
2. Determine the mass of the clay.
3. Observe and record any other characteristics of the clay.
4. Place the piece of dried potter's clay in the kiln. Follow your teacher's instructions for heating or firing the clay in the kiln.
5. Allow the clay to cool. Determine the mass of the fired clay. Trace the outline of the clay again. Observe and record any changes.

DATA AND OBSERVATIONS

TABLE 18-1

	Temperature of pump	
	Lower end	Upper end
Before pumping		
After pumping		

57

L2

Take Home Activity

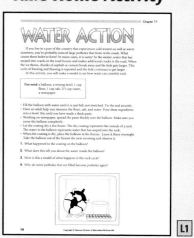

WATER ACTION Chapter 11

If you live in a part of the country that experiences cold winters as well as warm summers, you've probably noticed large potholes that form in the roads. What causes these holes to form? In many cases, it is water! In the winter, water that has seeped into cracks in the road freezes and makes additional cracks in the road. When the ice thaws, chunks of asphalt or cement break away and the hole gets larger. This cycle of freezing and thawing is repeated and the hole continues to get larger.

In this activity, you will make a model to see how water can crumble rock.

You need: a balloon, a mixing bowl, 1 cup flour, 1 cup salt, 2/3 cup water, a newspaper

- Fill the balloon with water until it is just full, not stretched. Tie the end securely.
- Have an adult help you measure the flour, salt, and water. Pour these ingredients into a bowl. Stir until you have made a thick paste.
- Working on newspaper, spread the paste thickly over the balloon. Make sure you cover the balloon completely.
- Let the coating dry a few hours. The dry coating represents the outside of a rock. The water in the balloon represents water that has seeped into the rock.
- When the coating is dry, place the balloon in the freezer. Leave it there overnight. Take the balloon out of the freezer the next morning and observe it.

1. What happened to the coating on the balloon?
2. What does this tell you about the water inside the balloon?
3. How is this a model of what happens in the rock cycle?
4. Why do some potholes that are filled become potholes again?

L1

*There may be more than one activity for this chapter.

Chapter 11 The Rock Cycle

REVIEW AND REINFORCEMENT

Study Guide*

Concept Mapping

Critical Thinking/ Problem Solving

ENRICHMENT AND APPLICATION

Integrating Sciences

Across the Curriculum

Technology and Society

ASSESSMENT

Multicultural Connection**

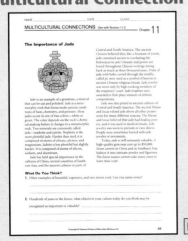

Performance Assessment

Review and Assessment

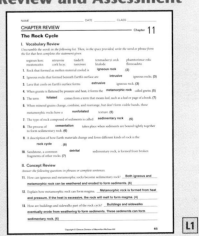

The Rock Cycle

THEME DEVELOPMENT

One theme that this chapter supports is stability and change. Although rocks may be stable for thousands or millions of years, they will usually change through Earth's forces into different types of rock. The theme of systems and interactions becomes evident as students learn about the rock cycle.

CHAPTER OVERVIEW

Students will study the way igneous, metamorphic, and sedimentary rocks are formed.

Tying to Previous Knowledge

Arrange students in small groups to brainstorm ways of identifying rocks. Students may recall from Chapter 10 that luster, hardness, and texture are used to identify minerals.

`COOP LEARN`

INTRODUCING THE CHAPTER

Have students look at the photograph on page 329 and describe the formations they see.

Uncovering Preconceptions

Students may think that new rock material is made. Explain that the amount of rock material on or within Earth is set and that new rock is formed out of old materials.

The ROCK Cycle

Did you ever wonder...

✓ **Where rocks come from?**

✓ **What's inside a rock?**

✓ **Why some rocks are smooth and rounded while others have jagged edges?**

✓ **Why there are so many different colors of rock?**

Science Journal

Before you begin to study about the rock cycle, think about these questions and answer them *in your Science Journal.* When you finish the chapter, compare your journal write-up with what you have learned.

Answers are on page 357.

T he Grand Canyon, in northwest Arizona, is 277 miles long, 1 mile deep, and 18 miles wide in some places. When viewed from the bottom near the river, people marvel at how small the river appears compared to the high wall of rock that rises above it. Isn't it amazing that this extraordinary canyon was formed little by little as the swiftly flowing Colorado River cut into many different rock layers? Here in the Grand Canyon, erosion has exposed many different types of rocks.

The Colorado River begins its journey in the Rocky Mountains, the largest mountain system in North America. Just as layers of rock in the Grand Canyon have been eroded, the peaks of the Rockies have been weathered and eroded to their present form.

▶ *Let's explore the great variety of Earth's rocks.*

The Grand Canyon, Arizona

328

`00:00` OUT OF TIME?

If time does not permit teaching the entire chapter, use the Chapter Overview on this page, Reviewing Main Ideas at the end of the chapter, and the Chapter 11 audiocassette to point out the main ideas of the chapter.

Learning Styles		
Kinesthetic	Activity, pp. 330, 338, 343; Find Out, pp. 339, 344; Investigate, pp. 348-349
Visual-Spatial	Explore, pp. 329, 330, 341; Visual Learning, pp. 331, 332, 340, 342, 345; Find Out, pp. 331, 346; Design Your Own Investigation, pp. 336-337; Activity, p. 339; Discussion, p. 347
Interpersonal	Activity, p. 352
Logical-Mathematical	Life Science Connection, pp. 350-351
Linguistic	Research, p. 332; Across the Curriculum, p. 333; Oral, p. 333; Visual Learning, pp. 334, 347, 350; A Closer Look, pp. 334-335; Multicultural Perspectives, pp. 336, 344; Activity, pp. 338, 342, 345; Discussion, pp. 340, 346, 350

LS

Explore! ACTIVITY

How are rocks different?

What To Do

1. Collect eight or nine different rocks from around your school and examine them closely.

2. In what ways are they the same? In what ways are they different? What characteristics could you use to sort them?

3. Try to sort them into three separate groups.

4. Record your answers and observations *in your Journal*.

Mexican agate

How are rocks different?

Time needed about 10–15 minutes

Materials rock samples, magnifying glass (optional)

Thinking Processes organizing information, classifying, thinking critically, comparing and contrasting, forming operational definitions

Purpose To classify rocks based on observed differences and similarities.

Preparation You can provide the rock samples or assign the collection of eight or nine rocks as homework.

Teaching the Activity

Students might group their rocks according to size. Explain that each rock was probably a piece of a larger rock and that there is no way of knowing how big the original rock was. L1

Expected Outcomes

Students should be able to sort rocks by characteristics such as color, shape, or texture.

Answers to Questions

Rocks are all relatively hard. The rocks will most likely differ in color, texture, and luster. Characteristics may vary but should include color, texture, and luster.

✔ Assessment

Performance Give students other rocks and have them classify these rocks into the groups they have made. Use the Performance Task Assessment List for Making Observations and Inferences in **PASC**, p. 17. L1

COOP LEARN

ASSESSMENT PLANNER

PORTFOLIO
Refer to page 359 for suggested items that students might select for their portfolios.

PERFORMANCE ASSESSMENT
Process, pp. 344, 346, 349
Skillbuilder, p. 335
Explore! Activities, pp. 329, 330, 341
Find Out! Activities pp. 331, 339, 344, 346
Investigate!, pp. 336–337, 348–349

CONTENT ASSESSMENT
Check Your Understanding, pp. 338, 342, 352
Reviewing Main Ideas, p. 357
Chapter Review, pp. 358–359

GROUP ASSESSMENT
Opportunities for group assessment occur with Cooperative Learning Strategies.

PREPARATION

Planning the Lesson

Refer to the Chapter Organizer on pages 328A–D.

Concepts Developed

In this section, students are introduced to igneous rocks and discover the relationship of igneous rocks to minerals.

1 MOTIVATE

Bellringer

Before presenting the lesson, display **Section Focus Transparency 31** on an overhead projector. Assign the accompanying **Focus Activity** worksheet. **L1**

LEP

Activity Students can explore some differences between igneous rocks. Materials needed are pumice and granite, a plastic container, and water.

Give students same-size samples of pumice and granite and ask them which sample they think will float. Have students drop both samples into the container of water. Pumice will float because it contains many air pockets. Ask students why the pumice floats. **L1**

Explore!

What makes a rock unique?

Time needed about 30–45 minutes

Materials pink and gray granite samples; mineral samples including quartz, feldspar, hornblende, and biotite; a rock sample that is not granite; magnifying lens

Thinking Processes organizing information, classifying, thinking critically, observing and inferring, comparing and contrasting

---

11-1 Igneous Rocks

How Do Igneous Rocks Form?

As viewed from the bottom of the Grand Canyon, rocks form shelves, steep slopes, and sharp cliffs. You may see variations in the rocks along the sidewalk as you are walking to school. You might pick up an unusual rock and wonder why it looks different from most of the other rocks nearby. While most of the rocks are flat and dull, this one is rounded and has shiny black and white pieces in it. You put the interesting rock in your pocket and decide that you'll ask your science teacher about it.

What exactly should you ask your teacher? You might begin by asking, "Why are rocks different from one another?" and "Is this a rock or a mineral?" You would probably also ask, "What kind of rock is this?"

Explore! ACTIVITY

What makes a rock unique?

You've noticed that rocks can be found with many shapes, colors, and textures. These characteristics can be used to classify and name rocks. Examine granite and some mineral samples to explore how.

What To Do

1. Use a magnifying lens to examine the granite and the mineral samples. *In your Journal*, compare and contrast the small fragments in the rock and the mineral samples.

2. Suppose you were asked to assemble granite using the mineral samples. Which minerals would you use?

3. Now examine a rock that isn't granite. Can you identify any of the minerals that make it up? Are they the same minerals you found in granite?

4. *In your Journal*, define the term *rock* based on your observations.

Purpose To compare mineral samples and an igneous rock and infer the composition of an igneous rock.

Teaching the Activity

As students examine their samples of granite, lead them to look at and describe characteristics like the shapes, sizes, and colors of the crystals. Have them note which crystals are shiny, which are rough, and which have other qualities. **L1** **COOP LEARN**

Answers to Questions

1. Answers will vary.

2. Pink granite is mostly composed of pink feldspar and quartz, and the gray granite of hornblende, quartz, white or gray feldspar, and biotite.

3. Answers will vary.

4. A rock is a mixture of minerals.

Figure 11-1

A The grains of the various minerals that make up granite are large enough to identify with the unaided eye.

Mica is found in small amounts.

Quartz is one of the major components.

Feldspar is a major component.

Hornblende is found in small amounts.

B Because the minerals in granite are interlocked with one another, granite does not weather easily. Where have you seen granite used?

In the Explore activity, you discovered that a **rock** is a mixture of one or more minerals. In addition to minerals, rocks can also be a mixture of mineraloids, glass, or organic particles. Often several different minerals are mixed together, which gives a rock its color and texture. In granite, the mineral pieces are large enough to be seen without a microscope. Look at **Figure 11-1**. It shows a photograph of granite and its components.

The photo helps you see that a rock can be made up of different minerals, but how do minerals combine to form granite and other types of rock? The following activity will help you see how minerals crystallize to form rock.

Find Out! ACTIVITY

What happens as a mineral cools?

What To Do

1. Place a small piece of salol (phenyl salicylate) on a microscope slide.

2. Wearing a thermal glove, set the slide on a hot plate for a few seconds—just until the salol melts.

3. Then quickly observe the slide with a magnifying glass or under a microscope.

4. Record your observations *in your Journal.*

Conclude and Apply

1. What is happening to the salol?

2. How is this similar to what happens when an igneous rock forms?

Tying to Previous Knowledge

Have students recall the various ways used to identify a mineral. Also have them review how minerals form and the relationship between crystal size and cooling rate. In this section, students will learn why intrusive igneous rocks contain larger crystals than extrusive igneous rocks.

Visual Learning

Figure 11-1 Ask students to compare and contrast the four minerals shown in the illustration. **Where have you seen granite used?** *Answers may include buildings, fireplaces, monuments, and sidewalks.*

GLENCOE TECHNOLOGY

 CD-ROM

Science Interactions Course 2, CD-ROM

Chapter 11

Find Out!

What happens as a mineral cools?

Time needed 10 minutes

Materials salol (phenyl salicylate), microscope slide, thermal glove, hot plate, microscope or magnifying glass

Thinking Processes thinking critically, observing, recognizing cause and effect, representing and applying data, making models

Purpose To model a geologic process.

Teaching the Activity

Discussion Begin by reviewing how the magma beneath Earth's crust melts and have students explain how the cooling salol is similar to cooling magma. Guide students in relating each part of this activity to the formation of an igneous rock. [L1]

Expected Outcome

Students should observe the salol crystallizing as it cools.

Conclude and Apply

1. It is crystallizing.

2. Both the salol and magma become solid as they cool.

✔ Assessment

Content Ask students to trace the sequence in the formation of an igneous rock.

In the Find Out activity, the salol behaved like lava from a volcano or like magma that is trapped below Earth's surface. As the lava or magma cools, it becomes solid, similar to the way fudge candy hardens as it cools. Many crystals of various minerals form from the cooling of lava or magma. The crystals grow together and form solid igneous rock. An **igneous rock** is a rock that formed as molten material cooled.

Figure 11-2

A *Intrusive igneous rock* Although formed deep within Earth, it is not uncommon to find intrusive igneous rock on Earth's surface. Forces in Earth, such as compression and tension, push some intrusive igneous rock to the surface. Some other intrusive igneous rock is exposed when erosion removes the rock and soil above it.

B All intrusive igneous rocks, including diorite shown here, as well as granite and gabbro, form slowly. Do you expect these rocks to have small or large mineral grains? Why?

■ Intrusive Igneous Rocks

When igneous rocks are formed by magma that cools beneath Earth's surface, they are called **intrusive** igneous rocks. Intrusive rocks are found at Earth's surface when rock and soil that once covered them is removed by erosion. They may also be found at the surface when forces in Earth, such as compression or tension, push them to Earth's surface. **Figure 11-2** shows you an example of an intrusive igneous rock and where it might form.

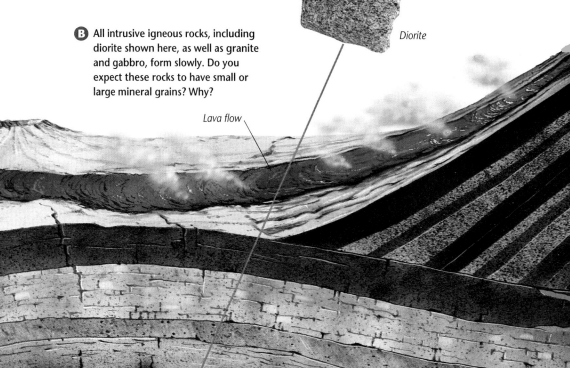

Diorite

Lava flow

Magma (trapped)

332

■ Extrusive Igneous Rocks

Igneous rocks formed by lava that cools on Earth's surface are called **extrusive** igneous rocks. This lava is exposed to air and moisture, and it cools quickly. Study **Figure 11-2** to learn more about how extrusive igneous rocks form. Observe the size of the minerals of the extrusive igneous rocks. Compare this with the size of the minerals in intrusive igneous rocks.

C *Extrusive igneous rock* Because lava is exposed to cooling air and moisture, minerals in extrusive igneous rock form quickly and are much smaller than minerals in intrusive rock. Most extrusive igneous rock minerals can be seen only through a microscope.

Rhyolite

D Rhyolite and andesite are examples of extrusive igneous rock.

Andesite

Magma

333

Across the Curriculum
Daily Life

Ask students if they have ever seen a science fiction movie where a monster or person picks up a huge rock. Ask why pumice is sometimes used as a prop in these movies. *Because it appears heavy but can actually be picked up easily because of all the air spaces it contains.* Have students brainstorm other uses for pumice then research how it is actually used in industry or the home. **L1**

GLENCOE TECHNOLOGY

 Videodisc
STVS: Earth & Space
Disc 3, Side 2
Fibers From Rocks (Ch. 11)

Exploring Lake Superior's Bottom (Ch. 15)

Meeting Individual Needs

Learning Disabled Provide samples of igneous rocks and the minerals that compose them. Have students match the mineral samples to the rock that contains them.

Different rocks can be given to different groups of students for examination and mineral composition verification. Each group can report its findings to the class. **LEP** **COOP LEARN**

ENRICHMENT

Oral Assign various volcanic regions on Earth to individual students. Then have these students prepare an oral report to the class on the landforms and rocks found in these areas. Use the Performance Task Assessment List for Oral Presentation in **PASC**, p. 71. **L1** **P**

Content Background

More than 90 percent of Earth's crust is made of igneous rock. It may take some magma trapped below Earth's surface hundreds of years to cool two or three degrees. As a result, some intrusive igneous rocks may take thousands of years to form. Six groups of minerals, in different combinations, compose most igneous rocks: olivine, quartz, amphiboles, pyroxenes, micas, and feldspars. The most common coarse-grained igneous rock is granite, which composes most of the continental crust. Basalt is the most common fine-grained igneous rock. Most of the oceanic crust is composed of basalt.

Basalt, the most common extrusive igneous rock, is shown in **Figure 11-3**. Basalt is a common rock of the Hawaiian Islands. The photograph also shows an unusual beach in the Hawaiian Islands that has black sand. Study **Figure 11-3** and try to figure out why the Hawaiian Islands have black sand beaches.

Figure 11-3

A In the picture on the left is a black sand beach in the Hawaiian Islands. Recall what you know about weathering, erosion, and igneous rocks. Where does the black sand come from?

B Below is a sample of basalt. Notice that it is made of small, dark mineral grains.

Basalt

Visual Learning

Figure 11-3 Ask students to describe the sand at beaches they have visited. Then have them compare and contrast the sand from the different beaches according to color, texture, size of grain. Lead them to infer that the type of sand on a beach depends on the rock from which it was formed and the weathering and erosion process the rock underwent. **Where does the black sand come from?** *As the basalt weathered and eroded, dark particles of the rock were transported and deposited on this beach.* L1

Natural Glass

Natural glass forms when thick, slow-flowing lava cools rapidly. One kind of lava that typically forms natural glass is rhyolitic lava. When rhyolitic lava cools on the surface of Earth, it forms crystals. The size of the crystals determines the texture of the igneous rock.

Some scientists compare rhyolitic lava to cold honey. The lava is a thick liquid that flows very slowly. Sometimes rhyolitic lava cools so quickly, crystals do not have time to form. The result is a smooth, glossy volcanic glass called obsidian.

Obsidian

Obsidian is natural glass, sometimes called supercooled liquid. Supercooled liquids form solids with no internal crystalline structure. Window glass is considered supercooled liquid. It is smooth and glossy in texture. The difference between window glass and obsidian is that window glass contains no impurities, so it's clear in color.

Obsidian is usually black in color. The presence of iron oxide will turn obsidian red.

Purpose

A Closer Look reinforces Section 11-1 by explaining in detail how obsidian, an extrusive igneous rock, is formed.

Content Background

Obsidian is classified as a dark-colored, glassy extrusive igneous rock.

Some igneous rocks can have two or more different-sized crystals. This type of rock is called porphyry and is formed when molten material begins cooling slowly deep in Earth but is then carried to the surface where the remainder of the molten material cools quickly. When this happens, the rock contains large crystals surrounded by fine-grained crystals or amorphous rock.

Teaching Strategies

Do the following demonstration in a well-ventilated classroom or fume hood. **CAUTION:** Do not allow students to breathe in the fumes. Melt some napthaline (moth balls) in a beaker over a bunsen burner.

Classifying Igneous Rocks

Unless we're near an erupting volcano, we can't actually observe the formation of igneous rocks to see whether they were created below or above Earth's surface. Yet we can identify igneous rocks as either intrusive or extrusive by the size of their crystals. Rocks that have large, visible crystals are called coarse-grained rocks. Rocks that have crystals so small that we cannot see them are referred to as fine-grained rocks. Coarse-grained, fine-grained, glassy, and porous are examples of rock textures.

Igneous rocks can also be identified and grouped by their overall color. Light-colored rocks must be composed of light-colored minerals, such as quartz and some feldspars. Dark-colored rocks are composed of minerals, such as pyroxenes and amphiboles.

Comparing and Contrasting
Examine samples of obsidian, pumice, and gabbro. In what ways do obsidian and pumice differ from gabbro, another igneous rock? In what ways are they similar? If you need help, refer to the **Skill Handbook** on page 642.

Obsidian and pumice have no mineral grains. The obsidian is smooth and shiny, while the pumice has small pits and holes. Gabbro, on the other hand, is coarse with large mineral grains. All three rocks are igneous rocks. **L1**

Other colors of obsidian are rare. Obsidian is found in abundance where volcanoes erupted in areas where cooling is rapid.

Obsidian for Tools

When obsidian fractures, it breaks into shards that are angular and sharp like a broken windowpane. For this reason, ancient cultures found obsidian useful for arrowheads and knives.

Scientists estimate that humans began making their own glass about 6000 years ago. Some scientists theorize that human-made glass was discovered by accident, when rocks and sand were melted by fire, creating a product similar to nature's own glass. Today's glass is made in a similar way. Sand, soda (sodium oxide) and lime (calcium oxide) are mixed and melted in a furnace.

What Do You Think?

Today, obsidian is often used in surgical instruments. Describe how the properties of obsidian make this rock useful for surgery.

Pour some of the melted liquid over the surface of an index card. Pour the rest into an aluminum foil cup. Ask students what the liquid poured out over the card and the liquid in the aluminum cup represent. *That over the card is like lava that cools on the surface of Earth. That at the cup bottom represents intrusive rock.* Then ask them to predict what size crystals will form when each cools. *Smaller crystals should form on the card than in the cup.*

Answers to
What do you think?
Answers should include that obsidian has sharp edges and can be shattered into small, fine fragments.

Going Further ▌▌▌▌▌▌►
Divide students into small groups. Provide a sample of glass and obsidian to each group. Have the students examine the samples and discuss their similarities and differences. Then discuss the various uses

of glass and relate them to its properties. Finally, have them imagine that they live in a culture that does not know how to make glass and discuss how they might use their sample of obsidian. Ask each group to summarize their discussions in three written paragraphs. Use the Performance Task Assessment List for Writing in Science in **PASC**, p. 87. **COOP LEARN** **P**

11-1 Igneous Rocks

Preparation

Purpose To design and carry out an investigation to classify igneous rocks according to their characteristics.

Process Skills communicating, making and using tables, forming operational definitions, forming a hypothesis, observing and inferring, separating and controlling variables, classifying, interpreting data

Time Required one class period to plan and begin the experiment; one-half class period to complete the experiment and summarize results

Materials igneous rock samples, including obsidian (A), basalt (B), granite (C), diorite (D), pumice (E), rhyolite (F); andesite (G), gabbro (H), scoria (I); hand lens

Possible Hypotheses Igneous rocks can be classified by several characteristics, including general color (light or dark) based on their composition. They can also be classified by their rate of cooling (large visible crystals cooled slowly; small crystals cooled quickly).

Plan the Experiment

Process Reinforcement Remind students that intrusive igneous rocks have large crystals and extrusive igneous rocks have very small crystals or no crystals at all. Also remind them that basaltic igneous rocks tend to be dark in color, and granitic igneous rocks tend to be light.

Possible Procedures Separate the rock samples into different-colored piles and classify the rocks according to color, or separate the rock samples into piles based on individual crystal size and classify the rock samples based on crystal size.

DESIGN YOUR OWN
INVESTIGATION

Igneous Rocks

Igneous rocks form from lava that cools quickly at or near Earth's surface or from magma that cools slowly deep inside Earth. The color of an igneous rock depends upon the type of minerals found in the lava or magma. If you wanted to classify an igneous rock, what would you need to determine about the way the rock formed?

Preparation

Problem
How would you classify igneous rocks?

Form a Hypothesis
Based on what you know about igneous rocks, form a hypothesis about how their characteristics can be used to classify them.

Objectives
- Determine which characteristics of igneous rocks are useful for classification purposes.
- Observe differences in the rocks' crystal sizes, textures, and colors.
- Infer how the rocks formed, based on the observed characteristics.

Materials
igneous rock samples
hand lens

Table 11-1

Rock Sample	Texture	Color	Minerals Visible
A. obsidian	glassy	black	none—cooled too fast
B. basalt	fine	black	feldspar, pyroxene
C. granite	coarse	varies	quartz, feldspar, micas, hornblende
D. diorite	coarse	salt & pepper	feldspar, hornblende, mica, little quartz
E. pumice	glassy, porous	usually gray	none—see A
F. rhyolite	fine	gray, yellow, red	as in C
G. andesite	fine	salt & pepper	as in D
H. gabbro	coarse	dark gray, green-black	feldspar, pyroxene, hornblende, olivine
I. scoria	frothy	as in H, possibly red	none—see A

DESIGN YOUR OWN
INVESTIGATION

Plan the Experiment

1. As a group, agree upon a way to test your hypothesis. Write down what you will do at each step of your test.

2. List the characteristics you think will be useful in classifying the rock samples. For texture, refer to the terms regarding texture on page 335.

3. How will you summarize your observations? If you need a data table, design one now *in your Science Journal.*

Check the Plan

1. Which characteristics will you use to classify your rock samples?

2. Review your data table. Have you included all the characteristics that you will observe?

3. Make sure your teacher approves your experiment before you proceed.

4. Carry out your experiment. Record your observations.

Analyze and Conclude

1. **Classify** Based on your observations, identify your igneous rock samples. Did your classification system help you to identify the rocks?

2. **Interpret Data** How does obsidian differ from most other igneous rocks?

3. **Observe and Infer** What minerals might have caused the various colors found in your rocks?

4. **Identify** Which samples are intrusive? Extrusive?

5. **Recognize Cause and Effect** Why do igneous rocks of the same composition sometimes have different-sized grains?

Going Further

Examine your igneous rock samples again. Classify them as basaltic, granitic, or andesitic.

Program Resources

Science Discovery Activities, 11-1
Activity Masters, pp. 47–48, Design Your Own Investigation 11-1

3 ASSESS

Check for Understanding

For the Apply question in the Check Your Understanding, have students make a table to compare granite and rhyolite. Table headings might be color, texture, minerals contained, and intrusive/extrusive. Use the Classroom Assessment List for Group Work in **PASC**, p. 97. L1

COOP LEARN

Reteach

Activity Have students construct their own igneous rocks with clay.

Direct them to roll many small balls in different colors of clay to represent the mineral crystals. They can then put these balls together to form a model of an intrusive igneous rock. By using even smaller balls of clay, they can form an extrusive igneous rock. **LEP** L1

Extension

Activity Have students who have mastered the concepts in this section collect igneous rocks and mount them on index cards. On each card, have students identify the rock as intrusive, extrusive, fine-grained, or coarse-grained. Also have students list some of the minerals each sample contains. L3

4 CLOSE

Activity

Have students make a table as shown.

	Intrusive	Extrusive
Place of formation	beneath surface	at surface
Crystal size	large	small
Rate of cooling	slow	fast
Texture	coarse	smooth

Direct students to fill in the appropriate information based on their understanding of this section. L1

Color	Intrusive (course texture)		Extrusive (fine texture)	
Dark	Gabbro		Basalt	Scoria
Intermediate	Diorite		Andesite	
Light	Granite		Rhyolite	Pumice

Figure 11-4

Granite, rhyolite, gabbro, and basalt are individual types of igneous rocks. Use **Figure 11-4** above to describe their characteristics. There are also some igneous rocks that are classified as intermediate in color. They are neither dark nor light in overall color. Diorite and andesite are examples of these intermediate rocks.

Igneous rocks are the most abundant type of rock on Earth. They've been classified to make them easier to identify and to study. By studying all types of rocks, geologists and other scientists have been able to hypothesize how Earth formed. They have been able to determine how mountains such as the Rockies were formed from an upheaval of Earth's crust, and how the rock layers of the Grand Canyon region have accumulated and eroded over millions of years.

check your UNDERSTANDING

1. What is the difference between a rock and a mineral?
2. How do igneous rocks form?
3. Describe the differences between intrusive and extrusive igneous rocks.
4. **Apply** How are granite and rhyolite similar? How are granite and rhyolite different?

check your UNDERSTANDING

1. A rock is made up of one or more minerals.
2. Igneous rocks form when magma or lava cools to form rocks.
3. Intrusive rocks form slowly beneath Earth's surface and have large mineral crystals. Extrusive rocks form on Earth's surface where molten materials cool quickly. Extrusive rocks have a fine-grained texture because their mineral crystals are smaller.
4. Both are light-colored, igneous rocks. Granite is an intrusive igneous rock with a coarse texture because of its large mineral grains. Rhyolite is an extrusive igneous rock with a fine, small-grained texture.

Metamorphic Rocks

How Do Metamorphic Rocks Form?

Suppose you discovered that the rock you found on your way to school is an igneous rock. It formed when crystals of one or more minerals grew together as magma or lava cooled. Will the minerals in the rock remain unchanged forever? You know that weathering can change rocks, but are there other ways that rocks can change?

To understand a different way that rocks can change, think about how the contents of your lunch bag might change after it's been in your locker all day. The apple you packed has been resting on your sandwich and cream-filled cake since early this morning. The heat in your locker has turned the cake into a gooey mess. The pressure from the apple has flattened your sandwich. In the following activity, observe what changes occur when pressure is similarly applied to a model of rock layers.

Section Objectives

- Explain how metamorphic rock is formed.
- Identify and classify metamorphic rocks.

Key Terms

metamorphic rock
foliated
nonfoliated

Find Out! ACTIVITY

What can happen to a rock when it is exposed to pressure?

To find out, you'll first need to make a crayon rock using four to six of each color of crayon—red, green, blue, and yellow.

What To Do

1. Use a pencil sharpener to make a pile of crayon shavings on a sheet of aluminum foil. They will represent different minerals in an igneous rock.

2. Fold the edges of the foil toward the middle to enclose the shavings within a rectangular packet.

3. Gently flatten the packet by squeezing it between your palms.

4. Now, unfold the foil packet and examine your rock.

5. Return the crayon rock into its foil packet and see what happens when you squeeze it between two boards using a vise or two C-clamps.

Conclude and Apply

1. How did the crayon rock change?

2. Recall that the individual shavings represent mineral crystals. What do you think would happen to the crayon minerals if heat were applied?

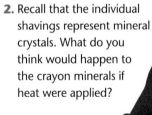

Find Out!

What can happen to a rock when it is exposed to pressure?

Time needed   15–20 minutes

Materials   crayons (4 to 6 of 4 colors per student), pencil sharpener, aluminum foil, vise or two C-clamps, two boards

Thinking Processes observing, inferring, recognizing cause and effect, forming a hypothesis

Purpose To illustrate how pressure can change the appearance of a material. L1

Expected Outcome

Students should find that the more pressure they exert, the more tightly bound together and integrated the crayon shavings are.

Conclude and Apply

1. The shavings became more compact and the individual shavings seemed to blend together.

PREPARATION

Planning the Lesson

Refer to the Chapter Organizer on pages 328A–D.

Concepts Developed

When rocks are buried deep within Earth, tremendous heat, great pressure, and chemical reactions can cause them to change into different rocks with different textures and structures. Students learn that these are metamorphic rocks.

1 MOTIVATE

Bellringer

Before presenting the lesson, display **Section Focus Transparency 32** on an overhead projector. Assign the accompanying **Focus Activity** worksheet. L1
LEP

Activity This activity will show students how metamorphic changes can be caused by heat. Materials needed are a toaster, slice of bread, and a knife. Toast a slice of bread. Cut the toast in half and have students compare the color and texture of the outer surface with its untoasted center. L1

2. They would melt, run together, and recombine.

✔ Assessment

Portfolio Have students make annotated drawings of their materials before, during, and after pressure has been applied. Students should relate their "rock" to the process that forms metamorphic rock. Use the Performance Task Assessment List for Scientific Drawing in **PASC**, p. 55. L1
P

---

2 TEACH

Tying to Previous Knowledge

Show students a sample of granite. Using the characteristics of igneous rocks from Section 11-1, have them describe the granite. Then show them a sample of the metamorphic rock gneiss, which can be formed from granite. Ask them to describe any similarities they see between the gneiss and the granite samples.

Visual Learning

Figure 11-5 Call on students' experiences to reinforce the concepts described in the captions. For example, ask if students have made meatballs or snowballs. Ask volunteers to relate the effects of the pressure in their hands and the compaction of the material they were using to the pressure on rock beneath Earth's surface. **How has pressure on the igneous rock affected the mineral grains?** *The pressure flattens and aligns the mineral grains in the rocks.*

Figure 11-6 To understand the sequence of events when the sedimentary rock becomes gneiss, have students study the drawings as they read the captions.

Theme Connection Metamorphic rock is formed when rock is changed by heat or pressure or both. In the Find Out activity on page 339, students exerted pressure on a crayon rock and observed the change. This theme of stability and change is more evident in the Explore activity on page 341.

Figure 11-5

Granite

Heat and pressure

Gneiss

A The weight of overlying rock layers causes pressure on formations of igneous rock deep beneath Earth's surface.

B Look at the diagram above. How has pressure on the igneous rock affected the mineral grains?

C The altered rock is now a metamorphic rock. The gneiss above was produced from granite by this process.

You've seen what happens to minerals in rocks when they're exposed to increases in pressure. **Metamorphic rock** forms when rock is changed by heat or pressure or both. **Figure 11-5** shows what can happen when pressure is applied to the igneous rock granite. The minerals in granite are flattened and form the metamorphic rock gneiss (NICE). What occurs in Earth to change these rocks?

Rocks beneath Earth's surface are under great pressure from overlying rock layers. They are also exposed to

Figure 11-6

Some forms of gneiss originate from igneous rock and result from the process shown above. Other forms of gneiss begin as sedimentary rock and pass through several stages before becoming gneiss.

A *Shale* When the weight of overlying layers exerts pressure on mud containing clay, the weight forces water out of the mud and presses the clay layers together forming solid rock, called shale. Microscopic grains of clay minerals are evenly distributed in shale.

B *Slate* Forces within Earth expose some shale to heat and pressure. The heat and pressure cause the clay minerals in the shale to separate into distinct layers. Once this change takes place, the rock is called slate. The mineral grains in slate are barely visible.

340 Chapter 11 The Rock Cycle

Meeting Individual Needs

Visually Impaired Have students handle the samples of igneous and metamorphic rocks. Have them describe the differences in textures between these two rock types. As they handle the rocks, describe to them the different ways these rocks are formed.

ENRICHMENT

Discussion In 1987, authorities in Great Britain decided to rope off Stonehenge. It is believed that ancient people may have used the structure as a calendar. The large rocks are sandstone. Many tourists visit Stonehenge each year and some have vandalized the monument. Ask the class if these types of monuments should be roped off to guard against vandals or be left open so all can appreciate them close up.

heat from magma. If the heat and pressure are great enough, the rocks melt and magma forms. If the heat and pressure are not great enough to melt the rocks, the mineral grains in the rock may change in size or shape. To better understand how metamorphic rocks can form, do the Explore activity below and then study **Figure 11-6**.

Explore! ACTIVITY

From what do metamorphic rocks form?

Metamorphic rocks can also form from rocks other than igneous rocks. In the following activity, you'll compare metamorphic rocks with nonmetamorphic rocks.

What To Do
Your teacher will provide you with samples of four metamorphic rocks and four nonmetamorphic rocks. Each of the metamorphic rocks formed from one of the nonmetamorphic rocks.

1. For each metamorphic rock, determine which nonmetamorphic rock it might be related to.

2. *In your Journal*, list the characteristics of the four pairs of rocks.

C *Schist* Continuing heat and pressure on slate may cause mineral grains to grow larger or new minerals to form. If this change takes place, the resulting rock is called schist. Mineral grains in schist are much larger than those in slate.

D *Gneiss* As temperature and pressure continue to increase, the minerals will separate into bands and the mineral grains will become large enough to easily identify. When these changes take place, the schist will have changed into gneiss.

11-2 Metamorphic Rocks **341**

Program Resources

Study Guide, p. 38
Laboratory Manual, pp. 57–58, Metamorphic Process [L2]
Multicultural Connections, p. 25, The Importance of Jade [L1]
Science Discovery Activities, 11-2
Section Focus Transparency 32

3 ASSESS

Check for Understanding

To answer the Apply question, have students review what happens to the minerals in shale when enough heat and pressure are applied. Use the Classroom Assessment List for Group Work in **PASC,** p. 97.

Reteach

Activity To help students remember how metamorphic rocks are formed, have them research the origin of the term *metamorphic*. (*Meta* means "change," *morph* means "form".)
L1 LEP

Extension

Research Have students research various metamorphic rocks and their uses. L3

4 CLOSE

Activity

Have students write a paragraph describing how metamorphic rock is formed. Then have them list several examples of metamorphic rock. L1

Classifying Metamorphic Rocks

Figure 11-7

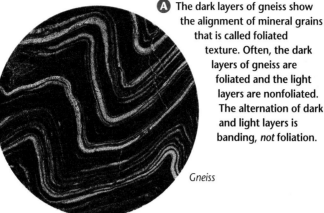

A The dark layers of gneiss show the alignment of mineral grains that is called foliated texture. Often, the dark layers of gneiss are foliated and the light layers are nonfoliated. The alternation of dark and light layers is banding, *not* foliation.

Gneiss

B Compare this photograph of marble with the one of gneiss above. How do the two rocks differ? Marble is an example of a metamorphic rock with a nonfoliated texture.

Marble

In any of the samples you observed in the Explore activity, did the mineral grains flatten and line up in parallel bands? Metamorphic rocks with this kind of **foliated** texture form when minerals in the original rock flatten under pressure.

The metamorphic rock slate forms from the sedimentary rock shale. Under heat and pressure, the minerals in shale become so tightly compacted that water can't pass between them. Slate is easily separated along its foliation layers.

In some metamorphic rocks, the mineral grains change, combine, and rearrange, but they don't form visible bands. This process produces a **nonfoliated** texture. Such rocks don't separate easily into layers. Instead, they fracture into pieces of random size and shape.

So far, we've discovered how two types of rock are formed. Next we'll observe how sedimentary rocks are formed and how some igneous and metamorphic rocks are formed from them. The next section will complete our investigation of different kinds of rock.

check your UNDERSTANDING

1. How do igneous rock and metamorphic rock differ?

2. By what characteristics are metamorphic rocks classified?

3. Apply Slate is a metamorphic rock that is sometimes used as building material for roofs. What properties make slate particularly useful for this purpose?

check your UNDERSTANDING

1. Igneous rock forms from molten material. Metamorphic rock forms when heat and pressure are applied to existing rock.

2. texture, which is either foliated or nonfoliated, and minerals

3. The tightly compacted minerals in slate do not allow water to pass between them. Therefore, slate helps protect a roof from water damage.

Sedimentary Rocks

How Do Sedimentary Rocks Form?

So far, you've explored two major types of rocks: igneous and metamorphic. In this section, you'll learn about a third type of rock, which is composed of sediments. This rock is called sedimentary rock. Where do sedimentary rocks come from?

You may recall that weathering and erosion are two major processes that change Earth's surface. Weathering breaks rocks or remains of plants and animals into smaller pieces called sediments. Sediments are transported to new locations by the agents of erosion—water, wind, ice, and gravity. Under certain conditions, deposited sediments recombine to form a solid rock called **sedimentary rock**. How do deposited sediments form rock?

Think of an area where layer after layer of sediments are deposited. The pressure from the upper layers pushes down on the lower layers. The sediments compress and form rock. This process is called compaction, and is shown in **Figure 11-8**. How else can sediments form rock? The activity on the next page will help you find out.

Figure 11-8

Compaction During the process of compaction, pressure from overlying layers pushes the sediment layers together. The amount of sediment stays the same, but the grains become more tightly packed.

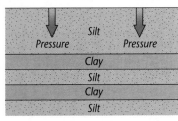

A Sediment accumulates in layers.

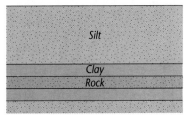

B Pressure from above squeezes the lower layers.

C As pressure continues, the lower layers continue to compact and eventually form rock.

Sandstone

Figure 11-9

The size of the grains in sediment and the kind of sediments determine the type of sedimentary rock that will form. Sandstone is formed from grains of sand up to 2 millimeters across.

Section Objectives

- Explain how sedimentary rock is formed.
- Identify and classify sedimentary rocks.
- Use a diagram of the rock cycle to explain how rocks form and change.

Key Terms

sedimentary rock
rock cycle

Program Resources

Study Guide, p. 39
Laboratory Manual, pp. 59–60, Concretions **L1**
Critical Thinking/Problem Solving, p. 19, Rocks All Around You **L2**
Teaching Transparency 22 **L1**
Section Focus Transparency 33

PREPARATION

Planning the Lesson

Refer to the Chapter Organizer on pages 328A–D.

Concepts Developed

In this section, students will investigate sedimentary rocks and how they form.

1 MOTIVATE

Bellringer

Before presenting the lesson, display **Section Focus Transparency 33** on an overhead projector. Assign the accompanying **Focus Activity** worksheet. **L1** **LEP**

Activity In this activity, students will observe how different-sized sediments settle in layers.

Materials needed are a mixture of sand, soil, and pebbles; graduated glass cylinders; and water. Have students work in groups. Each group fills a graduated cylinder halfway with water. Then have students slowly pour in the mixture of sand, soil, and pebbles. As the heavier pebbles settle first, the sand second, and the soil last, students observe layers of sediments forming. **COOP LEARN** **L1** **LEP**

2 TEACH

Tying to Previous Knowledge

The weathering and erosion of rocks forms sediments that may eventually help form sedimentary rocks. Ask students where these rocks must be located in order for weathering and erosion to occur. *on Earth's surface*

Find Out! ACTIVITY

How can sediments become cemented together?

What To Do

1. Fill a paper cup with sand.

2. In a second cup, mix one part white glue and one part water.

3. Poke small holes in the sand-filled cup large enough for the glue solution to drain through, but not large enough for much sand to run out.

4. Suspend the paper cup with the sand over a bowl.

5. Pour the glue solution into the cup and allow it to drain through the sand for several days. Tear away the paper.

6. Record your observations *in your Journal*.

Conclude and Apply

1. How is this block of sand similar to a sedimentary rock?

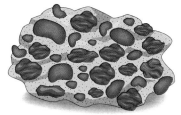

As you saw in the Find Out activity, cementation is also an important rock-forming process. To get a better idea of how cementation occurs in nature and helps to form sedimentary rocks, study **Figure 11-10**.

Figure 11-10

Cementation Cementation is another process that binds sediments tightly to form sedimentary rock.

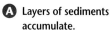
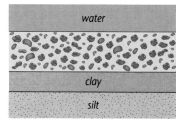
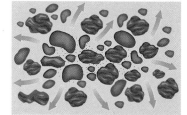

A Layers of sediments accumulate.

B Water is squeezed out. The mineral deposits that were dissolved in the water remain and crystallize.

C The crystallization of the minerals cements the sediments together to form rock.

Multicultural Perspectives

Inca Stonework

Stone has been an important building material since prehistoric times. The Incas ruled much of western South America from about 1438 until the Spanish Conquest in 1532. They were excellent stone masons (workers who build with stone). Although they used only stone and bronze tools and lacked use of the wheel, they shaped and transported huge stones over long distances. They cut the boulders so perfectly that they required no mortar. Have students research examples of Inca stonework and make a bulletin board displaying information about the stone used.

Classifying Sedimentary Rocks

Sedimentary rocks can be composed of any type of weathered and eroded rock material and sometimes even particles from plants and animals. To classify sedimentary rocks, you must look at the sediment they contain, as well as the way in which the rocks were formed.

■ Detrital Sedimentary Rocks

Detrital sedimentary rocks are made of the broken fragments of other rocks. These sediments, which are the solid products of weathering, have been compacted and cemented together.

In some detrital rock, the sediments are large and well rounded. In others, the sediments are large but have sharp angles.

Although not a rock, concrete is made from pebbles and sand grains that have been cemented together. Look at **Figure 11-11.** Notice how similar the concrete sidewalk looks to naturally occurring detrital rock.

Shale is a detrital sedimentary rock that requires little cementation to hold its particles together. Its sediments are clay-sized minerals, which are even smaller than sand-sized particles. Clay-sized sediments can be compacted together by pressure from overlying layers.

Sandstones are another very common detrital sedimentary rock. Look at **Figure 11-12** to learn more about sandstones.

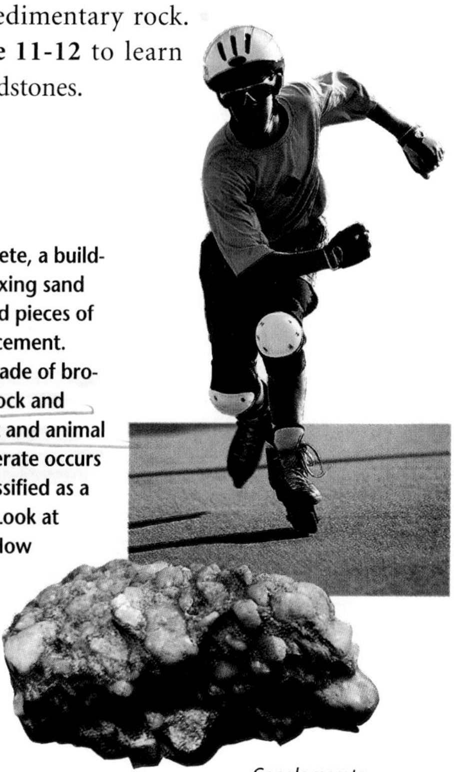

Figure 11-11

People make concrete, a building material, by mixing sand grains, pebbles, and pieces of crushed rock with cement. Conglomerate is made of broken fragments of rock and sometimes of plant and animal remains. Conglomerate occurs naturally and is classified as a sedimentary rock. Look at the photographs. How are concrete and conglomerate alike?

Conglomerate

Sandstone

Figure 11-12

Sandstone is a detrital sedimentary rock formed from sand-sized sediments, usually grains of the minerals quartz and feldspar. The sandstone rock formations shown here were formed from sand deposited in layers in the desert. Judging from the shape of the formations, what was the most likely agent of erosion that shaped the rock?

Visual Learning

Figure 11-11 Ask students to describe where they have seen concrete used. *Answers may include sidewalks, swimming pools, building walls, and stairways.* **How are concrete and conglomerate alike?** *They are similar in appearance, both are made of visible-sized pieces of rock. Conglomerate occurs naturally and is a rock. Concrete is human-made and is not a rock.*

Figure 11-12 Ask students to compare and contrast this photo to the photo of the Grand Canyon on page 328. Then ask them to compare and contrast the appearance of rock eroded by water and rock eroded by wind.

Judging from the shape of the formations (of sandstone) what was the most likely agent of erosion that shaped the rock? *Wind.*

Theme Connection The theme that this section supports is stability and change. Rocks may remain stable and unchanged for thousands of years, but when exposed to agents of change such as weathering and erosion, they may change into other forms of rock.

Activity Tell students that in many parts of the United States, large areas of sedimentary rock are found at Earth's surface. Have them research their own area's geologic history to find out about the presence of sedimentary rock. L2

Meeting Individual Needs

Learning Disabled, Visually Impaired Allow students to handle various samples of sediments: sand, pebbles, clay. Then have them handle the sedimentary rocks that can form from each of these sediments: sandstone, conglomerate, shale. Have a student volunteer explain the connection between the sediment and the sedimentary rock that they can form. LEP

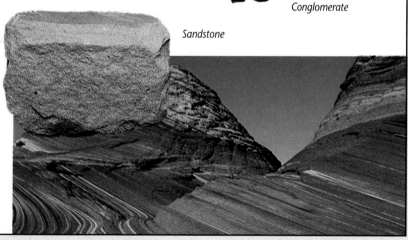

Discussion Mount Augustus located in western Australia is 8 kilometers long and 3 kilometers wide and made entirely of conglomerate. Conglomerate rock was first called puddingstone because it resembled the lumpy appearance of plum pudding from England. What would have caused the rock to have this appearance? *sediments of different sizes forming a rock*

Find Out!

What can happen to dissolved minerals?

Time needed about 10–15 minutes

Materials alum, water, teaspoon, tablespoon, shallow pan

Thinking Processes thinking critically, observing and inferring, recognizing cause and effect

Purpose To observe how a chemical sedimentary rock might form.

Preparation To speed up this activity, heat the saltwater solution to evaporate the water more quickly. Watch glasses are ideal for evaporating this type of solution. Use them if available instead of a shallow pan.

Teaching the Activity

Discussion Be sure to tell students that they are observing the formation of a chemical sedimentary rock on a very small scale. This process actually occurs in large lakes and seas and it may take hundreds or thousands of years for the water to evaporate and leave behind salt. [L1]

Expected Outcome

Students should see salt residue at the bottom or sides of the pan.

Conclude and Apply

1. Salt remains behind and forms a rock.

DID YOU KNOW?

Many of the earliest trade routes between continents were established for the purpose of trading salt. Gold and salt were considered of equal value—salt was even used as currency in early China.

■ **Chemical Sedimentary Rocks**

The sediments of detrital sedimentary rocks originate from weathering and are transported as solid particles.

But what happens when weathering causes some of the minerals in rocks to dissolve? Find out what kind of rocks might result from evaporation.

Find Out! ACTIVITY

What can happen to dissolved minerals?

What To Do

1. Dissolve 2 grams of alum in 20 mL of water to make a solution.

2. Pour the solution into a shallow pan and allow it to evaporate. Now look at the bottom of the pan.

Conclude and Apply

1. What do you think happens when ocean water evaporates? Answer the question *in your Journal.*

As you've just observed, some layers of sediment come from minerals that were once chemical compounds dissolved in solution. Chemical compounds become concentrated when the water in seas or lakes evaporates.

These mineral layers can form chemical sedimentary rocks. Rock salt is an example.

Rock salt can form when halite is carried in solution in ocean or lake water. As the water evaporates or

Figure 11-13

Ⓐ Limestone is an example of sedimentary rock. Limestone makes an excellent building stone because it can be carved easily. The Sphinx and many pyramids in Egypt are made of limestone.

Limestone

✔ **Assessment**

Process Ask students to formulate a model of how the process they observed in the Find Out! activity can lead to the formation of rocks in lakes and seas. Use the Performance Task Assessment List for Formulating a Hypothesis in **PASC,** p. 21. **COOP LEARN**

Program Resources

Making Connections:
Technology & Society, p. 25, Lunar Rocks [L2]
Multicultural Connections, p. 26, Shen Kua-Geologist in Early China [L1]
Science Discovery Activities, 11-3
Teaching Transparency 21

other conditions change, the concentration of elements increases until the point of saturation is reached. The halite precipitates onto the ocean or lake floor, forming rock salt.

■ Organic Sedimentary Rocks

Other sedimentary rocks have large amounts of the remains of once-living things, also known as fossils. They are classified as organic sedimentary rocks. One of the most common organic sedimentary rocks is fossil-rich limestone. It is made of the mineral calcite. Fossil-rich limestone

consists mostly of the remains of once-living (aquatic) organisms, together with calcite. Look at **Figure 11-13** to see examples of limestone and how it has been used.

Ocean animals, such as mussels and snails, make their shells from the mineral calcite and a few other minerals. When the animals die, their shells accumulate on the ocean floor. When these shells are compacted and cemented together, layers of sedimentary rock are formed.

Connect to...
Life Science

One of the most common fossils in ancient organic rocks is the brachiopod. Find out what a brachiopod looks like and draw a picture of it. Are any brachiopods living today? If so, where do they live?

Visual Learning

Figure 11-13 Tell students that the limestone used for statues like the Sphinx is an organically built sedimentary rock. Ask students to speculate what characteristics of limestone make it popular for carving statues. *Possible answers might include resistance to weathering and the ability to be carved easily.*

Connect to . . .

Life Science

Brachiopod means "arm foot." As students draw their pictures of brachiopods, ask them to think about why these creatures were given this name. There will be no single "right" drawing for students to create, since the rounded shells that covered brachiopods differed in size and shape. There were thousands of species of brachiopods, with about 200 still found today.

B The white cliffs of Dover are thick layers of soft, fine-grained, white organic limestone, commonly called chalk. These chalk cliffs are made mostly of tiny shells and crystals of the mineral calcite.

The Sphinx in Egypt

The white cliffs of Dover in England

There are many types of limestone. This example is called oolitic limestone.

ENRICHMENT

Discussion Show the class an igneous rock sample. Ask students to hypothesize if this rock has always been an igneous rock. If so, ask them to hypothesize how it came to be formed. If not, ask them to describe how this rock may have been formed from a metamorphic or sedimentary rock. L2

INVESTIGATE!

11-2 Very Sedimentary

Planning the Activity

TECH PREP

Time needed 25–30 minutes

Materials See student activity; rock samples should include: (a) calcite, (b) sandstone, (c) shale, (d) conglomerate, (e) rock salt.

Purpose To observe characteristics of sedimentary rocks in order to identify them.

Process Skills thinking critically, observing and inferring, recognizing cause and effect, forming operational definitions

Preparation Prepare the 5% HCl solution by adding 50 g of concentrated HCl into 950 mL of distilled water. Put 5% HCl solution in dropper bottles for student safety.

Teaching the Activity

Demonstration You may want to demonstrate the use of HCl on calcite to the class before students try it. [L2]

Process Reinforcement Ask students how they might test for the presence of a mineral in a rock if they could not see the mineral. Looking for a chemical reaction, such as that produced by HCl on calcite, is one possible answer.

Safety Remind students to note the Caution notice in the student text.

Possible Hypotheses After students test the HCl on the sample of calcite, ask them to form a hypothesis about the effects of HCl on the other rock samples. Possible hypotheses include that HCl will fizz when applied to any rock containing calcite.

Very Sedimentary

Now that you've explored three different types of sedimentary rocks, you can use your knowledge and power of observation to identify the unknown sedimentary rocks you'll examine in the following Investigate activity. In this activity, you'll identify and classify sedimentary rocks.

Problem

How can you classify sedimentary rocks?

Materials

dropper	water
hand lens	goggles
paper towels	apron
sedimentary rock samples	5% hydrochloric acid (HCl)
	calcite sample

Safety Precautions

Acid can cause burns and damage clothes; handle it with care. Report any spills to your teacher.

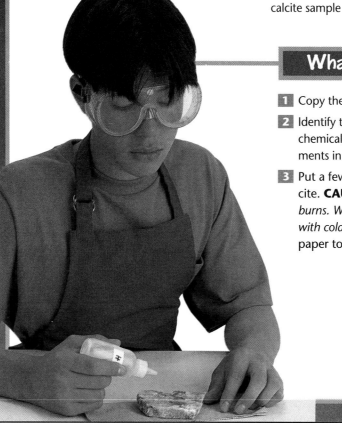

What To Do

1 Copy the data table *into your Journal.*

2 Identify the sediments in each sample as detrital, chemical, or organic. *Classify* the size of the sediments in the rocks.

3 Put a few drops of HCl on your sample of calcite. **CAUTION**: *HCl is an acid and can cause burns. Wear goggles and an apron. Rinse any spill with cold water. Observe what happens. Use a paper towel to remove remaining acid.*

Program Resources

Activity Masters, pp. 49–50, Investigate 11-2

Take Home Activities, p. 18, Water Action [L1]

Critical Thinking/Problem Solving, p. 5, Flex Your Brain

skip

After identifying the ancient sea organisms in this rock, geologists were able to determine that the rock is about 200 million years old.

Sample data

Data and Observations

Sample	Sediment Size	Observations	Minerals Present	Detrital, Chemical, or Organic	Rock Name
A	silt	fizzes in HCl	calcite	organic	limestone
B	sand		quartz, feldspar, hematite	detrital	sandstone
C	clay		kaolinite, feldspar	detrital	shale
D	sand & pebble		pebbles composed of any mineral or rock	detrital	conglomerate
E	varies		halite	chemical	rock salt

4 Put a few drops of HCl on each rock sample. Which samples contain calcite?

5 Describe any minerals present.

6 *Classify* your samples. Then, identify each rock.

Analyzing

1. How did examining the sediments help you to identify the detrital rocks?

2. Why did you test the rocks with HCl?

Concluding and Applying

3. Contrast detrital rocks with chemical and organic rocks. How do they differ?

4. **Going Further** Acid dissolved some minerals in the rock. Explain why this is a weathering process.

Discussion Before students do this activity, review with them the various ways that sedimentary rocks are classified. Make sure students understand what information should be written in the table.

Expected Outcomes

Outcomes will vary depending on the samples used. However, students should see that some sedimentary rocks contain calcite because of the bubbling action caused by the HCl.

Answers to Analyzing/ Concluding and Applying

1. Detrital rocks are identified by the kinds of sediments they contain.

2. to see if any contained calcite

3. Detrital rocks have easily identifiable sediments in them. Chemical sedimentary rocks precipitate from solution. Organic sedimentary rocks are formed from plant and animal remains. Detrital rocks are made of broken fragments of other rocks.

4. The material holding the sediments together was chemically dissolved by the acid solution. This is an example of chemical weathering.

✔ Assessment

Process Have students work in small groups to write the text for a brief field guide to sedimentary rocks in which they describe different sedimentary rocks and list characteristics for identification. Use the Performance Task Assessment List for Booklet or Pamphlet in **PASC**, p. 57. COOP LEARN P

Meeting Individual Needs

Visually Impaired Have students work with a fully sighted partner to do the Investigate activity. As the students perform each phase of the activity, have them discuss what each one has observed. Together they can record the data and observations.

Figure 11-14

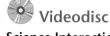

The Rock Cycle

Figure 11-14 shows how weathering, erosion, compaction, and cementation lead to the formation of sedimentary rock. The processes by which Earth materials change to form different kinds of rocks make up the **rock cycle**. The illustration below shows just one part of the rock cycle.

A Weathering loosens and breaks down rock. Some rock pieces fall into moving water, such as a river.

Life Science CONNECTION

Fossil of a mammal-like reptile

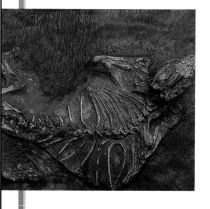

Using Rocks and Fossils to Tell Time

Studying sedimentary rock, such as that exposed in the Grand Canyon, can yield a wealth of information about what has occurred in Earth's past. Geologists use features of the rocks themselves and fossils within the rocks to help understand the history of Earth.

Telling Time with Fossils

Fossils help in many ways, but one of the most important is that they help geologists tell time. Because fossils are the remains or traces of once-living things, they record the features of those living things. You may be aware that living things evolve, or change over time.

Fossils record the changes in characteristics of living things and are therefore physical

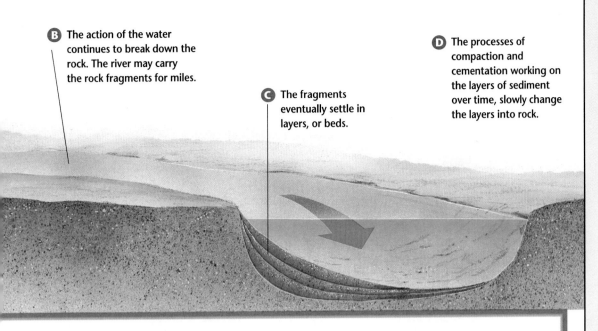

B The action of the water continues to break down the rock. The river may carry the rock fragments for miles.

C The fragments eventually settle in layers, or beds.

D The processes of compaction and cementation working on the layers of sediment over time, slowly change the layers into rock.

Content Background

Coal is an organic sedimentary rock. It is formed from trees and other plants that lived in swampy areas millions of years ago. These plants died and their remains were deposited in shallow, stagnant water. The water slowed down the normal processes of decay, the plants were only partially decomposed, and the decaying material became enriched in carbon. This formed layers of peat, a soft, fibrous material. As more organic plant remains accumulated, increased pressure caused the peat to become compacted into lignite or brown coal. More pressure and compaction caused the lignite to become bituminous or soft coal.

evidence of how plants and animals have changed over time. Careful study of fossils can help a scientist determine how animals preserved as fossils were related—whether one was an ancestor or descendant of another.

The Geologic Time Scale

The fossil record shows us how life has evolved during Earth's history. Geologists have used the fossil evidence found in rock layers to construct what is called the geologic time scale. It is simply time divisions with names for the history of Earth.

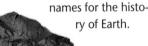

Trilobite

The geologic time scale is based on the fossil evidence of plants and animals. For example, the trilobite shown here is found in rocks that are between 544-505 million years old. This time in Earth's history is known as the Cambrian period.

What Do You Think?

If you found two fossils that were very similar, but one had a feature that indicated that it had evolved from the other, which one would be the older of the two? If you found a fossil in a layer of flat sedimentary rock and then another fossil in the flat sedimentary rock above the first one, which one probably lived *before* the other?

11-3 Sedimentary Rocks **351**

---

tion of Earth can move older layers on top of more recent layers, and how movement along fault lines can break the continuity of layers. Allow students to discuss how using fossils that geologists are able to link to a particular section of time would help them understand the history of rock layers in a certain area of Earth. **L2**

Answers to
What Do You Think?
The fossil with the "advancement" probably was formed after the fossil without it.
The fossil in the lower layer of rock probably was formed before the fossil in the higher layer.

Going Further ▸
Have students sequence the sections of geologic time by having them work in small groups to research the organisms

and geologic events of each section of geologic time. Then ask them to create murals portraying some of the distinctive plants and animals of "Cenozoic Park" or "Cambrian Park." If it is desirable to have fewer students working together in groups, encourage them to work with shorter sections of geologic time. Students may consult a book like *Life Story* by Virginia Lee Burton for details of each era and period. **L1**

3 ASSESS

Check for Understanding

For the Apply question, have students identify what type of rocks granite and slate are. *igneous; metamorphic* Have them describe the path through the rock cycle each could have taken to become part of a sedimentary rock. Use the Performance Task Assessment List for Group Work in **PASC**, p. 97. **L1** **P**

Reteach

Demonstration Pass around samples of shale, sandstone, and conglomerate. Then place the rocks on a cement slab behind a safety screen. Crush the rocks with a hammer. Have students examine the sediments that composed each rock. Have students relate the size of the particles to how the rocks were formed. **CAUTION:** *Wear safety goggles while performing this demonstration.* **L1**

Extension

Activity Have students use reference books to list the chemical composition of several sedimentary rocks. **L3**

4 CLOSE

Activity

Call on one student to pick a group of rocks. Call on another to explain which processes might act on rocks of that group and cause them to change into other rocks. Call on a third student to explain what might happen to these new rocks. Continue until students understand that Earth's rocks undergo continuous changes via the rock cycle. **L1**

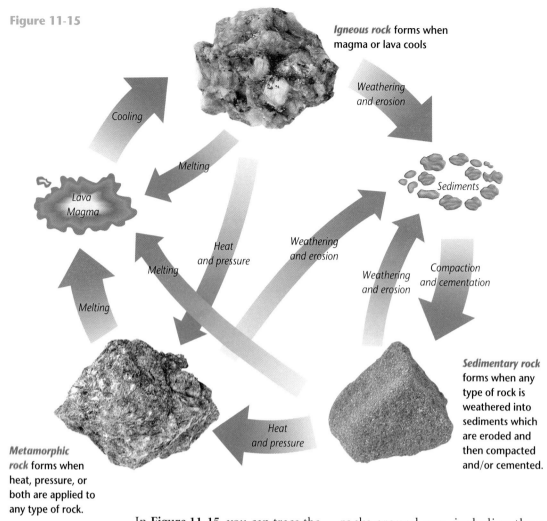

Figure 11-15

Igneous rock forms when magma or lava cools

Cooling

Weathering and erosion

Melting

Sediments

Lava Magma

Heat and pressure

Weathering and erosion

Melting

Weathering and erosion

Compaction and cementation

Melting

Metamorphic rock forms when heat, pressure, or both are applied to any type of rock.

Heat and pressure

Sedimentary rock forms when any type of rock is weathered into sediments which are eroded and then compacted and/or cemented.

In **Figure 11-15**, you can trace the formation of different types of rock. All the rocks you've learned about in this chapter formed through the processes of this cycle. And all the rocks around you, including those used to make buildings, monuments, and even sidewalks, are part of the rock cycle. They are all in a constant state of change.

check your UNDERSTANDING

1. How is sedimentary rock formed?
2. Why can limestone be classified as either a chemical or an organic sedimentary rock?
3. Explain how limestone can change into several other rocks in the rock cycle.
4. **Apply** How can particles or fragments of both granite and slate be found in the same detrital rock?

check your UNDERSTANDING

1. Broken particles of rocks, plants, and animals that have weathered and eroded are cemented together.

2. Limestone can be formed from either dissolved calcite or from shells and hard parts of calcareous animals.

3. Limestone, a sedimentary rock, can change into marble when it is subjected to intense heat. Or it can form into another sedimentary rock by being weathered, eroded, and redeposited.

4. The rock pieces could have originated in totally different places, yet have been weathered and transported to a new location where they became cemented together.

Leisure Connection

Collecting Rocks

You can learn about the history of Earth, understand the work of geologists, and uncover fossils while exploring your environment—just by collecting rocks.

Equipment

You'll need a few tools. A geologist's hammer and chisel are available at a hardware or camping store. A sturdy knapsack will carry tools and any specimens you collect. A pocketknife is handy for scraping rocks. To be safe, make certain to have on sturdy boots, a hard hat and safety goggles.

Maps

Most important for the serious collector is a pocket guidebook for collecting and a map of the area in which you'll be collecting, preferably a topographic map. Look for rocks where natural forces have uncovered them. Streams and cliffs are good places to look. You can also look in places where human activity has exposed rock, such as railroad cuts and quarries.

Fossils

You may be lucky enough to uncover fossils to add to your collection. Look for places where sedimentary rock is exposed. Chip away carefully at the rock and try to keep the fossil intact. Fossil collecting is more difficult than finding rocks. But if you become familiar with the rock formations, and are patient in your search, fossil collecting is an exciting way to make your own discoveries.

You Try It!

Marble, granite, and limestone are just some of the common rock materials used for buildings in most towns and cities. If you live in an urban area and can't find a natural place to collect rocks, use a magnifying lens to examine the materials on buildings around you. You can also determine the geological history of your town by learning which rock materials are from the region, and which are imported. Old stone buildings may even have fossils "built" right into them!

Purpose
The Leisure Connection reinforces all sections of this chapter by encouraging students to collect and identify various rocks in their community.

Content Background
The rocks found in a region can be used to piece together a history of its formation. For example, the discovery of igneous rocks in an area can indicate a volcanic origin. Sedimentary rocks might indicate that the area was once underwater. Fossils may be found in this type of rock and could be used to piece together some of the life history of the region. Metamorphic rocks might indicate great temperature and pressure changes within Earth's crust.

Teaching Strategies Obtain a rock collection that your students can examine. A collection belonging to a fellow student, member of the school staff, or local personality could further motivate and encourage students to start a rock collection of their own. You might also demonstrate a rock polisher which can be used to polish collected rocks in order to make jewelry.

Going Further ⅢⅢⅢ➡
Have students build a classroom rock collection by working in groups of three or four. Organize a field trip so that students can search for rocks in a promising area such as a quarry, or provide each group with several rock samples. Using a rock guide and a hand lens, have students try to identify their rocks. Then have them mount the rocks on cards, label them, and place them in the area designated for the collection. Conduct a field trip to a local museum to view rock, mineral, and gem collections. Students can write to the Museum of Natural History, New York, NY, to obtain more information about rock collections. Use the Performance Task Assessment List for Display in **PASC,** p. 63. **L1** **COOP LEARN**

Science and Society

Science and Society

Who Owns the Rocks?

Let's say you lean over your back fence and pick up an interesting rock from your friend's yard. Does this rock belong to you or to your friend? Did you know that there are laws to decide questions like this?

Black Gold

Let's say you own a house and the property on which your house is built. You decide to build a swimming pool in your backyard. While digging a hole for your pool, you discover oil right there in your own backyard! The newspaper publishes a story on your lucky find, but you don't really know anything about the oil business. Then the phone starts to ring. People are offering you money in return for the rights to the oil in your yard. What would you do?

Your Rights

In the United States, the law says that if you own the land, you own anything buried beneath that land. If you live in an urban area, chances are that as a landowner you have restricted rights, meaning you can't, for example, just stick an oil well in the middle of a busy city. But many landowners do own the land, the rocks beneath the land, and any mineral deposits found there. Actually, the law says that, in theory, as a landowner, you own your property all the way to the center of Earth.

You could let someone pay you to drill for oil in your backyard. What you are doing is selling the mineral rights to your land. The person who buys your mineral rights can drill for oil, but that person has only bought the rights to the minerals, not to your land.

EXPAND your view

Law of Capture

As a landowner, you own your land, and the rock formations underneath the ground. But you don't automatically own all the resources—such as gas, oil, and water—on the land or embedded within the rock layers beneath the surface. That's because such resources may shift in position, even when they are located underground. If your neighbor started drilling an oil well and was able to drain the oil from your land through his well, that is legal. It is legal because of the law of capture. This law says that if you stay on your own land but can capture such a resource, it now belongs to you.

The Controversy

The law of capture has become more controversial in this country in recent years because of the growing shortage of fresh water. It is legal for someone to drain water from an underground source on someone else's land. Such an action, however, may leave others without groundwater, and some people don't think this is fair. The law of capture is also an issue now because some oil companies have discovered ways to drain the oil from miles of underground oil fields.

Our government owns thousands of acres of land across the country. When it comes to the question of who should have the right to search rock formations for mineral deposits in these areas, there is disagreement. Environmental groups are opposed to mining and drilling in these areas because of possible environmental damage. Corporations fight for the chance to search for mineral deposits on those properties.

Science Journal

The government is considering selling off more of its mineral rights. The oil industry is in favor of more land becoming available for them to use. But some people are opposed to the idea because some of the methods used to extract oil from the ground have caused damage to the surface land.

Environmentalists want the government to keep certain wilderness areas pure and unspoiled. They want the government to find alternative energy sources.

In your Science Journal discuss how you think the government should respond to the oil industry's and environmentalists' requests.

EXPAND your view

What Do You Think?

Accept all answers that the students can justify.

Going Further ▐▐▐▐▶

Have students work in groups of three or four to hypothesize a scenario in which the discovery of a mineral or oil deposit might cause problems or conflicts. For example, one scenario might be the discovery of oil near a reef that is protected by the U.S. government. If the oil were extracted, (etc.). . . . Have them formulate models for safe and less damaging ways of extracting the deposit from Earth. Use the Performance Task Assessment List for Group Work in **PASC**, p. 97.
COOP LEARN

Where are the Flaming Cliffs?

Purpose

Igneous rocks are formed from molten material, and metamorphic rocks are formed when rocks undergo intense heat and pressure. Both processes often destroy any fossil remains in the rock. The processes that form sedimentary rocks, however, are conducive to fossil preservation. These SciFacts expand on the discussion on sedimentary rocks in Section 11-3 by giving students an example of a sedimentary rock formation, the Flaming Cliffs in north central China, that was the site of a great paleontological discovery.

Content Background

Until Andrews's discovery, paleontologists were divided over whether dinosaurs laid eggs, like reptiles, or gave birth to live young, like mammals. The large number of bones discovered at the site—which included individual specimens of different sizes and ages—allowed dinosaur growth patterns to be studied for the first time in depth. In addition, paleontologists were able to confirm that the fossils included specimens of both sexes—another first for research purposes.

Discussion

Ask students to discuss how the compaction and cementation of sedimentary rocks is related to the mold-and-cast method of fossil formation. Lead them to understand that when the hard parts of an organism are encased and buried in sediment, compaction and cementation turn the sediment into rock. Pores in the rock let water and air reach the organism's remains, decaying them and leaving behind a cavity in the rock called a mold. Later, other sediments may fill in the cavity, harden into rock, and produce a cast of the original organism, called a fossil.

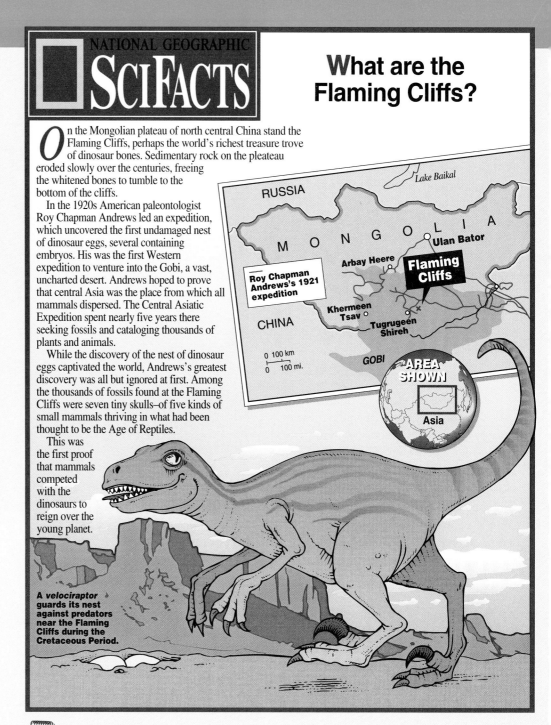

What are the Flaming Cliffs?

On the Mongolian plateau of north central China stand the Flaming Cliffs, perhaps the world's richest treasure trove of dinosaur bones. Sedimentary rock on the pleateau eroded slowly over the centuries, freeing the whitened bones to tumble to the bottom of the cliffs.

In the 1920s American paleontologist Roy Chapman Andrews led an expedition, which uncovered the first undamaged nest of dinosaur eggs, several containing embryos. His was the first Western expedition to venture into the Gobi, a vast, uncharted desert. Andrews hoped to prove that central Asia was the place from which all mammals dispersed. The Central Asiatic Expedition spent nearly five years there seeking fossils and cataloging thousands of plants and animals.

While the discovery of the nest of dinosaur eggs captivated the world, Andrews's greatest discovery was all but ignored at first. Among the thousands of fossils found at the Flaming Cliffs were seven tiny skulls–of five kinds of small mammals thriving in what had been thought to be the Age of Reptiles. This was the first proof that mammals competed with the dinosaurs to reign over the young planet.

A *velociraptor* guards its nest against predators near the Flaming Cliffs during the Cretaceous Period.

Map labels: Lake Baikal, RUSSIA, MONGOLIA, Ulan Bator, Arbay Heere, Flaming Cliffs, Roy Chapman Andrews's 1921 expedition, Khermeen Tsav, CHINA, Tugrugeen Shireh, 0 100 km, 0 100 mi., GOBI, AREA SHOWN, Asia

Science Journal

In your Science Journal, describe what the Age of Mammals was and suggest why it was given that name.

Science Journal

Students should mention that the Age of Mammals marked the end of the Mesozoic Era and the beginning of the Cenozoic Era. It is referred to as the Age of Mammals because mammals replaced dinosaurs as the dominant life-form on Earth.

Science Journal

Review the statements below about the big ideas presented in this chapter, and answer the questions. Then, re-read your answers to the Did You Ever Wonder questions at the beginning of the chapter. *In your Science Journal*, write a paragraph about how your understanding of the big ideas in the chapter has changed.

Heat and pressure

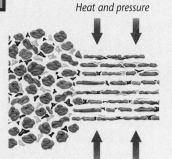

1 Intrusive igneous rocks form below Earth's surface and generally contain large mineral grains. Extrusive igneous rocks form on Earth's surface and generally have small mineral grains. *Explain how a volcano might cause both intrusive and extrusive igneous rocks.*

2 Pressure and heat can change minerals to form metamorphic rock. *Where on Earth is metamorphism most likely to occur: at the surface or deep within the crust? Why?*

3 Weathering breaks rocks and plant and animal remains into small sediments. *Explain the role of weathering in the formation of sedimentary rocks.*

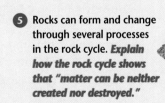

4 Sediments can be compacted, cemented, or precipitated out of solution to form sedimentary rock. *Describe the formation of rock salt. Is it an example of a detrital, organic, or chemical sedimentary rock? Why?*

5 Rocks can form and change through several processes in the rock cycle. *Explain how the rock cycle shows that "matter can be neither created nor destroyed."*

Have students look at the photographs and illustrations on this page. Direct them to read the statements to review the main ideas of the chapter.

Teaching Strategies

Divide the class into four groups. Have each of the groups use the photographs and illustrations to write the name of a rock type (igneous, sedimentary, or metamorphic) on a sheet of paper along with its characteristics. Have the fourth group make a list of the ways rocks form or change. Then have the first three groups fasten their papers to the chalkboard using magnetic strips. Have the fourth group make arrows between these papers and label them to complete the rock cycle diagram.
COOP LEARN

Answers to Questions

1. When the volcano erupts lava emerges and cools quickly forming extrusive igneous rocks. Meanwhile the underground magma chamber contains magma that cools slowly forming intrusive igneous rocks.

2. deep within the crust because temperatures and pressures are much higher there

3. Weathering is an important step in breaking rock apart so that it can be eroded and deposited to form rock.

4. Rock salt forms from the evaporation of seawater. Because it is a precipitate it is a chemical sedimentary rock.

5. All rocks come from other rocks or rock materials, therefore no new matter is involved.

GLENCOE TECHNOLOGY

MindJogger Videoquiz

Chapter 11 Have students work in groups as they play the Videoquiz game to review key chapter concepts.

Science Journal

Did you ever wonder...

• Igneous rocks or lava come from the solidification of magma. Metamorphic and sedimentary rocks can form from igneous rocks. (pp. 330, 339, and 343)

• Rocks are mixtures of minerals. (p. 330)

• Weathered rocks are often smoother and rounder than unweathered rocks. (p. 343)

• Rocks are different colors because of the different minerals they contain. (p. 330)

Project

Have students make a bulletin board display illustrating igneous, sedimentary, and metamorphic rocks. Each student should choose a different rock in each category and draw a picture of the rock. Students should attach a description of how the rock was formed and the minerals it usually contains to their drawings. Use the Performance Task Assessment List for Bulletin Board in **PASC,** p. 59. **COOP LEARN** **L1**

Using Key Science Terms

1. Foliated rocks have mineral grains that are flattened and line up in parallel bands. Nonfoliated rocks don't form any visible bands.

2. Intrusive igneous rocks cool slowly beneath Earth's surface; they have large mineral crystals. Extrusive igneous rocks cool quickly on Earth's surface; they have a fine-grained texture.

3. Igneous rocks form when magma cools and forms crystals. Metamorphic rocks form when pressure or heat changes the minerals in rocks. Sedimentary rocks form when weathered or eroded rock materials are joined through compaction and/or cementation.

4. Two or more minerals make up a rock.

5. The rock cycle is a series of events that changes rocks.

Understanding Ideas

1. igneous, because most rock is beneath Earth's surface and not weathered

2. Rocks are classified by how they form and by their structure. It is necessary to classify rocks to make them easier to study.

3. Sedimentary rocks are held together by compaction and cementation.

4. Heat and pressure are two processes that work together to form metamorphic rocks.

5. To form sedimentary rocks, you need sediments and a cementing agent.

6. Fossils are usually found in sedimentary rocks, because other rock types are either inhospitable to life or have undergone changes that may have destroyed fossils.

7. Sedimentary rocks are classified by their sediment type: chemical, organic, or detrital.

Developing Skills

1. See annotations.

Using Key Science Terms

extrusive
foliated
igneous rock
intrusive
metamorphic rock

nonfoliated
rock
rock cycle
sedimentary rock

Answer the following questions using what you know about science terms.

1. Compare and contrast foliated and non-foliated rocks.

2. Distinguish between intrusive igneous rocks and extrusive igneous rocks.

3. Compare and contrast igneous, metamorphic, and sedimentary rocks.

4. Describe the relationship between minerals and rocks.

5. What is the rock cycle?

Understanding Ideas

Answer the following questions in your Journal *using complete sentences.*

1. What is the most abundant type of rock? Why?

2. How are rocks classified? Why is classification of rocks necessary?

3. How are the particles that make up sedimentary rocks held together?

4. What processes work together to form metamorphic rocks?

5. What is needed in order to form sedimentary rocks?

6. In what kind of rocks will fossils normally be found? Why?

7. What characteristics are used to classify sedimentary rocks?

Developing Skills

Use your understanding of the concepts developed in each chapter to answer each of the following questions.

1. Concept Mapping Using the following processes, label each arrow to complete the concept map of the rock cycle: *weathering and erosion, compaction and cementation, heat and pressure, melting, cooling.*

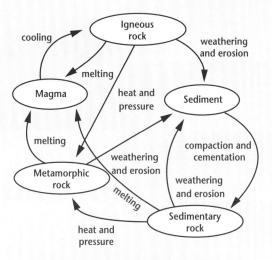

2. Observing and Inferring Infer what would have happened if you had allowed the salol in the Find Out activity on page 331 to cool in the refrigerator.

3. Forming a Hypothesis Refer to the rocks you observed in the Explore activity on page 341. Form a hypothesis stating why metamorphic rocks do not usually contain fossils.

4. Forming a Hypothesis Form a hypothesis explaining why some rock fragments in detrital sedimentary rocks are rounded.

2. The crystals would be small because the salol was cooled quickly.

3. Most rock with fossils that undergoes heat and pressure would not still contain the fossils.

4. Fragments may have been weathered before they were compacted or cemented into a sedimentary rock.

Critical Thinking

In your Journal, *answer each of the following questions.*

1. Describe some effects of heat and pressure on mineral crystals.
2. What might cause intrusive igneous rocks to appear at Earth's surface?
3. How could concrete become part of the rock cycle when it is not a natural rock?
4. Look at the two photographs of igneous rocks. Which is an extrusive igneous rock? Which is an intrusive igneous rock? Explain your answers.

5. Why are metamorphic processes difficult for scientists to study?
6. Why can both sedimentary and metamorphic rocks have bands of colors?

Problem Solving

Read the following problem and discuss your answers in a brief paragraph.

While on a class field trip in the city, you observed rocks used as building materials. In the city square, you noticed flowers arranged in a rock terrace, that is, layers of rock and soil with plants growing among the rock. The rocks were light-colored and contained many small fossils.

Soon your class left the square and entered a historic district with an assortment of buildings. The first building you noticed was light pink with small crystals of quartz that felt gritty to the touch. Continuing down the street, you observed another building with columns. This building was constructed of a light-colored, highly polished rock. Next, you saw a wooden building with a roof made of dark tiles. Some of the tiles had been broken off in layers.

Using what you've learned about rocks, name the rocks that you observed. What rocks or rock-like materials do people use for buildings and other structures?

CONNECTING IDEAS

Discuss each of the following in a brief paragraph.

1. **Theme—Systems and Interactions** What are the processes of the rock cycle that form different types of rocks?
2. **Theme—Scale and Structure** How do igneous rocks form with various colors and textures?
3. **Life Science** **Connection** Explain how fossils can help "tell time."
4. **Leisure Connection** How can rocks tell us about Earth's history?

Critical Thinking

1. Heat and pressure can melt and can change the size and shape of mineral crystals.
2. Faulting of Earth's surface, erosion, or human activity such as strip mining.
3. Concrete is made of rocks and is subject to the processes of the rock cycle.
4. The rock on the right is extrusive because the crystals are small. The rock on the left has large crystals. It is the intrusive rock.
5. Metamorphic processes usually take place beneath the surface of Earth.
6. Sedimentary rocks have bands of color because of different colors of sediments deposited in layers. Bands of color in metamorphic rocks can result from pressure that causes the minerals in the rock to flatten and line up in parallel bands.

Problem Solving

Limestone, granite, marble, slate, concrete, sandstone, and brick are the rocks described in the paragraphs.

Connecting Ideas

1. Erosion, heat, pressure, melting, and weathering are processes of the rock cycle.
2. The color of igneous rock is determined by the minerals of which it is made. The texture of an igneous rock is determined by the rate of cooling.
3. Fossils record the changes in characteristics of living things. By studying and identifying a fossil, scientists can often determine when the fossil was alive.
4. Accept reasonable answers. Rocks can reveal faulting and the presence of such things as volcanoes, oceans, vegetation, glaciers, etc.

✔ Assessment

Portfolio Review the portfolio options that are provided throughout the chapter. Encourage students to select one product that demonstrates their best work for the chapter. Have students explain what they learned and why they chose this example for placement into their portfolios.

Additional portfolio options can be found in the following **Teacher Classroom Resources:**

Making Connections: Integrating Sciences, p. 25
Multicultural Connections, pp. 25, 26
Making Connections: Across the Curriculum, p. 25
Concept Mapping, p. 19
Critical Thinking/Problem Solving, p. 19
Take Home Activities, p. 19
Laboratory Manual, pp. 57–58; 59–60
Performance Assessment P

Chapter Organizer

SECTION	OBJECTIVES	ACTIVITIES & FEATURES
Chapter Opener		**Explore!**, p. 361
12-1 Shore Zones (3 sessions; 1.5 blocks)	1. **Describe** how longshore currents form. 2. **Contrast** steep shore zones and flat shore zones. 3. **List** some origins of beach sand. **National Content Standards: (5-8) UCP2, UCP4, A1, B3, D1, F3-4**	**Explore!**, p. 363 **Design Your Own Investigation 12-1:** pp. 368–369 **Literature Connection**, p. 386
12-2 Humans Affect Shore Zones (2 sessions; 1 block)	1. **Relate** the ways in which human activities pollute shore zones. 2. **Describe** the effects of ocean pollution on sea life. **National Content Standards: (5-8) UCP2, UCP4, A1, C3-5, D1, F2**	**Skillbuilder:** p. 371 **Find Out!**, p. 373 **Life Science Connection**, pp. 372–373 **Science and Society**, pp. 384–385 **Teens in Science**, p. 386
12-3 The Ocean Floor (3 sessions; 1 block)	1. **Describe** some of the methods used to map the ocean floor. 2. **Name and describe** some features of the ocean floor. **National Content Standards: (5-8) UCP2, UCP4, A1, D1-2, E2, F3-5**	**Find Out!**, p. 377 **Investigate 12-2:** pp. 380–381 **A Closer Look**, pp. 382–383

ACTIVITY MATERIALS

EXPLORE!

p. 361 sandstone, sand
p. 363* fine sand, rectangular cake pan, water, 2 rulers

INVESTIGATE!

pp. 380–381* graph paper, blue and brown pencils

DESIGN YOUR OWN INVESTIGATION!

pp. 368–369* 3 different types of beach sand, stereomicroscope, magnet

FIND OUT!

p. 373* small rocks, shallow baking pan, water, 10 mL of vegetable oil, tongue depressors, cotton balls, detergent, paper toweling, feathers, measuring cup
p. 377* graph paper, shoe box with lid, soda straw, metric ruler, modeling clay, rubber bands

KEY TO TEACHING STRATEGIES

The following designations will help you decide which activities are appropriate for your students.

- **L1** Basic activities for all students
- **L2** Activities for average to above-average students
- **L3** Challenging activities for above-average students
- **LEP** Limited English Proficiency activities
- **COOP LEARN** Cooperative Learning activities for small group work
- **P** Student products that can be placed into a best-work portfolio
- 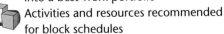 Activities and resources recommended for block schedules

Need Materials? Call Science Kit (1-800-828-7777).

[00:00] OUT OF TIME? We recommend that students do the activities with an asterisk.

Chapter 12 The Ocean Floor and Shore Zones

TEACHER CLASSROOM RESOURCES

Student Masters	Transparencies
Study Guide, p. 40 **Take Home Activities**, p. 19 **Activity Masters**, Design Your Own Investigation 12-1, pp. 51–52 **Critical Thinking/Problem Solving**, p. 5 **Science Discovery Activities**, 12-1 **Laboratory Manual**, pp. 61–64, Waves, Currents, and Coastal Features **Laboratory Manual**, pp. 65–68, Profile of a Coastline	**Section Focus Transparency 34**
Study Guide, p. 41 **Concept Mapping**, p. 20 **How It Works**, p. 15 **Making Connections: Integrating Sciences**, p. 27 **Critical Thinking/Problem Solving**, p. 20 **Multicultural Connections**, p. 27 **Making Connections: Across the Curriculum**, p. 27 **Science Discovery Activities**, 12-2	**Section Focus Transparency 35**
Study Guide, p. 42 **Multicultural Connections**, p. 28 **Activity Masters**, Investigate 12-2, pp. 53–54 **Making Connections: Technology & Society**, p. 27 **Science Discovery Activities**, 12-3	**Teaching Transparency 23**, Ocean Floor Features **Teaching Transparency 24**, Atoll Formation **Section Focus Transparency 36**

ASSESSMENT RESOURCES	TEACHING & TECHNOLOGY
Review and Assessment, pp. 71–76 **Performance Assessment**, Ch. 12 **PASC*** **MindJogger Videoquiz** **Alternate Assessment in the Science Classroom** **Computer Test Bank**	**Spanish Resources** **Cooperative Learning Resource Guide** **Lab and Safety Skills** **Science Interactions, Course 2, CD-ROM** **Computer Competency Activities**

*Performance Assessment in the Science Classroom

NATIONAL GEOGRAPHIC

Index to National Geographic Magazine

The following articles may be used for research relating to this chapter:

- "Life Without Light," by Ian R. McDonald and Charles Fisher, October 1996.
- "Our Polluted Runoff," by John G. Mitchell, February 1996.
- "Rebirth of a Deep-Sea Vent," by Richard A. Lutz and Rachel M. Haymon, November 1994.
- "Deep-Sea Geysers of the Atlantic," by Peter A. Rona, October 1992.

National Geographic Society Products Available From Glencoe

To order the following products for use with this chapter, contact your local Glencoe sales representative or call Glencoe at 1-800-334-7344:

- *GTV: Planetary Manager* (Videodisc)
- *NGS PictureShow Geology*, "Weathering and Erosion." (CD-ROM)

Additional National Geographic Society Products

To order the following products for use with this chapter, call the National Geographic Society at 1-800-368-2728:

- "Atlantic Ocean/Pacific Ocean" (Map Set)
- *Dive to the Edge of Creation* (Video)
- *Let's Explore a Seashore* (Video)
- *The Living Ocean* (Video)
- *Our Dynamic Earth* (Video)
- *What is the Earth made of?* (Video)

Teacher Classroom Resources

These are key components of the classroom resources package.

TEACHING AIDS

Section Focus Transparencies

34 SECTION FOCUS TRANSPARENCY

A DAY AT THE BEACH

Picture a sandy beach with dunes and a clear blue ocean. Sounds like a great vacation spot! The photo shows an area that had all of these things, over 180 million years ago!

1. What do you think caused the distinctive patterns seen in the canyon walls?
2. What mineral probably makes up the rock walls? What type of rock?

L1

35 SECTION FOCUS TRANSPARENCY

SHORELINE HABITAT

The photograph is an aerial view of Miami Beach in Florida. Other than people and a few palm trees, what lives along this shoreline? Very few of the natural inhabitants of the shoreline still remain along highly developed shorelines like this one in South Florida.

1. What other harmful effects do you think can occur along highly developed shore zones?
2. What can be done to preserve shore zone habitats?

L1

36 SECTION FOCUS TRANSPARENCY

BENEATH THE OCEAN FLOOR

This is a seismic section that uses sound waves similar to sonar to provide scientists with a two-dimensional picture of what the ocean floor and Earth's structure below the floor of the ocean looks like.

1. What do you think scientists explore for, using seismic sections like this?
2. What other methods can scientists use to explore the ocean floor?

L1

Teaching Transparencies

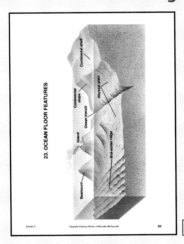

23. OCEAN FLOOR FEATURES

L1

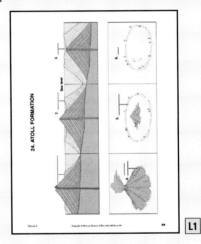

24. ATOLL FORMATION

L1

HANDS-ON LEARNING

Science Discovery Activity*

ACTIVITY 12-1

Sediment Sorting

Ocean floors near the coasts of continents are covered with deep layers of sediment. These materials have been carried to the sea by rivers and deposited into the ocean.

Getting Started
Sediments don't just pile up at the mouth of a river. When a river empties into the sea, the motion of the river water carries the sediments a good distance into the ocean water. As the sediments settle, they become sorted by size, shape, and weight. If you examined these sediments, what would you expect to find?

Hypothesize
Make a hypothesis about how sediments carried by a river are sorted when deposited in the ocean. Are the smallest and lightest sediments found closest to the mouth of the river or farthest out on the ocean floor?

Try It!
Your group will have the following:
- container of mixed sediments
- jar with screw-on lid
- stack of newspapers (about 5 cm) or book
- trough (to carry sediments)
- large pan
- water

1. Pour water into the large pan to a depth of about 5 cm. Lift one end of the pan and rest it on the stack of newspapers or book as shown in the drawing.

65

L1

Laboratory Manual*

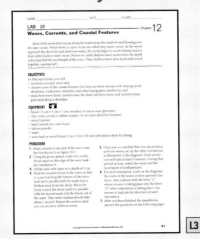

NAME _____ DATE _____ CLASS _____

LAB 20 Chapter **12**

Waves, Currents, and Coastal Features

Most of the waves that you see along the seashore are the result of wind blowing over the open ocean. Wind waves in open ocean are called deep water waves. As the waves approach the shoreline and shallower water, the waves begin to touch bottom and are then called shallow water waves. Waves are called shallow water waves when the depth is less than half the wavelength of the wave. These shallow water waves slow down and crowd together, causing surf.

OBJECTIVES
In this experiment, you will
- construct a simple wave tank.
- observe some of the coastal features that form as waves interact with sloping sandy shorelines, underwater obstacles, and other topographic landforms, and
- study how waves form currents near the shore and how waves and currents transport sand along a shoreline.

EQUIPMENT
- block (15 cm × 5 cm × 7 cm, wooden) to use as wave generator
- clay, rocks, wood or rubber stoppers (to simulate shoreline features)
- pencil (grease)
- sand (mixed sizes and clean)
- talcum powder
- water
- water tank or wood frame (1 m × 1 m × 10 cm) with plastic sheet for lining

PROCEDURE
1. Make a beach at one end of the water tank. See Simulation A in Figure 20-1.
2. Using the grease pencil, make two marks 10 cm apart on the edge of the wave tank. See Simulation A.
3. Fill the tank with water to a depth of 3 cm.
4. Hold the wooden block in the water so that it is just touching the bottom of the wave tank and is parallel with the mark that is farthest away from the shore. Move the block toward the shore until it is parallel with the second mark. Lift the block out of the water. This entire motion should take about 1 second. Repeat the motion until you can produce uniform waves.

5. Once you are satisfied that you can produce uniform waves, set up the other simulations as illustrated in the diagrams. Each simulation will take at least 3 minutes. During that period of time, watch the waves and the movement of sand particles.
6. For each simulation, mark on the diagrams the crests of the waves at their approach the shore. Also indicate with the letter "e" where erosion is taking place and the letter "d" where deposition is taking place. Use arrows to indicate the direction of sand movement.
7. After you have finished the simulations, answer the questions on the following pages.

61

L3

Take Home Activity

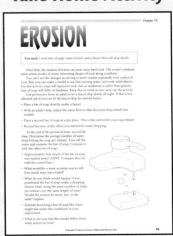

Chapter 12

EROSION

You need: 2 new bars of soap (same brand) and a faucet that will drip slowly

Over time, the motion of waves can wear away hard rock. The ocean's constant wave action results in many interesting shapes of rock along coastlines.

You can't see the changes occurring as water washes repeatedly over a piece of rock. But, you can make a model to see that moving water can erode solid objects. For this activity, soap will represent rock. Just as sandstone is softer than granite, bars of soap will differ in hardness. Keep this in mind as you carry out the activity.

Get permission from an adult to let a faucet drip slowly all night. If this is not allowed, ask if you can let the faucet drip for several hours.

- Place a bar of soap directly under a faucet.
- With an adult's help, adjust the water flow so that the water drips slowly but steadily.
- Place a second bar of soap in a dry place. This is the control for your experiment.
- Record the time of day when you started the water dripping.

At the end of the period of time, record the time. Determine the average number of water drops hitting the soap per minute. Turn off the water and examine the bar of soap. Compare it with the other bar of soap.

- Approximately how much of the bar of soap was washed away? (HINT: Compare this bar with the control bar.)
- What would be a more accurate way to tell how much mass was eroded?
- What do you think would happen if you positioned the bar of soap under a dripping shower head, using the same number of drops per minute over the same length of time? Would the erosion be more, less, or the same? Explain.
- Estimate how long a bar of soap like yours might last under the conditions in your experiment.
- What is one way that this model differs from water action on rock?

19

L1

*There may be more than one activity for this chapter.

Chapter 12 The Ocean Floor and Shore Zones

Study Guide*

Concept Mapping

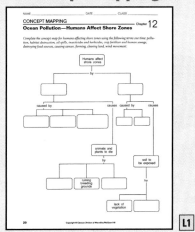

Critical Thinking/ Problem Solving

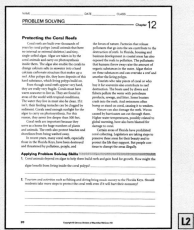

Integrating Sciences

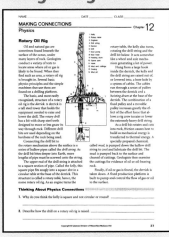

Across the Curriculum

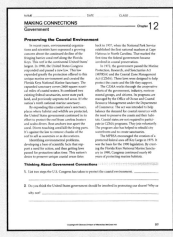

Technology and Society

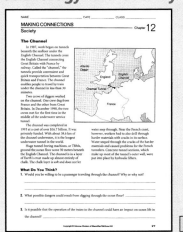

Multicultural Connection**

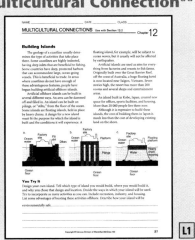

Performance Assessment

Review and Assessment

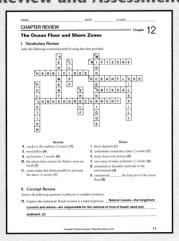

*One per section **Two per chapter

Chapter 12   The Ocean Floor and Shore Zones   **360D**

The Ocean Floor and Shore Zones

THEME DEVELOPMENT

The themes of this chapter are systems and interactions and scale and structure. Students discover how waves change rocks and other materials into sand and how currents move that sand to change the shoreline. The chapter describes the variety of geological structures on the ocean floor and discusses the interactions that cause pollution of ocean–shore systems.

CHAPTER OVERVIEW

Nearly three-fourths of Earth is covered with water and most of it is salty. The contact that this salt water makes with land is called a shore zone. Students will discover that shore zones can be steep, rocky, flat or sandy. Students will also see how human activities can adversely affect shore zones. Geological structures on the ocean floor are described.

Tying to Previous Knowledge

Ask students to think of examples of natural changes they have seen in the world around them. Examples might include erosion, snow blown into drifts, trees blown over in storms, or flooding.

INTRODUCING THE CHAPTER

Have students design an under-water city. Ask students to carefully consider what features such a city would require under these conditions.

⏱ 00:00 OUT OF TIME?

If time does not permit teaching the entire chapter, use the Chapter Overview on this page, Reviewing Main Ideas at the end of the chapter, and the Chapter 12 audiocassette to point out the main ideas of the chapter.

THE OCEAN FLOOR AND SHORE ZONES

Did you ever wonder...

✓ Where beach sand comes from?

✓ Why beaches are sometimes closed for health reasons?

✓ What land beneath the ocean looks like?

📝 **Science Journal**

Before you begin to study about the ocean floor, think about these questions and answer them *in your Science Journal.* When you finish this chapter, compare your journal write-up with what you have learned.

Answers are on page 387.

You've probably seen movies or read books about people living on the ocean floor. In your Journal, describe what an ocean city might be like. In which ocean would you build such a city? Would it matter on which part of the ocean floor you placed it?

Do you think of "oceans" when you hear the word Earth? If you're like many people, you think about forests, mountains, and other landforms. After all, the land is where we live and play. Very few people stop and think that land makes up only about one-fourth of Earth's surface.

Most of the water on Earth is found in the oceans. The oceans affect the climate in your area and around the world. People obtain food from ocean waters, and they use the oceans for recreation. People even misuse the oceans by dumping wastes into them.

▶ *In the following activity, you'll begin to discover why some shore zones are covered with sand.*

360

Learning Styles	Kinesthetic	Explore, p. 363; Activity, pp. 367, 383; Find Out, p. 373
	Visual-Spatial	Explore, p. 361; Demonstration, pp. 362, 374, 382; Visual Learning, pp. 363, 364, 365, 366, 371, 372, 376; Across the Curriculum, p. 366; Investigate, pp. 368-369, 380-381; Research, p. 375; Discussion, p. 376; Activity, p. 381; Project, p. 387
	Interpersonal	Activity, pp. 365, 370
	Logical-Mathematical	Across the Curriculum, pp. 364, 365; Find Out, p. 377
LS	**Linguistic**	Discussion, pp. 364, 367, 374, 375, 378; Multicultural Perspectives, pp. 365, 371; Research, p. 366; Across the Curriculum, pp. 371, 372; Life Science Connection, pp. 372-373; Visual Learning, pp. 374, 378; A Closer Look, pp. 382-383

Explore! ACTIVITY

From rock to sand. How can it happen?

When you think of oceans, the beach probably comes to mind. You may be familiar with the sand that's found on many shores, but do you know how the sand arrived there?

What To Do

1. Hold a piece of sandstone in one hand and some sand in the other.

2. Feel the differences and similarities between these two materials.

3. Now rub two pieces of sandstone together. Describe what happens.

4. How do the particles of sandstone compare with the sand? Hypothesize what forces in nature could possibly cause large rock surfaces to become as small as sand.

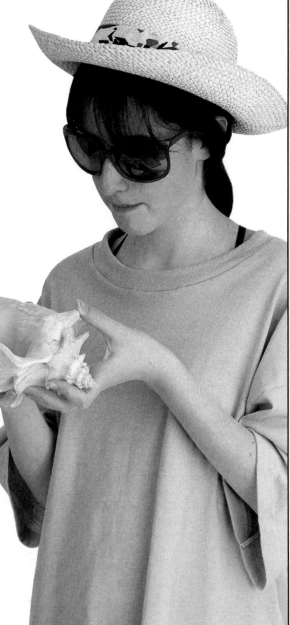

Explore!

From rock to sand. How can it happen?

Time needed 15–20 minutes

Materials pieces of sandstone, loose sand

Thinking Process observing and inferring

Purpose To demonstrate that sand comes from rock and other hard materials.

Preparation Obtain sand that is about the same size as the sand grains that compose the sandstone samples.

Teaching the Activity

Science Journal Suggest that students record their observations in a table in their journals. Students should easily see some relationship between the sand and the sandstone. Both are gritty, and the grain sizes and shapes are similar. However, the grains making up the sandstone are cemented while the grains in the sample of sand are loose. **L1**

Expected Outcomes

Students will observe that both feel gritty to the touch; grains in both samples will most likely be rounded.

Answer to Question

4. Students should compare size, roughness, and appearance. Forces might include those from moving water, wind, or ice.

✔ Assessment

Content Have students use hand lenses to identify minerals in the sand and sandstone samples. Then students can make a table comparing and contrasting the composition of the sand with that of the sandstone. Use the Performance Task Assessment List for Data Tables in **PASC,** p. 37.

ASSESSMENT PLANNER

PORTFOLIO
Refer to page 389 for suggested items that students might select for their portfolios.

PERFORMANCE ASSESSMENT
Process, pp. 363, 369
Skillbuilder, p. 371
Explore! Activities, pp. 361, 363
Find Out! Activity, pp. 373, 377
Investigate, pp. 368–369, 380–381

CONTENT ASSESSMENT
Oral, p. 373
Check Your Understanding, pp. 367, 375, 383
Reviewing Main Ideas, p. 387
Chapter Review, pp. 388–389

GROUP ASSESSMENT
Opportunities for group assessment occur with Cooperative Learning Strategies.

PREPARATION

Planning the Lesson

Refer to the Chapter Organizer on pages 360A-D.

Concepts Developed

This section describes how waves contribute to the rock cycle by breaking down rocks, coral, and shells, to form sand. Shores types are contrasted.

1 MOTIVATE

Bellringer

 Before presenting the lesson, display **Section Focus Transparency 34** on an overhead projector. Assign the accompanying **Focus Activity** worksheet. [L1]
[LEP]

Demonstration Demonstrate how wind causes waves by blowing across the surface of a pie plate half full of water. Have volunteers experiment to determine the relationship between the force of the "wind" and the height of the waves created. [L1]

2 TEACH

Tying to Previous Knowledge

Write on the chalkboard *erosion, deposition,* and *waves.* Have volunteers define the meaning of each term.

Section Objectives

■ Describe how longshore currents form.

■ Contrast steep shore zones and flat shore zones.

■ List some origins of beach sand.

Key Terms

longshore current

Changing Shore Zones

Imagine yourself sitting on a beach. You hear the gentle surf and watch gulls as they circle around a lighthouse, resting atop rocks and sand dunes yards away from the ocean's edge.

Now look at **Figures 12-1A** and **B**, which show sand disappearing from the beaches along the ocean. Structures cannot stand for very long without special supports along an eroding beach. The sand dunes that the buildings once stood upon are being carried away, bit by bit, by the ocean.

Shore zones constantly change. They change because waves and currents are constantly eroding and depositing sediments along the shore. In the next activity, you will discover how this happens.

Figure 12-1

Waves carry energy. When they hit land, waves release their energy against the shoreline and cause erosion. Much shoreline erosion takes place during storms when strong waves can make dramatic changes in a few hours.

A When the Cape Cod Lighthouse was first erected in 1797, the distance from the tower base to the cliff edge was 155m. By 1903 the forceful energy of waves against the shoreline eroded that distance to 95 m. In 1993 only 35 m of land stood between the tower and the edge of the cliff. If the structure is left as is, and the shore eroded at about the same rate, in what year might the sea claim this historic structure?

362

GLENCOE TECHNOLOGY

Software

Computer Competency Activities

Chapter 12

((**Program Resources**))

Study Guide, p. 40
Laboratory Manual, pp. 61–64, Waves, Currents, and Coastal Features [L3]
Take Home Activities, p. 19, Erosion [L1]
Science Discovery Activities, 12-1
Section Focus Transparency 34

Explore! ACTIVITY

How do waves affect the shoreline?

In this activity, you will determine whether the angle at which waves strike the shore makes any difference in their effects on the land.

What To Do

1. Pour enough fine sand into one end of a rectangular cake pan to form a small beach.
2. Add water to the pan until it is half full. At the end of the pan away from the sand, move a ruler back and forth on the surface of the water to create a series of waves that move directly parallel toward your beach. Record your observations *in your Journal.*
3. Now change the angle of your ruler to form waves that strike the beach at an angle.
4. In the sand, place a short ruler perpendicular to the beach, as shown.
5. Once again, create waves that strike the beach at an angle. Observe the shape of the beach near the ruler.

Conclude and Apply

1. How did the "groyne" that you constructed with the short ruler affect shore erosion?
2. Describe how building a groyne could protect a house on the shore zone.

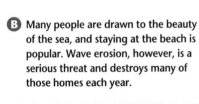

B Many people are drawn to the beauty of the sea, and staying at the beach is popular. Wave erosion, however, is a serious threat and destroys many of those homes each year.

Explore!

How do waves affect the shoreline?

Time needed 30 minutes

Materials fine sand, rectangular cake pans, rulers, water

Thinking Processes observing and inferring, recognizing cause and effect

Purpose To discover the effects of waves approaching a sandy shore from different angles.

Preparation Depending on the size of the pans used, 6-inch rulers or tongue depressors may be used to create the waves.

Teaching the Activity

The sand must be piled up to form a slope at one end of the pan, with the highest part above the level of the water. Rulers should be moved smoothly. Some experimentation will be needed to create a smooth wave action. Encourage students to maintain wave action in a given di-

Longshore Currents

As you continue with this section, you will be able to apply what happened in your model shoreline in the previous activity to actual events. When a low, gentle wave breaks on the shore, it moves sand onto the beach. At other times, as during a storm, waves remove more sand than they bring in, and the beach is eroded.

In most places, waves approach at a slight angle to the shore. Therefore, the sand is pushed along the beach. A **longshore current** is a flow of ocean water that runs close to the shore and parallel to it. **Figure 12-2** diagrams such a current for you.

Figure 12-2

Waves approaching the shoreline at an angle push water along the shore, causing what is called a longshore current.

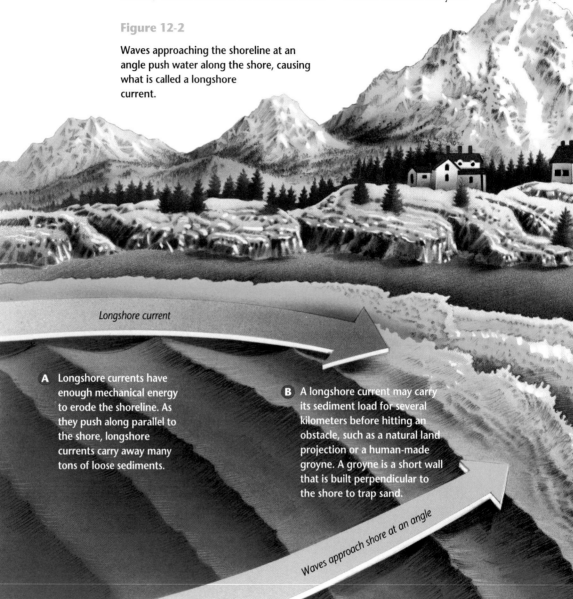

Longshore current

A Longshore currents have enough mechanical energy to erode the shoreline. As they push along parallel to the shore, longshore currents carry away many tons of loose sediments.

B A longshore current may carry its sediment load for several kilometers before hitting an obstacle, such as a natural land projection or a human-made groyne. A groyne is a short wall that is built perpendicular to the shore to trap sand.

Waves approach shore at an angle

Steep Shore Zones

All shore zones are affected by waves. Yet some shore zones are steep and rocky, with little if any beach area.

Rock fragments produced by the waves become sediment. But the constant pounding motion of the water in a steep shore zone prevents most of the sediment from settling at the base of the cliffs. Waves and longshore currents carry it away and deposit it in quieter waters somewhere along the coastline.

Figure 12-3

In steep shore zones, incoming waves smash into cliffs and rocks and give up their energy all at once. Sometimes the waves curl over and collapse on themselves, trapping and compressing air. The compressed air explodes, shooting ocean spray high into the air. The same violent pounding motion that erodes a steep shore zone also prevents most of the eroded sediment from settling there.

Down-drift side

Up-drift side

D Sediments deposited by a longshore current build up on the up-drift side of a groyne and enlarge the beach. But the groyne cuts off the normal supply of drift sand on its down-drift side. The eroded beach sediments are not replaced, and the shoreline erodes at a faster than normal rate. Would placing groynes close together or far apart best help solve this problem?

Groyne

When the longshore current hits an obstacle, the current's energy is transferred to the obstacle. With its energy gone, the current slows down and drops its sediment load.

365

Activity To simulate how far different-sized sediments are carried by waves, have students working in small groups perform this activity. Instruct students to fill a small glass jar three-quarters full of water. Then have them slowly add 1 tablespoon each of gravel, sand, silt, and clay to the jar. Tell them to put the lid on the jar and allow the particles to settle. Then have them shake the jar a few times and observe which sediment settles first. *Gravel settles first followed by sand, silt, and clay, respectively.* Have students hypothesize which sediments are deposited closest to the shore and which are carried the farthest. *Larger sediments are dropped closer to the shore while finer sediments are carried into deeper waters.* **L2** **COOP LEARN**

GLENCOE TECHNOLOGY

 CD-ROM

Science Interactions, Course 2, CD-ROM

Chapter 12

Across the Curriculum

Mathematics

Have students solve the following problems related to waves and beach erosion. Assume that an average "lifetime" is 75 years. **If 200 million waves hit a given rock in your lifetime, about how many waves will hit that rock in a year?** *2 666 667* **A week?** *51 282* **A day?** *7326* **An hour?** *305* Now assume that one hundred waves remove one milligram of rock. **About how much rock will be removed in an hour?** *3 milligrams* **In a day?** *73 milligrams* **A week?** *513 milligrams* **A year?** *26 667 milligrams* **75 years?** *2 000 000 milligrams*

Multicultural Perspectives

Native American Legends

Have students find several Native American legends that describe the relationship between Earth and its waters. Findings can be presented to the class as posters, oral presentations, or written reports. **L1** **LEP**

Visual Learning

Figure 12-3 **Would placing groynes close together or far apart best help solve this problem?** *Placing groynes close together allows the up-drift sediment catch of one groyne to reach the down-drift side of the preceding groyne.*

Flat Shore Zones

If you like a sandy beach, a flat shore zone is the place for you. Beaches are made of different materials. Some are made up of sand and stones, while others consist of shell fragments. You can find beach fragments ranging in size from fist-sized rocks to grains of sand almost as fine as powder.

Does it surprise you that most beach sand comes from sediment carried to oceans by rivers? Some sand is

Figure 12-4

Beaches extend as far inland as the tides and waves can deposit sediments. Beaches also extend some distance out below the surface of the water.

A Beaches are made of various materials. Some, like the beach pictured above, are mostly sand and small stones.

B Some beaches are mostly shell fragments and coral skeletons. The beach in Stuart, Florida, pictured below, is made up of shell fragments and sand.

formed as waves weather the rocks, shells, or coral found in the shore zone. Many beach sands contain a lot of quartz.

Other types of sand are made up of organic materials. Warm ocean waters, such as those of the Caribbean Sea, contain abundant marine life. The white beaches of a Caribbean island like Jamaica are made up of many fragments of seashell and coral.

Perhaps the next time you visit the seashore, or see scenes of the shore, you'll stop and think about the type of shore zone you're looking at and some of the factors that formed it. Keep in mind, however, that all shore zones are subject to constant change. In addition to the daily effects of winds, waves, and currents, humans have a great effect on shore zones. You will learn about some of these effects in the next section of this chapter.

C Sometimes you can find a wide variety of sediment fragments on the same beach. The beach in the St. Joseph Peninsula State Park, Panhandle Area, Florida has fragments that range in size from fist-sized rocks to powder-fine grains of sand. Which of the beaches shown on these pages would you most like to spend time on? Why?

check your UNDERSTANDING

1. How do longshore currents form?
2. Contrast the characteristics of a steep or rocky shore zone with those of a flat shore zone.
3. Are all beach sands alike? Explain your answer.

4. **Apply** At a nearby lakeshore, you notice that a long, low wall, just a little above the water's surface, is being built about 100 meters from shore. What effect might this construction have on the lake's shore?

check your UNDERSTANDING

1. Longshore currents are formed by waves moving at a slight angle toward the shore.
2. Steep shore zones are characterized by cliffs and features carved from the cliffs by waves. Flat shore zones are characterized by broad, flat, sandy beaches.
3. No, beach sands vary according to the source materials from which they formed.
4. Beach erosion should be reduced.

check your UNDERSTANDING

3 ASSESS

Check for Understanding

Assign the questions under Check Your Understanding. As you ask each question, have students jot down answers with words or a sketch.

Reteach

Discussion Ask students to imagine that someone put dishes into a washing machine instead of into a dishwasher. Students should discuss what would happen. Have them explain what would happen if the washing machine were rerun many times. *The dishes would be shattered. If the machine were run many times, the shattered dishes would break into smaller and smaller pieces.* Have students list the factors involved here: *The force of the water and the motion of the washing machine cause the dishes to crash against the walls of the machine and against one another. Continued action produces smaller and smaller pieces.* Have students compare this to changes along the shore.

Extension

Activity People use a variety of barriers to try to reduce the amount of sand that is removed from a beach. One such structure is a groyne, a rock wall that extends into the sea, perpendicular to the shoreline. Have students discover the effects of a groyne on changes in shorelines by repeating the Explore activity on page 363, using the stones to make a groyne. **L2**

4 CLOSE

Display pictures of flat and steep seacoasts. Suitable pictures can be found in photo books from the library, such as *America's Seashore Wonderlands* (National Geographic Society, 1985). For each picture, ask students to describe how a specific feature may have been formed.

12-1 Beach Sand

Preparation

Purpose To determine the characteristics of several samples of beach sand.

Process Skills classifying, comparing and contrasting, observing and inferring, interpreting data, measuring, communicating, making and using tables

Time Required 30–40 minutes

Materials samples of 3 different types of beach sand, stereomicroscopes, magnets

Possible Hypotheses Students may hypothesize that beach sand may be classified according to grain size, texture, color, luster, composition, and roundness.

Plan the Experiment

Process Reinforcement Most students should be able to easily observe color, grain sizes, and grain shapes. Some students might even be able to infer the composition of individual grains.

Possible Procedures First observe one of the sand samples using the stereomicroscope, and record all observations. Then place the sand grains from one sample in the middle of the circle of the sand gauge to determine the average size of the grains. Describe the texture and roundness of the sample. Determine if a magnet will attract grains from the sample. Then repeat these steps for the other two sand samples.

Teaching Strategies

Troubleshooting Caution students not to mix the sand samples. They will need to examine only a small quantity of sand. When they are finished with the sample, they should not try to put it back into its original container. Instead, they should dispose of the sand in a wastebasket or outdoors—not down a drain.

Beach Sand

You have learned how to identify different types of minerals and rocks by looking for certain distinctive characteristics. Can the composition of individual sand grains be used to identify and classify beach sands?

Preparation

Problem
What characteristics can be used to classify different samples of beach sand?

Form a Hypothesis
As a group, discuss what characteristics of beach sand would be most helpful for classification purposes. Then form a hypothesis about which of these characteristics you will observe to classify beach sand.

Objectives
- Determine which characteristics of beach sands can be used to identify different sands.
- Identify the rock and mineral compositions of different beach sands.

Materials
samples of 3 different
   types of beach sand
stereomicroscope
magnet

Meeting Individual Needs

Visually Impaired In the Investigation, visually impaired students can be responsible for step 2, determining the texture of the sand samples. For students with a moderate degree of impairment, provide extra large magnifying glasses.

DESIGN YOUR OWN
INVESTIGATION

Plan the Experiment

1 As a group, agree upon a way to test your hypothesis. Write down what you will do at each step of your test.

2 Make a list of the characteristics you will observe to classify the beach sands. You can use the sand gauge shown here to determine the average size of the grains. What other characteristics will you observe? How will you observe them?

3 How will you summarize your observations? If you need a data table, design one now *in your Science Journal.*

Check the Plan

Discuss and decide upon the following points and write them down.

1 Review the steps of your procedure. Can everyone in the group follow them?

2 Review your data table. If you need help, refer to Making and Using Tables in the *Skill Handbook.*

3 Make sure your teacher approves your experiment before you proceed.

4 Carry out your experiment. Record your observations.

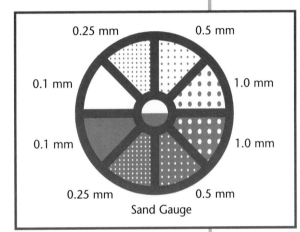

0.25 mm   0.5 mm
0.1 mm   1.0 mm
0.1 mm   1.0 mm
0.25 mm   0.5 mm
Sand Gauge

Analyze and Conclude

1. Classify What are some characteristics of beach sand? Did your observations support your hypothesis?

2. Observe and Infer Were the grains of a particular sample generally the same shape? Explain why this may or may not be the case.

3. Interpret Data Were all the sand particles made of rock, or were other materials found in your sand? Explain.

4. Draw a Conclusion Explain why you wouldn't find sand composed mostly of coral along a beach of Lake Superior.

Going Further

In Nome, Alaska, the beach sand contains gold that was mined during the late 1800s. How would you separate the gold from the rest of the sand particles? Where do you think the gold came from?

Science Journal Have students record their observations and the answers to all the questions posed in their journals. [L1]

Expected Outcome

Students should be able to recognize that sands vary in texture, color, the degree of roundness of the grains, and the size of the individual particles.

Analyze and Conclude

1. Characteristics include color, texture, roundness, grain size, luster, composition, and density. Also, some grains might be attracted to a magnet. Answers will vary depending on students' hypotheses.

2. Samples from different beaches may have different shapes, but grains from the same sample should be about the same shape if they are made of only one type of material. If the sample is a mixture of materials, the shapes of the grains will vary.

3. Answers will vary, but might include shell, rock, and mineral fragments.

4. Coral forms in warm, tropical, salty waters. Lake Superior is located too far north for coral formation, and is a freshwater lake.

✔ Assessment

Process Based on their observations, have students work in pairs to infer the source material for each sand sample. Use the Performance Task Assessment List for Group Work in **PASC,** p. 97.
COOP LEARN

Going Further

Gold has a higher specific gravity or density so it will be heavier. The gold was probably carried to the beaches by streams and rivers.

PREPARATION

Planning the Lesson

Refer to the Chapter Organizer on pages 360A-D.

Concepts Developed

The previous section described how ocean waves cause longshore currents and move sand from one place to another. This section looks at several types of pollution and their effects on shore zones.

1 MOTIVATE

Bellringer

 Before presenting the lesson, display **Section Focus Transparency 35** on an overhead projector. Assign the accompanying **Focus Activity** worksheet. [L1] [LEP]

Activity Have students working in pairs design signs that explain why a hypothetical beach is closed to the public. Reasons may include rough waters, sharks, an oil spill, high levels of pollution, and so on. [L1] [COOP LEARN] [P]

2 TEACH

Tying to Previous Knowledge

Refer students to Figure 12-5. Ask: **Do pollutants placed on a beach affect only that particular beach? Why or why not?** *No, just as longshore currents transport sand, they also carry pollutants.*

Theme Connection Have students determine from a map or globe that all oceans are interconnected. Thus, each ocean can be viewed as an individual system, or all oceans can be viewed as one large system.

Section Objectives

- Relate the ways in which human activities pollute shore zones.
- Describe the effects of ocean pollution on sea life.

Key Terms

pollution

Pollution

Imagine going to the beach shown in **Figure 12-5**. You notice a sign has been posted. It states that the beach is closed because waste materials have washed ashore. You may have heard of such an incident on the news. Every year, beaches around the world are spoiled by careless treatment of shore zones.

Shore zones are popular areas. In the United States, three out of every four persons live in a coastal state. Coastal cities and towns are active places. Commercial shipping and fishing are important industries, and factories often line the waterfronts.

Unfortunately, one side effect of these many human activities is

Figure 12-5

Ⓐ Pollution left behind by people along beaches is ugly and can be deadly to wildlife. Plastic objects such as bags, rings from drink six-packs, and toys are especially dangerous.

Ⓑ More than a million seabirds and 100 000 ocean mammals, such as seals and sea otters, are killed each year by plastic garbage in the ocean when they swallow or become entangled in the plastic.

370 Chapter 12 The Ocean Floor and Shore Zones

Program Resources

Study Guide, p. 41

Concept Mapping, p. 20, Ocean Pollution—Humans Affect Shore Zones [L1]

How It Works, p. 15, How Are Oil Spills Cleaned Up? [L2]

Making Connections: Integrating Sciences, p. 27, Rotary Oil Rig [L3]

Section Focus Transparency 35

---

pollution. **Pollution** is unwanted or harmful materials or effects in the environment. Pollution may range from a plastic cup left behind by a picnicker to heat released into the water by a factory or power plant.

Some pollution is just annoying and ugly, such as litter. Yet other types of pollution, such as wastes from factories, homes, and businesses, can cause great harm when they get into streams, lakes, or ocean water. These materials include toxic chemicals and metals from factories along with plas-

tic, paper, and garbage from homes and businesses. This refuse is dumped directly into the sea or in landfills near the shore. Medical wastes, such as used needles and plastic tubes, have washed up onto beaches, where they threaten humans and other animals with disease.

SKILLBUILDER

Recognizing Cause and Effect

You've learned how many human activities are polluting the ocean. As the human population grows, these activities will increase in size and number. Think of ways that you can help prevent pollution. Make a cause-and-effect chart listing the effects of your actions. If you need help, refer to the **Skill Handbook** on page 643.

Visual Learning

Figure 12-5 and 12-6 Students can use the figures to classify water pollutants in two groups—those that are litter but are not harmful versus those that are potentially harmful to life. Have students suggest who is responsible for each of the different pollutants.

Across the Curriculum

Daily Life/Language Arts

Have students find out about the problems associated with plastic pollutants by writing to the Center for Environmental Education, 1725 DeSales St. NW, Washington, DC 20036 and requesting a copy of *A Citizen's Guide to Plastics in the Ocean: More Than a Litter Problem.* With the information obtained about plastic pollutants, have students design posters that discourage people from polluting Earth's environment with plastics. Use the Performance Task Assessment List for Poster in **PASC,** p. 73. **P** **L1**

Figure 12-6

A Wastes from industry such as toxic chemicals, paper and plastics, and raw and treated sewage are dumped directly into the sea or in landfills near the shore.

B Medical wastes, such as used needles and plastic tubes, have washed up onto beaches, threatening humans and other animals with serious disease.

SKILLBUILDER

The chart might look something like this. Only two examples are given here, but students should list as many as possible. **L2**

Pollution Causes and Effects

Cause	Effects
People littering	Dirty beaches, animals sick from eating trash
Oil spills	Birds coated with oil get sick and die

Multicultural Perspectives

One Person Can Make a Difference!

Tell students that waste treatment facilities remove pollutants from sewage and other wastewater. In cities, these facilities often are located in low-income neighborhoods. L. Ann Rocker, an African American woman in New York City, found that the City Charter says that projects like garbage transfer stations and waste treatment facilities should be distributed equally among neighborhoods. Ms. Rocker has been working to make the city live up to its charter. Have students discuss why Ms. Rocker's stance is important. Interested students can find out the location of such facilities in their town or city. **L1**

Some of the most serious cases of ocean and shore zone pollution have been produced by oil spills. Oil spills are usually caused when offshore oil wells leak, oil tankers collide at sea, or accidents occur at oil refineries. Oil is very harmful to living things. It makes breathing and eating very difficult, causing many animals to suffocate or starve.

Figure 12-7

Oil kills animals because it makes breathing and eating difficult, causing many animals to suffocate or starve. Oil coats the feathers of birds, causing the feathers to tangle and stick together. When birds' feathers are affected by oil, the birds cannot fly, float, or keep warm. They soon die.

Life Science CONNECTION

Adaptation of Marine Life in the Rift Zones

It is always dark in the deepest parts of the ocean. There is no sunlight, and the temperature is only a few degrees above freezing. For many years, biologists believed that it was unlikely that any form of life could exist in such an environment. Without sunlight, plants would be unable to carry on photosynthesis. Without plants, there would be no food or oxygen.

Then, in 1979, marine scientists made a startling discovery. In a Pacific Ocean rift zone, 2500 meters under water, they found a thriving colony of bizarre sea creatures. Giant clams the size of footballs were piled on top of one another. Bright red worms, as thick as a person's wrist, were encased in white tubes more than two meters long. Pale, ghost-white crabs that had never been exposed to a ray of sunlight crawled along the rocks.

Ghost-white crabs

Giant clams

372 Chapter 12

Life Science CONNECTION

How do you clean up an oil spill?

Cleaning up an oil spill presents a problem because of the nature of oil. Some oil floats on the water's surface and forms a sticky coating on everything it touches. The following activity will help you appreciate the difficult job people have in cleaning up an oil spill.

What To Do

1. Place a few small rocks at one end of a shallow baking pan to represent a shoreline.

2. Pour water into the pan until it is half full. Next, pour 10 mL of vegetable oil into the water.

3. Gently slosh the water back and forth to make sure that the rocks are wet.

4. Try using the following materials to clean up the oil: tongue depressors, cotton balls, detergent, paper toweling, and feathers.

Conclude and Apply

1. Describe your efforts to remove the oil from the water and the rocks.

Living Conditions

All of these creatures were living in the vicinity of a hot-water vent—the chimneys of rock that spew forth water heated by magma beneath the ocean floor. The hot water warms the surrounding ocean to temperatures as high as 30° C. The animals cluster around the vent like people

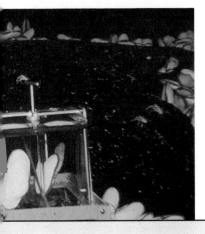

huddle around a camp fire on a cold winter night.

Although the vents provide heat, they don't provide sunlight. That left biologists puzzled. How could this community of creatures survive without plants, and how could plants grow without sunlight? In the deepest parts of the ocean, there is a substitute for sunlight. The vents spew forth a foul-smelling gas known as hydrogen sulfide. Through a process called chemosynthesis, some species of bacteria can convert hydrogen sulfide into food and oxygen. Then, deep-sea dwellers feed upon these bacteria, the way animals in shallow waters feed upon green plants.

Tube worms

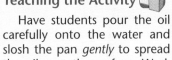
inter NET CONNECTION

ALVIN, the submersible that discovered abundant life in the rift zones, was developed at the Woods Hole Oceanographic Institute. Visit the World Wide Web to find out more about ALVIN's missions.

12-2 Humans Affect Shore Zones **373**

---

inter NET CONNECTION

A virtual History of ALVIN can be found at **http://dsogserv.whoi.edu/alvin/alvin_history/alvin_history.htm**

Going Further ⅢⅢⅢ▶
Divide the class into three groups to compare and contrast the creatures found

in different levels of the ocean. The first group can research creatures that live near the surface layers of the water; the second, those that inhabit depths of up to 1600 feet; and the third, those that live deep in the rift zones. Students should note the creatures' appearance, eating habits, and ability to produce biological light. Have students prepare oral presentations with visuals for the class. **COOP LEARN** **L1**

How do you clean up an oil spill?

TECH PREP

Time needed 30–45 minutes

Materials shallow baking pans, small rocks, water, vegetable oil, liquid measuring device, tongue depressors, cotton balls, detergent, paper toweling, feathers

Purpose To experience the difficulty of cleaning up an oil spill.

Preparation Distribute materials to groups of students. Provide a supply of cleanup materials in a central area. **COOP LEARN** **LEP**

Teaching the Activity

Have students pour the oil carefully onto the water and slosh the pan *gently* to spread the oil over the surface. Work on waterproof or covered surfaces and protect clothing from oil.

Science Journal Have students make a table to record their results, listing each material and its effectiveness in cleaning up the oil. **L1**

Expected Outcomes

Results will vary depending on the methods tried. Students should conclude that it is difficult to clean up an oil spill.

Conclude and Apply

1. The more absorbent materials did a better job of cleaning up the oil.

✔ Assessment

Oral Ask students what the oil does in the water. *Some floats.* Ask if this made the cleanup job easier or harder and how oil spills affect the ocean life. *Responses may include that because oil floats, it can sometimes be skimmed off or soaked up; but it also sticks to surfaces and can destroy habitats and poison plant and animal species.* Use the Performance Task Assessment List for Oral Presentations in **PASC**, p. 71.

NATIONAL GEOGRAPHIC SOCIETY

Videodisc

STV: Rain Forest

Herring gull with plastic six-pack holder

51849

Sea lion caught in fishing net

51848

3 ASSESS

Check for Understanding

Divide the class into three groups and assign each group one of the Check Your Understanding questions on page 375. Allow five minutes for the groups to come up with their answers. Then have each group share its answers with the class. **COOP LEARN**

Reteach

Demonstration Display four clear jars with the following: one with plain tap water, another with tap water and dissolved salt, a third with two ounces of vinegar and water, and a fourth with ammonia and water. Have students hypothesize if any of the jars contain polluted water. *Many students will incorrectly assume that because each jar contains clear water, none of the water is polluted.* Demonstrate how to gently fan the air above each jar so that the vapors can be smelled. Students should now conclude that some pollutants are not visible. L1

As you learned in the Find Out, oil pollution isn't easy to clean up. But at least you can see oil. Imagine how difficult it is to clean up harmful materials you can't see. Two examples of such materials are the substances used to kill insects (insecticides) and weeds (herbicides).

When the waters near shore zones become polluted, everyone and everything is affected. Food chains are disrupted, and the oxygen supply in the region is reduced.

Figure 12-8

Sewage Human sewage promotes the growth of some marine organisms, such as algae and plankton. When algae and plankton die, their remains are decomposed by bacteria, which use up large amounts of oxygen from the water. Fish and other organisms that use this oxygen die.

Landfills Rainwater penetrates landfills and picks up a variety of pollutants in dissolved form from the waste. The polluted water seeps into the soil and groundwater, and some makes its way to the ocean.

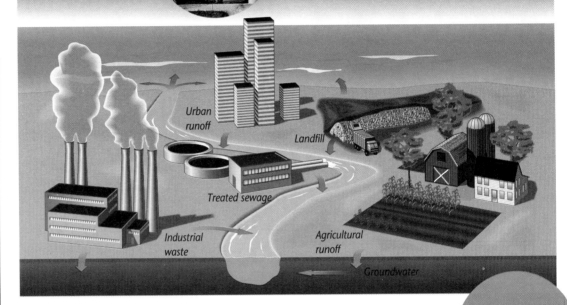

Urban runoff

Landfill

Treated sewage

Industrial waste

Agricultural runoff

Groundwater

Industrial waste The air over many large cities is continually polluted. When rain falls through polluted air, chemicals such as sulfuric acid are washed down upon Earth in the form of acid rain. The pollutants harm plant life and enter streams, rivers, and the ocean.

Agricultural runoff To increase their crop production, farmers use poisonous chemicals applied by crop dusters such as the one in the photograph to the right to kill pests. When rain falls on fields that have been treated with chemicals, the chemicals dissolve and mix with the rainwater. Some poisons seep into the soil and make their way into groundwater and finally to the ocean.

Habitat Destruction

The place where an organism lives is called its habitat. If one part of a habitat is altered or destroyed, all members are affected.

One way people destroy shore habitats is to fill them in. As populations in many flat shore zones grow, the need for more land becomes urgent. Huge areas of coastal wetlands are filled in with soil, rock, construction materials, and even garbage. The newly created land is used for buildings, roads, and airports, while thousands of acres of shore zone habitats have been destroyed.

Thus far in this chapter, you have learned about shore zones and the harmful effects humans can have on these important regions. But the shore zones make up only a small part of Earth's surface. Most of the solid Earth lies beneath the ocean waters, largely unexplored. Many people feel that the ocean basins will become more important as the world's population continues to increase. We certainly will look to the ocean floor for new sources of minerals. You'll learn about the ocean floor in the next section of this chapter.

Figure 12-9

Huge areas of coastal wetlands are filled in with soil, rock, construction materials, and even garbage. The newly created land is used for buildings, roads, and airports. But while making way for people and their needs, thousands of acres of shore zone habitats have been destroyed.

A The Back Bay area of Boston was once a marshy section of the Charles River. For years it was a favorite dumping ground for people living in Boston. In the mid-1800s the Bay was filled in.

B The marshlands and the homes and food of the animal species that lived there are gone. Over the years Boston has added 3000 acres of land by filling in the shallow coastal waters that surround it.

check your UNDERSTANDING

1. Describe some things an individual might do to reduce shore zone pollution.
2. Describe some of the effects of pollution on organisms in and near the shore zone.
3. **Apply** How can planting trees and grass help to preserve a shore zone habitat?

check your UNDERSTANDING

1. Activities that individuals can undertake include not littering, supporting community efforts that promote recycling, avoiding uprooting plants, planting grass and shrubs in unprotected soil.
2. Toxic chemicals can cause disease and death; fertilizers can reduce dissolved oxygen in the water; excessive sediment deposition can interfere with vital life processes of marine organisms; oil can cause disease.
3. Root systems of trees and grass help hold topsoil in place and keep it from washing into rivers and streams. This reduces the amount of sediment carried to the ocean.

PREPARATION

Planning the Lesson

Refer to the Chapter Organizer on pages 360A-D.

Concepts Developed

The ocean floor has mountains, valleys, and flat areas similar to those found on the continents. In most cases, methods of indirect measurement must be used to map the ocean floor.

1 MOTIVATE

Bellringer

 Before presenting the lesson, display **Section Focus Transparency 36** on an overhead projector. Assign the accompanying **Focus Activity** worksheet. L1

LEP

Discussion Students should recall that the major features of continents are mountains, valleys, and plains. Display a map of the ocean floor, if available. Students should recognize mountains, valleys, and plains. Students should suggest ways to figure out what the ocean floor looks like.

Visual Learning

Figure 12-10 This figure gives a pictorial history of recent developments in submersibles used to explore the ocean floor. Have students use Figure 12-10 to discuss how Jason was an improvement over Alvin and how the AUVs (Autonomous Underwater Vehicle) developed in the early 1990s are an improvement over Jason. *Jason is more cost effective than Alvin because it is unpeopled. However, the newer AUVs are not limited by a tether cable as Jason was.*

12-3 The Ocean Floor

Section Objectives

■ Describe some of the methods used to map the ocean floor.

■ Name and describe some features of the ocean floor.

Key Terms

continental shelf,
abyssal plain,
rift zone,
mid-ocean ridges

Exploring the Ocean Floor

How do we know what the ocean floor is like? Most of the information comes from making indirect measurements. Humans can't measure features of the ocean floor directly because they can't withstand the very high pressure exerted by the water at great depths. The following Find Out activity will help you to understand how indirect measurements are made. In the following activity, you will attempt to determine the shape of a surface without seeing it. Your teacher will provide you with a sealed shoe box having a series of numbered holes in a straight line along the lid of the box. You will insert a straw into each hole to determine the shape of the ocean-floor model on the bottom of the box. Once the shape has been determined, you will construct a graph to represent the shape of the model in the box.

Figure 12-10

Submersibles and semi-submersibles bring the ocean floor into view.

B In 1985, scientists aboard *Alvin* made a 2.5 hour descent to the ocean floor to explore the *Titanic*, a luxury liner that had lain unseen since it collided with an iceberg and sank in 1912. *Jason Jr.*, a sonar and camera equipped ROV shown below and to the right, explored the inside of the *Titanic* and sent images to the crew of *Alvin*.

A *Alvin*, shown above, is a people-operated submersible, first used in the 1970s. By the late 1980s, remotely operated vehicles (ROVs) with no people aboard were available.

376

Program Resources

Study Guide, p. 42
Teaching Transparency 23 L1
Science Discovery Activities, 12-3
Multicultural Connections, p. 28,
Berthel Carmichael—Research Mathematician L1
Section Focus Transparency 36

How can you determine the shape of something you can't see?

When exploring something you can't see, you often must use evidence based on indirect observations. This approach is especially true in science, when you have to consider such questions as what the inside of Earth is like or how deep the ocean is.

What To Do

1. First, construct a graph like the one shown here. Make the graph as long as the shoe box. Draw the vertical lines so they line up with the holes in the lid. Mark the vertical scale in centimeters as shown.

2. Place the straw beside a ruler and, starting with zero, mark the straw at 0.5-cm intervals. Label each mark.

3. Now make your measurements. Insert the zero end of the straw into hole 1 until it touches the ocean-floor model on the bottom of the box.

4. Record this measurement on your graph by placing a dot on line 1 at the depth measured.

5. Repeat Steps 3 and 4 for holes 2–10.

6. Connect the points on your graph to produce a side-view drawing of your ocean floor.

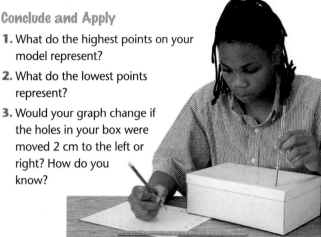

Data and Observations Sample data

Top of the box (sea level)

Conclude and Apply

1. What do the highest points on your model represent?

2. What do the lowest points represent?

3. Would your graph change if the holes in your box were moved 2 cm to the left or right? How do you know?

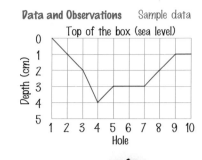

C All but the very top of the Plexiglass semi-submersible *Nemo*—invented in the late 1980s— floats below the surface. *Nemo* passengers get a clear view of a reef in the Red Sea, shown right.

How can you determine the shape of something you can't see?

Time needed 60 minutes

Materials small shoe box with lid, modeling clay, rubber bands, soda straws, graph paper, metric rulers

Thinking Processes making and using graphs, measuring in SI, observing, interpreting data

Purpose To make a profile of a surface that can't be seen.

Preparation To prepare a model, place a shoe box on its side. Punch ten holes, 3 cm apart, along a line down the middle of the lid of the box. The holes should be just large enough for a soda straw to be inserted. Number the holes 1 to 10 starting at the left. Use modeling clay to build a model of the ocean floor. The exact shape doesn't matter as long as there are variations in height. Secure the lid on the box with rubber bands.

Teaching the Activity

Demonstration If necessary, demonstrate to students how to mark a straw at 0.5-centimeter intervals as directed in the text. Demonstrate how to insert the straw vertically into the first hole, determine the measurement, and mark that measurement on the graph.

Troubleshooting Students must hold straws vertically for accurate measurements.

Science Journal Have students make their graphs in their journals. The depth scale on the graph will depend on the dimensions of the shoe box and the shape of the clay model. The depth at each hole must be recorded on the corresponding location on the vertical axis of the graph. **L1**

Expected Outcome

Students working with the same model should produce similar graphs.

Conclude and Apply

1. mountains

2. valleys or trenches

3. The graph would show the profile at this point and will probably be different. Have the students perform this with their boxes.

✔ Assessment

Performance Provide students with a map showing ocean floor topography. Have students construct a profile of an ocean basin floor. Then have students work in small groups to construct three-dimensional models of the basin floors. Profiles should be labeled with the name of the ocean. Use the Performance Task Assessment List for Model in **PASC**, p. 51. **COOP LEARN**
P

2 TEACH

Tying to Previous Knowledge

In Chapter 2, students learned about earthquakes and volcanoes. Ask students to recall some of the things they learned. Ask if they think earthquakes and volcanoes can occur under the ocean. Point out that they can, and they cause the same changes in Earth's surface as when they occur on continents.

Theme Connection To increase awareness of scale and structure, ask students to compare and contrast the topography of the ocean floor with that of the continents. *The structures—plains, valleys, mountains—are quite similar; they only differ in scale. The highest mountains, deepest valleys, and flattest plains on Earth are found on the ocean floor.*

Visual Learning

Figure 12-11 Using the information given in this figure, have students list adjectives that describe each feature of the ocean floor. For example, adjectives for the Continental Slope might be *steep, slanting,* or *sloping.* Create a table of students' responses on the chalkboard. Students may wish to copy the table in their journals for quick reference.

Inquiry Question Oceanographers use sound waves to determine the depth of the ocean because they know that underwater, sound travels about 1463 meters per second. Suppose a sound wave took 3 seconds to make its round trip. **How deep would the ocean be at that point?** *3 s/2 = 1.5 s; 1.5 s × 1463 m/s = 2194.5 m deep*

Ocean Floor Features

Figure 12-11

Early sailors obtained depth measurements in much the same way as you did in the previous Find Out—they lowered weighted ropes to the ocean floor. They then retrieved and measured the length of the submerged rope.

Continental shelf

Continental slope

Abyssal plain

A) Continental shelf and slope
As you leave the shore and move out under the water, the first part of the ocean floor is the continental shelf. Most sections of the continental shelf are flat and slant gently down and out into the ocean. At the outer edge, the bottom dips, forming what is called a continental slope.

B) Abyssal plains Ocean currents carry sediments from the continental shelves and slopes and deposit them on the ocean bottom. These deposits fill in valleys and depressions, creating flat surfaces called abyssal plains. The abyssal plains contain the flattest areas on Earth.

378   Chapter 12 The Ocean Floor and Shore Zones

ENRICHMENT

Discussion Discuss with students the importance of finding new uses for existing technologies. For example, sonar was invented in 1915 by a French scientist, Paul Langevin. Its original purpose was to help ships detect icebergs and avoid collisions. Sonar was not used to measure ocean depths until the 1920s. In 1925, a German research ship equipped with sonar discovered the Mid-Atlantic Ridge. Ask students to hypothesize as to other uses for sonar. Tell them that modern fishing fleets use sonar to locate schools of fish. Similar devices are used by biological research vessels to find whales.

Today, we use sonar, as shown in **Figure 12-11C**. We've discovered many features on the ocean floor. Some of these geologic structures are shown below. Read the captions of **Figure 12-11** to explore the wonders of the deep world of Earth's oceans.

C *Sonar* To map the ocean floor, oceanographers use sonar. They send a beam of sound waves toward the ocean floor. When the sound waves hit the bottom, they bounce back and are received by a recorder on the ship. The recorder measures the time it took for the waves to travel from the ship to the ocean floor and back.

E *Mid-ocean ridges and rift zones*
Regions where the seafloor is spreading or moving apart are called **rift zones**. In rift zones, magma from Earth's interior oozes onto the seafloor through the cracks formed by the gap between the plate boundaries. Alongside rift zones extend chains of underwater mountains called mid-ocean ridges. Active volcanoes are common along the mid-ocean ridges.

Reflected sound wave

Mid-ocean Ridge

Rift zone

Ocean trench

Abyssal plain

D *Ocean Trenches* Earth's crust, including the seafloor, is made up of a series of huge plates. Where these plates are converging or pushing into one another, one plate is pushed down under the other plate, creating a long, narrow valley called a trench. Trenches make up some of the deepest parts of the ocean. Active volcanoes and earthquakes are common along trenches.

379

Uncovering Preconceptions
Students may believe that the ocean bottom is flat and smooth. Before starting the Investigate activity on page 380, have students predict what the ocean bottom looks like across the mid-Atlantic. One way to do this would be to draw two possibilities on the chalkboard, one showing a flat bottom with a smooth curve up on each side, the other similar to the outline obtained with the data in the Investigate activity. Then have students vote for the one that they think is closest to what the ocean floor looks like.

Teacher F.Y.I.
Students may be interested to know that until 1977, only 5 percent of the ocean's floor had been charted. NASA's satellite *Seasat* measured the other 95 percent in only three months in 1978.

GLENCOE TECHNOLOGY

 Videodisc

The Infinite Voyage: To The Edge Of The Earth

Chapter 6
Exploring Life in the Canadian Arctic

12-2 Ocean Floor Profile

Planning the Activity

Time needed   30 minutes

Materials   graph paper, blue and brown pencils

Process Skills   making and using graphs, interpreting data

Purpose   To graph a set of data and use the graph to answer questions about the ocean floor.

Teaching the Activity

Discussion   Point out that students will work with actual data from the ocean floor. On a world map, show the 39° north latitude line from New Jersey to Portugal. The data were collected along this line.

Troubleshooting   Be sure that students realize that the graph is set up in a form that is different from what they may be used to. Usually, zero is at the bottom of the vertical axis. However, on this graph, it is at the top of the vertical axis. Students may become confused and reverse the data as they begin to plot the points. By walking around the room and checking students' progress, you may eliminate many problems.

Process Reinforcement   Have students compare this graph with the profile they made in the Find Out activity on page 377. *Both graphs plot depth on the vertical axis; thus the absolute values of the numbers increase as you move toward the x-axis.*

Science Journal   Have students attach their graphs in their journals and record their results.
L1

INVESTIGATE!

Ocean Floor Profile

In this activity, you will construct a profile, or side view, of the features of the ocean floor between New Jersey and Portugal. To make your profile, you will interpret a table of data that was collected by a depth-sounding technique similar to the sonar technique described earlier.

Problem
What does the ocean floor look like?

Materials
graph paper
blue and brown pencils

What To Do

1 Set up a graph as shown.

2 Examine the data listed in the table. This information was collected at 29 locations across the Atlantic Ocean. Each station was along the 39° north latitude line from New Jersey to Portugal.

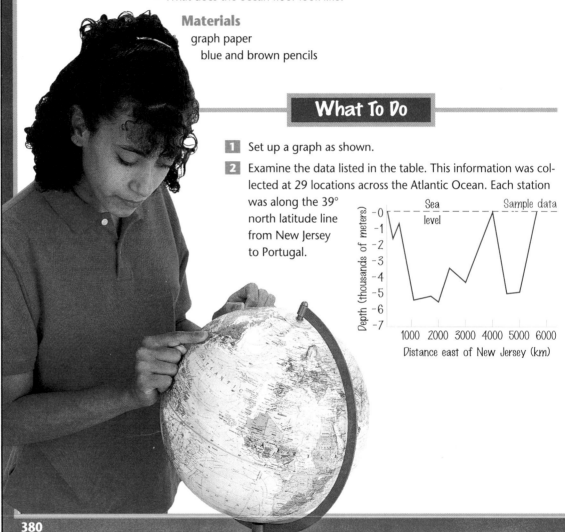

380

Program Resources

Activity Masters,   pp. 53–54, Investigate 12-2

Teaching Transparency 24 L1

Making Connections: Technology & Society,   p. 27, The Chunnel L1

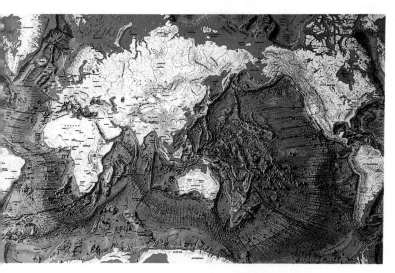

This map shows the mountains, valleys, and plains of Earth—those on dry land as well as those on the ocean floor.

Data and Observations

Station Number	Distance from New Jersey (km)	Depth to Ocean Floor (m)
1	0	0
2	160	165
3	200	1800
4	500	3500
5	800	4600
6	1050	5450
7	1450	5100
8	1800	5300
9	2000	5600
10	2300	4750
11	2400	3500
12	2600	3100
13	3000	4300
14	3200	3900
15	3450	3400
16	3550	2100
17	3600	1330
18	3700	1275
19	3950	1000
20	4000	0
21	4100	1800
22	4350	3650
23	4500	5100
24	5000	5000
25	5300	4200
26	5450	1800
27	5500	920
28	5600	180
29	5650	0

3 Plot each data point listed on the table. Then, connect the points with a line.

4 Color the ocean bottom brown and the water blue.

Analyzing

1. What ocean-floor features would you infer occur between 160 and 1050 km from the coast of New Jersey? Between 2000 and 4500 km? Between 5300 and 5600 km?

2. Would a profile taken across the Atlantic at 39° South latitude be similar to the one at 39° North latitude? Explain.

3. Would you see a similar profile for the Pacific ocean floor compared to the Atlantic? Explain.

Concluding and Applying

4. You have constructed a profile of the ocean floor along the 39° latitude line. If a profile is drawn to represent an accurate scale model of a feature, both the horizontal and vertical scales will be the same. What is the vertical scale of your profile? What is the horizontal scale?

5. **Going Further** Compare and contrast your profile with the ocean floor map on this page. How accurate do you think your profile is? Explain.

Expected Outcomes

The completed profile should show a ridge, a broad, deep area, a rise to an island, another deep area, and a rise to the shore of Portugal. If you did the Uncovering Preconceptions activity on TWE p. 379, return briefly to the results and make certain that everyone understands that the ocean bottom is *not* flat.

Answers to Analyzing/ Concluding and Applying

1. A continental slope occurs between 160 and 1050 km from New Jersey. The mid-Atlantic Ridge and a rift valley occur between 2000 and 4500 km from New Jersey. A continental slope lies between 5300 and 5600 km from New Jersey.

2. It would be similar but the continental shelves would be farther apart. (Refer to the map.)

3. No, refer to the map.

4. The vertical scale is 1 unit = 1000 meters. The horizontal scale is 1 unit = 1000 kilometers.

5. Going Further: The profile has tremendous vertical exaggeration, that is, slopes look a lot steeper than they really are. To make an accurate profile, the vertical and horizontal scales must be the same.

✔ Assessment

Performance Have students study a physiographic map of the United States. Have them determine elevations at various points along about the 40° latitude. Then have students make a graph of their data and compare it to the graph made in the Investigate activity. Use the Performance Task Assessment List for Graphs in **PASC**, p. 39.

ENRICHMENT

Activity On a map of the Atlantic Ocean, have students place a ruler along the 39° north latitude line from New Jersey to Portugal. Have them refer to their graph from the Investigate activity. Then ask: **What geographic feature is represented by measurement at Station Number 20?** *the Azores* L1

Demonstration To demonstrate a volcanic eruption, use some clay, baking soda, white vinegar colored with red food coloring, and a small plastic funnel. Invert the funnel over a loosely packed mound of baking soda. Cover the funnel with clay. Pour the vinegar into the funnel and have students describe the eruption.

Content Background

The volcanoes of the Hawaiian Islands are shield volcanoes made mostly of basalt. Mauna Loa is over 9000 meters tall; 5000 meters of this total is below sea level.

Figure 12-12

Some volcanic mountains build from magma bubbling from rift zones in the ocean floor and rise above the surface of the water. The islands near Iceland formed this way. Other islands, such as Hawaii, formed from seafloor volcanoes over hot spots in the ocean floor.

Purpose

A Closer Look explains the precious metal deposits being found at rift zones in the ocean floor.

Content Background

About 500 million tons of silver are dissolved in the ocean's water, along with an estimated five billion tons each of uranium and copper. Gold? Perhaps ten million tons! Despite the marked increase in coring, drilling, and dredging, the retrieval of precious metals has barely begun.

Teaching Strategy

Have students work in pairs and hold mock interviews between newscasters and undersea geologists. `COOP LEARN`

Debate

Have students debate whether it is worthwhile to try to recover the minerals that are at the bottom of the ocean. `COOP LEARN`

Mining Minerals at the Rift Zones

The year 1848 will always be remembered as the start of the Gold Rush. Hundreds of prospectors—people who search the ground for valuable minerals—headed out west to search for gold in the mountains of California. There may be another gold rush in the twenty-first century, except this time it will be to mine precious metals at the bottom of the ocean. Oceanographers say that a good place to look for metals will be at rift zones, where undersea surveys have already discovered huge amounts of such metals.

Where do they come from? It's a slow process, beginning when seawater passes through cracks in rocks at the rift zone. Beneath the ocean's crust, the seawater encounters magma. The hot seawater mixes with particles of metal in the magma and then rises back up to the surface. The seawater pours out of chimneys of rock standing up

Black plume at a rift zone

Answers to
What Do You Think?

Accept all answers that can be supported logically. Some students may suggest that the U.N. regulate the distribution of profits. Other students may say that the country that collects the metals should keep the profits.

Going Further ⫸

Have students work in small groups to draft their own Law of the Sea treaty, stating how they believe rights to the minerals in the ocean should be allocated. Urge students to create a treaty that all countries will want to sign. Students can write to the United Nations for a copy of the treaty:

United Nations Headquarters
United Nations Plaza
New York, NY 10017

`COOP LEARN` `L2`

In the activity just completed, you learned that the Atlantic Ocean floor has many interesting features. Now study the map shown in the Investigate. Down the middle of the Atlantic Ocean, you can see a rift zone.

Alongside the rift zones extend chains of underwater mountains called **mid-ocean ridges**. What other geologic features can you identify on the map?

At the beginning of this chapter, you were asked for ideas about living on the ocean floor. If you were asked to plan an ocean city, which part of the ocean floor would you choose?

to 30 meters tall. As it erupts, it forms black plumes. Then the dissolved metal particles in the plumes solidify and sink to the bottom of the ocean.

Mineral-rich nodules similar to those found at rift zones

Mineral Rights

Finding precious metals might not be nearly as troublesome as figuring out who has a right to own them. If a silver deposit is found in the middle of the Atlantic Ocean, which country has a right to claim it?

In an attempt to prevent such disputes, the United Nations drafted a treaty—The Law of the Sea. The treaty states that a nation has a right to minerals found within 370 kilometers of its shoreline. Any metals found beyond that boundary line are "the common heritage of humankind."

What Do You Think?

In 1988, 35 countries had signed the United Nations treaty. Yet, many industrialized countries, such as the United States and Japan, refused to sign. They feel that only countries that participate in undersea mining should have the right to claim any profits. Do you agree? What do you think would be the fairest way to distribute metals found at the bottom of the ocean?

12-3 The Ocean Floor **383**

Science and Society

Purpose

Science and Society reinforces the material on shore zones in Section 12-2 by describing some of the different ways these zones are eroded. Some ways people and nature interact with the ocean both to control and cause beach erosion are also discussed.

Content Background

Waves can change the shape of a beach in only a few hours. Temporary beach changes may be as great as 3 m in height and 50 m in width, for example. These shifts are especially noticeable between the seasons. In winter, when there are strong storms, the exposed beach becomes narrower and may even disappear in some places. In the spring, beaches begin to rebuild and, by the summer, they reach maximum development.

Big storms can cause even more dramatic effects on shore zones. The hurricane of 1938, for example, carved out depressions and piled up sandbars and sandbanks along the East Coast that would normally have taken centuries to form. A week after the storm, the U.S. Coast Guard issued a warning to sailors that its water-depth charts had been rendered useless and would have to be redrawn after new soundings were made all along the coast from Cape May, New Jersey, to Cape Cod, Massachusetts. Great Gull Island, off the tip of Long Island, lost six of its 18 acres. A family planning to sell 50 acres of beach land in Connecticut found they had only 2 acres left.

Science and Society

Beach Erosion

Imagine that you take a trip to the ocean. You walk down to the beach only to discover that there's no more sand. Everywhere you look, you see nothing but rocks and cobblestones. It's hard to believe that anything as large as a beach can simply vanish. Yet, in some places, longshore currents carry away tons of sand each year.

How does this happen? Waves and currents can cut away at the land that slopes toward the sea, eventually shaping the shoreline into steep cliffs. Land that is eroded in this way may form a flat underwater terrace. This type of formation permits the waves to pound directly at the base of the cliff. Finally, the overhanging portion above the notch may crumble, and more land is carried out to the sea.

Wave action and other currents can shape the shore in other dramatic ways. They can build up sandbars directly offshore. The sandbars then absorb much of the force of the waves, altering the effects of the waves' impact on the shore. Sandbars, once formed, may show continual and rapid change—eroding at one end and getting longer at the other as the sand particles are redistributed.

A fierce storm or hurricane may magnify the effects discussed above, by generating huge waves and strong currents. And the effects of a storm are further magnified if it happens to hit when tides are especially high. At such times, homes along the threatened beach may be flooded or destroyed. The day after such a storm, an onlooker can often notice a visible change in the shore's configuration from the day before.

Erosion Control

What would you do if you had a home by the beach, and you found out that your property was eroding away? You might want to build a groyne. Groynes are short walls built at right angles to the shore that trap sand being carried away by longshore currents.

Going Further ⫸

Have students work in teams to brainstorm ways that beach erosion could be harmful to ocean organisms. Suggest that students begin by listing different ocean creatures. For information, students might want to write to a nearby aquarium. Use the Performance Task Assessment List for Group Work in **PASC,** p. 97. COOP LEARN

Sometimes, however, human efforts to stop beach erosion can be detrimental. The problem with groynes is that they work too well. They capture so much sand that beaches down-current from the wall are deprived of sand. You would trap sand at your own beach, but meanwhile sand would continue to be eroded from the next beach area down-current.

Some property owners try to minimize the effects of shoreline erosion simply by constructing more groynes similar to the one in the photograph above. On the New Jersey shore more than 300 such walls have already been built. Seen from above, such beaches take on the appearance of a wavy line, with alternating peaks and valleys.

Other communities have decided to stop building groynes altogether. They try to prevent sand erosion through a technique known as beach nourishment. Truckloads of sand are dumped on the beach to replace sand that has been washed away. Unfortunately, beach nourishment is only a temporary solution, and it can be very expensive. The cost can be greater than one million dollars per kilometer per year to restore sand on a beach.

Other Shore Problems

Beach nourishment can also pose a serious hazard for the environment. In Miami, for instance, coarse quartz sand was replaced with a muddier, softer sand. When the waves broke upon the shore, they picked up a lot of mud, and the water became thick and cloudy. This natural pollution killed many coral reefs.

Another important cause of beach erosion is destruction of coastal vegetation. Plant cover stabilizes sand dunes and helps prevent the loss of beach sand or dirt. Off-road bikers or too many recreational hikers may harm the vegetation. So can real estate development and road building.

The twentieth century has seen more erosion than any other period in recorded history. This is largely because of all the construction that takes place on our beaches. Bulldozers clear away tons of sand every time someone decides to build a hotel, restaurant, or highway. This speeds the process of erosion.

Science Journal
Some people think that our government should pass laws that would place restrictions on beachfront construction. *In your Science Journal* discuss other ways you can think of to halt the process of beach erosion.

Content Background
Groynes extend 30–40 m into the sea on pebble beaches, and 60–200 m on sand beaches. Usually, the groynes are arranged depending on the length and angle of the waves during the strongest storm. As a rule, these spaces are equal to one to two groyne lengths. This kind of shore protection is used extensively along the Caucasus and the Crimea coasts of the Black Sea, as well as along the Atlantic coast of the United States.

Teaching Strategy
Have students form "news teams" to give on-the-spot reports of beach devastation caused by a recent storm or hurricane. **COOP LEARN**

Discussion
Have students discuss what methods they favor to preserve beaches from erosion. In addition to supporting active intervention, remind students that they can elect not to take action at all. Have students justify their stance based on what they learned in this excursion.

Science Journal
Accept all opinions that can be supported. Students might suggest that erosion could be halted with fences, brick walls, and banning people from seriously eroded beaches.

Going Further ⫸
Students may wish to write to organizations that work to control beach erosion. Students who live near the ocean can find out which methods of controlling erosion are best suited to their area; those who live inland can research an area they have visited. Possible sources of information include:

Environmental Protection Agency/Public Affairs
401 M St.
Washington, DC 20420
(202) 260-4361

U.S. Army Corps of Engineers
20 Massachusetts Ave., NW
Washington, DC 20314-1000
(202) 272-0660

Use the Performance Task Assessment List for Letter in **PASC,** p. 67. **COOP LEARN**

Teens in SCIENCE

Purpose

Teens in Science reinforces Section 12-2 by showing how humans can affect shore zones in a positive way. The selection explains how volunteer rescue teams help save marine animals.

Content Background

Whales travel in groups called pods, hence the name of the rescue group in this excursion. Creatures belonging to the order *Cetacea* that are more than 5 m long are generally referred to as whales, whereas smaller species are called dolphins or porpoises.

Teaching Strategy

Have students discuss what traits a person would need to be a successful member of the Pod Squad.

Answers to What Do You Think?

Answers will vary but may include: There's always something exciting to see; there's always some action going on.

Literature Connection

Purpose

Literature Connection extends Section 12-1, in which the characteristics of shore zones were presented. The Strombus snail's story describes the actions of waves along the shore.

Answers to What Do You Think?

Some may say they would have thrown the shell back into the sea before the snail died. Others may say that the snail would not have recovered, so they would have kept the shell.

Teens in SCIENCE

Rescue Team Is All Wet

Imagine that your telephone has just rung. It's the Pod Squad alerting you to another rescue. Fifteen-year-old Marc D'Anto is a member of a volunteer whale rescue team in Key Largo, Florida. He works with marine biologists and veterinarians.

"To me, the ocean is very exciting. It's a whole new world. Like a real-life video game," Marc says. "I guess Trish and Alex, two pilot whales, are the biggest rescues I've been involved in. Trish weighed close to 1200 pounds and Alex nearly 700 pounds. When we found them, they were floating in shallow water."

The rescuers used a crane and specially made harness to lift the whales to safety. They transported them in a moving van to a nearby canal. "The whales had to be given antibiotics to combat pneumonia. The medication made them sleepy. If we

didn't watch them closely, they would sink."

Marc and other volunteers stayed in the water 24 hours a day to keep the whales afloat. After months of care, both whales were strong enough to be set free.

What Do You Think?

Marc describes the ocean as a "real-life video game". What do you think he means by this?

Literature Connection

Shell collecting is a hobby enjoyed by many people. They like to search the seashore at low tide for shells that wash up at high tide. Empty shells are called dead shells. But sometimes shells are found that are still "alive"—an animal is attached to it and lives inside.

A Real Find

"The Strombus", by Latino author Sylvia C. Peña, is the story of a father and daughter who find a shell on the beach after a hurricane. Read the story to learn how the strombus inside the shell feels in the ocean during the hurricane and after being washed ashore. Also learn how finding a prized strombus shell, and then realizing the animal is still alive, affects the girl.

What Do You Think?

What do you think about the girl's behavior? How would you have reacted to finding the strombus?

Going Further ⫸

Students should work in small groups to find out about additional efforts to save whales and other ocean creatures. Possible topics include threats facing marine animals, such as ocean pollution, gill and drift net problems, and fishing. For information, students can contact:

Center for Marine Conservation
1725 DeSales St., NW
Washington, DC 20036
(202) 429-5609

The Cousteau Society
870 Greenbriar Circle Suite 402
Chesapeake, VA 23320
(804) 523-9335

Students can use their information to prepare public service announcements, write pamphlets, make posters, or create other displays. Select the appropriate Performance Task Assessment List from **PASC.** COOP LEARN **P**

Science Journal

Review the statements below about the big ideas presented in this chapter, and answer the questions. Then, re-read your answers to the Did You Ever Wonder questions at the beginning of the chapter. *In your Science Journal,* write a paragraph about how your understanding of the big ideas in the chapter has changed.

1 Land areas in contact with the ocean are called shore zones. *Compare the characteristics of a steep shore zone, as seen in the picture to the right, to a flat shore zone.*

2 Currents play a major role in shaping shore zones. *Describe how longshore currents move sand and sediment.*

4 Most of what we know about the ocean floor has come from indirect measurement. *What are some of the ways we indirectly measure the ocean floor?*

3 Human activities affect life in a shore zone. *How would an oil spill, as seen in the photo above, affect life in a shore zone?*

Have students look at the photographs on this page. Direct them to read the statements to review the main ideas in this chapter.

Teaching Strategy

Have students work in pairs to answer the questions posed within the text for each photograph. Have them record their answers, along with the questions, in their journals.

Answers to Questions

1. It often is bounded by cliffs and steep rocky bluffs; a flat shore zone has broad, sandy beaches.

2. Longshore currents move sand and sediment parallel to the shore.

3. Animals and plants can become covered with the oil; many birds and fish die when shore zones are polluted with oil.

4. The ocean is too deep for direct observations. Indirect observations include: sonar, ROV's and satellites.

GLENCOE TECHNOLOGY

MindJogger Videoquiz

Chapter 12 Have students work in groups as they play the Videoquiz game to review key chapter concepts.

Science Journal

Did you ever wonder...

• Most beach sand is made of sediment carried to the shore by rivers. Some beach sand is made of shells or coral that originated at the shore. (pp. 366–367)

• Beaches are sometimes closed because of pollution. (p. 370)

• The ocean floor has mountains, valleys, and flat areas. (pp. 378–379)

Project

Have students display pictures of shore zones found in magazines or travel brochures. If they have ever traveled to a shore, these pictures might include family photographs or postcards. Students could also make drawings of shore zones. Students should use these pictures to construct a display contrasting steep and flat shore zones. Use the Performance Task Assessment List for Display in **PASC,** p. 63.

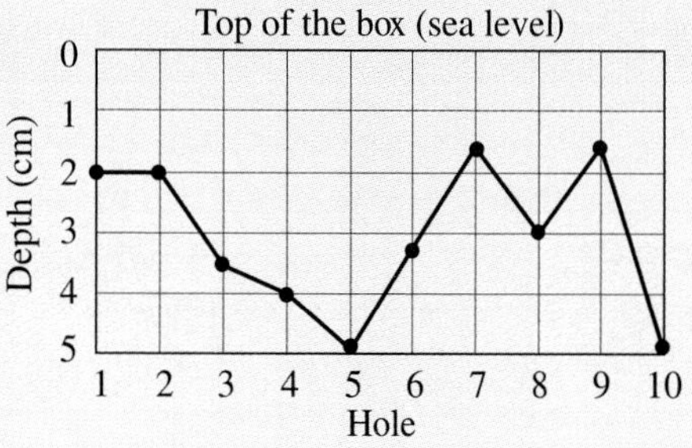

chapter 12
CHAPTER REVIEW

Using Key Science Terms

Students' paragraphs will vary, but should include all six key terms and reflect an understanding of the scientific concepts presented in this chapter.

Understanding Ideas

1. Rock fragments in waves act like chisels to erode rocky shore zones.

2. Gentle waves have less energy than storm waves, so in a given amount of time their effects on the beach are less noticeable than those of a storm wave.

3. Pollution is unwanted or harmful materials or effects in the environment. Examples include toxic chemicals from factories, sewage, as well as plastic, paper, and garbage from homes and businesses.

4. The continental shelf is an extension of the continent; the abyssal plain is in the deeper ocean out from the shelf.

5. Alongside rift zones in the oceans are chains of underwater mountains called mid-ocean ridges.

6. Beach sand varies due to source rock and the energy of the waves along the beach, among other factors.

7. Rainwater carries pesticides to rivers, which in turn carry them to the ocean.

8. wave action and longshore currents

9. Trenches are formed where converging plates move together and rift valleys are formed where plates are moving away from each other.

Developing Skills

1. See student page for answers.

2. See graph.

3. Features of the ocean floor are very similar to those on the continents. There appear to be high mountains, valleys, and flat areas on the ocean floor.

4. The detergent broke the oil into smaller particles. Using detergents to clean up oil spills is not necessarily environmentally

Using Key Science Terms

abyssal plain	mid-ocean ridges
continental shelf	pollution
longshore current	rift zone

Using the key science terms, write a brief paragraph describing your understanding of the ocean floor and shore zones.

Understanding Ideas

Answer the following questions in your Journal using complete sentences.

1. Explain how the action of waves against the rock produces interesting and beautiful rock formations in a steep shore zone.

2. What is the difference between the effect of a low gentle wave and a large storm wave on the shore?

3. Define pollution and give examples of several kinds of pollution.

4. What is the difference between the continental shelf and the abyssal plain?

5. Compare the location of rift zones and mid-ocean ridges.

6. Why does beach sand from different beaches vary in size, color and composition?

7. How would a pesticide used on a field far from the coast pollute ocean water?

8. What causes erosion and deposition of beach sand?

9. How does a trench compare to a rift zone?

Developing Skills

Use your understanding of the concepts developed in the chapter to answer each of the following questions.

1. **Concept Mapping** Complete the concept map of shore zones.

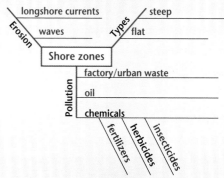

2. **Making and Using Graphs** Using the following information, construct a line graph similar to the one on page 380.

Hole #	Depth (cm)	Hole #	Depth (cm)
1	2 cm	6	3.5 cm
2	2 cm	7	1.5 cm
3	3.5 cm	8	3 cm
4	4 cm	9	1.5 cm
5	5 cm	10	5 cm

3. **Comparing and Contrasting** If the above graph represents a profile of the ocean floor, what can you say about those features as compared to land features?

4. **Observing and Inferring** Refer back to the Find Out activity on page 373. Explain what happened when detergent was used to remove the oil. Would this be an environmentally safe way to remove the oil? Explain why or why not.

safe because substances in the detergents may harm the organisms in the environment.

DATA AND OBSERVATIONS
Top of the box (sea level)

[Line graph: x-axis labeled "Hole" from 1 to 10; y-axis labeled "Depth (cm)" from 0 to 5.]

Critical Thinking

In your Journal, *answer each of the following questions.*

1. Specialized submarines called submersibles have been developed that can carry people safely to some of the deepest parts of the ocean, where they can make direct observations of the ocean floor. Why don't scientists use these machines to map the ocean floor?

2. The distance "as the crow flies" from North Point to South City is 15 km. The table shows how much coastline lay between the two cities in the years 1894, 1924, 1954, and 1984. What could account for the difference? What do you think the distance might be in 2004?

Distance Between South City and North Point	
Straight Line Distance	15 km
1894	27 km
1924	26 km
1954	22 km
1984	19 km

3. Phosphates are chemicals that contain phosphorous, which is a plant nutrient. Why is it a good idea to use phosphate-free laundry detergents?

Problem Solving

Read the following problem and discuss your answers in a brief paragraph.

Scientists use sound waves produced by sonar to map the ocean floor.

1. In using these waves, what two factors must they know in order to calculate how deep the floor is in a particular place?

2. Sound travels through ocean water at an average speed of about 1500 meters per second. If a sound wave takes 4 seconds to travel from the machine to the ocean floor and back, how deep is the ocean floor at that location?

CONNECTING IDEAS

Discuss each of the following in a brief paragraph.

1. **Theme—Scale and Structure** What kind of rock would most likely be found on the ocean floor near the continental slope? What kind might be found along a rift zone? Explain your answers.

2. **Theme—Systems and Interactions** Explain how the wastewater that leaves your home might pollute a shore zone.

3. **Theme—Systems and Interactions** How might a steep shore zone become a flat zone over time?

4. **Life Science Connection** On what parts of the ocean floor is there proba- bly an adequate amount of sunlight for ocean plants to photosynthesize? What parts may be so deep that only chemosynthesis is possible?

5. **Science and Society** What effect do groynes have on longshore currents and their movement of beach sediment?

✔ Assessment

Portfolio Review the portfolio options that are provided throughout the chapter. Encourage students to select one product that demonstrates their best work for the chapter. Have students explain what they learned and why they chose this example for placement into their portfolios.

Additional portfolio options can be found in the following **Teacher Classroom Resources:**

Multicultural Connections, pp. 27, 28

Making Connections: Technology and Society, p. 27

Concept Mapping, p. 20

Critical Thinking/Problem Solving, p. 20

Making Connections: Integrating Sciences, p. 27

Take Home Activities, p. 19

Making Connections: Across the Curriculum, p. 27

Laboratory Manual, pp. 61–68

Performance Assessment P

Critical Thinking

1. Due to immense pressures and lack of air and light on the ocean floor, these machines can only be used for short periods of time to study very limited areas.

2. Ocean currents are eroding the coastline between the two places. A likely prediction is that the distance could be less than 19 km but not less than 15 km.

3. Wastewater containing laundry detergents with phosphorous will eventually reach shallow waters. The phosphorous might allow algae and plankton to grow rapidly leading to the depletion of oxygen from the water.

Problem Solving

1. Scientists must know the speed of the sound waves and how much time it takes them to travel back to the source.

2. (1500 m/s × 4 s)/2 = 3 000 m

Connecting Ideas

1. Sedimentary rocks are most likely found near the continental slope as ocean debris collects. Igneous rocks are found along a rift zone.

2. After the water is treated, it is usually discharged into a river or the ocean. If the wastewater is not treated properly, it can cause pollution.

3. The rocks can be eroded by waves and bits of rocks carried by the waves. If currents do not transport this sediment, a flat shore zone may form.

4. The continental shelf receives enough sunlight to allow photosynthesis. The abyssal plain and deep ocean trenches receive no sunlight, making chemosynthesis the life-supporting process.

5. Groynes provide barriers to the currents, which causes sediment to be deposited on the updrift side of the groyne.

Chapter Organizer

SECTION	OBJECTIVES	ACTIVITIES & FEATURES
Chapter Opener		**Explore!**, p. 391
13-1 The Electricity You Use (2 sessions, 1 block)	**1. Trace** the source of the energy that runs appliances. **2. Describe** what water and steam do in an electric power plant. **3. Relate** the role of generators in producing electricity. **National Content Standards: (5-8) UCP2, B3, D1-2, E2, F5**	**Find Out!**, p. 393 **Skillbuilder:** p. 394
13-2 Fossil Fuels (3 sessions, 1.5 blocks)	**1. Compare and contrast** three different fossil fuels. **2. Trace** the steps in the formation of fossil fuels. **National Content Standards: (5-8) UCP2, UCP4, A1, B3, D1-2, F2**	**Explore!**, p. 396 **Design Your Own Investigation 13-1:** pp. 398-399 **Explore!**, p. 400 **Science and Society**, p. 415
13-3 Resources and Pollution (2 sessions, 1 block)	**1. Classify** energy resources as either renewable or nonrenewable. **2. Discuss** the environmental effects of burning fossil fuels. **National Content Standards: (5-8) B3, D1-2, F2, F4-5**	**A Closer Look**, pp. 404–405
13-4 Alternative Energy Resources (2 sessions, 1 block)	**1. Describe** alternative sources of energy. **2. Differentiate** among the ways alternative energy resources are used to produce electricity. **National Content Standards: (5-8) UCP2, UCP5, A1, B3, D1, D3, F2, F4**	**Explore!**, p. 407 **Investigate 13-2:** pp. 410-411 **Physics Connection**, pp. 412–413 **Technology Connection**, p. 416 **How It Works**, p. 417

ACTIVITY MATERIALS

EXPLORE!

p. 396* piece of coal, hand lens, pencil
p. 400 water, salad oil, jar with lid
p. 407* 400-mL beaker, water, pinwheel, sink or bucket

DESIGN YOUR OWN INVESTIGATION

pp. 398-399* clear plastic bottle with spray pump, 1 to 2 cups of small, clean pebbles, clear plastic tubing, 100-mL graduated cylinder, 100 mL vegetable oil, 50 mL cold water, 100 mL hot water, liquid detergent

INVESTIGATE!

pp. 410-411* 400 mL water, hot plate, coffee can painted black, newspaper, clear plastic, rubber bands, white glue, bowl, 250-mL beakers (3), black plastic, clear plastic shoe box, thin plastic foam sheets, tape, aluminum foil, 2 thermometers

FIND OUT!

p. 393* teakettle, water, burner, pinwheel, thermal mitt, goggles

KEY TO TEACHING STRATEGIES

The following designations will help you decide which activities are appropriate for your students.

- **L1** Basic activities for all students
- **L2** Activities for average to above-average students
- **L3** Challenging activities for above-average students
- **LEP** Limited English Proficiency activities
- **COOP LEARN** Cooperative Learning activities for small group work
- **P** Student products that can be placed into a best-work portfolio
- 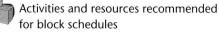 Activities and resources recommended for block schedules

Need Materials? Call Science Kit (1-800-828-7777).

OUT OF TIME? We recommend that students do the activities with an asterisk.

Chapter 13 Energy Resources

TEACHER CLASSROOM RESOURCES

Student Masters	Transparencies
Study Guide, p. 43 **Making Connections: Technology & Society**, p. 29	**Section Focus Transparency 37**
Study Guide, p. 44 **Critical Thinking/Problem Solving**, p. 21 **Multicultural Connections**, p. 29 **Science Discovery Activities**, **13-1, 13-2** **Activity Masters**, Design Your Own Investigation 13-1, pp. 55–56	**Teaching Transparency 25**, Petroleum Usage **Section Focus Transparency 38**
Study Guide, p. 45 **Making Connections: Across the Curriculum**, p. 29	**Section Focus Transparency 39**
Study Guide, p. 46 **Multicultural Connections**, p. 30 **Take Home Activities**, p. 20 **Making Connections: Integrating Sciences**, p. 29 **Concept Mapping**, p. 21 **Science Discovery Activities**, **13-3** **Laboratory Manual**, pp. 69–72, Acid Rain **Laboratory Manual**, pp. 73–76, Solar Energy Application **Laboratory Manual**, pp. 77–80, Wind Power **Activity Masters**, Investigate 13-2, pp. 57–58	**Teaching Transparency 26**, Nuclear Power Plant **Section Focus Transparency 40**

ASSESSMENT RESOURCES	TEACHING & TECHNOLOGY
Review and Assessment, pp. 77–82 **Performance Assessment**, Ch. 13 **PASC*** **MindJogger Videoquiz** **Alternate Assessment in the Science Classroom** **Computer Test Bank**	**Spanish Resources** **Cooperative Learning Resource Guide** **Lab and Safety Skills** **Science Interactions, Course 2, CD-ROM** **Computer Competency Activities**

*Performance Assessment in the Science Classroom

NATIONAL GEOGRAPHIC TEACHER'S CORNER

Index to National Geographic Magazine

The following articles may be used for research relating to this chapter:

- "In the Heart of Appalachia," by Jeannie Ralston, February 1993.
- "A Comeback for Nuclear Power?" by Peter Miller, August 1991.
- "Living With Radiation," by Charles E. Cobb, Jr., April 1989.
- "The Quest for Oil," by Fred Hapgood, August 1989.
- "Wrestlin' for a Livin' With King Coal," by Michael E. Long, June 1983.

- "Energy: Facing Up to the Problem, Getting Down to Solutions," A Special Edition, February 1981.

National Geographic Society Products Available From Glencoe

To order the following products for use with this chapter, contact your local Glencoe sales representative or call Glencoe at 1-800-334-7344:

- *Eye on the Environment: Nuclear Power* (Poster)
- *GTV: Planetary Manager* (Videodisc)
- *STV: Electricity and Simple Machines* (Videodisc)

Additional National Geographic Society Products

To order the following products for use with this chapter, call the National Geographic Society at 1-800-368-2728:

- *Solar Energy* (Kids Network Curriculum Unit)
- *Energy: The Fuels and Man* (Video)
- *What Energy Means* (Video)

Teacher Classroom Resources

These are key components of the classroom resources package.

Teaching Transparencies

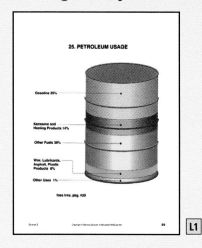

25. PETROLEUM USAGE

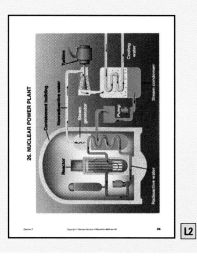

26. NUCLEAR POWER PLANT

Section Focus Transparencies

37 SECTION FOCUS TRANSPARENCY

EARLY EXPERIMENTS WITH ELECTRICITY

THE FAR SIDE   By GARY LARSON

Early but unsuccessful practical jokes

1. Where do modern humans normally obtain electricity?
2. Why must Norg shake hands to "get" the joke?

38 SECTION FOCUS TRANSPARENCY

BLOWOUT!

An offshore oil platform during a blowout and fire is a terrifying sight. The exploration for fossil fuels is risky business. Major oil well blowouts can send millions of dollars worth of oil and gas up in smoke, as well as causing tremendous environmental damage.

1. What energy source is used in your area to heat homes and generate electricity?
2. List several advantages and disadvantages of fossil fuels.
3. Why is exploring for fossil fuels worth the risk?

39 SECTION FOCUS TRANSPARENCY

WHERE'S THE FIRE DEPARTMENT?

Call the fire department, the house is on fire! Or is it? No, it's the ground that's on fire. The smoke and gases seen escaping here are coming from burning abandoned underground coal mines in western Pennsylvania.

1. Why do you think this type of fire is difficult to put out?
2. What other environmental problems can occur from the burning of fossil fuels?

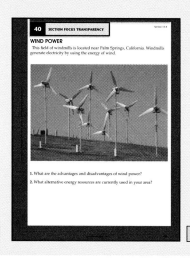

40 SECTION FOCUS TRANSPARENCY

WIND POWER

This field of windmills is located near Palm Springs, California. Windmills generate electricity by using the energy of wind.

1. What are the advantages and disadvantages of wind power?
2. What alternative energy resources are currently used in your area?

Science Discovery Activity*

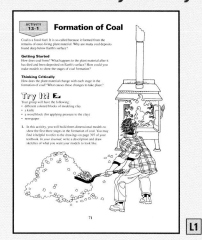

ACTIVITY 13-1   **Formation of Coal**

Coal is a fossil fuel. It is so-called because it formed from the remains of once-living plant material. Why are many coal deposits found deep below Earth's surface?

Getting Started
How does coal form? What happens to the plant material after it has died and been deposited on Earth's surface? How could you make models to show the stages of coal formation?

Thinking Critically
How does the plant material change with each stage in the formation of coal? What causes these changes to take place?

Try It!
Your group will have the following:
- different colored blocks of modeling clay
- a knife
- a wood block (for applying pressure to the clay)
- newspaper

1. In this activity, you will build three-dimensional models to show the first three stages in the formation of coal. You may find it helpful to refer to the drawings on page 397 of your textbook. In your Journal, write a description and draw sketches of what you want your models to look like.

Laboratory Manual*

NAME _____ DATE _____ CLASS _____

LAB 22   Chapter 13
Acid Rain

The major products formed from burning fossil fuels such as coal and gasoline are carbon dioxide and water. However, nitrogen dioxide and sulfur dioxide are also formed. These gases dissolve in precipitation to form acid rain. Acid rain is one of the most harmful forms of pollution.

When acid rain falls on a pond or lake, the acidity of the water increases. The rise in acidity is usually harmful to organisms in the water. If the acidity becomes too high, all living things in the water will die. The pond or lake is then considered "dead."

OBJECTIVES
- generate a gas that represents acid rain,
- observe the reaction of this gas with water, and
- demonstrate how the gas can spread from one location to another.

EQUIPMENT
- apron
- calcium carbonate
- forceps
- goggles
- hydrochloric acid solution
- microplate (96-well)
- paper towel
- paper (white)

- pipet (plastic microtip)
- sandwich bag (sealable, plastic)
- scissors
- solution (universal indicator)
- straw
- watch or clock
- water (distilled)

CAUTION: The hydrochloric acid solution is corrosive. The universal indicator solution can cause stains. Avoid contacting these solutions with your skin or clothing.

PROCEDURE
1. Wear an apron and goggles.
2. Place the microplate on a flat surface.
3. Using the plastic microtip pipet, completely fill all the wells except A1, A12, D6, H1, and H12 with distilled water.
4. Use a paper towel to wipe away any water on the surface of the microplate.
5. Using the microtip pipet, add 1 drop of the indicator solution to each well containing water. Rinse the pipet with distilled water.
6. Use the forceps to add a small lump of calcium carbonate to well D6.
7. Using the scissors to cut four 1-cm lengths of soda straw. Insert one length of soda straw in each of the wells A1, A12, H1, and H12 as shown in Figure 22-1. Cut a 0.5-cm length of soda straw and place it in well D6.

8. Carefully place the microplate into the plastic sandwich bag and seal the bag. Place the bag on the piece of white paper.
9. Using the scissors, punch a small hole in the plastic bag directly over well D6.
10. Fill the microtip pipet one-fourth full with the hydrochloric acid solution.
11. Slip the tip of the pipet through the hole above well D6. Direct the stem of the pipet into the soda straw in well D6.
12. Add 4 drops of hydrochloric acid to the well. Observe the surrounding wells.
13. After 30 seconds, note any color changes in the surrounding wells. Record a color change in the solution in a well by marking a positive sign (+) to the corresponding well in the microplate shown in Figure 22-2.

Take Home Activity

Chapter 13

SOLAR HEATER

You need: thin cardboard from cereal boxes, aluminum foil—2 feet, scissors, glue, paper fastener, hole punch

Light from the sun provides heat that we can use. You can make a small solar heater to warm your hands or to slightly warm up some food.

- Cut out the pattern on this page. Trace and cut ten shapes from the cardboard and ten shapes from the aluminum foil using the pattern. Have an adult help you use the pattern to glue each piece of aluminum foil, shiny side up, to each piece of cardboard. After the glue is dry, punch a hole through each piece as shown on the pattern.
- Put the ten pieces in a stack, shiny side up, so that the holes are lined up. Join them together with a paper fastener. Have an adult help you use the pattern to spread the shapes into a circle, connecting the notches. Curve them upward slightly into the shape of a bowl.
- Try your solar heater on a sunny day. See how long it takes to warm your hands in front of it. Try this at different times of day to see if different amounts of sunshine affect the heat produced.
- Demonstrate your solar heater as you describe to a friend what is happening. Try heating different things in front of your heater: a small cup of water, a piece of bread, a rock, a piece of metal.

*There may be more than one activity for this chapter.

Chapter 13 Energy Resources

REVIEW AND REINFORCEMENT

Study Guide*

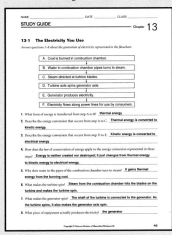

Concept Mapping

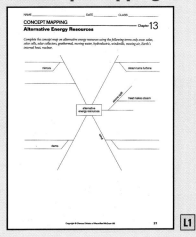

Critical Thinking/ Problem Solving

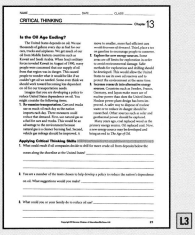

ENRICHMENT AND APPLICATION

Integrating Sciences

Across the Curriculum

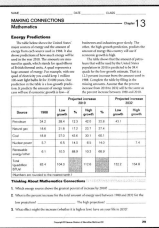

Technology and Society

ASSESSMENT

Multicultural Connection**

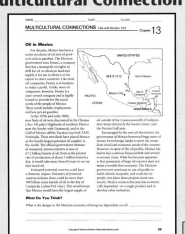

Performance Assessment

Review and Assessment

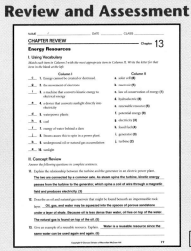

*One per section **Two per chapter

Energy Resources

THEME DEVELOPMENT

The themes that this chapter supports are systems and interactions and scale and structure.

CHAPTER OVERVIEW

In this chapter, students will study the role of water and steam in an electric power plant and learn how turbines and generators produce electricity. Students will be introduced to the formation and retrieval of fossil fuels and the environmental effects of burning fossil fuels.

Tying to Previous Knowledge

Students can infer how they use energy in a day by creating a list of ten things they do. Beside each item, have them write where they think the energy comes from for doing each activity.

INTRODUCING THE CHAPTER

Have students identify what is in the photograph on page 390. *windmills and a furnace* Ask them what the objects might be used for. *Grinding wheat, pumping water, using solar and wind energy.*

Uncovering Preconceptions

Students may think that electricity comes to their homes from large batteries, similar to those in cars. Explain to students that electric generating plants produce electricity and send it out to homes through wires.

ENERGY RESOURCES

Did you ever wonder...

✓ **Where the electricity in your home comes from?**

✓ **How a power plant produces electricity?**

✓ **If we'll ever run out of electricity?**

Science Journal

Before you begin to study energy resources and conservation, think about these questions and answer them *in your Science Journal*. When you finish the chapter, compare your journal write-up with what you have learned.

Answers are on page 418.

R emember the last time you were without electricity? Perhaps you were camping. What did you do for light at night? How did you cook?

Perhaps your city experienced a power outage. How long did the power failure last? The longer it lasted, the more likely it was to have serious effects, such as food defrosting inside freezers or accidents resulting from traffic lights that didn't work. When the power returned, you and others may have breathed a sigh of relief. Traffic flow could resume normally. Food would not spoil in freezers.

▶ **In this chapter, you'll read the story behind the flick of a switch that brings us electricity. You'll read about power plants. Day and night, power plants provide the electricity that many people have become totally dependent upon.**

A solar furnace in Odeillo, France

00:00 OUT OF TIME?

If time does not permit teaching the entire chapter, use the Chapter Overview on this page, Reviewing Main Ideas at the end of the chapter, and the Chapter 13 audiocassette to point out the main ideas of the chapter.

Learning Styles		
Kinesthetic	Find Out, p. 393; Activity, p. 394; Investigate, pp. 398-399, 410-411; Explore, p. 407	
Visual-Spatial	Demonstration, pp. 394, 408; Visual Learning, pp. 394, 397, 403, 405, 406, 412, 414; Explore, pp. 396, 400; Across the Curriculum, p. 397	
Interpersonal	Debate, p. 404; Activity, p. 406; Discussion, p. 414	
Logical-Mathematical	A Closer Look, pp. 404-405	
Linguistic	Explore, p. 391; Activity, pp. 392, 404; Research, p. 395; Discussion, pp. 397, 401, 402, 404, 412; Multicultural Perspectives, pp. 403, 408; Physics Connection, pp. 412-413; Across the Curriculum, p. 413; Science at Home, p. 418	

LS

Explore! ACTIVITY

How important is electricity to you?

Take a survey to see how electricity is important in your daily life.

What To Do

1. List how you use electricity in your daily activities.

2. Classify the items and activities as things you need (refrigerator), things that make life easier (blow dryer), or things you have or do for fun (TV set).

3. If you didn't have electricity, how would you supply the things you need? What changes would this cause in your lifestyle? Record your answers and observations *in your Journal.*

Explore!

How important is electricity to you?

TECH PREP

Time needed
20–25 minutes

Materials No special materials are required for this activity.

Thinking Processes organizing information, classifying, practicing scientific methods, observing, representing and applying data, predicting

Purpose To infer how electricity is used in a day and hypothesize how life would change without electricity.

Teaching the Activity

Discussion Have students identify essential electrical devices and justify their choices. Extract three or four answers from volunteers. Then ask the class to identify a few "fun" electrical devices and discuss how essential they are. **L1**

Science Journal In their journals, have students write a story about a morning without electricity, from awakening to arriving at school.

Expected Outcome

Students should identify both essential and nonessential electrical appliances in their homes.

Answers to Questions

Answers will vary. Make sure that students can explain their reasoning.

✔ Assessment

Process Working in cooperative groups, have students classify the electrical devices in the school according to whether they are essential. Students could then make a poster encouraging energy conservation in the school. Use the Performance Task Assessment List for Poster in **PASC,** p. 73.

ASSESSMENT PLANNER

PORTFOLIO
Refer to page 420 for suggested items that students might select for their portfolios.

PERFORMANCE ASSESSMENT
Process, pp. 391, 399, 411
Skillbuilder, p. 394
Explore! Activities, pp. 391, 396, 400, 407
Find Out! Activity, p. 393
Investigate!, pp. 398–399, 410–411

CONTENT ASSESSMENT
Oral, p. 393
Check Your Understanding, pp. 395, 401, 406, 414
Reviewing Main Ideas, p. 418
Chapter Review, pp. 419–420

GROUP ASSESSMENT
Opportunities for group assessment occur with Cooperative Learning Strategies.

PREPARATION

Planning the Lesson

Refer to the Chapter Organizer on pages 390A–D.

Concepts Developed

Students will learn how electricity is produced in power plants and transmitted to homes.

1 MOTIVATE

Bellringer

 Before presenting the lesson, display **Section Focus Transparency 37** on an overhead projector. Assign the accompanying **Focus Activity** worksheet. L1

LEP

Activity Draw a vertical line to divide the chalkboard in half. Direct individual students to come to the chalkboard and write on one side of the line the name of an appliance they listed in the Explore activity that uses household electricity. On the other side of the line, have students list the appliances that run on batteries. Electricity plays an important role in our lives. To underscore the idea that electricity is a useful and important commodity, challenge students to create a list of 20 items they use every day that do not use electricity. Stress to students that battery-powered items cannot be included in this list. Many students will be hard-pressed to think of 20 items. L1

Section Objectives

- Trace the source of the energy that runs appliances.
- Describe what water and steam do in an electric power plant.
- Relate the role of generators in producing electricity.

Key Terms

generator

The Source of Electricity

When you wake up in the morning and reach for a lamp or wall switch, you take it for granted the light will come on. You may turn a number of appliances on and off as you get ready for school without thinking about the electricity you're using.

Where does the electricity in your home come from? Imagine jumping through an electrical outlet to see for yourself. Once behind the outlet, you find that you're inside electrical wires. Study **Figure 13-1** to discover where electricity comes from and how it gets to your home.

Figure 13-1

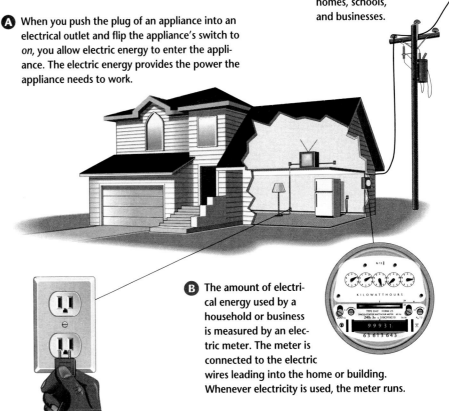

A When you push the plug of an appliance into an electrical outlet and flip the appliance's switch to *on*, you allow electric energy to enter the appliance. The electric energy provides the power the appliance needs to work.

C The wires that supply electricity to the outlets all come together at a fuse box or a breaker box. A transformer, often on a utility pole outside the home, converts high-energy electricity into lower-energy electricity that can be used by homes, schools, and businesses.

B The amount of electrical energy used by a household or business is measured by an electric meter. The meter is connected to the electric wires leading into the home or building. Whenever electricity is used, the meter runs.

392

Program Resources

Study Guide, p. 43

Making Connections: Technology & Society, p. 29, Hydroelectric Power, Past or Future? L1

Section Focus Transparency 37

ENRICHMENT

Activity So that students may better understand the impact of electrical devices on lifestyle, have them research and then write about a day in the life of a person who lives without electricity in his or her home. Then students can speculate on how life would change for the person if devices were more available. L2

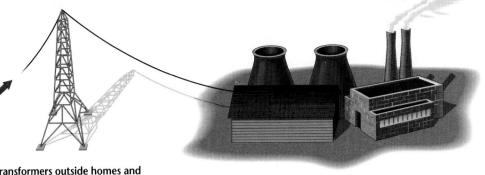

 D Transformers outside homes and buildings are connected to wires that eventually lead into even larger lines called transmission lines. Transmission lines carry electric current in a much greater strength.

E Transmission lines receive electricity from the electric power plant where electricity is generated.

After your journey through the outlet to the power plant, you've finally arrived at the source of the electricity. What goes on there? Do this Find Out activity to discover one way electricity is generated.

  Find Out! ACTIVITY

How can thermal energy be converted to kinetic energy?

How can thermal energy make things move? You know that if you get your hand too near a heat source, you move rapidly to get away. However, that's not the way it works with mechanical objects.

What To Do

1. Fill a teakettle about three-fourths full with cool water and place it on a burner.

2. Turn on the burner. What happens to the water as energy from the burner is transferred to the kettle? Is the transferred energy still thermal energy?

3. When steam begins coming out of the kettle's opening, hold a pinwheel in the path of the steam and observe. **CAUTION:** *Steam causes severe*

burns. Do not hold your arm, hand, or the pinwheel too near the steam.

Conclude and Apply

1. What happens to the pinwheel?

2. Explain why the following statement is true: The demonstration you have just done shows that energy changes forms.

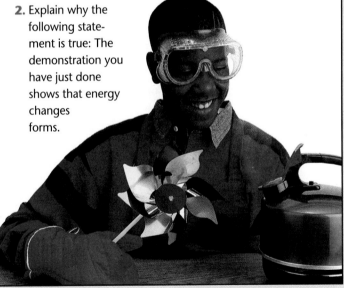

Find Out!

How can thermal energy be converted to kinetic energy?

Time needed 15 minutes

Materials hot plate, teakettle, pinwheel, mitt or other hand covering, goggles, water

Thinking Processes thinking critically, recognizing cause and effect, observing

Purpose To observe that thermal energy can be changed to mechanical energy.

Preparation Have all the equipment ready and be sure that the kettles do not leak.

Teaching the Activity

This activity will require waiting time, so you may want to fill kettles with hot tap water. Have students record their observations while they wait, or give them an energy word puzzle to complete. **L1**

Safety Be sure the students point the steam nozzle away from themselves and others.

2 TEACH

Tying to Previous Knowledge
Review potential energy, kinetic energy, and the law of conservation of energy (Chapter 4).

GLENCOE TECHNOLOGY

CD-ROM
Science Interactions, Course 2, CD-ROM
Chapter 13

Science Journal Have students write their responses to the Conclude and Apply questions in their journals. **L1**

Expected Outcome
Students will learn that thermal energy can be changed to mechanical energy.

Answers to Questions
2. The water boils. Yes, it is still thermal energy.

Conclude and Apply
1. The pinwheel spins.
2. Thermal energy from steam is partly converted into kinetic (mechanical) energy when it comes into contact with the blades of the pinwheel.

✔ Assessment

Oral Ask students to explain how thermal energy can be converted to mechanical energy. Have them give examples of machines or devices that run through this energy conversion. Encourage them to come up with additional examples besides the pinwheel. (Students may mention steam engines, moving trains, ships, etc.) Students could then make a bulletin board showing the examples that they identify. Use the Performance Task Assessment List for Bulletin Board in **PASC**, p. 59.

Energy Changes Forms at a Power Plant

Recall that in Chapter 4 you learned about the law of conservation of energy. According to the law of conservation of energy, energy can change form, but it cannot be created or destroyed. The total amount of energy in a system does not change. You worked with the law of conservation of energy as you completed the activity with the teakettle and the pinwheel. You neither created nor destroyed energy. You simply caused it to change form. First you transferred thermal energy from the burner to the water. Steam from the boiling water acted on the blades of your pinwheel, changing thermal energy to mechanical energy.

■ **Generating Energy**

Electricity can be generated at power plants using the same principles of energy transfer and change as your teakettle and pinwheel. **Figure 13-2** shows some of the equipment at a power plant and how it works.

As you can see, the main steps for generating electricity include boiling water to produce steam, and passing the steam through the blades of a turbine, which causes the axle of a generator to turn. A **generator** is any machine that converts mechanical energy to electrical energy. Most

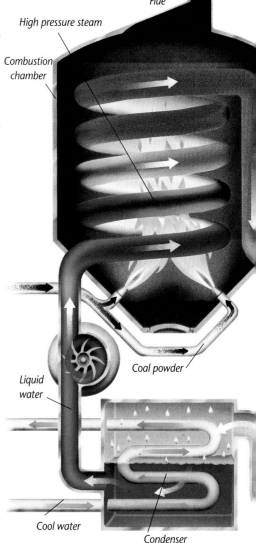

Figure 13-2

Ⓐ The generation of electricity begins with thermal energy. The power plant shown here burns coal in a combustion chamber. The burning coal creates thermal energy in the combustion chamber, which causes the water inside the pipes to turn into steam.

Flue
High pressure steam
Combustion chamber
Coal powder
Liquid water
Cool water
Condenser

394 Chapter 13 Energy Resources

---

generators consist of a coil of wires that rotates within a magnetic field. When the magnetic field of the magnet interacts with the wires, it causes electrons within the wires to move. This movement of electrons is electricity.

You know how important electricity is to our way of life. You found that out when you listed all the activities and appliances that use electricity. A world without electricity might be interesting to imagine in a story. What changes might you need to make in the way you are living if you did not have electricity in your school or community?

B The steam created in the combustion chamber is directed at the turbine. The pressure of the steam turns the blades of the turbine. The blades are connected to a shaft. When the blades turn, the shaft turns. The turning of the shaft is mechanical energy produced by the action of steam on the turbine. What part of the teakettle and pinwheel activity is the turbine like?

C The turning turbine shaft is connected to an electric generator. A simple generator contains a coil of wire that spins through a magnetic field. This rotation of the coil generates electricity. The electricity travels out of the coil and into electrical wires.

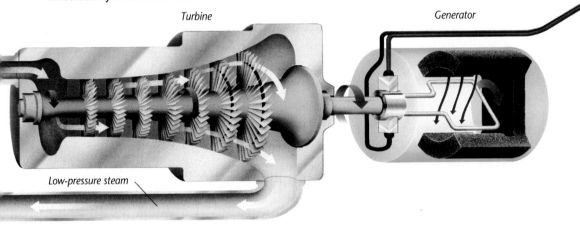

Turbine

Generator

Low-pressure steam

check your UNDERSTANDING

1. How does a blow dryer or other appliance you might use every day at home get the energy to operate?
2. Use the terms water, steam, mechanical energy, and electrical energy to discuss how steam helps generate electricity.
3. What is the relationship between a turbine and a generator in an electric power plant?
4. **Apply** Do you think it would be possible to generate electricity without coal? Explain.

check your UNDERSTANDING

1. Electricity generated at a power plant flows through wires into outlets in homes.
2. Water turns into steam. Thermal energy from the steam is converted into mechanical energy when it turns the turbine's blades. The turbine produces mechanical energy to run the generator that produces electrical energy.
3. The spinning turbine shaft is connected to the generator which makes electricity.
4. Yes; as long as there is kinetic energy to drive the generator. The kinetic energy would not have to come from burning coal.

Visual Learning

Figure 13-2 What part of the teakettle and pinwheel activity is the turbine like? Steam from the teakettle is like the low-pressure steam from the combustion chamber. The turbine is like the pinwheel turned by the steam.

3 ASSESS

Check for Understanding

To help students answer the Apply question, ask them if coal is the only source of thermal energy. *no* Use the Performance Task Assessment List for Group Work in **PASC**, p. 97.

Reteach

Project Have students make flash cards, each with one step in producing electricity as outlined in Section 13-1. Tell students to shuffle the cards and place them on a desktop. Have students arrange the steps in order from left to right in front of them. Check to be sure they are correct. Repeat the steps until students have memorized them. **L1**

Extension

Research Students who have mastered the concepts in this section can research the development of electric power from the experiments of scientists from Michael Faraday through Thomas Edison. **L3**

4 CLOSE

With books closed, have students draw a diagram showing how energy changes form at a power plant and produces electricity. Look for the following steps: burning coal produces thermal energy, thermal energy is transferred to water, water is turned to steam, steam drives turbine, turbine spins generator, electricity is produced. **L1**

13-2

PREPARATION

Planning the Lesson

Refer to the Chapter Organizer on pages 390A–D.

Concepts Developed

Students will learn about the formation of coal, oil, and natural gas and that they are fossil fuels formed long ago in Earth's history.

1 MOTIVATE

Bellringer

Before presenting the lesson, display **Section Focus Transparency 38** on an overhead projector. Assign the accompanying **Focus Activity** worksheet. L1

LEP

Explore!

What does a piece of coal look like?

Time needed 15 to 20 minutes

Materials lignite or bituminous coal, hand lens

Thinking Processes thinking critically, observing and inferring, organizing information, classifying, practicing scientific methods, forming a hypothesis

Purpose To examine coal and interpret its origin.

Preparation Cover the desk or countertops with old newspaper.

Teaching the Activity

Troubleshooting Be sure that samples have discernable fossils.

Science Journal In their journals, have students draw the piece of coal that they observe in this activity. L1

 13-2 **Fossil Fuels**

Section Objectives

- Compare and contrast three different fossil fuels.
- Trace the steps in the formation of fossil fuels.

Key Terms

fossil fuel

The Energy in Coal

A power plant can use any one of a number of different resources for the energy needed to operate turbines and generators. Some power plants, for instance, generate electricity from the force of rushing water or blowing winds. Others generate electricity by burning oil or natural gas. In the United States, more than 57 percent of the electric power plants rely on burning coal. That makes coal very important.

But what is coal made of? How can something like coal generate energy? Do the Explore activity below to begin to investigate the nature of coal.

Explore! ACTIVITY

What does a piece of coal look like?

Where does coal come from? A careful examination of a coal sample may give you some clues.

What To Do

1. Examine a piece of coal under a hand lens. What do you see? What color is the coal?
2. Describe its luster. Are there layers in the sample? Try running your finger or pencil tip along one of them.
3. Now see if you can detect any fossils in the sample. A fossil is any evidence of past life. It might be the imprint of either plant or animal life. If your coal sample has any fossils, decide whether they are remains of plants or of animals.
4. Record your observations and answers to the questions *in your Journal.*

Your examination of coal may have given you some clues as to how coal forms. Coal is a sedimentary rock. The distinguishing feature of coal, however, is that the sediments that form it are the remains of once-living organisms.

Most coal is made up of the remains of plants that captured energy from the sun to make food and to grow. These plant remains retain some of that solar energy in a changed form—chemical potential energy.

396 Chapter 13 Energy Resources

Expected Outcomes

Students will learn that coal is rock-like and contains fossils.

Answers to Questions

1. The coal is black and shiny. Soft brown coal looks brown and earthy.

2. Luster is shiny. Yes, there are layers in the sample.

3. Most fossils in coal are plants.

✔ Assessment

Portfolio Have students write a paragraph describing the coal sample that they examined. Students could then work in cooperative groups to research and write about where in the United States coal is found and mined. Use the Performance Task Assessment List for Writer's Guide to Nonfiction in **PASC**, p. 85. P

Formation of Coal

Much of the coal mined in the United States began forming more than 300 million years ago. Coal started as the remains of plants that died in swampy regions. As time went by, more plants died and sediment covered and compressed the plants. **Figure 13-3** on this page will give you a clearer idea of how the coal mined today formed.

Connect to...
Physics

Trace the energy conversions from the formation of coal from plants to the production of electricity at a coal-burning power plant.

Figure 13-3

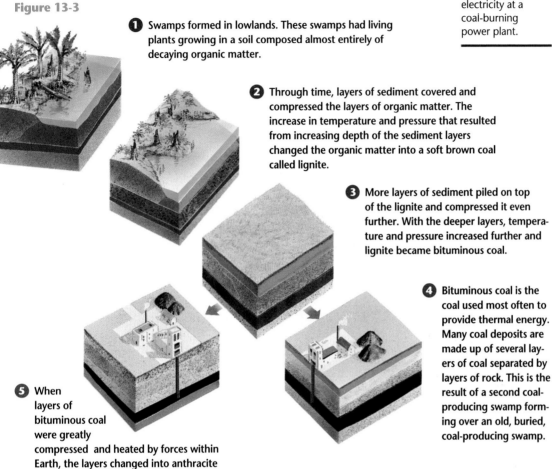

1 Swamps formed in lowlands. These swamps had living plants growing in a soil composed almost entirely of decaying organic matter.

2 Through time, layers of sediment covered and compressed the layers of organic matter. The increase in temperature and pressure that resulted from increasing depth of the sediment layers changed the organic matter into a soft brown coal called lignite.

3 More layers of sediment piled on top of the lignite and compressed it even further. With the deeper layers, temperature and pressure increased further and lignite became bituminous coal.

4 Bituminous coal is the coal used most often to provide thermal energy. Many coal deposits are made up of several layers of coal separated by layers of rock. This is the result of a second coal-producing swamp forming over an old, buried, coal-producing swamp.

5 When layers of bituminous coal were greatly compressed and heated by forces within Earth, the layers changed into anthracite coal, the hardest of all coals.

Because of its link with fossil plants from the past, coal is one of several different kinds of fossil fuels. A **fossil fuel** is the remains of ancient plants or animals that you can burn today to produce thermal energy. Other fossil fuels are oil and natural gas. Do the Investigation that follows to discover some difficulties with recovering another type of fossil fuel.

13-2   Fossil Fuels   **397**

13-1 Extracting Oil

Preparation

Purpose To determine which kind of rock material—sand, gravel, or a mixture—is the easiest to extract oil from.

Process Skills practicing scientific methods, separating and controlling variables, thinking critically, observing and inferring, recognizing cause and effect, organizing information, making and using tables, measuring in SI, predicting

Time Required 1 class period

Materials See reduced student text.

Possible Hypotheses Students may predict that any one of the three kinds of rock will yield the most oil.

The Experiment

Troubleshooting This is a messy lab. Students should wear old clothes and lab aprons. Have students save old spray bottles and bring them in. Make certain that the tubing in the spray bottles goes all the way to the bottom and is inserted into the rock material. Make certain that bottles are clean or have students wash the bottles and the spraying mechanism in liquid detergent.

Process Reinforcement Ask students to identify the variables in this investigation. *gravel, sand, gravel and sand mixed*

Possible Procedure For each experiment, start with a clean spray bottle. Use the same amount of rock in each experiment based on weight or volume. Put the measured amount of gravel, sand, or mixture into the bottle and then add 100 mL of vegetable oil. For each variable, spray out as much oil as possible and measure it in a graduated cylinder as per the photo.

Extracting Oil

In the section following this lab, you'll learn how oil is formed and trapped in rock. In this lab, you will be an oil driller and find out how to extract oil from the ground. The purpose of this experiment is to see what kind of rock yields the most oil.

Preparation

Problem

What kind of material—sand, gravel, or a mixture—yields the most oil?

Form a Hypothesis

Have your group agree on a testable hypothesis about extracting oil.

Objectives

• Compare how easy it is to extract oil from sand, gravel, or a mixture of the two.
• Infer from your investigation what kind of rock yields the most oil.

Materials

clear plastic bottle with spray pump
plastic tubing
100-mL graduated cylinder
100 mL vegetable oil
liquid detergent
sand
clean gravel

Safety

Do not let your "oil well" spray other students.

Teaching Strategies

Troubleshooting Students may obtain very little oil from spraying but it is important to record these data. Students will need more oil to start each experiment with 100 mL.

Discussion Discuss with students that some rocks are harder to extract oil from than others.

Science Journal Have students design data tables and record observations and data in their science journal. [L1]

Meeting Individual Needs

Learning Disabled Show students that oil on water can be broken up by detergent. Have students fill a beaker three quarters full of water. Tell them to add a small drop of vegetable oil to the water. What does the oil do? *floats* Have students drop a small amount of detergent in the middle of the oil "slick." What does it do? *The oil clears out in a circle toward the edges of the beaker.* What does detergent do to oil? *breaks it up* [L2]

DESIGN YOUR OWN
INVESTIGATION

Plan the Experiment

1 Examine the photos and the materials provided. Also look at the concluding questions. Then as a group, decide which materials will be used to test the hypothesis.

2 Write a step-by-step procedure to test the hypothesis. In each test, what will stay the same? What will vary? What will you be comparing? Repeat the procedure to test the sand, gravel, and the mixture.

3 Between tests in your experiment, it is probably a good idea to clean out the sprayer by running detergent water through it to remove the "oil."

Check the Plan

1 How will you mix your oil into the rock?

2 How many tests will you run?

3 How will you clean your oil well between tests?

4 How will you measure how much oil is extracted from each type of rock?

5 How will you record your data?

6 Before you proceed, make certain your teacher approves your plan.

7 Carry out your investigation. Record your observations in a data table *in your Science Journal.*

This huge oil gusher on May 28, 1923, helped to bring much of the oil boom activity to West Texas. It took 21 months to drill the well, located in Reagan County, Texas, before oil flowed.

Analyze and Conclude

1. Measure in SI How many milliliters of oil did you extract from each kind of rock?

2. Compare and Contrast Which rock yielded the most oil—the gravel, the sand, or the mixture of sand and gravel?

3. Draw a Conclusion Does the data from your experiment support your hypothesis? Which rock type would oil drillers prefer to drill into?

Going Further

Predict what would happen if you added hot water, cold water, or detergent to the oil. Would this increase the amount of oil extracted?

Program Resources

Study Guide, p. 44

Activity Masters, pp. 55–56, Design Your Own Investigation 13-1

Critical Thinking/Problem Solving, p. 21, Is the Oil Age Ending? [L3]

Multicultural Connections, p. 29, Oil in Mexico [L1]

Science Discovery Activities, 13-1, 13-2

Teaching Transparency 25 [L1]

Section Focus Transparency 38

Expected Outcome

Students will learn that pumping alone may not bring oil easily to the surface and that some materials are difficult to extract oil from. Gravel is easier to extract oil from than the other two materials.

Analyze and Conclude

1. Answers will vary.

2. Gravel yielded the most oil.

3. Answers will vary depending on predictions. Drillers would prefer loose gravel materials from which to extract oil.

✔ Assessment

Process Students can work in small groups to model a pressurized oil well. To do this, students fill a medium-sized balloon with water, tie it off, and place it in a plastic grocery bag. Then they cut a 0.5-m piece of plastic tubing on an angle and insert the angled end into the bag, but not into the balloon. Then they submerge the bag with the balloon in it into a sink full of water to drive air out of the bag. Do not let water enter the bag. With the air driven out, tie shut the top of the bag with string without squeezing off the tubing. Then puncture the balloon with the end of the tubing and unclamp the tubing over a sink. There will be a "gusher." Students can draw a cartoon to show how the "gusher" occurred. Use the Performance Task Assessment List for Cartoon/Comic Book in **PASC,** p. 61. **COOP LEARN** [P]

Going Further

The oil may mix slightly with hot water so more oil will be pumped out. Detergent will break up the oil and allow it to be pumped out more easily. Cold water and oil do not mix. Cold water can actually force the oil level up above the bottom of the drill.

Oil and Natural Gas

You probably remember that the other fossil fuels are oil and natural gas. What makes these two energy resources fossil fuels? Both formed from the remains of ancient organisms, and most major deposits formed millions of years ago. Unlike coal, however, which came from land plants, most oil and natural gas formed from plants and animals that lived in ancient shallow oceans.

As ancient sea organisms died, their remains settled on the ocean floor. Most organisms decayed. But in some places, they were buried under thick layers of sand and mud. Just as layers of decaying plants were compressed during the formation of coal, so the layers of sediment and partly decayed organisms were compacted over time. Slowly, chemical reactions changed the organisms into oil and natural gas. How are these fuels found within the ground now? The next activity will help answer that question.

 ACTIVITY

How are water and oil found in natural settings?

You may already be aware of how oil and water interact with each other. But what does that tell you about how they're likely to be found in natural settings?

What To Do

1. *In your Journal*, write your prediction of how oil and water might be layered if they are found within the same beds of rock. Which would you be able to pump out of the ground first?

2. Pour equal amounts of water and vegetable oil into a small bottle or jar. Shake the container.

3. Describe the appearance of the mixture. Observe what happens after the container is allowed to rest for a few minutes. What does the oil and water mixture do?

4. What does this tell you about how oil and water interact? What does this model show you about how oil and water are likely to be found in natural settings? Did this model support your prediction?

5. Record your observations and answers *in your Journal*.

Figure 13-4

A A petroleum-bearing rock must be porous (have holes in it) and be permeable (the holes must be connected) so that fluids such as oil and water can flow through it. Sandstone is porous and permeable, while shale is porous but not permeable. Oil and other fluids can flow through sandstone but not through shale.

B The oil and water gather in a pool, in a porous rock layer, under the impermeable rock layer. Because oil is less dense than water, oil lies on top of the water. If natural gas is present, it lies on top of the oil.

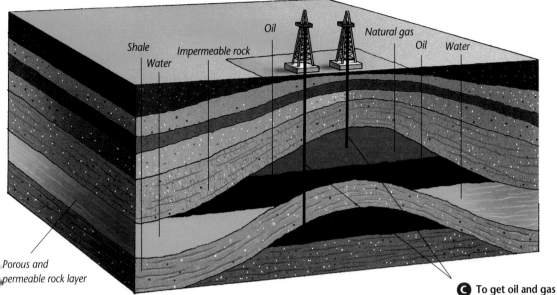

Porous and permeable rock layer

C To get oil and gas from the ground, engineers must drill through the rock layers to reach the reservoir.

■ Oil and Natural Gas in Natural Settings

Oil, gas, and water may get squeezed into the spaces of porous sandstone. If there's a space for oil and gas to accumulate, an oil or gas reservoir is formed. See how oil, gas, and water are arranged in **Figure 13-4**. Which is the least dense? How can you tell?

Coal, oil, and natural gas are all important resources for thermal energy and also for generating electricity. But you've seen that fossil fuels can be difficult to obtain. Fossil fuels can also create problems with the environment in a number of different ways. You'll find out about some of these problems next.

check your UNDERSTANDING

1. In what ways are fossil fuels used in the United States?
2. How is the formation of coal like the formation of oil and natural gas? How is it

different?
3. **Apply** Natural gas is often taken from the same locations where oil is drilled. Why do you think this is true?

check your UNDERSTANDING

1. coal for fuel in electric power plants; oil for gasoline to fuel cars and other vehicles; natural gas for cooking and heating
2. All were formed from living organisms. Coal formed mainly on land from plant remains; oil and natural gas form mainly in shallow oceans from the remains of sea

organisms, largely protists.
3. Where underground space is available, natural gas may have formed between oil deposits and the overlying layer of impermeable rock.

3 ASSESS

Check for Understanding

As they complete the Apply question have students discuss the ways coal and oil are formed. Use the Performance Task Assessment List for Group Work in **PASC**, p. 97.

Reteach

Discussion Have each student list three facts about coal, oil and gas. Then have students work with a partner to compare lists. Lead a discussion to create a master list on the chalkboard. Limit each heading (oil, gas, coal) to 5 or 6 facts. **COOP LEARN** **L1**

Extension

Discussion Have students who have mastered the concepts in this section discuss what life would be like if coal or oil deposits had never formed. Have them speculate as to what would take the place of the fuels. **L2**

4 CLOSE

Discussion

Have students brainstorm and discuss the ways coal, gas, and oil are used. Have them give specific examples for each. **L1**

PREPARATION

Planning the Lesson

Refer to the Chapter Organizer on pages 390A–D.

Concepts Developed

Students will learn about renewable and nonrenewable energy resources and the effects of burning fossil fuels.

1 MOTIVATE

Bellringer

 Before presenting the lesson, display **Section Focus Transparency 39** on an overhead projector. Assign the accompanying **Focus Activity** worksheet. L1

LEP

Discussion Ask students to make a list of items they would not want to see used up. Direct a discussion of what life would be like if the items on the list were actually used up. L1

2 TEACH

Tying to Previous Knowledge

Ask students to define the word *renew*. Ask them to make a list of things they can renew, such as library books, magazine subscriptions, movie rentals, etc. Tell students that fossil fuels are *nonrenewable*. Have them speculate on the meaning of the word.

Theme Connection

The theme that this section supports is energy. Using fossil fuels to create energy causes pollution. It takes energy to clean up pollution. The world faces a future shortage of oil.

Section Objectives

- Classify energy resources as either renewable or nonrenewable.
- Discuss the environmental effects of burning fossil fuels.

Key Terms

renewable resources
nonrenewable
   resources

Fossil Fuel: How Much and How Long?

The processes that caused coal to form continue today in some places on Earth. The same is true for the formation of oil and natural gas. This doesn't mean that we'll be able to find all the fossil fuel we need. Fossil fuels are natural resources. A natural resource is anything that occurs naturally that people use. Some natural resources occur in almost endless supplies. Others, such as fossil fuels, do not.

■ Renewable Resources

Natural resources that can be replaced by natural processes in fewer than 100 years are **renewable resources**. For example, think about the sun. It will continue to shine for millions of years. Energy from the sun will continue to warm the planet and provide light for growing plants. Study **Figure 13-5** to learn more about renewable and reusable resources.

Figure 13-5

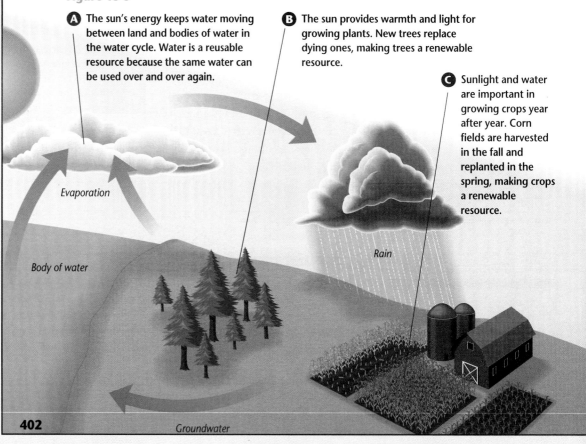

A The sun's energy keeps water moving between land and bodies of water in the water cycle. Water is a reusable resource because the same water can be used over and over again.

B The sun provides warmth and light for growing plants. New trees replace dying ones, making trees a renewable resource.

C Sunlight and water are important in growing crops year after year. Corn fields are harvested in the fall and replanted in the spring, making crops a renewable resource.

Evaporation

Body of water

Rain

Groundwater

402

Program Resources

Study Guide, p. 45
Making Connections: Across the Curriculum, p. 29, Energy Predictions L3
Section Focus Transparency 39

Meeting Individual Needs

Learning Disabled Tell students that paper is made out of wood. Have them discuss the merits and problems of recycling. Emphasize that recycling paper saves trees. Have students generate a list of all the useful things that come from trees. Ask them what could replace these things. When they say plastic, remind them that most plastics are derived from petroleum. **LEP**

■ Nonrenewable Resources

Resources that people use up much faster than nature can replace them are **nonrenewable resources**. A nonrenewable resource is one that may be replaced only over a very long time, or it may never be replaced.

Take coal as an example. It took millions of years for decaying plants to change into a seam of coal. An electric power plant burns the amount of coal in a small seam in a single day. Plants dying in today's swamps won't become coal for millions of years. Our supplies of coal, therefore, are limited. Even so, we may be able to rely longer on coal than on any other fossil fuel.

The United States has a great supply of bituminous coal; however, more than 430 billion tons of it are still in the ground. The United States recovers and burns over 900 million tons of coal each year. Of that 900 million tons, electric power plants burn more than 750 million tons.

Even at this rate of usage, the United States would still have enough coal to last hundreds of years.

Supplies of other fossil fuels are much more limited. The United States and some other countries of the world still have reserves of oil and natural gas. Reserves are places people know of where natural resources are deposited. According to some estimates, all the world's oil reserves could be emptied within the next 25 years if people continue to use oil at current rates.

World Energy Production and Consumption

World Energy Production

- North America
- Eastern Europe and Russia
- Asia and Pacific Rim
- Middle East
- Western Europe
- Africa
- Central America

World Energy Consumption

Figure 13-6

A The map below shows the locations of the known oil, gas and coal reserves. Judging from locations marked on the map, what can you infer about distribution of these resources?

B The pie charts above compare the amount of world energy produced with the amount of energy consumed. Which area has the greatest difference between amount produced and amount consumed? Which area uses the most energy?

Key

 Coal reserves

 Oil and natural gas reserves

13-3 Resources and Pollution **403**

Multicultural Perspectives

The Real McCoy

During the industrial revolution, steam-driven engines and turbines became important sources of power. An African-American inventor, Elijah McCoy (circa 1844–1928) patented many inventions related to steam engines. Born in Canada, McCoy moved to Michigan where he lived and worked for most of his life. During the 1870s, he patented several inventions that lubricated steam engines and steam cylinders. One of these inventions was a self-lubricating device which fed oil into industrial machinery. McCoy's invention reduced the time machines had to be shut down so that moving parts could be oiled. Have students research more about McCoy's inventions and their importance to industry.

Discussion Point out to students that volcanic eruptions also produce pollution. Dust, poison gas, and steam are vented into the atmosphere during an eruption. Ask students to list sources of air pollution caused by human activities. Have them come to the chalkboard and write their ideas. *Answers may include: vehicle fumes, incinerators, burning of fossil fuels for heat or to power factories.*

Debate Students can compare and contrast two sides to the air pollution problem. Side one will take the stance that pollution is a necessary outcome if economic growth is to continue. Side two will argue that air pollution is unacceptable no matter what the gain. Divide the class into groups of four. Give each group a position, then give them time to prepare to defend the position.

These questions should follow the debate. Can we have progress without air pollution? Is the level of air pollution this country has now acceptable? What might be the consequences of laws that absolutely prohibit all air pollution?

Activity Students can identify major sources of air pollution in their area. They can contact a local source and request to have data sent to them that shows air pollution levels over a one-month period. Students can graph the data to see if there are any patterns. If cars cause the most pollution, then weekday readings may be higher than weekend readings. L2

Burning Fossil Fuels: Costs to Be Paid

DID YOU KNOW?

In one year, burning the coal to light one 100-watt light bulb for 12 hours a day creates more than 900 pounds of carbon dioxide and 8 pounds of sulfur dioxide.

Fossil fuels are available in limited supplies and there may be scarcities in the future. But air pollution is a threat connected with fossil fuels right now.

Most of the air pollution traced to oil comes from burning gasoline, kerosene, and other oil-based fuels used in cars, trucks, and other motor vehicles. Because electric power plants burn so much coal, most of the air pollution traced to coal comes from them. Burning coal gives off nitrous oxide, which is made of nitrogen and oxygen, and carbon dioxide, which is made of carbon and oxygen.

Figure 13-7

A Earth reradiates some heat from the sun back into space in the form of infrared rays. Burning fossil fuels releases gases that absorb infrared rays. Too many of these gases in the atmosphere could work like a greenhouse roof and prevent heat from escaping. Earth's temperature could rise to an uncomfortable and dangerous level.

B Burning high-sulfur coal releases large amounts of the gas sulfur dioxide, which when combined with water vapor in the air, forms acid rain. Acid rain damages plant and animal life as well as buildings.

Electric Expense

The electric company charges your family for the electric energy you use. The company sells energy in units called kilowatt hours (kWh). A kilowatt hour is equal to 1000 watts of power used for one hour.

For example, one 100-watt light bulb left burning for ten hours uses one kilowatt hour of electric energy. If that bulb burns day and night for a month, it uses 72 kilowatt hours of energy.

This electric utility meter is being read electronically. Information from the meter is downloaded to a small hand-held computer which then sends the reading to a larger computer at the electric company.

Reading the Meter

To figure out your monthly bill, an electric company employee comes to your home and reads the electric meter. The meter might have dials like those on the one shown. It works something like a clock. The hands on the dials move to

Purpose

A Closer Look demonstrates to students that there are costs in generating electricity for which the consumer must pay. The content relates the theme of energy by introducing an electricity measurement unit, the kilowatt hour, and explaining the reading of the electric meter.

Content Background

Some students may think electricity is free because they have never seen an electric bill. Many costs go into the price the consumer must pay for electricity. Some areas tax electric consumption, adding to the cost. Some electric companies add a surcharge for their fuel consumption during winter and summer months. Others have a higher daytime rate and a lower nighttime rate. Some electric companies offer customers a discount for using most of their electricity at off-peak hours.

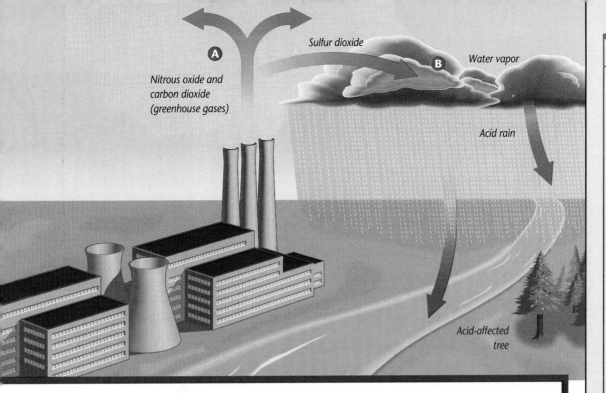

A — Nitrous oxide and carbon dioxide (greenhouse gases)

Sulfur dioxide

B — Water vapor

Acid rain

Acid-affected tree

Visual Learning

Figure 13-7 Help students recall what they learned in Chapter 9: Acid rain, which causes metals to corrode and pollutes soil and water, also forms when nitrogen oxides from car and truck exhausts react with water in the air. Students can formulate ways in which acid emissions from motor vehicles and from power plants can be reduced.

The following demonstration may help students better understand the potential effects of acid rain. Show students one effect of sulfuric acid on a substance. Materials needed are dilute sulfuric acid, nylon from women's hose, a large test tube, goggles, and an apron.

Cut a piece of nylon about 5 cm × 5 cm. Roll it tightly and push it down into the test tube. Pour enough sulfuric acid to cover the nylon. Place it in a test-tube rack and have students observe from a distance. Sulfuric acid is a component of acid rain.

measure the amount of electricity used.

Read the meter from left to right. If the needle is between numbers, read the smaller number. This meter reads 18 432 kilowatt hours. It read 17 268 kilowatt hours last month.

The difference between the two readings (18 432 kWh – 17 268 kWh = 1164 kWh) is how much electric energy was used this past month.

In this particular city, using one kilowatt hour of electricity costs 10.658 cents. Multiplying the cost of one kilowatt hour by the number of kilowatt hours used (10.658 cents per kWh × 1164 kWh) gives the amount of the bill—$124.06.

 USING MATH

What can you and your family do to reduce your electric bill by $10.00 a month? Do you think you can reduce your bill by $25.00 a month? How?

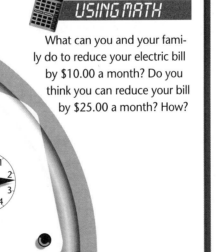

Rr 13 $\frac{8}{9}$

KILOWATTHOURS

2A

2A

405

GLENCOE TECHNOLOGY

Software
Computer Competency Activities
Chapter 13

Teaching Strategies

Contact your local electric company and ask for information on the various charges for electricity. Share this information with the class.

USING MATH

Answers will vary but should include shutting off lights and turning off appliances such as televisions and stereos when not in use.

Going Further ▐▐▐▐▐▶

Students can investigate electricity consumption and costs at your school. Ask a building service worker to accompany a group of students to the school electric meter so that students may read it once in the morning and once at the end of the school day and record their readings. Find out the cost of one kilowatt hour of electricity. Then students can calculate the cost of electricity used that day. Continue this process for one week, having different groups of students read the meter. Based on their findings, students can project the cost of the school's electricity for one month. Have students suggest ways the school could cut down on electricity consumption. Use the Performance Task Assessment List for Group Work in **PASC**, p. 97. **COOP LEARN**

3 ASSESS

Check for Understanding

As the students answer the Apply question have them write a list of nonrenewable resources and a list of renewable resources.

Reteach

Discussion Ask students what would happen if we used twice as many trees as we planted. Students should be able to list things such as: the building industry would have to turn to other materials; people who rely on wood for fuel would have to change fuels or get cold in the winter; people who cut the wood would lose jobs; paper would begin to become scarce and expensive; birds and animals that live in or around trees might become extinct. L1

Extension

Research Have students who have mastered the concepts in this section research the price of automobile fuel or home heating fuel for the past ten years. They should find the average price for each year and create a line graph. L3

4 CLOSE

Activity

Have student pairs make drawings with slogans urging people to stop polluting the atmosphere by suggesting ways to reduce the consumption of fossil fuels. L1

Figure 13-8

Smokestack scrubbers remove the pollutants from industrial smoke.

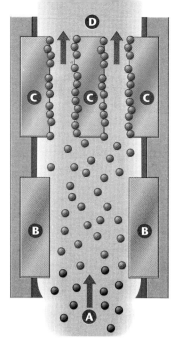

Ⓐ A fan blows the polluted smoke past electrically charged plates.

Ⓑ The plates give the particles of pollution a positive electric charge.

Ⓒ The smoke with its positively charged particles of pollution moves past negatively charged plates. The positively charged particles are attracted by and held to the negatively charged plates.

Ⓓ The smoke, now stripped of its pollutants, is released through the smokestack.

Carbon dioxide and nitrous oxide are both greenhouse gases. They get that name because they keep solar heat from escaping from Earth, much as a greenhouse retains solar heat inside. An accumulation of greenhouse gases may contribute to global warming. You will study more about greenhouse gases and global warming in Chapter 15.

Some coal contains large amounts of sulfur. When high-sulfur coal burns, sulfur dioxide is produced. The sulfur dioxide combines with water vapor in the air to form tiny droplets of acid. These droplets gather together to form acid rain, which is a severe pollution problem in some industrial areas. **Figure 13-8** shows one device that helps remove some pollutants.

Only time will tell what the future of our fossil fuel resources will be. We can be certain, however, that we will always need energy. What qualities would the perfect energy source have? It should be clean, it should be plentiful, and it should be renewable. In the next section, you'll read about several alternatives to fossil fuels that meet these requirements.

check your UNDERSTANDING

1. What makes coal a nonrenewable resource?
2. Why might the burning of fossil fuels be considered a costly practice?
3. **Apply** List several problems that result from burning fossil fuels. What alternatives to fossil fuels can you suggest that would help do away with some of these problems? How can you reduce consumption of fossil fuels?

check your UNDERSTANDING

1. Only a certain amount now lies in natural deposits. It will take millions of years for more coal to form.

2. Air pollution harms people and the environment; technology to reduce air pollution from fossil fuels costs money.

3. Air pollution, acid rain, acid deposition. Students may suggest using waterwheels and windmills. Student answers will vary, look for logical support for answers.

Alternative Energy Resources

If Not Fossil Fuels, What?

Imagine that you're responsible for developing the power source for a new city. You know you'll have to build a power plant to generate the electricity. You decide not to use fossil fuels as your energy source. But where do you go from there? What alternative sources of energy might you use? The following activity may help you develop some ideas.

Explore! ACTIVITY

How many different methods can you use to make your pinwheel spin?

You know that one way to generate electricity is to cause turbines to spin. Use a pinwheel as a model of a turbine and see how many ways you can make it spin.

What To Do

1. Demonstrate some of the ways you are already aware of.

2. Try this one. Fill a 400-mL beaker with water.

3. Hold the pinwheel over a sink, a bucket, or a pan.

4. Pour the water on the pinwheel, being careful that the water falls into the sink, bucket, or pan. Did you get the pinwheel to turn? Record your observations and answers *in your Journal.*

Program Resources

Study Guide, p. 46
Concept Mapping, p. 21, Alternative Energy Sources, L1
Making Connections: Integrating Sciences, p. 29, Leapfrogging Fossil Fuel Technologies with Biomass L2
Science Discovery Activities, 13-3
Section Focus Transparency 40

Section Objectives

■ Describe alternative sources of energy.
■ Differentiate among the ways alternative energy resources are used to produce electricity.

Key Terms

hydroelectric
solar cell

13-4

PREPARATION

Planning the Lesson

Refer to the Chapter Organizer on pages 390A–D.

Concepts Developed

In this section, students will learn about alternatives to fossil fuel use for electric energy production.

Explore!

How many different methods can you use to make your pinwheel spin?

Time needed 10 minutes

Materials pinwheel, 400-mL beaker, water, sink or bucket

Thinking Processes thinking critically, observing and inferring, recognizing cause and effect

Purpose To experiment to find out that moving water exerts force that can power a turbine (pinwheel).

Teaching the Activity

Tell students that they are demonstrating how a hydroelectric generating station operates. L1

Science Journal Have students make a diagram showing their observations and include lines to show the force that made the pinwheel spin.

✔ Assessment

Content Have students identify the energy changes that occur as the water is poured on the pinwheel. *Potential energy changes to kinetic energy.* Students could then create a device that uses falling water to make something move. Use the Performance Task Assessment List for Invention in **PASC**, p. 45.
COOP LEARN

1 MOTIVATE

Bellringer

 Before presenting the lesson, display **Section Focus Transparency 40** on an overhead projector. Assign the accompanying **Focus Activity** worksheet. [L1]

LEP

Demonstration

Materials needed are a 1.5-volt battery, a 1.5-volt solar cell, and a small electric motor (this apparatus can be purchased as a kit). Connect the battery to the motor using two wires. Show students that the motor operates. Disconnect the battery and connect the solar cell. With the solar cell pointed at a light source, show students that the motor operates.

Put an index card over the face of the solar cell and the motor will stop. Ask students what the major drawback is to using solar cells. *They only work in the light.* [L1]

2 TEACH

Tying to Previous Knowledge

Ask students what happens when they wear dark clothes on a sunny day. *They get hot or warm, depending on the season.* Elicit from students that they are experiencing warming by direct action of the sun.

Theme Connection The theme that this section supports is energy. This section covers alternative resources for energy production. All forms of energy production use some form of energy to produce another form of energy. The most abundant energy source is solar energy. Alternative energy sources will, in the long run, save money and the environment.

Hydroelectric Energy

Now you've seen that there is more than one way to get a pinwheel to spin. The same must be true of turbines in electric power plants. You might be pleased to discover that a number of alternative sources of energy are already at work generating electricity. One of them is waterpower. Power plants that use waterpower to generate electricity are called **hydroelectric** plants.

People in southern Canada and the eastern United States use the river waters at Niagara Falls to generate electricity for a number of large cities. In other places, where there are no natural waterfalls, people have built concrete dams to produce

Figure 13-9

A Hydroelectric power plants take advantage of the potential energy of water stored behind a dam.

B The stored water is released through gateways into pipes near the base of the dam that lead to the blades of turbines. Because of the weight of the reservoir water above it, the water entering the gateways is under great pressure as it falls to the turbines.

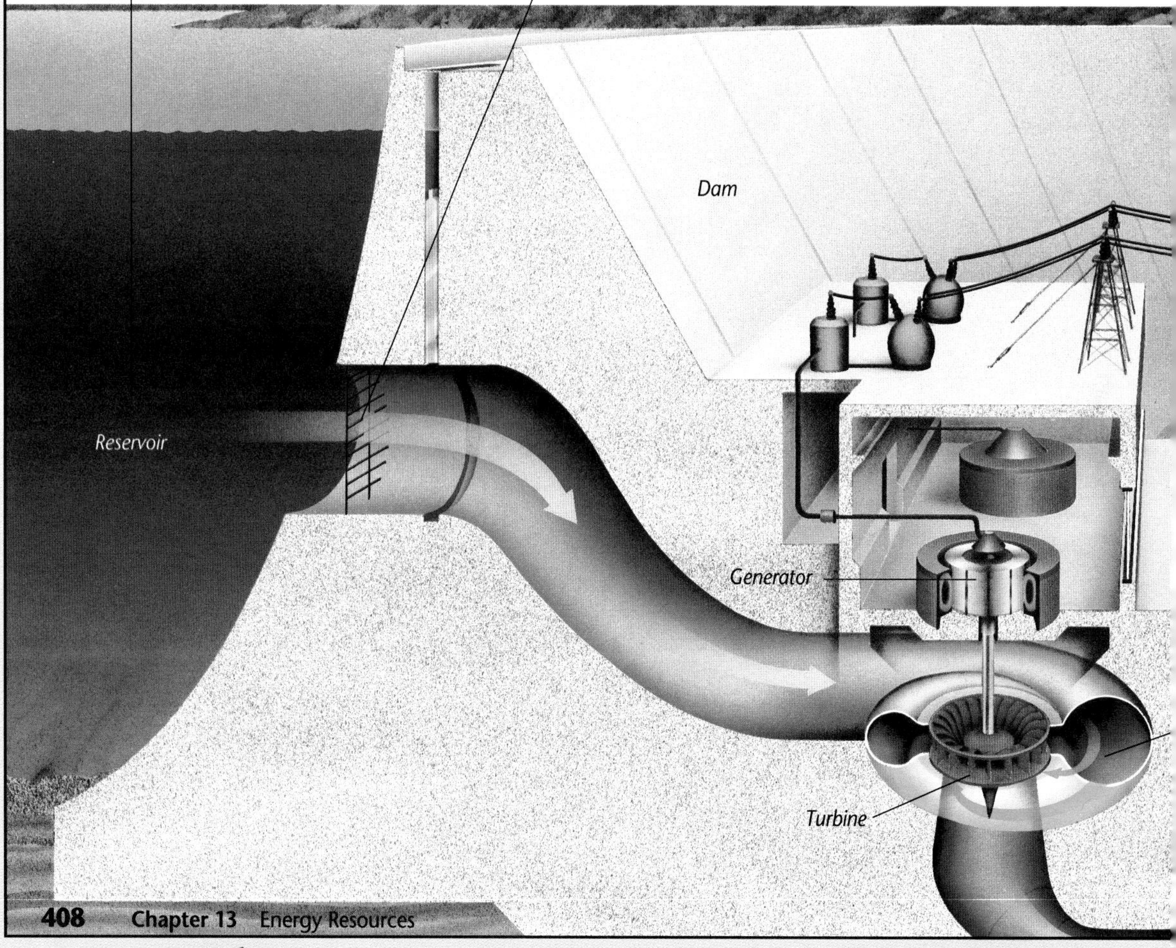

Dam

Reservoir

Generator

Turbine

408 Chapter 13 Energy Resources

Multicultural Perspectives

Early Use of Fossil Fuels

The Chinese probably burned both petroleum and natural gas for fuel and light as early as the fourth century BCE. Documents from the second century AD record that the Chinese searched for and drilled deep "fire wells" for natural gas. They constructed bamboo pipelines to transport the gas to other locations. Have students do research to find out when natural gas was first used in the United States.

hydroelectric power. The Shasta Dam, in northern California's Sacramento River, is the tallest structure of its type in the world. What happens to the waters of the Sacramento River behind the dam?

The river water that backs up behind a dam makes up the hydroelectric plant's reservoir. Most reservoirs are the size of lakes. Lake Shasta extends 56 kilometers up the Sacramento River. Now look at **Figure 13-9** of the dam and power plant to find out how a hydroelectric plant generates electricity.

Hydroelectric power is a clean, renewable source of energy. However, the dams that must sometimes be constructed to provide hydroelectric power create lakes that flood and destroy many different ecosystems.

About 6.5 percent of the electricity used in the United States comes from hydroelectric power plants. Hydroelectric power is one source that may meet future energy needs. Do the activity that follows to investigate another possible source.

Figure 13-10

The High Aswan Dam in Egypt, pictured below, is one of the world's largest hydroelectric plants. It gets its power from the Nile River.

Power lines

C The pressure of the water turns the turbines that drive the electrical generators in the plant.

D Power plant operators open or close the gateways in the dam to control when the generators will run and how much electric power they will produce. When does the water have the most potential energy? The most kinetic energy?

13-4 Alternative Energy Resources 409

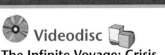

13-2 Warming Race

Planning the Activity

Time needed 1 class period to build solar collector; 5–10 minutes three times during the day to conduct the experiments; 20–30 minutes to organize the data

Purpose To design and construct a solar device to warm water and to compare and contrast it with an electric hot plate.

Process Skills making models, comparing and contrasting, collecting and interpreting data, making and using tables, forming a hypothesis

Materials Have students bring in newspapers to use for insulation. A variety of packing materials, including "bubble" sheets, could also be used as insulators.

Preparation Arrange a place in the classroom where each group's solar collector can be stored until students have conducted their warming experiments.

Teaching the Activity

Process Reinforcement Be sure that students set up their data tables correctly. Have them explain how the data that they collected influenced their understanding of solar energy.

Possible Hypotheses Students may hypothesize that the viability of using solar energy for heat depends partly on the efficiency of the solar energy collector and also depends on the availability of sunlight.

Possible Procedures Students should position the device so that solar energy both hits the water directly and is reflected onto the water. To achieve this, students could cut the short end from one side of the shoe box and line the remaining "insides" with foil. Students could divide the water among three beakers, place the beakers in the coffee can, and set the coffee can in the box (open side up), then cover the open end of the box with clear plastic, and position the device in bright sunlight.

Discussion You may wish to help students make plans for constructing their devices by holding a preliminary discussion on the available materials. Have students identify which colors absorb and reflect light, which materials are insulators, etc. Students can sketch or describe their devices before starting construction.

Science Journal Have students sketch their devices and record their hypotheses in their journals.

INVESTIGATE!

Warming Race

Solar energy seems like a perfect solution to our energy problems. Solar energy is nonpolluting, readily available and constantly renewed on Earth's surface. However, solar energy has a number of drawbacks. **In your Journal,** *make a list of some drawbacks that you can think of. Then, do this activity to explore one drawback you may not have thought of.*

Problem
Which will warm water better—solar energy or electrical energy?

Materials
400 mL water	three 250-mL beakers
hot plate	black plastic
coffee can, painted black	clear plastic shoe box
newspaper (for insulation)	thin plastic foam sheets (for insulation)
clear plastic	tape
rubber bands	aluminum foil
white glue	2 thermometers
bowl	

Safety Precautions

Be cautious around all sources of heat.

What To Do

1 Construct a device that uses the sun's energy to warm water. You will not use all of the materials, and you may find you do not need to use very many of them.

Program Resources

Laboratory Manual, pp. 69–72, Acid Rain L2 ; pp. 73–76, Solar Energy Application L3 ; pp. 77–80, Wind Power L2

Activity Masters, pp. 57-58, Investigate 13-2

2 At three different times of the day—morning, afternoon, and evening—place your solar heater in the sun to warm 200 mL of water for five minutes. Make certain that you always start with water at the same temperature.

3 *In your Journal*, record the temperature of the water every 30 seconds.

4 Warm 200 mL of water in a beaker on a hot plate for five minutes.

5 *In your Journal*, record the temperature every 30 seconds.

Mana La, the solar-powered car pictured above, is on a test run in Hawaii. Notice the solar collecting panels, which wrap around the top and sides of the vehicle.

Analyzing

1. *In your Journal*, make a table that summarizes the results of all four warming experiments.

2. Graph the results of all of the warming experiments—temperature on the vertical axis, time on the horizontal axis.

3. How do the warming experiments in the solar device compare to each other? Did water reach a higher temperature more quickly at one time of the day than at another?

4. How did the solar energy-device experiments compare to the electric device? With which device did water reach a higher temperature?

Concluding and Applying

5. Expand your graph to predict how long it would take for water warmed by solar energy to reach the same temperature as the water warmed by electric energy.

6. What might be one of the drawbacks to using solar energy to warm water?

7. **Going Further** Redesign your solar device to improve its ability to warm water and try heating 200 mL of water again. Compare your results to the previous experiment's results.

Expected Outcome

Students will find that the solar collector heats water to the greatest temperature at midday, when the sun is shining most brightly. They will observe that the hot plate heats water faster and brings it to a higher temperature than does the solar collector.

Answers to Analyzing/ Concluding and Applying

1,2. Answers will vary depending on the type of solar device constructed, the season of the year, and the degree of insulation. Tables and graphs should show that solar energy warms the water more slowly and that temperatures reached in the solar collector will not be as high as those on the hot plate.

3. Water reached a higher temperature more quickly in the morning and afternoon, when sunlight was more direct.

4. The electric device was quicker and it brought the water to a higher temperature.

5. Check student graphs for accuracy. Students may see that there are not enough hours of daylight for water to be warmed to the same temperature.

6. The greater amount of time that it takes for solar energy to warm water might be a drawback. Solar water heaters only warm water when the sun is shining.

7. Answers will vary. Students might improve their device by taping thin plastic foam sheets (insulators) to the outside of the shoe box.

✔ **Assessment**

Process Students can compare and contrast the collectors that different groups made. Then students can identify features on each that helped and hindered collection of solar energy. Students can then construct and display improved collectors in the lunchroom or another public area. Use the Performance Task Assessment List for Display in **PASC,** p. 63.
COOP LEARN

Figure 13-11 Students can infer the purpose of each component in the solar collector illustrated and then explain how it works. You may wish to explain that a heat exchanger is a device that allows the warm liquid in the pipes to warm the air that will be circulated through the house. Students may find it helpful to picture the heat exchanger as a pair of interleaved fins. Hot liquid passes through one set of fins while cold air passes through the other. As the hot liquid gives off thermal energy to the cold air, the liquid cools and the air is heated.

Inquiry Questions In the future, a large array of solar panels might be placed in space to collect solar energy. The energy would then be transmitted to Earth in the form of microwaves. How would this method of collecting solar energy eliminate some of the problems we have collecting solar energy from Earth's surface? *Weather, seasons, and latitude will not affect the collection of energy from space.*

Discussion Students can suggest how they could use solar energy at home. Students may decide to raise curtains and blinds in the windows to let more sunlight enter in cold weather. Tea can be made by placing a jar of water with a tea bag in it in the sun to "brew."

Solar Energy

Figure 13-11

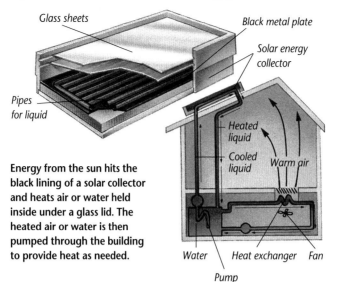

Glass sheets — Black metal plate

Solar energy collector

Pipes for liquid

Heated liquid

Cooled liquid — Warm air

Water — Heat exchanger — Fan

Pump

Energy from the sun hits the black lining of a solar collector and heats air or water held inside under a glass lid. The heated air or water is then pumped through the building to provide heat as needed.

Suppose your new city could find a single source for all its energy needs. This source might be the sun. The sun sends as much energy to Earth's surface in 40 minutes as humans use all around the world in one year. Energy from the sun, called solar energy, is clean and renewable. Look at **Figure 13-11** to see one way in which solar energy can be used.

Another way to use the sun's energy is to concentrate it with mirrors. For example, on pages 390 and 391 is a picture of a towering structure of flat mirrors that stands just outside the town of

Physics
CONNECTION

The containment tower at the Palo Verde, Arizona, nuclear power plant

Promises and Problems

Nearly 20 percent of the electricity in the United States comes from nuclear energy. One kind of nuclear energy is released when atoms of uranium are split apart. The particles move apart at high speed. They collide with other particles and give off thermal energy. The thermal energy causes water to boil, which makes steam to drive a turbine. In this way, electricity is produced.

The Advantages

One kilogram of uranium releases a million times more energy than one kilogram of fossil fuel. Nuclear energy does not release pollutants, such as CO_2 and sulfur emissions, as coal-burning power plants do.

The Disadvantages

However, nuclear energy has its disadvantages. Uranium releases radioactive particles. Therefore, the fuel in nuclear reactors must be handled with extreme care. Even though nuclear power plants are designed to contain radiation, accidents can and have occurred.

Physics
CONNECTION

Purpose
This excursion extends the discussion of alternative energy resources presented in Section 13-4.

Content Background
Of all the methods of electricity generation, nuclear-powered generating plants are the most controversial. Nuclear power, when compared with coal-burning power plants, is a clean power source. The waste is the problem. In addition to the concern

of how to contain nuclear waste safely, there is the problem of the water used to cool the reactor. This water is hot, and if it is returned to the environment, it will warm the body of water into which it flows. This may disrupt life cycles of aquatic organisms. Long-term problems of safety and waste disposal have caused nuclear power plants to be expensive.

Teaching Strategies
Have students conduct a debate.

Odeillo, France. The mirrors are positioned to focus energy from the sun on one part of the tower. In that tower is a high-temperature laboratory where temperatures can reach 3300° C.

■ Solar Cells

There are other uses for solar energy on a much smaller scale. For example, you may have used a solar calculator. A **solar cell** converts light from the sun into electricity, which powers the calculator. In one type of solar cell, thin layers of silicon, a hard dark-colored element, are sandwiched together and attached to tiny wires. As light strikes the different layers, it causes an electrical current to flow.

Solar cells are efficient because they generate electricity in just one step. Compare that with the number of steps required to produce energy from coal at an electric power plant that uses coal as an energy source.

Solar energy seems like a perfect energy source, doesn't it? But there are some drawbacks. Solar energy is available only when the sun is shining. Solar collectors and cells work less efficiently on cloudy days. They don't work at all at night. And during winter, when days are short, the collectors and cells generate less energy. Finally, although silicon is cheap and plentiful, solar cells are still expensive to make.

Solutions to Some of the Problems?

Even the most safely run nuclear power plant produces radioactive waste. Currently, there is no entirely safe way to store these wastes, which remain dangerous for thousands or millions of years.

Some government officials and scientists have suggested putting the waste in storage areas deep underground, in rural areas away from cities and towns. But others have said there is no way to make sure the storage area won't leak in the thousands of years the waste will remain dangerous.

Some people have suggested that the use of nuclear power should not be expanded until the serious problems connected with it are solved. Others think that nuclear power should continue to be developed and that safety and environmental problems can be solved as they occur.

Use the World Wide Web to find out how the design of the reactor contributed to the nuclear accident in 1986 at Chernobyl.

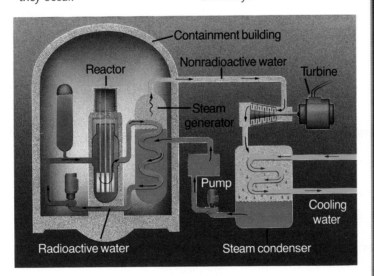

Containment building
Nonradioactive water
Reactor
Turbine
Steam generator
Pump
Cooling water
Radioactive water
Steam condenser

Present this situation. The local power company must decide whether to build a nuclear power plant near our town. It will provide jobs for hundreds of people and bring in outside construction workers to spend their money in town. Have one team represent citizens opposed to this plan and the other, citizens in favor of it. COOP LEARN

Chernobyl 10-year Anniversary information can be found at **http://www.uilondon. org/chernidx.html/**

Going Further ⫸
Have some pairs of students research nuclear fission reactions and how they are used to produce power. Students can report their findings in a written report or diagram. Use the Performance Task Assessment List for Group Work in **PASC**, p. 97. COOP LEARN

Inquiry **Question** What are some of the advantages of using wind energy? *Wind is nonpolluting, free, produces no waste, and causes no environmental harm.*

Visual Learning

Figure 13-12 Have students contrast a conventional and Darrieus windmill. Invite them to communicate the advantages of the Darrieus windmill in generating electricity.

3 ASSESS

Check for Understanding

To help students answer the Apply question have them think about the amount of sunlight that their area receives. Use the Performance Task Assessment List for Group Work in **PASC,** p. 97.

Reteach

Project Have students make a table by listing solar, wind, and water power in one column, and advantages and disadvantages of each in succeeding columns. L1

Extension

Project Have students plan a town where all the energy needs are met by alternative energy sources. Tell them to list as many buildings as they can in their town and the energy sources used to heat and cool them. L3

4 CLOSE

Discussion

Divide the class into groups of three and have the groups brainstorm the best possible alternative energy sources for their area. Have the groups defend their choices in a discussion. L1

Wind Energy

DID YOU KNOW?

The earliest windmills were probably built in Iran in the 600s. They were used to grind grain. These windmills had sail-shaped arms that revolved around a vertical axis.

What source of energy makes a kite soar overhead or a sailboat move on a quiet lake? It's wind power, an energy source that has been used for thousands of years. Egyptians used wind power to propel their sailing barges along the Nile River more than 5000 years ago.

One way to take advantage of the energy in wind is windmills. Today, people use windmills to generate electricity on a large scale. Examine **Figure 13-12** to see a windmill that generates electricity.

Wind energy is another clean and renewable alternative energy source. Like waterpower, it can be used only at certain locations. Winds must reach speeds of about 32 kilometers per hour to turn the blades of windmills. Also, if the winds are not blowing steadily, no electricity is generated.

Now that you've studied energy resources, you know that wise use and planning of resources can supply us with electricity for as long as we need it. As time goes on, we'll continue to use electricity, but the energy resources that generate it will be different from those we now use.

Figure 13-12

The power of the wind is often harnessed by windmills. The turning windmill blades are connected to a turbine, which drives an electrical generator.

A The Darrieus rotor in the photograph is a windmill with a vertical axis. Unlike traditional windmills, the Darrieus can operate in winds from any direction.

B Because its blades do not rise and fall but spin around, gravity does not slow the rotation of the Darrieus's blades. The Darrieus is also easier to maintain because all of its mechanical parts are located on the ground.

check your UNDERSTANDING

1. List the alternative energy resources that are renewable. What makes each of these renewable?

2. Which alternative energy source is not renewable? Explain.

3. **Apply** Do you think solar energy could serve well as an alternative energy resource where you live? What conditions would be needed in an area that wanted to use solar power?

check your UNDERSTANDING

1. Hydroelectric: Earth's total water supply never changes. Solar: The sun will last as long as life on Earth. Wind: There will always be wind on Earth's surface.

2. Nuclear energy; uranium cannot be replaced by natural processes within 100 years.

3. In most regions, solar power has potential because overall there is enough sunlight during the year to power solar batteries. Some coastal communities, where fog and rain are common, may be less well suited.

Science and Society

Using Coal Resources

Some coal lies within 200 feet of Earth's surface. To mine this coal, tons of soil and rock lying above the coal must be moved out of the way. It's a simple but dirty process.

First, bulldozers clear off an area and make it level. Then, many small holes are drilled in the rock and soil above the coal. The holes are filled with explosives. The explosives are set off and the rock is shattered. Next, earth movers, some as tall as a 20-story building, clear away the rock and soil, shoving it into massive heaps. The exposed coal is scooped up and loaded into huge trucks. This process is called strip mining.

The Cost of Strip Mining

For over 100 years, huge expanses of land were scarred by strip mining.

Once the coal was removed, the mined areas, looking like landscapes from some barren planet, were abandoned. Abandoned strip mines were more than just an eyesore. Serious environmental problems, including erosion and acid mine drainage, resulted.

Reclaiming the Land

Since 1978, companies that strip-mine in the United States have been required by law to restore the land they disturb. In this process, the open coal pit is leveled, and the rock and soil that were removed to expose the coal are placed back in the leveled pit. Topsoil is replaced and replanted.

The restored area looks better than the abandoned pits, but it isn't as good as new. The process of succession is slow to replace all the natural diversity the ecosystem had before strip mining destroyed it.

Below, a pit mine

Below, a reclaimed strip mine

You Try It

Write a short letter to your local newspaper that outlines some of the problems from mining and burning coal. Suggest ways in which people can conserve energy so that less coal is used.

Science and Society

Purpose Science and Society reinforces Section 13-2 on fossil fuels by explaining how coal is extracted in strip mining. It also presents problems involved with strip mining.

Content Background Processes called coal gasification and liquefaction treat coal with high heat, oxygen, and hydrogen to produce a mixture of gases, including methane or liquid petroleum. Methane can be separated from the mixture for use in homes. The other gases are low grade and can be used in industry, but are too expensive to transport. The liquid products produced by liquefaction also need extensive cleaning and refining to be usable. Processing plants to extract and refine the products cost upward of $4 billion. In this regard, the costs far outweigh the benefits. The United States has huge energy requirements and an abundance of coal to meet many of those needs. However, coal will not cheaply replace oil.

Teaching Strategies Tell students that mining disasters and black lung disease have had a serious effect on coal miners. Black lung disease is caused by breathing coal dust. It is fatal, and long-term health care must be provided for its victims. Have students discuss who should pay the costs of this long-term care, the taxpayers or the mining companies. Steer the discussion in the direction of a debate.

Debate Have students debate whether or not the United States should seek a clean way to use coal rather than find alternate energy sources.

Answers to You Try It
Student responses will vary but should outline dangers of coal use. Students may feel that alternatives to coal should be developed, especially since coal is a nonrenewable resource.

Going Further ▌▌▌▌▶
Divide the class into five groups. Assign one of the following topics to each group to research: (1) the formation of coal in ancient swamps; (2) coal mining techniques; (3) the hazards of coal mining; (4) the uses of coal and its by-products; and (5) production and consumption of coal in the world by continent. Have each group prepare a report and poster on their findings. For further information, students can call: United Mine Workers of America 1-800-843-8109. Use the Performance Task Assessment List for Group Work in **PASC**, p. 97. **COOP LEARN**

Technology
Connection

Purpose

This Technology Connection extends Section 13-4 on alternative energy resources by explaining how people can tap into the geothermal energy within Earth's crust. It also presents the drawbacks of geothermal energy.

Content Background

Geothermal energy is plentiful where volcanic activity is near the surface. Geysers are a product of geothermal energy. Old Faithful in Yellowstone National Park is a familiar name to many students. Water from the surrounding area seeps down through cracks in the ground and comes into contact with heated rock. Steam pressure builds in about 65 minutes and the geyser erupts. It is an extremely predictable occurrence. Tapping steam power like this and forcing it across a turbine produces electricity.

Teaching Strategies
Have students compare and contrast the relative pros and cons of geothermal energy in areas where people have direct access to local geysers and in areas where technology must be used to exploit this alternative energy resource.

Science Journal

Student answers will vary. However the plant would disrupt the natural beauty of the park and it would also endanger local ecosystems.

Technology
Connection

Geothermal Energy

An erupting volcano or geyser produces vast amounts of thermal energy. Can we tap into that energy source? If we can, not only will we conserve the fossil fuels we would have burned otherwise, but we might even reduce the volcano's or geyser's destructiveness by giving it another outlet.

Natural Water Heaters

People in Iceland have used hot water and steam from local geysers for hundreds of years to heat their homes and to wash laundry and dishes. Those of us who live farther from naturally boiling springs need to use technology.

Engineers drill a deep well into Earth's crust until they reach a layer of heated rock that surrounds a magma chamber. If the rock is porous, it may already hold water in the form of steam under great pressure. When the drillers tap into the reservoir, steam and hot water rise to Earth's surface where they can be used to turn the blades of turbines. This energy source is called geothermal ("Earth heat") energy. The picture on this page shows a geothermal power plant in Iceland.

Geothermal power plant west of Reykjavik, Iceland

Limitations on Geothermal Sources

All rocks everywhere get hotter as you go deeper into Earth's crust. Then why don't we use geothermal energy everywhere? For one thing, geothermal energy must be found near Earth's surface. Drilling deep wells is expensive and causes pollution.

Geothermal wells release hydrogen sulfide and sulfur-dioxide gas, which are poisonous. Expensive industrial scrubbers can keep these gases from polluting the atmosphere.

Another drawback is the water. It may contain toxic elements such as ammonia, boron, mercury, or radioactive elements. A geothermal power plant has to plan for safe disposal of cooled wastewater.

Science Journal
Yellowstone National Park is a large wilderness area in Wyoming that has numerous geysers and hot springs—sources of geothermal energy. *In your Science Journal,* discuss how you would feel about a geothermal power plant being built on the border of Yellowstone National Park. How might the plant affect the park?

Going Further ⫸
Students can work in small groups to research the use of geothermal water for heating in places such as Reykjavik, Iceland; Villeneuve la Garenne, France; or Boise, Idaho. Have groups write a report of their findings and present it orally, along with graphic aids, to the class.

Students could begin their research by reading Chapters 3 and 4 in *Tapping Earth's Heat,* by Patricia Lauber. Garrard Publishing Company, 1978. Use the Performance Task Assessment List for Writing in Science in **PASC,** p. 87.

HOW IT
WORKS

Tidewater Power Plants

It's a simple concept. Find a bay where high tide and low tide vary by more than 10 meters (33 feet). Build a dam across the bay. When the gates are open and the tide is coming in, seawater fills a reservoir behind the dam. Install turbines in the dam. Now, every time the tide comes in or goes out, the turbines are turning and you're making electricity. It's nonpolluting, renewable, and—after the initial construction costs are met—free. The diagram on this page is an example of a tidal power plant that has been operating in France.

So Why Aren't Tidal Power Plants More Common?

The answer to that question is in the fact that the operation of tidal power plants isn't quite that simple.

First, there aren't many bays with such a large tidal range. And when you find one, the shape of the bay may be wrong for storing a large volume of seawater on the landward side.

Second, tidal power plants don't generate much electricity. The water is not under great pressure, as it is at the foot of the huge dams that generate hydroelectric

power. Also, of course, the turbines turn (and therefore generate electricity) only part of each day. Still water means still turbines. And unfortunately, we have not yet created a good way to store electricity in large quantities. So an industry or a community that wanted to use tidal power would need other reliable power sources.

What Do You Think?

What other problems can you think of that must be solved before tidal power plants are common? What about such a plant's environmental impact on its own intertidal zones and on currents hundreds of miles down the coast? Discuss the future of tidal power as a group.

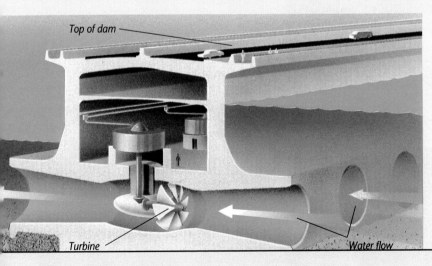

Top of dam

Turbine

Water flow

the requirements for tidewater power. *Examples include the Bay of Fundy in Canada, the Severn Estuary in Great Britain, The Rance Estuary in France, Cook Inlet in Alaska, and the Gulf of California in Mexico.* Students could consult tide tables of various coasts around the world. Tide tables are published by the National Oceanic and Atmospheric Administration of the U.S. Department of Commerce. ⬛L2

Have students look at the diagram and picture on this page. Direct them to read the statements to review the main ideas of this chapter.

Teaching Strategies

Direct students to pretend they are energy locked up in the coal and oil shown in the diagram. Have students describe their journey from the beginning of their life cycle (living things), to lighting a light bulb in a house.

Answers to Questions

1. A generator is a coil of wire that rotates inside a magnetic field and thus produces energy.

2. Low-sulfur coal can be used instead of high-sulfur coal. Smokestack scrubbers can be used to remove some of the pollutants.

3. In hydroelectric plants, the turbines spin because of the direct action of the water on the turbines. In fossil-fuel burning plants, steam is produced by heating water and the steam is pumped through the turbines, causing them to turn.

Science Journal

Review the statements below about the big ideas presented in this chapter, and answer the questions. Then, re-read your answers to the Did You Ever Wonder questions at the beginning of the chapter. *In your Science Journal,* write a paragraph about how your understanding of the big ideas in the chapter has changed.

1 Wires transport electricity from generators to our homes, schools, and other places that need electric power. Many power plants have steam turbine generators that rely on coal for fuel. *Describe how a generator produces electricity.*

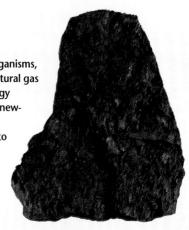

2 Derived from ancient living organisms, the fossil fuels coal, oil, and natural gas are our most widely used energy resources. They are also nonrenewable. Burning coal and oil has released harmful pollutants into the air. *What steps can be taken to prevent some of the damaging effects of burning fossil fuels?*

3 As fossil fuel supplies dwindle, alternative energy resources become increasingly important. Some of those already being used are hydroelectric, nuclear, solar, wind, and geothermal power. *How is the way hydroelectric plants cause turbines to spin different from the way fossil fuel-burning plants cause turbines to spin?*

Science at Home

Have students keep track of energy use in their homes. Tell them to assess the ways they might be able to save energy. Have students present their ideas in the form of an oral report to the class.

Science Journal

Did you ever wonder...
• Electricity used in homes comes from an electric power plant. (pp. 392–393)
• A power plant uses a generator to convert kinetic energy into electrical energy. (pp. 394–396)
• Eventually all the fossil fuels will be used up, and we will not be able to generate electricity with them. However, if we explore and use alternatives to fossil fuels, the electricity supply is in no danger. (pp. 402–414)

GLENCOE TECHNOLOGY

MindJogger Videoquiz

Chapter 13 Have students work in groups as they play the Videoquiz game to review key chapter concepts.

Using Key Science Terms

fossil fuel   nonrenewable resource

generator   renewable resource

hydroelectric   solar cell

Answer the following questions using what you know about the science terms.

1. Compare and contrast geothermal power and hydroelectric power.
2. What are fossil fuels and what makes them different from one another?
3. Draw a generator and describe how it works.
4. Compare and contrast nonrenewable and renewable resources.
5. How does a solar cell produce energy?

Understanding Ideas

Answer the following questions in your Journal using complete sentences.

1. Why is a corn crop considered a renewable resource?
2. What are greenhouse gases? Why are they called greenhouse gases?
3. How does power production by nuclear plants help reduce air pollution?
4. How does a fossil fuel-burning plant contribute to acid rain?
5. Why are fossil fuels used more often than alternative energy resources?

Developing Skills

Use your understanding of the concepts developed in this chapter to answer each of the following questions.

1. **Concept Mapping** Create a sequence concept map of the formation of coal. You may use the format below or modify it to use in your map.

Initiating event

Swamps produce layers of organic matter.

Event 1

Sediment covers organic matter.

Event 2

Compression of organic matter forms lignite.

Event 3

Compression of lignite forms bituminous coal.

Final outcome

Coal may be used to produce thermal energy.

2. **Sequencing** Outline the steps in the generation and transportation of electricity to your school.
3. **Hypothesizing** Refer to the Explore activity on page 400. Hypothesize what can happen to wildlife as a result of an oil spill.
4. **Designing an Experiment to Test a Hypothesis** Refer to the Explore activity on page 407. Hypothesize what other methods can be used to spin the pinwheel. Design an experiment to test your hypothesis.

Critical Thinking

Use your understanding of the concepts developed in this chapter to answer each of the following questions.

1. You find a fossil in a chunk of coal. Is it more likely to be a coral or a twig? Explain your answer.
2. What geographic limitations do solar, wind, geothermal, hydroelectric, and tidal energy resources have?

Using Key Science Terms

1. Geothermal energy uses Earth's heat. Hydroelectric energy is produced by falling or flowing water.
2. Coal, oil, and natural gas are fossil fuels. Coal is formed from once-living plants in swampy areas, and oil and natural gas are formed primarily of the remains of ancient sea organisms.
3. A generator is a coil of wire that rotates inside a magnetic field to produce electricity.
4. Renewable resources can be replaced quickly. Nonrenewable resources can only be replaced over a long period of time.
5. A solar cell generates electricity by producing an electric flow of energy from the sun.

Understanding Ideas

1. Because, it can be replanted and reharvested every year.
2. Nitrogen oxide and carbon dioxide are greenhouse gases. They collect in the atmosphere and keep solar energy from escaping.
3. Nuclear power plants do not rely upon burning fossil fuels to produce energy. They have no emissions of greenhouse gases.
4. Coal-burning plants release sulfur dioxide which combines with water to form acid rain.
5. Currently in most regions, fossil fuels are less expensive to use than alternative energy resources.

Developing Skills

1. See student page.
2. 1) Coal is burned to release thermal energy. 2) Thermal energy heats water changing it to steam. 3) Steam flows through turbines. 4) Turbines turn and drive the generator. 5) The generator produces electricity, which flows through wires to the school.
3. Oil spills can seriously harm or kill wildlife.
4. Students should suggest methods using the sun or the wind.

Program Resources

Review and Assessment, pp. 77–82 L1
Performance Assessment, Ch. 13 L2
PASC
Alternate Assessment in the Science Classroom
Computer Test Bank L1

Critical Thinking

1. It will be a twig. Coal forms in swampy areas where twigs are abundant.

2. Hydroelectric plants must be near running water. Solar energy requires abundant sunshine. Wind power is efficient only where wind speeds are a steady 32 km per hour. Geothermal energy can be tapped only in areas with underground volcanic activity. Tidal energy requires being near the ocean where tidal range is great.

3. We are now making the greatest use of fossil fuels ever. Oil usage has reached its peak and will fall off before the use of other fossil fuels does.

4. Reduce rate of consumption by conservation of resources.

Problem Solving

Answers will vary, but students should base their answers on the use of windmills and running water to run a turbine.

Connecting Ideas

1. Burning coal produces thermal energy that changes into mechanical energy. A generator converts mechanical energy into electricity.

2. Burning oil releases soot and waste gases. Oil spilled in the ocean coats the ocean surface; some sinks and affects bottom dwellers.

3. The sun is the ultimate source of energy; it provided the light that plants used to grow.

4. Earth's thermal energy sources are near the surface in these areas.

5. If high sulfur coal is exposed to rain it might produce acid runoff. If this water flows into local ponds and streams it makes them more acidic. This might cause the death of plants and animals.

6. Tidal energy stations are expensive. Plants and animals living nearby will be affected because of water flow change. There are few areas with the tidal range necessary to make them work.

3. The diagram below shows one geologist's idea of the place fossil fuels will occupy in human history. What statement is the geologist making about the use of fossil fuels today? What statement is she making about oil in particular? Do you agree or disagree with the geologist's view of the future of fossil fuels?

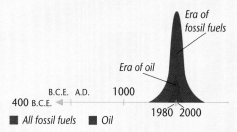

4. Imagine that there's a natural resource called a barghopper. Using barghoppers at the rate of 5000 a year, you've now used up all of the existing barghoppers. Nature still produces barghoppers at the rate of 2000 a year. What change do you make to change the barghopper from a nonrenewable to a renewable resource?

Problem Solving

Read the following problem and discuss your answers in a brief paragraph.

Dmetri lives with his family in a cabin atop a mountain. The cabin is above the tree line, and is so far from the nearest city that there's no electric service and no natural gas. Much of the year, the mountain is covered with snow. During the spring and summer, the snow melts, forming rapidly moving meltwater streams in the nearby valleys.

Consider sources of electricity that could be used in Dmetri's home. List suggestions for Dmetri, telling him what to do to get electricity for his cabin throughout the year.

CONNECTING IDEAS

Discuss each of the following in a brief paragraph.

1. Theme—Energy How can coal be used to generate electricity?

2. Theme—Systems and Interactions How is oil related to air pollution? How is it related to pollution of the ocean?

3. Theme—Energy Why is the sun an important source of energy in the fossil fuels used today?

4. Technology Connection Why is geothermal energy usually most readily available in regions of frequent volcanic activity?

5. Science and Society What environmental effects would you expect from exposing high-sulfur coal to rain? What might happen to local ecosystems if strip mining allowed drainage of water into local streams, lakes, and ponds? Explain your answer.

6. How Does It Work? Why aren't there many tidal energy stations operating today?

✔ Assessment

Portfolio Review the portfolio options that are provided throughout the chapter. Encourage students to select one product that demonstrates their best work for the chapter. Have students explain what they learned and why they chose this example for placement into their portfolios.

Additional portfolio options can be found in the following **Teacher Classroom Resources:**

Making Connections: Integrating Sciences, p. 29
Multicultural Connections, pp. 29, 30
Making Connections: Across the Curriculum, p. 29
Concept Mapping, p. 21
Critical Thinking/Problem Solving, p. 21
Take Home Activities, p. 20
Laboratory Manual, pp. 69–72; 73–76; 77–80
Performance Assessment P

Earth Materials and Resources

In this unit, you learned how to distinguish among metals, nonmetals, and metalloids. You compared minerals based on properties, such as hardness, luster, and cleavage. You classified rocks as sedimentary, igneous, and metamorphic. You learned how some of Earth's materials can be used as sources of energy.

You also investigated the structure of Earth's oceans and learned about features of Earth's shorelines and the ocean basin floor.

Try the exercises and activity that follow—they will challenge you to use and apply some of the ideas you learned in this unit.

CONNECTING IDEAS

1. Explain why sedimentary rocks will likely form on ocean floors. What types of natural resources would you expect to find within the layers of sedimentary rocks on the ocean floor? Explain your answer.
2. Arrange groups a through d in order from the simplest to the most complex of Earth's structures: a) Rocky Mountains and the Mid-Ocean Ridge; b) silicon, aluminum, and oxygen; c) granite and basalt; and d) quartz, feldspar and mica. Classify each group according to the type of natural resource it represents.

Exploring Further ACTIVITY

Granite: Looking between the grains

What To Do

1. Pretend you could reduce your size to fit between the particles in a piece of granite.
2. Write a story about what you might see in your travels through the rock.
3. Describe the different particle sizes, elements, minerals, crystals, gas molecules, liquids, colors, lusters, or any other properties you might see.

Earth Materials and Resources

THEME DEVELOPMENT

The two themes supported in this unit are systems and interactions and energy. The unit explored how the rock cycle showed the interaction of force and pressure. In addition, students learned how coal containing stored energy can be used by people to perform work.

Connecting Ideas
Answers

1. Sedimentary rocks will likely form on the ocean floors because these rocks form through the accumulation and consolidation of sediments. Oil and natural gas might be found in these rock layers.

2. The correct order is: b, d, c, a. The simplest structures are elements, followed in complexity by minerals, rocks, and rock formations such as mountains.

Exploring Further

Granite: Looking between the grains

Purpose To apply and communicate knowledge of the properties of minerals and rocks.

Background Information

Materials Allow students to examine samples of granite if possible.

Troubleshooting If students have trouble getting started, encourage them first to draw or describe aloud what they might see in their travels through the rock.

Answers to Questions

Students' responses will vary. Stories should creatively relate what students have learned about minerals and rocks. Encourage students to use all five senses (touch, taste, smell, sound, sight) as they describe the different particle sizes, elements, and so forth.

✔ Assessment

Portfolio Have students compile their stories into a class booklet. One possible title is *Looking Between the Grains*. Use the Performance Task Assessment List for Booklet in **PASC**, p. 57. **COOP LEARN** **P**

Air: Molecules in Motion

UNIT OVERVIEW

UNIT FOCUS

In Unit 4, students will learn that the structure of atoms affects how they react to form molecules. They will learn how some elements combine to form gases and observe how gases are affected by temperature and pressure. In addition, students will communicate the importance of gases in the respiratory systems of living things.

THEME DEVELOPMENT

This unit develops the themes of energy and systems and interactions. Energy is central to all the atomic, molecular, and biological processes discussed in Unit 4. Students will communicate how energy allows the body to synthesize biochemical molecules. The input and output of the respiratory system is an example of systems and interactions. A knowledge of gases will help students communicate how the body uses oxygen, a gas in air, to combine with food, releasing energy.

Connections to Other Units

The information on gases presented in this unit will help students understand how cells and simple organisms function, one of the topics of Unit 5. In addition, a firm grounding in gases, atoms, and molecules will better enable students to grasp the material on chemical reactions in Unit 5.

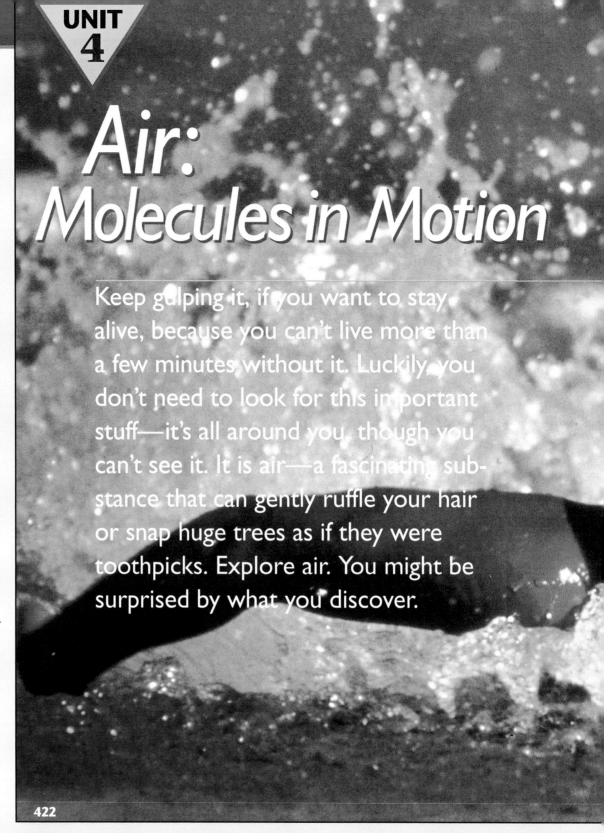

UNIT
4

Air: Molecules in Motion

Keep gulping it, if you want to stay alive, because you can't live more than a few minutes without it. Luckily, you don't need to look for this important stuff—it's all around you, though you can't see it. It is air—a fascinating substance that can gently ruffle your hair or snap huge trees as if they were toothpicks. Explore air. You might be surprised by what you discover.

422

GLENCOE TECHNOLOGY

 Videodisc

Use the *Science Interactions, Course 2* **Integrated Science Videodisc** lesson, *Behavior of Gases: Up, Up, and Away,* after Chapter 14 and Chapter 15.

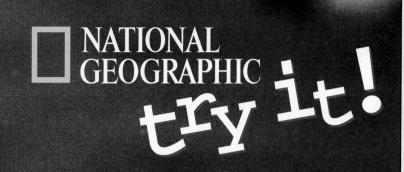

NATIONAL GEOGRAPHIC try it!

Can you place a paper towel in an aquarium full of water and still keep it dry?

What To Do

1. Half-fill an aquarium with water. Crumple a paper towel and put it in a baby food jar so the towel won't fall out when you turn the jar over. What do you think will happen if you put the upside-down jar in the aquarium? Write your prediction *in your Science Journal*, then test it.

2. What will happen if you half-fill the jar with water, place a plastic coffee can lid over the jar, and turn it upside down in the aquarium? Record your prediction, then test it.

3. Now, add 30 mL of water to a soft drink can and heat until the water boils. What will happen if you use a thermal mitt to turn the can upside down over the aquarium, placing the rim of the can just below the level of the water? Write your prediction, then test it.

423

GETTING STARTED
Discussion Some questions you may wish to ask your students are:

1. What is air made of? Students may know that there is oxygen in the air but may not know about the other gases that make up air (nitrogen, carbon dioxide, trace amounts of others). They may not realize that each of these gases has mass and exerts pressure.

2. When you squeeze a balloon, why does the pressure increase? Expect students to speculate on relationships among pressure, force, and area without coming up with the fact that pressure equals force divided by area.

3. Why do we need to breathe? Most students will not know that cells need oxygen for the process of respiration. Oxygen combines with glucose to release stored energy from glucose.

Responses to these questions will help you determine misconceptions that students have.

3. As the air in the can cools, its pressure decreases, and water moves into the can.

✔ Assessment

Oral Ask students to explain why water did not move into the can at first. If students have difficulty, guide them to observe that the can was partly filled with boiling water and mostly filled with air. Then ask students what changed as they held the can in the water. Challenge them to formulate a hypothesis about how temperature affects air pressure.

Try It!

Purpose Students observe that air exerts pressure.

Background Information Air pressure is a result of the energy of the molecules of the gases that make up air. As these molecules bump into objects, they exert pressure.

Materials aquarium, paper towel, baby food jar, plastic coffee can lid, empty soft drink can, thermal mitt, hot plate

Troubleshooting In Step 3, have students heat the can on a hot plate. Make sure students use a thermal mitt to handle the can, which will be hot.

Answers to Questions
1. The towel stays dry because air pressure inside the jar keeps the water out.
2. Air pressure pushes against the lid, holding it in place.

Chapter Organizer

SECTION	OBJECTIVES	ACTIVITIES & FEATURES
Chapter Opener		**Explore!**, p. 425
14-1 How Do Gases Behave? (4 sessions; 2 blocks)	1. **Identify** the gas phase of matter by its properties. 2. **Find** the relationships involving pressure, volume, and temperature of a gas. **National Content Standards: (5-8) UCP1-2, A1-2, B1, E2, G1, G3**	**Explore!**, p. 426 **Skillbuilder:** p. 428 **Investigate 14-1:** pp. 430-431 **Find Out!**, p. 434 **Skillbuilder:** p. 435 **A Closer Look**, pp. 428-429 **Earth Science Connection**, pp. 432-433 **Leisure Connection**, p. 449 **History Connection**, p. 447
14-2 What Are Gases Made Of? (2 sessions; 1 block)	1. **Relate** how the behavior of gases can be explained by a particle theory of matter. 2. **Describe** evidence for an atomic theory of matter. **National Content Standards: (5-8) UCP1-2, A1-2, B1, G1, G3**	**Explore!**, p. 436 **Explore!**, p. 437 **How Do We Know?**, p. 439 **Technology Connection**, p. 450
14-3 What Is the Atomic Theory of Matter? (2 sessions; 1 block)	1. **Describe** evidence that pure substances are made up of identical particles. 2. **Specify** what is meant by element, compound, and atom. **National Content Standards: (5-8) UCP1-2, A1-2, B1, D1, F5, G1, G3**	**Find Out!**, p. 440 **Design Your Own Investigation 14-2:** pp. 442-443 **Health Connection**, p. 448

ACTIVITY MATERIALS

EXPLORE!

p. 425 balloon, boxes of various sizes
p. 426* chair, table, odor source
p. 436 balloon, soda bottle or flask, drinking straw
p. 437* flashlight, ice cubes

INVESTIGATE!

pp. 430–431 large air cylinder with piston or plastic syringe, petroleum jelly, various weights, such as bricks or books

DESIGN YOUR OWN INVESTIGATION

pp. 442–443 pennies, nickels, dimes, quarters, balance, rubber cement

FIND OUT!

p. 434* capillary tube, thermometer, crushed ice, water, deep beaker, beaker tongs, ruler, hot plate
p. 440* 250-mL beaker, concentrated washing soda solution, 2 test tubes, 2 electrodes, 4 alligator clamps, power supply, ruler

KEY TO TEACHING STRATEGIES

The following designations will help you decide which activities are appropriate for your students.

- **L1** Basic activities for all students
- **L2** Activities for average to above-average students
- **L3** Challenging activities for above-average students
- **LEP** Limited English Proficiency activities
- **COOP LEARN** Cooperative Learning activities for small group work
- **P** Student products that can be placed into a best-work portfolio
- 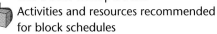 Activities and resources recommended for block schedules

Need Materials? Call Science Kit (1-800-828-7777).

[00:00] **OUT OF TIME?** We recommend that students do the activities with an asterisk.

Chapter 14 Gases, Atoms, and Molecules

TEACHER CLASSROOM RESOURCES

Student Masters	Transparencies
Study Guide, p. 47 **Take Home Activities,** p. 22 **How It Works,** p. 16 **Activity Masters,** Investigate 14-1, pp. 59–60 **Multicultural Connections,** p. 31 **Concept Mapping,** p. 22 **Science Discovery Activities, 14-1** **Laboratory Manual,** pp. 81–86, The Behavior of Gases	**Teaching Transparency 27,** Charles's Law **Section Focus Transparency 41**
Study Guide, p. 48 **Multicultural Connections,** p. 32 **Making Connections: Across the Curriculum,** p. 31 **Science Discovery Activities, 14-2, 14-3** **Laboratory Manual,** pp. 87–88, Air **Laboratory Manual,** pp. 89–90, Air Pressure	**Section Focus Transparency 42**
Study Guide, p. 49 **Critical Thinking/Problem Solving,** p. 22 **Making Connections: Integrating Sciences,** p. 31 **Making Connections: Technology & Society,** p. 31 **Activity Masters,** Design Your Own Investigation 14-2, pp. 61–62	**Teaching Transparency 28,** Law of Definite Proportions **Section Focus Transparency 43**

ASSESSMENT RESOURCES	TEACHING & TECHNOLOGY
Review and Assessment, pp. 83–88 **Performance Assessment,** Ch. 14 **PASC*** **MindJogger Videoquiz** **Alternate Assessment in the Science Classroom** **Computer Test Bank**	**Spanish Resources** **Cooperative Learning Resource Guide** **Lab and Safety Skills** **Science Interactions, Course 2, CD-ROM** **Computer Competency Activities**

*Performance Assessment in the Science Classroom

NATIONAL GEOGRAPHIC TEACHER'S CORNER

Index to National Geographic Magazine

The following articles may be used for research relating to this chapter:

- "Worlds Within the Atom," by John Boslough, May 1985.

GLENCOE TECHNOLOGY

The following multimedia resources are available from Glencoe.

Glencoe Physical Science Interactive Videodisc
Behavior of Gases

Physical Science CD-ROM

National Geographic Society Series
STV: Water
Newton's Apple: Physical Sciences

Teacher Classroom Resources

These are key components of the classroom resources package.

TEACHING AIDS

Section Focus Transparencies

41 SECTION FOCUS TRANSPARENCY Section 14-1

TIRE PRESSURE

Automobile manufacturers provide information about the correct cold inflation pressure for the tires on the automobile. Tires are "cold" when the car has been sitting for at least three hours or has been driven no more than a mile during that time.

1. Why should tire pressure be checked when the tires are cold?

2. Suppose a driver decides to check tire pressure right after a two-hour trip. Should the driver release air from the tires to achieve the recommended pressures? Give a reason for your answer.

3. Offer a suggestion as to why the recommended inflation pressures for front and rear tires are different.

L1

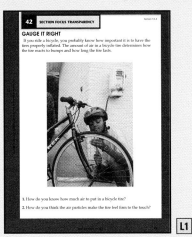

42 SECTION FOCUS TRANSPARENCY Section 14-2

GAUGE IT RIGHT

If you ride a bicycle, you probably know how important it is to have the tires properly inflated. The amount of air in a bicycle tire determines how the tire reacts to bumps and how long the tire lasts.

1. How do you know how much air to put in a bicycle tire?

2. How do you think the air particles make the tire feel firm to the touch?

L1

43 SECTION FOCUS TRANSPARENCY Section 14-3

CHANGE OF SUBSTANCE

When you bake a cake, does the finished product look like all the ingredients? Some chemical change has taken place, forming new substances that look different from the materials that formed them. Similar changes also occur in other chemical changes, such as when sodium and chlorine react to form common table salt, sodium chloride.

Sodium

Sodium chloride

Chlorine

1. Describe the appearance of the sodium and the chlorine. How do they differ from the appearance of the sodium chloride?

2. Sodium and chlorine react in a one-to-one ratio. Do you think that equal masses of these elements react with each other? Explain your answer.

L1

Teaching Transparencies

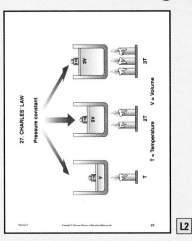

27. CHARLES' LAW
Pressure constant

$V = Volume$
$T = Temperature$

27   L2

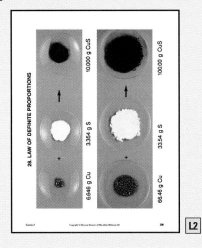

28. LAW OF DEFINITE PROPORTIONS

10.000 g CuS   3.354 g S   6.646 g Cu

100.00 g CuS   33.54 g S   66.46 g Cu

28   L2

HANDS-ON LEARNING

Science Discovery Activity*

ACTIVITY 14-1   Bubble and Boyle

Do you know that some deep-sea salvagers use air to raise sunken cargo ships? How do trapped air bubbles change the buoyancy of a sunken ship? How do you think you can use a bubble of air to lift a submerged object? Do you think you can use the bubble to lower a submerged object?

Getting Started
Place an empty medicine dropper in a container of water. What happens to the dropper? Try pushing down on the dropper. Is the dropper empty? How do you know? Now completely fill the dropper with water and place it in the container. What happens to the dropper? How does the volume of air in the dropper affect its buoyancy? How does a change in the volume of air in the dropper affect its buoyancy?

Partially fill a medicine dropper with water and turn it upside down. Tightly cover the tip of the dropper with your fingertip. Now gently squeeze the bulb. What happens to the volume of the bulb as you squeeze it? What happens to the volume of air?

Hypothesize
Think about a dropper floating just beneath the surface of a container of water. Will a change in atmospheric pressure affect the buoyancy of the dropper? Will an increase in air pressure cause the dropper to change its position? Write a hypothesis about how a change in pressure will affect the buoyancy of a dropper floating near the surface of water.

Try It!
You will have the following materials to use:
• 100-mL graduated cylinder • water • rubber bands
• medicine dropper • plastic wrap

76   L1

Laboratory Manual*

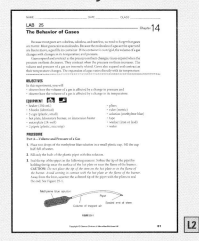

NAME _____ DATE _____ CLASS _____

LAB   25   Chapter 14
The Behavior of Gases

Because most gases are colorless, odorless, and tasteless, we tend to forget that gases are matter. Most gases exist as molecules. Because the molecules of gas are far apart and are free to move, a gas fills its container. If the container is not rigid, the volume of a gas changes with changes in its temperature and pressure.

Gases expand and contract as the pressure on them changes. Gases expand when the pressure on them decreases. They contract when the pressure on them increases. The volume and pressure of a gas are inversely related. Gases also expand and contract as their temperature changes. The expansion of a gas varies directly with its temperature.

OBJECTIVES
In this experiment, you will
• observe how the volume of a gas is affected by a change in pressure and
• observe how the volume of a gas is affected by a change in its temperature.

EQUIPMENT
• beaker (250-mL)   • pliers
• 5 books (identical)   • ruler (metric)
• 5 cups (plastic, small)   • solution (methylene blue)
• hot plate, laboratory burner, or immersion heater   • tape
• microplate (24-well)   • washer (iron or lead)
• 2 pipets (plastic, microtip)   • water

PROCEDURE
Part A—Volume and Pressure of a Gas
1. Place two drops of the methylene blue solution in a small plastic cup. Fill the cup half-full of water.
2. Fill only the bulb of the plastic pipet with this solution.
3. Seal the tip of the pipet in the following manner. Soften the tip of the pipet by holding the tip near the surface of the hot plate or near the flame of the burner. CAUTION: Do not place the tip of the stem on the hot plate or in the flame of the burner. Avoid coming in contact with the hot plate or the flame of the burner. Away from the heat, squeeze the softened tip of the pipet with the pliers to seal the end. See Figure 25-1.

Methylene blue solution
Pipet
Sealed end of stem
Column of trapped air
FIGURE 25-1

81   L2

Take Home Activity

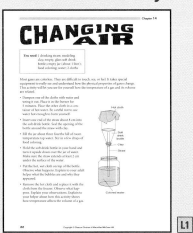

Chapter 14

CHANGING AIR

You need: 1 drinking straw; modeling clay; empty, glass soft drink bottle; empty jar (about 1 liter); food coloring; water; 2 cloths

Most gases are colorless. They are difficult to touch, see, or feel. It takes special equipment to really see and understand how the physical properties of gases change. This activity will let you see for yourself how the temperature of a gas and its volume are related.

• Dampen one of the cloths with water and wring it out. Place it in the freezer for 5 minutes. Place the other cloth in a container of hot water. Be careful not to use water hot enough to burn yourself.

• Insert one end of the straw about 4 cm into the soft drink bottle. Seal the opening of the bottle around the straw with clay.

• Fill the jar about three fourths full of room temperature tap water. Stir in a few drops of food coloring.

• Hold the soft drink bottle in your hand and turn it upside down over the jar of water. Insert the straw extends at least 2 cm under the surface of the water.

• Put the hot, wet cloth on top of the bottle. Observe what happens. Explain to your adult helper what the bubbles are and why they appeared.

• Remove the hot cloth and replace it with the cloth from the freezer. Observe what happens. Explain your observations. Explain to your helper about how this activity shows how temperature affects the volume of a gas.

22   L1

424C   Chapter 14   Gases, Atoms, and Molecules   *There may be more than one activity for this chapter.

Chapter 14 Gases, Atoms, and Molecules

Study Guide*

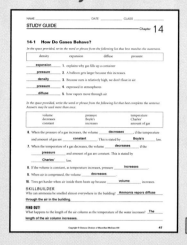

Concept Mapping

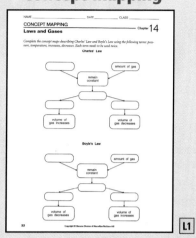

Critical Thinking/ Problem Solving

Integrating Sciences

Across the Curriculum

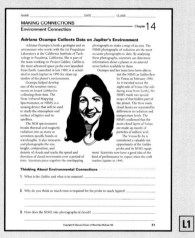

Technology and Society

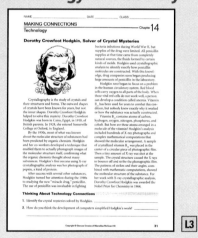

Multicultural Connection**

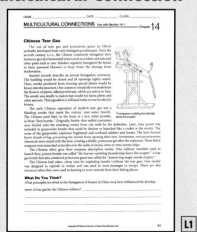

Performance Assessment

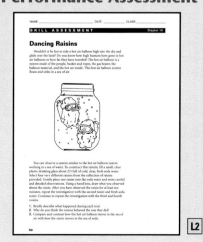

Review and Assessment

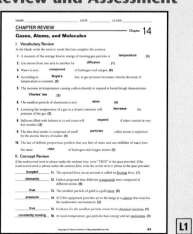

Gases, Atoms, and Molecules

THEME DEVELOPMENT

Gas behavior can be understood on a large scale by accounting for the interactions of its particles on a submicroscopic scale. Structure is highlighted in the atomic theory of matter, in which the structure of a compound is shown to be a combination of different types of atoms.

CHAPTER OVERVIEW

In this chapter, students identify properties of gases and observe the relationships between pressure and volume, volume and temperature, and pressure and temperature. Students communicate the atomic theory of matter.

Tying to Previous Knowledge

Have students list all the properties of gases that they can think of. *Possible responses include: invisible, light, airy, have no shape, are compressible, and spread out.*

INTRODUCING THE CHAPTER

Have students look at the photograph on page 425 and tell where the air in a tire is stored. Have them formulate a model of how air pressure in the tire changes when the tire is pumped up or goes flat.

Uncovering Preconceptions

Students may think that gases have no real "substance" or mass. The Explore activity on page 436 proves that this is not the case. The force of the air prevents the balloon from being inflated. Only after the straw allows an escape route for trapped air can the balloon be inflated.

GASES, ATOMS AND MOLECULES

Did you ever wonder...

✓ **Why you sometimes see remnants of exploded car tires at the side of the road?**

✓ **What's in a balloon?**

✓ **Why a hairspray can feels cold when you use it?**

Science Journal

Before you begin to study gases, atoms, and molecules, think about these questions and answer them *in your Science Journal*. When you finish the chapter, compare your journal write-up with what you have learned.

Answers are on page 451.

I magine you are casually riding your bike through the neighborhood. In the alley behind one building, you cannot avoid some broken glass. You hear a loud "Thwop!" and your front tire is flat. As you walk your bike home, you see your older sister at the gas station. She is putting air in her car's tires. You remember that she said the tires squealed when she turned corners and that they probably needed more air. She looks up and sees your plight, and you gratefully accept her offer to drive you home. As you round the corner toward home, you notice the car's tires no longer squeal. You begin to wonder what made the difference.

▶ **In the next activity, you'll explore a useful and unique property of gases.**

424

Learning Styles	Kinesthetic	Explore, pp. 425, 426, 436; Activity, pp. 427, 433, 441, 445; Investigate, pp. 430-431; Find Out, p. 440
	Visual-Spatial	Visual Learning, pp. 427, 429, 434, 441; Activity, p. 441; Explore, p. 437
	Interpersonal	Activity, p. 446
	Logical-Mathematical	A Closer Look, pp. 428-429; Visual Learning, pp. 432, 438, 444, 446; Find Out, p. 434; Investigate, pp. 442-443; Multicultural Perspectives, p. 445; Project, p. 451
LS	**Linguistic**	Earth Science Connection, pp. 432-433; Discussion, pp. 437, 439; Multicultural Perspectives, p. 438

Does air exert pressure?

What To Do

1. Obtain a partially inflated balloon and a box that's smaller than the balloon. Predict *in your Journal* whether the balloon will fit in the box.

2. Try to close the lid. Does the lid stay closed without help?

3. Repeat Steps 1 and 2 using various sizes of boxes. How does the size of the box relate to how easy it is to close the lid?

PREPARATION

14-1

Planning the Lesson

Refer to the Chapter Organizer on pages 424A–D.

Concepts Developed

This section explores the behavior of gases under varying conditions. The inverse relationship between pressure and volume in a gas (Boyle's law) is observed in an experiment. Next, the students find that the Kelvin temperature and volume of a gas are directly proportional (Charles's law). The direct relationship between Kelvin temperature and pressure is also presented.

Explore!

Do gases move?

Time needed   5 minutes

Materials   open classroom space; chair for each student; table; odor source

Thinking Processes thinking critically, observing and inferring, practicing scientific methods, forming a hypothesis

Purpose   To investigate the movement of a gas.

Preparation   Try to minimize air currents in the classroom.

Teaching the Activity

Science Journal   Have students hypothesize about the nature of the odor in the Explore activity and how it traveled through the room.

14-1 How Do Gases Behave?

Section Objectives
- Identify the gas phase of matter by its properties.
- Find the relationships involving pressure, volume, and temperature of a gas.

Key Terms
Boyle's law
Charles's law

Identifying Properties of Gases

You know that matter exists in three common physical phases— solid, liquid, and gas. Each phase has its own characteristic properties and behaviors. In this section, you will study gases.

Explore! ACTIVITY

Do gases move?

What To Do

1. With your classmates, arrange chairs facing away from a small table or desk like four equally spaced spokes of a wheel. Everybody except the experimenter should sit in one of the chairs.

2. The experimenter then will put on the small table something that has a strong odor such as a cut orange or an open bottle of perfume. The sitting students are to raise their hands as soon as they can smell whatever is on the table. Which hands are raised first? How long is it before everyone can smell the odor? In what direction does the odor travel?

It's possible that vapors from the thing you smelled moved without any help from a fan or any other outside source. **Figure 14-1** reviews this characteristic of gases as well as a few others.

Expected Outcome

Depending on air currents, one side of the room probably will detect the odor first. It should be detectable to all soon thereafter. A reasonable hypothesis for students to make is that the smell travels as a "cloud."

Answers to Questions

Answers will vary depending upon the air currents in the room.

✔ Assessment

Oral   Ask students to give evidence to support the hypothesis that gases move. Students could then draw their own cartoon showing how gases move. Use the Performance Task Assessment List for Cartoon/Comic Book in **PASC,** page 61.
P

Figure 14-1

 Expansion The shape and volume of a body of gas are determined by the shape and volume of its container. Examples of gas containers include lightbulbs, tires, air mattresses, and balloons.

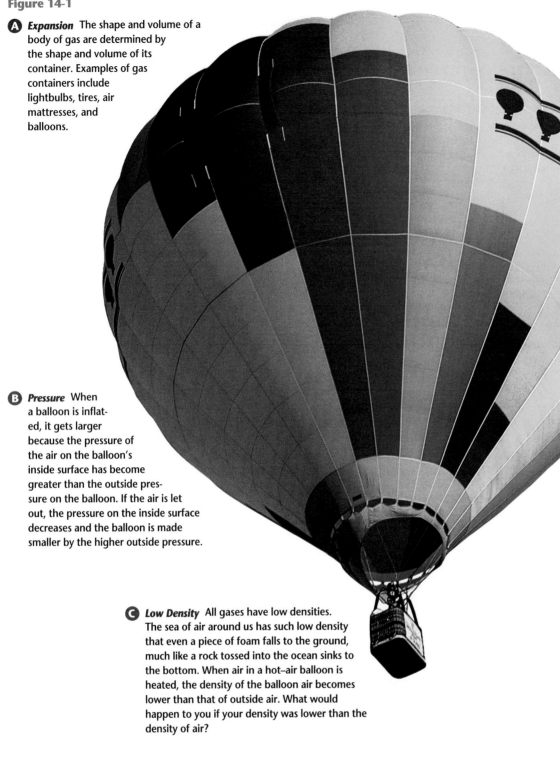

B **Pressure** When a balloon is inflated, it gets larger because the pressure of the air on the balloon's inside surface has become greater than the outside pressure on the balloon. If the air is let out, the pressure on the inside surface decreases and the balloon is made smaller by the higher outside pressure.

C **Low Density** All gases have low densities. The sea of air around us has such low density that even a piece of foam falls to the ground, much like a rock tossed into the ocean sinks to the bottom. When air in a hot–air balloon is heated, the density of the balloon air becomes lower than that of outside air. What would happen to you if your density was lower than the density of air?

14-1 How Do Gases Behave? **427**

Meeting Individual Needs

Learning Disabled Students might not be familiar with such words and terms as the ones below. Be on the lookout for unfamiliar terms and be prepared to provide definitions and real-life examples if necessary.

- expansion
- air mattress
- density
- wheel spokes
- ammonia
- inflate

1 MOTIVATE

Bellringer

Before presenting the lesson, display **Section Focus Transparency 41** on an overhead projector. Assign the accompanying **Focus Activity** worksheet. L1

LEP

Activity This activity allows students to explore the effect of temperature on air pressure.

Materials needed are tennis balls; buckets of ice water; buckets of hot water; meterstick.

Provide each pair of students with two wet tennis balls. Have students drop each ball from the same height. Then have them place one ball in ice water and the other in hot water. After several minutes, have them drop the balls again and compare the height to which the balls bounced in both drops. *The "hot" ball bounces much higher.* Explain that heating the ball increased the air pressure inside. L1 COOP LEARN

2 TEACH

Tying to Previous Knowledge

Review the three physical phases of matter—solid, liquid, and gas. The information in this chapter is applicable to all matter because under some conditions, all matter will exist in its gas phase.

Visual Learning

Figure 14-1 Bring a balloon to class. Students can explain the air pressure changes that occur as you inflate and deflate the balloon. **What would happen to you if your density was lower than the density of air?** *You would rise off the ground into the air.*

14-1 How Do Gases Behave? **427**

The ammonia gas that is dissolved in the water of the cleaner comes out of solution or is released into the air as the water evaporates. L1

Theme Connection The theme that this section supports is systems and interactions. Because gases are not understood at the level of the individual particle as yet, they must be envisioned as systems of interacting particles. Some of the characteristics of the individual gas particles can be inferred from their macroscopic behavior as a system. For example, during the Explore activity, students smell the gas diffusing to occupy the space available in the classroom.

Uncovering Preconceptions Most students will realize that the liquid and gas phases of matter are interchangeable depending on temperature, but they are probably not aware that pressure alone can change a gas to a liquid. Placing a gas under great pressure will cause it to liquify. When the pressure is removed, the liquid will revert back to gaseous form.

GLENCOE TECHNOLOGY

Videodisc

STVS: Chemistry

Disc 2, Side 1
Sand Blasting with Dry Ice (Ch. 8)

Relating Gas Pressure and Volume

SKILLBUILDER

Observing and Inferring

As you leave your apartment building, you notice a neighbor washing her apartment door with a cleaner containing ammonia. An hour later, you walk into the building and you can smell ammonia cleaner almost everywhere. The smell is strongest near your neighbor's door. State an inference to explain your observations. If you need help, refer to the **Skill Handbook** on page 642.

If you've ever inflated a balloon or a bicycle tire, you may have already noticed a very interesting property of gases. When you pump air into a bicycle tire, you are compressing it. That means you are squeezing the air so it takes up less space. In other words, you are reducing its volume. Unlike water and other liquids, gases can be compressed. As you squeeze or compress a gas such as air, you increase the pressure of the gas. You noticed this when you were trying to stuff a balloon into different boxes in the Explore activity. The smaller the box you attempted to put it in, the harder the gas in the balloon pushed back as you tried to close the lid. The pressure got higher as the volume got smaller. Pressure and volume are two important characteristics of gases. **Figure 14-2** describes other characteristics of gases that are important to understand.

The Density of a Gas

We move through a gas, air, all the time. Rarely are we aware of its presence. We know that gases exist and that they are made of atoms and molecules. How can we measure other properties of something that is generally invisible?

You Try It!

The purpose of this activity is to compare the density of dry ice, which is solid carbon dioxide, with that of gaseous carbon dioxide. Flatten a tall kitchen trash bag so there is no air in it. You will use a cube-shaped piece of dry ice so you can measure its dimensions. **CAUTION:** *Dry ice is very cold and must be handled with extreme care because it could freeze your skin. Do not touch dry ice with your fingers. Always use tongs when handling the dry ice.* Quickly use a balance to find the mass of the dry ice cube. Then, measure its dimensions, and immediately place the dry ice in the large plastic bag. Seal the bag tightly as soon as you put the dry ice in. Try not to let any gaseous carbon dioxide escape.

Purpose

A Closer Look reinforces Section 14-1 by providing an activity that allows students to see that a gas has density.

Teaching Strategy

Obtain two balloons, one filled with helium and the other with air. Ask students to predict which will rise or sink in the air. Display the balloons and let them go. Be sure to hold their strings. Ask why each behaves the way it does. *Helium is less dense than air, so the helium-filled balloon rises. But air is compressed in a balloon, forcing more mass into a smaller space. So the air in the balloon is denser than open air, and the balloon sinks.*

Answers to You Try It!

The density of dry ice (solid carbon dioxide) is 1.51 g/mL at −56.6°C. The density of gaseous carbon dioxide is 0.00184

Figure 14-2

When describing a gas, there are four variables to consider: amount, pressure, volume, and temperature. To find how these four variables are related, choose two at a time to test. While keeping all the other variables constant, make changes in one of the two test variables and measure its effect on the other. This process of controlling the number of variables is called a controlled experiment.

A *Amount* is the number of gas particles.

B *Volume*, expressed in cubic units, is the amount of space the gas takes up.

C *Temperature* of gas can be measured on several different scales.

D *Pressure*, often expressed in atmospheres (atm), is related to the force the gas exerts on the objects it touches. The force is equal to the pressure times the area of the object. One atm is equal to 14.7 pounds per square inch.

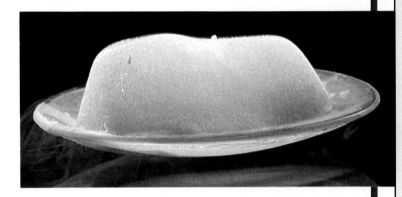

Inquiry Question An undersea diver complained that her air tank was only "half full." The tank company assured her that *all* of its tanks were always completely full of gas. Can both parties be technically correct in this dispute? How? *Yes. Gas expands to fit any volume. So while the diver had only half the quantity of the gas she needed, the tank was nevertheless "filled."*

Visual Learning

Figure 14-2 Students can use real balloons and observe the effects of changing one condition at a time or students can call upon previous experiences to hypothesize the effect of changing each condition. Inform students that one atmosphere (1 atm) is about equal to the amount of pressure that the atmospheric air exerts on us under normal conditions.

GLENCOE TECHNOLOGY

CD-ROM

Science Interactions, Course 2, CD-ROM

Chapter 14

Once the dry ice is sealed in the bag, you may begin your calculations of the volume and density of the dry ice. The volume is equal to the length times the width times the height, or $V = l \times w \times h$. The density is the mass divided by the volume, or $D = M/V$.

While you are doing your calculations, the dry ice will be changing into a gas inside the plastic bag, so the bag will appear to be partially inflated. When you are sure there is no longer any solid dry ice in the bag, put the bag into a shape that is as box-like as possible. When it is in this shape, measure its dimensions as well as you can. Then, calculate the volume and the density. Since

the bag is not perfectly box-like, you will not be able to get measurements that are extremely accurate. However, with patience and care, you will get a close approximation. Because little carbon dioxide was allowed to escape, the mass of the dry ice will be about the same as the mass of

the gaseous carbon dioxide.

How does the volume of the gaseous carbon dioxide compare with the volume of the dry ice? Is the number of molecules the same in the dry ice and in the gaseous carbon dioxide? Why do the molecules in gaseous carbon dioxide take up more space?

---

g/mL at 20°C. The volume of gaseous carbon dioxide is greater than the volume of solid carbon dioxide. The molecules in gaseous carbon dioxide have more energy than the molecules of solid carbon dioxide, so the gaseous molecules are not as strongly attracted to each other; they tend to diffuse evenly throughout whatever container they occupy.

Going Further ▮▮▮▮▮

Light a candle, cut the corner off the

bag from You Try It in which CO_2 is trapped, hold the bag above the lighted candle, and carefully pour the CO_2 gas over the flame. After students observe the flame going out, they can hypothesize whether CO_2 gas is more or less dense than air and what use for CO_2 gas this experiment suggests. *CO_2 gas is denser than air. The flame goes out because CO_2 won't support combustion and surrounds the candle, preventing oxygen from reaching it. CO_2*

gas can be used as a fire extinguisher. Use the Performance Task Assessment List for Formulating a Hypothesis in **PASC,** p. 21.
COOP LEARN

Planning the Activity

TECH PREP

Time needed 30 minutes

Purpose To investigate the relationship between pressure and the volume of a gas.

Process Skills making and using graphs, observing and inferring, comparing and contrasting, recognizing cause and effect, interpreting data

Materials large air cylinder with piston or plastic syringe; petroleum jelly; 4 weights, such as bricks or books

Preparation Make sure the weights you use are of equal size and stack well. Books, bricks, or wooden blocks would work well here.

Teaching the Activity

Possible Procedure Students will probably place a variety of weights on the cylinder plunger. They should perform multiple trials with each weight.

Troubleshooting If groups have difficulty planning their investigation, suggest that they experiment with different numbers of weights. Stress that they should wait between trials to allow the temperature inside the piston to come to equilibrium with the room temperature.

Process Reinforcement Ask students to identify the cause and effect in this investigation. Be sure that students clearly describe or illustrate the procedure they plan to follow.

Possible Hypotheses Students may hypothesize that increasing the pressure on a gas will decrease its volume based on their experiences with squeezing a balloon.

Science Journal Have students record their hypotheses and results in their journals. They should include an illustration of their experimental design.

Pressure and Volume

Have you ever wondered why a balloon pops when you step on it? Maybe on television you've seen a scuba diver's bubbles get bigger as the bubbles near the surface. Why do these things happen? In this activity, you'll discover how the pressure and volume of a gas are related.

Problem
How are the pressure and volume of a gas related?

Materials
large air cylinder with piston or plastic syringe
petroleum jelly

various weights, such as bricks or books

Safety Precautions

What To Do

Work with your group to *hypothesize* how the pressure and volume of a gas are related. Plan a way to test your hypothesis using an air cylinder. Here are some things that your group should do when using the air cylinder to help you in your investigation.

1 Remove the cap from the air cylinder. Lightly lubricate the plunger of the piston with petroleum jelly. Insert the plunger into the cylinder. Make sure the plunger is snug, but moves easily.

2 Pull the plunger back so that the cylinder contains 30 mL of air. Replace the cap.

Meeting Individual Needs

Visually Impaired Modify the Investigate by setting up a "shunt" tube on the side of the piston. Attach a balloon to the end of this tube. When weight (pressure) is added, the balloon will fill in proportion to the pressure that was applied to the piston. The students can *feel* how much the balloon is inflated to understand the relationship between pressure and volume. **LEP**

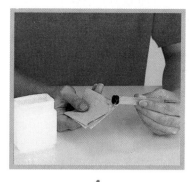

A

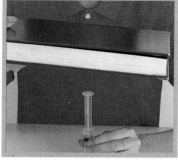

B

Sample data

Data and Observations

No. of Weights	Volume of Air			Average Volume
	Trial 1	Trial 2	Trial 3	
1	20	19.5	20.5	20 mL
2	9.5	10	10	10 mL
3	7	6.5	7	7 mL
4	4.5	5	5	5 mL

If you moisten two plunger cups and push the ends together, there will be almost no air left in the cups. The difference between the cup pressure and the outside air pressure will create a force so strong that you cannot pull them apart. Try it!

3 Set a weight on the cylinder's platform. Twist the plunger a bit to keep it from sticking. Read the volume of air in mL. Remember that pressure can raise the temperature of a gas, so wait a minute or so for the temperature to return to room temperature.

Show your experimental plan to your teacher. If you are advised to revise your plan, be sure to check with your teacher before you begin. Carry out your plan. Record your data in a table *in your Journal*.

Analyzing

1. *Identify the variables* in your experiment. How were they controlled?

2. Determine the effect of increasing pressure on the volume of a gas.

Concluding and Applying

3. Going Further Graph your data, putting pressure on the horizontal axis and volume on the vertical axis. *Interpret* your graph to find the relationship between pressure and volume of an enclosed gas.

Figure 14-3

The diagram below demonstrates Boyle's law. As gas pressure increases, volume decreases, as long as the temperature is constant.

1P

P equals pressure

2P

3P

3.0 liters
25°C

1.5 liters
25°C

1 liter
25°C

In the Investigate, you found the same result that scientists discovered over 300 years ago. They found that the pressure and volume of a gas are inversely related. What does that mean? It means that when one goes up, the other goes down. For this relationship to be true, the amount of gas in the system and the temperature of the system must be held constant. This relationship is known as Boyle's law. **Boyle's law** can be stated thus: The volume of a certain amount of gas is inversely proportional to the pressure, if the temperature remains constant. The law is illustrated in **Figure 14-3**.

You now have a better understanding of how pressure and volume are related. More important, you have a sense of what it means to carry out a controlled experiment. Boyle's law tells how pressure and volume are related, but it does not tell why. You will learn why later.

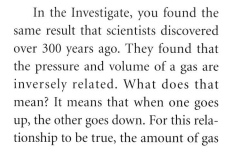

Earth Science CONNECTION

Research Giants

Have you ever heard of an aerostat? An aerostat is an aircraft that is supported by the buoyancy of a gas that is less dense than air.

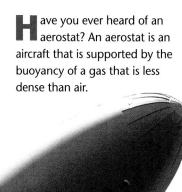

Aerostats include blimps and dirigibles. The simplest kind is the balloon, because it has no means of propulsion or steering. In other words, a balloon just goes wherever the movement of the air takes it.

Basically, a balloon is a large bag that is filled with a gas that is lighter, or less dense, than air, for example, hot air or helium. This large bag displaces a lot of air, and has an upward force on it—called a buoyant force—that is equal to the weight of the displaced air.

These giant balloons can lift a load of instruments higher than a jetliner but lower than an orbiting satellite.

Effect of Temperature on Gas Pressure and Volume

What happens to the tires on your bike in hot weather? They probably get harder. What's going on inside the tire? What happens when you put a balloon over a heat source, like a lamp? It swells and eventually explodes, doesn't it?

What goes on in the bike tire and balloon is the same thing that goes on in the tires on cars and trucks that travel roads and highways. You know that there is air inside the tires. What happens to that air as the tire pounds against the road? The tire heats up. As the tire heats up, so does the air inside the tire. If it heats up too much, the tire may explode. How could this happen if no additional air were pumped into the tire? Can you think of any other examples where you have observed that a gas changes with a change in temperature? How is the volume of a gas affected by a change in its temperature if pressure is held constant? Try this next activity to find out what happens.

Connect to...
Life Science

When you breathe, air rushes into and is expelled from your lungs. Explain breathing to your classmates in terms of volume and pressure in the chest cavity and in the air around you.

The volume of helium needed to lift the load could be equal to the volume of a small house on the ground. Once the balloon reaches the altitude it's designed for, the volume of the gas could be the volume of many houses.

What kind of information can be obtained by balloons? Balloons carry many different kinds of instruments. Some have telescopes for viewing objects in space. Others collect and analyze the gases in the atmosphere. Still others measure and record temperature, pressure, and the amounts and kinds of radiation from space. Ground-based radar can track balloons to find out about the speed and direction of high-altitude winds.

Information collected in these studies helps scientists understand the makeup and behavior of the gases in our atmosphere. As we learn more about this important part of the environment, we will be able to predict the atmosphere's behavior and to know how we can protect and preserve it.

What Do You Think?

What would be an advantage of taking pictures through telescopes on high-altitude balloons, rather than through ground-based telescopes? What are the disadvantages? In what ways are research balloons different from the multicolored balloon shown on page 427?

433

How does the volume of a gas depend on its temperature?

Find Out!

How does the volume of a gas depend on its temperature?

Time needed   30 minutes

Materials capillary tube; thermometer; crushed ice and water; deep beaker; ruler; heat source; beaker tongs

Thinking Processes measuring, making and using graphs, observing and inferring, comparing and contrasting, forming a hypothesis

Purpose To investigate the relationship between the temperature and volume of a gas.

Teaching the Activity

Troubleshooting Touch one end of the capillary tube to oil, just enough to allow a small amount of oil into the tube to form a plug.

Science Journal Have students record their observations in their journals. [L1]

Expected Outcome

Students should find that as the temperature increases, the volume also increases.

Conclude and Apply

1. A straight line graph should result.
2. It increases. The volume of a fixed amount of gas increases as temperature increases.
3. Pressure was held constant by not changing the force pressing on the top of the capillary tube. The amount of gas was held constant by using a sealed-off tube.

✔ Assessment

Process Students can design and carry out an experiment to show the effects of heat and cold on an inflated balloon. For example, students might hypothesize that putting an inflated balloon in a freezer will cause its volume to decrease, while putting the same balloon in direct sunlight on a hot day will cause it to expand. Use the Performance Task Assessment List for Designing an Experiment in **PASC,** p. 23. **COOP LEARN** **P**

Sample data

Data and Observations

Temp (°C)				
Volume represented by length of air columns	Results will vary, but students should find that as temperature increases, the length of the air column increases.			

What To Do

1. Copy the data table *into your Journal.* Obtain a thermometer and plugged capillary tube from your teacher.
2. Prepare a mixture of crushed ice and water in a deep beaker. Place the thermometer and tube into this beaker. Be sure the air column is below the water level. Wait until the temperature becomes constant and bring the thermometer close to the side of the beaker. Read and record the temperature.
3. Place a ruler in the water next to the tube and record the length of the air column.
4. Replace the ice water with tap water and slowly heat the water. Measure the length of the air column at various temperatures as the water heats to boiling. Record the temperatures and column lengths in the data table.

Conclude and Apply

1. We can assume that the diameter of the tube is constant, so that as the length increases, the gas volume increases proportionally. Because we're looking for a relationship between temperature and volume, let's use the length of the column instead of the volume to make our calculations easier. Plot a graph with volume (column length) on the vertical axis and temperature on the horizontal axis.
2. As the temperature increases, what happens to the volume? What is the relationship between the temperature and volume of a gas?
3. How did you hold the variables of pressure and amount of gas constant in this activity?

You have discovered that the relationship between the volume of a gas and its pressure is inverse: as one goes up, the other goes down. You have seen that the relationship of the volume and temperature of a gas is direct: as one goes up, so does the other, and vice versa. If you could repeat the Find Out activity, measur-ing pressure instead of volume, you would find that the relationship between the temperature and pressure of a gas is direct: as one goes up, the other goes up too, and vice versa. These relationships are shown in **Table 14-1.** These relationships were found by careful experiments. In each case, two variables were held constant

434 Chapter 14 Gases, Atoms, and Molecules

Visual Learning

Figure 14-4 Is the relationship between pressure and temperature inversely proportional? *No, it is direct: as one goes up, so does the other, and vice versa.*

Pressure constant

Table 14-1 summarizes the relationships among the variables in a gas. Is the relationship between pressure and temperature inversely proportional?

Figure 14-4

This diagram demonstrates Charles's Law. If the pressure on a gas remains constant and the temperature of the gas is increased, the volume of the gas also increases. An everyday example of this law is a cake baking in a hot oven. The increase in temperature causes the bubbles of gas in the batter to expand—the cake rises.

Table 14-1

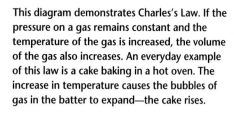

Boyle's Law	P↑	V↓
	P↓	V↑
Charles's Law	T↑	V↑
	T↓	V↓
Pressure and Temperature	P↑	T↑
	P↓	T↓

and the other two were allowed to change. The results tell how the gas behaves. But they do not tell why. That will require more investigation.

The relationship between the temperature and volume of a gas is called Charles's law. **Charles's law** can be stated thus: Gases increase or decrease their volume as the temperature rises and falls, provided pressure and amount of gas are held constant. Charles's law is shown in **Figure 14-4**.

If instead of allowing the volume to change, you had just a solid, immovable plug in the capillary tube, what would have happened when you heated the air column?

You have learned some ways in which gases behave, but you haven't explored why they behave that way. In the next section, you will discover what's behind the behavior of gases.

SKILLBUILDER

Making and Using Tables

Use **Table 14-1** to answer the following question. If you need help, refer to the **Skill Handbook** on page 639. What would happen to the volume of a gas if you doubled the pressure and doubled the temperature?

3 ASSESS

Check for Understanding

Assign questions 1-3 of Check Your Understanding. As you discuss the Apply question, demonstrate the effect of heating a balloon.

Reteach

Discussion This chart shows the consequences of changing one of the three variables.

Pressure Up	V-Down	T-Up
Pressure Down	V-Up	T-Down
Volume Up	P-Down	T-Up
Volume Down	P-Up	T-Down
Temperature Up	P-Up	V-Up
Temperature Down	P-Down	V-Down

Ask students to give examples of each relationship. L1

4 CLOSE

Have students create a poster illustrating the relationship between pressure and volume or temperature and volume. P

check your UNDERSTANDING

1. Give an example for each of the characteristics of a gas.
2. When a basketball bounces, the temperature of the ball increases. What happens to the temperature of the air in the ball?
3. **Apply** One way to fill up a balloon is to blow air into it—increase the inside air pressure. Can you think of another way to expand a balloon?

14-1   How Do Gases Behave?   **435**

check your UNDERSTANDING

1. A cake rising is an example of volume increasing with temperature. A piston in an engine shows that pressure increases when volume decreases. Heating a balloon shows that pressure increases in proportion to temperature.

2. The pressure of the gas on the inside walls of the ball increases. The temperature of the air also increases.

3. Another way to expand a balloon is to heat it up or decrease the pressure outside the balloon.

PREPARATION

Planning the Lesson

Refer to the Chapter Organizer on pages 424A–D.

Concepts Developed

In this section students first do an experiment that demonstrates that air takes up space. Next, they observe the movement of dust particles, which is caused indirectly by the movement of air. Finally, evidence for gases being composed of small particles—the atomic theory of matter—is presented.

1 MOTIVATE

Bellringer

 Before presenting the lesson, display **Section Focus Transparency 42** on an overhead projector. Assign the accompanying **Focus Activity** worksheet. L1
LEP

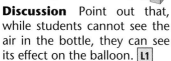

Can you see air?

Time needed 10 minutes

Materials balloon; empty soda bottle or flask; straw

Thinking Processes observing and inferring, recognizing cause and effect

Purpose To observe indirect evidence of the presence of air.

Teaching the Activity

Discussion Point out that, while students cannot see the air in the bottle, they can see its effect on the balloon. L1

Science Journal Have students draw a picture or cartoon that shows the function of the straw in this activity. L1

What Are Gases Made Of?

Section Objectives

- Relate how the behavior of gases can be explained by a particle theory of matter.
- Describe evidence for an atomic theory of matter.

Key Terms

atomic theory of matter

Describing Gases

You've seen huge trailer trucks on the road supported by nothing but the air in the tires. You know air is supporting all that weight. But it's hard to imagine how. Using a model can help.

Explore! ACTIVITY

Can you see air?

No. But you can see what air can do, such as bend trees and blow dust around.

What To Do

1. Push a deflated balloon into an empty soda bottle or flask and stretch the open end of the balloon back over the bottle's mouth, as shown in the picture. Try to blow up the balloon. What happens?

2. Remove the open end of the balloon from the bottle's mouth but keep the balloon in the bottle. Try to blow up the balloon now. What happens?

3. Insert a straw into the bottle between the balloon and the glass. Now, try to blow up the balloon in the bottle. What happens?

No matter how hard you huff and puff, you cannot blow up the balloon—until the straw is put in. But why? Even though you cannot see it, there is air inside the bottle. That air is keeping the balloon from inflating. How can air do this? Well, imagine the bottle filled with sand. What would you have to do to inflate the balloon? You would have to remove some of the sand. Think of the air as tiny sand particles—so small you can't see them. As you tried to inflate the balloon, you squeezed the particles closer and closer together. Like sand, it gets harder and harder to squeeze the particles in air closer together.

The idea that a gas, such as air, is made of particles can explain some of the other properties of gases you have

---

Expected Outcome

The air in the bottle will prevent the balloon from inflating unless it is given an escape route through the straw.

Answers to Questions

1. It can't be done.
2. The balloon is still impossible to blow up.
3. The balloon inflates.

✓ Assessment

Content Have students write a story or poem, based on their observations, titled "Invisible Air." Students can then recite their poems or read their stories to the class. Use the Performance Task Assessment List for Oral Presentation in **PASC**, page 71. COOP LEARN P

Figure 14-5

The arrangement, size, and movement of gas particles determine the properties of the gas.

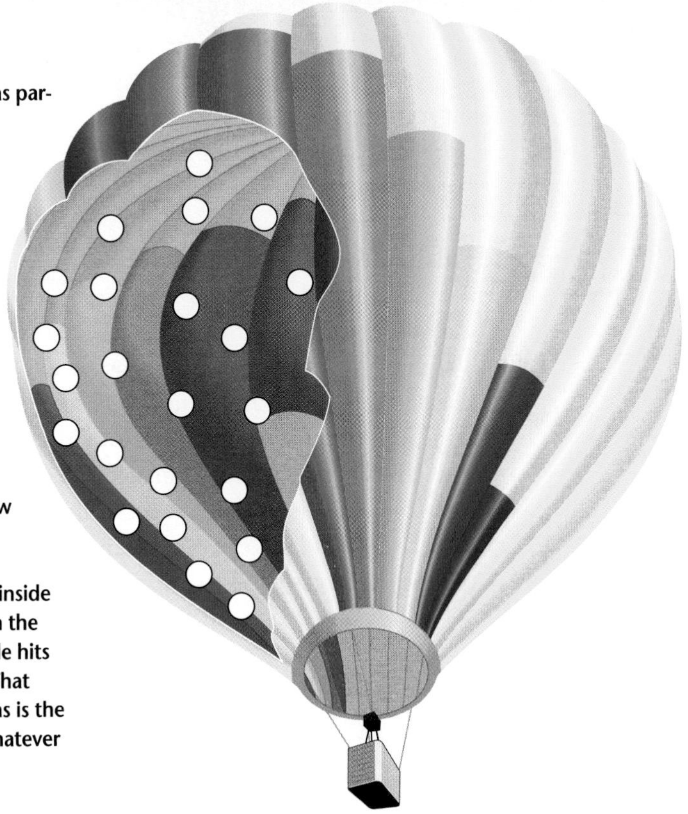

A Gas particles are always moving. When air is forced into a deflated balloon, the particles of air spread out and the balloon inflates. The air, like all gases, expands to fill the container it is in.

B Density is the mass of a material within a certain volume. Substances with large, tightly packed particles have a lot of mass per volume and a high density. All gas particles have a lot of space between them, giving gases low density.

C As the particles of hot air move around inside the balloon, they constantly collide with the walls of the balloon. Every time a particle hits something, the particle exerts a force. That force is pressure. The pressure of any gas is the result of its particles pushing against whatever they touch.

already observed. Look at **Figure 14-5** to see how expansion, low density, and pressure can be explained by particles.

■ Explaining Pressure and Temperature

What if gas particles were in motion? Moving gas particles could bang into the walls of a container and cause pressure. The kinetic energy of moving particles is measurable. In fact, the temperature of a gas is proportional to the average kinetic energy of the gas particles. So as temperature increases, the kinetic energy of the gas increases.

 Explore! ACTIVITY

How do gas particles move?

If gas particles are too small to see, how can anyone see them moving? Try this activity to find out.

What To Do

1. Light a flashlight in a darkened room.

Watch the dust specks in the beam of light. Do the specks ever change speed?

2. Set a tray of ice cubes in the light beam and watch the dance of the dust specks now. What happens?

14-2   What Are Gases Made Of?   **437**

Program Resources

Study Guide, p. 48

Laboratory Manual, pp. 87–88, Air L1 ; pp. 89–90, Air Pressure L2

Multicultural Connections, p. 32, Designing Away Sonic Boom

Making Connections: Across the Curriculum, p. 31, Adriana Ocampo Collects Data on Jupiter's Environment L1

Science Discovery Activities, 14-2, 14-3

Section Focus Transparency 42

Discussion Have students name and discuss the characteristics of air, such as how it moves and what it can do. Write the key points on the chalkboard.

2 TEACH

Tying to Previous Knowledge

The behavior of gases described in the previous section leads directly into this question: What is a gas composed of that makes it behave as it does?

Explore!

How do gas particles move?

Time needed 10 minutes

Materials flashlight, ice cubes

Thinking Processes observing and inferring, recognizing cause and effect

Purpose To observe indirect evidence of the movement of gas particles.

Teaching the Activity

Science Journal Have students write what their observations of dust lead them to infer about gas particles.

Expected Outcome

The ice will cause a change in the direction of the specks. Students should conclude that the particles are somehow affected by the ice.

Answers to Questions

1. not if the air in the room is still

2. The dust specks change direction.

✔ **Assessment**

Performance Divide the class into several cooperative groups. Have each group model how gas particles move by making a drawing, choreographing a dance, etc. Students could then present their model to the class. Use the Performance Task Assessment List for Group Work in **PASC,** page 97. COOP LEARN

Inquiry Question A clown finds that when she fills her balloons at sea level and brings them to a show at a mountain ski lodge, they burst in the mountain air. Can you explain this? *The air pressure at the higher altitude is reduced; now the pressure difference between the inside air and outside is too great for the rubber to withstand and the balloon bursts.*

GLENCOE TECHNOLOGY

 Videodisc

The Infinite Voyage: Unseen Worlds

Chapter 9
Studying the Basic Building Blocks: The Atom

Chapter 10
The Scanning Tunneling Microscope: Observing Atomic Particles

Chapter 11
The Fermi Lab Accelerator: Splitting the Atom

Figure 14-6

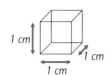

Billions of gas particles are in one cubic centimeter of space.

The dust specks you saw were never still. Although dust is not a gas, it seems reasonable to assume that gas particles in the air around the dust are also moving.

The idea that gas particles are in motion could help explain how a gas expands to fill a space. You've probably observed that shortly after a bottle of perfume is opened, the fragrance seems to fill the room, too. The gas particles of the perfume spread rapidly. This suggests that the particles of a gas are in motion. It also suggests that there are relatively large spaces between the gas particles.

So you've seen evidence that tiny gas particles have energy and are constantly moving very rapidly. This can also be used to explain pressure. The particles continually collide with the walls of any container they are in. As each particle of gas strikes the container wall, it exerts a push, or force—like that caused by a baseball hitting a wall. There are billions of gas particles in even a cubic centimeter of space. The push of that many particles adds up to quite a force being exerted.

The air in the soda bottle illustrated this. The air particles in the bottle exerted enough force in their collisions with the surface of the balloon to keep the balloon from inflating in the bottle.

Figure 14-7

A Air does not have to be trapped in a container to exert pressure. The air particles in the atmosphere constantly collide with one another and with different surfaces. Air particles exert pressure on you, on your books, on the floor and ceiling and walls of your classroom—on everything.

B Air exerts pressure on light things, such as feathers, and heavy things, such as rocks. Does air exert a greater pressure on rocks than on feathers?

Multicultural Perspectives

The First Gas Mask
Garrett A. Morgan, the son of a former slave, was born in 1875 in Kentucky. After 6 years of schooling, he moved to Cincinnati, started a sewing machine repair business, and began work on numerous inventions. In 1912, he tried to create a breathing device that could be used to help people survive exposure to noxious gases and smoke. His invention became the world's first gas mask, consisting of a safety hood connected to a long tube which cooled and filtered incoming air. Have students find out about other inventions of Morgan. *Morgan also was the inventor of the traffic light.*

The Atomic Theory of Matter

It seems that particles could be used to explain what you have observed and to accurately predict the behavior of matter. This is one reason that scientists have theorized that such particles exist and make up all gases, as well as solids and liquids. In fact, they make up your body. What are these particles called? Atoms.

The theory that matter is composed of small particles called atoms is the **atomic theory of matter**. It was first suggested by a Greek named Democritus. An expanded theory building on Democritus' idea of the atom was put forth by the chemist John Dalton. His theory will be explained as the chapter continues.

In the next section, you will examine more about atoms and some of the evidence that has led to the atomic theory of matter.

How Do We Know?

Thinking about atoms

You might wonder how someone living 2000 years ago and without the benefit of experimental testing that we do today could imagine something like the atom.

Early thinkers had to rely on their experience—with rain, snow, and wind; heat and cold; salt water and fresh water; and the lives of animals and plants around them. Lucretius, a Roman poet, helped us understand how Democritus reasoned in these lines about atoms.

… Their nature lies beyond our range of sense,

… Especially since things we can see, often

Conceal their movements, too, when at a distance.

Take grazing sheep on a hill, you know they move,

… Yet all this, far away, is just a blur,

A whiteness resting on a hill of green …

These ancient philosophers knew that even things they could see would disappear when at a great distance. They reasoned that matter could appear as solid as the flock of sheep on the hillside, if it were made of very tiny particles. Like the sheep, these particles could be moving, and yet that movement would be invisible.

check your UNDERSTANDING

1. How does a particle theory of matter explain diffusion?
2. How does a particle theory of matter explain gas pressure?
3. What evidence do scientists have that gases are made of particles?
4. **Apply** When weather balloons are sent up from Earth, only a small amount of gas is added to the balloon. As the balloon rises, it expands until it appears full. Use the particle theory to explain why this happens to a sealed balloon.

check your UNDERSTANDING

1. Particles move at rapid speeds in all directions. If they encounter resistance, they stop. Otherwise they spread until their relative density is uniform.
2. The particles push on the walls of whatever they are contained in to create pressure.
3. The way gas diffuses, exerts pressure, and inhabits volume are all evidence that a gas is made of particles.
4. As the atmospheric pressure decreases, there is less force pushing on the balloon, so the gas particles inside the balloon are free to push out on the rubber, filling the balloon to capacity.

---

3 ASSESS

Check for Understanding

Assign questions 1–4 of Check Your Understanding. As an extension of the Apply question, invite a hot-air balloonist to visit your class to explain how the flight of a hot-air balloon is controlled.

Reteach

Demonstration Demonstrate the effect of pressure, volume, and temperature on gas particles with polystyrene balls and an empty plastic jug. Put several polystyrene balls in an empty jug. Show pressure to be the force of the balls on the jug walls as you swirl the jug. Simulate an increase in temperature by swirling the jug faster. Simulate changes in volume by changing to a larger or smaller jug. L1 LEP

Extension

Activity As an independent home activity, direct students to trap some air in a resealable freezer bag; place the bag in the freezer; then warm the bag up with a hair dryer. Have the students record their observations. L3

4 CLOSE

Discussion

Have students compare and contrast gases to what they know about liquids and solids. **Is the relationship between pressure and volume true also for a liquid?** *No; liquids are not compressible* **Is the relationship between temperature and volume true for both solids and liquids?** *Both expand to lesser degrees when heated.* L1

PREPARATION

Planning the Lesson

Refer to the Chapter Organizer on pages 424A–D.

Concepts Developed

The atomic theory of matter is developed in this section. The critical part of the atomic theory is presented in an activity that simulates the way atoms combine in fixed, small, whole number ratios.

Find Out!

How do hydrogen and oxygen make up water?

Time Needed   30 minutes

Materials 250-mL beaker, washing soda solution, two test tubes, electrodes, alligator clamps, DC power supply (battery), ruler

Thinking Processes observing and inferring, interpreting data

Purpose To investigate the relationship between elements in a compound.

Preparation Any DC power source can be used for this experiment. For a 5% washing soda solution, dissolve 150 g washing soda (sodium carbonate) in 2850 mL of distilled water. Plain water may be used instead of washing soda solution.

Teaching the Activity

Safety Tell students to be careful when using electricity.

Discussion Call students' attention to the bubbles that start rising in the test tubes when the power supply is connected. Point out that electricity is being used to break down water into its chemical "ingredients." This is how chemists know that water is a compound—it can be broken down into hydrogen and oxygen. L1

14-3

What Is the Atomic Theory of Matter?

Section Objectives

■ Describe evidence that pure substances are made up of identical particles.

■ Specify what is meant by element, compound, and atom.

Key Terms

atom

Pure Substances—Elements and Compounds

You have seen evidence that there are pure substances. Aluminum metal has certain properties. If you cut aluminum into smaller and smaller pieces by ordinary means, you can never find a piece of this metal that differs in its properties from the larger piece. Aluminum is an example of an element.

But aluminum can combine with other elements to form entirely new substances. Is there some pattern in the way elements combine when they form other substances? Let's find out.

Find Out! ACTIVITY

How do hydrogen and oxygen make up water?

You know that hydrogen and oxygen are gases and that they are elements. How much of each gas is needed to make water?

What To Do

1. Partially fill a 250-mL beaker with concentrated washing soda solution. Overfill two test tubes with the same solution. Insert an electrode completely into each test tube.

2. Holding your thumb over the mouth of one test tube, invert it into the beaker. When you let go, there should be no bubble at the top of the tube. Repeat with the second test tube. **CAUTION:** *Wash your hands thoroughly after touching the solution.*

3. Connect the electrodes with alligator clamps to a DC power supply. What happens?

4. Let the setup run for at least 15 minutes. Use a ruler to measure the amounts of gas collected in the tubes. Record these measurements.

Conclude and Apply

1. What do you notice about the relationship between the volumes of gas in the test tubes?

Science Journal In their journals, have students write the statement, "Water is not an element." Have them give written evidence, based on their observations, to support the statement.

Expected Outcomes

The amount of hydrogen in one test tube should be approximately twice the amount of oxygen in the other. Students should conclude that the ratio of hydrogen to oxygen in water is 2 to 1.

Conclude and Apply

1. There is twice as much hydrogen as oxygen.

✔ Assessment

Process Have students interpret several illustrations of substances to infer whether they are elements. For example, you may show illustrations of gold, helium, and table salt. Students could then create a display showing various elements. Use the Performance Task Assessment List for Display in **PASC,** page 63. COOP LEARN

The gases you collected in your test tubes were hydrogen and oxygen. The proportion should have been very close to, if not exactly, twice as much volume of hydrogen gas as oxygen. Experiments like this show that when you break up water, you always end up with the same volume ratio of hydrogen to oxygen.

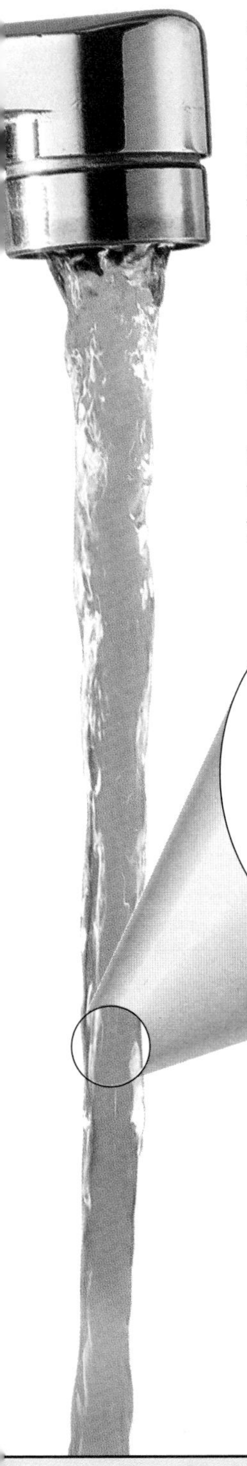

Why do you think you ended up with two gases when you separated the water? **Figure 14-8** shows that when you keep dividing a drop of water, you eventually enter a microscopic world of molecules, elements, and atoms.

As you see, the molecules of water can be broken down into different atoms. **Atoms** are the smallest particle of an element. If you had to "build" some water, you would need to use atoms as your building material. Just as a carpenter needs different building materials to construct a house, you need different atoms to "build" water. You'll need two atoms of the element hydrogen and one atom of the element oxygen to build one molecule of water. However, you'll need to build a lot of molecules to end up with even a tiny drop of water!

DID YOU KNOW?

Atoms are so small that 150 billion of them would fit on the period at the end of this sentence.

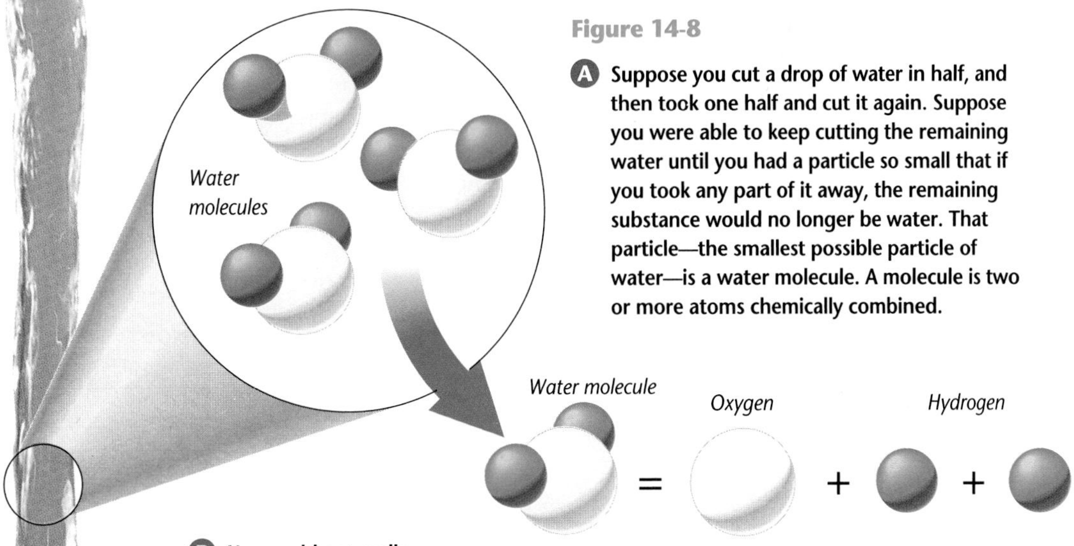

Figure 14-8

Water molecules

A Suppose you cut a drop of water in half, and then took one half and cut it again. Suppose you were able to keep cutting the remaining water until you had a particle so small that if you took any part of it away, the remaining substance would no longer be water. That particle—the smallest possible particle of water—is a water molecule. A molecule is two or more atoms chemically combined.

Water molecule   *Oxygen*   *Hydrogen*

B You could not really break up a molecule of water by cutting, but electric current can separate the molecule into the elements hydrogen and oxygen. Water is therefore a compound of these two elements.

C Elements and compounds are both pure substances. Elements cannot be divided into any other pure substance by ordinary chemical means, but compounds can.

D The substances into which compounds can be broken are often elements, but may be other compounds. Hydrogen peroxide, for example, is a compound that can be broken down into oxygen—an element—and water, a compound.

14-3   What Is the Atomic Theory of Matter?   **441**

14-2 An Atomic Model

Preparation

Purpose To investigate the relationship between the numbers and masses of substances that combine with one another.

Process Skills observing and inferring, comparing and contrasting, interpreting data, predicting

Time Required 30 minutes

Materials You may want to have the students use their own coins.

Possible Hypotheses Students will probably predict that the mass of each group of combination coins will be the same as the sum of that many individual coins. Some students may predict that the combinations will have a slightly greater mass because of the mass of rubber cement.

Plan the Experiment

Process Reinforcement Encourage students to approach the activity as if the coins were unknown quantities in a black box. They know the aggregate weight of any groups of coins, but they do not know the weight of single coins. Point out that chemists analyzing gases were presented with similar data and determined that individual units of different sizes were combining with one another.

Possible Procedures Weigh a predetermined number of one kind of coin, and do the same for a different coin. Then, glue the coins together and weigh them as combinations or "coin compounds." Compare the weight of the coin compounds and the weight of a comparable number of loose coins of the same denomination.

An Atomic Model

When you look at water you cannot tell that it is made up of two atoms of hydrogen and one atom of oxygen. How do we know that matter is really made of atoms? One way to tell is by making a model. In this experiment you will use a model to test a prediction based on the atomic theory.

Preparation

Problem
What is the relationship of the masses of atoms when they combine?

Form a Hypothesis
Is mass affected when a compound is made? Will the mass of separate elements be the same as the mass of the compound?

Objectives
- Compare the sum of masses of elements separately with the mass of the compound formed.
- Calculate ratios of masses.

Possible Materials
pennies
nickels
dimes
quarters
rubber cement
balance

Safety Precautions

Keep the rubber cement capped when not in use; the fumes can be dangerous.

Program Resources

Activity Masters, pp. 61–62, Design Your Own Investigation 14-2

Making Connections: Technology & Society, p. 31, Dorothy Crowfoot Hodgkin, Solver of Crystal Mysteries **L3**

DESIGN YOUR OWN
INVESTIGATION

Plan the Experiment

1 What will you weigh as separate coins? What "coin compounds" will you make with coins and the rubber cement?

2 Will you have a data table to record your results?

3 What kind of mathematical operations will you perform to make your comparisons?

Check the Plan

1 Why do you need to use the same balance, coins, and procedure for each weighing?

2 Have your teacher check your plan before you begin your experiment.

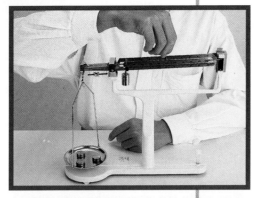

Analyze and Conclude

1. Calculate How does the mass of your "coin compounds" compare with the sum of the masses of the individual coins? What is the ratio of the total mass of the one type of coin in your compound compare with the total mass of the other type of coin?

2. Hypothesize Hypothesize the ratio of the mass of 100 of your first coin to 100 of your second coin. What

would the mass of 100 units of this "coin compound" be?

3. Infer Infer how you expect the masses of elements in a compound to relate.

4. Predict What would be the ratio of two hydrogen atoms to one oxygen atom? What would the ratio of hydrogen to oxygen be in a water molecule? in a liter of water?

Going Further

If you had made fifteen 7-cent combinations, what would be the mass of the fifteen 7-cent combinations?

Meeting Individual Needs

Learning Disabled Encourage these students to keep a list of the definitions of terms used in this section so they can differentiate between such terms as *molecule* and *compound*. Point out that such terms have subtle differences, so they should study their definitions carefully.

Teaching Strategies

Troubleshooting Be sure students do not expect weight ratios to reflect monetary value; i.e., nickels should not weigh five times as much as pennies.

Science Journal Have students record their observations in their journals. They can also draw pictures in their journals to illustrate what they did in this activity.

Expected Outcomes

Whether students weigh 1000 of each type of coin or 15, the weight ratio between the two coins will remain the same. Students should conclude that distinct items of fixed weight are combining in this experiment.

Analyze and Conclude

1. The sum of the masses of individual coins is the same as that of the coin compound. The ratios will depend on the combinations used.

2. The ratio would stay the same and the mass would be 100 times the mass of a single "coin compound."

3. The ratio of masses of elements in a particular compound will always be the same.

4. The mass ratio for all three cases would be 1 to 8.

✔ Assessment

Process Ask students to hypothesize about making different combinations than the one they did in the experiment. Ask them how the coins separately would compare to them together. Students could then write an article for the science and health section of a newspaper explaining the use of models to tell that matter is made up of atoms. Use the Performance Task Assessment List for Newspaper Article in **PASC,** page 69.
COOP LEARN

Going Further

The mass would be the same as the mass of 15 nickels plus the mass of 30 pennies.

Content Background

Chemists make the distinction between the weight of molecules and the number of molecules in a sample with the mole concept. A mole is simply the name for a specific number of molecules, in the same way that a dozen is a name for twelve. The number a mole stands for is huge—6.02×10^{23}. One mole of helium, for example, weighs exactly 4 grams, while one mole of carbon weighs 12 grams, and so on, in complete accord with the periodic table.

How Do We Know that Matter Is Made of Atoms?

Only recently has technology been available that can get even close to seeing a single atom. So how could anyone tell that matter was made of atoms? In the past, scientists had to rely on other experiments to investigate this. The main evidence for the existence of atoms came from chemical reactions.

■ The Ratios Remain the Same

When you mix a solution, such as a powdered fruit drink, you can make it as strong or weak as you want. You can add a little more or less sugar to suit your tastes, or add more water to weaken the flavor. Without concern for taste, you can add as much of any ingredient you desire, up to the saturation point.

Mixing ingredients is not the same as chemically combining them, though. Think back to the Investigate activity you just finished. In order to make a 6-cent combination, how many pennies and nickels did you need? Could you use more nickels and still get a 6-cent combination? If matter is made of atoms, we would expect them to behave similarly to the coin

Figure 14-9

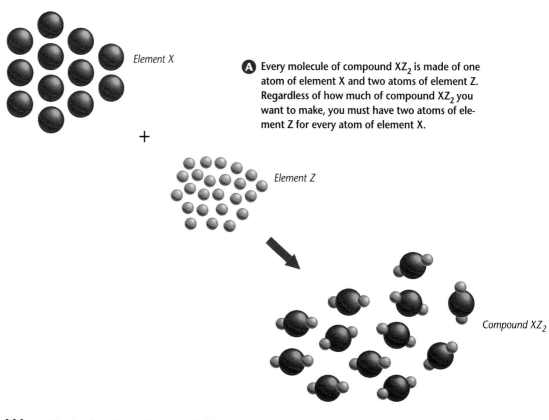

Element X

+

Element Z

A Every molecule of compound XZ$_2$ is made of one atom of element X and two atoms of element Z. Regardless of how much of compound XZ$_2$ you want to make, you must have two atoms of element Z for every atom of element X.

Compound XZ$_2$

Meeting Individual Needs

Gifted Students might enjoy the challenge of figuring out formulas for compounds, given the ratios of the elements that they are composed of. For example, in benzene, the ratio of weight of hydrogen to carbon is 1 to 12. **What is the formula for benzene?** *1 carbon atom for every hydrogen atom.* In iron oxide, or rust, the ratio of weight of iron to oxygen is 2.33 to 1. **What is the formula for iron oxide?** *2 iron atoms for every 3 oxygen atoms.* **L3**

Visual Learning

Figure 14-9 Help students relate compound XZ$_2$ to water, H$_2$O. A molecule of water always contains two atoms of hydrogen and one atom of oxygen. A molecule of XZ$_2$ always contains one atom of X and two atoms of Z. Ask students to give analogies of situations where relationships have a constant ratio. For example, in building a car, if you have one chassis and six wheels, you can only use four wheels and two will be left over.

model you explored. We would expect the ratios of elements needed to make a compound to be fixed, just like you needed one penny for every nickel to make a 6-cent combination.

To make ordinary table salt, for example, you need one atom of chlorine for every atom of sodium. If you use more sodium, you won't get more salt; you'll just end up with extra sodium. Look at **Figure 14-9** to see how this works.

If we can't see atoms, how do we know this happens? Scientists count on the fact that if matter is made of atoms, then each atom must have a certain mass. That means if you have two piles of an element, and one pile has twice as much mass as the other,

then that pile has twice as many atoms. Mass is something we can measure fairly easily.

So if you add 1 g of sodium to 1.54 g of chlorine, then you will end up with 2.54 g of salt. However, if you add 1.5 g of sodium to 1.54 g of chlorine, you would still get only 2.54 g of salt, but you'd have 0.5 g of sodium left over.

Every compound is made with an exact ratio of elements. For example, you need 2.67 g of oxygen for every 1 g of carbon to make carbon dioxide, never more, never less. No matter how much of a compound you have, the ratio of masses of the elements that make up that compound is always the same.

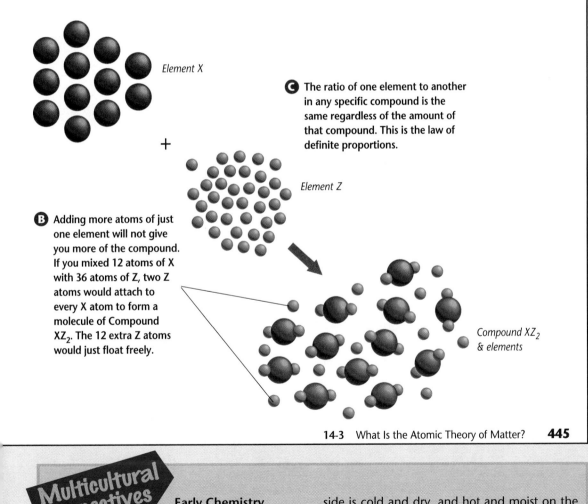

Element X

+

B Adding more atoms of just one element will not give you more of the compound. If you mixed 12 atoms of X with 36 atoms of Z, two Z atoms would attach to every X atom to form a molecule of Compound XZ_2. The 12 extra Z atoms would just float freely.

Element Z

C The ratio of one element to another in any specific compound is the same regardless of the amount of that compound. This is the law of definite proportions.

Compound XZ_2 & elements

14-3 What Is the Atomic Theory of Matter? **445**

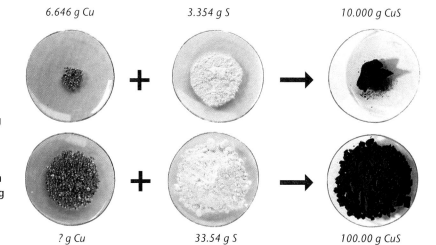

Visual Learning

Figure 14-10 Help students write proportions to express the relationship between the weight of copper and sulfur in copper sulfide. **How much copper would you need to combine with 33.54 g of sulfur to make 100.00 g of CuS?** *66.46 g of copper*

3 ASSESS

Check for Understanding

Have students work with partners to answer questions 1–4 of Check Your Understanding. Use the Performance Task Assessment List for Group Work in **PASC**, p. 97.

Reteach

Modeling Use paper models of elements labeled with their atomic weights to show how atoms combine in simple, whole number combinations. For example, label carbon atoms **12** for their atomic weight and label oxygen atoms **16**. Using this model, students can easily see that a combination of 2 oxygens and 1 carbon in carbon dioxide gives a weight ratio of 32:12, or 2.67:1, as mentioned in the text. Repeat this process with other compounds. L1 LEP

4 CLOSE

Activity

Have students imagine a group has challenged the atomic theory of matter. Have them draft a short presentation (including models, diagrams, etc.) that would persuade an objective audience that the atomic theory of matter is correct. L1 P

■ Who Developed the Atomic Theory?

When John Dalton analyzed the masses of elements in the early 1800s, he could see the same kind of ratios you observed in the Investigate. He created the model to account for the ratios he observed. This model is called the atomic theory of matter.

Dalton proposed that different elements would be composed of different atoms. Just as pennies are different from nickels, atoms of helium are different from atoms of sodium, and just as pennies are all the same, atoms of helium are the same, too. The atoms of one element would differ in mass from the atoms of another element, just like the mass of a penny is different from the mass of a nickel.

The atomic theory of matter can be summarized in the following statements. Try to explain each in terms of the model you used in the Investigate activity.

1. Matter consists of atoms.
2. All atoms of a given element are identical.
3. Different elements have different atoms.
4. Atoms maintain their mass in chemical reactions.

Figure 14-10

Copper sulfide (CuS) is a compound made by combining copper with sulfur. For every gram of sulfur, you must add 1.98 grams of copper. If you combined 3.354 g of sulfur with 6.646 g of copper, you would have 10.000 g of copper sulfide.

How much copper would you need to combine with 33.54 g of sulfur to make 100.00 g of CuS?

6.646 g Cu   3.354 g S   10.000 g CuS

? g Cu   33.54 g S   100.00 g CuS

check your UNDERSTANDING

1. What evidence do you have that every water molecule is alike?
2. Table salt is made up of sodium and chlorine. Is it an element or a compound? Why?
3. A 9-g sample of distilled water in Los Angeles contains 8 grams of oxygen and 1 gram of hydrogen. How many grams of hydrogen would a 90-g sample of distilled water in London contain? Explain.
4. **Apply** Draw 4 particles of hydrogen and 1 particle of carbon. Draw 1 particle of methane (made up of 1 carbon and 4 hydrogens). Which of the particles that you drew is a molecule? How do you know?

check your UNDERSTANDING

1. Based on the electrolysis experiment, students can infer that a water molecule is made up of two hydrogen atoms and one oxygen atom. Any other combination would not give water.

2. It is a compound because it is made up of more than one element.

3. 10 g, because the weight ratio of oxygen to hydrogen is always the same for water—8 to 1.

4. Methane; methane is a compound, while the other five particles are just atoms of elements.

The Gas Laws

Does it seem strange that some of the earliest experiments were done on gases? Most gases are, after all, invisible. How did these early scientists even know they were there?

One of the most important characteristics that a scientist possesses is curiosity. If you hold a glass upside down and push it into a container of water, what happens? If you are a curious person, you might begin to wonder what is in the glass that won't allow the water to enter.

Robert Boyle was very curious about nature. The relationship between pressure and volume that you know as Boyle's law was first published in 1662. But Boyle didn't spend his life working with gases. He went on to study blood circulation, water expansion, color, electricity, the bending of light in transparent objects, as well as the way sound travels in air.

But a scientist doesn't always spend his time doing fancy experiments. Jacques Charles, a French physicist, was another scientist who was interested in how and why things happened. When the French Academy of Science, where Charles worked, decided to experiment with balloon flight, Jacques

Charles was given the job of making the hydrogen gas and filling the balloon. He was so fascinated by the idea that, within a month of the first flight, Charles and his brother built their own balloon and flew for 90 minutes. Charles took along several scientific instruments, but managed to show only that air pressure decreased as the balloon rose.

Even though his interest and research in the behavior of gases continued, Charles never published his experiments. However, he explained them to the French chemist, Joseph Gay-Lussac. Gay-Lussac repeated the experiments and published the results in 1802. For this reason, Charles's law is sometimes called Gay-Lussac's law. Gay-Lussac also published results on the pressure-temperature relationship of a gas when the volume is constant. This law completes the circle of laws relating pressure, volume, and temperature of gases.

What Do You Think?

If you were doing an experiment and something unexpected happened, would you just assume you'd done something wrong or would you try to figure out why the unexpected occurred? Why is curiosity an important quality for a scientist to possess?

Curiosity propels scientists to keep asking questions and to learn from their mistakes.

Going Further ▸

Have student groups make time-line posters showing the historical progress of learning about gases. They should include key dates for discoveries of gases and laws and major gas-related events, such as the 1937 *Hindenburg* disaster. Use the Performance Task Assessment List for Poster in **PASC**, p. 73. **COOP LEARN**

Purpose

The History Connection provides background on Robert Boyle and Jacques Charles, scientists whose laws regarding the behavior of gases are discussed in Section 14-1.

Content Background

The Scientific Revolution of the 1600s brought important research by Galileo, Newton, Boyle, and others. During this century, gas was given its name—from the Greek *chaos*, meaning "space," because a gas fills space—and gas studies began. In the 1700s, scientists first isolated hydrogen, oxygen, nitrogen, and carbon dioxide for study. After Charles's law was published, Italian scientist Amadeo Avogadro discovered that a standard volume of any gas at the same temperature and pressure contains the same number of particles. The number of particles was later determined to be 602 billion trillion (6.02×10^{23}) and was designated Avogadro's number. All gases—oxygen, chlorine, methane, and so on—obey the gas laws of pressure, temperature, and volume.

Teaching Strategies

Ask students to write three things that they are curious about in science, in school, or at home. Then have them write a way that they think they could investigate each item to learn more about it. Hold a class discussion on the role of curiosity in learning. Emphasize that if people were not curious, they would learn nothing.

Answers to

What Do You Think?

If you are a scientist, your curiosity will drive you to find out why the unexpected occurred.

Health CONNECTION

Purpose

Health Connection reinforces the principles of substances combining presented in Section 14-3 and the concepts of the pressure of a gas presented in Sections 14-1 and 14-2 by describing the breathing of divers underwater.

Content Background

In space travel, we must carry our environment's gases, pressure, and temperature with us. The same is true of underwater activities. Scuba gear provides the right gases at the right pressure, and insulated wet suits help maintain body temperature. Scuba divers now explore shipwrecks, seek treasure, go sport diving and spear fishing, take underwater photographs, and do archaeological research.

Teaching Strategy

Float a cork in a large container or sink of water. Hold a clear cup upside down over the cork and slowly push the cup straight down into the water until it touches bottom. Slowly bring it back to the surface. Ask: **If this water were the ocean, if the cup were the size of a car, and if you were on a raft like the cork, could you descend to the ocean floor in this way?** *Yes, until the oxygen is used up and if the ocean isn't very deep because water pressure would compress the air at great depths. Oceanographers call this a diving bell.*

Answers to

What Do You Think?

At 100 m, pressure is 11 atm, a crushing force that could damage our bodies. At greater depths, the pressure is even worse. So below 100 m, a diving vessel is used to carry our normal 1-atm environment with us.

Health CONNECTION

Breathing Underwater

Unlike fish, a person's body has no way to draw oxygen from water. We must take our environment with us when we travel in the depths of the oceans or the vacuum of space. As you know, the pressure on your body due to the mass of air around you is one atmosphere. This is as if a weight of almost 15 pounds is pressing on every square inch of your body. Underwater, the pressure increases due to the added mass of water above you. Every 10 meters adds 1 atmosphere pressure. At a depth of 30 meters, the total pressure on your body would be 4 atm. That's nearly 60 pounds of pressure per square inch!

Pushing Back

The muscles controlling your lungs and diaphragm evolved to work in 1 atm pressure. At a depth of 40 meters, the pressure on your chest would make it impossible to inflate your lungs to breathe, even if you had a supply of air available. In 1943, Jacques-Yves Cousteau and Emile Gagnan invented the SCUBA (Self-Contained Underwater Breathing Apparatus). SCUBA equipment provides air to the lungs at a pressure that matches the underwater environment. Therefore, you can breathe comfortably.

If the pressure in your lungs is increased to match the outside pressure, what happens when you swim toward the surface? Boyle's law tells us that if the amount of a gas remains constant, as the pressure decreases, volume increases. If you went from an outside pressure of 5 atm to a pressure of 1 atm without exhaling, the air would expand to five times

Going Further ▐▐▐▐▐▶

Tell students to think of people exploring underwater as organisms who are living in an environment to which they are not adapted. Students can name other examples of organisms in different environments and what they need to survive. Examples include fish on a mountaintop, algae in the desert, a polar bear in Florida, a parrot at the South Pole, and a pine tree on the ocean floor. Students can use their examples to create humorous posters or cartoons for a class display. Use the Performance Task Assessment List for Group Work in **PASC,** p. 97.

the volume, certainly enough to rupture your lungs.

The Bends

When you breathe air under increased pressure for 30 minutes or longer, more nitrogen from the air dissolves in your blood than normal. As you move up to the surface, the nitrogen becomes less soluble and comes out of your blood as bubbles. This is much like what happens when you release the cap from a soda bottle and you see bubbles of carbon dioxide gas rising in the liquid.

If you rise to the surface too quickly, these bubbles of nitrogen form in joints and muscles and cause pain. If they form in the spinal cord, brain, or lungs, they can cause death. This effect is called "the bends" or decompression sickness. If a diver rises slowly, decreasing the pressure gradually, the nitrogen can be released through the lungs.

What Do You Think?

If SCUBA equipment can provide air at a pressure equal to that underwater, why do people still have to use enclosed diving vessels at depths below 100 m?

Cooking Under Pressure

Sometimes, foods are cooked by boiling them in water. The time it takes for them to cook varies depending on factors such as the temperature of the boiling water. But water does not always boil at the same temperature.

The temperature at which water boils depends on the pressure surrounding the water. As the molecules in the air bounce against the surface of the liquid, they exert pressure. The water molecules throughout the liquid must have enough energy to overcome that pressure before they change state from liquid to gas.

Air pressure varies with changes in altitude—the greater the altitude, the less the pressure. For example, the air pressure in Miami, which is very near sea level, is greater than the pressure in Denver, the mile-high city.

Therefore, water boils at a slightly higher temperature in Miami than in Denver. And so, a cook in Denver must increase cooking times for everything from boiled eggs to baked cakes.

Science Journal

A pressure cooker increases the pressure in the pot. What does that do to the temperature at which water boils? Will foods cook more quickly or more slowly in a pressure cooker? Answer these questions *in your Science Journal.*

Purpose

The Leisure Connection applies the gas laws described in Section 14-1, showing the relation between pressure, temperature, and volume to cooking procedures.

Content Background

Cooking in an open pot allows the water to vaporize into steam (boil) around 100°C. Adding more heat will not raise the temperature of the water above 100°C. A liquid boils when its internal vapor pressure equals atmospheric pressure. A pressure cooker increases the air pressure above the liquid, so the liquid must reach a higher temperature before its vapor pressure equals the air pressure. As a result, added heat can drive the temperature of water above 100°C.

Teaching Strategy

Ask students about their experiences and observations when cooking and boiling water. Help them visualize the "battle zone" at the surface of hot water, where air and water molecules constantly bounce against one another. Few water molecules can escape to move as water vapor among the air molecules until the boiling temperature is reached. Then, many water molecules acquire enough energy to move among the air molecules.

Science Journal

The increased pressure in the cooker raises the temperature at which water boils. Foods will cook faster.

Going Further ⫸

Have students work in pairs to think of other examples of pressure cookers, both natural and manufactured. Heating-system boilers, geysers, and volcanoes are examples. For each example, have students consider the energy source, the pressure, what the cooker is made of (metal, rock), how strong the cooker is, whether it could explode from the pressure, and how to prevent explosion. Have each pair share one example with the class. **COOP LEARN**

Technology
Connection

Technology
Connection

What Is a Vacuum?

Purpose

The Technology Connection extends Sections 14-1 and 14-2 by explaining how air pressure and a partial vacuum can move something.

Content Background

Vacuum is the absence of matter. This manner of definition of a vacuum is analogous to defining darkness and silence as the absence of light and sound. In 1677, the Dutch philosopher Spinoza wrote that "nature abhors a vacuum." He meant that a vacuum is unstable, because matter, especially gas, fills every open space. People create partial vacuums to do useful work. We can use a vacuum to move something; in a vacuum cleaner, it lets air carry dirt into a bag. Vacuums are good insulators, with few molecules to conduct heat or electricity.

Teaching Strategy

Guide students to understand what Spinoza meant by "nature abhors a vacuum." Ask: **When you move something, what happens to the space it used to occupy?** *Something rushes in to fill it.* **If you scoop water from a bathtub, what fills the hole?** *the remaining water* **If you dig soil, what rushes in to fill the void?** *air* **When you walk, what fills the space behind you?** *air* **When you open a vacuum-packed food container, what do you hear?** *the air swooshing in*

Answer to
You Try It!

Removing air from inside the straw creates a partial vacuum, allowing the atmospheric pressure outside the straw to collapse it.

Going Further

Present small groups with one or more of these jobs and ask them

V acuum cleaners, vacuum-sealed containers, vacuum-packed foods—just what is a vacuum?

A vacuum is a space that has no matter in it. There is no such thing as a complete vacuum because no one has ever been able to remove all the air molecules within a given space. Even in the near-vacuum of outer space, it is estimated that there are about 100 molecules of matter in every cubic meter of space.

Why would anyone want to create a vacuum in the first place? Have you ever used a straw to drink from a glass or can? When you draw on the straw, you remove some of the air inside the straw. This produces a lower pressure inside the straw than in the air outside—a partial vacuum. The greater air pressure of the air on the surface of the liquid pushes the liquid up the straw.

How else are partial vacuums used? A vacuum conducts heat poorly, so it's a good

insulator. This property is used in devices such as insulating bottles. These bottles contain a double-walled glass container that has had the air removed from between the walls. Thermal energy can't pass through this space from either direction, so the liquid inside stays hot or cold.

You may recall when you studied sound that sound is a result of vibrations of molecules. Because there are no molecules—or atoms—in a vacuum, it does not conduct sound.

Light bulbs contain a partial vacuum and nitrogen or argon gas, which is why a light bulb will pop when it breaks. Since there is little oxygen in the light bulb, the burning filament lasts longer than it would if surrounded by air. You will learn more about this in a later chapter.

One of your favorite uses for a vacuum may be the television. The television tube works because there are very few air molecules in the tube. The beam inside the television tube would never get to the screen if it had to travel through air molecules.

You Try It!

Roll a wax-coated paper straw between your fingers a few times to remove the stiffness. Be sure to leave the straw open. Now press your finger tightly over one end so that no air can enter. Put the other end of the straw in your mouth and suck on it. What happens? Why did this happen?

450 Chapter 14 Gases, Atoms, and Molecules

to draw and/or write how they could use a vacuum to accomplish it: (1) Soundproof a room. *Enclose it in a vacuum;* (2) Tightly seal a food container lid. *Pump air from the container, creating a vacuum that lets outside air pressure seal the lid;* (3) Move small cylinders of chemical samples among several laboratories in a building. *Use a network of tubing with a vacuum pump;* (4) Keep a can of orange juice frozen in the sunshine. *Insulate it in a vacuum;* and (5) Continually move sand from a quarry into a glass factory. *Use tub-*

ing and a vacuum pump, as a vacuum cleaner. Have groups share their solutions with the class. Use the Performance Task Assessment List for Group Work in **PASC,** p. 97. `COOP LEARN`

Science Journal

Review the statements below about the big ideas presented in this chapter, and answer the questions. Then, re-read your answers to the Did You Ever Wonder questions at the beginning of the chapter. *In your Science Journal*, write a paragraph about how your understanding of the big ideas in the chapter has changed.

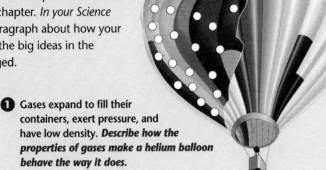

1 Gases expand to fill their containers, exert pressure, and have low density. *Describe how the properties of gases make a helium balloon behave the way it does.*

1P

3.0 liters
25°C

2P

1.5 liters
25°C

3P

1 liter
25°C

2 The temperature, volume, pressure, and amount of a gas are all related. As one changes, so do the others. *Describe how the temperature, volume, amount, and pressure of a gas change as you pump up a bicycle tire.*

3 Elements are made up of particles called atoms. Compounds are made up of atoms that are chemically combined. *What characteristics of gases prove that they are made of particles?*

4 The atomic theory states that matter consists of atoms, all atoms of an element are identical, and different elements have different atoms. *Why is the ratio of masses of elements in a compound always the same?*

Pressure constant

Use the diagrams to review the main ideas presented in the chapter.

Teaching Strategies

Pose questions like the ones below to the class. Then have students make up their own gas law or atomic theory problems.

1. The pressure on a gas in a closed container increases. What effect will this have on:

(a) volume? *decreases*

(b) number of particles? *same*

2. The number of particles of a gas in a rigid, closed container increases. What effect will this have on:

(a) volume? *same*

(b) pressure? *increases*

Answers to Questions

1. The helium gas fills the irregular shape of the balloon; the gas expands, causing the balloon to become firm; the gas is low in density, even lighter than air, which causes the balloon to rise.

2. Pumping up a tire increases the amount of air in the tire, which in turn increases the temperature. If the volume stays the same, then the pressure increases. But if the volume increases, then the pressure increases less.

3. Pressure. They expand to fill a container.

4. All atoms of a single element have the same weight, and atoms always combine in the same proportions to form a specific compound.

Science Journal

Did you ever wonder...

• A foreign object such as a nail can rip a hole in the tire. The rapidly escaping compressed air can cause the hole to enlarge tearing the tire apart. (pp. 432–433)

• Molecules or atoms of the gas that fills the balloon. (p. 427)

• As the pressure decreases as a result of being sprayed out of the can, the temperature also decreases. (p. 434)

Project

Students can use a barometer to measure air pressure and chart how the atmosphere changes over the course of time. Students might want to develop a scale to quantify the changes of atmospheric pressure and chart the changes on a graph. They can take notes on the weather to see if pressure and weather can be correlated in some way. Then they can obtain official barometer readings to corroborate those taken in class. Use the Performance Task Assessment List for Group Work in **PASC,** p. 97. L2 P

GLENCOE TECHNOLOGY

MindJogger Videoquiz

Chapter 14 Have students work in groups as they play the Videoquiz game to review key chapter concepts.

Chapter 14 Gases, Atoms, and Molecules **451**

Using Key Science Terms

1. An atom is the smallest particle of an element.

2. All matter is composed of small particles called atoms.

3. Boyle's law states that the volume of a certain amount of gas is inversely proportional to the pressure if the temperature remains constant. That is, if there is an increase in volume there will be a decrease in pressure.

4. Charles's law states that gases increase or decrease their volume at the same rate that Kelvin temperature changes, if the amount of gas and the pressure remain constant. That is, an increase in temperature will cause an increase in volume of a certain amount of gas.

Understanding Ideas

1. The trapped gas in the bag was subjected to great pressure as your hand hit it. The bag could not withstand this pressure, so it broke open.

2. Three times the starting pressure; four times less

3. A gas expands to fill its container, exerts pressure, and has low density.

Developing Skills

1. The student group toward which the air is flowing should react first. The other groups should not react for awhile because the air is flowing away from them.

2. See reduced student page for completed concept map.

3. The temperature of the gas is no longer constant. The volume of the gas will increase because the temperature of the gas will increase. The average volume will be more for the same amount of weight than in the first set of trials.

4. The dust specks will speed up their movement when placed over hot water.

Using Key Science Terms

atom	Boyle's law
atomic theory of matter	Charles's law

In your own words, explain:
   1. an atom
   2. the atomic theory of matter
   3. Boyle's law
   4. Charles's law

Understanding Ideas

Answer the following questions in your Journal *using complete sentences.*

   1. Suppose you blow up a brown paper lunch bag, you hit it hard, and it explodes. Use what you have learned in this chapter to explain why the bag exploded.

   2. How much pressure increase is required to cut the volume of a gas in a balloon to one-third of the original? How much would the pressure of a gas change if its volume were four times greater?

   3. What part(s) of the atomic theory of matter help(s) explain why the airship that takes overhead pictures of football games stays inflated?

Developing Skills

Use your understanding of the concepts developed in this chapter to answer each of the following questions.

   1. Recognizing Cause and Effect Repeat the Explore activity on page 426 with the experimenter using a small hand-held fan to direct the odor toward one group of

students. How do these results compare with the results taken when the air in the classroom was as still as possible?

2. Concept Mapping Complete the concept map of gases.

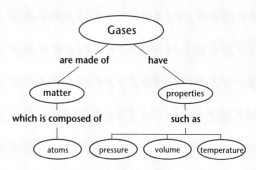

3. Separating and Controlling Variables Repeat the Investigation on pages 430-431 without waiting between trials. Which variable have you changed? How do these results compare with the results taken after waiting between trials?

4. Observing and Inferring Repeat the Explore activity on page 437, placing the flashlight over very hot water. What happens to the movement of the dust specks you see in the beam of light?

Critical Thinking

In your Journal, *answer each of the following questions.*

   1. You have a weak spot in the wall of your bicycle tire. Is your tire more likely to tear on a hot day or a cold day?

   2. Compare the piece of bread and the piece of volcanic rock shown in the picture. How may their formations have been similar?

Program Resources

Review and Assessment, pp. 83–88, L1

Performance Assessment, Ch. 14 L2

PASC

Alternate Assessment in the Science Classroom

Computer Test Bank L1

3. The pressure on a gas in a closed container was increased. What can you say about the volume of the gas?

4. You are taking a bouquet of inflated balloons to a friend in the hospital on a winter day with the air temperature near freezing. You blew the balloons up in your house, but by the time you got to the hospital, they appeared shriveled and smaller. Why did this happen? What will happen when you take the balloons into the hospital?

Problem Solving

Read the following problem and discuss your answer in a brief paragraph.

Juan and Sam were camping and a rainstorm during the night got their tent bottom wet. The next day was warm and bright, so Sam left his inflatable air mattress in the sun to dry while they went hiking. When they returned, the mattress had a jagged hole in it and all the air was gone.

Sam thought that an animal had ripped the mattress. How did Juan explain what happened using what he learned in this chapter?

Critical Thinking

1. Pressure increases with temperature. On a hot day, the air inside a tire will expand and more likely tear the tire.

2. Both were formed when air bubbles inside were trapped, then expanded after being heated; the material around the bubbles later solidified.

3. If the pressure is increased, the volume would decrease because the distance between particles would decrease.

4. The cold air lowered the temperature of the air inside the balloons, decreasing their volume so the balloons shrank. The warm hospital air will gradually warm the air in the balloons, and they will expand.

Problem Solving

Thermal energy from the sun heated the gas inside the mattress. It expanded more than the mattress could hold and caused the mattress to "explode."

Connecting Ideas

1. Its temperature is rising.

2. The gases, warmed by the body's heat, expand. They cannot expand much in the enclosed space so the pressure increases and pushes harder on the intestinal walls causing pain.

3. Aerostats can lift instruments high into the atmosphere to perform studies and collect information about gases in our atmosphere.

4. Increasing the pressure in a container increases the boiling temperature of the liquid. This allows the food to reach a higher temperature than it normally would, cooking it faster.

5. Sucking the air out of the straw reduces the pressure inside the straw. Meanwhile, the pressure is still high on the liquid outside of the straw. The difference in pressure moves the liquid up through the straw.

CONNECTING IDEAS

Discuss each of the following in a brief paragraph.

1. Theme—Systems and Interactions On TV, the weather map shows a large air mass moving into your area. Its volume is increasing, but its pressure is staying the same. What is happening to its temperature?

2. Theme—Energy Explain how gas from digesting food in the intestines can cause a stomachache.

3. Earth Science Connection How do aerostats help scientists learn about gases in the atmosphere?

4. Leisure Connection Describe how cooking with a pressure cooker might speed up the process of cooking.

5. Technology Connection Describe how you are able to drink liquids through a straw.

✔ Assessment

Portfolio Review the portfolio options that are provided throughout the chapter. Encourage students to select one product that demonstrates their best work for the chapter. Have students explain what they learned and why they chose this example for placement into their portfolios.

Additional portfolio options can be found in the following **Teacher Classroom Resources: Multicultural Connections,** pp. 31, 32

Making Connections: Integrating Sciences, p. 31

Making Connections: Across the Curriculum, p. 31

Concept Mapping, p. 22

Critical Thinking/Problem Solving, p. 22

Take Home Activities, p. 22

Laboratory Manual, pp. 81–90

Performance Assessment P

Chapter Organizer

SECTION	OBJECTIVES	ACTIVITIES & FEATURES
Chapter Opener		**Explore!**, p. 455
15-1 So This Is the Atmosphere (3 sessions, 1.5 blocks)	**1. Describe** the composition of the atmosphere. **2. Discuss** ways people affect air composition. **National Content Standards: (5-8) UCP2, A1, D1, F1-2, F4**	**Find Out!**, p. 457 **Investigate 15-1:** pp. 458–459 **Skillbuilder:** p. 461 **Life Science Connection,** pp. 460–461
15-2 Structure of the Atmosphere (3 sessions, 1.5 blocks)	**1. Describe** the structure of Earth's atmosphere. **2. Explain** what causes atmospheric pressure. **National Content Standards: (5-8) UCP1-2, A1-2, D1**	**Explore!**, p. 464 **Design Your Own Investigation 15-2:** pp. 466–467 **Science and Society,** pp. 476–477
15-3 The Air and the Sun (2 sessions, 1.5 blocks)	**1. Explain** what causes wind. **2. Describe** Earth's wind systems. **National Content Standards: (5-8) UCP2-3, A1, B1, B3, D1, D3, F2-3**	**Explore!**, p. 470 **Find Out!**, p. 472 **A Closer Look,** pp. 470–471 **Teens in Society,** p. 478

ACTIVITY MATERIALS

EXPLORE!

p. 455 deflated ball, pan balance, bicycle pump, inflation needle
p. 464* 4 textbooks, modeling clay, waxed paper
p. 470* clear plastic storage box and lid, soil, stiff cardboard, thermometer, heat lamp

INVESTIGATE!

pp. 458–459* measuring cup with mL gradations, metric ruler, 2 rubber bands, 2 pencils, white vinegar, test-tube stand, test-tube clamp, test tube, paper towels, steel wool, water, tongs, scissors, beaker

DESIGN YOUR OWN INVESTIGATION

pp. 466–467* small coffee can, drinking straw, rubber balloon, heavy paper (28 cm × 21.5 cm), transparent tape, scissors, metric ruler, rubber band

FIND OUT!

p. 457* petroleum jelly, 4 plastic lids, permanent marker or grease pencil, magnifying glass, microscope slides, microscope
p. 472* 2 small paper bags, string, meterstick, ring stand, lamp with 100-watt bulb

KEY TO TEACHING STRATEGIES

The following designations will help you decide which activities are appropriate for your students.

- **L1** Basic activities for all students
- **L2** Activities for average to above-average students
- **L3** Challenging activities for above-average students
- **LEP** Limited English Proficiency activities
- **COOP LEARN** Cooperative Learning activities for small group work
- **P** Student products that can be placed into a best-work portfolio
- 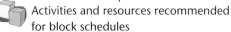 Activities and resources recommended for block schedules

Need Materials? Call Science Kit (1-800-828-7777).

⏱ **OUT OF TIME?** We recommend that students do the activities with an asterisk.

Chapter 15 The Air Around You

Student Masters	Transparencies
Study Guide, p. 50 **Activity Masters**, Investigate 15-1, pp. 63–64 **Making Connections: Across the Curriculum**, p. 33 **Making Connections: Technology & Society**, p. 33 **Laboratory Manual**, pp. 91–92, Radiant Energy and Climate	**Section Focus Transparency 44**
Study Guide, p. 51 **Activity Masters**, Design Your Own Investigation 15-2, pp. 65–66 **Take Home Activities**, p. 23 **Science Discovery Activities**, **15-1, 15-2, 15-3** **Laboratory Manual**, pp. 93–96, Smoke Pollution	**Section Focus Transparency 45**
Study Guide, p. 52 **Concept Mapping**, pp. 5, 23 **Critical Thinking/Problem Solving**, p. 23 **Multicultural Connections**, pp. 33, 34 **Making Connections: Integrating Sciences**, p. 33 **Science Discovery Activities**, **15-3** **Laboratory Manual**, pp. 97–98, Temperature of the Air **Laboratory Manual**, pp. 99–100, Air in Motion	**Teaching Transparency 29**, Major Air Circulation **Teaching Transparency 30**, The Coriolis-go-round **Section Focus Transparency 46**

ASSESSMENT RESOURCES	TEACHING & TECHNOLOGY
Review and Assessment, pp. 89–94 **Performance Assessment**, Ch. 15 **PASC*** **MindJogger Videoquiz** **Alternate Assessment in the Science Classroom** **Computer Test Bank**	**Spanish Resources** **Cooperative Learning Resource Guide** **Lab and Safety Skills** **Science Interactions, Course 2, CD-ROM** **Computer Competency Activities**

*Performance Assessment in the Science Classroom

NATIONAL GEOGRAPHIC TEACHER'S CORNER

Index to National Geographic Magazine	National Geographic Society Products Available From Glencoe	Additional National Geographic Society Products
The following articles may be used for research relating to this chapter: • "Is Our World Warming?" by Samuel W. Matthews, October 1990. • "Air: An Atmosphere of Uncertainty," by Noel Grove, April 1987.	To order the following products for use with this chapter, contact your local Glencoe sales representative or call Glencoe at 1-800-334-7344: • *Eye on the Environment: Ozone Layer* (Poster) • *GTV: Planetary Manager* (Videodisc) • *STV: Atmosphere* (Videodisc)	To order the following products for use with this chapter, call the National Geographic Society at 1-800-368-2728: • *Front Line of Discovery: Science on the Brink of Tomorrow* (Book) • *Atmosphere: On the Air* (Video) • *Investigating Global Warming* (Video) • *Ozone: Protecting the Invisible Shield* (Video)

Teacher Classroom Resources

These are key components of the classroom resources package.

Section Focus Transparencies

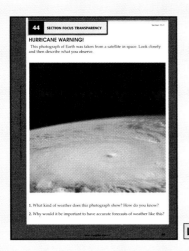

Teaching Transparencies

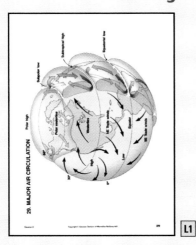

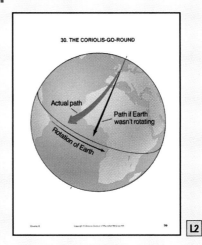

Science Discovery Activity*

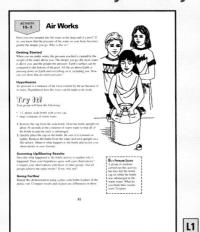

Laboratory Manual*

Take Home Activity

Chapter 15 The Air Around You

REVIEW AND REINFORCEMENT

Study Guide*

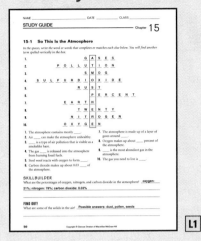

Concept Mapping

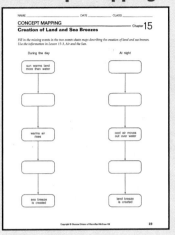

Critical Thinking/ Problem Solving

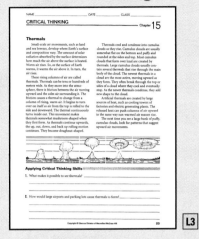

ENRICHMENT AND APPLICATION

Integrating Sciences

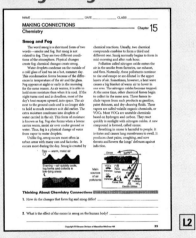

Across the Curriculum

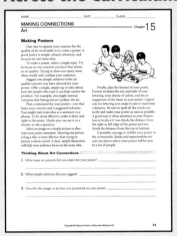

Technology and Society

Multicultural Connection**

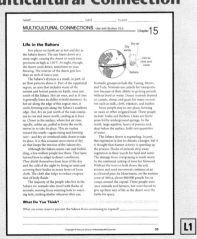

ASSESSMENT

Performance Assessment

Review and Assessment

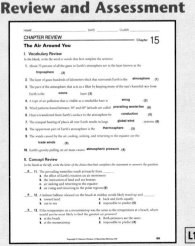

*One per section **Two per chapter

Chapter 15   The Air Around You   **454D**

CHAPTER 15

The Air Around You

THEME DEVELOPMENT

Earth's atmosphere is a system of gases that can be affected by human activities. The theme of systems and interactions becomes evident when students perform the Find Out activity on page 457.

CHAPTER OVERVIEW

The composition of Earth's atmosphere is presented first. The structure of the atmosphere is then described. Finally, the transfer of the sun's energy through the atmosphere and how this causes winds on Earth is described.

Tying to Previous Knowledge

It is important for students to realize that sunlight can be converted to thermal energy.

INTRODUCING THE CHAPTER

Before reading page 454, have students hypothesize why the hot-air balloons shown on the page remain afloat. *Accept all reasonable responses.*

Uncovering Preconceptions

Because it is invisible, many students think that air does not have any mass. Have students make fans out of paper and wave them near their faces. Ask students to describe what they feel. Help them conclude that if they feel the air, it must have mass.

THE AIR AROUND YOU

Did you ever wonder...

✓ **What the air is made of?**

✓ **Where Earth's air ends and outer space begins?**

✓ **Why the wind blows?**

Science Journal

Before you begin to study the air around you, think about these questions and answer them *in your Science Journal.* When you finish the chapter, compare your journal entry with what you've learned.

Answers are on page 479.

A rainbow of colors soars overhead. This is a balloon meet. It's your first, and you're quite excited.

"How do they stay afloat?" you wonder. "How can they turn or go up and down without any machinery or wings?"

You remember that last weekend the meet was scheduled but called off because of very high winds. Earlier today, you overheard two of the balloon pilots saying that the air was perfect for a great flight.

You know it's air that enables a balloon to soar through the sky. What properties of air can support the weight of the balloon and the pilot? What properties of air cause the balloon to change altitude or direction?

In this chapter, you will study the properties of air that enable balloons to fly as you learn about one part of Earth, its atmosphere.

▶ ***In the following activity, you'll begin your exploration of Earth's atmosphere and its properties.***

454

00:00 OUT OF TIME?

If time does not permit teaching the entire chapter, use the Chapter Overview on this page, Reviewing Main Ideas at the end of the chapter, and the Chapter 15 audiocassette to point out the main ideas of the chapter.

Learning Styles		
Kinesthetic	Explore, pp. 455, 464, 470; Activity, pp. 463, 475; Design Your Own Investigation, pp. 466-467; Find Out, p. 472; Multicultural Perspectives, p. 474
Visual-Spatial	Find Out, p. 457; Visual Learning, pp. 460, 463, 469, 471, 472, 474; Discussion, p. 461; Activity, pp. 465, 471; Across the Curriculum, p. 474
Interpersonal	Discussion, p. 456; Debate, p. 462; Across the Curriculum, p. 473; Project, p. 479
Logical-Mathematical	Investigate, pp. 458-459; Across the Curriculum, p. 469; Activity, p. 472
LS **Linguistic**	Life Science Connection, pp. 460-461; Discussion, pp. 461, 468; Across the Curriculum, p. 463; Multicultural Perspectives, p. 464; Research, p. 469; A Closer Look, pp. 470-471

Explore! ACTIVITY

Does air have mass?

What To Do

1. Use a pan balance to determine the mass of a completely deflated ball. Record your measurement *in your Journal.*

2. Use a bicycle pump to inflate the ball to its recommended maximum pressure. Predict what you think the mass of the inflated ball is.

3. Use the pan balance to determine the mass of the air-filled ball.

4. Were there any changes in mass? If so, what were they?

5. Finally, *in your Journal,* use your observations to explain any conclusions that you make.

ASSESSMENT PLANNER

PORTFOLIO
Refer to page 481 for suggested items that students might select for their portfolios.

PERFORMANCE ASSESSMENT
Process, pp. 457, 467
Skillbuilder, p. 461
Explore! Activities pp. 455, 464, 470
Find Out! Activities pp. 457, 472
Investigate, pp. 458–459, 466–467

CONTENT ASSESSMENT
Oral, pp. 464, 472
Check Your Understanding, pp. 461, 465, 475
Reviewing Main Ideas, p. 479
Chapter Review, pp. 480–481

GROUP ASSESSMENT
Opportunities for group assessment occur with Cooperative Learning Strategies.

PREPARATION

Planning the Lesson

Refer to the Chapter Organizer on pages 454A–D.

Concepts Developed

This section relates to Chapter 6, Thermal Energy, and Chapter 14, Gases, Atoms, and Molecules. Students will be able to describe the composition of the atmosphere and discuss the ways people affect air composition.

1 MOTIVATE

Bellringer

Before presenting the lesson, display **Section Focus Transparency 44** on an overhead projector. Assign the accompanying **Focus Activity** worksheet. L1

LEP

Discussion To reinforce communication skills, ask students if they think air pollution is a problem around the world. Many may think that the problem exists only in large cities or only in the United States. Inform them that some people living in other parts of the world have respiratory problems caused by sulfurous smog. Smog can be so bad that governments issue anti-pollution masks for people to wear. L2

2 TEACH

GLENCOE TECHNOLOGY

CD-ROM

Science Interactions, Course 2, CD-ROM

Chapter 15

15-1 So This Is the Atmosphere

Section Objectives
- Describe the composition of the atmosphere.
- Discuss ways people affect air composition.

Key Terms

atmosphere
smog

Figure 15-1

The atmosphere is the layer of gases hundreds of kilometers thick that surrounds Earth.

What's in the Atmosphere?

It's early morning, and you're getting ready for work. Before you leave, you decide to check the weather report coming over the computer screen. The smog is bad today. Once again, you will need to wear your air-filter mask. Pollution in the atmosphere has increased the temperature, and it could reach 104°F today. You will have to wear clothes designed to keep you cool. The ozone layer is thinner than it was when you were young, so you will have to use strong sunblock lotion to protect your skin from the sun. You sigh and remember attending a balloon rally when you were a teenager. It was on a beautiful day before the atmosphere became so polluted.

Could this be your future? It's one possible future you could face. Your life depends on the air you breathe and the condition of the atmosphere in which you live. The **atmosphere** is the layer of gases hundreds of kilometers thick that surrounds Earth.

Program Resources

Study Guide, p. 50
Laboratory Manual, pp. 91–92, Radiant Energy and Climate L2
Making Connections: Across the Curriculum, p. 33, Making Posters LEP
Section Focus Transparency 44

When you think of the atmosphere, you probably think about oxygen, wind, or the air you inhale or exhale. Yet the atmosphere contains more than just gases.

You may be surprised to learn that you breathe in more than you think! What's in the air other than the gases you need to stay alive? Let's find out.

Find Out! ACTIVITY

What solids are in the air around you?

What To Do

1. Smear thin layers of petroleum jelly onto four plastic lids.

2. Place the lids in four outdoor locations around your home and school.

3. After one week, collect the lids. Be sure to mark each one to indicate the location from which it came.

4. Examine each lid with a magnifying glass. *In your Journal,* record the materials you can identify.

5. Sort the solids—large pieces of dust, plant pieces, seeds, insect parts, and so on—taken from each sample site.

6. Place the materials on microscope slides and examine each slide with a microscope.

Conclude and Apply

1. Which of the solids collected could have been a result of human activity?

2. If you collected seeds, do any of them suggest how some plants disperse their seeds?

3. Do you think any of the materials might be harmful to people? Explain your answer. (HINT: Think about people you may know who have allergies.)

From the activity, you may have discovered that some things are more abundant than others in the air. Together, gases, solids, and liquids make up the atmosphere we live in. Each time you take a breath, you breathe in a mixture of these substances. Yet, you're probably most concerned with the oxygen compo-

nent of Earth's atmosphere. After all, you know that you need oxygen to stay alive. But how much of the air is made of oxygen? It may be less than you think. Other important gas components found in the atmosphere include nitrogen, methane, and carbon dioxide. In the following Investigate, you'll find the answer.

15-1   So This Is the Atmosphere   **457**

Tying to Previous Knowledge

Ask students to describe what happens when they leave a steel object outside for a long period of time. They should know that steel rusts when exposed to moisture in the air. Explain that rust is oxidized iron. Show students a sample of red hematite and a rusty iron nail. Point out that it is the oxygen that causes iron to rust. Have students compare the two samples you are holding.

Theme Connection The theme of scale and structure is apparent in this section as students study the composition of the atmosphere. They will find out the amount of oxygen and other gases in the atmosphere.

clude and Apply in their journals. L1

Expected Outcome

Students will observe the particulate matter in the air from natural and human activities.

Conclude and Apply

1. Possible answers include paper particles, wood or soot, metal shavings.

2. by moving air or wind

3. Wood, soot, and pollens could be unhealthy if they are inhaled.

✔ Assessment

Process Have groups of two or three students conduct a survey about the effects of air pollutants on humans. Ask groups to generate five questions to ask people who suffer from allergies or diseases caused by air pollutants. Students should survey at least 15 people, tally their results, and describe any trends they find in the data. Use the Performance Task Assessment List for Surveys and Graph of Results in **PASC**, p. 35. **COOP LEARN**

Find Out!

What solids are in the air around you?

Time needed   10 minutes setup time, 1 week wait time, 50 minutes follow-up time

Materials   4 plastic lids, permanent marker or grease pencil, petroleum jelly, magnifying glass, microscope slides, microscope

Thinking Processes   organizing information, classifying, observing and inferring

Purpose   To collect, examine, and classify par-

ticulate matter that has settled out of the air.

Teaching the Activity

Have students predict the solids they might find based on human activities in the area. Have a volunteer record students' predictions on the chalkboard for comparison after they have completed the activity. **COOP LEARN** L1

Science Journal   Suggest that students make a chart to help them organize their findings. Have students record their responses to Con-

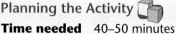

15-1 How Much Oxygen Is in the Air?

Planning the Activity

Time needed 40–50 minutes

Purpose This activity will show how much oxygen is in the air.

Process Skills organizing information, making and using tables, observing and inferring, measuring in SI, practicing scientific method, separating and controlling variables, analyzing data, predicting

Materials See student text page.

Preparation Be sure all materials are available.

Teaching the Activity

Review percentage calculations with the class.

Process Reinforcement Be sure students set up their data tables correctly, as they will be using their data for later calculations. If students are having difficulty, go over the mechanics of using tables with them.

Discussion Tell students that oxygen in the air is very reactive with other substances. The vinegar cleans the steel wool of impurities such as oils, exposing the raw iron (steel) to oxygen.

Activity Have students graph the results of this activity, plotting water level in millimeters on the vertical axis and the time on the horizontal axis. L2

Science Journal In addition to recording their data in their journals, help students write a paragraph describing the reaction that occurred in the test tube. *Oxygen is a very reactive element. The vinegar removes any substances on the steel wool, thus promoting oxidation.* L1

INVESTIGATE!

How Much Oxygen Is in the Air?

Steel wool, which is mostly iron, reacts with the oxygen in air. The steel wool will combine with oxygen in a test tube to form rust. This reaction will continue until all of the oxygen has been used. In this activity, you will determine how much oxygen was used, and calculate the percentage of oxygen in the air.

Problem

How can the amount of oxygen in the air be measured?

Safety Precautions

Materials

measuring cup with mL gradations	2 pencils	test-tube clamp	2 rubber bands
	tongs	test tube	metric ruler
	white vinegar	paper towels	beaker
scissors	test-tube stand	water	steel wool

What To Do

1 Copy the data table *into your Journal.* Then, measure the length of the test tube in millimeters, and record it *in your Journal* as well.

2 Mix 30 mL white vinegar with 20 mL of water in the beaker.

3 Unroll a bale of steel wool. Cut a strip that is 2 cm wide and 20 cm long (see photo **A**). Soak the steel wool in the vinegar solution for 1 minute.

4 Using tongs, remove the steel wool from the vinegar solution. Use the tongs to stretch out the strip of steel wool, and dry it thoroughly between two paper towels.

Meeting Individual Needs

Physically Challenged Some students with physical limitations may not be able to manipulate the equipment used in this activity. Allow those students to observe and record the information gathered during this experiment.

A   **B**

5 Pour out the vinegar solution, rinse the beaker, and fill it about 2/3 full of water.

6 Using two pencils, push the steel wool into the bottom 2/3 of the test tube, keeping it as loose as possible (see photo **B**).

7 Use rubber bands to attach the ruler to the test tube. Position the tube such that its open end is at about the 0 mm mark.

8 Turn over the test tube and ruler and insert it into the beaker so the opening of the tube is just below the surface of the water. Then, attach the tube to the stand, using the clamp. Readjust the ruler so that the 0 mm mark is at the water line.

9 Observe and record the level of the water in the test tube, in millimeters, every 2 minutes until the water level stops changing.

Sample data

Data and Observations

Time Elapsed (Minutes)	Water Level (mm)
2	4
4	5
6	8
8	10

Analyzing

1. Use this formula to calculate the percentage of oxygen in the air.

$$\frac{\text{final water level (mm)}}{\text{tube length (mm)}} \times 100 = \% \text{ oxygen}$$

2. About what percentage of the air in the test tube was oxygen?

Concluding and Applying

3. Why was it important to stretch the steel wool and pack it loosely before inserting it in the water?

4. Based on your observation, do you think the air is mostly oxygen?

5. **Going Further** What do you predict will happen to the steel wool after you remove it from the test tube and expose it to air?

Expected Outcomes

Students will realize that air is made partly of oxygen and that what they observed was a chemical reaction that used about 21 percent of the gas trapped in the inverted test tube.

Answers to Analyzing/ Concluding and Applying

2. about 21 percent

3. to increase the surface area available for the reaction to take place

4. No; air is mostly some other gas (nitrogen).

5. It will continue to rust by reacting with the air.

✓ Assessment

Performance Have students, working in small groups, design and carry out an experiment to determine how other metals react with oxygen. For example, students might compare and contrast how steel, copper, and aluminum react with oxygen by putting objects made from these substances outside for several weeks and noting any changes in the metals. Have students write a brief report detailing their procedures, results, and conclusions. Select the appropriate Performance Task Assessment List in **PASC.** COOP LEARN P

Program Resources

Activity Masters, pp. 63–64, Investigate 15-1

Making Connections: Technology & Society, p. 33, Lichens—Tiny Pollution Labs L2

Science Discovery Activities, 15-1

Gases in Our Atmosphere

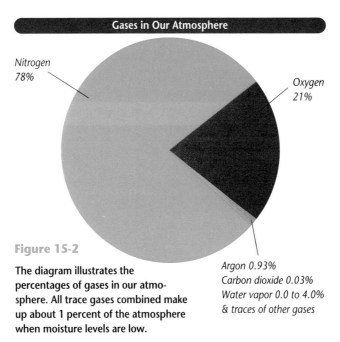

Nitrogen 78%

Oxygen 21%

Argon 0.93%
Carbon dioxide 0.03%
Water vapor 0.0 to 4.0%
& traces of other gases

Figure 15-2

The diagram illustrates the percentages of gases in our atmosphere. All trace gases combined make up about 1 percent of the atmosphere when moisture levels are low.

From the Investigate activity, you found that oxygen makes up about 20 percent of the gases in the test tube. The gases in the tube are the same as the gases in the atmosphere. Therefore, you've shown that the atmosphere is only about 20 percent oxygen. What gases make up the other 80 percent?

Figure 15-2 shows the percentages of gases that occur naturally in the atmosphere. However, human activities can increase the amount of certain gases while decreasing others.

Burning fossil fuels releases gases into the atmosphere. Sulfur dioxide and nitrous oxides are among them. When these gases build up in the

Life Science CONNECTION

Smog and Our Health

Athletes who train in big cities have to get up early—often before 5 A.M.—for their daily run. Later on, smog levels are too high. Breathing in dirty air during such strenuous activity would be dangerous for the athletes' bodies, as smog is harmful to the heart as well as the lungs.

Too Much Ozone

What exactly makes the air dirty? Two of the worst gases we can breathe in are ozone and carbon monoxide. The ozone layer high up in the atmosphere is crucial to our existence. Ozone, formed from pollutants released into the air we breathe, is a serious pollutant. Breathing in ozone is irritating, causing a burning sensation in your nose and throat. It can give you a headache, make your eyes sting and, if levels are high enough, blur your vision. Ozone can also damage lung tissue, reducing your ability to fight infection, and lead to diseases like pneumonia, chest colds, and bronchitis.

atmosphere, they can result in smog. **Smog** is a type of air pollution that is visible as a smokelike haze. Breathing smog is unhealthy for everyone. Smog can irritate your eyes and damage the tissues of your lungs, making them more susceptible to disease. Smog is also harmful to plants, because it prevents plants from absorbing the carbon dioxide they need.

SKILLBUILDER

Making and Using Graphs

Use the graph in **Figure 15-2** to help you answer the following questions. If you need help, refer to the **Skill Handbook** on page 640.
1. What is the most abundant gas in the atmosphere? What percentage of the total volume of gases is it?
2. When added together, what percentage of the total volume of gases are nitrogen, oxygen, and carbon dioxide?

check your UNDERSTANDING
1. Which gas in the atmosphere is most important to people? Why?
2. Why can driving a car be harmful to the atmosphere?
3. **Apply** Why do you think smog is more likely to form over cities than rural areas?

Cars and Smog

Carbon monoxide, a gas produced when gasoline is burned in a car engine, is absorbed by red blood cells when we breathe in. Red blood cells are supposed to be absorbing oxygen, and are less efficient at this when carbon monoxide gets in the way, so the body receives less oxygen than it needs. Decreased oxygen levels can cause heart trouble and chest pains. This is why, on days when pollution levels are extremely high, the elderly, young children, and people with heart or chest problems are warned to stay indoors and rest.

Low levels of oxygen in the brain can impair coordination

and motor functions. When we exercise, we need even more oxygen than usual to keep our bodies functioning. That's why it's dangerous to exercise when smog levels are high.

interNET CONNECTION

Use the World Wide Web to find out about daily pollution and weather conditions in the San Francisco Bay area for a six-month period. What are some safer transportation alternatives if air quality is unhealthy on a particular day?

15-1 So This Is the Atmosphere **461**

SKILLBUILDER
1. nitrogen; 78 percent
2. 99.03 percent [L2] LEP

GLENCOE TECHNOLOGY

 Software
Computer Competency Activities
Chapter 15

3 ASSESS

Check for Understanding

Prior to answering the Apply question, have students compare and contrast characteristics of cities. They can then pick out those characteristics that could contribute to smog.

Reteach

Discussion Tell students that a jar contains a different color sand for each gas in the atmosphere. Ask them if the sand would appear layered or multicolored. Have students explain why. *Gases are mixed in the atmosphere.* Have volunteers make such a model of the gases in the atmosphere. [L1]

Extension

Research Have students research the impact of air pollution on their town or city. Direct students to list the major sources of pollution in the region and brainstorm possible solutions. [L3]

---

a small piece of aluminum foil and insert it in the center of the terrarium, being careful not to let the stick touch water. Cover the terrarium with a glass or plastic lid. Light the incense and let it burn for a few minutes every hour for a week. Have students observe and record the general appearance of the plants in their journals each day for up to one month. [L2]

check your UNDERSTANDING

1. oxygen, for life processes
2. Burning gasoline produces harmful smog-producing gases.
3. Cities have more people and factories, so human activities that require burning fuels are greater.

4 CLOSE

Discussion

Have students discuss the roles of individuals, industry, and government in improving air quality. Create a class chart from students' responses. [L1]

PREPARATION

Planning the Lesson

Refer to the Chapter Organizer on pages 454A–D.

Concepts Developed

Section 15-2 deals with the structure of the atmosphere and the cause of atmospheric pressure. Connections can be made to Chapter 1, Forces and Pressure, and Chapter 14, Gases, Atoms, and Molecules.

1 MOTIVATE

Bellringer

 Before presenting the lesson, display **Section Focus Transparency 45** on an overhead projector. Assign the accompanying **Focus Activity** worksheet. L1

LEP

Debate Explain to students that the ozone layer protects people from the harmful effects of ultraviolet rays from the sun. These rays cause skin to burn when exposed to the sun. Point out that the gas used in air conditioners (Freon) sometimes escapes and destroys ozone. Have students debate the human consequences and possible remedies of ozone depletion. L1

2 TEACH

Tying to Previous Knowledge

Ask students if they ever suffered from an earache when riding in an airplane or felt fullness in their ears when going up a mountain or in an elevator. Point out that as the altitude increases, atmospheric pressure decreases, creating an imbalance between the inside and outside of the eardrum. Yawning helps to equalize the pressure.

Section Objectives

- Describe the structure of Earth's atmosphere.
- Explain what causes atmospheric pressure.

Key Terms

troposphere
ozone

Figure 15-3

Almost all of the clouds, from fluffy, white puffs to gray sheets of storm clouds, appear in the troposphere.

Layers of the Atmosphere

Return for a moment to your future. After breakfast, you leave your home to travel downtown. As you wait with others for public transportation, you join in a discussion about the atmosphere's ozone. An outer layer of ozone shields Earth's life-forms from the sun's harmful rays. You recall that the loss of ozone protection was already a concern when you were a child. Air pollution from human activities in the last half of the 1900s was at the heart of the concern. But what exactly is the ozone and where is it found in the atmosphere?

The atmosphere changes as you move away from Earth's surface. Some layers contain gases that easily absorb energy from the sun, while others do not. As a result, some layers are warmer than others. Based on the temperature differences, we can divide the atmosphere into five layers.

■ Troposphere

You live in the **troposphere**, the layer closest to the ground. The troposphere contains 75 percent of all the gases in the atmosphere as well as dust, ice, and liquid water. Raging thunderstorms, sizzling heat, numbing cold, and all other kinds of weather occur in this layer.

■ Stratosphere

Just above the troposphere is the stratosphere. It is within the stratosphere that the ozone layer is found. Use **Figure 15-4** to locate the ozone layer in the atmosphere. About how far above Earth's surface does it lie? **Ozone** is a gas that absorbs some of the harmful radiation from the sun. As a layer in the atmosphere, ozone acts much like a filter. The ozone layer filters many of the sun's harmful rays, keeping them from reaching the troposphere. Thinning areas and holes in the ozone layer let harmful rays pass on through.

■ Mesosphere and Thermosphere

Beyond the stratosphere are the mesosphere, the thermosphere, and the exosphere. The mesosphere and the thermosphere are useful for radio transmission. The exosphere is the uppermost part of Earth's atmosphere. Beyond it is outer space. Unlike states or countries with defined boundaries, the upper part of the atmosphere has no special ending point. The air just becomes less and less dense until there is no air at all. Where there is no atmosphere at all, outer space begins.

462 Chapter 15 The Air Around You

Program Resources

Study Guide, p. 51
Laboratory Manual, pp. 93–96, Smoke Pollution L1
Take Home Activities, p. 23, Air Pressure in Action L1
Section Focus Transparency 45

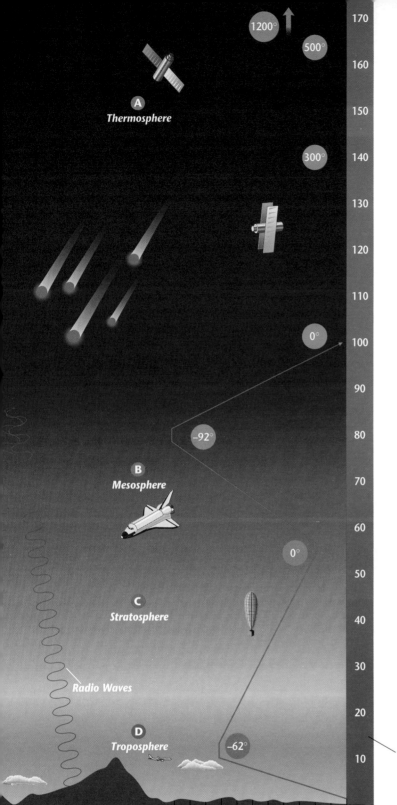

Figure 15-4

The division of Earth's atmosphere into layers is based on temperature differences.

A *Thermosphere* Only a small portion of the thermosphere is shown. This layer extends beyond 500 km in altitude. Beyond the thermosphere is the exosphere, and beyond that is outer space. Temperatures in the thermosphere and exosphere increase rapidly, and may reach 1200° C.

B *Mesosphere* Obstacles on the Earth's surface block radio waves and make long distance transmission at this level impossible. The solution has been to transmit in the obstacle-free mesosphere. The boundary between the mesosphere and thermosphere, called the ionosphere, reflects radio waves back to Earth's surface.

C *Stratosphere* Earth's protective layer of ozone lies within the stratosphere. Weather balloons are flown here and some high-flying jet aircraft travel within the lower stratosphere.

D *Troposphere* The troposphere is the lowest layer of Earth's atmosphere. This is the layer in which we live. It is here where clouds form and weather occurs.

Altitude (km)

463

Visual Learning

Figure 15-4 To compare and contrast the layers of Earth's atmosphere, have students study the figure. Stress that each layer of the atmosphere has unique characteristics. Ask students to use the figure to determine the altitude of the troposphere (*up to about 15 km above Earth's surface*) and locate the ozone layer (*in the stratosphere, about 40 km above Earth's surface*). Point out that as a component of the stratosphere, the ozone layer is several kilometers thick. Students may incorrectly infer from Figure 15-4 that there is an abundance of ozone molecules because the molecules are widely dispersed throughout the stratosphere. Explain that if all the ozone molecules were brought down to sea level, where pressure is greater, the ozone layer would be less than 3 mm thick.

NATIONAL GEOGRAPHIC SOCIETY

Videodisc

STV: Atmosphere

Layers of atmosphere; artwork

38686

Ozone layer; artwork

38687

ENRICHMENT

Activity Have students show that gases can exert pressure. Pairs of students need a flask with a two-hole stopper to fit, glass tubing, and water.

Have students bend a 15-cm glass tube 90° and let it cool. Insert another glass tube into one hole of the two-hole stopper. Push it through the hole so that when the stopper is placed on the flask, the tube comes to within 0.5 cm of the bottom. Insert the bent tube in the other hole and push it through the same as the first tube. Fill the flask with water and insert the stopper.

Tell students to slowly blow through the bent tube until most of the water is gone. Students should conclude that the gas replaced the water because the gas from their lungs had greater pressure. L3

Is atmospheric pressure the same in all layers of the atmosphere?

Time needed 15–20 minutes

Materials modeling clay, waxed paper, 4 textbooks

Thinking Processes observing and inferring, comparing and contrasting, making models

Purpose To observe the effects of varying amounts of pressure on a substance.

Preparation Be sure the clay is pliable. Use an ice cream scoop to measure equal amounts of clay. If you are using plasticine clay, be sure students know to work it with their hands to soften it. If you are using wet clay, be sure to keep it moist. Since four textbooks are needed, have groups of four students work together. COOP LEARN

Teaching the Activity

Troubleshooting Put a piece of the waxed paper on desktops to make cleanup easier. Observe to make sure students are not pressing down on the books. L1

Science Journal After students have read the procedure, have them predict what might happen to the balls of clay and record these predictions in their journals. L1

Expected Outcome

Students will observe that the layers of clay get successively thinner toward the bottom of the pile because of the additional pressure.

✔ Assessment

Oral/Content Ask students to identify and label the layer of the atmosphere represented by each book in their models. Use the Performance Task Assessment List for Model in **PASC**, p. 51.

Atmospheric Pressure

Why do you suppose clouds are found only near Earth's surface, in the troposphere? Clouds are made of water, which has mass. You know that gravity pulls anything with mass toward Earth's surface. So, you would expect to find most of the water near Earth, rather than in the mesosphere or thermosphere.

In the Explore activity on page 455, you discovered that air also has mass. Earth's gravity pulls the air toward the ground the same as it does water, a stone, or you. Air has weight too, and this weight causes air molecules to push together, producing pressure on each other. We call this air pressure.

Explore! ACTIVITY

Is atmospheric pressure the same in all layers of the atmosphere?

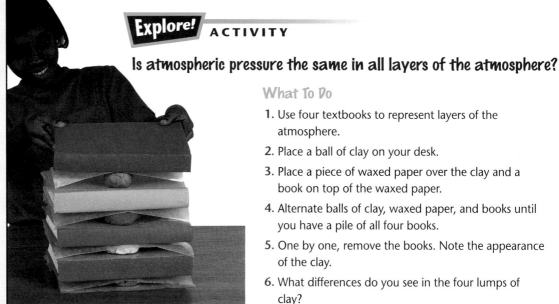

What To Do

1. Use four textbooks to represent layers of the atmosphere.
2. Place a ball of clay on your desk.
3. Place a piece of waxed paper over the clay and a book on top of the waxed paper.
4. Alternate balls of clay, waxed paper, and books until you have a pile of all four books.
5. One by one, remove the books. Note the appearance of the clay.
6. What differences do you see in the four lumps of clay?
7. How would you account for these differences?

Think of your desktop as Earth's surface. The closer the air was to the surface, the more compressed the air became under the weight of the layers pushing down from above. On Earth, air pressure tends to be greatest at sea level. There is more air pressing down from above at sea level than on a mountaintop.

Air pressure doesn't change only if you go up or down in altitude. It can change from hour to hour, right where you live.

You've probably heard weather forecasters use the terms "high-pressure system" or "low-pressure system." You've seen these marked on weather maps as "H" or "L."

464 Chapter 15 The Air Around You

Meeting Individual Needs

Gifted Tell students that before ozone was present in the atmosphere, much ultraviolet radiation reached Earth's surface. If life was present, it would have had to exist deep in the ocean where it was protected from the harsh ultraviolet radiation. Have teams debate whether life would have evolved in the same way if ozone had never formed in the atmosphere. COOP LEARN L3

Multicultural Perspectives

Native American Legends

Interested students can read some of the Native American legends about Earth's air. (Michael J. Caduto and Joseph Bruchac. *Keepers of the Earth*. Golden, CO: Fulcrum, Inc., 1989) Have students paraphrase the stories and discuss their meanings with the rest of the class. L2

Figure 15-5

Scientists and weather forecasters use a type of barometer called an aneroid barograph to measure and record changes in air pressure. A change in air pressure usually indicates a coming change in weather, as seen in the bottom left photo. By analyzing the changes in air pressure, weather forecasters can predict what the weather will be.

A The measuring part of an aneroid barograph (a type of barometer) is a metal chamber from which almost all of the air has been removed. When air pressure goes up, the chamber contracts. When air pressure goes down, the chamber expands.

B Levers transfer the movement of the chamber to a pen—the recording part of the instrument. The pen moves up and down as the chamber moves in and out.

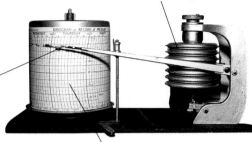

C The pen's up-and-down movements draw a line on a paper attached to a very slowly turning cylinder. One complete turn of the cylinder might record pressure changes for a week.

The "L" marks the center of a large mass of air with low pressure. Often, a low-pressure system moving into your area means clouds, precipitation, and storms. A high-pressure system often means clear weather.

But how do the weather forecasters on television know whether the air pressure is going up or down? As you can see in the figure above, an instrument called a barometer can be used to measure air pressure. The activity on pages 466-467 will help you understand how a barometer works.

check your
UNDERSTANDING

1. Tell which layer has each of these characteristics—(a) a raging thunderstorm, (b) a blurred boundary with outer space, (c) a protective ozone layer.

2. What is the pushing together of air molecules called?

3. **Apply** Would you expect air pressure to be greater in the thermosphere than in the stratosphere? Why?

15-2 Structure of the Atmosphere **465**

check your
UNDERSTANDING

1. **a.** troposphere; **b.** exosphere; **c.** stratosphere

2. air pressure

3. No; the weight of upper layers on the stratosphere would make its pressure greater than that in the thermosphere.

Check for Understanding

To help answer the Apply question, have students explain why air exerts pressure. *Atoms and molecules in the atmosphere are pushed together because of the mass of the air above them. Because the molecules become more densely packed, they collide more often. These collisions cause air to exert pressure.*

Reteach

Activity To demonstrate that air exerts pressure, obtain a jar large enough that a student can stick his or her hand into it easily. Also obtain a plastic bag with an opening large enough to fit over the rim of the jar. Secure the plastic bag to the rim of the jar. Be sure the bag fits tightly. Have a student make a fist and try to push the plastic bag gently into the jar. Air in the jar will resist the student's attempts to push on the bag. [L1]

Extension

Activity Have students who have mastered the concepts of this section put a barometer inside a large, sturdy, clear plastic bag. Tell them to inflate the bag with a bicycle pump and watch the reaction of the barometer. Have students record their observations and draw conclusions. [L3]

4 CLOSE

Display a barometer in the classroom. Have students hypothesize what would happen to the barometer's readings if it were placed on a different floor of the school or on a higher floor in a large office building. Provide students with a chance to observe the barometer each day and observe pressure differences. [L1]

15-2 Atmospheric Pressure

Preparation

Purpose To see how a barometer changes with the weather.

Process Skills making and using tables, thinking critically, observing and inferring, measuring in SI, making models

Time Required 45 minutes and then a few minutes each day to check and record barometric measurements

Materials See reduced student text.

Possible Hypotheses Students should predict that a change in pressure makes the barometer react and that high pressure brings clear weather while low pressure brings clouds and precipitation.

The Experiment

Troubleshooting Remind students that once the can is sealed, it should not be reopened. If the barometer is constructed during long periods of high pressure or low pressure, the indicator straw may not move up or down.

Process Reinforcement Have students compare their data with a barometer in the classroom. If you do not have a barometer, have students obtain local weather information and make a note of the barometric pressure.

Possible Procedures Stretch the balloon over the can and anchor it with a rubber band. Make a gauge by attaching the drinking straw to the balloon and attaching a strip of paper to the side of the can. Be sure to mark where the straw indicates on the strip and above the mark write *high* and below it *low*.

Teaching Strategies

Troubleshooting Make certain the straw does not touch the strip of paper. It should be free to move up and down.

Atmospheric Pressure

When high- and low-pressure systems, as shown on TV weather maps, move into your area, do you notice the increase or decrease in pressure? You probably don't. Our bodies are not sensitive enough to detect small changes in air pressure. But you can construct a barometer that detects air-pressure changes.

Preparation

Problem
How does a barometer change with the weather?

Form a Hypothesis
As a group, hypothesize how a barometer will react to high- and low-pressure systems, and predict what kind of weather will accompany changes in the barometer.

Objectives
• Make a barometer.
• Observe changes in atmospheric pressure.

Materials
small coffee can
scissors
transparent tape
large rubber balloon
construction paper
rubber band
drinking straw
12 bean seeds

Safety Precautions

Be careful when using scissors.

Program Resources

Activity Masters, pp. 65–66, Design Your Own Investigation 15-2
Science Discovery Activities, 15-2

Meeting Individual Needs

Limited English Proficiency Students whose second language is English will benefit from a demonstration of how to construct their instruments as a volunteer reads aloud the procedures used in this Investigation. **LEP**

DESIGN YOUR OWN
INVESTIGATION

Plan the Experiment

1 With the help of the photos and the information on pages 462-465, construct a barometer with the materials provided.

2 A barometer needs a pointer and a gauge to show the changes in air pressure. As you construct your barometer, think about what material will be most sensitive to changes in air pressure.

3 Review your hypothesis and write a plan to test your hypothesis using the barometer.

4 Design a data table to record the barometer readings. For how many days will you record information about changes in air pressure?

5 Information about air pressure and weather is available from TV and newspapers. How can this help your investigation?

Check the Plan

1 Where will you keep the barometer?

2 Will you record your information at the same time every day?

3 How will you use the information from TV or newspapers?

4 Make certain that your teacher approves your plan before you begin.

5 Carry out your investigation. Record your observations.

Analyze and Conclude

1. Explain how your barometer works.

2. **Compare and Contrast** If the atmospheric pressure changed during your investigaton, what weather occurred on the days of change? Did this support your hypothesis?

3. **Observe** What was the reading on your barometer on cloudy, rainy, or snowy days? What was the reading on sunny, clear days?

4. If you made a graph of the changes shown by your barometer, what would it look like? Would it be a straight line or something else? Use the barometric readings from the TV or newspaper to construct an actual graph.

5. **Hypothesize** What might happen if your barometer were placed on a mountaintop? What might happen if it were held in the stratosphere? Explain.

Going Further

If the temperatures at sea level and on a mountaintop are the same, will the air pressure be the same or different? Explain.

Discussion Review with students the idea of atmospheric pressure using the photo on page 464. Then discuss that heavier air creates more force that the sensitive balloon will record.

Science Journal Have students prepare data tables in their journals. Have them prepare enough room to record the data from their barometers for two weeks or however long this experiment goes on. **L1**

Analyze and Conclude

1. The barometer works on the sensitivity of the balloon to changes in the force pushing down on it. Air presses down on the balloon on the can. When air pressure is high, the balloon is depressed and the straw points toward high. When air pressure is low, less pressure is applied to the balloon and the straw points to low.

2. Answers will vary but low pressure should be followed by clouds and precipitation. High pressure brings clear days.

3. Rainy days should have a low pressure reading. Sunny, clear days should have a high pressure reading.

4. This will depend on the weather but the graph should be constructed with barometric readings on the vertical axis and days of the experiment on the horizontal axis.

5. The pressures will be lower both on the mountaintop and in the stratosphere. The air gets less dense as altitude increases.

✔ Assessment

Process Have students observe their barometers in an air-conditioned room for a period of time, graph their data, and then write a brief explanation of their results. There should be no variation in the readings. Use the Performance Task Assessment List for Written Summary of a Graph in **PASC**, p. 41. **P**

L1 **COOP LEARN**

Going Further ⫸

Air pressure would be greater at sea level. At like temperatures, atmospheric pressure decreases as altitude increases.

15-3

The Air and the Sun

PREPARATION

Planning the Lesson

Refer to the Chapter Organizer on pages 454A–D.

Concepts Developed

This section develops the concept that air exerts pressure. The differences in pressure cause wind. Students will learn how wind is generated and how moving air forms wind systems.

1 MOTIVATE

Bellringer

 Before presenting the lesson, display **Section Focus Transparency 46** on an overhead projector. Assign the accompanying **Focus Activity** worksheet. L1

LEP

Discussion
Ask students the following questions. **Why does a metal spoon get hot if you leave it in a pot on top of a hot burner? Why does it get hot inside a closed car on a hot summer day? Why is the attic very warm in the springtime while the ground floor is cool?** Answers will lead to a discussion of energy transfer. Inform students that they will learn that the answers to the questions are conduction, radiation, and convection, respectively.

2 TEACH

Tying to Previous Knowledge

All students have seen and/or felt the effects of the wind. Ask them to share any personal experiences with wind.

Section Objectives
■ Explain what causes wind.
■ Describe Earth's wind systems.

Key Terms
trade winds
prevailing westerlies

Energy from the Sun

The future workday ends, and you return to your home. You place your dinner in the microwave oven. While waiting for your dinner to cook, you flick on the news scanner and request to see the news about the planetary atmospheric program. The space agency runs the program that is studying the atmospheres of Venus and Mars. The scientists hope to find solutions to Earth's atmospheric problems through the study. The atmosphere on Mars is too thin to support life. As **Figure 15-6** shows, Mars can't hold much of the energy that radiates from the sun. As a result, Mars is a very cold, lifeless planet. Venus is a very hot, lifeless planet. Its atmosphere is so dense that most of the energy coming from the sun changes to heat, is trapped in the atmosphere, and can't escape. On Earth, there's a delicate balance between the amount of radiation that's trapped and the amount that escapes.

Earth's atmosphere and the sun interact to provide an environment that can support life. How do they do this? The sun is the source of most energy on Earth. Energy is transferred from the sun to Earth and heat is transferred from one part of Earth to another.

Figure 15-6

The sun radiates energy to the planets revolving around it. How that radiated energy affects a specific planet depends on that planet's atmosphere.

Ⓐ Most radiation entering Venus's atmosphere is trapped by thick gases and clouds.

Program Resources

Study Guide, p. 52
Laboratory Manual, pp. 97–98, Temperature of the Air L2
Concept Mapping, p. 23
Critical Thinking/Problem Solving, p. 23, Thermals L3
Multicultural Connections, p. 34, El Niño L1 ; p. 33, Life in the Sahara L1

Making Connections, Integrating Sciences, p. 33, Smog and Fog L2
Science Discovery Activities, 15-3
Teaching Transparency 29 L1
Section Focus Transparency 46

You may recall that there are three ways to transfer heat from one object to another—radiation, conduction, and convection. These three processes are also at work in the atmosphere.

■ Heat Transfer Through Radiation

Radiation from the sun travels through space on its way to Earth. As you can see in **Figure 15-7**, some of this radiation is reflected back out into space. However, about 70 percent of it is absorbed.

The radiation from the sun is trapped in Earth's atmosphere in much the same way as heat is trapped in a greenhouse. You can demonstrate this warming effect by doing the next Explore activity.

Figure 15-7

Ⓐ The sun is the source of most energy on Earth.

Ⓑ Three different things happen to the energy Earth receives from the sun. Some energy is reflected back into space, some is absorbed by the atmosphere, and some is absorbed by land and water surfaces.

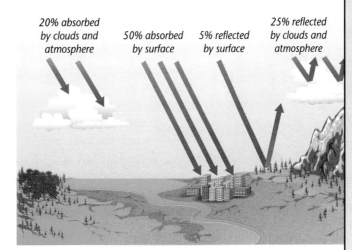

20% absorbed by clouds and atmosphere

50% absorbed by surface

5% reflected by surface

25% reflected by clouds and atmosphere

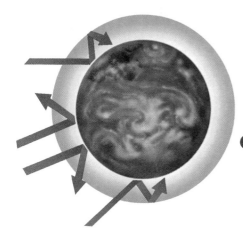

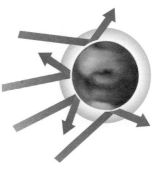

Ⓑ Earth's atmosphere creates a delicate balance between energy received and energy lost.

Ⓒ On Mars, a thin atmosphere allows much radiation to escape.

15-3 The Air and the Sun   **469**

Theme Connection The theme that this section supports is energy. Energy in the atmosphere is constantly being moved from one place to another. The net effect of the transfer of thermal energy in the air is wind.

Visual Learning

Figure 15-7 Use the following questions to help students understand what happens when solar energy reaches Earth. **How much of the sun's energy is absorbed by Earth's surface?** *50 percent* **How much of the solar energy is reflected back into space?** *30 percent* **How much is absorbed by Earth's atmosphere?** *20 percent*

Across the Curriculum

Math

Earth is about 150 000 000 km from the sun. The radiation coming from the sun travels at 300 000 km/second. Have students determine about how long it takes for the radiation leaving the sun to reach Earth.

$$\frac{150\ 000\ 000\ km}{300\ 000\ km/s} = 500\ s\ or\ 8.3\ min$$

GLENCOE TECHNOLOGY

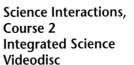 **Videodisc**

Science Interactions, Course 2 Integrated Science Videodisc

Lesson 4

STVS: Physics

Disc 1, Side 2
Wind Engineering (Ch. 7)

How does a greenhouse trap heat?

Time needed
30–40 minutes

Materials clear plastic box with clear lid, soil, stiff cardboard, thermometer, heat lamp

Thinking Processes recognizing cause and effect, representing and applying data, observing and inferring, making models

Purpose To recognize the cause and effect of the heat buildup in a model greenhouse.

Teaching the Activity

Have the students work in pairs. L1 COOP LEARN

Expected Outcome

Students will learn that radiation can enter a system, be converted to heat, and get trapped.

Answer to Question

The temperature was higher in the closed box.

✔ Assessment

Performance Have students design and carry out an experiment in which they measure the temperature of the soil and an equal amount of water every five minutes when the soil and water are exposed to sunlight. Use the Performance Task Assessment List for Designing an Experiment in **PASC**, p. 23.

Explore! ACTIVITY

How does a greenhouse trap heat?

To demonstrate how heat is trapped in Earth's atmosphere, you will need a clear plastic storage box and lid.

What To Do

1. Fill the bottom of the storage box with about 3 cm of soil.
2. Insert a stiff piece of cardboard into the soil at about half the length of the box.
3. Use the cardboard as a prop for a thermometer. The bulb of the thermometer should be facing up.
4. Overhead, about 30 cm from the box, place a heat lamp.
5. Do not cover the box with the lid. Turn on the heat lamp and wait 10 minutes.
6. Turn off the heat lamp and read the thermometer. Record the temperature *in your Journal.*
7. Allow the box to return to room temperature. Then, put the lid on it.
8. Turn on the heat lamp and once again wait 10 minutes.
9. Turn off the heat lamp and read the thermometer again. Record the temperature. In which situation was the temperature higher?

Giant tree-fern from the Atlantic rain forest in southern Brazil

Global Warming and the Greenhouse Effect

The greenhouse effect, or the trapping of heat by our atmosphere, has become a serious environmental concern. Some people fear that average temperatures on Earth are rising at an abnormal rate.

It's normal for Earth's average temperatures to rise and fall over time. However, some believe that humans may be rapidly changing the temperatures. Such global warming might melt enough of the polar ice to raise sea levels everywhere. Weather patterns also might shift.

A Greenhouse Without the Green

Carbon dioxide is one of the main gases that causes the greenhouse effect. Pollution caused by human activities adds carbon dioxide to the atmosphere. The worldwide loss of trees and plants has an effect because trees and plants remove carbon dioxide from the air and give off oxygen. Destruction of forest lands has a double impact—less carbon dioxide is taken out of the

470

Purpose

A Closer Look expands the discussion of the greenhouse effect by showing a relationship between increases in carbon dioxide levels in the atmosphere and temperatures on Earth.

Content Background

Scientists have yet to prove conclusively that global warming is actually occurring. Some scientists blame recent rises in temperature on El Niño, a warming of the surface of the Pacific Ocean off the west coast of South America. This is caused by an unexplained weakening in the Pacific high pressure system. Unusual warming trends begin to spread throughout the eastern Pacific equatorial region and as far north as Alaska. This anomaly lasts for one to two years and can occur in four- to twelve-year cycles.

Teaching Strategy

Have students list examples of the

■ Conduction Transfer

Heat is also transferred from Earth's surface to the atmosphere by conduction. Conduction is the transfer of heat through a material from a higher temperature to a lower temperature. Have you ever left a metal spoon in a hot pan? If you have, you know that the spoon handle becomes hot. This is an example of conduction.

The atmosphere is not a good conductor of heat. Only air that directly touches hot surfaces, like a hot road, becomes heated by conduction. As the air moves over Earth's surface, it picks up heat from the surface.

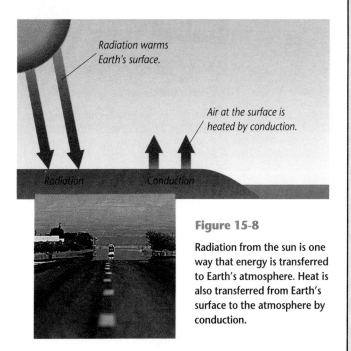

Radiation warms Earth's surface.

Air at the surface is heated by conduction.

Radiation   Conduction

Figure 15-8

Radiation from the sun is one way that energy is transferred to Earth's atmosphere. Heat is also transferred from Earth's surface to the atmosphere by conduction.

Activity Have students experiment with heat conduction using a metal iced tea spoon, paraffin, ice, a metal bell, and a steel ball smaller than the bowl of the spoon.

Have students use thermal gloves to heat the spoon for about 5 to 10 seconds in a Bunsen burner flame. Turn off the burner. Place a small piece of paraffin in the spoon and let the melting wax just about fill the spoon. Put the ball in the wax. Place the bottom of the spoon on a piece of ice to cool. Turn the spoon so that the ball is facing down and secure the tip of the handle to a ring stand. Place the bell directly below the ball so that when the ball drops, it hits the bell. With the burner, heat the spoon handle near the ring stand. After a minute or so, the ball will drop and ring the bell. Ask students how the wax melted. *conduction of heat through the spoon* L1

Visual Learning

Figure 15-8 Have a volunteer use the figure to explain the transfer of heat through conduction, as the class follows along. L1

NATIONAL GEOGRAPHIC SOCIETY

Videodisc

STV: Atmosphere

Smog: Chicago, Illinois

38685

atmosphere, and more is put into the atmosphere when the forests are burned.

Although global warming is not completely understood, many people feel we should take action to slow it down. What ways can you think of to reduce levels of carbon dioxide?

What Do You Think?

What would happen if global warming were to continue? Would our planet come to resemble another nearby planet of our solar system? Which one?

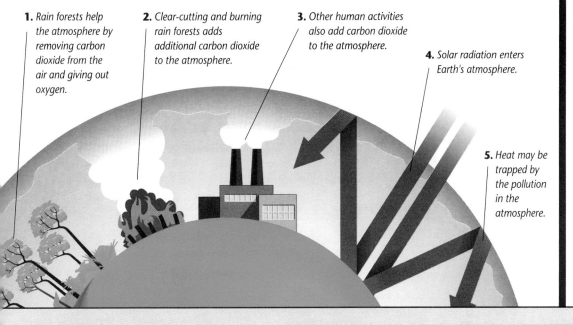

1. *Rain forests help the atmosphere by removing carbon dioxide from the air and giving out oxygen.*

2. *Clear-cutting and burning rain forests adds additional carbon dioxide to the atmosphere.*

3. *Other human activities also add carbon dioxide to the atmosphere.*

4. *Solar radiation enters Earth's atmosphere.*

5. *Heat may be trapped by the pollution in the atmosphere.*

greenhouse effect. For each, have them explain the heat source and how the heat is trapped and held. For example, cars heating up inside.

Answers to
What Do You Think?

If global warming continues, Earth could hypothetically become like its neighbor Venus.

Going Further ⫸

Have students discuss what a world without trees would be like. Ask students how oxygen would be replaced in the atmosphere and what types of building materials could be substituted for wood. Then direct partners to write a story about a world with no trees. Stories may include information on how the trees were destroyed, the effects of the destruction,

what trees were originally used for, how these products were replaced, and stories told by older people who were alive when trees existed. **COOP LEARN** L1

Visual Learning

Figure 15-9 As students study heat transfer through convection in the figure, ask the question posed: **What do we call such movement of air?** *wind*

Find Out!

Why does hot air move?

Time needed 15–20 minutes

Materials 2 small paper bags, meter stick, string, ring stand, lamp with 100-watt bulb

Thinking Processes observing and inferring, recognizing cause and effect, separating and controlling variables, forming a hypothesis

Purpose To observe one effect of convection.

Teaching the Activity

Have students work in pairs. `COOP LEARN`

Science Journal Have students identify the variables and controls in this experiment in their journals. `L1`

Expected Outcome

Students will observe that heated air rises.

Conclude and Apply

1. The heated air is less dense, thus the bag containing it is forced aloft by the cooler, denser air surrounding the bag.

2. The light bulb provided the energy necessary to heat air in the bag.

✔ Assessment

Oral Have students work in small groups to prepare a script and perform a puppet show that demonstrates why hot air rises. Use the Performance Task Assessment List for Skit in **PASC,** p. 75. `COOP LEARN`

Find Out! ACTIVITY

Why does hot air move?

What To Do

1. Tie two small paper bags filled with air closed with string. Hang the bags from the ends of a meter stick with more string.

2. Suspend the meter stick on a ring stand and adjust the strings and meter stick until the meter stick is balanced.

3. Place a lamp with a 100-watt bulb below one of the bags. **CAUTION:** *Make sure the bulb is at least 25 cm from the bag so that it does not catch on fire.*

4. Stand back about 1 meter from the bags. Remain very still as you observe what happens. Try not to stir up any air currents.

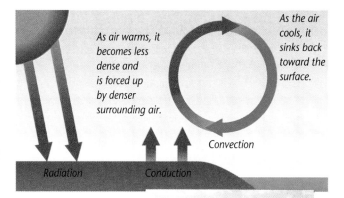

As air warms, it becomes less dense and is forced up by denser surrounding air.

As the air cools, it sinks back toward the surface.

Radiation   Conduction   Convection

Figure 15-9

This up-and-down cycle sets up a convection current of moving air as seen in the photo on the right. What do we call such movement of air?

Conclude and Apply

1. What happened to the suspended bags?

2. Explain the purpose of the lamp in this experiment.

■ Convection Transfer

When the air inside the bag over the lamp was heated, it became less dense and had less air pressure than the surrounding room air. The pressure in the room air pushed the heated bag upward. As a result, the meter stick tilted. Convection causes movement of air masses in the atmosphere in a similar manner. Convection is the transfer of heat by the movement of an air mass from one place to another. Warm air in the atmosphere is less dense and has less air pressure than cooler air. The cooler, denser air forces the warmer air upward. In the atmosphere, as the warm air is forced upward, it begins to cool and becomes more dense. It can then sink back down to the surface where the newly warmed air is being forced upward. This up-and-down cycle sets up a convection current of moving air we call wind.

ENRICHMENT

Activity Have students measure the temperature in the room at different heights and compare the temperatures to discover any differences. Students can tape a thermometer to a meter stick to get the readings from the ceiling. The activity can be performed in a gym at 1-meter height differences using metersticks taped together. Calibrated thermometers are a must in this activity.

Global Winds

Look at how the sun's radiation strikes different places on Earth, shown in **Figure 15-10**. Where are the sun's rays more direct, at the North Pole or at the equator? Because the equator receives so much solar radiation, it is usually warmer than any other place. The unequal heating of places all over Earth results in large, global wind systems. The trade winds and the prevailing westerlies are examples of wind systems that can have a great effect on global climate.

■ Trade Winds

Hotter air over the equatorial area rises and creates low pressure. Staying aloft, the heated air moves toward the poles. Cooler air from the polar regions moves toward the low pressure at the equator. The winds caused by the air sinking and returning to the equator are called **trade winds**.

The trade winds are warm and steady. In the Northern Hemisphere, early ship captains were able to use these winds to help them sail southwest and to explore the Americas. In the Southern Hemisphere, the trade winds would help a sailing ship glide northwest. Even today, airplane pilots use the trade winds to help save fuel. If you travel from Miami, Florida, to Ecuador in a jet, the pilot might ride the trade winds to increase the plane's speed and save fuel.

Figure 15-10

A Near the poles, the sun's rays are spread out more than at the equator. So, equal amounts of energy don't heat equally—each square meter of land at the poles receives less energy than each square meter at the equator.

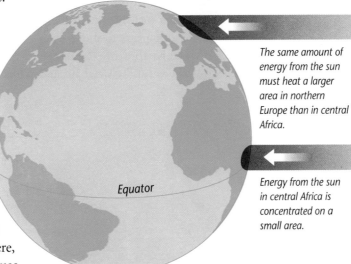

The same amount of energy from the sun must heat a larger area in northern Europe than in central Africa.

Energy from the sun in central Africa is concentrated on a small area.

Equator

B Whether near the poles or near the equator, humans have adapted to living in regions where temperatures vary as a result of large global wind systems.

473

■ Prevailing Westerlies

Earth's rotation on its axis affects other air movements. Some of the warm air traveling away from the equator does not cool enough to sink back to the surface. It continues to move toward the North and South poles. At the same time, cold, polar air is moving along the surface of the land toward the equator. Earth's rotation prevents either one of these masses from moving in a straight southerly or northerly direction.

The rotation of Earth deflects air from its north or south path. Between 30° and 60° latitude, the wind is deflected to the east. In the northern hemisphere, it appears to move from a southwestern to a northeastern direction. The winds between 30° and 60° latitude are called the **prevailing westerlies**. The northern prevailing westerlies are responsible for much of the weather movement in the United States and Canada.

■ Local Winds

Within the global patterns, smaller wind systems also exist. Whether you enjoy a bright, sunny day or a cold, rainy one often depends on the wind systems in your local area.

Do you live by a large lake or the sea? If you do, you have probably enjoyed days and nights at the beach. Have you ever noticed that during the daytime a cool breeze seems to blow gently toward the land from the water? Then at nighttime, the cooling breeze blows from the land toward

Figure 15-11

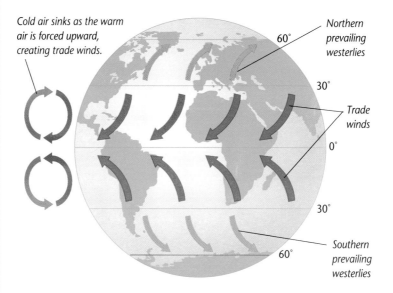

Cold air sinks as the warm air is forced upward, creating trade winds.

Northern prevailing westerlies

60°

30°

Trade winds

0°

30°

Southern prevailing westerlies

60°

A Warmer air over the equatorial area is less dense and is forced upward, creating low pressure. The warmer air moves toward the poles.

B Cooler air from the polar regions sinks and moves toward the lower pressure at the equator. The air sinking and returning to the equator creates the trade winds.

The northern prevailing westerlies are responsible for much of the weather movement in the United States and Canada.

Airplane pilots often tap into the energy of the trade winds. If you travel from Miami, Florida, to Ecuador in a jet, the pilot might ride the trade winds to increase the plane's speed and save fuel.

Multicultural Perspectives

Energy from Wind

Humans have harnessed wind power for millennia. Wind power has pumped water, driven sailing ships, and generated electricity. Historian Robert Temple in his book *The Genius of China* called the Chinese "the greatest sailors in history." By the second century A.D., they had developed sophisticated sails, masts, rudders, and other items for sailing. A document from the year A.D. 260 describes Chinese sailing vessels with as many as seven masts. Thanks to wind power, the Chinese became great explorers and traders more than 1000 years before the western "age of exploration." Have students make simple sailboats and hold a competition to gauge the effectiveness of different sail designs. [L2]

Figure 15-12

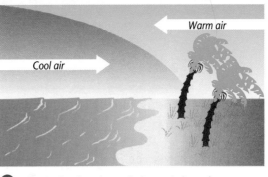

A During the daytime, the water and land absorb most of the sun's radiation. Over the land, much heat is radiated back up into the air above the land. Over the water, the air remains cool.

B Later in the day, the cool, dense air from the sea begins to move on land, forcing the warm air over the land upward. A sea breeze begins to blow from the water to the land.

C At night, the heat that had been trapped by the water begins to escape. The air over the sea is heated. Over the land, very little heat is given off, and it is cooler than the sea.

D During the nighttime, the cool air from the land moves toward the sea, forcing the warmer air over the sea upward.

the water. Can you explain why this happens?

Land and sea breezes are only one example of the many wind patterns that can create the weather in local communities. No matter what the locality, however, the air around us makes Earth a special planet in our solar system—the only planet known to have a life-supporting atmosphere.

check your UNDERSTANDING

1. Describe three ways heat is transferred in the atmosphere.
2. Why do weather patterns tend to move from west to east across the United States? Do they move west to east all over the world?
3. **Apply** Why does a cool breeze come from the water during the day and from the land at night?

check your UNDERSTANDING

1. through radiation, conduction, and convection
2. Prevailing westerlies blow air masses from west to east. The prevailing westerlies cause air masses to move from west to east across the United States and other parts of the Northern Hemisphere. No, all locations around the world do not have weather patterns moving west to east.
3. During the day, the water is cooler than the land. The cooler, denser air over the water forces the warmer air upward over the land. At night, the air over the water is warmer than that over the land so the wind direction is reversed.

Content Background

Content Background

Specific heat is the amount of thermal energy it takes to raise the temperature of one gram of a material 1°C. Given the same amount of solar energy, water is slower to heat than land. So during the daytime, warm air is forced upward off the land and is replaced by cooler air from the ocean, creating a sea breeze. Water also loses heat more slowly than land. At night, the air over water is warmed and forced upward by cooler air from the land, thus creating a land breeze.

3 ASSESS

Check for Understanding

Have students draw diagrams to support their answers for questions 1 and 2 of Check Your Understanding.

Reteach

Activity Have students list examples of conduction, convection, and radiation. Then have students build mini-greenhouses from a split oatmeal box, and sprout some plant seeds in it. As a control, students can plant seeds in an open container. Tell them to leave a closable hole in one end large enough to insert a thermometer for periodic temperature checks. Keep the mini-greenhouses in a sunny spot until the plants sprout. L1

Extension

Research Have students who have mastered the concepts in this section research and report on some local winds, such as the Santa Ana, mistral, chinook, sirocco, and bora. L2

4 CLOSE

Have students compare and contrast land and sea breezes. Be sure the answers indicate an understanding of convection. Have students write the ways conduction, convection, and radiation are used in heating systems. L1

Science and Society

Purpose Science and Society expands the discussion begun in Section 15-2 of the ongoing destruction of one layer of Earth's stratosphere.

Content Background Before ozone was present in Earth's atmosphere, much ultraviolet radiation reached Earth's surface. As ozone began to build up in the atmosphere, the planet became more hospitable to life forms.

The holes in the ozone layer are created, in part, by abundant ice crystals and a circular wind pattern above Antarctica. The ice crystals seem to promote CFC destruction of ozone. The pattern of ozone dissipation is strongest in September and October. As it weakens near the end of November, the ozone hole closes slowly. It has been found that a 1 percent ozone depletion causes a 2 percent rise in incoming ultraviolet radiation.

Teaching Strategy Discuss how students could adjust their lifestyles to reduce the use of CFCs. **What items could they live without? What activities would they have to cut down on? Which items could be made using other materials?**

Science and Society

The Disappearing Ozone Layer

As you'll remember from reading this chapter, the ozone layer is part of the stratosphere—it lies between 10 and 50 kilometers above your head. This layer of ozone gas absorbs some of the sun's radiation that is harmful to living things. The harmful radiation that manages to pass through the ozone layer makes you tan or sunburn—and too much of it can cause skin cancer, as well as other health problems. Currently, as many as 27,000 Americans develop skin cancer every year, and as many as 6000 die from it per year.

To Tan or Not to Tan?

What would happen if something happened to the ozone layer and more of the sun's dangerous radiation reached Earth? There could be a sharp increase in rates of skin cancer in humans and other animals. People who sunbathe would have to be more careful too, as exposure could lead to a very painful sunburn, cancer, or death.

It's a frightening thought—but the ozone layer is already starting to develop holes. In 1986, scientists discovered two holes in the ozone layer—a small hole over the North Pole, and a much larger hole over Antarctica. Since that time, the holes have disappeared, then reappeared at certain times of year. No one is quite sure what's causing the holes, or why they open and close. But most scientists do agree that the holes are gradually getting bigger, and that the ozone layer has become thinner all around Earth.

Chlorofluorocarbons

One of the possible causes of the holes is a group of chemicals called chlorofluorocarbons (CFCs), which destroy ozone. CFCs are used in automobile air conditioners, in some aerosol sprays, in refrigerators, bicycle seats, foam cups, polystyrene egg cartons, and

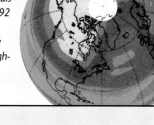

The Nimbus 7 satellite, as seen on the next page, is used in mapping global ozone distribution.

The two views of the Northern Hemisphere show ozone distribution for February 1992 at the top and February 1993 at the bottom. The darker red indicates a higher ozone concentration.

foam packaging for fast-food containers. We used to think CFCs were ideal for consumer products—they're nonflammable, nontoxic, and decompose very slowly. CFCs enter the atmosphere when these products are manufactured and used.

CFC gases rise slowly from Earth to the ozone layer, where the sun's rays are very powerful. The CFCs break down in the intense ultraviolet light and release chlorine. It is the free chlorine gas that destroys ozone.

You may have heard that some major fast-food chains converted from using foam packaging to cardboard hamburger containers, in an effort to stop using CFCs. Several countries, including the United States, have agreed to phase out CFCs that deplete the ozone layer by the year 2000.

Many people think that despite these laws and promises, not enough is being done to get rid of CFCs. It can take up to 150 years for CFCs to decompose, and a further 15 years for the gas to rise to the stratosphere. Some scientists say that even if we stopped using CFCs today, the ozone layer would continue to thin for the next 100 years.

How Can You Help?

We feel we can't change laws or influence the policies of international companies. But politicians and businesses can be affected by the way we act. If people stopped buying products containing CFCs, the companies that make them would need to find an alternative—and that process is already starting. Imagine you do some grocery shopping for your family, and you have to buy a box of eggs. If you buy eggs packed in a recycled paper container, instead of in a foam

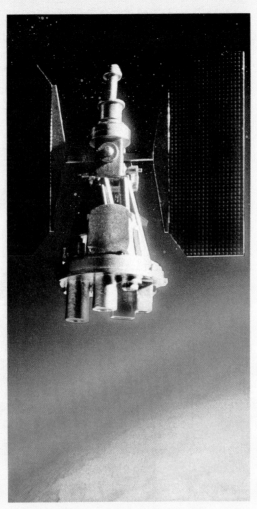

container that contains CFCs, you've made a decision that will help the environment.

Science Journal

In your Science Journal discuss what conveniences you would be willing to give up if you knew they were destroying the ozone layer.

Debate

Tell students that in 1990, nations around the world agreed to phase out the majority of chemicals that cause ozone depletion by the year 2000. Divide students into teams to debate whether or not this time frame is acceptable, given the fact that the problem is potentially deadly to Earth's inhabitants. L2

Activity

Have students work in small groups to compile a list of products that contain CFCs. Tell them to check the labels on all aerosol cans they can find in their homes. Further, have them list any appliances that use CFCs. Have the class create a list of alternative products that don't use CFCs and that they would buy to help save the ozone layer from destruction. **COOP LEARN** L1

Science Journal

Students' responses will vary but might include using pump sprays rather than aerosol sprays, which often contain CFCs; buying only food items packed in paper; and using motor vehicles less frequently to cut down on air pollution, in general.

Going Further ⟶

Have pairs of students research skin cancer and its prevention. Tell students to find out about the causes and treatment of skin cancer, as well as the need for early detection and self-examination for this serious disease. To request information on skin cancer, students can call The American Cancer Society toll free at 1-800-227-2345. Have students prepare oral reports for the class. Use the Performance Task Assessment List for Oral Presentation in **PASC**, p. 71. **COOP LEARN**

Teens in SCIENCE

Purpose

Teens In Science explains how convection, a concept discussed in Section 15-3, can be applied to make hot-air ballooning possible.

Content Background

The earliest known balloon flight was made in 1709 in Portugal. In a recent crossing of the Pacific Ocean from Japan to Alaska, one hot-air balloon attained a speed of 382 kilometers per hour. The highest recorded hot-air balloon flight was made in 1988, when a balloon attained an altitude of 19.8 kilometers.

Hot-air balloons use convection to rise through more dense, cooler air. A large propane burner is mounted under the opening of the balloon. The balloon begins its inflation on its side. Flaming propane gas is directed toward the opening of the balloon. Strong gusts of heated air go into the balloon, which begins to expand. The hotter the inner air of the balloon becomes, the more it expands and inflates the balloon. Finally, the balloon becomes upright. When it is in this position, the air is heated even more. Soon the air is warm enough that the balloon becomes much less dense than the air around it. Like any mass of air warmed at Earth's surface, it is forced upward by surrounding, denser air, dragging its gondola (basket) under it.

Teaching Strategy

Have students design and build model hot-air balloons from tissue paper. They can work in design teams of three or four. Each team can demonstrate how their model works for the class.

COOP LEARN

Teens in SCIENCE

Flying High and Loving It

Have you ever let go of the string holding a helium-filled balloon? Did your balloon shoot straight up like a rocket, or did it seem to drift in the direction of the wind? Having trouble remembering? Just ask 18-year-old commercial hot-air balloon pilot, David Bair. He ought to know.

Earning His Wings

David took his first ride in a hot-air balloon at the age of four. His hometown of Albuquerque, New Mexico is host to an annual festival called the International Hot-Air Balloon Fiesta. "My family met a pilot who had no crew. We volunteered to help out."

But not all the flights have been smooth. "Navigating is tricky. You don't steer a balloon as you would an airplane."

The Science of Flight

Why is navigating a hot-air balloon so difficult? As you know, if the temperature of a gas is increased, air molecules become more active and they begin to move away from each other. When the heat source in the balloon is turned on, air in the balloon heats up. Molecular action in the balloon makes air in the balloon less dense than the surrounding air, and the balloon is pushed upward by the colder air. To go down, the heating source is simply turned off. Air in the balloon becomes more dense, and the balloon sinks. Balloon pilots can control the up and down movements of the balloon.

"To move side to side, you have to depend on wind currents. This past summer, I got caught in a severe wind. Even though I kept the balloon at a stable temperature, this wind current tossed my balloon at a speed of 400 feet a minute. Normally, ballooning is very safe, but weather is part of what makes ballooning so exciting."

Going Further ⅢⅢⅢ▶

Use the following activity to demonstrate to students that air temperature decreases with altitude. Use a minimum/maximum thermometer, some helium balloons, and 300 meters of monofilament. A kite string reel will make the takeup of the monofilament easier. This activity should be done on a still, sunny day so that winds are not a factor. Attach several large helium balloons together with a short piece of string. Tie the thermometer apparatus and monofilament to an ac-cessible place on the connector string. Hold the monofilament firmly as you let the balloons and the thermometer float upward. Let the apparatus stay in place for several minutes; then reel it in. The thermometer should show a drop in temperature of about 3°C. Based on this information, have students determine the average drop in temperature per 100 meters. *about 1°C* Also ask students to determine the temperature at various heights up to 3000 meters. ☐2

Science Journal

Review the statements below about the big ideas presented in this chapter, and answer the questions. Then, re-read your answers to the Did You Ever Wonder questions at the beginning of the chapter. *In your Science Journal,* write a paragraph about how your understanding of the big ideas in the chapter has changed.

1 Three major components make up the atmosphere. *Describe these three components.*

2 The atmosphere is made up of several layers. *What are these layers and why is each important?*

3 Heat transfer occurs three ways. *How do each of these three ways affect the atmosphere?*

4 Two of the global wind systems are the trade winds and prevailing westerlies. *Where are these two wind systems located?*

Science Journal

Did you ever wonder...
- Air is a mixture of gases; the most abundant are nitrogen (78 percent) and oxygen (21 percent). (p. 460)
- Outer space is the region outside Earth's atmosphere where there is no air. (p. 462)
- Winds occur because of unequal heating of Earth's surface. (p. 473)

Project

Divide the class into three groups. Have the first group make a list of the gases found in the atmosphere. Have the second group identify uses for each of these gases. Have the third group determine whether the gas is a pollutant. Make a wall chart on which the groups can enter their information. Use the Performance Task Assessment List for Group Work in **PASC,** p. 97.

COOP LEARN

chapter 15
REVIEWING MAIN IDEAS

Have students use the photographs and illustrations on this page to review the main ideas presented in this chapter.

Teaching Strategy

Divide the class into groups. Assign each group one of the questions. Students can share their responses aloud or write them on the chalkboard.

Answers to Questions

1. The atmosphere is composed of solids, liquids, and gases. Two of these gases are oxygen and nitrogen. Some of the solids in the atmosphere are dust and pollen. Liquid water is also found in the atmosphere.

2. The layer closest to Earth is the troposphere, where our weather occurs. The next layer is the stratosphere, which contains the protective ozone layer. The next layer is the mesosphere, used for radio wave transmission. Above the mesosphere lies the thermosphere and exosphere, where satellites orbit Earth.

3. The air and water in the atmosphere are warmed directly by radiation from the sun. This radiation also warms Earth's surface, which in turn warms the air near it by conduction. Warm air near the surface is less dense than cooler air, so this warm air is forced up away from Earth's surface, and cooler air sinks to replace it. This warming process is called convection.

4. The trade winds are located between 0° and 30° north latitude, and between 0° and 30° south latitude. The prevailing westerlies are located between 30° and 60° south latitude and 30° and 60° north latitude.

GLENCOE TECHNOLOGY

 MindJogger Videoquiz

Chapter 15 Have students work in groups as they play the Videoquiz game to review key chapter concepts.

Using Key Science Terms

1. *Atmos* means "vapor"; *sphere* means "the place or extent of action"; the definition of *atmosphere* is "the blanket of air that surrounds and protects Earth." *Tropo* means "to change"; the troposphere is where weather occurs.

2. The prevailing westerlies occur between 30° and 60° latitude and blow to the west. The trade winds occur 30° north and south of the equator. Their movements are not in the same directions.

3. Ozone absorbs harmful radiation from the sun. Smog is air pollution. As ozone is destroyed, more radiation enters the atmosphere. As smog increases, less carbon dioxide is absorbed by plants, and, instead, remains in the atmosphere. This increase in atmospheric carbon dioxide accentuates the greenhouse effect.

Understanding Ideas

1. Radiation is the transfer of thermal energy through space. Conduction is heat moving from one material to another. Convection is due to density differences.

2. Air pressure is higher closer to the surface of Earth. The air is compressed under the weight of the layers pushing down from above.

3. Deforestation increases the greenhouse effect. Burning gives off carbon dioxide, and then there are fewer trees to absorb carbon dioxide.

4. Cooler air has higher pressure than warmer air. Cool air, with high pressure, will move into an area of warm air, forcing the warm air upward.

5. The surfaces absorb heat from the sun. The air immediately above the surface is heated through convection and conduction.

Developing Skills

1. See student page.

U sing Key Science Terms

atmosphere	smog
ozone	trade winds
prevailing westerlies	troposphere

1. Look up the origin of the words atmosphere and troposphere. What are the meanings of the main parts of the terms? Explain why the words are appropriate.

2. Explain the differences between the prevailing westerlies and trade winds.

3. Explain the relationship between ozone, the greenhouse effect, and smog.

U nderstanding Ideas

Answer the following questions in your Journal using complete sentences.

1. Discuss the differences in heat transfer by radiation, conduction, and convection.

2. Is air pressure greater or less closer to Earth's surface than at higher altitudes? Explain your answer.

3. Does deforestation increase or decrease the greenhouse effect? How?

4. How do temperature and pressure work together to cause winds?

5. On a clear, sunny day, why is the temperature significantly warmer for baseball players on the field or tennis players on the court than for the fans in the stands?

D eveloping Skills

Use your understanding of the concepts developed in this chapter to answer each of the following questions.

2. As air is pumped into the deflated ball, the books will rise. This shows that air under pressure can move things.

3. Results will vary depending on the locations chosen.

4. The thermometer in the dark soil should register a higher temperature.

---

1. Concept Mapping Complete the following concept map of air.

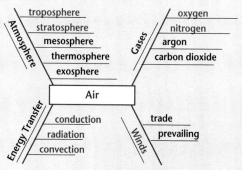

2. Predicting Deflate the ball used in the Explore activity on page 455 and put two heavy books on top of it. Predict what will happen if you pump air into the ball. Now try it and see if your prediction was accurate. In what way can this property of air be useful?

3. Predicting Prepare the plastic lids from the Find Out activity on page 457 with fresh petroleum jelly and place in the same areas as used before, but either much higher or much lower. Predict what you might find in the lids after this trial. Are there more or fewer particles in the lids? Why?

4. Forming a Hypothesis Prepare the greenhouse used in the Find Out activity on page 470 using 1/2 dark soil and 1/2 white sand. Repeat the activity using a thermometer for each side of the box. Hypothesize whether there will be a difference in the temperature readings. Did your observations support your hypothesis?

Program Resources

Review and Assessment, pp. 89–94 L1
Performance Assessment, Ch. 15 L2
PASC
Alternative Assessment in the Science Classroom
Computer Test Bank L1

Critical Thinking

In your Journal, *answer each of the following questions.*

1. Why do you think European traders took one route to the Americas and a different route on the return voyage?

2. Study the figures shown. Where would you have to be to feel a breeze during the daytime? During the nighttime? Is this a global or local pattern?

Valley breeze — Warm air

Mountain breeze — Cool air

3. Jet airliners have pressurized cabins. This means that a nearly constant air pressure is maintained. Why do you think this is necessary?

4. Holes in the ozone layer have been found near the poles. Why is this dangerous?

5. From what you have learned in this chapter, explain why sailboarders and hang gliders are usually found near the water's edge or shoreline.

6. Based on your understanding of the atmosphere, why do you think helium balloons float?

Problem Solving

Read the following problem and discuss your answers in a brief paragraph.

Alicia and her friends are hiking in the Peruvian Andes. They are experienced hikers, but Alicia has never hiked at such a high altitude. They begin their hike at the base of a mountain. By noon, they are three kilometers higher. Alicia begins having trouble keeping up with the others. She feels weak and is gasping for air.

1. What do you think is happening to Alicia? Why?

2. What can she do to help herself?

CONNECTING IDEAS

Discuss each of the following in a brief paragraph.

1. **Theme—Scale and Structure** Relate Newton's law of inertia to the movement of air.

2. **Theme—Energy** Why do you think Denver, Colorado banned most wood-burning fireplaces?

3. **Theme—Systems and Interactions** Would you expect atmospheric pressure to be greater at the poles or at the equator? Why?

4. **A Closer Look** How are the Amazon Rain Forest, carbon dioxide, and global warming related?

5. **Life Science Connection** What factors about a city and its location might make it an unhealthful place for people with breathing disorders?

Critical Thinking

1. They used one route to catch winds that carried them west toward the Americas and a different route to catch winds that carried them east back to Europe.

2. mountainside; valley; local

3. Air at high altitudes is too thin to breathe.

4. Harmful radiation from the sun can get into the troposphere.

5. During the day, warm air rises over the land and sinks over the cooler ocean water, creating updrafts over land (good for hang gliders) and breezes over water (good for windsurfing).

6. Helium is less dense than air.

Problem Solving

1. Alicia is not getting enough oxygen in the less dense air of the higher altitude.

2. She should sit down and rest awhile so her body isn't using as much oxygen or move to a lower elevation.

Connecting Ideas

1. Air masses will tend to stay still or in motion until acted on by another air mass.

2. Fireplaces emit carbon dioxide, soot, and other pollutants into the air and contribute to smog.

3. Atmospheric pressure is greater at the poles because cold air is denser than warm air.

4. The Amazon rain forest is being burned on a large scale. The burning wood causes an increased carbon dioxide buildup in the atmosphere. Carbon dioxide is a greenhouse gas. It holds heat in the atmosphere, causing global warming.

5. Cities burn fossil fuels to generate electricity. People use motor vehicles, which also burn fossil fuels, to get around in cities. If a city is located in an area of minimum winds or stagnant air, pollutants build up and remain in place.

✔ Assessment

Portfolio Review the portfolio options that are provided throughout the chapter. Encourage students to select one product that demonstrates their best work for the chapter. Have students explain what they learned and why they chose this example for placement into their portfolios.

Additional portfolio options can be found in the following **Teacher Classroom Resources:**
Multicultural Connections, pp. 33, 34

Making Connections: Technology and Society, p. 33
Concept Mapping, p. 23
Critical Thinking/Problem Solving, p. 23
Making Connections: Integrating Sciences, p. 33
Take Home Activities, p. 23
Making Connections: Across the Curriculum, p. 33
Laboratory Manual, pp. 91–100
Performance Assessment, Ch. 15

Chapter Organizer

SECTION	OBJECTIVES	ACTIVITIES & FEATURES
Chapter Opener		**Explore!**, p. 483
16-1 How Do You Breathe? (3 sessions; 2 blocks)	**1. Compare** how different organisms take in oxygen. **2. Trace** the pathway of air into and out of the lungs. **3. Describe** the pressure changes that occur within the chest cavity when you breathe. **National Content Standards: (5-8) UCP1-2, UCP5, A1, C1, C3, C5**	**Explore!**, p. 485 **Explore!**, p. 488 **Investigate 16-1:** pp. 490–491 **Explore!**, p. 492 **A Closer Look,** pp. 486–487
16-2 The Air You Breathe (3 sessions; 1.5 blocks)	**1. Compare** air that is inhaled with air that is exhaled. **2. Explain** cellular respiration and its relationship to gas exchange in your body. **National Content Standards: (5-8) UCP1-2, A1-2, B3, C1, C3, F1, G1, G3**	**Find Out!**, p. 495 **Design Your Own Investigation 16-2:** pp. 498–499 **Physics Connection,** pp. 496–497 **Technology Connection,** p. 506
16-3 Disorders of the Respiratory System (2 sessions; 1 block)	**1. Discuss** respiratory disorders and their causes. **2. Determine** how to keep your lungs healthy. **National Content Standards: (5-8) UCP2, A1-2, C1, C3, E2, F1, F5**	**Explore!**, p. 502 **Skillbuilder:** p. 503 **Science and Society,** p. 505 **SciFacts,** p. 507

ACTIVITY MATERIALS

EXPLORE!

p. 483 clock or watch with second hand
p. 485* goldfish in aquarium
p. 488* No special materials required.
p. 492 metric tape measure
p. 502* No special materials required.

INVESTIGATE!

pp. 490–491 round balloon, metric ruler

DESIGN YOUR OWN INVESTIGATION

pp. 498–499 clock or watch with second hand, 2 drinking straws, 200 mL bromothymol blue solution, 400-mL beakers (2), graduated cylinder

FIND OUT!

p. 495* towel, mirror

KEY TO TEACHING STRATEGIES

The following designations will help you decide which activities are appropriate for your students.

- **L1** Basic activities for all students
- **L2** Activities for average to above-average students
- **L3** Challenging activities for above-average students
- **LEP** Limited English Proficiency activities
- **COOP LEARN** Cooperative Learning activities for small group work
- **P** Student products that can be placed into a best-work portfolio
- Activities and resources recommended for block schedules

Need Materials? Call Science Kit (1-800-828-7777).

00:00 **OUT OF TIME?** We recommend that students do the activities with an asterisk.

Chapter 16 Breathing

TEACHER CLASSROOM RESOURCES

Student Masters	Transparencies
Study Guide, p. 53 **Take Home Activities**, p. 24 **How It Works**, p. 17 **Critical Thinking/Problem Solving**, p. 5 **Activity Masters**, Investigate 16-1, pp. 67–68 **Concept Mapping**, p. 24 **Science Discovery Activities**, 16-1	**Teaching Transparency 31**, Respiratory System **Teaching Transparency 32**, Comparing Respiratory Systems **Section Focus Transparency 47**
Study Guide, p. 54 **Multicultural Connections**, p. 35 **Making Connections: Integrating Sciences**, p. 35 **Making Connections: Across the Curriculum**, p. 35 **Activity Masters**, Design Your Own Investigation 16-2, pp. 69–70 **Science Discovery Activities**, 16-2 **Laboratory Manual**, pp. 101–102, Carbon Dioxide and Respiration	**Section Focus Transparency 48**
Study Guide, p. 55 **Critical Thinking/Problem Solving**, p. 24 **Making Connections: Technology & Society**, p. 35 **Multicultural Connections**, p. 36 **Science Discovery Activities**, 16-3	**Section Focus Transparency 49**

ASSESSMENT RESOURCES	TEACHING & TECHNOLOGY
Review and Assessment, pp. 95–100 **Performance Assessment**, Ch. 16 **PASC*** **MindJogger Videoquiz** **Alternate Assessment in the Science Classroom** **Computer Test Bank**	**Spanish Resources** **Cooperative Learning Resource Guide** **Lab and Safety Skills** **Science Interactions, Course 2, CD-ROM** **Computer Competency Activities**

*Performance Assessment in the Science Classroom

NATIONAL GEOGRAPHIC TEACHER'S CORNER

National Geographic Society Products Available From Glencoe

To order the following products for use with this chapter, contact your local Glencoe sales representative or call Glencoe at 1-800-334-7344:

- *Newton's Apple Life Sciences* (Videodisc)
- *STV: Human Body Series,* "Respiratory, Circulatory, Digestive Systems" (Videodisc)

Additional National Geographic Society Products

To order the following products for use with this chapter, call the National Geographic Society at 1-800-368-2728:

- *The Incredible Machine* (Book)
- *Are We Getting Enough Oxygen?* (Kids Network Curriculum Unit)
- *The Incredible Human Machine* (Video)
- *Man: The Incredible Machine* (Video)
- *Your Body Series,* "Circulatory and Respiratory Systems" (Video)

Teacher Classroom Resources

These are key components of the classroom resources package.

TEACHING AIDS

Section Focus Transparencies

47 SECTION FOCUS TRANSPARENCY

WATER WORLD

Scuba diving requires special equipment that allows a person to breathe while his or her body is in the water.

1. What special equipment is required for scuba diving? Why is this equipment needed?

2. How do scuba divers and fish differ in the way they get oxygen?

L1

48 SECTION FOCUS TRANSPARENCY

VIOLENT BREATHS OF NATURE

Hurricanes and tornadoes are violent reminders of nature's strength. Hurricanes, also called typhoons, are by far the largest storms in terms of area covered.

1. A hurricane is a visible reminder that air is always around us. What is air?

2. Why is air important?

L1

49 SECTION FOCUS TRANSPARENCY

THANK YOU FOR NOT SMOKING

You see these signs in a lot of places these days. There is a current trend to prevent smoking in nearly all public buildings . . . and in some private homes.

1. Why do you think it is necessary for this "No Smoking" sign to be posted in this restaurant?

2. Do you think such signs are necessary?

L1

Teaching Transparencies

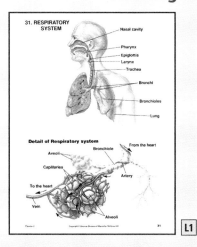

31. RESPIRATORY SYSTEM

L1

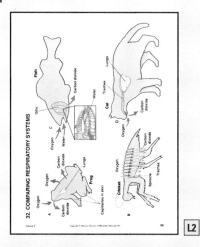

32. COMPARING RESPIRATORY SYSTEMS

L2

HANDS-ON LEARNING

Science Discovery Activity*

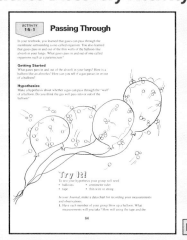

ACTIVITY 16-1 Passing Through

L1

Laboratory Manual*

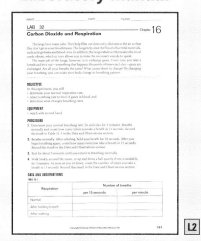

LAB 32 — Chapter 16

Carbon Dioxide and Respiration

L2

Take Home Activity

BETTER BREATHING

L1

*There may be more than one activity for this chapter.

Chapter 16 Breathing

Study Guide*

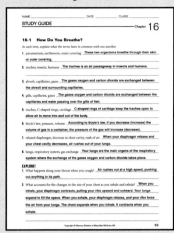

Concept Mapping

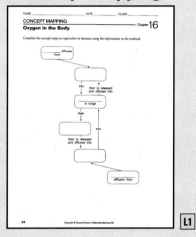

Critical Thinking/ Problem Solving

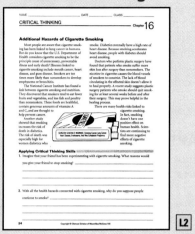

Integrating Sciences

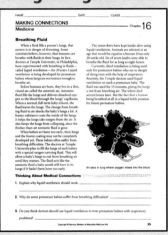

Across the Curriculum

Technology and Society

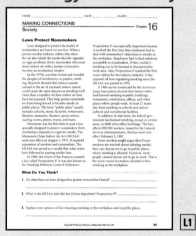

Multicultural Connection**

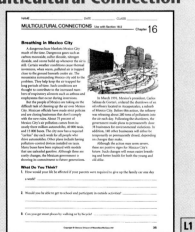

Performance Assessment

Review and Assessment

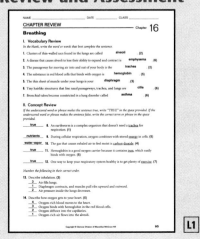

*One per section **Two per chapter

Chapter 16   Breathing   **482D**

Breathing

THEME DEVELOPMENT

Disruptions that can occur to the respiratory system illustrate the interactions between systems. For example, due to a buildup of carbon dioxide in the bloodstream, respiratory distress forces the heart to work harder. Thus problems in the respiratory system cause problems in the circulatory system.

CHAPTER OVERVIEW

Students will study the main features and functions of the respiratory systems of one-celled organisms, fishes, insects, and humans.

Tying to Previous Knowledge

The respiratory system provides oxygen to cells in tissues where energy is released. Help students compare this to the burning of fuel in an engine. The carburetor supplies oxygen for the burning of gasoline, which provides energy to move the car. Waste gases are emitted through the tailpipe.

Uncovering Preconceptions

Build a simple model to illustrate the function of the epiglottis. Three cardboard tubes in an inverted Y shape can be used to represent the pharynx, trachea, and esophagus. Drop a small object down the pharynx tube. Students should see that the object can drop through either tube. The epiglottis ensures that "food" goes down the esophagus.

Breathing

Did you ever wonder...

✓ **Why you can see your breath on a cold day?**

✓ **Why you breathe faster when you exercise?**

✓ **Why a doctor thumps on your chest when you're having an examination?**

Science Journal

Before you begin to study the respiratory system, think about these questions and answer them *in your Science Journal.* When you finish the chapter, compare your journal write-up with what you have learned.

Answers are on page 508.

482

I t's a rainy day. You're relaxing by reading a magazine while your dog is asleep at your feet. As you finish a page, you become aware of a big sigh and the dog's rhythmic breathing. In and out, in and out, he breathes deeply and slowly. Occasionally, he even snores. You're aware of the dog's breathing, but he isn't. Come to think of it, most of the time you're probably not aware of your own breathing. You don't have to be. From the moment you're born until the moment you die, air enters and leaves your body automatically. You don't have to think about it.

Breathing is actually one step in supplying your body with the oxygen it needs. In this chapter, you will discover how and why you and some of the organisms around you breathe. You'll also find out what happens to the air you breathe in. Finally, you will learn about some disorders of the respiratory system.

▶ **In the activity on the next page, explore your breathing patterns and compare them with those of your classmates.**

OUT OF TIME?

`00:00`

If time does not permit teaching the entire chapter, use the Chapter Overview on this page, Reviewing Main Ideas at the end of the chapter, and the Chapter 16 audiocassette to point out the main ideas of the chapter.

Learning Styles		
Kinesthetic	Explore, pp. 488, 492; Investigate, pp. 490-491, 498-499; Visual Learning, p. 492; Find Out, p. 495; Demonstration, p. 501
Visual-Spatial	Explore, p. 485; Visual Learning, pp. 487, 488, 489, 496, 497, 503; Demonstration, p. 494; Across the Curriculum, p. 504
Interpersonal	A Closer Look, pp. 486-487
Intrapersonal	Physics Connection, pp. 496-497; Enrichment, p. 495
Logical-Mathematical	Multicultural Perspectives, p. 485
Linguistic	Across the Curriculum, p. 487; Visual Learning, p. 493; Enrichment, p. 503
Auditory-Musical	Explore, pp. 483, 502

LS

Explore! ACTIVITY

What happens when you breathe?

The act of breathing is just doing what comes naturally for most of us. Breathing is actually a complex activity that requires coordination between your respiratory system and your brain (telling your respiratory system what to do). What exactly happens to your respiratory system when you breathe? Try this activity to explore your breathing and how often it takes place.

What To Do

1. Put your hand on your chest. Notice your breathing. Feel your chest move up and down.

2. Take a deep breath. What happens to your rib cage? In which directions does it move?

3. Count your breathing rate for one minute. How does your breathing rate compare with the rates of your classmates?

4. *In your Journal*, write a paragraph describing your observations. What explanation can you offer for why breathing rates vary?

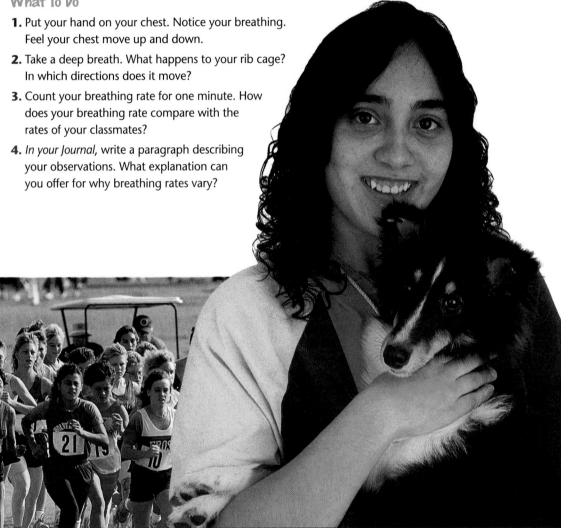

How Do You Breathe?

PREPARATION

Planning the Lesson

Refer to the Chapter Organizer on pages 482A–D.

Concepts Developed

Students are introduced to the respiratory system by comparing and contrasting how single-celled organisms, fish, and humans take in oxygen and release carbon dioxide. The mechanics of breathing are illustrated: from the pathway of air into the lungs to the inhalation and exhalation of atmospheric gases.

1 MOTIVATE

Bellringer

 Before presenting the lesson, display **Section Focus Transparency 47** on an overhead projector. Assign the accompanying **Focus Activity** worksheet. L1

LEP

Discussion Ask students to describe times that they could not breathe normally. Perhaps they held their breath while swimming underwater or were stuck in a smoky area. Discuss what would have happened if they had been unable to resume breathing quickly.

2 TEACH

Tying to Previous Knowledge

Discuss with students what they learned in Chapter 3, *Circulation*. Remind them that oxygen is carried into the body by the respiratory system. The exchange of gases between the circulatory system and the respiratory system takes place in the capillaries.

Section Objectives

- Compare how different organisms take in oxygen.
- Trace the pathway of air into and out of the lungs.
- Describe the pressure changes that occur within the chest cavity when you breathe.

Key Terms

trachea, gills, lungs, alveoli, diaphragm

How Some Organisms Take In Oxygen

You and a dog have similar respiratory systems. However, while most organisms need oxygen, they don't obtain it the same way.

■ One-celled Organisms

One-celled organisms take in oxygen directly from their watery environment. The figure below shows how oxygen and carbon dioxide move into and out of a paramecium.

■ Complex Organisms

Most complex organisms have specific body structures to take in oxygen and release carbon dioxide. The

Figure 16-1

Ⓐ One-celled organisms, such as a paramecium, take in oxygen directly from their watery environment. The outer surface of a paramecium is covered by a membrane. Oxygen in the water diffuses into a paramecium's body through the membrane. Likewise, carbon dioxide wastes move out of the body through the membrane.

Ⓑ Most complex organisms have specific body structures to take in oxygen and release carbon dioxide. Earthworms have capillaries in their skin where carbon dioxide is exchanged for oxygen. An earthworm takes in oxygen from moist soil and releases carbon dioxide waste through its skin.

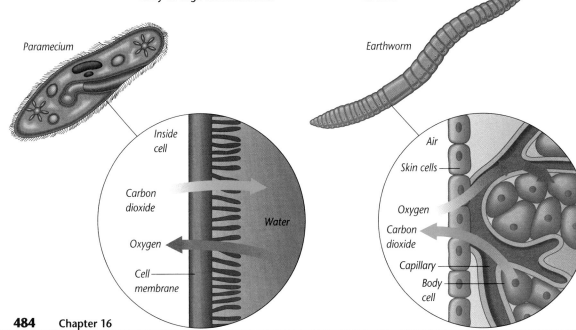

Paramecium

Inside cell

Carbon dioxide

Oxygen

Cell membrane

Water

Earthworm

Air

Skin cells

Oxygen

Carbon dioxide

Capillary

Body cell

484 Chapter 16

Program Resources

Study Guide, p. 53
Teaching Transparency 31 L1
Teaching Transparency 32 L2
Take Home Activities, p. 24, Better Breathing L1
Section Focus Transparency 47

figures on these two pages show how grasshoppers and earthworms take in oxygen. The grasshopper's respiratory system is more complex than the earthworm's. The grasshopper has a system of tubes, each called a trachea.

The **trachea** is a passageway through which air travels into and out of the body. But not all complex organisms possess a trachea. In the next activity, you will explore how fish obtain oxygen.

 Explore! ACTIVITY

What can you learn by watching a goldfish?

What To Do

1. Observe goldfish in an aquarium. Watch the overall behavior of the fish for several minutes.

2. Note any body parts that move. Which body parts moved while you watched the fish?

3. Did any of the fish's body movements seem to be related to each other?

4. *In your Journal*, describe the fish's movements and explain how they are related to the fish's body functions.

Grasshopper

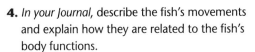

Trachea
Exoskeleton
Oxygen
Carbon dioxide
Air

C Unlike the paramecium and earthworm, oxygen doesn't diffuse through an insect's outer body covering. Grasshoppers and other insects take in oxygen through tiny openings along their sides. Each opening connects to a tube called a trachea.

16-1 How Do You Breathe? **485**

16-1 How Do You Breathe? **485**

Theme Connection The theme that this section supports is energy. Our cells require oxygen in order to utilize nutrients via cell respiration. Cell respiration produces the energy for all cellular activities. Chemical energy is transformed into mechanical energy and heat. Without oxygen, we would lack the energy to move, speak, and think. In short, we could not live for more than a few minutes without a constant supply of oxygen.

GLENCOE TECHNOLOGY

 CD-ROM

Science Interactions, Course 2, CD-ROM

Chapter 16

NATIONAL GEOGRAPHIC SOCIETY

 **Videodisc**

STV: Human Body Vol. 1

Circulatory and Respiration Systems
Unit 1, Side 1
Need for Oxygen

7795-9506

■ Gills

Fish have body structures that extract oxygen from the water. When you observed the goldfish in the last activity, you probably noticed the flaps on either side of its head. Beneath these flaps are gills. **Gills** are respiratory structures of some aquatic animals through which oxygen is removed from water.

When a fish opens its mouth, water flows in at the same time as a cover, or flap, over the gills closes. When the mouth closes, the water moves through the mouth, over the gills, and out past the flap that is now opened.

Each gill is made of several spongy structures called gill filaments. Because the filaments are feathery in nature, they provide increased surface area for water to pass over. Tiny capillaries extend throughout each filament. When the water passes over the filaments, oxygen diffuses from the water into the fish's blood as it travels through the capillaries. This oxygen-rich blood is then delivered to all parts of the fish's body. At the same time, carbon dioxide moves from the capillaries of the gills out into the water. The water then flows out of the fish through the opened gill covers. **Figure 16-2** shows a fish's respiratory system.

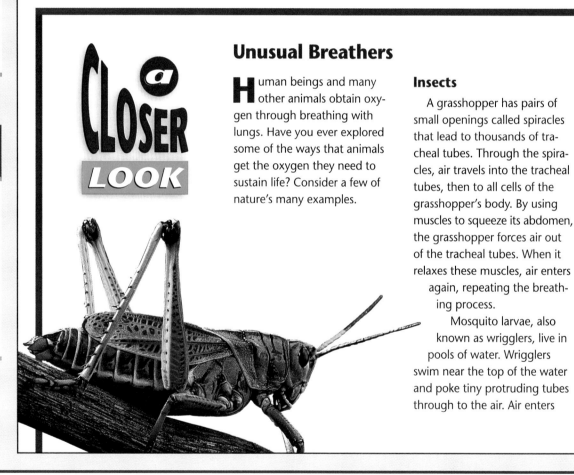

Unusual Breathers

Human beings and many other animals obtain oxygen through breathing with lungs. Have you ever explored some of the ways that animals get the oxygen they need to sustain life? Consider a few of nature's many examples.

Insects

A grasshopper has pairs of small openings called spiracles that lead to thousands of tracheal tubes. Through the spiracles, air travels into the tracheal tubes, then to all cells of the grasshopper's body. By using muscles to squeeze its abdomen, the grasshopper forces air out of the tracheal tubes. When it relaxes these muscles, air enters again, repeating the breathing process.

Mosquito larvae, also known as wrigglers, live in pools of water. Wrigglers swim near the top of the water and poke tiny protruding tubes through to the air. Air enters

Purpose

A Closer Look reinforces Section 16-1 by describing some unusual ways that organisms take in oxygen.

Content Background

Animals with spiracles usually have two pairs in the thorax and eight in the abdomen. The spiracles open and close periodically to prevent excess water loss through evaporation.

There are four pairs of book lungs in

scorpions and up to two in spiders. Regardless of the number, each lung is enclosed in a cavity that opens to the outside by a small slit.

Teaching Strategy

Invite pairs of students to read the selection together and then discuss which organism has the most interesting method of breathing and why. `COOP LEARN`

Activity Have students make drawings,

Figure 16-2

The main respiratory organs of fish are their gills. The gills contain capillaries.

A When a fish opens its mouth, water flows in. At the same time, a flap over the gills closes.

B When the mouth closes, the flap over the gills opens. Water moves through the mouth, over the gills, and out past the open flap.

C As water moves over the gills, oxygen diffuses from the water into the gill capillaries.

D At the same time, carbon dioxide, a waste gas carried by the circulatory system, moves from the capillaries out into the water.

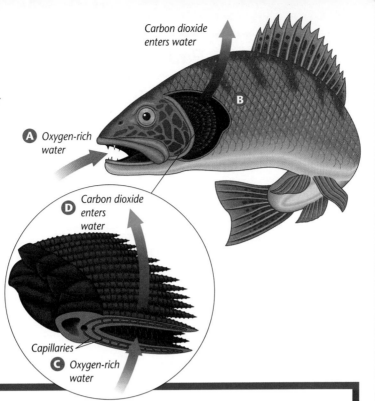

Carbon dioxide enters water

B

A Oxygen-rich water

D Carbon dioxide enters water

Capillaries

C Oxygen-rich water

Visual Learning

Figure 16-2 Help students use Figure 16-2 to trace the sequence that takes place when fish extract dissolved oxygen from water. Help them to visualize what it is like to have water flushing over the gill filaments. Have them suggest why gills with many blood vessels are an advantage.

Across the Curriculum

Language Arts

The word *respiration* derives from the Latin word *spirare*, which means "to breathe." Ask students to list as many words as they can think of that end with *-spiration*. Words such as *aspiration, inspiration, perspiration, transpiration,* and *expiration* all have common meanings that we may not associate with breathing. Have students look up these words in the dictionary to find their common meanings and how they might relate to breathing. **LEP** **L1**

these tubes and moves to tracheal tubes throughout their bodies.

Some aquatic beetles carry extra oxygen in large bubbles within their thick hairs. When beetles are underwater, oxygen passes from these bubbles to the tracheal tubes leading to all parts of their bodies. The oxygen bubbles also give them added buoyancy for traveling up to the water's surface.

Spiders

Spiders and scorpions have book lungs connected to tracheal tubes. Book lungs look a lot like gills, and they work in a similar way, removing oxygen from air instead of water. Book lungs are a series of thin "plates" full of blood vessels that catch and carry oxygen throughout the animal's body. The European water spider carries bubbles of air within its book lungs to bell-shaped webs that it builds under water. It uses these webs to store oxygen for future use.

Science Journal

Imagine that you are writing a science fiction novel about a creature with an unusual way of getting oxygen. Write a description *in your Science Journal* of what this animal looks like and exactly how it breathes.

16-1 How Do You Breathe? **487**

or models of some of the breathing mechanics they read about in this selection: spiracles, bubble breathers, or book lungs. Models should be accompanied by the name of the breathing method and a brief description of its function. **L3**

Science Journal

Answers will vary according to the creature students have selected. All students should have described the animal's appearance and method of breathing in detail.

Going Further ▌▌▌▌▌▶

Birds are also unusual breathers. Invite students to work in teams to find out how birds obtain oxygen. Have students describe the appearance of avian lungs: **How does air pass through the lungs? Are there additional air sacs? If so, how are they placed and what is their function?** Students can present their work as brief oral reports with illustrations. Use the Performance Task Assessment List for Oral Presentation in **PASC**, p. 71. **COOP LEARN**

Visual Learning

Figure 16-3 Display a three-dimensional model of the respiratory system. To reinforce sequencing skills, have volunteers trace the pathway of air through the model as others describe it. **When are you most likely to breathe through your mouth?** *when a lot of oxygen is needed, such as when doing aerobic exercises* **Why do you think it's important not to talk while eating?** *so food doesn't enter the trachea* L1

Explore!

What is your trachea like?

Time needed 5–10 minutes

Materials There are no special materials or preparation required for this activity.

Thinking Processes observing and inferring, recognizing cause and effect, forming operational definitions

Purpose To investigate features of the trachea.

Teaching the Activity

You may want to demonstrate the movements in this activity before the students begin.

Science Journal Suggest that students draw diagrams to enhance their written observations in their journals. L1

Expected Outcome

The students should feel the features of the trachea and hypothesize what these features do.

Answers to Questions

1. Students should feel firm rings stacked on top of one another.

2. These rings keep the trachea open so that air can pass through.

3. When you cough, you inhale first and then quickly release the air with great force. A cough can dislodge food or other obstructions in the trachea.

4. Description should summarize information in answers 1-3.

✓ Assessment

Performance Have small groups of students use their observations from this Explore activity and Figure 16-3 to make three-dimensional models of the trachea. Use the Performance Task Assessment List for Model in **PASC**, p. 51. COOP LEARN P

Your Pathway for Air

Your body has its own structures through which it receives oxygen and expels carbon dioxide wastes. These structures make up your respiratory system. The major parts of your respiratory system are shown in **Figure 16-3**. Refer to this diagram as you follow the path of air from your nose to your lungs.

Figure 16-3

Ⓐ Air enters your body through your nostrils and sometimes your mouth. When are you most likely to breathe through your mouth?

Ⓑ As the air moves past the tissues in your nasal cavity, your body tissues transfer heat and moisture to the air.

Ⓒ The warmed, moistened air moves to the pharynx, a passageway at the back of your nose and mouth, and then on to the larynx, or voice box.

Ⓓ A protective flap of tissue called the epiglottis covers the top of the larynx. When you breathe, the epiglottis is open. When you swallow, the epiglottis closes so that food or liquid moves toward your esophagus and not toward your lungs. Why do you think it's important not to talk while eating?

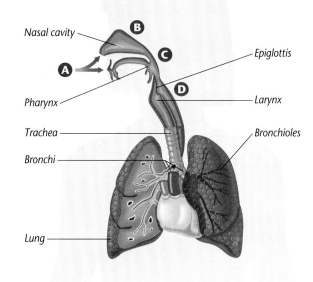

Explore! ACTIVITY

What is your trachea like?

What To Do

1. Place your fingers on the front of your neck and gently move them up and down. What do you feel?

2. Turn your head from side to side and continue breathing. How does the structure in your neck help keep you alive?

3. Now, gently cough while keeping your fingers on your neck. What happens when you cough?

4. *In your Journal*, describe the structure and possible function of the structure you felt in your neck.

Program Resources

How It Works, p. 17, What Are Hiccups? L1

Science Discovery Activities, 16-1

When you ran your fingers up and down your throat, you felt the top of your trachea. Remember, an insect's trachea carries air in and out of its body. Your trachea is also a passageway for air moving into and out of your body. The rings you felt are C-shaped rings of cartilage that keep the trachea open. What advantage is there to having a trachea that is open all the time?

■ Lungs

At the lower end of the trachea are two short branches called bronchial tubes through which air moves into the lungs. The **lungs**, which are located in the chest cavity, are the main organs of your respiratory system. Here in the lungs, the exchange of oxygen and carbon dioxide takes place.

Within your lungs, the bronchial tubes branch into increasingly smaller and smaller passageways. At the ends of the narrowest tubes are clusters of tiny, thin-walled sacs called **alveoli**, which are shown in **Figure 16-4**. You can get an idea of what a mass of alveoli looks like if you imagine a tight cluster of grapes. Your lungs contain millions of alveoli that are surrounded by capillaries. Just as oxygen and carbon dioxide pass between the gills and capillaries in fish, the exchange of these gases takes place between the alveoli and capillaries in your body. Oxygen and carbon dioxide pass easily, or diffuse, through the thin walls of the alveoli. In addition, the large number of alveoli provides a huge surface area for gases to be exchanged.

Connect to...
Physics

The surface area of your lungs is about 20 times as great as that of your skin. Prepare a talk that tells what this means about how well adapted the body is to take in oxygen.

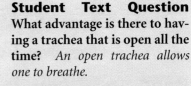

Content Background
The term *breathing* refers to the process by which air moves in and out of the lungs. *Respiration* refers to the exchange of oxygen and carbon dioxide in the lungs (external respiration) or between the blood and body cells (internal respiration).

Connect to . . .
Physics

Students should be able to find answers for this Connect to . . . feature in basic library references.

Answer: *Increased surface area allows for increased passage of molecules back and forth. If we had to rely on our skin as a breathing organ, we would not be able to obtain enough oxygen, nor get rid of wastes efficiently. Another factor is that volume increases more rapidly than surface area. Thus, at larger volumes, an organism's surface area will not be sufficient for moving molecules into and out of the body.*

Figure 16-4

Trachea

Alveoli

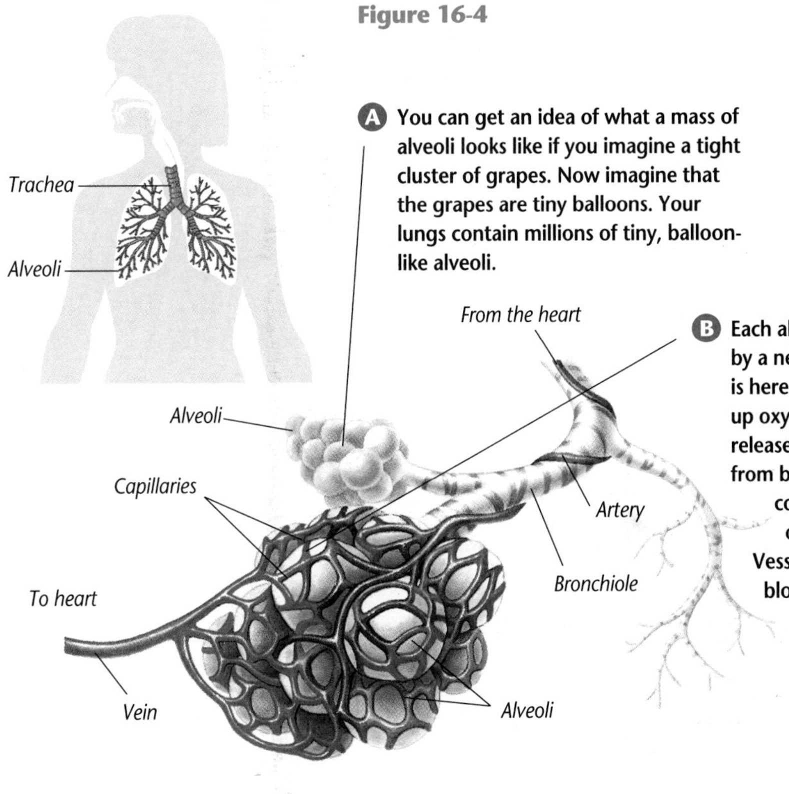

(A) You can get an idea of what a mass of alveoli looks like if you imagine a tight cluster of grapes. Now imagine that the grapes are tiny balloons. Your lungs contain millions of tiny, balloon-like alveoli.

From the heart

(B) Each alveolus is surrounded by a network of capillaries. It is here that your blood picks up oxygen from the air, and releases carbon dioxide waste from body cells. Vessels colored red carry oxygen-rich blood. Vessels colored blue carry blood with less oxygen.

Alveoli

Capillaries

To heart

Artery

Bronchiole

Vein

Alveoli

16-1   How Do You Breathe?   **489**

Visual Learning

Figure 16-4 To help students visualize the lung, display a large cluster of grapes in a net bag. Have students observe how the net fits around the grapes. Relate this to the description of the lungs in Figure 16-4. Take the grapes out of the net and use the main branch and smaller branches down to the cluster as an analogy for the trachea, bronchioles and alveoli. **LEP**

Planning the Activity

Time Needed
20–25 minutes

Materials round balloon, metric ruler

Process Skills making and using tables, making and using graphs, measuring in SI, practicing scientific methods, interpreting data, forming operational definitions

Purpose To measure the vital capacity of the lungs.

Teaching the Activity

Demonstration Show students how to conduct this activity. They must try to inhale and then exhale as much air as possible to measure vital capacity. Some students may complain that the balloon has not inflated at all. Explain that it has, but vital capacity is low.

Troubleshooting The size of the balloon could possibly affect the results. Be sure all students have balloons of similar capacities.

Process Reinforcement To help students review reading a line graph, call out several balloon diameters from the graph on page 491 and have students supply the lung capacities by reading the vertical axis.

Science Journal Students will probably record lower average values than those listed in the table as sample data. [L1]

INVESTIGATE!

Take a Deep Breath

Vital capacity is the largest amount of air your lungs expel after taking the deepest breath you can. In this activity, you will find your vital capacity.

Problem
What is your vital capacity?

Materials
round balloon   metric ruler

What To Do

1 Copy the data table *into your Journal.*

2 Stretch a balloon several times. Take as deep a breath as you can. Exhale into the balloon as much air as possible. Pinch the balloon closed.

3 Measure the diameter of the balloon in centimeters as shown (see photo **A**). Record the data.

4 Repeat Steps 2 and 3 four more times.

Sample data

Data and Observations

Trial	Diameter in Centimeters	Vital Capacity in Cubic Centimeters
1	18	3000
2	17.5	2750
3	18	3000
4	18	3000
5	17	2500
Total	88.5	14 250
Average	17.7	2850

Program Resources

Activity Masters, pp. 67–68, Investigate 16-1

Critical Thinking/Problem Solving, p. 5, Flex Your Brain

Concept Mapping, p. 24, Oxygen in the Body [L1]

Meeting Individual Needs

Physically Challenged Prior to beginning this Investigate, check with the school nurse to find out whether or not your physically challenged students are able to participate. If participation is limited, have the physically challenged students be responsible for measuring the balloons.

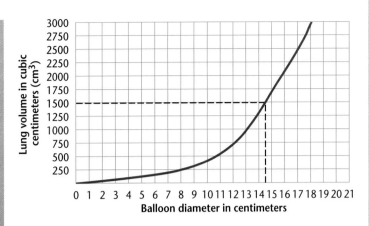

A

5 Vital capacity is expressed in cubic centimeters. To calculate vital capacity, find the balloon diameter on the horizontal axis of the graph. Follow this number up to the red line and move across to the corresponding capacity. The dashed line on the graph shows an example of how to find vital capacity.

6 Record your vital capacity for each trial.

7 Calculate and record your average vital capacity.

Analyzing

1. What is your vital capacity?

2. Were there differences in the diameters of the balloon during the five trials?

3. How does your average compare with the averages of other class members?

Concluding and Applying

4. How could you improve the accuracy of this activity?

5. Can you infer how this activity could be used to find people who might have a lung disease?

6. **Going Further** Tidal volume is the amount of air that you exhale after drawing a normal breath. Design an experiment to find your tidal volume.

How Air Moves In and Out

You learned in Chapter 14 how gases act under certain conditions. Air is a mixture of gases. It follows the same laws in your lungs as it would in a laboratory. For example, you know that according to Boyle's law, if you decrease the volume of gas in a container, the pressure of the gas will increase. You can demonstrate Boyle's law by squeezing an empty plastic bottle. Air rushes out when you squeeze the bottle, as illustrated in **Figure 16-5**.

Figure 16-5

Ⓐ If you hold your hand over an open plastic bottle, you feel no flow of air. Why?

Ⓑ Squeezing the bottle increases the pressure inside the bottle. The air inside the bottle has a higher pressure than the air outside the bottle.

Ⓒ As you release your grip, the air pressure inside the bottle drops below the air pressure outside the bottle.

Explore! ACTIVITY

How does your chest size change when you breathe?

What To Do

1. Inhale and use a metric measuring tape to find the size of your own chest. When measuring, place the tape around your chest and directly under your armpits.

2. Measure the size of your chest when you exhale.

3. *In your Journal*, compare your chest size when inhaling and exhaling. Explain what you think caused the differences in the measurements.

■ The Diaphragm

The size of your chest changes as air moves in and out of your lungs. Like hands on a plastic bottle, something in your chest cavity exerts pressure or relieves pressure on your lungs. These pressure changes are caused by your **diaphragm**, a thin sheet of muscle under your lungs. Follow the events in **Figure 16-6** to see how the pressure changes caused by the diaphragm allow you to inhale and exhale.

You've seen that air pressure is important in helping you to breathe. What happens when you go to the

Figure 16-6

Inhaling

A When you inhale, your diaphragm contracts and muscles pull your ribs upward and outward. These actions increase the size of your chest cavity. Your lungs expand to fill the space.

B Because there is now more room for the air in your lungs, the pressure of that air decreases. The air pressure in your lungs becomes lower than the air pressure outside your body. The higher-pressured outside air rushes in through your nose to fill the low-pressured lungs.

Exhaling

C When you exhale, your diaphragm relaxes and moves upward. The muscles attached to your ribs also relax. These two actions reduce the size of your chest cavity.

D Because there is now less room for the air in your lungs, the air pressure in your lungs increases. The air pressure in your lungs becomes higher than the air pressure outside your body. The high-pressure air rushes out of your lungs, like it did from the squeezed bottle.

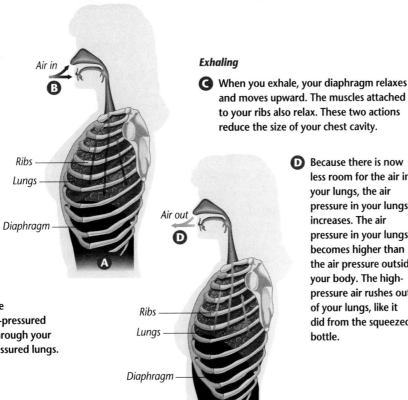

Air in

B

Ribs

Lungs

Diaphragm

A

Air out

D

Ribs

Lungs

Diaphragm

C

Visual Learning

Figure 16-6 Help students summarize the process of exhaling and inhaling by completing these sentences.
• When exhaling, air rushes *out*, the diaphragm *relaxes*, and the chest cavity *decreases*.
• When inhaling, air rushes *in*, the diaphragm *contracts*, and the chest cavity *increases*.

3 ASSESS

Check for Understanding

Have students check their answers to the Apply question by blowing out a candle and describing what they felt occur in their chests. [L1]

Reteach

Demonstration Use a lung demonstration apparatus to illustrate to students how the downward movement of the diaphragm reduces air pressure within the chest cavity.

Extension

Research Have students research the way singers practice breathing in order to have greater breath control while singing. [L3]

4 CLOSE

Demonstration

To demonstrate how the diaphragm contracts and relaxes, use a sports water bottle and masking tape. Have a volunteer remove the plastic straw from the bottle and cover the hole with a piece of tape. Instruct the volunteer to squeeze the bottle and describe what happens. *The air rushes out the small air hole.* Allow the bottle to resume its normal shape and have students listen as the air rushes back in. Ask students which state of the bottle represents the diaphragm contracting and which represents the diaphragm relaxing. [L1]

---

mountains where the air pressure is lower? You probably find it harder to breathe.

People who live at high altitudes have physical traits that help them breathe at the lower air pressure. The Aymara and Quechua Indians of the Andes Mountains have barrel-shaped chests, strong diaphragms, and large lungs. They also have more capillaries around the alveoli, larger hearts, and more red blood cells. All of these traits deliver oxygen to their body cells and tissues more efficiently.

check your UNDERSTANDING

1. Compare how a paramecium and a fish obtain oxygen.
2. Draw a diagram of the respiratory system indicating the direction of air flow in and out.
3. Explain how air pressure relates to inhalation and exhalation.
4. **Apply** Describe what occurs inside your chest cavity when you blow out the candles on a birthday cake.

check your UNDERSTANDING

1. Oxygen diffuses through a paramecium's thin outer covering, oxygen diffuses from gill filaments into capillaries in fish.
2. See Figures 16-3 and 16-6.
3. Exhalation—pressure inside the lungs is greater than outside, and gases in the lungs are pushed out; inhalation—air pressure inside the lungs is lower than air pressure outside the body, and air rushes into the lungs.
4. Diaphragm contracts and flattens; muscles between ribs contract, causing rib cage to expand; lungs fill with air. Blowing out, ribs move closer. Increased pressure on lungs forces air out.

PREPARATION

Planning the Lesson

Refer to the Chapter Organizer on pages 482A–D.

Concepts Developed

Students will explore the composition of the air that is inhaled and exhaled and how the circulatory system interacts with the respiratory system to transport oxygen and carbon dioxide throughout the body.

1 MOTIVATE

Bellringer

 Before presenting the lesson, display **Section Focus Transparency 48** on an overhead projector. Assign the accompanying **Focus Activity** worksheet. L1

Demonstration Put a sugar cube or a small pile of sugar crystals on a piece of aluminum foil. To make the sugar ignite easily, add a sprinkle of powdered charcoal to the corner of the cube or on top of the pile. Then light with a match. **Use caution** because the sugar will melt and boil. Extinguish the flame. Lead a discussion of the necessity of oxygen for the oxidation of sugar within cells. Explain that the respiratory system provides oxygen for this chemical activity that releases energy for cell functions.

GLENCOE TECHNOLOGY

Videodisc

STVS: Human Biology

Disc 7, Side 2
Nicotine and the Lungs (Ch. 15)

Section Objectives

■ Compare air that is inhaled with air that is exhaled.

■ Explain cellular respiration and its relationship to gas exchange in your body.

Key Terms

*hemoglobin,
cellular respiration*

How Oxygen Gets To All Parts Of Your Body

Because only 21 percent of the air you inhale is oxygen, your body needs to move a steady supply of air into and out of your lungs. You learned the path air travels into your lungs, but how does your body remove the oxygen it needs?

■ **Hemoglobin**

Your blood contains an oxygen-binding substance called **hemoglobin**. Hemoglobin is found in the red blood cells and contains iron, which easily bonds with oxygen. **Figures 16-7** and **16-8** show how oxygen moves into the

Trachea
Bronchi
Bronchioles
Alveoli

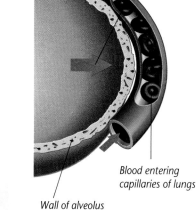

Oxygen

Blood entering capillaries of lungs

Wall of alveolus

Figure 16-7

Ⓐ As blood moves through your capillaries around the alveoli, oxygen diffuses from the alveoli into the capillaries and then into your red blood cells.

Ⓑ Once inside the red blood cells, oxygen binds with molecules of hemoglobin.

Ⓒ After leaving your lungs, the oxygen-rich blood moves to your heart, which pumps blood to the rest of your body.

Carbon dioxide

Blood leaving capillaries of lungs

Figure 16-8

Ⓐ The hemoglobin and oxygen remain bonded as the red blood cells move throughout your body.

Ⓑ As the blood passes body cells that have low amounts of oxygen, the oxygen in the red blood cells is released by the hemoglobin and diffuses from the red blood cells into individual body cells.

Ⓒ At the same time, carbon dioxide wastes diffuse from your body cells into the blood in your capillaries. The carbon dioxide then diffuses from the capillaries into the alveoli and is exhaled from the lungs.

494 Chapter 16 Breathing

Program Resources

Study Guide, p. 54

Multicultural Connections, p. 35, Breathing in Mexico City L1

Making Connections: Integrating Sciences, p. 35, Breathing Fluid L2

Making Connections: Across the Curriculum, p. 35, Growing Up Above 3000 Meters L2

Section Focus Transparency 48

What makes air vital?

In the 1700s, a British chemist, Joseph Priestley, discovered that a mouse couldn't live in a container in which a candle had previously been burned. He reasoned that some substance in the air was destroyed when the candle burned. He also discovered that if he put a mint plant into the container for eight or nine days, and then returned a live mouse to the container, it lived. The substance necessary for life had returned. It was later to be called oxygen.

red blood cells, where it bonds with the hemoglobin. Because of hemoglobin, your blood is able to carry oxygen throughout your body, where it is needed. The heart pumps this oxygen-rich blood to all parts of your body. Then, the oxygen is released by the hemoglobin and taken in by body cells. As oxygen is being released to your body cells, carbon dioxide is being taken up by the blood. About 30 percent of the carbon dioxide attaches itself to passing hemoglobin. The rest travels back to your lungs in plasma, the watery substance in blood.

Once the blood is back in the lungs, the carbon dioxide diffuses from the plasma and red blood cells into the alveoli. Your lungs release this waste gas from your body whenever you exhale.

Find Out! ACTIVITY

What other gas do you exhale?

It's tempting to think of the air you breathe in and out as being made up only of oxygen. Try this experiment to find out if that is true.

What To Do

1. Use a towel to wipe off your hands.
2. Hold the palm of your hand up to your mouth and exhale into it.
3. Feel the palm with your other hand.
4. Then, hold a mirror up to your mouth and breathe onto it. Observe what happens to the mirror.

Conclude and Apply

1. How did your hand feel after you breathed into it?
2. What did you see on the mirror?
3. *In your Journal,* summarize your observations and infer what other gas is released from your body when you exhale.

2 TEACH

Tying to Previous Knowledge

Remind students that bodies, like machines, require fuel to run. Students know that cars need gas and that they give off exhaust. The body also requires fuel to function and must get rid of its waste gases. The burning of fuel requires oxygen; one waste product is carbon dioxide.

Find Out!

What other gas do you exhale?

Time needed 5–10 minutes

Materials towel, mirror

Thinking Processes thinking critically, observing and inferring

Purpose To observe that water vapor is exhaled during respiration.

Teaching the Activity

Students could complete this activity independently.

Science Journal Students' summaries should include that gases are exhaled when they breathe. L1

Expected Outcomes

Students may feel their hands warm up at first. The moisture in the palm should be felt with the other hand.

Conclude and Apply

1. It felt wet and warm.
2. water droplets
3. Students should note that water vapor was released during exhalation.

✔ Assessment

Performance Have students find out the average composition of the air they breathe and make a pie graph displaying these values. *Most air is about 20 percent oxygen, 74 percent nitrogen, and about 6 percent water vapor and other trace gases.* Use the Performance Task Assessment List for Graph from Data in **PASC**, p. 39. P

Program Resources

Laboratory Manual, pp. 101–102, Carbon Dioxide and Respiration LEP

ENRICHMENT

Research Have students research the bends. This condition occurs in deep-sea diving or high-altitude flying when a person goes too rapidly from an abnormal to a normal atmospheric pressure. Have students find out what causes the feeling of cramps in this condition and how it is treated. L3

Theme Connection The theme that this chapter supports is energy. As Figure 16-9 shows, a consistent mixture of gases in the atmosphere is inhaled while a different percentage of gases is exhaled. The extraction of oxygen in the body that occurs during respiration allows us to break down the food needed to provide our cells with energy.

Student Text Question

What happens to cause this change in composition of the air you breathe? *Oxygen is removed from the air and taken into the body. Carbon dioxide is removed from the body and released to the air.*

Content Background

Impulses from the nervous system regulate the rate of breathing when too much carbon dioxide has collected in the bloodstream. This is a homeostatic function, one of the many in the human body.

Comparing Air Inhaled With Air Exhaled

You discovered in the Find Out activity that water vapor is one gas that is evident when you exhale. Where does this water vapor come from? Tissues lining the respiratory system are very moist. When you inhale atmospheric air, it is immediately exposed to these moist tissues, and it becomes more humid. But the amount of water vapor in air is not the only thing to change when you breathe. The amount of carbon dioxide and oxygen gases changes as well, as you can see in **Figure 16-9**. What happens to cause this change in composition of the air you breathe?

Figure 16-9

Composition of the Air You Inhale and Exhale

Inhaled air
78.62% 0.5% 0.04% 20.84%

Exhaled air
74.5% 6.2% 3.6% 15.7%

■ Nitrogen
■ Water vapor
■ Carbon dioxide
■ Oxygen

The air you inhale and the air you exhale are different in composition from one another.

Physics CONNECTION

Your Larynx

What kinds of sounds can you make with your voice? You may be able to sing or to imitate the sound of a door creaking, a dog growling, or a bass drum booming. You may have noticed that when you get excited, the pitch of your voice goes up or that when you're depressed, the pitch goes down. Your larynx makes a wide range of sounds possible.

When you speak or sing, air rushing out of your lungs as you exhale vibrates your vocal cords to produce sound. Muscles in your throat control the length and shape of the cords. Loose, relaxed vocal cords produce a low sound. Tightened, stretched cords make higher pitches.

When you sing, air moving across your vocal cords causes them to vibrate.

Building a Vocal Cord Model

Try building a model of the vocal cords. Find a metal, plastic, or cardboard tube one inch

Purpose

This Physics Connection provides additional information about the human respiratory system by explaining how the larynx functions.

Content Background

Human speech requires both raw and finished sounds. The former are produced by the larynx; the latter, by the mouth, teeth, tongue, palate, and facial muscles.

The larynx, about 4 cm long, is shaped like a triangular box. It is made of nine cartilages of which the most prominent is the thyroid cartilage, or "Adam's apple." These cartilages are linked by muscles and ligaments. Lying on either side of the opening of the larynx (the glottis), are the two folds of tissue called the vocal cords. They are attached to the thyroid cartilage in front and to a pair of smaller cartilages in back. When a person sings or talks, air passes against the vocal cords and causes them to

Cellular Respiration

Think about what is needed before machines can run or microwave ovens can cook. Don't these objects need energy to work? Your body needs energy to operate, too. Your body gets this energy from a chemical reaction that happens within individual cells. This process is called cellular respiration. During **cellular respiration**, oxygen combines with stored nutrients in cells to release energy, carbon dioxide, and water. Your respiratory system supplies the oxygen and removes the carbon dioxide produced in this process.

Figure 16-10

The process of respiration is always happening, but the rate at which it happens varies with the types of activities that you do. The more energy you need to carry out an activity, like swimming, the greater your rate of respiration. As your respiration rate increases, your need for oxygen increases. During which activity do you think your need for oxygen would be greater: swimming or walking?

Teacher F.Y.I.

Tell students that during one minute while the body is at rest, approximately 200 mL of oxygen are used by body cells; an equal amount of carbon dioxide is produced.

Visual Learning

Figure 16-10 During which activity do you think your need for oxygen would be greater: swimming or walking? *Swimming requires a higher rate of breathing and cellular respiration than walking and therefore would require more oxygen.*

in diameter. Find a rubber stopper for one end of this tube, fitted with a smaller glass or metal tube through its center.

Connect this smaller tube to an air pump. A foot pump, like the one you might use to inflate a bike tire, would be a good choice. Cut two pieces of rubber from toy balloons. Stretch these pieces across opposite sides of the open end of the tube, securing them with tape or rubber bands. Rubber bands may be better, as they will allow you to make alterations more easily. Be sure to leave plenty of width in the pieces of rubber for further alteration.

As you pump air through the tube, the rubber sheets should vibrate—producing sound.

Loosen or tighten the sheets as you wish. Looser sheets should produce a low pitch. Tighter sheets, with the rubber pieces stretched thinner, will produce a higher pitch.

If you have easy access to the materials, construct more than one of these models. You may be able to combine different sounds, or even to play a tune, alternating use of the models or by sounding them at the same times.

You Try It!

See if you can match the pitch of your own speaking or singing voice with that of your model vocal cords. Also,

take note of the effects of different quantities of air on the pitches and volumes of sounds from the model.

Across the Curriculum

Health

Tell students that as recently as one hundred years ago, people believed that the night air carried diseases, and so they slept with the windows shut tight. Today we know that this is not so. Ask students to suggest reasons why. *Fresh air not only makes us feel better, but also keeps up the supply of oxygen.*

497

---

vibrate. Changing the frequency of the vibrations changes the tone of the sounds.

Teaching Strategy

Have students work with a partner to practice tonal variations including whispering (speaking without using their vocal cords). **COOP LEARN** **L1**

Demonstration

Have students demonstrate their vocal cord models' tonal range for the class.

Activity

Invite students to learn more about common speech disorders, such as lisps, cluttering, stuttering (stammering). **What causes these disorders? What treatments are available?** Have students share their findings with the class.

Going Further

Students can work in small groups to find out how ventriloquists make dummies and puppets speak. After students gather

information, have them practice this technique of speaking while retracting their tongues and moving only the tip. Remind them to distract the audience with the puppet. When students feel comfortable with the technique, invite volunteers to demonstrate their new-found skills for the class. **COOP LEARN**

16-2 Breathing and Carbon Dioxide

Preparation

Purpose To design an experiment to observe the effect of exercise on the amount of carbon dioxide exhaled by the lungs.

Process Skills organizing information, making and using tables, comparing and contrasting, designing an experiment, separating and controlling variables, observing, interpreting data

Time Required 45 minutes

Materials See reduced student text. Make the bromothymol blue solution by adding 2 to 6 mL of bromothymol blue to 200 mL of distilled water. A more concentrated solution will increase the time needed to change the color of the solution.

Safety Before any student does any exercise, make sure he or she has no medical problem that the exercise would complicate. Also, caution students not to inhale the solution.

Possible Hypotheses Students may hypothesize that exercise will increase the amount of carbon dioxide exhaled by the lungs. Some students may hypothesize that exercise will have no effect.

The Experiment

Process Reinforcement
Check students' journal entries before they begin to see that they have designed a logical and valid experiment, i.e., they are testing only one variable.

Possible Procedures
Procedures will vary but should be along the following lines: Label beakers "A" and "B." Measure 50 mL bromothymol blue solution with a graduated cylinder and pour into beaker A. Repeat for beaker B. Record the time it takes for the color change to occur while exhaling through the straw into the solution in beaker A. Continue to exhale until the solution turns yellow. Record the time in the data table. Run in place for three minutes and repeat the process

Breathing and Carbon Dioxide

You may have noticed that when you exercise or play sports, your breathing rate changes. Is the amount of carbon dioxide you exhale also related to your body's level of activity? You can test for the presence of carbon dioxide in your breath by using chemicals that change color in the presence of carbon dioxide.

Preparation

Problem
How does exercise affect the amount of carbon dioxide exhaled by the lungs?

Form a Hypothesis
Will exercise affect the amount of carbon dioxide you exhale? How?

Objectives
- Predict the effect of exercise on the amount of carbon dioxide exhaled.
- Recognize the cause and effect relationship between exercise and amount of carbon dioxide exhaled.

Materials
clock or watch with second hand
200 mL bromothymol blue solution
2 drinking straws
400-mL beakers (2)
graduated cylinder

Safety

Be careful not to inhale or swallow bromothymol blue solution. Inform your teacher of any medical conditions that prevent you from exercising.

498

DESIGN YOUR OWN
INVESTIGATION

Plan the Experiment

1 Bromothymol blue solution turns green and then yellow as carbon dioxide is added to it. How will you use this information as you design your experiment?

2 You may want to look at the amount of time it takes for the color change to occur under different circumstances. What will you want to keep constant? What will change?

3 Write out your plan and design a data table *in your Science Journal.*

Check the Plan

1 Would the amount of solution make a difference in your tests? Would it matter who was doing the exercise and exhaling for each test?

2 Have your teacher check and approve your plan before you begin. Carry out your experiment and note any changes you make as you go.

Analyze and Conclude

1. Analyze Were you able to determine a difference in the amount of carbon dioxide exhaled before and after exercise? How can you tell?

2. Infer Infer why exercising causes you to breathe faster than usual.

3. Compare Look back at your hypothesis and compare your results with your predictions. What is going on within the cell during exercise that explains the results?

Going Further

Design a concept map on the computer that describes the process that goes on in the cell during exercise.

 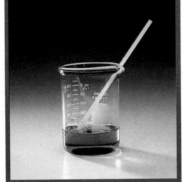 

16-2   The Air You Breathe   **499**

Meeting Individual Needs

Behaviorally Disordered Provide behaviorally disordered students a clearly structured environment for this Investigate. Keep groups of students working at the same pace through each of the steps. Stress the importance of carefully pouring the correct amount of the solution into each beaker and how too much or too little solution will affect the color change.

Program Resources

Activity Masters, pp. 69–70, Design Your Own Investigation 16-2
Science Discovery Activities, **16-2**

---

of exhaling into the solution in beaker B. Stop when the color of the solution in beaker B matches that in beaker A. Record this time in the data table. **L2**

Teaching Strategies

Troubleshooting Remind the students to blow quickly into the bromothymol blue solution after exercising.

Science Journal Have students record their hypotheses, data tables, and results in their journals. **L1**

Expected Outcomes

Students should find that it takes less time for the solution in beaker B to change color after they have exercised than did the solution in beaker A.

Analyze and Conclude

1. Yes. It took longer for the solution to change color before exercising because there was less carbon dioxide in the exhaled breath.

2. Exercising causes you to breathe faster to increase the amount of oxygen supplied and carbon dioxide removed from the body.

3. Oxygen combines with stored nutrients in the cells to release carbon dioxide, energy, and water.

✔ Assessment

Process Have small groups of students conduct an experiment to test the relationship between age and the amount of carbon dioxide exhaled. Students should test at least five volunteers. Make sure that participants have a doctor's permission to exercise. Students can make graphs displaying their data and results. Use the Performance Task Assessment List for Graph from Data in **PASC,** p. 39. **COOP LEARN** **P**

Going Further ⦀⦀⊪

1st box—Cell with nutrients; 2nd box off to side with arrow to first—oxygen taken in through exercise; 3rd, 4th, and 5th boxes as outflows from 1st—carbon dioxide, energy, and water are released.

16-2   The Air You Breathe   **499**

3 ASSESS

Check for Understanding

Direct students to the Check Your Understanding questions. To answer the Apply question, students must recall that it is the beating of the heart that pumps the blood through the bloodstream and that oxygen and carbon dioxide are transferred between the bloodstream and the lungs.

Reteach

Discussion Discuss with students why an aerator is used in an aquarium. Because fish take in oxygen from the water through their gills, oxygen must be dissolved in the water they "breathe." Ask them whether they think the level of carbon dioxide in the water would increase or decrease if fish were added to a tank of water.

Extension

Scientific Drawing Have students who have mastered the concepts in this section draw two pictures of human lungs. In the first, ask students to label the gases and their concentrations to represent air that is inhaled. In the second, students' labels should represent the composition of exhaled air. L3

4 CLOSE

Discussion

In the next section, students will learn about respiratory diseases. You can lead into this topic by having students discuss how healthy lungs relate to the amount of oxygen that can be delivered to the cells.

Figure 16-11

A When you run, your breathing becomes more rapid and deep to supply the added oxygen your increased rate of respiration requires. An increase in the rate of respiration also results in an increase in the amount of carbon dioxide produced by your body cells.

B When running, and immediately afterward, you exhale more carbon dioxide than usual. Eventually, as you rest, your rate of respiration slows down and your breathing rate returns to its before-exercise rate.

In the Investigate, you explored the relationship between physical activity and the amount of carbon dioxide exhaled. You probably discovered that increased activity caused an increase in the amount of carbon dioxide your body gives off. You know that carbon dioxide is a waste product of cellular respiration. How, then, are breathing and cellular respiration related?

You've learned that breathing—a function of the respiratory system—gets oxygen and carbon dioxide into and out of your body. The circulatory system moves these two gases around within your body. The cells use oxygen and give off carbon dioxide during cellular respiration, which is how your body obtains energy.

Your respiratory system is very efficient, especially when you have clean, fresh air to breathe. What happens if the air is not so fresh and clean? What if a disease prevents the efficient transfer of oxygen to your lungs and body cells? In the next section, you will learn how these problems affect your respiratory system.

check your UNDERSTANDING

1. How does the air you inhale differ from the air you exhale?
2. What role does hemoglobin play in the transfer of oxygen between your lungs and body cells?
3. Why is the amount of water vapor higher in air you exhale than in air you inhale?
4. **Apply** Explain why both your heart rate and breathing rate increase with exercise.

check your UNDERSTANDING

1. Air you inhale has more oxygen and less carbon dioxide than the air you exhale.
2. Iron in the hemoglobin of red blood cells bonds with the oxygen and carries it to body cells that need oxygen for respiration.
3. Water is a waste product of respiration. Exhaling it as water vapor is a way to get rid of it.
4. The increased heart rate pumps blood through your body faster. As a result, more oxygen gets into the bloodstream, and more carbon dioxide is released from the bloodstream into the lungs, where it can be exhaled.

16-3 Disorders of the Respiratory System

Keeping Your Lungs Clean

As you learned, air is a mixture of gases, but it also contains particles of dirt, pollen, dust, and smoke. These pollutants can damage your respiratory system and interrupt the flow of oxygen to your body's cells. Every year thousands of people in the United States die from diseases related to smoking and air pollution.

Your body has some defenses against the particles that mix with the different gases in air. When you inhale, these particles become stuck in a moist lining in the trachea and lungs. This lining is covered with tiny hairlike structures called **cilia**, shown in **Figure 16-12**. Cilia beat in an upward direction, causing a current that carries the particles to the throat, where they are swallowed and disposed of by acid in the stomach. What do you think happens if your cilia stop working?

■ Lung Disease

When inhaled air contains large amounts of dust, pollen, smoke, or smog particles, cilia lining the respiratory system can be affected. Smoke from cigarettes, for example, temporarily paralyzes cilia, preventing them from performing their sweeping jobs. Particles not swept out by cilia

usually reach the alveoli, where they are engulfed by white blood cells. White blood cells help prevent infections by consuming both dirt and bacteria. However, some substances such as asbestos, a material used for insulation, can't be consumed by white blood cells, and the substance remains in the lungs. In the following activity, you can determine how a classmate's lungs sound.

Figure 16-12

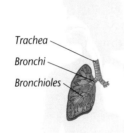

Trachea
Bronchi
Bronchioles

Cilia

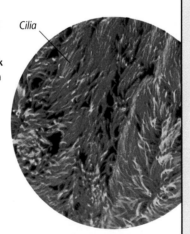

Ⓐ The linings of your nasal passageways, trachea, and lungs are covered with tiny hairlike structures called cilia. Cilia are covered with mucus. They beat toward your pharynx and cause the mucus to move, like a flowing sheet, in that direction.

Ⓑ Dust and other particles stick to the mucus and move with it toward the pharynx as fast as one centimeter per minute. At the pharynx, the particles and mucus are swallowed and broken down by acids in your stomach.

Program Resources

Study Guide, p. 55

Critical Thinking/Problem Solving, p. 24, Additional Hazards of Cigarette Smoking L2

Multicultural Connections, p. 36, Fighting Tuberculosis Among the Navajo L1

Making Connections: Technology & Society, p. 35, Laws Protect Nonsmokers L1

Science Discovery Activities, 16-3

Section Focus Transparency 49

PREPARATION

Planning the Lesson

Refer to the Chapter Organizer on pages 482A–D.

Concepts Developed

In this section, respiratory dysfunctions are presented. Normal breathing depends on healthy lung tissue and the proper balance of gases in the atmosphere. Environmental conditions such as pollution and unhealthy practices, such as tobacco smoking, disrupt this delicate balance.

1 MOTIVATE

Bellringer

Before presenting the lesson, display **Section Focus Transparency 49** on an overhead projector. Assign the accompanying **Focus Activity** worksheet. L1
LEP

Demonstration Use this activity to model our cells' need for oxygen in order to produce energy. Materials needed are a candle, matches, glass cup or jar.

Place the candle on a heat-resistant surface and light it. Then cover the candle with the glass. The flame should extinguish quickly. Discuss with students why the flame went out and whether a larger glass would put out the flame more or less rapidly. Explain that oxygen was necessary to the burning process. Lead students to conclude that just as the flame was extinguished when the oxygen supply was depleted, any disruption in the supply of oxygen will harm the ability of the cells to release energy. L1

Section Objectives
■ Discuss respiratory disorders and their causes.
■ Determine how to keep your lungs healthy.

Key Terms
cilia, asthma, cystic fibrosis, emphysema, lung cancer

Explore!

What is percussing?

Time needed 5–10 minutes

Materials No special materials or preparations are required for this activity.

Thinking Processes observing and inferring, forming operational definitions

Purpose To use the technique of percussing, a technique doctors use to check patients' lungs.

Teaching the Activity

Have students work in same-sex pairs. **COOP LEARN**

Troubleshooting Be sure not to let students diagnose each other.

Science Journal Students' observations will probably all describe a clear, hollow sound.
L1

Expected Outcome

All students' lungs should sound hollow. If a student's lungs sound unusual, check with a school nurse as to follow-up procedure.

Answer to Questions

3. If the lungs are healthy, they will sound hollow. In a patient with pneumonia, the lungs will sound solid or filled with fluid.

✔ Assessment

Content Have your students draw cartoons that explain what percussing is and why doctors do it. Use the Performance Task Assessment List for Cartoon/Comic Book in **PASC**, p. 61. **P**

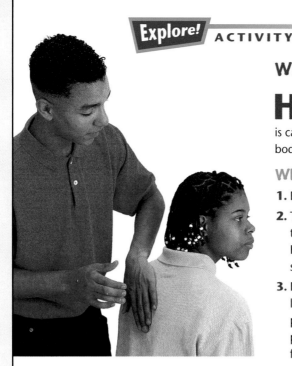

Explore! ACTIVITY

What is percussing?

Have you ever gone to a doctor and had him or her thump on your back? This procedure is called percussing. A doctor can tell whether a body part is solid or air-filled by percussing.

What To Do

1. Put one hand flat on your partner's back.
2. Tap the third finger of that hand with the three middle fingers of your other hand. Healthy lungs should make a clear, hollow sound.
3. Pneumonia patients have fluid around their lungs. *In your Journal*, infer how percussing can help a doctor diagnose if a patient's lungs are healthy or filled with fluid.

■ Asthma

Did you ever see a person who was having a difficult time breathing? That person may have been having an asthma attack. **Asthma** is a disorder of the lungs in which there may be shortness of breath, wheezing, or coughing. When a person has an asthma attack, the bronchial tubes become constricted very quickly. As a result, the flow of air to the lungs is reduced. Asthma is often an allergic reaction. An attack can be caused by breathing certain substances, such as plant pollen. Stress and eating certain foods also have been related to the onset of asthma attacks.

■ Cystic Fibrosis

What color are your eyes? Your hair? These traits were passed from your parents to you. They're not harmful—they're what make you you. However, harmful traits also can be passed from parents to their children. One of these traits is a disease called cystic fibrosis (CF). If both parents carry this trait in their genetic makeup, they have a 25 percent chance of passing the trait to a child.

Cystic fibrosis affects the respiratory system by blocking the air passages. Fluid that lines the lungs and air passages thickens. This thickened fluid builds up, blocking the flow of air and causing lung damage. A person with CF coughs, wheezes, suffers from frequent lung infections, and usually dies at a young age. Although some people with CF live into adulthood, there is presently no cure for the disease.

2 TEACH

Tying to Previous Knowledge

Have students recall what they know about dangerous substances in cigarette smoke and the exhaust of gasoline-powered vehicles, which cause respiratory illnesses.

Meeting Individual Needs

Learning Disabled Students may have difficulty distinguishing the various respiratory diseases such as asthma and emphysema. Have groups of students create a chart showing the causes of each disease and its effects. Then lead a discussion on ways these diseases can be prevented. **COOP LEARN**

■ Emphysema

Smoking has been shown to cause severe damage to lungs. One disease that is closely linked with smoking is emphysema. **Emphysema** is a disease that occurs when air passageways or alveoli lose their ability to expand and contract. When a person has emphysema, air becomes trapped in the alveoli. Eventually the alveoli stretch and rupture. As a result, the overall surface area of the lungs is decreased. The lungs become scarred, and less oxygen moves into the bloodstream. The amount of oxygen carried by the blood decreases while the amount of carbon dioxide increases, resulting in a shortness of breath. Some people affected with emphysema can't blow out a match or walk up a flight of stairs. Because the heart works harder to supply oxygen to body cells, people who have emphysema often develop heart problems as well.

Recognizing Cause and Effect

Infer what would happen in your body if you were near a volcano that had erupted and the carbon dioxide level of the surrounding air had risen sharply. If you need help, refer to the **Skill Handbook** on page 643.

Theme Connection The disruptions that occur to the respiratory system described in this lesson illustrate the theme of systems and interactions. For example, when the alveoli of a person with emphysema are destroyed, the lungs' ability to provide oxygen to the bloodstream is weakened. The buildup of carbon dioxide adds to the problem as the heart is forced to work harder. Thus, a problem in the respiratory system causes problems in the circulatory system as well.

Content Background

Scientists at the Los Alamos National Laboratory in New Mexico discovered that exercising in polluted environments can be dangerous to your health. Nitrogen dioxide, found in cigarette smoke and automobile exhaust, is a chemical that is harmful if inhaled. Rats that exercised after being exposed to nitrogen dioxide suffered five times more lung damage than rats that did not exercise. The implication is that exercising in a polluted environment is more harmful than not exercising at all.

SKILLBUILDER

Students should infer that inhaling the ash dust would have a similar effect as breathing the dust, pollen, and dirt from polluted air, interrupting the ability of the respiratory system to provide oxygen to the bloodstream. In addition, the heart would be forced to work harder to rid the body of the excess carbon dioxide.

Figure 16-13

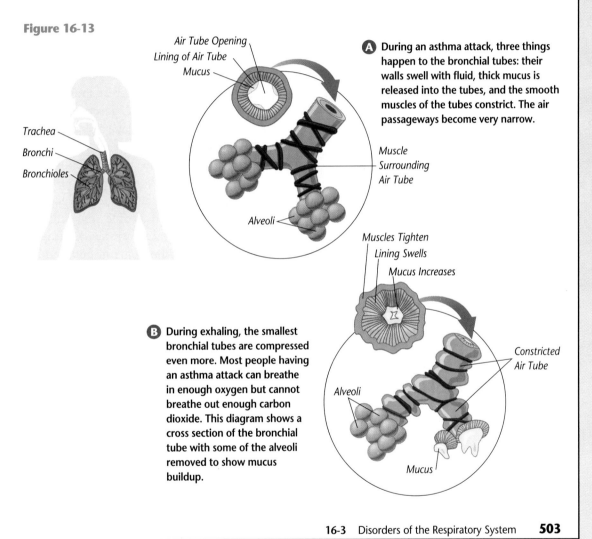

A During an asthma attack, three things happen to the bronchial tubes: their walls swell with fluid, thick mucus is released into the tubes, and the smooth muscles of the tubes constrict. The air passageways become very narrow.

B During exhaling, the smallest bronchial tubes are compressed even more. Most people having an asthma attack can breathe in enough oxygen but cannot breathe out enough carbon dioxide. This diagram shows a cross section of the bronchial tube with some of the alveoli removed to show mucus buildup.

ENRICHMENT

Activity Have the class determine the profile of a smoker versus a nonsmoker. Have students create a survey that individuals can take anonymously. The survey should include questions regarding age, sex, and if parents smoke. Do friends smoke? Do they participate in sports? Students should brainstorm as many questions as possible. Have students tally the results, analyze for trends, and report results to the class. **COOP LEARN**

Visual Learning

Figure 16-13 Have students use the diagrams to identify the problem areas in the bronchial tubes during an asthma attack.

GLENCOE TECHNOLOGY

 Software

Computer Competency Activities
Chapter 16

■ **Lung Cancer**

If you were asked which cancer caused the most deaths among men and women in the United States, would you know that the answer is lung cancer? When cilia are damaged, the lungs lose a defense against disease

Figure 16-14

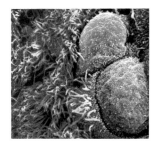

A Healthy lungs have clear bronchial tubes through which air can freely pass. When healthy, the tiny alveoli of the lungs are able to take in a full supply of oxygen and give up their full load of carbon dioxide.

B Cancer cells interfere with the normal function of all lung cells. Because cancer cells grow faster than normal cells, they soon outnumber normal cells. Soon, normal cells weaken and are no longer able to carry out the activities that keep lung tissue healthy.

C The cilia in healthy lungs beat continually and strongly to keep the mucus sheet moving toward the pharynx. When cancer invades the lung, the cilia are weakened.

and **lung cancer** can develop. Inhaling the tar in cigarette smoke is the greatest contributing factor to lung cancer. Tar is a black, sticky substance that builds up on the linings of the smoker's mouth, throat, and lungs.

Carbon monoxide is another poisonous substance found in cigarette smoke. You might be familiar with this compound, because it's one of the gases found in car exhaust. When a smoker inhales, carbon monoxide enters the respiratory system. Here, it interferes with the binding of oxygen to hemoglobin. This happens because carbon monoxide binds more easily and more firmly to hemoglobin than oxygen does. Thus, the smoker's body cells receive less oxygen than they need. This puts a great strain on the smoker's heart. Cigarette smoking damages the circulatory system as well as the respiratory system.

Most living things we know of would die without oxygen. Your respiratory system takes in oxygen and gets rid of carbon dioxide. You can keep your respiratory system healthy by avoiding smoking and polluted air. Also, regular exercise helps you increase your body's ability to use oxygen. This makes your breathing more efficient.

check your UNDERSTANDING

1. What two diseases of the respiratory system are linked to smoking?
2. What happens during an asthma attack?
3. How does air quality affect respiration?

4. **Apply** Why are people with respiratory disorders warned to stay indoors on days when air pollution is severe or pollen counts are higher than normal?

504   Chapter 16   Breathing

check your UNDERSTANDING

1. emphysema and lung cancer
2. The bronchial tubes constrict quickly, making it difficult for the person to breathe.
3. Pollutants such as smoke, dust, and pollen can damage the lungs in several ways. They can cause the bronchial tubes to constrict or stretch and rupture the alveoli, coat the respiratory structures with tar, or interfere with the ability of hemoglobin to carry oxygen to the body's cells.

4. Air pollution and pollen irritate the lungs and can cause damage to the lungs as outlined in the answer to question 3.

Science and Society

Cigarette Ads—Are They a Crime?

A dvertising is meant to make you want to buy the product being advertised. But what if the product can be shown to injure the user? Should our government protect us from harming ourselves?

Past and Future Cigarette Ad Bans

The Public Health Cigarette Smoking Act of 1970 banned all advertising of cigarettes on radio and TV in the United States.

One result of the Act was to turn the tobacco companies' attention to the rest of the world. As smoking declines in the U.S., Canada, and Western Europe, markets are expanding elsewhere. In the 1970s and 1980s, global tobacco use rose 75 percent and is still rising. For example, in China, cigarette consumption has risen 500 percent since 1965.

In 1996, the United States announced regulations to curb the marketing and sale of tobacco products to young people after the government classified nicotine as an addictive drug. These regulations bar the sale of cigarettes from vending machines and self-service displays that are accessible to people under 18. Also, free samples and tobacco advertising on billboards within 1000 feet of schools and playgrounds are prohibited.

The Tobacco Industry

Since 1966, cigarette packages and ads have warned that users were risking their health. That's the same year that one very large tobacco company announced that it was diversifying, or buying into, other less controversial fields. Was it trying to protect its image? Many large tobacco companies now own subsidiaries that make and sell such diverse products as cookies,

crackers, canned fruits and vegetables, flour, frozen foods, and even television programs. Millions of people get their paychecks from diversified tobacco companies. If cigarette ads are banned, what would be the effect on the tobacco companies and the companies they own? Could people lose their jobs as the result of such a ban? If, as some people believe, banning tobacco ads would not affect people's decision to smoke, what would be the point?

What Do You Think?

In a small-group discussion, list some effects you would expect if all cigarette ads are banned. Try to decide whether you are for or against such an ad ban.

EXPAND
your view

Technology
Connection

Technology
Connection

Garrett A. Morgan: Gas Mask Inventor

Purpose
The Technology Connection extends the concepts of Chapter 16 by profiling Garrett A. Morgan, the inventor of the gas mask.

Content Background
The National Safety Device Company was set up to manufacture the hood, with Garrett as the general manager. For three months he urged African Americans to buy stock in the company for $10 a share, but his efforts were unsuccessful. A month later the stock sold for $100 a share; two years later, it was up to $250 a share but it was sold out. On October 22, 1914, the company staged a spectacular exhibit around the Morgan helmet. They filled a canvas tent with thick smoke from burning sulphur, formaldehyde, and manure. A volunteer wearing the helmet stayed in the suffocating tent for a full twenty minutes and experienced no ill effects.

After Morgan's dramatic rescue of the men trapped in the tunnel, a group of prominent Cleveland citizens gave him a solid gold, diamond-studded medal inscribed, "To Garrett A. Morgan, our most honored and bravest citizen." Morgan set up his own company to manufacture and sell the hood.

Teaching Strategy
Have students work in pairs to read the selection and then make a diagram showing how Morgan's "Safety Hood" operated. Suggest that students use different colored pencils or markers to show the inward and outward air flow.

COOP LEARN L2

Demonstration
After students read the selection, invite a firefighter to come to class and demonstrate the masks and other respiration equipment currently in use.

African American inventor Garrett A. Morgan (1877-1963) developed "The Safety Hood" in 1912—the predecessor of the gas mask. A patent was granted to Morgan in 1914 for a device consisting of a hood placed over the head of the user. A tube from the hood featured an inlet opening for air, with the tube long enough to enter a layer of air underneath dense smoke or gas. The tube could be placed beyond the reach of gas fumes and dust, and through it pure air could be furnished to the user. The lower end of the tube was lined with an absorbent material, such as a sponge, that was moistened with water before use. This lining prevented smoke and dust from penetrating the tube and cooled the outside air entering the tube. A separate tube contained a valve for exhaled air.

A Life Saver
The original intent of Morgan's invention was to allow firefighters to enter fires

without suffocating from smoke and gases. Morgan had an opportunity to personally prove the value of his invention following a tunnel explosion in the Cleveland, Ohio, Waterworks. Morgan, his brother, and two volunteers saved several

men trapped in the smoke and gas-filled tunnel under Lake Erie from almost certain suffocation. The men entered the burning tunnel wearing Safety Hoods and carried the trapped workers to safety. After 1914, many fire departments across the country were using the Morgan Safety Hood to save lives and property.

During World War I, Morgan's Safety Hood was improved and used as a gas mask by the United States Army. Thousands of lives were saved during this war thanks to Garrett Morgan's invention.

Science Journal
Research the type of "gas masks" used by firefighters today in resource books at the library. Write a report *in your Science Journal* on how firefighters' respiration equipment has changed since the time of Morgan's Safety Hood.

Activity
Have students make a diagram tracing an escape route from their homes in case of fire. Students can begin by making a scale drawing of their home, showing all doors and windows. Then they can trace the best route from each location.

Going Further ⫸
Invite students to learn more about other African American inventors and scientists. Students can work in teams to research and report. Possible inventors or scientists include: Percy Lavon Julian, Lloyd A. Hall, Ernest Everett Just, Daniel Hale Williams, Louis Tompkins Wright, Charles Richard Drew, Lewis Howard Latimer, Benjamin Banneker, and George Washington Carver. Students may wish to display the results of their work in a prominent place in the school.

COOP LEARN

NATIONAL GEOGRAPHIC
SciFacts

What's causing Mexico City's smog?

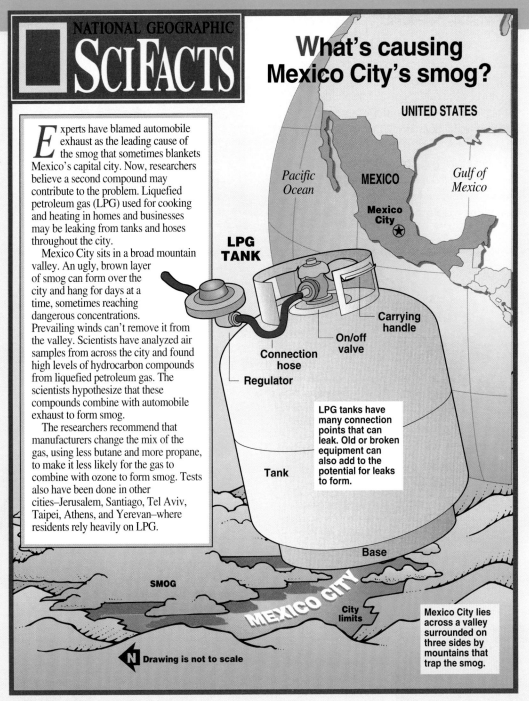

E xperts have blamed automobile exhaust as the leading cause of the smog that sometimes blankets Mexico's capital city. Now, researchers believe a second compound may contribute to the problem. Liquefied petroleum gas (LPG) used for cooking and heating in homes and businesses may be leaking from tanks and hoses throughout the city.

Mexico City sits in a broad mountain valley. An ugly, brown layer of smog can form over the city and hang for days at a time, sometimes reaching dangerous concentrations. Prevailing winds can't remove it from the valley. Scientists have analyzed air samples from across the city and found high levels of hydrocarbon compounds from liquefied petroleum gas. The scientists hypothesize that these compounds combine with automobile exhaust to form smog.

The researchers recommend that manufacturers change the mix of the gas, using less butane and more propane, to make it less likely for the gas to combine with ozone to form smog. Tests also have been done in other cities—Jerusalem, Santiago, Tel Aviv, Taipei, Athens, and Yerevan—where residents rely heavily on LPG.

UNITED STATES

Pacific Ocean

MEXICO

Gulf of Mexico

Mexico City ★

LPG TANK

Carrying handle

On/off valve

Connection hose

Regulator

LPG tanks have many connection points that can leak. Old or broken equipment can also add to the potential for leaks to form.

Tank

Base

SMOG

MEXICO CITY

City limits

Mexico City lies across a valley surrounded on three sides by mountains that trap the smog.

N Drawing is not to scale

Science Journal

In your Science Journal, list other areas of the world with serious pollution problems. Suggest reasons for these pollution sites. What steps are being taken to solve some of these problems?

Science Journal

Answers will vary, but most major cities, such as Los Angeles and Tokyo, are plagued by air pollution problems. Smog—both photochemical and sulfurous—and acid rain are caused by the burning of fossil fuels in factories, vehicles, and homes. Efforts to remedy the problem include reducing car emissions, burning low-sulfur coal, carpooling, and switching to renewable energy sources, such as solar power.

NATIONAL GEOGRAPHIC
SciFacts

What's causing Mexico City's smog?

Purpose
These SciFacts detail Mexico City's continuing efforts to control its air pollution problem. Smog formed by automobile exhaust and liquefied petroleum gas often blankets the city for days. Smog presents a serious health risk to humans, causing damage to lungs and other organs. As such, this feature relates to the discussion on disorders of the respiratory system found in Section 16-3.

Content Background
The effects of air pollution on the human body vary from minor to severe. Compounds found in smog cause the eyes to water and sting. If conditions are bad enough, vision may be blurred. Ozone irritates the nose and throat, reducing the ability of the lungs to fight infection. This, in turn, makes people more susceptible to diseases such as pneumonia and asthma. Carbon monoxide that is absorbed by red blood cells reduces the heart's ability to transport oxygen throughout the body. When oxygen levels in the brain are reduced, motor functions and coordination are impaired. At high levels, air pollution can cause death; about 60 000 deaths each year in the United States are related to air pollution.

Activity Invite a local Environmental Protection Agency representative to the class to discuss local pollution problems. Have the official brainstorm with students ways that they can reduce environmental concerns.

Have students use the photographs, illustrations, and questions on this page to review the main ideas presented in this chapter.

Teaching Strategy

Divide the class into small groups. Have each group make a model of the respiratory system. Students can design two- or three-dimensional models using papier-mâché, clay, poster paper, tracing paper, or transparencies as materials. The models must include the nasal cavity, pharynx, epiglottis, larynx, esophagus, trachea, bronchi, and lungs. After the groups have finished their models, have them share their work with the class. **L2**

COOP LEARN

Answers to Questions

1. Paramecium–membrane; earthworm–capillaries near skin; fish–gills; insects–spiracles, trachea; spiders–book lungs; humans–lungs.

2. Oxygen and carbon dioxide diffuse through the thin walls of alveoli. Oxygen is picked up by the capillaries surrounding the alveoli, which in turn, release carbon dioxide as waste from the body.

3. During cellular respiration, oxygen combines with stored nutrients in the body's cells to produce carbon dioxide and water. During this reaction, energy is released.

4. Some effects of reduced oxygen availability are difficulty in breathing and diseases of or damage to the respiratory system.

GLENCOE TECHNOLOGY

MindJogger Videoquiz

Chapter 16 Have students work in groups as they play the Videoquiz game to review key chapter concepts.

Science Journal

Review the statements below about the big ideas presented in this chapter, and answer the questions. Then, re-read your answers to the Did You Ever Wonder questions at the beginning of the chapter. *In your Science Journal*, write a paragraph about how your understanding of the big ideas in the chapter has changed.

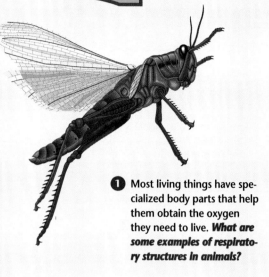

1 Most living things have specialized body parts that help them obtain the oxygen they need to live. *What are some examples of respiratory structures in animals?*

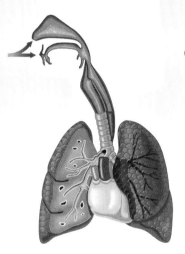

2 The lungs are the main organ of your respiratory system. *How does gas exchange take place in the lungs?*

3 Cellular respiration occurs in the body's cells when oxygen combines with food to release energy and produce carbon dioxide and water. *How is cellular respiration important to your body?*

4 Lung diseases and disorders reduce the amount of oxygen that can be transported to the body's cells. *What are some of the effects of reduced oxygen availability?*

Science at Home

Ask students to use the technique from the Investigate activity on pages 490–491 with family members and friends. Have students make a bar graph of their data with vital capacity on the vertical axis and height in centimeters on the horizontal axis. Have them attempt to identify factors that cause participants to differ in their vital capacity, such as gender, weight, age, athleticism, and so on. **L1**

Science Journal

Did you ever wonder...

• Water is a waste product of respiration and thus is released as water vapor when you exhale. (p. 496)

• An increase in the rate of breathing supplies more oxygen to the bloodstream. (p. 500)

• A clear, hollow sound means that your lungs are not filled with fluid. (p. 502)

Using Key Science Terms

alveoli
emphysema
asthma
gill
cellular respiration
hemoglobin
cilia
lung
cystic fibrosis
lung cancer
diaphragm
trachea

For each set of terms below, choose the one that does not belong and explain why it does not belong.

1. gills, alveoli, diaphragm, trachea
2. asthma, cellular respiration, emphysema, lung cancer
3. gills, lungs, hemoglobin
4. cystic fibrosis, lung cancer, emphysema
5. Look up hemo (or hem) and globin in the dictionary. Explain why hemoglobin is an appropriate name for the substance it represents.

Understanding Ideas

Answer the following questions in your Journal *using complete sentences.*

1. In what ways are gill filaments similar to alveoli?
2. Where is hemoglobin found and what are its functions?
3. How do asthma, emphysema, and lung cancer impair the performance of the respiratory system?
4. How do the circulatory system and respiratory system work together?
5. What is the difference between breathing and cellular respiration?

Developing Skills

Use your understanding of the concepts developed in this chapter to answer each of the following questions.

1. **Concept Mapping** Complete the concept map of breathing.

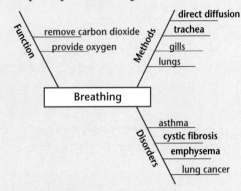

2. **Predicting** Predict what will happen when you repeat the Explore activity on page 483 after running in place for 20 seconds. How does the new number compare with the new numbers of your classmates?
3. **Recognizing Cause and Effect** After performing the Investigate on page 498, hold your nose and use a straw to breathe for 30 seconds. How do you feel? Would you want to run or walk anywhere using the straw to breathe? Why?
4. **Observing and Inferring** After performing the Explore activity on page 488, keep your fingers on your trachea and swallow. What movement did you feel? Recalling the text, what happens when you swallow?

Using Key Science Terms

1. Fish have gills. Humans breathe with the other structures.
2. Cellular respiration is a chemical process. All others are lung diseases.
3. Hemoglobin is a molecule. The others are respiratory organs.
4. Cystic fibrosis is inherited; smoking causes the other two diseases.
5. *Hem(o)*–blood; *globin*–colorless protein. *Hemoglobin* is an oxygen-binding protein in red cells.

Understanding Ideas

1. Both structures allow diffusion of gases. The feathery nature of the gills and the large number of alveoli provide increased surface area for gas exchange.
2. Hemoglobin in red blood cells contains iron that bonds with oxygen and releases it to body cells.
3. Asthma causes the bronchial tubes to constrict. In emphysema the alveoli lose their elasticity. In both cases, oxygen uptake is impaired. In lung cancer defenses against disease are reduced because the cilia of the lungs have been damaged.
4. The respiratory system brings oxygen into the body and the circulatory system transports it to body cells.
5. Breathing is the inhaling and exhaling of air. Cellular respiration is a chemical reaction that releases energy, carbon dioxide, and water.

Developing Skills

1. See reduced student page.
2. The number of breaths taken will increase for all students.
3. Breathing this way is similar to how people with emphysema or having an asthma attack feel.
4. The epiglottis closes so food does not enter the trachea.

Critical Thinking

1. When you inhale, air pressure is greater outside the body. The pressure is greater inside the chest cavity when you exhale.

2. Cilia move trapped dirt and dust particles to the throat where they can be swallowed and removed from the respiratory system.

3. The surface area of the lungs is greatly increased by having many sacs.

Problem Solving

1. Answers will vary depending on a student's height, sex, and athletic involvement.

2. Answers will most likely indicate less than the average adult.

3. Exercise can strengthen the muscles involved in breathing, thus increasing the amount of air that enters the lungs.

4. A person who exercises regularly will have greater vital capacity, and be more fit.

Connecting Ideas

1. Your breathing rate slows down because you are less active. Your body requires less oxygen.

2. Smoke contains a lot of CO_2 and CO and no oxygen.

3. As air passes the vocal cords, sounds are produced. Low sounds are produced when air rushes past loose vocal cords, and high sounds are produced when air rushes past tight vocal cords.

4. Answers may include that smoking cigarettes makes a person appear more sophisticated, makes a person appear to fit in with the group, makes a person more attractive.

5. Insects such as the grasshopper do not have lungs. The grasshopper does have a respiratory system with many trachea, though, through which air moves into and out of the body.

Critical Thinking

In your Journal, *answer each of the following questions.*

1. Explain where the air pressure is greater, inside the chest cavity or outside the body, when you inhale and when you exhale.

2. Why is it an advantage to have cilia in your respiratory system?

3. What is the advantage of the lungs having many masses of air sacs instead of two large sacs?

Problem Solving

Read the following problem and discuss your answers in a brief paragraph.

People who are physically active have a larger lung capacity than those who are less active. The largest possible amount of air that can be exhaled after drawing a deep breath is called the vital capacity. A relationship exists between a person's height and vital capacity. The average adult male's vital capacity is 5000 cm³, while that of an adult female is 4000 cm³.

1. To find your calculated vital capacity, multiply your height (in centimeters) by one of the following factors:

20 for females

25 for males

22 for female athletes

29 for male athletes

2. How does your calculated vital capacity compare with the average adult of your sex? What could account for any differences in the figures?

3. Explain how exercise can increase one's vital capacity.

4. How can information about vital capacity be used to evaluate physical fitness?

CONNECTING IDEAS

Discuss each of the following in a brief paragraph.

1. Theme—Stability and Change What happens to your breathing rate when you are sleeping? Why?

2. Theme—Systems and Interactions People trapped in fires often die of smoke inhalation rather than from burns. How is smoke inhalation fatal?

3. Physics Connection Briefly describe how the larynx produces low and high sounds.

4. Science and Society Cigarette ad writers try to convince you that using their product will bring high-performance cars, exotic vacations, and the perfect mate into your life. How do ad writers aim their product at you, the adolescent?

5. A Closer Look How is a grasshopper's respiratory system different from yours? How is it similar?

✔ Assessment

Portfolio Review the portfolio options that are provided throughout the chapter. Encourage students to select one product that demonstrates their best work for the chapter. Have students explain what they learned and why they chose this example for placement into their portfolios.

Additional portfolio options can be found in the following **Teacher Classroom Resources: Making Connections: Integrating Sciences,** p. 35

Multicultural Connections, pp. 35, 36

Making Connections: Across the Curriculum, p. 35

Concept Mapping, p. 24

Critical Thinking/Problem Solving, p. 24

Take Home Activities, p. 24

Laboratory Manual, pp. 101–102

Performance Assessment **P**

Air: Molecules in Motion

In this unit, you investigated how the structure of an element's atoms determines how the element is classified and how it reacts with other elements to form molecules, such as the gases in Earth's atmosphere.

You also saw that your respiratory system interacts with the gases in Earth's atmosphere by taking in oxygen when you breathe in, and releasing carbon dioxide when you breathe out. The oxygen taken into your body during cellular respiration is very important because it allows cells to carry on life processes.

Try the exercises and activity that follow—they will challenge you to use and apply some of the ideas you learned in this unit.

CONNECTING IDEAS

1. The surface area of the palm of your hand is about 15 square inches. Air pressure is about 15 pounds per square inch. Calculate the amount of air pressure on your hand. Explain why you do not feel this pressure on your hand.

2. What are the most common gases in Earth's atmosphere? Which of these does the human body use? Describe several examples of how human activities affect the concentration of gases in Earth's atmosphere. How do you think human respiration might change if Earth's atmosphere were all oxygen?

Exploring Further ACTIVITY

What causes air to move?

What To Do

1. Draw a diagram of the respiratory system.

2. Use an arrow to show where cold air is heated before it reaches the lungs. What method of heat transfer is involved?

3. *In your Journal*, describe the process that enables your lungs to take in gases from the atmosphere. How is this process related to the process that causes winds?

Air: Molecules in Motion

THEME DEVELOPMENT

Unit 4 explored the themes of energy and systems and interactions. Respiration demonstrates how the body can extract energy from its environment. The formation of molecules shows the role of interactions in the creation of matter.

Connections to Other Units

Circulation in Unit 1 and energy in Unit 2 provided the foundation for respiration in this unit. Gases, atoms, and molecules presented in this unit developed from the discussion of elements in Unit 3.

Connecting Ideas
Answers

1. 225 pounds of force; this pressure is balanced by an equal amount of pressure on the other side of the hand.

2. The four most common gases in Earth's atmosphere are nitrogen (78%), oxygen (21%), argon (1%), and carbon dioxide (.04%). All the cells in the human body require oxygen. Activities such as burning fuel affect the concentration of gases in the atmosphere. If our atmosphere were all oxygen, the rate of human respiration and the size of our lungs might change.

Exploring Further What causes air to move?

Purpose To describe differences in air pressure that result in breathing and cause winds.

Materials Provide different colors of pencils for students' drawings.

Preparation Have students review the diagram of the respiratory system in Figure 16-3 on page 488.

Answers to Questions

2. Cold air is heated in the nasal passages. Heat transfer takes place due to radiation and convection.

3. The act of breathing and winds are the result of air moving from areas of high pressure to low pressure.

✔ Assessment

Process Cut a piece of cardboard the size of a 5 or 10 gallon aquarium. Cut holes in the cardboard so that the glass chimneys from two hurricane lamps will fit snugly. Place a lighted candle in the aquarium so that it will be directly below one of the chimneys. Place the cardboard assembly over the aquarium. Light a newspaper roll and place over the chimney without the candle. Have students record their observations and explain the process that occurs. Use the Performance Task Assessment List for Making Observations and Inferences in **PASC**, p. 17.

Life at the Cellular Level

UNIT OVERVIEW

UNIT FOCUS

In Unit 5, students will observe the structures of different cells and the functions of these structures. They will learn about the basic chemical reactions that occur in cells and why these reactions are essential to life.

THEME DEVELOPMENT

Systems and interactions and energy are the two themes developed in this unit. The first theme is illustrated by the cell and the way it changes and reproduces. The cell is the unit of living matter upon which living systems are based. Students will learn how chemical interactions drive cells and enable them to carry out their functions. The flow of energy through cells drives metabolism—growth and development. The flow of energy through ecosystems determines how organisms interact through all levels of a community.

Connections to Other Units

Unit 5 provides a perspective of life at the cellular level, which relates to the concepts about molecules presented in Unit 4.

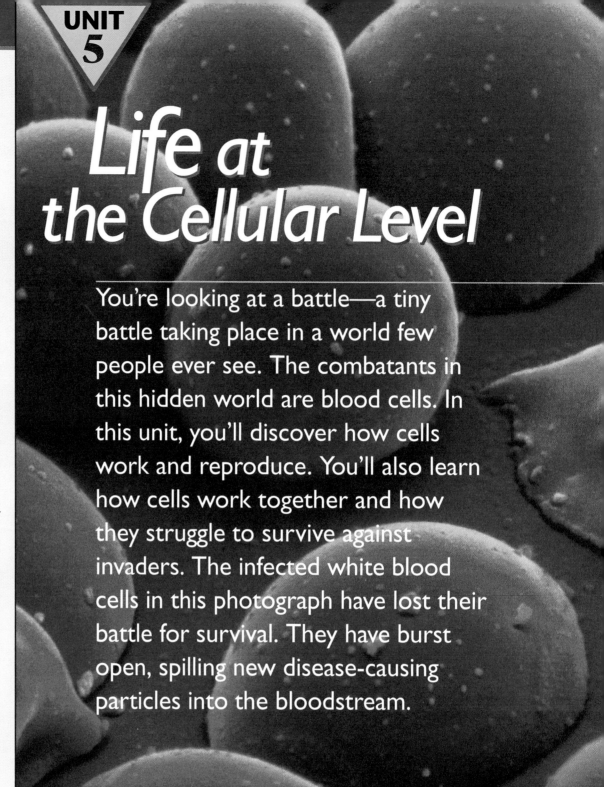

UNIT
5

Life at the Cellular Level

You're looking at a battle—a tiny battle taking place in a world few people ever see. The combatants in this hidden world are blood cells. In this unit, you'll discover how cells work and reproduce. You'll also learn how cells work together and how they struggle to survive against invaders. The infected white blood cells in this photograph have lost their battle for survival. They have burst open, spilling new disease-causing particles into the bloodstream.

512

GLENCOE TECHNOLOGY

 Videodisc

Use the *Science Interactions, Course 2* **Integrated Science Videodisc** lesson, *Photosynthesis and Cellular Respiration: Earth Recycles,* after Chapter 19 of this unit.

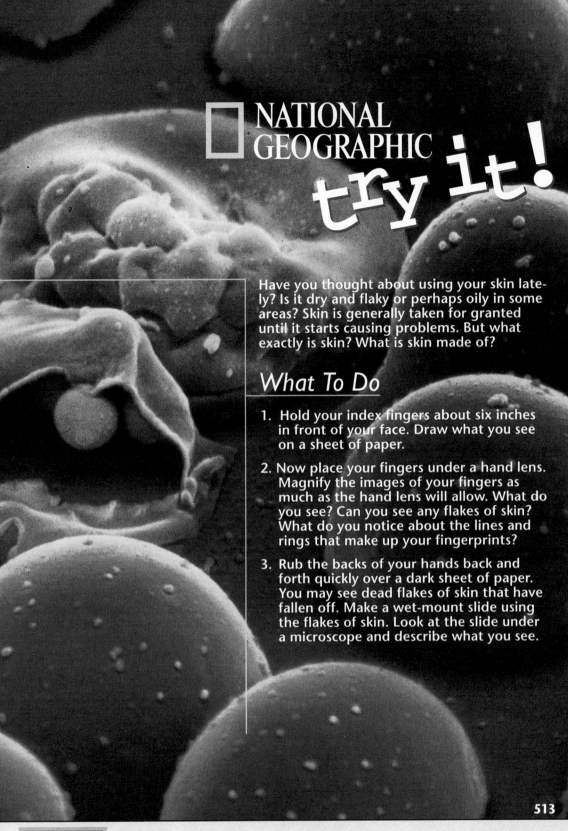

NATIONAL GEOGRAPHIC
try it!

Have you thought about using your skin lately? Is it dry and flaky or perhaps oily in some areas? Skin is generally taken for granted until it starts causing problems. But what exactly is skin? What is skin made of?

What To Do

1. Hold your index fingers about six inches in front of your face. Draw what you see on a sheet of paper.

2. Now place your fingers under a hand lens. Magnify the images of your fingers as much as the hand lens will allow. What do you see? Can you see any flakes of skin? What do you notice about the lines and rings that make up your fingerprints?

3. Rub the backs of your hands back and forth quickly over a dark sheet of paper. You may see dead flakes of skin that have fallen off. Make a wet-mount slide using the flakes of skin. Look at the slide under a microscope and describe what you see.

513

Discussion Some questions you may want to ask your students are:

1. How does a microscope allow the user to magnify something hundreds of times its natural size? Most students will not be able to explain that light passing through one or more lenses can produce an enlarged image of the object being viewed and the power of the lenses determines the degree of magnification.

2. If all living things are made up of similar units called cells, why doesn't every living thing look exactly alike? Most students will have some ideas but few will be able to formalize that each organism has unique DNA that determines the traits of its different types of cells.

3. Where in the cell is the information that directs the functions of the cell? Most students will not know that the answer is the nucleus.

Responses to these questions will help you determine misconceptions that students have.

depths of complexity in living matter. At this point, students will not be able to see skin cells clearly. Have them look at a prepared slide of an epidermal cell instead.

✔ Assessment

Oral Have students compare and contrast the differences in what they see using their unaided eyes, the hand lens, and the microscope. Ask them what a more powerful microscope might reveal. Then ask if the most powerful microscope is always the most desirable tool. Ask students how seeing things at different levels of magnification is useful in different situations. Use the Performance Task Assessment List for Oral Presentation in **PASC,** p. 71.

Try It!

Purpose Students will observe the way in which lenses allow closer examination of cell tissue than does the unaided eye.

Background Information Like all living tissue, skin is made up of cells. Tissues, in turn, make organs which make up organisms. Cellular levels of organization will become apparent to students as they look at other examples of living cellular material in this unit.

Materials hand lens, microscope, dark paper, slide, coverslip

Troubleshooting Some students may not be able to distinguish skin cells in the flakes of skin from their hands. Provide a prepared slide of an epidermal cell for students to examine.

Answers to Questions

This activity will guide students to see that even familiar tissue such as skin can reveal greater

Chapter Organizer

SECTION	OBJECTIVES	ACTIVITIES & FEATURES
Chapter Opener		**Explore!**, p. 515
17-1 The World of Cells (3 sessions; 1.5 blocks)	1. **Identify** cells as structures common to all living things. 2. **Conclude** that different cells usually have different functions. 3. **Draw** conclusions about why most cells are small. **National Content Standards: (5-8) UCP2, A1-2, C1-2, E2, G2-3**	**Find Out!**, p. 516 **Find Out!**, p. 518 **Find Out!**, p. 519 **Find Out!**, p. 520 **Design Your Own Investigation 17-1:** pp. 522–523 **A Closer Look,** pp. 520–521
17-2 The Inside Story of Cells (2 sessions; 1 block)	1. **Identify** the parts of a typical cell. 2. **Describe** the jobs of cell parts. 3. **Compare and contrast** plant and animal cells. **National Content Standards: (5-8) UCP2, A1, C1, C3**	**Explore!**, p. 525 **Find Out!**, p. 529 **Find Out!**, p. 530 **Explore!**, p. 532 **Skillbuilder:** p. 533
17-3 When One Cell Becomes Two (3 sessions; 1.5 blocks)	1. **Describe** the process of mitosis and its end products. 2. **Give examples** of instances where cell reproduction takes place. **National Content Standards: (5-8) UCP2, UCP4, A1, C1-3, E2, F1, F5**	**Find Out!**, pp. 536–537 **Investigate 17-2:** pp. 540–541 **Physics Connection,** pp. 536–537 **Technology Connection,** p. 543 **Science and Society,** pp. 544–545 **Health Connection,** p. 546

ACTIVITY MATERIALS

EXPLORE!

p. 525 prepared gelatin; resealable sandwich bag

p. 532 clear food container, plastic resealable sandwich bag, prepared gelatin

INVESTIGATE!

pp. 540–541* 5 corn seedlings, permanent marker, ruler, paper towels, 5 plastic bags, labels

DESIGN YOUR OWN INVESTIGATION

pp. 522–523* newspaper, scissors, slides, microscope, metric ruler, prepared slides of frog skin, drawing compass, onion skin, tomato skin, flower petals, calculator

FIND OUT!

p. 516 talcum powder, salt, Elodea, prepared slide of frog blood, microscope, slides, coverslips

p. 518* prepared slides of guard cells, human cheek cells, yeast cells; microscope

p. 519* human cheek cell slide, Elodea cells slide, microscope, drawing compass

p. 520 large and small clear containers, water, potassium permanganate crystals, plastic spoons

p. 529* fresh red onion, slide, coverslip, water, microscope, dropper

p. 530* white onion, onion root tip slide, slide, coverslip, microscope, iodine, paper towels

pp. 536–537* prepared slide of an onion root tip, microscope

KEY TO TEACHING STRATEGIES

The following designations will help you decide which activities are appropriate for your students.

- **L1** Basic activities for all students
- **L2** Activities for average to above-average students
- **L3** Challenging activities for above-average students
- **LEP** Limited English Proficiency activities
- **COOP LEARN** Cooperative Learning activities for small group work
- **P** Student products that can be placed into a best-work portfolio
- Activities and resources recommended for block schedules

Need Materials? Call Science Kit (1-800-828-7777).

[00:00] OUT OF TIME? We recommend that students do the activities with an asterisk.

Chapter 17 Basic Units of Life

TEACHER CLASSROOM RESOURCES

Student Masters	Transparencies
Study Guide, p. 56 **Activity Masters,** Design Your Own Investigation 17-1, pp. 71–72 **Laboratory Manual,** pp. 103–106, Observing Cells	**Section Focus Transparency 50**
Study Guide, p. 57 **Critical Thinking/Problem Solving,** p. 5 **How It Works,** p. 18 **Making Connections: Technology & Society,** p. 37 **Making Connections: Across the Curriculum,** p. 37 **Multicultural Connections,** p. 37	**Teaching Transparency 33,** Animal and Plant Cells **Teaching Transparency 34,** Cells: Structure and Function **Section Focus Transparency 51**
Study Guide, p. 58 **Take Home Activities,** p. 26 **Concept Mapping,** p. 25 **Activity Masters,** Investigate 17-2, pp. 73–74 **Making Connections: Integrating Sciences,** p. 37 **Critical Thinking/Problem Solving,** p. 25 **Science Discovery Activities, 17-1, 17-2, 17-3** **Multicultural Connections,** p. 38	**Section Focus Transparency 52**

ASSESSMENT RESOURCES	TEACHING & TECHNOLOGY
Review and Assessment, pp. 101–106 **Performance Assessment,** Ch. 17 **PASC*** **MindJogger Videoquiz** **Alternate Assessment in the Science Classroom** **Computer Test Bank**	**Spanish Resources** **Cooperative Learning Resource Guide** **Lab and Safety Skills** **Science Interactions, Course 2, CD-ROM** **Computer Competency Activities**

*Performance Assessment in the Science Classroom

NATIONAL GEOGRAPHIC TEACHER'S CORNER

Index to National Geographic Magazine

The following articles may be used for research relating to this chapter:

- "DNA Profiling: The New Science of Identity," by Cassandra Franklin-Barbajosa, May 1992.
- "A New Kind of Kinship," by Joel L. Swerdlow, September 1991.
- "The Awesome Worlds Within a Cell," by Rick Gore, September 1976.

National Geographic Society Products Available From Glencoe

To order the following products for use with this chapter, contact your local Glencoe sales representative or call Glencoe at 1-800-334-7344:

- *STV: The Cell* (Videodisc)
- *Newton's Apple Life Sciences* (Videodisc)

Additional National Geographic Society Products

To order the following products for use with this chapter, call the National Geographic Society at 1-800-368-2728:

- *Everyday Science Explained* (Book)
- *Biotechnology* (Video)
- *DNA: Laboratory of Life* (Video)
- *Discovering the Cell* (Video)

Teacher Classroom Resources

These are key components of the classroom resources package.

TEACHING AIDS

Section Focus Transparencies

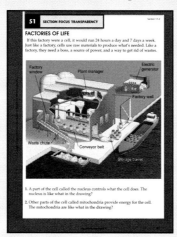

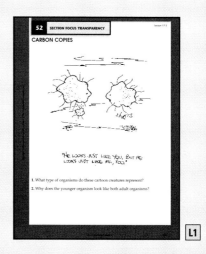

Teaching Transparencies

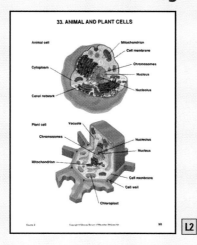

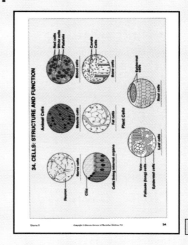

HANDS-ON LEARNING

Science Discovery Activity*

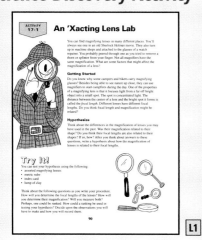

Laboratory Manual*

Take Home Activity

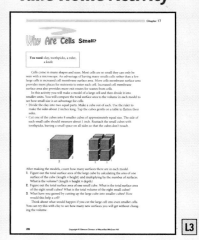

*There may be more than one activity for this chapter.

Chapter 17 Basic Units of Life

REVIEW AND REINFORCEMENT

Study Guide*

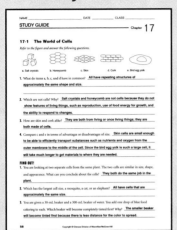

Concept Mapping

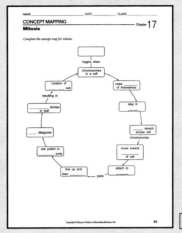

Critical Thinking/ Problem Solving

ENRICHMENT AND APPLICATION

Integrating Sciences

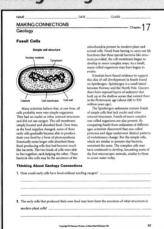

Across the Curriculum

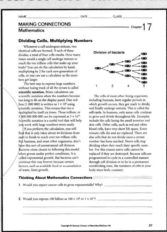

Technology and Society

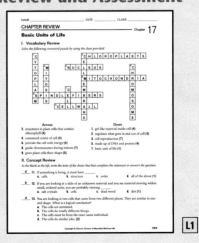

Multicultural Connection**

ASSESSMENT

Performance Assessment

Review and Assessment

Basic Units of Life

THEME DEVELOPMENT

The themes that are supported by this chapter are scale and structure and energy. Cells are the basic unit of all living things. Cells may be similar in scale, but they can vary greatly in structure, especially if they have specific functions, such as communication or movement. All cells need energy for metabolic functions.

CHAPTER OVERVIEW

Students will study the basic cell structures. They will also learn about mitosis, the process by which cells reproduce.

Tying to Previous Knowledge

Students are aware of the existence of cells from their study of bones and muscles in Chapter 7 and the nervous system in Chapter 8. As they study this chapter, students will learn more about how cells are alike and how they are different.

INTRODUCING THE CHAPTER

Have students look at the photographs of the humpback whale and the bees on page 514. Ask them what these organisms have in common. Help students to understand that all living things have in common the fact that they are made up of cells.

BASIC UNITS OF LIFE

Did you ever wonder...

- ✓ **How your body replaces the skin on a scraped knee?**
- ✓ **Why leaves are usually green?**
- ✓ **What makes a wooden bat strong?**

You'll find the answers to these questions as you read this chapter.

Science Journal

Before you begin to study about the basic units of life, think about these questions and answer them *in your Science Journal*. When you finish the chapter, compare your journal write-up with what you have learned.

Answers are on page 547.

Chipmunk

Did you know that you began life as a single cell? That first cell then divided into two. Those two became four. The dividing continued on and on until that original cell had become about two trillion cells by the time you were born nine months later!

You're not the only thing made up of cells. All living things—from whales to chipmunks to blades of grass—are made up of these remarkable structures. Cells are so small that most of them can't be seen without the help of a microscope.

This chapter will take you on a voyage into the inner world of the basic unit of life—the cell. You will see how vital cells are to small organisms, as well as to large ones.

▶ **In the activity on the next page, explore a model of living things.**

Humpback whale

514

Learning Styles		
	Kinesthetic	Explore, pp. 525, 532; Enrichment, p. 528
	Visual-Spatial	Explore, p. 515; Activity, pp. 517, 524, 533; Find Out, pp. 516, 518, 519, 520, 529, 530, 536; Visual Learning, pp. 517, 518, 521, 524, 527, 528, 531, 532, 535, 538; Enrichment, p. 523; Demonstration, pp. 527, 530
	Logical-Mathematical	A Closer Look, pp. 520-521; Investigate, pp. 522-523, 540-541; Physics Connection, pp. 536-537
	Linguistic	Across the Curriculum, p. 528; Science at Home, p. 547
LS	**Intrapersonal**	Multicultural Perspectives, p. 531

Explore! ACTIVITY

What are living things made of?

Nearly everything in our world is made up of many smaller parts.

What To Do

1. Take a look at the honeycomb in the photograph. What does the whole comb look like?

2. *In your Journal,* describe the individual structures that make up the honeycomb.

In many ways, the structures that form the honeycomb can serve as a model of the basic structure that makes up all living things.

Honeybees

ASSESSMENT PLANNER

PORTFOLIO
Refer to page 549 for suggested items that students might select for their portfolios.

PERFORMANCE ASSESSMENT
Process, pp. 515, 518, 523, 525, 529
Skillbuilder, p. 533
Explore! Activities, pp. 515, 525, 532
Find Out! Activities, pp. 516, 518, 519, 520, 529, 530, 536–537
Investigate, pp. 522–523, 540–541

CONTENT ASSESSMENT
Oral, pp. 519, 537
Check Your Understanding, pp. 524, 533, 542
Reviewing Main Ideas, p. 547
Chapter Review, pp. 548–549

GROUP ASSESSMENT
Opportunities for group assessment occur with Cooperative Learning Strategies.

Explore!

What are living things made of?

Time needed 5 minutes

Thinking Process observing, forming operational definitions

Purpose To observe an object made up of small, uniform units, or cells.

Teaching the Activity

Discussion Ask students to suggest names for the units of which the honeycomb is made. *Possible responses include boxes, units, or compartments. Some students may be aware that they are called cells.*

Science Journal
Suggest that students include a sketch of the honeycomb in their journals. **L1**

Expected Outcome
Students will observe that the honeycomb is made up of individual units.

Answers to Questions
1. Answers will vary, but students may compare the honeycomb to a grid.
2. They are six-sided units (hexagons).

✔ Assessment

Process Ask students to list other real-life objects that have a structure like that of a honeycomb. Have students draw diagrams of each to support their thinking. *Possible answers include apartment buildings and storage units.* Use the Performance Task Assessment List for Making Observations and Inferences in **PASC**, p. 17.

PREPARATION

Planning the Lesson

Refer to the Chapter Organizer on pages 514A–D.

Concepts Developed

All living things are made of cells. Cells vary in shape and in some details of their structure. Different kinds of cells in a single organism differ in structure and function. Cells in most organisms are generally small.

1 MOTIVATE

Bellringer

 Before presenting the lesson, display **Section Focus Transparency 50** on an overhead projector. Assign the accompanying **Focus Activity** worksheet. L1
LEP

Find Out!

What is everything made of?

Time needed   30 minutes

Materials   microscope, slides, coverslips, talcum powder, salt, Elodea, prepared stained slide of frog blood

Thinking Process observing, forming operational definitions

Purpose   To observe that living things are made up of small units.

Teaching the Activity

Science Journal Have students sketch the crystals and cells they observe. L1

Expected Outcome

Students will infer that living things are made up of small units.

◆ 17-1 ◆ **The World of Cells**

Section Objectives

■ Identify cells as structures common to all living things.
■ Conclude that different cells usually have different functions.
■ Draw conclusions about why most cells are small.

Key Terms

cell

What Is a Cell?

When you first looked at the honeycomb in the Explore activity photograph, what did you notice about it? You may have noticed that it looked very organized because it is made of many small units that were about the same general size and shape. It was also easy to see each unit, wasn't it?

Living things, or organisms, are also made up of many small units. But, in most organisms, these units are so small in size, they aren't easy to see using only your eyes. You need a tool called a microscope to be able to see what these very small structures look like.

Find Out! ACTIVITY

What is everything made of?

What To Do

1. Look at several different types of materials to find out what they are made of.

2. Make separate slides of talcum powder, salt, and Elodea (an aquarium plant). Using the directions on page 619, make a wet-mount slide of a single Elodea leaf. Your teacher will give you a prepared stained slide of frog blood.

3. Carefully follow the directions given on page 619, on how to use a microscope. Look at each slide first under low, then high power on the microscope.

Conclude and Apply

1. Describe the appearance of each sample. How do they differ?

2. Which of the samples looks as if it is made up of many small organized units?

Conclude and Apply

1. Talcum powder and salt look like crystals. Frog blood is made up of oval structures. The plant is made up of units with green objects inside.

2. the plant and the frog blood

✔ Assessment

Content Have students classify talcum powder, salt, Elodea, and frog's blood as living or nonliving. Students can explain their classification in a cartoon panel(s) that they create. *Elodea and frog's blood are living or were once living; talcum powder and salt are nonliving.* Use the Performance Task Assessment List for Cartoon/Comic Book in **PASC**, p. 61.

Of the samples you just looked at, you probably saw small organized units in the Elodea and blood and maybe even in the salt, but not in the talcum powder. Would you say that any of these samples were alive? Or were they alive at one time? How could you know? You would have to put them to some sort of test, wouldn't you? In order to know if these units were from a living thing, you would have to know if they came from some-thing that had cells. You would also need to know if the organism could grow, reproduce, use food for energy, and respond to changes. These are features of living things. The salt and talcum powder you used were in the form of crystals. Although crystals grow, they do not show the rest of the features of living things. The units you saw in the Elodea and blood samples, however, are cells. A **cell** is the basic unit of life in all living things.

Connect to...
Earth Science

Many non-living things, such as icicles, crystals, and sand dunes, appear to grow. Prepare a poster that shows the processes involved in the growth of living things and the "growth" of nonliving things.

Figure 17-1

Ability to grow is a feature of living things. Salt crystals, like those used on pretzels, grow, but they do not show the rest of the features of living things. Is salt a living thing?

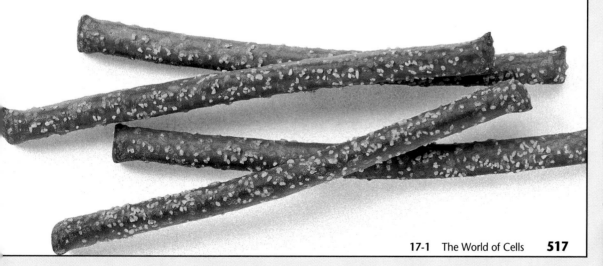

17-1   The World of Cells   **517**

Activity   Using children's interlocking plastic building blocks, have volunteers make two different structures. Ask students how the same blocks were able to make two different structures. *The structure depends on how you arrange the blocks.* [L1]

2 TEACH

Tying to Previous Knowledge

In Chapter 7, students learned about muscles. In Chapter 8, students learned about the nervous system. Have students recall the structure of muscle and nerve cells.

How Do We Know?

The cell theory brought together two fields of biology. Schwann was a zoologist and Schleiden was a botanist. What makes the cell theory important is the recognition of the cell as a fundamental structural component of all living things.

Connect to . . .
Earth Science

Answer: Accept all reasonable responses. Living things grow larger by production of more cells or cells enlarge through metabolism. Nonliving things "grow" by addition of two or more of the same materials, but not by dividing.

NATIONAL GEOGRAPHIC SOCIETY

   **Videodisc**

STV: The Cell
Viewing the Cell
Unit 1
Viewing the Cell

00373-11798

What do the traits of these cells help tell you about their jobs? *The nerve cell looks like a wire for carrying messages; the plant cell looks like a tube for carrying materials.*

Visual Learning

Figure 17-2 To compare and contrast the traits of cells, have students describe the pictured nerve cell and plant cell. Ask: **Judging from the shape of the plant cell shown, which job do you think it does?** *tubular shape suggests that this cell carries water throughout the plant*

GLENCOE TECHNOLOGY

CD-ROM

Science Interactions, Course 2, CD-ROM

Chapter 17

Find Out!

Do all cells look the same?

Time needed 15 minutes

Materials microscope, prepared slides of leaf guard cells, human cheek cells, and yeast cells

Thinking Processes observing, comparing and contrasting, forming operational definitions

Purpose To compare and contrast cell shapes and structures.

Preparation Prepared slides can be obtained from biological supply houses.

Teaching the Activity

Troubleshooting If students are not sure of what to look for, refer them to the photograph showing a pair of guard cells.

Science Journal Have students make drawings and record results in their journals. [L1]

DID YOU KNOW?

A single yeast cell is an organism all by itself, but your body is made up of more than 10 trillion (10 000 000 000 000) cells, none of which can exist alone for very long.

Figure 17-2

The shape of a cell is often related to the job it does.

Nerve cell

A Notice that the nerve cell has extensions that look like electric wires. The job of nerve cells is to pass messages in the form of chemical impulses from nerve cell to nerve cell throughout the body.

Are All Cells Alike?

With the help of a microscope, you've been able to see the cells that make up Elodea and frog blood. Did these two types of cells look like each other? Are the cells that make up the flowers on a rosebush the same as the cells that make up the wings of a butterfly?

Find Out! ACTIVITY

Do all cells look the same?

In this activity, you will look at cells from different types of organisms.

What To Do

1. You will need prepared slides of guard cells on the surface of a leaf, human cheek cells, and yeast cells. These slides may have been stained with different colors so that the parts of the cells can be seen more easily.

2. Place each slide on the microscope stage and focus with the low-power objective in place. *In your Journal*, draw what you see on each slide and label your drawing.

Conclude and Apply

1. How are the cells alike?

2. How do they differ?

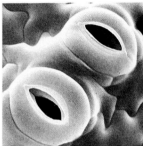

Guard cells

You've just observed an important fact about cells—cells come in different sizes and shapes. Your body contains many different kinds of cells. The shape of a cell may tell you something about the unique job of each cell. **Figure 17-2** shows you the unique shapes of nerve cells and plant stem cells. What do the traits of these cells help tell you about its job?

Plant cell

B Various plant cells have different jobs. Two plant-cell jobs are to carry water throughout the plant and to collect sunlight and change it into food. Judging from the shape of the plant cell shown, which job do you think it does?

518 Chapter 17 Basic Units of Life

Expected Outcome

Students will infer that cells can differ in shape and structure.

Conclude and Apply

1. All cells seem to have a definite outer boundary and various small structures inside them.

2. Cells vary in shape, size, color, and contents.

✔ Assessment

Process Have students, working in groups, use their observations to try to determine the function of the three types of cells they have drawn in their journals. Ask students to list several reasons to support their conclusions. Select the appropriate Performance Task Assessment List in **PASC.** COOP LEARN

How Big Is a Cell?

Think back to any organisms you have seen on your way to school. Some of them, such as trees, dogs, or your classmates, are large. Others, like blades of grass, a mosquito, or a caterpillar, are quite small. Are their respective cells large and small as well?

Find Out! ACTIVITY

Does the size of a living organism tell you anything about the size of its cells?

What To Do

1. Your teacher will give you two slides. One slide has human cheek cells. The other slide is of Elodea cells.

2. Look at the cells with a microscope using low power for each slide. Notice the size and shape of each.

3. You must use the same power each time to be able to compare the cells. Draw two circles the exact same size *in your Journal*.

4. Then, draw each cell exactly as you see it in relation to the circle. Are the cells on each slide about the same size?

Conclude and Apply

1. What does the size of an organism tell you about the size of its cells?

The two types of cells you just looked at were about the same size. Yet one slide contained cells from a human being, while the other slide contained cells from a small plant. Bigger organisms do not have bigger cells. They just have a larger number of cells.

Is there some advantage to a cell being so small? Why, for instance, aren't you made up of just one large cell instead of trillions of tiny cells? Let's think for a minute about objects, their sizes, and distance.

Figure 17-3

What makes you bigger than a toad? Both you and a toad are made of cells. Are human cells bigger than toad cells, or do humans just have more cells? The answer is number, not size. Most of your cells are not any larger than the toad's cells. You just have a lot more of them.

Meeting Individual Needs

Learning Disabled, Physically Challenged
If students are unable to draw to scale, have them count the number of cells that extend across the center of the field and record these numbers. Help students compare the two numbers. Cells that are smaller will have more cells visible across the field.

Find Out!

What is the relationship between the size of an object and the distance to its center?

Time needed 20–25 minutes

Materials small clear container, large clear container, water, potassium permanganate crystals, plastic spoons

Thinking Processes observing and inferring, comparing and contrasting

Purpose To observe that the larger the object, the greater the distance to its center.

Teaching the Activity

Safety Do not allow students to handle potassium permanganate, which is poisonous and a skin irritant.

Troubleshooting Try to use the same size of crystal in each container of water.

Science Journal Suggest that students make a two-column table in their journals to record their results. L1

Expected Outcome

The potassium permanganate spreads faster in the small beaker.

Conclude and Apply

1. The larger the area, the greater the amount of time needed for a material to spread.

✔ Assessment

Content Ask students to draw diagrams that explain the advantages smaller cells have over larger cells. *Students' diagrams should show that it takes less time for materials to spread through smaller cells.* Use the Performance Task Assessment List for Scientific Drawing in **PASC**, p. 55.

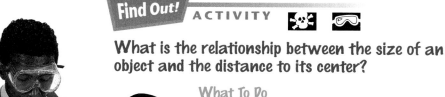

Find Out! ACTIVITY

What is the relationship between the size of an object and the distance to its center?

What To Do

1. Obtain small and large clear containers.
2. Fill each container with water that is the same temperature.
3. Your teacher will place a potassium permanganate crystal in the smaller container. At the same time, your teacher will also place a potassium permanganate crystal in the larger container.
4. *In your Journal*, note the amount of time it takes for the potassium permanganate to spread throughout the water in each container. Be sure to wash your hands if any solution touches them.

Conclude and Apply

1. What is the relationship between the size of a substance and the amount of time needed for a material to spread through a substance?

Why Cells Divide

H ow does cell size in other living things compare? Surprisingly, a tiny mouse and a gigantic elephant have something in common with each other and with you. The cells in mice, elephants, and humans are all about the same size.

More than ten trillion cells make up your body. As you can imagine, each cell is very small. If you could line up 1000 of those cells, they would total less than 2 centimeters in length—only about the width of a thumbnail.

Why Small?

Whatever the size of the whole organism, the cell remains small. That's because it is important for cells to have as much surface area as possible. Through this surface area, cells absorb needed materials and give off wastes. A larger surface area allows more of this activity to go on. In addition, the smaller cell size means that the incoming materials have to travel a shorter distance to reach the center of the cell.

Surface to Volume

The diagram on the next page shows how surface area is related to volume. The first cube has a volume of 1 cubic centimeter (height × width × depth). Its surface area is 6

📟 Purpose

A Closer Look extends Section 17-1's discussion of cell size by giving a mathematical explanation for why cells must be small. The surface-to-volume ratio of a cube is used as an analogy for a cell.

TECH PREP

Teaching Strategies

Ask students if they have ever seen or heard about movies with titles such as *The Blob* that depict giant, single-celled

The greater the size of the container of water, the longer it took for the potassium permanganate to spread through it. As the size of an object increases, so does the distance from its sides to its center. Suppose the containers you observed were cells. Materials might travel from the center of the smaller cell to its edges in a shorter length of time than materials traveling similarly in a larger cell. The ability of a cell to function well depends on the efficient flow of materials around it and into and out of it. Since materials travel at the same rate, it appears that materials may be supplied more efficiently in small cells than in larger ones.

Figure 17-4

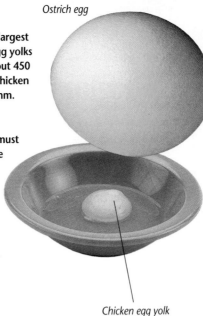
Ostrich egg

Chicken egg yolk

Ⓐ Bird egg yolks are the largest known cells. Ostrich egg yolks have a diameter of about 450 mm. The diameter of chicken egg yolks is about 45 mm.

Ⓑ Because cell materials must travel further, they take longer to get to the cell part where they are needed in large cells than in small cells. Which cells—large or small—appear to be most efficient? Why?

Visual Learning

Figure 17-4 Have a volunteer read the captions aloud and answer the question posed: **Which cells—large or small—appear to be most efficient? Why?** *Small cells; since materials must travel farther in large cells before they are available for work, large cells will be less efficient than small cells.*

Theme Connection As students learn about differences in cells, they are exposed to the theme of scale and structure. Although cells are generally similar in scale, as students learned in the Find Out activity on page 519, they vary in structure.

Discuss ways cells in a body exhibit division of labor and specialization. For example, refer students to Figure 17-2. Some nerve cells have very long processes, which may indicate their function in communication. Ask if these cells look very strong. *no* Then ask students if they think nerve cells could move body parts. *no* Ask students what cells move parts of the body. *muscle cells*

square centimeters (height × width × 6 sides of the cube). Comparing the surface area to the volume produces a ratio of 6/1, or 6.

Using the same math on the second cube, which is 4 centimeters on each side, yields a surface area of 4 × 4 × 6, or 96, and a volume of 4 × 4 × 4, or 64. Thus, this cube has a smaller surface-to-volume ratio—94/64, or about 1.5. Even though it is larger, the second cube has a relatively smaller surface-to-volume ratio than the first cube.

Now look what happens when the larger cube is divided into 64 small cubes. The large cube still has the same volume, 64, but it now has a total surface area of 384 (1 × 1 × 6 × 64 cubes), and its surface-to-volume ratio increases from 1.5 to 6. The divided cube has a much greater surface area.

To get an idea of relative surface area, pretend that the large cube and the individual parts of the divided cube are gift boxes. Suppose that you are wrapping each gift box in fancy paper. You'd need a lot more paper to wrap all of the small boxes than to wrap the one large box. However, you could store either the one large box or all of the small boxes in the same space because their volumes are equal.

Volume	1	64	64
Surface area	6	96	384
Surface-to-volume ratio	6	1.5	6

What Do You Think?

Keeping in mind surface-to-volume ratio, why do you think it would take less time to digest food that is well chewed?

NATIONAL GEOGRAPHIC SOCIETY

◉ **Videodisc**

STV: Human Body Vol. 3

Bacteria (green) on skin, magnified 8000×

48854

monsters. Most students will be familiar with such movies. Ask if such creatures can exist. Most students will answer in the negative. Have students read the feature and determine a scientific reason why such creatures do not exist. Have students read the feature and answer the question at the end.

Answer to
What Do You Think?
Well-chewed food has a greater surface-to-volume ratio than food that is less well-chewed.

Going Further ▸
Have groups of students discuss the ways in which a large surface-to-volume ratio would aid processes such as digestion, excretion, and transport of nutrients and other materials. Use the Performance Task Assessment List for Group Work in **PASC**, p. 97. COOP LEARN L3

17-1 Exploring Cell Size

Preparation

Purpose To observe cells to compare their sizes.

Process Skills observing and inferring, comparing and contrasting

Time Required 40 minutes

Materials See reduced student text. In addition to the listed materials, students will need drawing compasses and calculators. Prepared slides can be purchased from biological supply companies.

Possible Hypotheses Most students will predict that the period is larger.

The Experiment

Process Reinforcement To review what students know, observed, and learned about cell size, ask them to explain why cells are small. Small cells allow materials to move through the body of the cell more quickly. Have students recall the cells they have already observed under magnification. Students may wish to refer to their Science Journals where they drew diagrams. Ask students to predict how the size of any of those cells will compare with the size of a period. Then tell them that they will see just how small cells are—by measuring and comparing.

Possible Procedures Observations made through the microscope will be at different magnifications. Draw circles 100 mm in diameter and copy what is seen in the field of view in the circle. Measure the diameter of the object being viewed. For low power, multiply by 0.015, and for high power, multiply by 0.0035. This will bring all of the measurements into the same scale. A proportion comparing the actual diameter of the object to actual diameter of the field (either 1.5 mm or 0.35 mm) and diameter of the drawn object to diameter of the drawn field (100 mm) is being used.

Exploring Cell Size

Measuring an object as small as a cell can be difficult. However, if you can compare things that can be measured, you can come up with a fairly accurate measurement for something so tiny. For example, if you measure a worm and find out it is 4 cm long and then find an insect about one-fourth as long, some simple math will tell you that the insect is 1 cm long. You will apply this to a cell and the period at the end of a sentence.

Preparation

Problem
Both cells and the periods at ends of sentences are extremely small. What method can you use to measure them and to compare their size?

Form a Hypothesis
Make a hypothesis as to whether the period or the cell will be larger.

Objectives
• Observe cells to compare their sizes.

Materials
prepared slide of period from newspaper
prepared slides of frog skin
microscope   onion skin
paper and pencil   tomato skin
metric ruler   drawing compass
flower petals   calculator

Safety Precautions

Data and Observations			Sample Data
Type of cell	Drawing measurement	Multiply by	Actual diameter of object viewed
Period	35 mm	0.015	0.525 mm
Frog skin	7 mm	0.0035	0.025 mm
Flower petal	7–43 mm	0.0035	0.025–0.15 mm

Program Resources

Activity Masters, pp. 71–72, Design Your Own Investigation 17-1 L3

DESIGN YOUR OWN
INVESTIGATION

Plan the Experiment

1 Copy the data table into your Science Journal and decide which cells you will view.

2 We know that the field of view for the microscope on low power is 15 mm. *In your Science Journal,* draw a 100-mm circle to represent that field of view. Then draw the period in that circle as you see it through the microscope on low power.

3 How will you compare the size of the period in your 100-mm drawing to the size of the period in real life?

4 Repeat the experiment with the frog skin cell on high power. The field of view for the microscope on high power is 0.35 mm.

5 How will you use your drawing to determine the actual cell size?

6 Once again, repeat the experiment with another cell type.

Check the Plan

1 Why is the multiplication factor different for the two trials?

2 What part of the cell will you measure?

3 Make sure your teacher has approved your plan before you begin.

4 Do the experiment.

Analyze and Conclude

1. Compare and Contrast Which type of cell is largest? Is the period smaller or larger than the cells?

2. Sequence Sequence the cells in order from smallest to largest. How many times larger is the largest than the smallest?

3. Infer What can you infer about the size of cells?

17-1   The World of Cells   **523**

Teaching Strategies

Troubleshooting Remind students to set their compasses for the radius (50 mm), not the diameter of 100 mm.

Science Journal Have students record their predictions about the size of a cell for comparison with their actual measurements. Students may wish to record their data tables on graph paper and attach them to their Science Journals.

Expected Outcome

Students will conclude that cells are smaller than a period in a newspaper.

Analyze and Conclude

1. Answers will vary some. The frog cell will not be the largest. The period is larger than the cells.

2. Answers will vary; check against data.

3. The sizes of cells can vary, but most cells are about the same size.

Going Further

What differences would be needed in order for a scientist to examine the sizes of the inner structures of a cell?

✔ Assessment

Process The point of a pin measures about 0.20 mm in diameter. Most cells are about 0.01 mm to 0.02 mm in diameter. Have students use their data to calculate about how many frog skin cells could fit on this pinpoint. *Answer: about 8* Select the Performance Task Assessment List for Using Math in Science in **PASC,** p. 29.

Going Further

A much stronger microscope with greater magnification would be needed.

Visual Learning

Figure 17-5 To compare the two types of electron microscopes, have students answer the question posed: **If you wanted to learn about the texture of an eyelash, which microscope would you use?** *the SEM*

3 ASSESS

Assign the questions under Check Your Understanding. Allow students to test their predictions for the Apply question by doing the experiment.

Reteach

Discussion Ask students to recall cells from the lining of a human cheek and the cells from the skin of a leaf. Ask how the shapes are similar. *They are broad and flat.* Then ask what the function of these cells is. *They cover an area, as all "skin" tissues do.* Ask how the structure of a covering material is related to its function. *By being broad and flat, they cover more area.*

Extension

Activity Have students who have mastered the concepts presented in this section observe prepared slides of human blood. Explain that there are different types of white cells. Have students draw red blood cells, platelets, and several kinds of white blood cells. L3 P

4 CLOSE

Writing in Science

Have students write a brief paragraph to answer the following question. **Is there an advantage for a cell to be small?** Students point out that materials move more efficiently in a small cell. Use the Performance Task Assessment List for Writing in Science in **PASC**, p. 87. L1 P

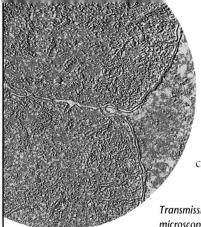

Transmission electron microscope (TEM) image

In the Investigate activity, you used a microscope to observe cells. Due to the invention and improvement of the microscope over the centuries, scientists are able to compare the cells of various organisms. They have discovered that while cells may vary in size and function, all cells have the same basic structure. Knowledge about cells was greatly advanced due to the development of the electron microscope. **Figure 17-5** shows the images of two different types of electron microscopes. In the next section, you will learn more about the features of cells.

Figure 17-5

Scanning electron microscope (SEM) image

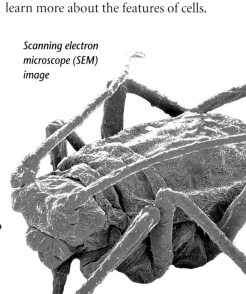

A Development of the electron microscope greatly advanced the study of cells. The transmission electron microscope (TEM) is used to study the inside parts of cells, like the rat liver cell above. The TEM uses an electron gun to produce an electron image of the specimen. That image is then converted to a visual image, which people can understand.

B The scanning electron microscope (SEM) is used to study the details on the surfaces of objects, such as this insect. SEMs form and display an image of the specimen. If you wanted to learn about the texture of an eyelash, which microscope would you use?

check your UNDERSTANDING

1. How could you determine whether a green patch found on an orange is living or nonliving?
2. You are given two slides labeled "Rabbit cells." You examine them briefly and observe that they differ somewhat from each other in shape. Explain why two cells from the same animal might have different shapes.
3. Through which type of cell, a chicken egg yolk or an ostrich egg yolk, might material move more efficiently? Explain.
4. **Apply** In the Find Out activity on page 520, you compared the amount of time it took for potassium permanganate to spread throughout water in containers of different sizes. While the size of the containers differed, the temperature of the water was the same. Now suppose you repeated the activity but filled the larger container with hot water and the smaller container with cool water. Would your results be the same or would they be different? Explain your answer.

check your UNDERSTANDING

1. Examine the patch under a microscope to see whether it contained cells. If cells are present, it is (or once was) living.
2. They probably came from different parts of the body and have different functions.
3. The chicken egg, because it is smaller.
4. In this activity, two factors will affect the rate at which spreading takes place—water temperature and size of the container. The potassium permanganate will spread faster in the hot water than in cool water. However, the distance the chemical must travel is still greater in the large container.

17-2 **The Inside Story of Cells**

The Parts of a Cell

You've learned that all living things are composed of the same basic units—cells. As you've explored the world of cells, you've discovered that cells differ from one another in size, shape, and function. Even with these differences, however, most cells share some common traits. An understanding of these features will help you understand how a cell does its job.

Each different type of cell in your body has a specific job to do. Nerve cells transmit impulses. Muscle cells contract and cause bones to move.

A cell and its activities might be compared to a business that operates 24 hours a day, making different products. It operates inside a building. Only materials that are needed to make specific products are brought into the building. Finished products and waste products are then moved out onto loading docks to be carried away. A cell performs similar functions to that of a business, and it also has a barrier that encloses it. Try this next activity to see what that barrier is like.

Section Objectives

- Identify the parts of a typical cell.
- Describe the jobs of cell parts.
- Compare and contrast plant and animal cells.

Key Terms

cell membrane, cytoplasm, nucleus, chromosomes, mitochondria, cell wall, chloroplasts

Explore! ACTIVITY

What holds a cell together?

What To Do

1. Make a model of a cell using semisolid gelatin and a clear, plastic, resealable sandwich bag.

2. Fill the bag with the gelatin and close it.

3. Gently poke the center of the bag. *In your Journal*, describe what happens to the gelatin inside. Do the bag and its contents have a definite shape? Can you change the shape of the bag? Does the shape stay changed? What helps keep the shape of the bag?

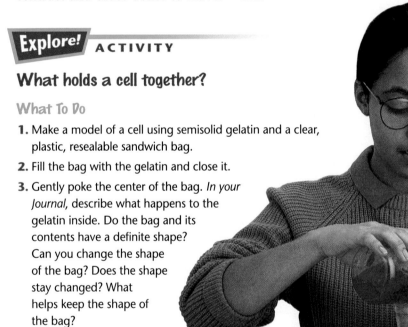

Explore!

What holds a cell together?

Time needed   5 minutes

Materials   prepared gelatin, plastic resealable sandwich bag

Thinking Processes   observing and inferring, making models

Purpose   To simulate the function of the cell membrane.

Preparation   Prepare unflavored gelatin or gelatin dessert using 1¼ the amount of water specified on the package.

Teaching the Activity

Troubleshooting   Be sure that bags are completely sealed. Collect and store the bags in a refrigerator for the Explore activity on page 532.

Science Journal   Have students answer each of the questions in step 3 in their journals. L1

17-2

PREPARATION

Planning the Lesson

Refer to the Chapter Organizer on pages 514A–D.

Concepts Developed

Students will learn about the structures that make up cells. Students will also see that plant cells differ from animal cells. Animal cells lack the cell wall and chloroplasts found in plant cells.

1 MOTIVATE

Bellringer

Before presenting the lesson, display **Section Focus Transparency 51** on an overhead projector. Assign the accompanying **Focus Activity** worksheet. L1
LEP

Expected Outcome

Students will see that the shape of the bag and its contents are changeable.

Answers to Questions

3. The bag and its contents do not have a definite shape; the shape of the bag can be changed, but it will not necessarily stay changed. The pressure of the gelatin on the bag gives it some firmness and helps keep the shape of the bag.

✔ Assessment

Process   Ask students to hypothesize how their results would have differed if the gelatin had been in (a) a liquid state or (b) a solid state. Have students draw diagrams to illustrate their hypotheses. Select the appropriate Performance Task Assessment List in **PASC**.

2 TEACH

Tying to Previous Knowledge

To compare and contrast cells, ask students to think back to the human cheek cells and the *Elodea* cells that they observed. **Did students notice a difference between the edge of an animal cell and the edge of a plant cell?** *The edges of the plant cell may look thicker than that of the animal cell.* Ask students to compare the overall shape of animal cells and plant cells. *Plant cells seem to be more rectangular in shape than animal cells.* Students will see that the cell wall accounts for these observed differences.

Theme Connection As they read this section and look at Figures 17-6 and 17-7, students will see evidence of the theme of scale and structure. Although cells differ in some of the details of their structures, they are made of the same components. Each component has a structure of its own, which helps it accomplish its particular function within the cell.

NATIONAL GEOGRAPHIC SOCIETY

Videodisc

STV: The Cell

Parts of the Cell
Unit 2
Parts of the Cell

12694-27234

The cell is the basic unit of structure and function in all living things. It is the basic building block of organisms. Among the many-celled organisms, there are two basic cell types—the animal cell and the plant cell. Although these two types of cells share many common structures, there are a few exceptions. The following diagrams will help you to identify the parts of typical cells. Each cell part will then be discussed, comparing their respective jobs within the cell.

Figure 17-6

ANIMAL CELL

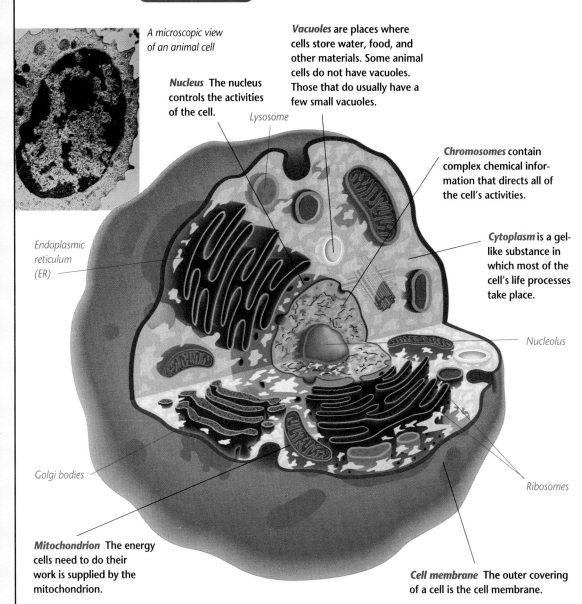

A microscopic view of an animal cell

Vacuoles are places where cells store water, food, and other materials. Some animal cells do not have vacuoles. Those that do usually have a few small vacuoles.

Nucleus The nucleus controls the activities of the cell.

Lysosome

Chromosomes contain complex chemical information that directs all of the cell's activities.

Cytoplasm is a gel-like substance in which most of the cell's life processes take place.

Endoplasmic reticulum (ER)

Nucleolus

Golgi bodies

Ribosomes

Mitochondrion The energy cells need to do their work is supplied by the mitochondrion.

Cell membrane The outer covering of a cell is the cell membrane.

Meeting Individual Needs

Visually Impaired Obtain three-dimensional models of cells and allow students to touch the parts of the cell as you identify them. Keep the models at hand so that students can refer to them as often as necessary. You can also use hard-boiled eggs and carefully cut the white around the long axis (without cutting into the yolk). When the cut white is pulled away, students can feel the round yolk and equate that to the nucleus of a cell.

Learning Disabled, Behaviorally Disordered As students work through the section, they should refer to Figures 17-6 and 17-7 as each new cell part is introduced. To minimize searching back through pages, provide each student with a small removable adhesive tag to mark this page. Students will use the diagram more often if they can find it quickly. As an alternative, you may wish to have students make their own drawings of cells for reference.

Figure 17-7

PLANT CELL

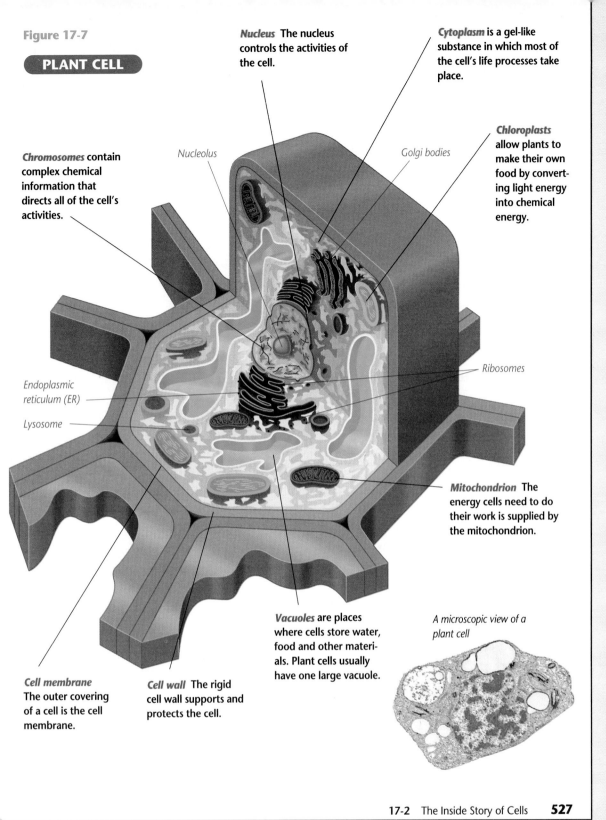

Chromosomes contain complex chemical information that directs all of the cell's activities.

Nucleolus

Nucleus The nucleus controls the activities of the cell.

Cytoplasm is a gel-like substance in which most of the cell's life processes take place.

Golgi bodies

Chloroplasts allow plants to make their own food by converting light energy into chemical energy.

Ribosomes

Endoplasmic reticulum (ER)

Lysosome

Mitochondrion The energy cells need to do their work is supplied by the mitochondrion.

Cell membrane The outer covering of a cell is the cell membrane.

Cell wall The rigid cell wall supports and protects the cell.

Vacuoles are places where cells store water, food and other materials. Plant cells usually have one large vacuole.

A microscopic view of a plant cell

17-2   The Inside Story of Cells   **527**

17-2   The Inside Story of Cells   **527**

Visual Learning

Figures 17-6 and 17-7
Ask students to compare the shapes of the animal and plant cells. *The animal cell is round and the plant cell is more rectangular.* Point out that these are generalizations of plant and animal cells and do not represent any one type of cell. Have students recall cells that they have already seen and compare them to the cells shown here. For example, students may comment that blood and cheek cells look thinner than the animal cell in this figure. Accept all reasonable comparisons.

Have students compare the computer art of the animal and plant cells with the microscope views of "real" animal and plant cells. Have students identify some of the structures visible within the real cells.

Note that additional "teacher-only" labels have been added to the cell diagrams for the purpose of information and discussion.

Demonstration If models are available, use them as you discuss the parts of cells with the class. Have students locate parts, such as the nucleus, in the diagram and then on the model. [L1]

Uncovering Preconceptions
Students often have the idea that cells are flat, not three-dimensional. This idea is reinforced by microscope studies, because the microscope has such a shallow depth of field and prepared slides are usually made of very thin sections of tissues. Use three-dimensional models, as well as the diagrams in this section, to help students see that cells have depth.

Flex Your Brain Use the Flex Your Brain activity to have students explore MEMBRANES. L1

Across the Curriculum

Language Arts

Greek and Latin words are used in naming cell parts. Help students pronounce the names of the cell parts in the cell diagrams. Have students make a list of the cell parts and use a dictionary to find the origins of the words and their meanings. L2

Visual Learning

Figures 17-8 and 17-9 To reinforce the parts of cells, have students copy the diagrams of the plant and animal cells (shown in the figures) in their student journals. Then have students label the cell membranes and the cytoplasm for each. Students can add to these diagrams as they learn more about the parts of the cell in this section. L1

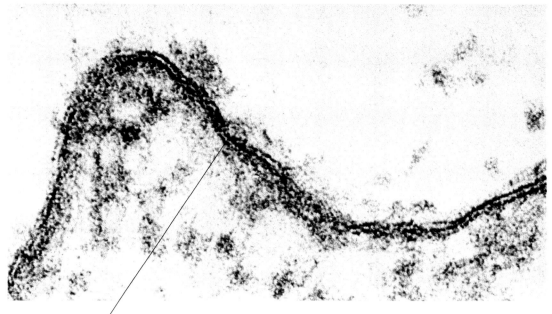

Figure 17-8

A The cell membrane contains the cell and is partly responsible for its shape. Because the membrane is flexible, it allows the shape of the cell to change under pressure. Food and oxygen enter the cell through the cell membrane. Water and other products made by the cell exit the membrane.

B A cell can be compared to a factory. Products going into a factory are regulated by the receiving department. Products leaving the building are regulated by the shipping department. Both of these jobs in a cell are carried out by the cell membrane.

Cell membrane

Animal cell   *Plant cell*

■ Cell Membranes

In the model of a cell you made in the Explore activity, a plastic bag represented the outer covering or barrier of a cell. Most cells are surrounded by an outer covering called the cell membrane. The **cell membrane** is a flexible structure that forms the outer boundary of the cell. You can't see a cell membrane using a regular light microscope. However, by using chemical tests and the electron microscope, scientists have found that a cell

membrane is a double-layered structure that surrounds the contents of the cell. In the Explore activity, you made a model of a cell using a plastic sandwich bag and gelatin. The part of the model that represented the cell membrane was only a single-layered structure. How could you have more accurately represented the structure of a cell membrane?

Figure 17-8 shows the cell membrane of a cell and tells about the important jobs it performs.

528 Chapter 17 Basic Units of Life

⟨**Program Resources**⟩

Study Guide, p. 57
Critical Thinking/Problem Solving, p. 5, Flex Your Brain
Teaching Transparency 33 L2
Making Connections: Across the Curriculum, p. 37, Dividing Cells, Multiplying Numbers L1
Section Focus Transparency 51

ENRICHMENT

Activity To help students formulate models encourage them to construct models of cells, using readily available materials. For example, a variety of shapes of pasta can be used to represent many cell structures. Display students' models in the classroom and have students refer to them as review aids during class discussions. L1

■ Cytoplasm

Think back to the photograph of the honeycomb you examined at the beginning of this chapter in the Explore activity. What filled each part of the comb? Honey, of course. What about cells? Are they filled with anything? In the Find Out activity that follows, you will discover the answer to this question.

What's inside cells?

What To Do

1. Take a layer of skin from a fresh red onion bulb. Bend it so that you can peel off a single, paper-thin layer from the inside.

2. Prepare a wet-mount slide of this layer of cells.

3. Observe the sample under low power and make a drawing of what you see. If you watch long enough, you may be able to see something moving around inside each rectangular cell.

Conclude and Apply

1. How would you describe the material (cytoplasm) from looking at these cells?

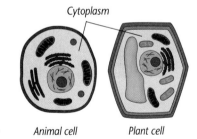

The liquid, found in both plant and animal cells, is where many of the cell's activities take place. **Cytoplasm** is a gel-like material inside the cell membrane. **Figure 17-9** shows where cytoplasm is located within the cell.

Quite a large portion of the cell you observed is made up of colorless cytoplasm. Cytoplasm contains a large amount of water, but it also contains chemicals and cell structures that carry out life processes for the cell. Some structures found in the onion cell have a small amount of color. The large red storage area in the onion cell is called a vacuole. The gel-like cytoplasm constantly moves around the structures within the cell. What use does this movement probably serve for the cell?

Figure 17-9

Just as air surrounds the workers in a factory, cytoplasm surrounds the internal structures of the cell. Cytoplasm is a soft, gel-like substance in which most of the cell's life processes take place. Like air, cytoplasm is constantly moving.

Cytoplasm

Cytoplasm

Animal cell   Plant cell

17-2   The Inside Story of Cells   **529**

Meeting Individual Needs

Behaviorally Disordered, Physically Challenged   Some students may not have the patience or dexterity to peel the onion for the Find Out activity. To help them, peel the onion and place it on the slide that they have prepared. Depending on the individual's dexterity, you may have to assist with adding the coverslip, too.

Demonstration Show students a photograph of a building that has an atrium-style lobby with footbridges that cross to connect different parts of the building. If photographs are not available, ask if students have been to such a building. Point out that the vacuole of a plant cell occupies much of the volume of a plant cell. The vacuole of a plant cell contains water, not air. Ask students if they saw any strands of cytoplasm passing through the central vacuole when they observed the onion cell in the Find Out activity on page 529. Compare these strands of cytoplasm with the footbridges that cross an atrium in a building.

Find Out!

Where is the cell's command center located?

Time needed 20 minutes

Materials slide, coverslip, onion, iodine stain, onion root tip slide, microscope, paper towels

Thinking Processes Comparing and contrasting

Purpose To locate and identify the nucleus of a cell.

Preparation To prepare iodine stain, mix 0.15 g potassium iodide with 100 mL water to which 0.03 g iodine has been added. Note: Both potassium iodide and iodine are poisonous—wash hands if contact is made.

Teaching the Activity

Troubleshooting To stain cells, hold a piece of paper towel at one side of the coverslip and place a drop of stain on the slide next to the opposite edge of the coverslip. This will draw the stain under the coverslip.

Science Journal Have students sketch their observations

■ Nucleus

Factories generally have a manager who directs everyday business for the company from a central office. A cell also has a command center that controls its activities. Just where is this center? What does it look like and what is it called?

Find Out! ACTIVITY

Where is the cell's command center located?

What To Do

1. Observe a layer of onion skin again, this time using a white onion.

2. Make a wet-mount slide of onion skin and look at it first under low power, then under high power.

3. With your teacher's help, let a small drop of iodine seep under the coverslip. **CAUTION:** *Iodine is poisonous. Wash your hands to remove any iodine that gets on your skin.* Look for a large round structure in the cytoplasm that takes on color. Draw what you observe *in your Journal.*

4. Now look at a prepared slide of an onion root tip under high power.

Conclude and Apply

1. What structure(s) in the cytoplasm became colored by iodine?

2. Did you see any movement in the living tissue?

3. How did your observations compare with the prepared, stained slide?

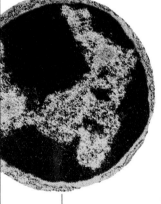

Nucleus

Figure 17-10

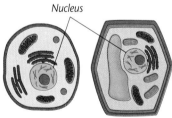

The nucleus contains genetic blueprints for operation of the cell. The nucleus has its own structures including the chromosomes.

Nucleus

Animal cell   *Plant cell*

When you saw a large sphere in the cells you examined, you were looking at the nucleus of the cell. In many cells, the nucleus is the largest structure you can see in the cytoplasm. The **nucleus** of a cell is its command center—the structure that directs all the activities of the cell. It contains complex chemical information that directs the cell's activities, including its ability to reproduce. This material inside the nucleus is separated from the cytoplasm by a thin membrane.

---

of the prepared slide so they can compare it with their sketch of the living tissue. [L1]

Conclude and Apply

1. the nucleus

2. Answers will vary, but students may observe movement in the cytoplasm.

3. The nucleus of the prepared slide will probably have a different color stain, which may show more detail than the iodine stain. There was no observed movement in the prepared slide.

✔ Assessment

Performance Have students look back at Figures 17-6 and 17-7 to identify the cell's nucleus, cytoplasm, and cell membrane. Have pairs of students prepare a poster or other display identifying parts of the cell. Select the appropriate Performance Task Assessment List for Poster in **PASC**, p. 73 COOP LEARN P

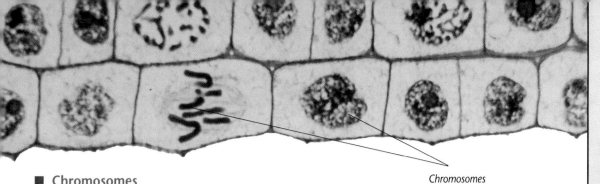

Chromosomes

■ Chromosomes

Chromosomes are threadlike structures made up of proteins and DNA, the molecules that control the activities of the cell. Look at **Figure 17-11**, and identify the nucleus and the chromosomes.

Chromosomes aren't visible all the time. When a cell is not reproducing, the nucleus looks grainy. The best time you can observe chromosomes is when a cell is dividing. What does this tell you about the cells you observed in the onion root tip?

■ Mitochondria

Almost any factory uses some type of machine to do work. The energy needed for these machines to run is supplied by a power plant located nearby. Cells do work, so they require energy, too. Inside each cell are structures that enable the cell to release energy obtained through food digestion.

Look at **Figure 17-12**. **Mitochondria** are the power plants of a cell, which release energy needed for cell activities. Inside these round to rod-shaped structures, molecules from food digestion are broken down to release energy that can be used for the activities of the cell.

Figure 17-11

Chromosomes take up most of the space of the nucleus. These long strands of material contain complex chemical information that controls all of the cell's activities including its ability to reproduce. Chromosomes are usually only visible when a cell is dividing.

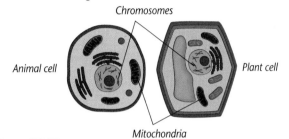

Chromosomes

Animal cell

Plant cell

Mitochondria

Figure 17-12

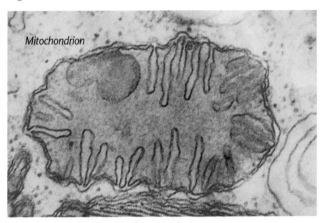

Mitochondrion

Power companies supply the energy for a factory to carry out its work. Mitochondria supply the energy for cells to do their work. Some cells use more energy than others because they are more active. Muscle cells have more mitochondria than do the cells that produce fingernails. Why do you think that is so?

Inquiry Questions What would happen if the nucleus of a cell became damaged in some way? *The cell would no longer function correctly because the nucleus controls all the activities of the cell.* Where do you think most of the materials that are needed by the cell are located? *in the cytoplasm* L3

Content Background

Not all cells have nuclei. The red blood cells of mammals have nuclei when they are first formed in the bone marrow, but they lose their nuclei before entering circulation. White blood cells, on the other hand, retain their nuclei. There are several kinds of white blood cells, which can be identified by the shapes of their nuclei.

Visual Learning

Figure 17-12 After students have studied Figure 17-12, ask the question posed: **Muscle cells have more mitochondria than do the cells that produce fingernails. Why do you think that is so?** *Muscle cells exert force to move, therefore, muscle cells work harder and require more energy than do the cells that produce fingernails.*

NATIONAL GEOGRAPHIC SOCIETY

◉ **Videodisc**

STV: Human Body Vol. 3

Chromosome

48757

Multicultural Perspectives

A History of the Cell
Have students prepare a time line of the discovery of cell structures, using encyclopedias or reference books. Students should include dates, discoveries, and the names and origins of the scientists involved. Provide an area for students to display their timelines.

Program Resources

Multicultural Connections, p. 37, A Pioneer in Cell Science L1
Teaching Transparency 34 L2
How It Works, p. 18, How Does a Cut Heal? L2
Making Connections: Technology & Society, p. 37, Selling Cells L1

Plant Cell Adaptations

In your trip through the structure of a cell, you have observed the cell membrane, cytoplasm, nucleus, chromosomes, and mitochondria of different cells. Animal cells contain all these structures. Plant cells, however, contain some additional cell parts. In the next activity, you will make a model of a plant cell to learn about these cell parts.

Explore! ACTIVITY

How do plant and animal cells differ?

What To Do

1. Take a clear food container and a plastic bag filled with semisolid gelatin.

2. Place the bag inside the container, pressing gently so that the plastic bag fits snugly up against the sides of the container.

3. *In your Journal*, record how the shape of the bag now compares with the shape of the container. If the bag were placed in a different type of container, would its shape change?

■ Cell Walls

In your model, the plastic container represents an adaptation found in plant cells, fungi, and bacteria. This structure, the outermost rim of the cell, is called the cell wall. The **cell wall**, shown in **Figure 17-13**, is a rigid structure located outside the cell membrane that supports and protects the cell. Cell walls from dead plants are used as wood. If you play baseball, cell walls may have helped you "muscle" a ball over the pitcher's head. The wooden bat you may have used is made up of the dead cells from an ash tree. Cell walls remain strong even though the contents of the cells are no longer there. Just imagine what would happen if your baseball bat had been made from animal cells whose only outer covering was a flexible cell membrane.

Figure 17-13

Cell walls are much thicker than cell membranes and are made of different substances in different organisms.

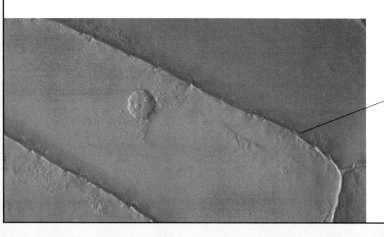

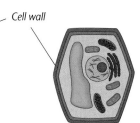

Tiny openings, or pores, in the cell wall permit substances to pass through. In your model of the cell, which part represented the cell wall?

Cell wall

Plant cell

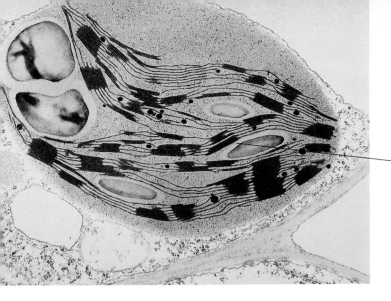

Figure 17-14

Chloroplasts are structures that contain chlorophyll, a green pigment that allows plants to make their own food by converting light energy into chemical energy in the form of a sugar called glucose.

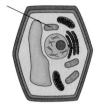

Chloroplast

Plant cell

■ Chloroplasts

You already know that plant cells have a structural adaptation that animal cells do not, namely, a cell wall. But when you studied plant and animal cells under the microscope and compared their structures, you may have made another interesting discovery. Green plant cells have another structure that animal cells don't, chloroplasts.

Chloroplasts are small structures that contain chlorophyll, a green pigment that allows plants to make their

own food. Most chloroplasts are located in the leaves of a plant. It is the green color of the chlorophyll that makes the leaves green.

You have now completed your journey through the major parts of plant and animal cells. The structures in these cells carry out certain life processes. In the next section, you will learn more about one of these processes—reproduction.

SKILLBUILDER

Making and Using Tables
Make a table that lists the parts of a cell and each of their jobs. If you need help, refer to the **Skill Handbook** on page 639.

check your UNDERSTANDING

1. Compare the job of a cell membrane with that of a cell wall.
2. What cell parts are found in green plant cells?
3. Describe the relationship that appears to exist between the job of a cell and the

number of mitochondria it contains.
4. What cell parts are more clearly visible when a cell is dividing?
5. **Apply** Suppose a disease destroyed all the chloroplasts in a green plant. Explain what would happen to the plant and why.

17-2   The Inside Story of Cells   **533**

check your UNDERSTANDING

1. The cell membrane maintains the chemical balance of a cell by controlling what enters and exits the cell. The cell wall supports the cell.
2. cell walls and chloroplasts
3. It appears that cells with more work to do have more mitochondria.

4. chromosomes
5. The plant would not be able to produce its own food and would die.

PREPARATION

Planning the Lesson

Refer to the Chapter Organizer on pages 514A–D.

Concepts Developed

As students investigated the size of a cell in Section 2, they set the stage for what will be discussed in this section. When a cell becomes too large to function efficiently, it divides. Cell division, or mitosis, produces new cells for growth and repair of the organism.

1 MOTIVATE

Bellringer

Before presenting the lesson, display **Section Focus Transparency 52** on an overhead projector. Assign the accompanying **Focus Activity** worksheet. L1

LEP

Ask students to bring in baby pictures and pictures of themselves in first or second grade. Identify the pictures by number and place them on a bulletin board. Then have students identify the person in each picture. When all pictures have been identified, discuss how students have grown and changed. L1

2 TEACH

Tying to Previous Knowledge

In Section 17-2, students learned that the difference between a large organism and a small one is not the size of the cells but the number of cells. Ask students how a growing organism gets more cells. *It makes them.* Point out that students will find out how this is accomplished in this section.

When One Cell Becomes Two

Section Objectives

■ Describe the process of mitosis and its end products.
■ Give examples of instances where cell reproduction takes place.

Key Terms

mitosis

Change and Growth

What happens to the tiny green shoots that, in spite of traffic, push through the cracks in playgrounds and parking lots? They often grow tall and strong, and produce roots that are hard to pull out. Puppies grow too, maturing into full-grown adult dogs. A green and black banded caterpillar sealed inside a pale green cocoon emerges as an orange and black monarch butterfly. All living things change and grow, often right before your eyes. Are you also changing?

Besides growing taller, you can find other evidence that the cells in your body are increasing in number. When you cut yourself, you see

Figure 17-15

2 WEEKS
Piglet weighs 4.5 kg (10 lb)

5 WEEKS
Pig weighs 11.5-13.5 kg (25-30 lb)

NEWBORN
Newborn weighs 1.8 kg (4 lb)

A As pigs, humans, and other animals develop, many of their body cells increase in number, causing the animal to grow gradually larger.

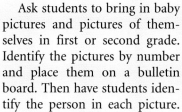

Program Resources

Study Guide, p. 58
Concept Mapping, p. 25, Mitosis L1
Take Home Activities, p. 26, Why Are Cells Small? L3
Making Connections: Integrating Sciences, p. 37, Fossil Cells L2
Science Discovery Activities, 17-1, 17-2, 17-3
Section Focus Transparency 52

---

dramatic evidence of cell reproduction. A cut is a break in your skin. Have you ever scraped your knee on the ground? As you know, when a cut occurs, blood initially flows through the opening. But in time, the tear in your skin is no longer visible.

Why? Your body actually repairs itself by sealing off the flow of blood and then producing new skin cells. The new cells fill the break in your skin as the dead cells are replaced. Cuts heal as new cells are produced.

GLENCOE TECHNOLOGY

 Software
Computer Competency Activities
Chapter 17

6 WEEKS

Pig weighs 13.5-15.75 kg (30-35 lb)

B When many animals reach the adult stage, cells in some body tissues continue to reproduce and replace old ones that wear out. Other cells, such as nerve and bone cells, become more specialized and are not replaced as easily, if at all.

ADULT

Pig weighs 95-105 kg (210-230 lb)

535

Visual Learning

Figure 17-15 Have a volunteer read aloud captions A and B as students trace the development of the pig. Ask students to relate cell reproduction to growth. *As the total number of cells increases, the body grows larger.*

Theme Connection This section offers students evidence of the theme of stability and change. Cell division brings about changes in the size or shape of an organism. However, the mechanism of cell division provides stability, as cells generally produce more cells of the same kind.

Discussion Ask students to identify areas of the body that are subject to a lot of wear and tear. Students will probably mention skin. Point out that skin cells are constantly rubbing off at the surface and being replaced from below. The cells at the bottom layer of the skin divide often.

GLENCOE TECHNOLOGY

 Videotape

The Secret of Life

Use the videotape *On the Brink: Portraits of Modern Science* to explore the factors that control cell growth.

ENRICHMENT

Discussion When a wound does not heal evenly, a scar forms. Students may have scars from childhood accidents or surgery. Ask students if plants ever get scars. Students may be surprised to learn that they do. If there are any trees in the schoolyard, they may have had initials carved in them. If there are any, show these to students and point out that trees do not heal from such wounds for a long time.

Find Out!

When are chromosomes visible?

Time needed 20 minutes

Materials microscope, prepared slide of onion root tip

Thinking Processes observing and inferring, comparing and contrasting

Purpose To locate and identify cells that are in the process of dividing.

Preparation Prepared slides of onion root tips can be purchased from biological supply houses.

Teaching the Activity

Troubleshooting To help students find the area where cell division can be seen, have them look at the slide without using the microscope. Have students locate the tip of the root and the cap of dead cells that covers it. Tell students to search the area just behind the root cap.

Science Journal Have students make their drawings and answer the Conclude and Apply questions in their journals.
L1

Expected Outcome

Students will observe that many of the cells in the root tip are dividing.

Conclude and Apply

1. The drawing of the dividing cell shows chromosomes. The drawing of the cell that is not dividing does not show chromosomes.

2. The presence of visible chromosomes is evidence of cell division.

An Introduction to Cell Reproduction

In order to understand how your body makes new cells, you first need to review the features of a nucleus and chromosomes. In the following Find Out activity, you will take a closer look at these cell structures.

Find Out! ACTIVITY

When are chromosomes visible?

Chromosomes are generally visible only when a cell is undergoing reproduction. If you can see a cell's chromosomes, then the cell is probably reproducing.

What To Do

1. Examine a prepared slide of an onion root tip under both low and high power. Look at cells that are undergoing reproduction.

Physics CONNECTION

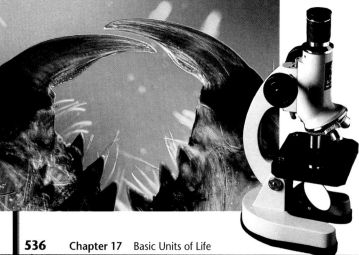
A magnified view of a spider's fangs

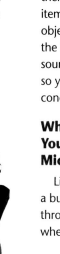

Light Microscope

It's easy to think of a microscope as a super magnifying glass, but a microscope is much more complex. Unlike a magnifying glass, which has only one lens, a microscope usually has at least three lenses and sometimes more. First, there is the lens closest to your eye, called the eyepiece. Then, there is the lens closest to the item you want to see, called the objective lens. Finally, there is the lens closest to the light source that illuminates the item so you can see it, called the condenser.

What Happens When You Look into a Microscope?

Light, from either a mirror or a built-in light bulb, passes through the condenser lens where it is intensified and

Physics CONNECTION

Purpose

Physics Connection extends the activities of Chapter 17 by describing how a light microscope works. This excursion also focuses on scale and structure by discussing how the lenses of a light microscope work together to expand the scale of an object.

Content Background

In 1590, a Dutch maker of reading glasses, Zacharias Janssen, put two magni-

fying glasses together in a tube to create the first crude compound microscope. In the mid 1600s, Anton Van Leeuwenhoek, another Dutch scientist, made a simple microscope with a tiny glass bead for a lens.

Teaching Strategy

Early scientists used various tools to help view objects. Have students look at a newspaper through the curved side of an empty glass, the flat bottom of an empty glass, a test tube filled with water and tightly stop-

2. Make a drawing *in your Journal* of a cell that's not reproducing. Then make a drawing of a cell that is reproducing.

Conclude and Apply

1. How do your drawings differ?

2. What structures indicated that the cell was undergoing reproduction?

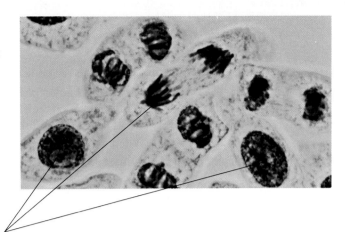

Onion root tip cell's chromosomes

Did you see chromosomes in the onion root tip cell? Chromosomes are threadlike structures located in the nucleus of the cell. In most cells, chromosomes play an important role in the reproduction of cells.

focused on the specimen you are looking at. The light then passes through the specimen and is collected by the objective lens, which shapes the light to form a magnified image of the specimen. That light image is then gathered by the eyepiece lens, which magnifies it again. Finally, the light carries the image into your eye, where it is projected on the layer at the back of your eye called the retina. As a result, you see an onion cell, a skin cell, bacteria, or whatever else you may be looking at through the microscope.

Magnification

Most microscopes actually have more than one objective lens. Usually, they have three objective lenses with different powers of magnification—10×, 40×, and 100×—meaning they enlarge the image of an item to 10 times, 40 times, or 100 times its natural size. Some microscopes also have changeable eyepiece lenses with different powers of magnification. If you multiply the power of the objective lens by the power of the eyepiece, you get the total magnification power of the microscope. For example, if your microscope has an eyepiece of 10 and objective lenses of 10×, 40×, and 100×, it can

magnify items from 100 to 1000 times their natural size!

What Do You Think?

The image produced by a microscope is backwards, like the reflection in a mirror, and upside down. For example, if you move a specimen slide to the right, the image you see through the eyepiece will look like it is moving to the left. Or if you move the slide down (toward you), the image will look like it is moving up (away from you). Why does this happen? Can you think of a way to correct the visual image you see so that it operates the same way as the real item?

17-3 When One Cell Becomes Two **537**

pered, and a magnifying glass. Ask them to compare their observations. **What might early scientists have learned by using such tools?**

Answers to
What Do You Think?

1. The convex lenses of the microscope invert the image.

2. Reflecting the image off a mirror will position it so that it matches the object being viewed; or

adding another lens to the microscope.

Going Further ⅢⅢⅢ➤

Stereoscopic light microscopes give a three-dimensional view of an object and are used to look at the surface of thick structures that light can't pass through. Challenge students to describe investigations that would require a stereoscopic microscope rather than a compound light microscope.

How Body Cells Reproduce

Have you ever watched a magician at work? Objects seem to disappear or reappear with sleight of hand. Such tricks can even make one object appear to become two. When a cell reproduces, one cell becomes two identical cells. It's not magic, however—it's mitosis.

Mitosis is the process by which the nucleus of a cell divides to produce two nuclei, each with the same type and number of chromosomes

Figure 17-16

STAGE 1

Mitosis begins with the chromosomes becoming fully visible. Each chromosome makes a copy of itself. The identical chromosomes remain joined together as a pair. At this point, the nuclear membrane begins to disappear and threadlike spindle fibers form. The chromosome pairs attach to the spindle fibers.

STAGE 2

The spindle fibers move all of the joined chromosome pairs to the center of the cell. The chromosome pairs line up along the middle of the cell.

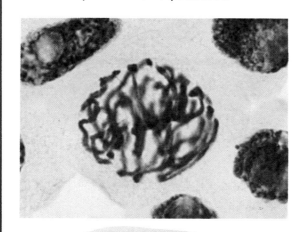

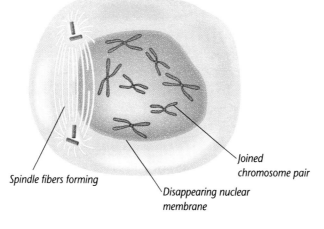

Spindle fibers forming

Disappearing nuclear membrane

Joined chromosome pair

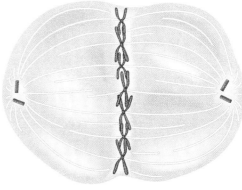

Chromosomes line up along cell center

538   Chapter 17   Basic Units of Life

Program Resources

Critical Thinking/Problem Solving, p. 25, When Cell Growth Gets Out of Control L3

Multicultural Connections, p. 38, Can a Disease Be Beneficial? L1

that the parent cell had. After the nucleus divides, the cytoplasm also usually separates. Follow the steps in **Figure 17-16** as this process is described.

Cell reproduction by mitosis has been taking place in your body from the moment you were conceived, and it continues even now. As you read this page, many cells in your body are dividing through mitosis. What evidence do you have that this is true?

All of your body cells reproduce by the same process—mitosis. But, the rate at which mitosis occurs in different types of cells may vary. In the next activity, you will observe the rate of mitosis in a young plant root.

Content Background

Mitosis is part of the cell cycle of growth and division. Interphase is the longest stage and is the part of the cycle during which the cell grows and duplicates the chromosomes. Prophase is the first stage of division, in which the nuclear membrane breaks down and the chromosomes become short and thick. Metaphase is the next stage, in which chromosomes line up across the middle of the cell. During anaphase, the double-stranded chromosomes separate. During telophase, the chromosomes move to opposite ends of the cell. Finally, the cytoplasm divides and each new cell is in interphase again. This is called cytokinesis.

Inquiry Question The nucleus is normally held together by a membrane. The membrane breaks down at the start of mitosis. Why do you think this breakdown occurs? *to allow the chromosomes to move from the nucleus to opposite ends of the cell*

STAGE 3

The chromosome pairs split apart. The spindle fibers seem to guide or pull the members of each chromosome pair to opposite ends of the cell.

STAGE 4

The spindle fibers disappear and the cytoplasm divides in half. The result is two new cells—each identical to the original cell. The process of mitosis has been called the dance of the chromosomes. Why do you think it might be called that?

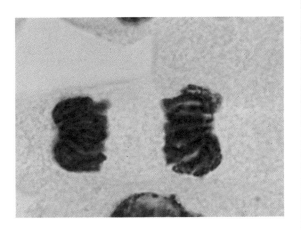

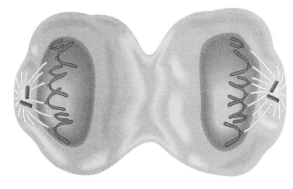

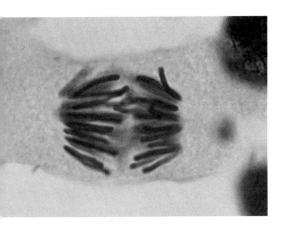

Chromosomes and their copies

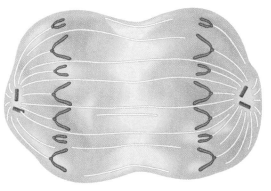

Cytoplasm divides in half

NATIONAL GEOGRAPHIC SOCIETY

 Videodisc

STV: The Cell
How Cells Reproduce
Unit 3

27458-45320

STV: Human Body Vol. 3
Cell dividing 1

48722

ENRICHMENT

Discussion To extend the discussion about mitosis, ask students if they know what cancer is. Students may know that this is a condition in which parts of the body grow in abnormal ways, forming growths that are called tumors. Ask students what could cause such growths? *mitosis* Point out that cancer is caused when mitosis gets out of control. Cells continue to divide and grow, even if there is no room for them. Stress that many tumors are benign (not cancerous), and that they should not assume that someone has cancer because he or she has a tumor. However, they should know that any tumor should be brought to the attention of a doctor.

Planning the Activity

Time needed 30 minutes each day for 3 consecutive days

Materials See student activity. Corn seeds are available from plant stores or farm supply houses.

Process Skills measuring in SI, observing and inferring, interpreting data, forming a hypothesis, predicting

Purpose To determine the rate of growth of a young root.

Preparation Begin sprouting corn seeds a few days before the activity is scheduled. Soak the seeds overnight in water, then place them between damp paper towels. Place the paper towels in plastic bags. Check each day to see if seeds have sprouted.

Teaching the Activity

Troubleshooting Try to match seedlings so that each group of students gets seedlings with roots of the same length. Discard any seedlings that show signs of mold.

Process Reinforcement To reinforce the importance of cell reproduction to growth, have students draw diagrams showing the stages of mitosis.

Science Journal Have students make a data table in their journals to record all their data and results. Students may also wish to graph the growth of the roots. L1

INVESTIGATE!

48-Hour Cell Reproduction

You have learned that cells reproduce through the process of mitosis. In the following activity, you will observe the average growth of young growing roots as an indirect measurement of how many new cells are formed during a 48-hour period.

Problem
What is the average growth rate of young, growing roots over a 48-hour period?

Materials
5 young corn seedlings	metric ruler
	5 plastic bags
permanent marking pen (not water-soluble)	labels
	paper towels

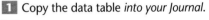

What To Do

1 Copy the data table *into your Journal*.

2 Label the five plastic bags (see photo **A**). Write "Seedling 1" on the first bag, and write "Seedling 2" on the second plastic bag. Continue labeling until all five bags have been identified.

3 Obtain five young corn seedlings from your teacher. On the first seedling, referred to as "Seedling 1", locate the growing root (see photo **B**).

4 Use the marking pen to then place a dot on the root 10 mm from the tip end. *Record* this measurement in the data table under "Original Length."

5 Wrap the seedling in a moist paper towel and place carefully into the plastic bag labeled "Seedling 1" (see photo **C**).

6 Repeat Steps 4 through 6 with the other four seedlings. Record each length in the table and place each wrapped seedling into the appropriate plastic bag.

Multicultural Perspectives

Hopi Plants

Discuss the fact that plant growth is vital to farming cultures. The Hopi people of the American Southwest, for example, have farmed successfully in near-desert conditions for 1000 years by breeding plants that have long roots and grow quickly. The Hopi place plants far apart so each one gets enough water. Have interested students do research to find out what types of plants the Hopi people grow and report back to the class. L2

A

B

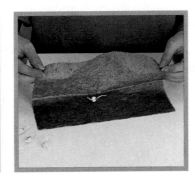
C

7 After 48 hours, *measure* the length of each root from its tip to the dot you made. Record measurements in the table under "Final Length."

8 Subtract the original length from the final length. Record this measurement in the table under "Growth of Root."

9 Calculate the average growth of a root tip. Record this number in the table under "Average Growth of Roots."

Sample data

Data and Observations

Seedling	Original Length (mm)	Final Length (mm)	Growth of Root (mm)	Average Growth of Roots (mm)
1	10	16	6	
2	10	14	4	
3	10	16	6	5
4	10	17	7	
5	10	12	2	

Analyzing

1. What evidence do you have that mitosis occurred in the corn roots?

2. Water is a nutrient needed by all living things. How did you supply the seedlings with the water they need to exist?

3. How much, on average, did the roots grow in 48 hours?

Concluding and Applying

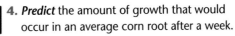
4. *Predict* the amount of growth that would occur in an average corn root after a week.

5. Suppose you found that the length of the seedlings had not changed during the 48-hour period. *Hypothesize* about some possible explanations for this lack of growth.

6. **Going Further** Using similar steps, design an experiment to determine the effect of temperature on cell reproduction. Show your design to your teacher. If you are advised to revise your plan, be sure to check with your teacher again. Check for all safety concerns. Carry out your plan.

Program Resources

Activity Masters, pp. 73–74, Investigate 17-2 [L2]

As students answer the Apply question, they may assume that hair is made up of cells. Point out that hair is a protein product of cells, as are fingernails. Help students infer that the products of mitosis are not the strands of hair, but the cells that produce the hair.

Reteach

Making Inferences Have students use their journals to re-call the Find Out activity on page 536 and the Investigate activity on pages 540–541. Ask if all the cells in the root tip were dividing. In the Find Out, students saw that dividing cells were located only near the tip. In the Investigate, students saw that the dot did not get pushed downward. From this, they can infer that there was no division behind the dot. L1

Extension

Research Have students who have mastered the concepts in this section do research to learn more about cancer. Students can do library research or contact a health-care professional for information on cancer. Students might also interview a representative from the local unit of the American Cancer Society. Have students prepare an oral presentation with visuals to present to the class. L2

4 CLOSE

Discussion

Ask students if they think the rate of mitosis is constant in humans, and ask them to cite evidence for their answers. *No, there are growth spurts and growth plateaus.* Discuss with students the uneven nature of human growth, but remember that many students at this age are sensitive about their heights. Some students may be experiencing growth spurts while others are at plateaus.

Cells That Reproduce Rapidly

Figure 17-17

In the Investigate activity, you observed growth in the corn seedlings' roots. This growth occurred due to mitosis, a process that occurs in most living cells. Through mitosis, the corn plants produced new cells that were identical to their parent cell.

Why do root cells of young plants divide through mitosis so rapidly? **Figure 17-17** may help you answer this question.

Once a new cell forms, how does it stay alive? How does a cell maintain itself within its environment? In Chapter 19, these questions will be answered as you discover how cells obtain the materials they need for life.

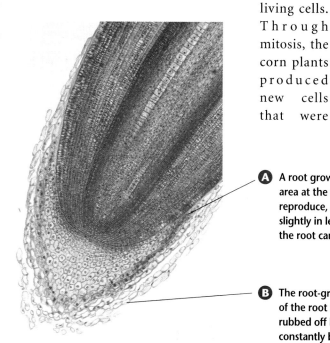

A A root grows longer because cells in a particular area at the end of the root are able to rapidly reproduce, and the new cells are able to grow slightly in length. If the growth area is damaged, the root cannot grow longer.

B The root-growth area is protected at the very tip of the root by a root cap. Root cap cells are easily rubbed off by soil particles, but these cells are constantly being replaced from within. Plant root cells, like human skin cells, are an example of cells with a rapid rate of mitosis.

check your UNDERSTANDING

1. How do the end products of mitosis compare with the original cell?
2. You know that each new nucleus produced by mitosis contains the same number of chromosomes as the original cell. A cell from the body of a frog contains a nucleus with 26 chromosomes. If one of these cells undergoes mitosis, how many chromosomes will be in each new cell produced? Explain your answer.
3. How could you tell whether or not a cell was undergoing the process of mitosis? Which specific structures are visible during this process?
4. **Apply** What cell process causes hair to constantly increase in length?

check your UNDERSTANDING

1. The two new cells produced are identical to the original cell.
2. Since the new cells are exact copies of the original, they will each contain 26 chromosomes.
3. You could look under a microscope to see if you could identify chromosomes. If you can see them, the cell is dividing.
4. Cells that make hair are constantly being produced by mitosis.

Technology Connection

Genetic Engineering

Today scientists are developing new and faster ways to improve herds and crops. The field is genetic engineering, which means the production of new genes by substituting or adding new genetic material to them. The changed genetic material is called "recombinant DNA".

How Does It Work?

Scientists decide on a goal—say, to protect a certain variety of crop plant from a certain kind of insect pest. First, they remove DNA from the bacteria that kills the pests. Then they coat tiny pieces of metal "bullets" with this bacterial DNA and shoot it into the crop plant's cells. As the treated plant cells reproduce, bacterial DNA becomes part of the new plant cells. When the pests feed on the products of these "new" plant cells, the pests die.

This technology can be used to improve and increase plant and animal production, size, resistance to disease, and other areas.

Genetic engineering can also be used to solve different types of problems. Mussels produce a sticky protein on their feet that helps them attach to rocks. Researchers have isolated the gene that controls the stickiness and have transferred the gene to bacteria and yeasts for mass production. The resulting glue works well in wet surroundings and is used in mouth, eye, and bone surgery.

 inter NET CONNECTION

Plants aren't the only living things undergoing genetic engineering. Damage to certain human genes can cause disease. Scientists are working to find those genes and developing genetic engineering techniques to repair the damage. This work is tied to a larger project—mapping

Blue mussels

the entire human genome. Information on the Human Genome Project is available on the World Wide Web from the Department of Energy. What is the Human Genome Project? What type of human genetic information is available to scientists now? How will that change in the future? How might this information be used?

Debate Divide the class into teams of debators. Pose this question for debate: Who should make decisions about genetic engineering—elected officials, scientists, voters, government agencies, or some other group or bodies? [L2]

Answers to
What Do You Think?

Students' responses will vary, but should be based on facts, rather than opinion alone. Students may write to the American Medical Association and the Food and Drug Administration for information on genetically engineered plants and animals.

Going Further ⮞
To find out about genetically engineered plants and animals that have already been created, students can contact the Food and Drug Administration, the agricultural department of a local university, or a state or local extension service. Groups of students should prepare a list of questions that they would like answered, such as "How many are already on the market?" Have students create booklets or pamphlets from the information they gather. Use the Performance Task Assessment List for Booklet or Pamphlet in **PASC**, p. 57. **COOP LEARN** **P**

EXPAND your view

Technology Connection

Purpose
This Technology Connection extends the information about cell reproduction in 17-3 by explaining how scientists can combine DNA from different organisms to create an organism with certain desirable properties. This feature gives students some insight into the issues involved.

Content Background Decoding DNA offers scientists the opportunity to create plants and animals with desirable qualities. It also may enable scientists to treat or prevent certain genetic diseases such as Alzheimer's disease and Huntington's disease.

Teaching Strategies Encourage students to weigh the potential advantages and risks of genetic engineering by discussing the questions posed at the end of this feature. Ask: **What factors might lead individuals or companies to abuse the potential of genetic engineering?** *Responses may include desire for fame, power, or profits.*

GLENCOE TECHNOLOGY

 Videodisc
The Infinite Voyage: The Geometry of Life
Chapter 7
Selective Breeding

inter NET CONNECTION

For more information on the Human Genome Project, refer to (http://www.er.doe.gov/production/oher/hug_top.html).

In Section 17-3, students learned that cells reproduce by mitosis. This **Science and Society** addresses the fact that the body does not continue to produce cells indefinitely. Modern technology has enabled people to take better care of their bodies, thereby increasing the average American life expectancy.

Content Background

During its lifetime, an organism undergoes two different types of biological changes—growth and aging. Growth refers to an increase in the size of the organism and the development of specialized structures. Aging involves a leveling off of cell production and a decrease in the body's efficiency.

Aging is commonly classified as either primary aging or secondary aging. Primary aging refers to the inevitable changes that occur in the structure of an organism. Secondary aging refers to structural changes that result from disease or accident.

Teaching Strategy The average American life expectancy is about 75 years. However, this rate does vary according to sex. The male life expectancy is about 71 years while that of females is 78 years. Discuss with your class possible reasons for this difference in life expectancy. Discuss stereotyping that is mentioned.

Activity

A society in which cell longevity, and therefore people, increases will bring about changes in consumer needs. Challenge students to invent a product or service that would answer such changes, for example, inventing a device that makes opening jars easier for people. L2

Science *and* Society **Our Aging Population**

How long can human beings live? Most gerontologists (scientists who study old age) think that the human body is designed to live no longer than 120 years. However, 110 years is probably the longest that anyone could hope to live—if he or she is extremely healthy and extremely lucky. Research in molecular biology has given some scientists reason to think we can extend the natural human life span to as long as 130 years! Nevertheless, our cells simply cannot continue to reproduce indefinitely. They wear out, and as a result, we get old and eventually die.

But even though we can't live forever, human beings in America are living longer than they ever have before. In 1900, the average American life expectancy was only 47 years. Today, life expectancy for the average American is 75 years. So, in less than one century, our life expectancy has increased by 28 years. That's pretty remarkable, considering that it took 2000 years for the average human life expectancy to increase from 25 years (the life expectancy when Julius Caesar was born in 100 B.C.E.) to 50 years.

When Are We Old?

Sixty-five may already be out-of-date as the dividing line between middle age and old age. After all, many older people don't begin to experience physical and mental decline until after age 75.

Why Are People in the United States Living Longer?

The main reason that people in the United States are living longer is that more people survive childhood. Before modern medicine changed the laws of nature with vaccines and antibiotics, many youngsters died of common childhood diseases such as measles and whooping cough. Now that the chances of dying young are much lower, the chances of living long are much higher due to better diets and health care.

Overall, our population is getting older. Fewer Americans are having children and

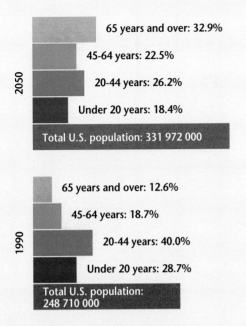

65 years and over: 32.9%

45-64 years: 22.5%

20-44 years: 26.2%

Under 20 years: 18.4%

Total U.S. population: 331 972 000

2050

65 years and over: 12.6%

45-64 years: 18.7%

20-44 years: 40.0%

Under 20 years: 28.7%

Total U.S. population: 248 710 000

1990

more of them are living longer. Since 1950, the number of Americans who are 65 and older has more than doubled, to reach 28 million. Of that 28 million older Americans, 2.6 million are over the age of 85—four times as many as in 1950.

Twelve out of every 100 Americans are 65 years old or older. The United States Census Bureau predicts that percentage will reach as high as 25 out of every 100 by the year 2035. By the year 2050, more than one-third of our population may be over 65, and only about one-fifth will be under 20!

One reason for the graying of America is the baby boom that followed World War II. Between the mid-1940s and the mid-1960s, 76 million children were born in the United States—increasing our population by one-third. Starting in 2010, the first of the baby boom generation will reach 65, and by 2030, all of the surviving baby boomers will be 65

to 85 years old. By 2050, there will be only five people of traditional working age for every four people who are past 65, the age of retirement.

What Are the Consequences of a Gray America?

As American society continues to age, the shift in our population will have far-reaching effects on our economy and our way of life. Some people fear such changes will be for the worse. For example, money that should be used to provide an education for the young will be used instead to provide expensive medical care for the elderly. Also, working-age people will have to pay incredibly high taxes so the government can afford to pay Social Security benefits to the retired.

On the other hand, some people see opportunity, not disaster, in the changes caused by our aging population. Today, many men and women in their "golden years" are healthy and alert, still active, and young in outlook if not in years.

As our society ages, we will need the contributions of our millions of older citizens. And with long lives ahead of them, they will need to stay active and involved.

Science Journal

In your Science Journal, discuss what changes might come about in American society as a result of the graying population. For example, do you think more people will work past the age of 65? What kinds of jobs might they have? Will they live with their adult children or perhaps with roommates or in senior citizen facilities? How will they get the medical care they need? Who will pay for it? As you look ahead, what effects do you think the aging of America will have on your life?

Content Background

Since 1900, the difference between the life expectancy of males and females has widened. In 1900, females were expected to live an average of 2 years more than males. Today, that difference has grown to more than 7 years. However, the difference between the life expectancy of white Americans and Americans of other races has decreased. In 1900, white Americans lived an average of 14 years more than other Americans. Today, the difference is about 4 years.

Activity

Have students interview a person from the "baby boomer" generation. Have students write questions on index cards that pertain to education, jobs, and the economy. Ask them to find out how the baby boomers affect the economy today. Have students tally the results of their surveys and share their findings with the class. [L2]

Science Journal

Responses should reflect changes in current attitudes and policies toward people over the age of 65. It is more likely that older Americans will work past the normal retirement age, possibly in part-time positions. There will be a greater need for senior citizen facilities and recreational programs geared for older people. In addition, there will be an increased demand for geriatric specialists.

Going Further ⅢⅢⅢ▶

Challenge students to identify at least ten ways that scientific investigation has contributed to an increase in life expectancy. Answers may include nutrition research, the development of exercise programs for people of various ages, and information that encourages people to change some habits, such as smoking. Have pairs of students create cartoon panels or illustrated lists of their responses. Have the class create a bulletin board entitled "Science Improves Life Expectancy."

COOP LEARN

EXPAND
your view

Health CONNECTION

Purpose

Health Connection reinforces Section 17.3 by explaining that damaged cells reproduce by mitosis just as healthy cells do. This excursion also focuses on scale and structure by describing how tumors affect body tissues.

Content Background

The parts of the body most often affected by cancer are the skin, the digestive organs, the lungs, and the female breasts. Interestingly enough, the prevalent body site affected by cancer varies from country to country (see *World Book Encyclopedia*). For example, stomach cancer is more common in Japan than in the United States. Yet, lung cancer is more common in the United States than in Japan. More Americans are affected by skin cancer than by any other form of the disease.

Teaching Strategy

GLENCOE TECHNOLOGY

Videodisc

STVS: Physics

Disc 1, Side 1
Diseased Cells and Lasers (Ch. 15)

STVS: Human Biology

Disc 7, Side 1
Diagnosing Disease with Glowing Cells (Ch. 13)

Develop the idea that the chief cause of skin cancer is exposure to ultraviolet rays from the sun. As a result, sunbathers and people who work outdoors are most likely to be affected by the disease. Discuss with your class what people are most likely affected by this and what precautions one could take to avoid exposure to the sun. Students will probably think of

Health CONNECTION

Skin Cell Mitosis and Cancer

Each kind of cell in our bodies that reproduces undergoes mitosis at its own particular rate. The rate at which a cell reproduces is part of the information stored in the nucleus of that kind of cell. Sometimes a cell is damaged, however, and some of the information stored in its nucleus becomes permanently changed. This change often affects how the cell grows and reproduces.

What Is a Tumor?

The original damaged cell and the damaged cells it produces through mitosis form a growing mass of tissue called a tumor. A tumor that is one centimeter across (about the size of a pea) can contain as many as one billion damaged cells!

Some tumors are benign, meaning the cells do not have the ability to invade other body tissues. Other tumors are malignant, meaning the cells are much more seriously damaged. They can invade other tissue and spread to other parts of the body. This is called cancer. Cancerous cells undergo mitosis in less than half the time of normal cells. Both kinds of tumors may grow very rapidly or very slowly, but most cancer tumors grow and spread quickly.

What Is Skin Cancer?

One of the most common kinds of cancer is skin cancer. The top photograph is a magnified section of normal skin. The bottom

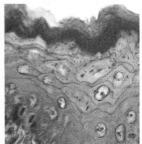

one shows a malignant tumor of the skin called a melanoma. Each year, more than 500 000 cases of skin cancer are reported!

The major cause of skin cancer is ultraviolet radiation from the sun. However, people can use proper sunscreen to prevent ultraviolet rays from damaging their skin cells.

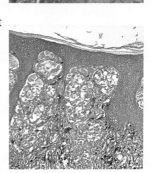

 USING MATH

Your skin cells normally reproduce themselves by mitosis every 14 days. If that's the case, how many of your skin cells are undergoing mitosis each day? Here are some facts to help you calculate the answer. *Your skin measures 1 900 000 mm². One skin cell measures 0.02 mm in diameter, so the area of one cell is 0.0004 mm².* Now make the following calculations.

1. How many skin cells are present on your body? (Need help? Divide the area of one cell into the area of all of your skin.)

2. How many skin cells undergo mitosis each day? (Need help? Divide the number of days needed to reproduce by mitosis into the number of skin cells.)

sunbathing and recreation as increasing exposure. Encourage them to think of occupations in which people are exposed to UV light for long periods of time. *fishers and construction workers*

 USING MATH

1. 4 750 000 000 skin cells
2. About 336 000 000 cells

Going Further ⟩⟩⟩⟩⟩

Have students use reference sources to learn more about the role humans are reputed to play in the depletion of the ozone layer. Student groups could make posters for the school to inform other students on how to protect themselves from the threat of skin cancer. Use the Performance Task Assessment List for Poster in PASC, p. 73. **COOP LEARN**

Science Journal

Review the statements below about the big ideas presented in this chapter, and answer the questions. Then, re-read your answers to the Did You Ever Wonder questions at the beginning of the chapter. *In your Science Journal,* write a paragraph about how your understanding of the big ideas in the chapter has changed.

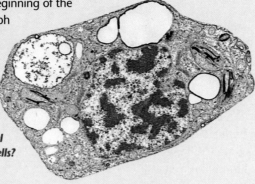

1 The cell is the basic unit of life for all living things. Most cells have a covering called a cell membrane. *What additional outer structure exists in plant cells?*

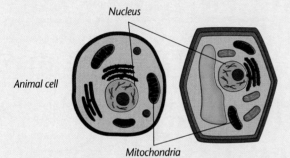

Nucleus

Animal cell

Mitochondria

Plant cell

2 A cell contains many structures that sustain its life. The nucleus directs the activities of the cell and contains information that controls the traits of an organism. Mitochondria release energy for the cell. *How do plant cells convert light energy into a more usable form of energy?*

3 Cells reproduce themselves through a process called mitosis. In mitosis, the nucleus of a cell divides so that each new cell has the same number of chromosomes as the parent cell. *What processes in the body give evidence that mitosis is taking place?*

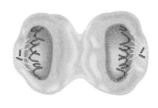

Science Journal

Did you ever wonder...
• Cells are replaced when remaining cells reproduce. (pp. 538–539)
• Green plant cells have chloroplasts which contain chlorophyll, a green pigment. (p. 533)
• Wooden bats are strong because the cell walls of the ash tree remain after the tree has died. (p. 532)

Science at Home

Have students list foods that come from plants. Ask students which items on the list have been changed greatly by the time they eat the food and which are likely to maintain their original microscopic structure. *Students should recognize that fresh vegetables and fruits retain their structure, while processed foods, such as flour, do not.* L1

Students will work with a partner to read the main ideas and answer the questions.

Teaching Strategies

Show students a photograph of a one-celled organism, such as an amoeba. Point out that the entire body of this organism is a single cell. Have students compare the structure of an amoeba to that of a human.

Answers to Questions

1. Plant cells have a rigid outer structure called a cell wall.

2. Plant cells convert light energy into chemical energy in the form of food in chloroplasts.

3. Growth and healing in the body are evidence that mitosis is occurring.

GLENCOE TECHNOLOGY

MindJogger Videoquiz

Chapter 17 Have students work in groups as they play the Videoquiz game to review key chapter concepts.

Using Key Science Terms

1. Both surround the cell.

2. Both are structures found in plant cells.

3. In mitosis, the chromosomes (the genetic material) duplicate and separate and move into two new cells.

4. The cell contains cytoplasm, in which mitochondria and chloroplasts are found.

5. The nucleus contains the cell's chromosomes.

Understanding Ideas

1. Living organisms grow, reproduce, use food for energy, and respond to stimuli.

2. (a) cell wall or cell membrane, (b) nucleus, (c) mitochondria

3. Animal cells don't contain chlorophyll and so must get their food from outside sources.

4. Growth and repair of damaged body parts are evidence.

5. The cell is very active and uses a lot of energy.

Developing Skills

1. See reduced student page.

2. Both contain a nucleus, mitochondria, cytoplasm, cell membranes, and chromosomes. Only plant cells contain a cell wall and chloroplast.

3. Refer to pages 538 and 539 for steps of mitosis.

4. Graphs will vary. Look for graphs using the *x*-axis for each seedling and the *y*-axis for the amount of growth.

U sing Key Science Terms

cell	cytoplasm
cell membrane	mitochondria
cell wall	mitosis
chloroplasts	nucleus
chromosomes	

For each set of terms below, explain the relationship that exists.

1. cell wall, cell membrane
2. chloroplasts, cell wall
3. chromosomes, mitosis
4. cytoplasm, mitochondria, chloroplasts
5. nucleus, chromosomes

U nderstanding Ideas

Answer the following questions in your Journal using complete sentences.

1. What characteristics determine if something is living or nonliving?

2. Which cell part is being compared with the following analogy?
   a. walls of a building from which a business operates
   b. manager who directs the business from a central office
   c. power plant that supplies energy to the business

3. Why can't animal cells make their own food?

4. How do you know that some of your body cells are reproducing?

5. What might a large number of mitochondria in a cell tell you about the cell's level of activity?

D eveloping Skills

Use your understanding of the concepts developed in this chapter to answer each of the following questions.

1. **Concept Mapping** Complete the following concept map of the basic units of life.

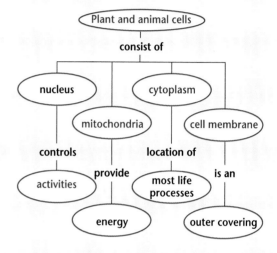

2. **Comparing and Contrasting** Compare and contrast plant and animal cells.

3. **Sequencing** Sequence the steps of mitosis.

4. **Making and Using Graphs** From the Investigate on pages 540-541, make a bar graph to represent the growth of each seedling.

C ritical Thinking

In your Journal, *answer each of the following questions.*

1. Why is it important that each new cell produced by mitosis has the same number of chromosomes as the parent?

2. The figure shows a cell that is reproducing by mitosis. Carefully examine what is occurring in the cell. Explain what will happen next in the cell as the mitosis process continues.

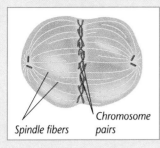

Spindle fibers   Chromosome pairs

3. Has the total number of cells in your body changed since your birth? Has the structure of your individual cells changed during your lifetime? Explain your answer.

Problem Solving

Read the following paragraph and discuss your answers in a brief paragraph.

Sharlene is investigating the growth rate of various types of cells. One of the organisms she is studying is *E. coli*, a one-celled bacterium. Sharlene observes that this bacterium can double its size in about 30 minutes if the environmental conditions are suitable. She also observes that once the cell has grown to twice its original size, it divides to form two new cells.

1. Why does *E. coli* divide when it grows to twice its size rather than continue to grow as a single cell?

2. If *E. coli* can divide at such a rapid rate, why doesn't the population of this bacterium outnumber all other kinds of organisms?

CONNECTING IDEAS

Discuss each of the following in a brief paragraph.

1. Theme—Energy How do chloroplasts produce food for plants?

2. Theme—Scale and Structure Explain the relationship between the size of a cell and the movement of materials through it.

3. Theme—Scale and Structure Cells by themselves tend to be spherical in shape. Cells in many-celled organisms often have regular shapes. How might you explain the difference?

4. Physics Connection What is the magnification power of a microscope equipped with a 10X eyepiece and a 100X objective lens? How has the development of the microscope helped scientists understand the structure of cells?

5. Health Connection How is mitosis changed when cancer is present?

Critical Thinking

1. If each cell did not have the correct set of chromosomes, the cell would lack some of the information needed to survive.

2. Each chromosome has made a copy of itself. Original and copy chromosomes are lining up on the spindle fibers in the middle of the cell. Next, the two strands will separate and one strand of each will move to opposite ends of the cell.

3. The total number of cells has increased. A sample response might be that some cells have become more specialized. Bone has replaced cartilage; some nerve cells have grown longer.

Problem Solving

1. It may be difficult for materials to move through the cytoplasm and supply all parts of the cell. Division will decrease the volume of each cell.

2. The death of many bacteria, either from consumption by white blood cells, lack of food, or buildup of waste chemicals, has an effect on the *E. coli* population.

Connecting Ideas

1. Chloroplasts convert light energy from the sun into chemical energy in the form of sugar called glucose.

2. It takes more time for materials to move through larger cells.

3. The regular shape is the result of cells pushing against one another.

4. 1000✕; Scientists can see structures inside cells.

5. Cancerous cells generally undergo mitosis in less than half the time of normal cells.

✔ Assessment

Portfolio Review the portfolio options that are provided throughout the chapter. Encourage students to select one product that demonstrates their best work for the chapter. Have students explain what they learned and why they chose this example for placement into their portfolios.

Additional portfolio options can be found in the following **Teacher Classroom Resources:**

Multicultural Connections, pp. 37, 38

Making Connections: Technology and Society, p. 37

Concept Mapping, p. 25

Critical Thinking/Problem Solving, p. 25

Making Connections: Integrating Sciences, p. 37

Take Home Activities, p. 26

Making Connections: Across the Curriculum, p. 37

Laboratory Manual, pp. 103–106

Performance Assessment P

Chapter Organizer

SECTION	OBJECTIVES	ACTIVITIES & FEATURES
Chapter Opener		**Explore!**, p. 551
18-1 How Does Matter Change Chemically? (1 session; .5 block)	1. **Describe** materials before and after chemical changes. 2. **Recognize** when chemical reactions have taken place. 3. **Identify and describe** several chemical reactions. **National Content Standards: (5-8) UCP2-3, A1, B1, C3, E1**	**Find Out!**, p. 552 **Find Out!**, pp. 554–555 **A Closer Look**, pp. 554–555 **Teens in Science**, p. 578 **SciFacts**, p. 577
18-2 Word Equations (3 sessions; 1.5 blocks)	1. **Distinguish** between reactants and products. 2. **Write** word equations. **National Content Standards: (5-8) UCP2, A1, B1, D1, D3**	**Explore!**, p. 557 **Investigate 18-1:** pp. 560–561
18-3 Chemical Reactions and Energy (3 sessions; 1.5 blocks)	1. **Relate** how energy is involved in chemical reactions. 2. **Differentiate** endothermic and exothermic reactions. **National Content Standards: (5-8) UCP2-3, A1, B1, B3, F1**	**Explore!**, p. 563 **Design Your Own Investigation 18-2:** pp. 568–569 **Skillbuilder:** p. 570 **Life Science Connection**, pp. 564–565
18-4 Speeding Up and Slowing Down Reactions (2 sessions; 1 block)	1. **Describe** how a catalyst affects a chemical reaction. 2. **Explain** how to control a chemical reaction with an inhibitor. **National Content Standards: (5-8) UCP2, A1, B1, C1, C3, E2, F1-2, F5**	**Find Out!**, p. 571 **Science and Society**, p. 575 **Health Connection**, p. 576

ACTIVITY MATERIALS

EXPLORE!

p. 551 no special materials required
p. 557 magnesium ribbon, white (distilled) vinegar, beaker
p. 563* rubber band, pencil, index card

INVESTIGATE!

pp. 560–561* 3 test tubes; test-tube rack; small pieces of copper, zinc and magnesium; graduated cylinder; wooden splint; 15 mL dilute hydrochloric acid; goggles; apron

DESIGN YOUR OWN INVESTIGATION

pp. 568–569* stopwatch, 3% hydrogen peroxide solution, thermometer, 25-mL graduated cylinder, 8 test tubes and rack, goggles, raw liver, raw potato, apron

FIND OUT!

p. 552* paper, matches, marshmallow, banana
pp. 554–555* colorless carbonated soft drink, limewater, drinking straw, clear glass
p. 571* goggles, apron, hydrogen peroxide, 2 test tubes, spoon or spatula, manganese dioxide, wooden splint, matches, beaker, water, hot plate, test-tube rack

KEY TO TEACHING STRATEGIES

The following designations will help you decide which activities are appropriate for your students.

- **L1** Basic activities for all students
- **L2** Activities for average to above-average students
- **L3** Challenging activities for above-average students
- **LEP** Limited English Proficiency activities
- **COOP LEARN** Cooperative Learning activities for small group work
- **P** Student products that can be placed into a best-work portfolio
- Activities and resources recommended for block schedules

Need Materials? Call Science Kit (1-800-828-7777).

[00:00] OUT OF TIME? We recommend that students do the activities with an asterisk.

*For adequate development of the concepts presented, we recommend that students do the activities with an asterisk.

Chapter 18 Chemical Reactions

TEACHER CLASSROOM RESOURCES

Student Masters	Transparencies
Study Guide, p. 59 **Take Home Activities**, p. 27 **Making Connections: Integrating Sciences**, p. 39 **Making Connections: Technology & Society**, p. 39 **Science Discovery Activities, 18-1** **Science Discovery Activities, 18-2** **Laboratory Manual**, pp. 107–108, Chemical Changes	**Section Focus Transparency 53**
Study Guide, p. 60 **Critical Thinking/Problem Solving**, p. 5 **Multicultural Connections**, p. 39 **Activity Masters**, Investigate 18-1, pp. 75–76 **Concept Mapping**, p. 26 **Laboratory Manual**, pp. 109–112, Chemical Reactions	**Teaching Transparency 35,** Chemical Reaction **Section Focus Transparency 54**
Study Guide, p. 61 **Critical Thinking/Problem Solving**, p. 26 **Activity Masters**, Design Your Own Investigation 18-2, pp. 77–78 **How It Works**, p. 19 **Making Connections: Across the Curriculum**, p. 39	**Teaching Transparency 36,** Photosynthesis **Section Focus Transparency 55**
Study Guide, p. 62 **Multicultural Connections**, p. 40 **Science Discovery Activities 18-3**	**Section Focus Transparency 56**

ASSESSMENT RESOURCES	TEACHING & TECHNOLOGY
Review and Assessment, pp. 107–112 **Performance Assessment**, Ch. 18 **PASC*** **MindJogger Videoquiz** **Alternate Assessment in the Science Classroom** **Computer Test Bank**	**Spanish Resources** **Cooperative Learning Resource Guide** **Lab and Safety Skills** **Science Interactions, Course 2, CD-ROM** **Computer Competency Activities**

*Performance Assessment in the Science Classroom

NATIONAL GEOGRAPHIC TEACHER'S CORNER

Index to National Geographic Magazine

The following articles may be used for research relating to this chapter:

- "Is Our World Warming?" by Samuel W. Matthews, October 1990.
- "Air: An Atmosphere of Uncertainty," by Noel Grove, April 1987.

Additional National Geographic Society Products

To order the following products for use with this chapter, call the National Geographic Society at 1-800-368-2728:

- *Everyday Science Explained* (Book)
- *What's Good to Eat: Foods the Body Needs* (Filmstrip)
- *Your Body Series,* "Digestive System" (Video)

GLENCOE TECHNOLOGY

The following multimedia resources are available from Glencoe.

The Infinite Voyage Series
The Future of the Past
Secrets from a Frozen World

National Geographic Society Series
GTV: Planetary Manager

Glencoe Physical Science Interactive Videodisc
Periodicity
Chemical Detectives

Physical Science CD-ROM

Teacher Classroom Resources

These are key components of the classroom resources package.

TEACHING AIDS

Teaching Transparencies

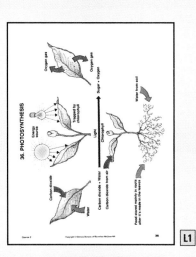

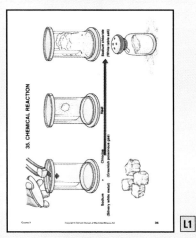

Section Focus Transparencies

53 SECTION FOCUS TRANSPARENCY

CHEMICAL CHANGES

In a chemical change, or chemical reaction, substances change into other substances. One example of such a chemical change is the reaction between iron and oxygen, forming the oxide of iron commonly called rust.

1. Describe the chemical changes you see in this scene.
2. Which of the changes shown are physical rather than chemical? How do you know?

54 SECTION FOCUS TRANSPARENCY

BUILDING BLOCKS

Have you ever built a model animal or car using interlocking blocks? You probably took it apart eventually to use the parts for something else. Some chemical reactions are similar to this.

1. How are some chemical reactions similar to building a model?
2. If building a molecule is shown as A + B → AB, how would you show taking a molecule apart?

55 SECTION FOCUS TRANSPARENCY

HOT AND COLD PACKS

Hot packs often use a chemical process to produce heat. When a specific chemical is dissolved in water, it releases heat. Cold packs contain different chemicals, some of which absorb heat when dissolved in water.

1. What are some practical uses for hot packs?
2. What are some practical uses for cold packs?
3. What safety precautions do you think should be taken with hot and cold packs?

56 SECTION FOCUS TRANSPARENCY

LOCK AND KEY

The advantage of locking something is that not just anyone can open it. The correct key must be used for the lock to be opened. Another key won't work.

1. What type of key will open the lock on a door?
2. How does this type of key differ from a key used for a suitcase?

HANDS-ON LEARNING

Science Discovery Activity*

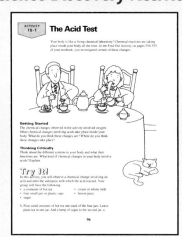

ACTIVITY 18-1 **The Acid Test**

Your body is like a living chemical laboratory! Chemical reactions are taking place inside your body all the time. In the Find Out Activity on pages 554-555 of your textbook, you investigated certain of these changes.

Getting Started
The chemical changes observed in the activity involved oxygen. Other chemical changes involving acids take place inside your body. What do you think these changes are? When do you think these changes take place?

Thinking Critically
Think about the different systems in your body and what their functions are. What kind of chemical changes in your body involve acids? Explain.

Try It!
In this activity, you will observe a chemical change involving an acid and infer the substance with which the acid reacted. Your group will have the following:

• a container of hot tea
• four small jars or plastic cups
• sugar
• cream or whole milk
• lemon juice

1. Pour small amounts of hot tea into each of the four jars. Leave plain tea in one jar. Add a lump of sugar to the second jar, a

Laboratory Manual*

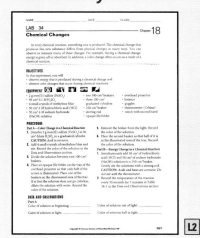

NAME _____ DATE _____ CLASS _____

LAB 34   Chapter 18
Chemical Changes

In every chemical reaction, something new is produced. The chemical change that produces this new substance differs from physical changes in many ways. You can observe or measure many of these changes. For example, during a chemical change, energy is given off or absorbed. In addition, a color change often occurs as a result of a chemical reaction.

OBJECTIVES
In this experiment, you will
• observe energy that is produced during a chemical change and
• observe color changes that occur during chemical reactions.

EQUIPMENT
• 2 g iron(II) sulfate (FeSO₄)
• 96 cm³ 0.1 M H₂SO₄
• 4 small crystals of methylene blue
• 50 cm³ 6 M hydrochloric acid (HCl)
• 50 cm³ 6 M sodium hydroxide (NaOH) solution
• two 100-cm³ beakers
• three 100-cm³ graduated cylinders
• 250 cm³ beaker
• stirring rod
• opaque file folder
• overhead projector
• balance
• goggles
• thermometer (Celsius)
• watch with second hand

PROCEDURE
Part A — Color Change in a Chemical Reaction
1. Dissolve 2 g iron(II) sulfate (FeSO₄) in 90 cm³ dilute H₂SO₄ in a graduated cylinder. CAUTION: Acid is corrosive.
2. Add 4 small crystals of methylene blue and stir. Record the color of the solution in the Data and Observations section.
3. Divide the solution between two 100-cm³ beakers.
4. Place an opaque file folder on the tray of the overhead projector so that only half of the screen is illuminated. Place one of the beakers on the illuminated area of the tray. If at first the solution does not go colorless, dilute the solution with water. Record the color of the solution.

5. Remove the beaker from the light. Record the color of the solution.
6. Place the second beaker so that half of it is in the illuminated area of the tray. Record the color of the solution.
Part B — Energy Change in a Chemical Reaction
1. Simultaneously add 50 cm³ of hydrochloric acid (HCl) and 50 cm³ of sodium hydroxide (NaOH) solution to a 250-cm³ beaker. Gently stir the solutions with a stirring rod. CAUTION: Acids and bases are corrosive. Do not stir with the thermometer.
2. Record the temperature of the reaction every 30 seconds for 5 minutes in Table 34-1 in the Data and Observations section.

DATA AND OBSERVATIONS
Part A
Color of solution at beginning: _____
Color of solution in light: _____
Color of solution out of light: _____
Color of solution half in light: _____

Take Home Activity

Chapter 18
Make a Fire Extinguisher

You need: a clean, wide-mouthed jar, vinegar, baking soda, a small candle, matches, and a small piece of modeling clay

You can perform some useful chemical reactions right in your own kitchen. For example, you can easily make a carbon dioxide gas fire extinguisher. Ask an adult to help you.

• Secure the candle upright in the clay.
• Light it with a match.
• Add enough baking soda to cover the bottom of the jar.
• Slowly pour a small amount of vinegar over the baking soda. Notice the foaming action that takes place.
• Hold the jar above the candle and tip it so that the gas, but not the liquid, pours out of the jar and onto the flame.
• Explain to the adult how you know a chemical reaction took place. Even though you couldn't see the carbon dioxide gas, how do you know it was in the jar?
• What happened when you "poured" the gas over the flame?

---

*There may be more than one activity for this chapter.

Chapter 18 Chemical Reactions

REVIEW AND REINFORCEMENT

Study Guide*

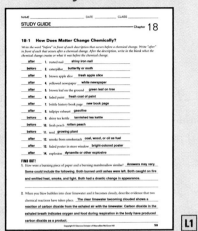

Concept Mapping

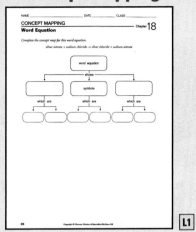

Critical Thinking/Problem Solving

ENRICHMENT AND APPLICATION

Integrating Sciences

Across the Curriculum

Technology and Society

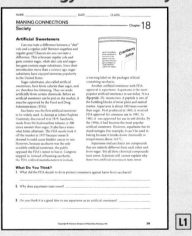

ASSESSMENT

Multicultural Connection**

Performance Assessment

Review and Assessment

Chemical Reactions

THEME DEVELOPMENT

The primary theme supported in this chapter is systems and interactions. Whether or not a reaction will occur usually depends on whether the reactants in the system are present in sufficient quantity and whether the interactions between reactants have enough energy to undergo a chemical change.

CHAPTER OVERVIEW

Students will be introduced to chemical reactions through examples of common reactions. Students will then learn to write word equations. Exothermic and endothermic reactions and catalysts and inhibitors will be studied.

Tying to Previous Knowledge

Students may be familiar with the *Hindenburg* dirigible disaster in 1937. What they will not realize is that this disaster demonstrates a *chemical reaction* between hydrogen gas and oxygen.

INTRODUCING THE CHAPTER

Have students look at the photograph and describe possible relationships that might occur between plants and their environment. *Answers might include reactions between sunlight and the plants' leaves, between the roots and the soil, and so on.*

Chemical Reactions

Did you ever wonder...

✓ **Why most cars built today must use unleaded gasoline?**

✓ **Why chemicals with strange-sounding names are in your food?**

✓ **Why your body is warm?**

✓ **Why some things rust and others don't?**

Science Journal

Before you begin to study about chemical reactions, think about these questions and answer them *in your Science Journal.* When you finish the chapter, compare your journal write-up with what you have learned.

Answers are on page 579.

550

I*magine what that tricycle once looked like and how it felt to ride it. Not long ago, the wheels probably glistened in the sun. Now much of the trike is coated with reddish-brown rust.*

Rusting is just one of many kinds of chemical changes that go on around you. A hamburger's cooking, flowers' growing and blooming, and the burning of gasoline in an automobile are all actions during which the properties of substances change.

▶ **In this chapter, you will find out how chemical changes take place.**

Learning Styles	**Kinesthetic**	Find Out, pp. 552, 554, 571; A Closer Look, pp. 554-555; Explore, p. 563; Project, p. 579
	Visual-Spatial	Explore, pp. 551, 557; Visual Learning, pp. 553, 556, 559, 566, 567, 570, 573; Demonstration, pp. 553, 559, 566, 572; Activity, p. 558; Across the Curriculum, p. 559; Investigate, pp. 560-561; Life Science Connection, pp. 564-565
	Logical-Mathematical	Investigate, pp. 568-569; Across the Curriculum, p. 573; Research, p. 573
LS	**Linguistic**	Science Journal, p. 573; Activity, p. 553; Discussion, pp. 556, 564, 565; Across the Curriculum, pp. 558, 564; Multicultural Perspectives, pp. 559, 567; Research, p. 567

Explore! ACTIVITY

What kinds of things rust?

Perhaps you have noticed that certain objects around you rust, and other objects don't ever rust at all. What is rust and why does it form?

What To Do

Record the following observations and the answers to the questions *in your Journal.*

1. Look for rust on objects at home, in school, and on your way to school. Make a list of these rusty objects.

2. Think about the items on your list. What do things that rust have in common?

3. Where does rust form? Do you notice more rust on objects that are indoors or outdoors? Why do you think this is so?

ASSESSMENT PLANNER

PORTFOLIO
Refer to page 581 for suggested items that students might select for their portfolios.

PERFORMANCE ASSESSMENT
Process, pp. 555, 557, 569
Skillbuilder, p. 570
Explore! Activities, pp. 551, 557, 563
Find Out! Activities, pp. 552, 554, 571
Investigate, pp. 560–561, 568–569

CONTENT ASSESSMENT
Oral, p. 561
Check Your Understanding, pp. 556, 562, 570, 574
Reviewing Main Ideas, p. 579
Chapter Review, pp. 580–581

GROUP ASSESSMENT
Opportunities for group assessment occur with Cooperative Learning Strategies.

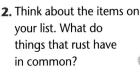

Explore!

What kinds of things rust?

TECH PREP

Time needed 30 minutes

Materials No special materials or preparation are required.

Thinking Processes observing and inferring, comparing and contrasting

Purpose To observe changes caused by rusting.

Teaching the Activity

Discussion Help students conclude that although the reactants in the reaction are iron and oxygen, water promotes the reaction, so most rust will be found in places exposed to moisture. **L1**

Science Journal Students' lists of rusty objects may include items such as tools, kitchen devices, bicycles, motor vehicles, and so on. **L1**

Expected Outcome

Students will discover that certain metal objects that are outside and exposed to moisture are most likely to be rusted.

Answers to Questions

2. They are metal, usually outside where they may get wet.

3. Rust forms on exposed iron surfaces. Students will probably notice more rust outdoors. Outdoor objects are more likely to be exposed to moisture.

✔ Assessment

Content Let students working in small groups share their list of rusty objects. Then ask each group to come to a consensus as to how and why rust forms. Have students create single-panel cartoons with captions that define rust. Use the Performance Task Assessment List for Cartoon in **PASC**, p. 61.
COOP LEARN **P**

18-1

PREPARATION

Planning the Lesson

Refer to the Chapter Organizer on pages 550A-D.

Concepts Developed

This section gives the students an overview of chemical reactions.

1 MOTIVATE

Bellringer

 Before presenting the lesson, display **Section Focus Transparency 53** on an overhead projector. Assign the accompanying **Focus Activity** worksheet. L1

LEP

Find Out!

Are all chemical changes alike?

Time needed 15 minutes

Materials paper, matches, marshmallow, banana

Thinking Processes comparing and contrasting, observing and inferring

Purpose To observe changes that take place as the result of chemical reactions.

Teaching the Activity

Science Journal To reinforce observational skills, make sure students include the *nature* of the changes involved in these reactions such as any change in color, texture, smell, size, and so on. L1

Expected Outcome

Students should conclude that chemical changes depend on the particular substance(s) reacting.

18-1 How Does Matter Change Chemically?

Section Objectives

- Describe materials before and after chemical changes.
- Recognize when chemical reactions have taken place.
- Identify and describe several chemical reactions.

Key Terms

chemical reaction

Observing Chemical Changes

You're already familiar with many changes around you. Melting is a common change. It is a physical change because the material that is changed stays the same material. Ice is frozen water, and when ice melts, it is still water.

If a different substance is produced by the change, the change is chemical. Burning wood turns to ashes. Rust forms on tools left in the rain. The paint on buildings fades. Plants and animals grow. How are these chemical changes similar? How are they different? This section will help you answer these questions.

Find Out! ACTIVITY

Are all chemical changes alike?

Chemical changes are all around you and you observe them every day, whether you realize it or not.

What To Do

You are going to observe several chemical changes.

1. Observe what happens when a piece of paper burns. *In your Journal*, list the changes that you and your classmates observe.

2. Watch what happens as a marshmallow burns. Again, list the changes you observe. Which changes are alike for the paper and the marshmallow? Which are different?

3. Next, observe how a banana changes over a period of a few days. Describe

any changes in the color, odor, and texture of the skin or the fruit inside.

Conclude and Apply

1. Compare and contrast the changes in the banana, the paper, and the marshmallow.

2. How do these changes differ from what you observe when you boil a pot of water?

552 Chapter 18 Chemical Reactions

Conclude and Apply

1. Paper and marshmallow both turned black; the paper got thin, and the marshmallow got "gooey." Meanwhile, the banana darkened and shriveled. All three changed in odor, the paper's change perhaps being the most drastic.

2. These changes are chemical because the substances become new substances, whereas boiling water is a physical change.

✔ Assessment

Content Have pairs of students list other chemical reactions that are similar to those observed in this activity. *Answers might include rotting fruit and vegetables, the burning of fossil fuels, cooking of various foods, and so on.* Use the Performance Task Assessment List for Group Work in **PASC**, p. 97. **COOP LEARN** L1

---

552 Chapter 18 Chemical Reactions

Chemical Reactions Around You

When a piece of paper burns, a banana ripens, or a bike rusts, the chemical and physical properties of the substance change. These changes may happen quickly or slowly. In each case, however, new substances are formed. A chemical reaction has taken place. A **chemical reaction** is a well-defined process that results in the formation of new substances having properties that are different from those of the original substances.

Have you ever noticed changes in a piece of newspaper? As is shown in **Figure 18-1**, the changes may be physical, but many are chemical changes caused by the reaction between substances in the air and in the paper. Whenever you see a cut apple turn brown, exhaust gases come out of a car's tailpipe, or a caterpillar change into a butterfly, you know that chemical reactions are taking place because new substances are formed.

Figure 18-1

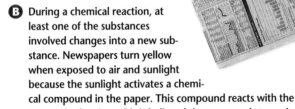

A Cutting, tearing, and crumpling change the newspaper's size, shape, and texture. These are physical changes and alter only the physical properties of the paper, not the paper itself. The newspaper *looks* different after you tear and crumple it, but it's still the same newspaper. No new substance is created.

B During a chemical reaction, at least one of the substances involved changes into a new substance. Newspapers turn yellow when exposed to air and sunlight because the sunlight activates a chemical compound in the paper. This compound reacts with the oxygen in the air, and it is believed that a new substance is formed. Because this new substance reflects yellow light and absorbs some other colors, the paper appears yellow.

C The burning of the paper is a chemical reaction because the oxygen in the air combines with substances in the newspaper. New gaseous and solid products and energy are produced. It is easy to see the changes in both the chemical and physical properties of a newspaper that has been burned.

18-1   How Does Matter Change Chemically?   **553**

Program Resources

Study Guide, p. 59

Laboratory Manual, pp. 107–108, Chemical Changes L2

Take Home Activities, p. 27, Make a Fire Extinguisher L1

Making Connections: Integrating Sciences, p. 39, Aspirin—Helpful or Harmful? L3

Making Connections: Technology & Society, p. 39, Artificial Sweeteners L1

Science Discovery Activities; 18-1, 18-2

Section Focus Transparency 53

Activity Have students guess which of the processes listed below involve a chemical reaction. *The correct answers are in italics.* water evaporating, *log burning, an acid contacting a base, corrosion on a battery,* boiling away a liquid, *an apple rotting, digesting food in your stomach,* paint peeling off a wall

2 TEACH

Theme Connection Chemical reactions are directly related to the theme of stability and change. As the result of a chemical reaction, the chemical and physical properties of the substances involved change. New substances that have specific properties different from those of the original substances are formed.

Visual Learning

Figure 18-1 Prior to reading the text that accompanies these photographs, have students hypothesize whether the changes that have occurred are chemical or physical. L1

Demonstration To compare a physical change and a chemical reaction, compare caramel with sugar water.

Materials needed are sugar, water, beaker, a small heavy frying pan, wooden spoons, and a heat source.

Make caramel by placing 1/2 cup sugar in a frying pan. Stir constantly as you heat the sugar over a low flame. When the sugar is straw-colored, remove the pan from the heat. At the same time, have a student stir 1/2 cup sugar and 1/2 cup water together in a beaker. Discuss differences between the changes. Students should cite indications that a new substance was formed when the caramel was made but that the sugar and water in the beaker are still the same. L1 LEP

In the last section, you saw that chemical reactions take place around you all the time. Would you be surprised to find out that chemical reactions take place inside your body every moment of your life?

Find Out! ACTIVITY

Is oxygen changed inside your body?

You know you breathe in oxygen from the air around you. But what happens to that oxygen?

What To Do

1. Colorless limewater turns cloudy or milky in the presence of carbon dioxide gas. Test this by adding a few drops of colorless carbonated soft drink to a small amount of colorless limewater.

Find Out!

Is oxygen changed inside your body?

Time needed 10 minutes

Materials limewater (saturated CaO solution) in a container; colorless, carbonated soft drink; clear glass; straw or tubing

Thinking Processes observing and inferring, recognizing cause and effect

Purpose To investigate a chemical reaction that takes place in the body.

Teaching the Activity

Explain that the soft drink is used as a control to show how a solution that is known to have carbon dioxide in it reacts with limewater.

Science Journal Suggest that students illustrate each step of the activity in their journals and then record their observations. L1

Expected Outcome

Students should conclude that some oxygen breathed in is converted to carbon dioxide.

Conclude and Apply

1. Bubble compressed air into the limewater.

2. Oxygen is breathed in, but carbon dioxide is breathed out.

3. Yes, the gas is changed to a different gas.

What We Breathe Can Destroy Bridges!

Of all the chemical reactions possible, corrosion probably receives more attention and costs more money than any other. Rust and other types of corrosion take their toll on bridges, overpasses, and ships. Corrosion occurs when metals, such as iron or steel, are affected by ordinary water and oxygen in the environment. If rain is acidic or if there are dissolved salts in the atmosphere, corrosion occurs even faster.

Rust Resisters

There are ways to protect metal from corrosion. Perhaps you've seen the overpasses along the expressway being painted. You may have coated your ice-skate blades or your bicycle chain with oil before you stored them for next season. Both of these methods of protection help to keep oxygen and water in the air from reaching the metal and causing corrosion.

Purpose

A Closer Look relates to Section 18-1 by presenting some common reactions with water and oxygen in the air—rust and corrosion of various metals and alloys.

Content Background

Corrosion is the forming of undesired chemical compounds on a metal from chemicals in the surrounding air or water. Rust (iron oxide) is the most familiar example. Tarnish (oxides and sulfides) on silver, brass, and copper is commonly seen. Methods of resisting corrosion include painting, oil-coating, plastic-coating, and galvanizing.

Teaching Strategy

Have each student observe the dull appearance of a penny that is several years old. Then have them vigorously rub a pencil eraser over the penny. Discuss what happened and why. *Students removed the tarnish—a copper oxide—and exposed the*

2. Observe what happens when the bubbles of carbon dioxide in the soft drink come in contact with the limewater.

3. Now gently blow out through a straw into a glass half-filled with colorless limewater. **CAUTION**: *Be sure not to take in any limewater.*

4. *In your Journal*, describe the changes you observe in the limewater as you continue to blow out through the straw.

Conclude and Apply

1. How could you be sure that limewater does not turn cloudy when oxygen is present?

2. What can you infer from your observations?

3. Has a chemical reaction taken place?

Oxygen combines with food you have taken in. Through a series of chemical reactions called respiration, new substances, carbon dioxide and water, and energy are produced. It is this carbon dioxide you exhaled into the limewater. The energy sustains life.

At the same time, digestion takes place. During digestion, food is broken down by chemical reactions.

You already know that the reaction of some metals with oxygen can produce an oxide. For instance, if aluminum is exposed to air, a thin coating of aluminum oxide is formed. Unlike iron oxide, which is loose and dusty, aluminum oxide consists of very tightly packed units. These units are so tightly packed that the aluminum oxide actually protects the metal beneath from further reactions.

Galvanizing

Another method for protecting metals from corrosion relies on the element zinc. Perhaps you've heard of galvanized steel, as is shown in the photo. It's often used for garbage cans,

plumbing, and gutters. Galvanized steel is steel coated with a layer of zinc by electroplating or by applying molten zinc. The zinc slowly forms a zinc oxide coating that protects the iron or steel beneath it.

Magnesium sheets bolted to bridge pillars and ship hulls protect in the same way. Both magnesium and zinc are allowed to corrode in order to protect the structural metal beneath.

You Try It!

Design a sculpture of iron or steel using discarded items you might have at home or might find in an alley or vacant lot. Then tell how you might protect it from corrosion.

---

unoxidized copper underneath. The oxide formed slowly as the copper pennies were exposed to oxygen in the atmosphere.

**Answer to
You Try It!**
Possible methods of protection include keeping it in a dry environment; covering it with paint, varnish, or shellac; or coating it with light oil periodically.

Going Further ⅢⅢ▶
Chemicals in the environment are important in corrosion. Have students explain the corrosion in these two scenarios:

• Recent Indiana winters had been severe, and lots of salt was used on the highways to melt snow and ice. Now cars in the area are corroding rapidly. Why? *The salt promotes a reaction between iron and oxygen.*

• Francesca polished her grandmother's silver spoons and brass candlesticks. But in a few days, they became tarnished. Why? *Pollutants in the air, especially sulfur compounds, reacted with the surface of the metals to form dull-looking compounds called tarnish.*

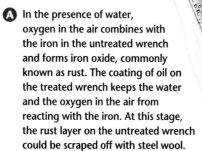
3 ASSESS

Check for Understanding

Before students answer the Apply question under Check Your Understanding, remind them of the teacher demonstration on p. 553 in which they compared caramel with sugar water.

Reteach

Discussion Focus on reactions as changes. Whenever substances are chemically combined, a permanent change will occur. The most important indication, however, is the formation of new products. Help students make a list of chemical reactions. L1

Extension

Activity Have students sprinkle table salt on a dark penny and then put a few drops of vinegar on the penny. Ask students to describe what happens. Explain that acid (vinegar) and salt react with the copper. L3

4 CLOSE

Discussion

Review that sugar dissolving in water is *not* a chemical reaction. Acid reacting with a base to form a salt *is* a chemical reaction. Ask students how both processes are the same. *Both involve change.* Then ask what is different about these changes. *In the chemical reaction, the starting substances became different substances.*

Describing Chemical Reactions

Remember how you described what happened to the marshmallow and the banana? You probably used words that told about changes in physical properties, such as color, shape, texture, and smell. How could you describe a rusting bicycle or a burning paper? When you say that paper burns and iron rusts, you are describing the chemical properties that change during chemical reactions. In **Figure 18-2**, you can clearly see what effect rusting had on the unprotected tool. The tool that was protected from rust still has its original properties.

How would you describe the chemical reactions that occur when a building is demolished by using explosives? It might appear that the substances haven't changed. You have brick or stone before and after the explosion. But what was the reaction that caused the building to collapse? While descriptive words might describe what an explosion looks or sounds like, these words don't tell us much about the chemical reaction that takes place. The next section will help you describe the chemical reactions that take place around you.

Figure 18-2

Two wrenches—one treated with a coating of oil and the other untreated—are left outside and forgotten. Rain collects in a puddle around the wrenches. The scene is set for chemical reaction.

A In the presence of water, oxygen in the air combines with the iron in the untreated wrench and forms iron oxide, commonly known as rust. The coating of oil on the treated wrench keeps the water and the oxygen in the air from reacting with the iron. At this stage, the rust layer on the untreated wrench could be scraped off with steel wool.

B Over time, the chemical reaction continues. More of the iron in the untreated wrench is changed into rust. What will eventually happen to the untreated wrench if it stays where it is? What will happen to the treated wrench?

check your UNDERSTANDING

1. List several ways that you can tell that a chemical change has taken place.
2. Name three chemical reactions that you read about in this section.
3. List two chemical changes not mentioned in this section. How did you decide that they were chemical rather than physical changes?
4. **Apply** When you dissolve sugar in water, does a chemical reaction take place? How do you know?

check your UNDERSTANDING

1. New properties appear; old properties disappear; sometimes an energy change is observed; it is not easily reversible.
2. Answers might include paper burning, marshmallow burning, a banana ripening.
3. Examples will vary. An example is CO_2 bubbles in rising bread dough. Reasons should include those found in the answer to question 1.
4. No; a new substance is not formed. The sweetness of the sugar is still present in the water, and unchanged sugar can be obtained by evaporating the water.

Writing Word Equations

How might you describe the chemical reactions in respiration? You could describe changes in the physical and chemical properties of the substances in words. Another way is to write a word equation. For example, you could use a plus sign (+) to mean *and* and an arrow (→) to mean *produces*. Then, the word equation for respiration would look like this:

oxygen + food → carbon dioxide + water

In any chemical reaction, the substances that you start with are the **reactants**. The new substances formed by the reaction are the **products**. In the above equation, oxygen and food are the reactants, and carbon dioxide and water are products. Can you describe what is happening in another chemical reaction?

Section Objectives

■ Distinguish between reactants and products.
■ Write word equations.

Key Terms

reactants
products

Explore! **ACTIVITY**

Can we describe any chemical reaction?

What To Do

You will observe a chemical reaction and describe what is happening.

1. Drop a piece of freshly sandpapered magnesium ribbon into a small amount of white (distilled) vinegar.

2. Observe carefully for several minutes. Record your observations *in your Journal.*

3. How do you know a reaction is taking place? How can you keep the reaction going?

4. What do you need to know to describe the reactants and products?

Let's take this reaction step by step to see what happened. Vinegar contains acetic acid, which contains hydrogen. When the magnesium was placed in the vinegar, it reacted with the acetic acid. Bubbles of hydrogen gas were released. This is the word equation for the reaction:

magnesium + acetic acid → magnesium acetate + hydrogen

What are the reactants in this reaction? What are the products?

This reaction has more than one reactant and more than one product. In another reaction, sugar, one reactant, can be heated until it breaks

---

PREPARATION

Planning the Lesson

Refer to the Chapter Organizer on pages 550A-D.

Concepts Developed

In this section, different kinds of reactions are examined and represented with word equations.

1 MOTIVATE

Bellringer

Before presenting the lesson, display **Section Focus Transparency 54** on an overhead projector. Assign the accompanying **Focus Activity** worksheet. [L1]

[LEP]

---

shown by the bubbles. [L1]

Expected Outcomes

Bubbles signify that a reaction is taking place. Adding more of the reactant that has been used up sustains the reaction. Students should conclude that the two react until one of them is used up.

Answers to Questions

3. Bubbles demonstrate a reaction is occurring. By adding more magnesium or more vinegar, the reaction will continue.

4. You need to know what substances reacted and what substances were present at the end of the reaction.

✔ **Assessment**

Process Have students work in pairs to identify the reactants and products in each of the chemical reactions presented in Section 18-1. Use the Performance Task Assessment List for Group Work in **PASC**, p. 97. [L2] **COOP LEARN**

Explore!

Can we describe any chemical reaction?

Time needed 10 minutes

Materials magnesium ribbon, white (distilled) vinegar, beaker

Thinking Processes observing and inferring, comparing and contrasting

Purpose To observe a chemical reaction.

Preparation Encourage students to experi-

ment with the relative amounts of magnesium and vinegar they need to use.

Teaching the Activity

Discussion Stress the idea that a reaction continues until one of the reactants is used up. That is why adding extra vinegar or magnesium causes the bubbling to resume—it supplies more raw material for the reaction.

Science Journal Students should observe that a gas is being liberated from the beaker as

2 Teach

Tying to Previous Knowledge

In the previous section, chemical reactions were introduced. In this section, students will learn how to describe any chemical reaction in terms of its reactants (original substances) and products (new substances). Discuss what an equation is—a statement in which one quantity is equivalent to another. Equations are used in this section to represent chemical reactions.

Theme Connection Stability and change are represented in each type of chemical reaction. For each set of reactants, if circumstances are the same, the same products are formed each time they react.

Flex Your Brain Have the students use the Flex Your Brain activity to explore CHEMICAL REACTIONS.

GLENCOE TECHNOLOGY

 Videodisc

The Infinite Voyage: Crisis in the Atmosphere

Chapter 1
Historical Aspects of the Greenhouse Effect and Fossil Air

Chapter 2
Studying Man-Made Carbon Dioxide

Chapter 5
Chlorofluorocarbons and Their Effect on the Ozone Layer

Chapter 6
NASA: Studying the Ozone

down into carbon and water, two products. When rust forms, iron and oxygen, two reactants, produce rust, one product. You can see that the reactions may vary in the number of reactants and the number of products. In general, any word equation can be written in the form:

$$reactant(s) \rightarrow product(s)$$

■ Acid—Base Reactions

Have you ever enjoyed an ice-cold glass of lemonade? Have you ever used a recipe that called for baking powder to make a cake rise? If so, you have used acids and bases.

In general, acids taste sour, react with metals, and usually contain hydrogen. Bases taste bitter, feel slippery, and usually are hydroxides.

Suppose you spill some household ammonia, ammonium hydroxide, on the kitchen counter. You wipe it up, but the counter still feels slippery. You try several different solutions, but you find that lemon juice removes the slippery feeling.

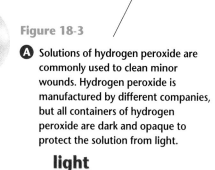

Figure 18-3

A Solutions of hydrogen peroxide are commonly used to clean minor wounds. Hydrogen peroxide is manufactured by different companies, but all containers of hydrogen peroxide are dark and opaque to protect the solution from light.

light

hydrogen peroxide → oxygen + water

B In the presence of light, hydrogen peroxide breaks down into oxygen and water. The word equation above describes this chemical reaction.

558 Chapter 18

Program Resources

Study Guide, p. 60
Critical Thinking/Problem Solving, p. 5, Flex Your Brain
Multicultural Connections, p. 39, Drs. Alvarez and Azencio: Two Kinds of Chemists [L1]
Section Focus Transparency 54

Across the Curriculum

Health

Ask students why they might apply an antiseptic to a cut. Antiseptics are compounds that are applied to living tissue to kill microorganisms that cause disease and/or infection. Most common antiseptics like hydrogen peroxide, benzoyl peroxide, or potassium permanganate disinfect by oxidation, a type of chemical reaction. Antiseptics also kill human cells. If possible, allow a volunteer to apply some H_2O_2 to a minor cut and observe the reaction.

What has happened? Acids, such as the citric acid found in lemon juice, and bases, such as ammonium hydroxide, react together to form a salt and water:

acid + base → salt + water

For this particular reaction,

citric acid + ammonium hydroxide → ammonium citrate + water.

The water is formed from the hydrogen from the acid and the hydroxide from the base. The parts of the acid and base that are left over form the salt.

■ Other Reactions

Chemical reactions occur constantly around you. Several of these are explored in **Figures 18-3–18-5**. Whether treating a cut, watching the blasting off of the space shuttle, or exploring a cave, chemical reactions affect and control your life.

carbonic acid + calcium carbonate → calcium hydrogen carbonate

Figure 18-4

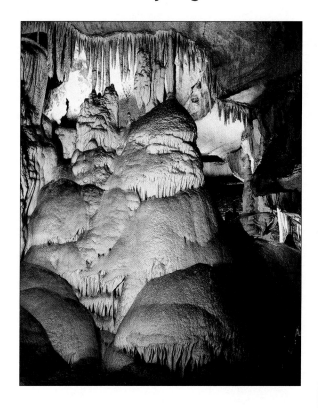

A Limestone caves are largely the result of chemical reactions. The calcium carbonate of the limestone reacts with acids such as the carbonic acid in rainwater that trickles down through the ground. The reaction produces calcium hydrogen carbonate. The calcium hydrogen carbonate dissolves in the water and washes away easily, leaving empty caverns in the limestone.

B The word equation above describes the most common chemical reaction that produces caves.

Figure 18-5

A The fuel system of the Space Shuttle includes separate tanks of liquid hydrogen and liquid oxygen. When mixed in correct proportion and ignited, the hydrogen and oxygen react to provide energy. The reaction also produces water, seen here as steam.

liquid hydrogen + liquid oxygen → water (steam) + energy

B The chemical reaction created by the fuel system is described in the word equation above.

Multicultural Perspectives

Ancient Metallurgists

Ask students to name places where steel is used. Students might mention automobiles, airplanes, buildings, tools, kitchen utensils, and plumbing fixtures. Explain that steel is an alloy of iron and other elements. Its strength depends on its carbon content. The more carbon, the more brittle the steel. The process of removing some of the carbon and turning the metal into more flexible steel was first described in the Chinese book *Huai Nan Tzu,* which dates to 120 B.C.E. Carbon reduction was accomplished by blowing oxygen over the cast iron to "oxygenate" it. The expertise of the Chinese metallurgists has since benefited all people from various cultures.

Visual Learning

Figures 18-3, 18-4, and 18-5 Before students study the captions, have them "read" the corresponding word equation and identify the reactants and products in each chemical reaction. In Figure 18-3, hydrogen peroxide is the reactant; oxygen and water are the products. In Figure 18-4, calcium carbonate (limestone) and the carbonic acid in groundwater are the reactants; calcium hydrogen carbonate is the product. In Figure 18-5, liquid hydrogen and liquid oxygen are the reactants; water vapor and energy are produced. L1

Across the Curriculum

Geography

Have students observe topographic maps of the area around Mammoth Cave in Kentucky. Help them identify the many sinkholes, caves, and depressions that formed when limestone reacted with rainwater. Maps can be obtained from either the United States Geological Survey or the Kentucky Geological Survey. L3 COOP LEARN

Demonstration Give students another opportunity to observe a chemical reaction.

Materials needed are table sugar, a test tube, and a Bunsen burner.

Place a small quantity of sugar in a test tube. Heat until the sugar burns. Black carbon will remain in the test tube. Water vapor and carbon dioxide gas will be released. As you discuss what happened, ask students to write a word equation that describes the reaction.

sugar + oxygen $\xrightarrow{\text{heat}}$ *water + carbon dioxide + carbon*

18-2 Word Equations 559

Planning the Activity

Time Needed 25 minutes

Materials 3 test tubes and rack; pieces of copper, zinc, and magnesium; 15 mL dilute hydrochloric acid; graduated cylinder; wooden splint; goggles; apron

Process Skills observing and inferring, comparing and contrasting, recognizing cause and effect

Purpose To compare reactivity of metals with acid.

Preparation For a 10% HCl solution, add 50 g concentrated HCl to 450 mL of distilled water.

Teaching the Activity

Discussion Have students describe what happened in each of the three test tubes. Which of the three elements was most reactive? L2

Safety You may want to have students wear rubber gloves as they work with hydrochloric acid.

Troubleshooting Trap some of the gas in the third test tube by holding a narrow-neck flask or large test tube over the tube. Keep the flask or test tube inverted so the lighter-than-air hydrogen does not float away. Be careful when introducing the glowing splint. Hydrogen gas can give a surprising pop.

Process Reinforcement
Have students compare and contrast the chemical reactions involved in this experiment. *All the reactions involved hydrochloric acid and a metal. No change occurred in the first tube. Both the zinc and magnesium reacted with the acid to form a gas.* L2

Science Journal Have students record their observations and the answers to the questions in their journals. L1

Describing Reactions

In this activity, you will compare the reaction of several metals with an acid to observe the results and describe what happens using word equations.

Problem
What metals will react with an acid?

Materials
3 test tubes
test-tube rack
goggles
graduated cylinder
pieces of copper,
    zinc, and magnesium

wooden splint
15 mL dilute
    hydrochloric acid
apron

Safety Precautions

Use caution when using acid. Wear lab aprons and goggles at all times.

 **What To Do**

1. Copy the data table *into your Journal.*

2. Put on goggles and an apron.

3. Set the three test tubes in a rack. Pour 5 mL of dilute hydrochloric acid, which contains hydrogen chloride, into each tube. **CAUTION:** *Handle acid with care. Immediately rinse away any spilled acid with plenty of water.*

4. Place a small piece of copper in the first tube, zinc in the second tube, and magnesium in the third tube.

5. Observe what happens in each tube and record your observations in the data table.

6. Have your teacher collect some of the gas from the test tube that contains the most bubbles. Bring a lighted splint near the gas. Record what you observe.

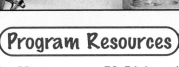

Program Resources

Activity Masters, pp. 75–76, Investigate 18-1

Concept Mapping, p. 26, Word Equation L1

Laboratory Manual, pp. 109–112, Chemical Reactions L2

Teaching Transparency 35

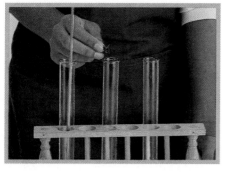

A

B

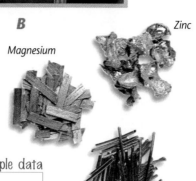

Magnesium

Zinc

Copper

Certain metals cause chemical reactions within your body that are essential for good heath. Copper helps prevent anemia, zinc helps convert nutrients into energy, and magnesium helps with respiration.

Sample data

Data and Observations

Tube	Substances Mixed	Observations
1	Hydrochloric Acid + Copper	no change
2	Hydrochloric Acid + Zinc	bubbles form
3	Hydrochloric Acid + Magnesium	most bubbles form, tube slightly warm

Analyzing

1. What evidence of chemical reaction did you observe? Which metals reacted with the acid?

2. Write a word equation for each reaction that took place.

3. When ignited, hydrogen explodes with a "pop." What can you infer about the identity of the gas produced? Why?

Concluding and Applying

4. Predict which of these metals would be least affected by acid rain if this metal were used to build statues.

5. **Going Further** The method of displacement in this Investigate is the most common way to prepare large amounts of hydrogen for use in the laboratory. Which of the metals you tested would be best for this purpose? Why?

Expected Outcomes

The zinc and magnesium will react with the acid. The gas collected over the magnesium will pop, indicating that it is hydrogen. Students should conclude that zinc and magnesium can react with hydrochloric acid to produce hydrogen gas.

Answers to Analyzing/ Concluding and Applying

1. bubbles; zinc, magnesium

2. zinc + hydrogen chloride → zinc chloride + hydrogen; magnesium + hydrogen chloride → hydrogen + magnesium chloride; there is no reaction with copper and the acid.

3. It is hydrogen because it reacted explosively with oxygen when the lighted splint was brought near it.

4. copper

5. Magnesium; its reaction was faster than that of the other metals.

✔ Assessment

Oral Obtain several varieties of the mineral calcite and have pairs of students observe the reaction that occurs when the mineral is exposed to hydrochloric acid. Have students describe and explain the reaction to their partner. Students should infer that a gas is being liberated as they observe the bubbles that form. Use the Performance Task Assessment List for Oral Presentation in **PASC**, p. 71.

COOP LEARN

Meeting Individual Needs

Physically Challenged Pair physically challenged students with students who are more able to manipulate the equipment used in this activity. Physically challenged students can be responsible for observing the reactions and recording their observations.

3 ASSESS

Check for Understanding

Have students check at home or at the grocery store to see what types of containers tomato sauce and orange juice are packaged in. Have them compare their findings to their answer to the Apply question. L2

Reteach

Activity Have students use building blocks to represent the reactants and products in a chemical reaction. By labeling the blocks with the names of the compound they represent, students can create a visual model of a chemical reaction. **LEP** L1

Extension

Discussion To help students classify chemical reactions, explain that a **synthesis reaction** takes place when two or more substances combine to form one new substance. A **decomposition reaction** takes place when one new substance is broken down into two or more new substances. Ask students to list as many examples as possible of each type of reaction. For example, *synthesis*—rust; *decomposition*—breakdown H_2O_2 into O_2 and H_2O.

4 CLOSE

Encourage students to talk about the various chemical reactions they have been studying. Then have students write a word equation to describe each chemical reaction covered in this section. L1 P

Reactions like those that you just investigated can be useful or destructive. For example, Figure **18-6** shows how ozone, a form of oxygen, is involved in both helpful and harmful chemical reactions. By being able to determine what products will be formed from certain reactions, we can predict the effects of some industrial processes on the environment or the effects of a new medicine on the body.

Review the word equations you studied in this section. Can you suggest other factors that play a part in determining whether chemical reactions will take place? In the next section, we'll look at the role played by energy in chemical reactions.

Figure 18-6

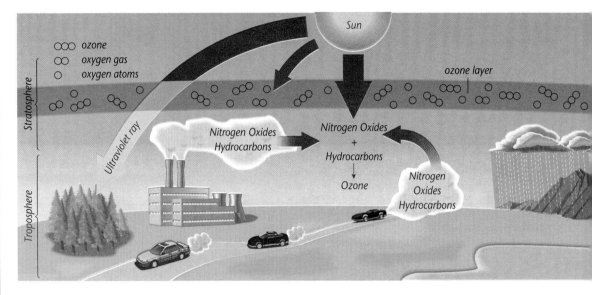

A The exhaust from cars and some industrial plants contains nitrogen oxides and hydrocarbons. These gases, in the presence of sunlight, may react with oxygen in the air to produce ozone. Ozone damages plants and reacts with body tissue in a harmful way.

B A layer of ozone is also created high in the stratosphere when ultraviolet rays from the sun react with regular oxygen molecules. The ozone layer absorbs most of the ultraviolet rays from the sun, which can cause sunburn and some forms of cancer.

1. Write a word equation to describe this chemical reaction: Iron and sulfur form iron sulfide.
2. Mercury(II) oxide breaks down when it is heated. Write a word equation for this reaction.
3. Mercury(II) oxide is a red, powdery substance. All substances that contain mercury, including mercury itself, are poisonous. Use this information to compare the properties of the reactants and products from Question 2.
4. **Apply** Many foods, such as tomato sauce, contain acids. What factors would you consider when choosing a container to cook or store such foods?

1. iron + sulfur → iron sulfide
2. mercury(II) oxide → mercury + oxygen
3. reactant: red, powdery, poisonous solid; products: mercury—silvery, liquid, poisonous metal; oxygen—colorless, gas, supports burning

4. Choose substances that would not react with the acids. Glass or a metal such as copper could be used because they do not react with these common acids.

Chemical Reactions and Energy

Energy in Reactions

Have you ever seen a fireworks display on the Fourth of July? Brilliant colors and the sounds of exploding rockets fill the night sky. Did you know that these sights and sounds are caused by a series of chemical reactions? Although you can't actually see

these reactions taking place, you can certainly observe the results! You've seen that chemical reactions produce new substances. In addition to new substances, light and sound are produced in the fireworks reaction. Where did this energy come from?

Section Objectives

■ Relate how energy is involved in chemical reactions.

■ Differentiate endothermic and exothermic reactions.

Key Terms

endothermic reaction
exothermic reaction

Explore! ACTIVITY

Does a rubber band have energy?

What To Do

1. Slip one end of a rubber band over a pencil.

2. Push the other end of the rubber band through a small hole in the center of an index card, as shown.

3. Hold the pencil flat against the back of the card.

4. Pull on the rubber band to stretch it a bit. Then, let it go and listen to the sound it makes.

5. Pull on the rubber band several more times. Each time, stretch the rubber band a little bit more.

6. Answer the following questions *in your Journal*. What happens to the sound? How do you know that a stretched rubber band has energy? Where does the energy come from?

563

Explore!

Does a rubber band have energy?

Time needed   10 minutes

Materials   rubber band, pencil, index card

Thinking Processes   observing and inferring, forming operational definitions

Purpose   To demonstrate the concept of potential energy.

Preparation   Any object to which the rubber

band can be hooked can replace the pencil.

Teaching the Activity

When stretched, the rubber band has a great deal of potential to move (potential energy) but almost no active (kinetic) energy. When let go, the situation reverses. [L1]

Science Journal   Have students record their observations in their journals. [L1]

PREPARATION

Planning the Lesson

Refer to the Chapter Organizer on pages 550A-D.

Concepts Developed

This section examines chemical reactions from an energy perspective. First, the potential energy of chemicals is discussed. Then, endothermic and exothermic reactions are introduced.

1 MOTIVATE

Bellringer

Before presenting the lesson, display **Section Focus Transparency 55** on an overhead projector. Assign the accompanying **Focus Activity** worksheet. [L1] **LEP**

Expected Outcome

Students should conclude that the more they stretch the rubber band, the more energy they give it.

Answers to Questions

6. The sound increases in volume. The band makes a sound when released. The energy comes from the pull of the hand.

✔ Assessment

Content   Have students list several activities or events from their daily lives that involve the release of energy. *Answers will vary, but may include burning fossil fuels and eating foods and using the resultant energy to do various activities.*

Ask the students to describe how they feel when they have a fever. Ask them why the body gets hot. Discuss the chemical reactions that occur and the making and breaking of chemical bonds associated with the release of energy. Some of this energy is released in the form of thermal energy, which raises the body temperature.

2 TEACH

Tying to Previous Knowledge

Review the concepts of energy and potential energy. Remind students of the Investigate activity they did in the previous section (pages 560–561). Ask them when the chemical potential energy was released. *It was released when a lighted splint was placed near the gas and the hydrogen reacted with the oxygen.*

Across the Curriculum

Psychology

Psychologists are able to study the function of the brain thanks to an unusual exothermic reaction involving formaldehyde, a common reagent, and catecholamines, an important group of brain chemicals. To locate catecholamines, psychologists expose a part of the brain to formaldehyde. The reaction that takes place *fluoresces*, or gives off light, showing exactly where catecholamine cells are located in the brain.

Chemical Energy

The stretched rubber band in the Explore activity absorbs the energy you put in when you pull on the rubber band. When you let go, the potential energy of the rubber band is released and transformed into other forms of energy. There is a similar energy change in chemical reactions.

The substances in the fireworks in the rockets contain chemical potential

Figure 18-7

Ⓐ Certain metals and metal salts each produce a flame that has a characteristic color. The substances take in energy from the flame and then release it in the form of light.

Ⓑ Because each of these substances always produces the same color flame, scientists use a flame test to identify some unknown substances.

Strontium chloride produces a red flame

Magnesium metal produces a white flame

Life Science CONNECTION

Edible Fuel

Like automobiles, our bodies have efficient engines. If they get the fuel they need, they have a better chance of running at peak performance. One essential nutrient that provides chemical energy is the carbohydrate.

Where Are Carbohydrates Found?

Carbohydrates come in many shapes and sizes. They're in foods such as apples, pears, grapes, oatmeal, spaghetti, bread, corn, potatoes, tortillas, and rice. All carbohydrates are made of carbon, oxygen, and hydrogen. All sugars and starches are carbohydrates.

When you eat foods with sugar, such as refined white sugar, your body can quickly use the sugar as fuel to provide needed energy. Inside your body's cells, the sugar combines with oxygen. This chemical reaction, called respiration, provides quick energy.

Life Science CONNECTION

Purpose
The Life Science Connection expands the discussion of exothermic reactions in Section 18-3 by describing an exothermic reaction that occurs in our bodies—respiration.

Content Background
The three basic nutrient types are carbohydrates, fats, and proteins. Because our bodies need more carbohydrates than fats and proteins, an ideal diet is about 55%

carbohydrates. Digestion is a reaction that breaks carbohydrates down into sugars.

Teaching Strategy
Ask students to name things that their bodies need energy for. List students' suggestions on the chalkboard. Students are likely to name only physical activities. If they do, remind them of the need for energy to keep vital organs operating and to generate heat to keep us warm. Point out that this energy comes from two sources,

energy that was stored when the substances were formed. When the rockets are ignited, chemical reactions take place. The chemical potential energy in the substances is released and changed into light, motion, and sound, as well as thermal energy.

Look back at the chemical reactions you studied in the last section. Notice how some of these reactions release energy while other reactions absorb energy.

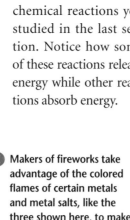

Copper chloride produces a blue flame

 Makers of fireworks take advantage of the colored flames of certain metals and metal salts, like the three shown here, to make colorful fireworks.

Discussion Ask students if they have used hot packs for sore muscles or to give off warmth while they engage in winter sports. Have them describe how they are used. If you have packs available, have students read the labels to see what chemicals the packs contain. If feasible allow students to activate a heat pack and feel the warmth generated.

GLENCOE TECHNOLOGY

 Videodisc

The Infinite Voyage: The Future of the Past

Chapter 1
Preserving Frescoes in Florence

Chapter 2
The Cologne Cathedral: Preserving Stained Glass

Chapter 3
Preserving Medieval Glass with a Chemical Lacquer

Chapter 4
The Statue of Liberty Restoration Project

Chapter 5
Winterthur Museum: New Cleaning Techniques for Old Paintings

Chapter 7
Restoring the Parthenon

Extended Energy

What can you eat if you need energy over a longer period of time? Strenuous activities require a lot of energy. If an athlete's muscles are to get the energy they need for a long race, runners must consume large amounts of carbohydrates. Athletes are counting on the starch in these foods to provide the fuel for respiration reactions throughout the event.

The starches in carbohydrates cannot provide quick energy because they are composed of several joined sugar molecules. There must be a chemical reaction to break apart the starch molecule. Then, the resulting sugar can be used by the cell for energy.

You Try It!

Test for starch in bread. Put on goggles and a lab apron. Add a few drops of iodine to two teaspoons of water and stir with a wooden splint. **CAUTION:** *Iodine is poisonous and stains. Do not taste food you have tested.*

Put a few drops of solution on a small piece of bread and observe what happens. Starch turns blue when tested with iodine.

Test some other foods, including spaghetti and apples, to see which contain starch.

about half from carbohydrates and the other half from fats. Both are carbon-hydrogen-oxygen compounds. Excess carbohydrates are changed to fats and stored by the body.

Answers to You Try It!
The test is positive for starch in bread.

Going Further ▸
To reinforce the chemical reactions our bodies use to digest carbohydrates, have students provide the missing word in each statement below.
- Digestion begins in the mouth where a chemical in _(saliva)_ breaks apart big starchy carbohydrate molecules into sugars in a _(chemical)_ reaction.
- In the stomach, strong acid slows or _(inhibits)_ carbohydrate digestion briefly.
- In the _(small intestines)_, another chemical reaction breaks apart starchy carbohy-

drates into smaller sugar molecules.
- These sugar molecules pass through the intestines' walls into the blood. When they reach the cells, they combine with oxygen in a(n) _(exothermic)_ reaction that releases energy.

Demonstration This activity allows students to observe an endothermic process.

You will need self-sealing plastic bags, ammonium chloride, goggles, and an apron.

Place the ammonium chloride in a large plastic bag. Inside this bag, also place a smaller bag full of water. Open the inner bag of water. Close the outer bag and shake. Ask students to feel the bag and describe what is happening. *The liquid is getting colder.* Explain that the materials in the bag took in heat from their environment, making the mixture colder than it was. Point out that bags like these are similar to the cold packs used by athletic trainers. **CAUTION:** *Avoid contact with ammonium chloride.* L1

Student Text Question
What kind of energy is absorbed in this reaction?
thermal energy (heat)

Visual Learning

Figure 18-8 Have students identify the reactants and the products in the word equation. *Aluminum oxide is the reactant; aluminum and oxygen are the products.*

Uncovering Preconceptions
Endothermic reactions, which absorb heat, are harder to grasp. Endothermic reactions actually take heat from the environment. The telltale sign of an endothermic reaction is that the chemicals get colder.

Connect to . . .

Life Science

Answer: ATP stores more energy

Connect to...
Life Science
ADP and ATP are important biological compounds that help supply the energy needed to run many reactions in your body. Which molecule, ADP or ATP, stores more energy?

Chemical Reactions that Absorb Energy

Exploding fireworks give off a lot of energy. However, some reactions require a lot of energy to get them started and keep them going. Does the following equation look familiar?

$$\text{carbon dioxide} + \text{water} \xrightarrow{\text{light}} \text{sugar} + \text{oxygen}$$

The word *light* over the arrow means that light is necessary for this reaction to occur. You would read this equation, "In the presence of light, carbon dioxide and water produce sugar and oxygen." Recall that this photosynthesis reaction takes place only in the light. Green plants absorb energy from sunlight. This energy is changed into the chemical energy in sugar and in oxygen. If you keep a green plant in the dark for too long, it will eventually die.

The photosynthesis reaction involves the absorbing of energy. A chemical reaction in which energy is absorbed as the reaction continues is called an **endothermic reaction**. The refining of aluminum metal from its ore is also an endothermic reaction, as shown in **Figure 18-8**. What kind of energy is absorbed in this reaction?

Figure 18-8

aluminum oxide + energy → aluminum + oxygen

Endothermic reaction An endothermic reaction makes it possible to extract aluminum—one of the major metals used to make the outside covering of airplanes—from its bauxite ore.

Bauxite ore

Bauxite is mostly aluminum oxide. Aluminum oxide is separated from the other substances in bauxite and melted with certain other substances such as cryolite. An electric current is passed through the mixture. The aluminum oxide absorbs the electrical energy and breaks down into oxygen and aluminum.

566 Chapter 18 Chemical Reactions

Program Resources

Study Guide, p. 61
Critical Thinking/Problem Solving, p. 26, The Nitrogen Cycle L2
Teaching Transparency 36
Section Focus Transparency 55

ENRICHMENT

Demonstration Repeat the procedure that was performed in the demonstration on this page, substituting calcium chloride for ammonium chloride. This will create a hot pack instead of a cold pack. Calcium chloride is a common salt used to melt ice. Ask students to explain why the bag got hot instead of cold. *An exothermic process took place this time.*

Chemical Reactions that Release Energy

Imagine that you've worked hard for several hours. When you eat, the chemical reactions of digestion release some of the chemical energy in food. This energy can be used by the body for movement or internal processes of the cells. The heat of your body is an example of chemical energy that has been changed to thermal energy.

A chemical reaction in which energy is released is called an **exothermic reaction**. The explosion of fireworks is an exothermic reaction that produces light, sound, and thermal energy. Think about the word equation for the burning of rocket fuel, as shown in **Figure 18-9**. You don't have to keep supplying energy for the reaction to continue.

In the activity that follows, you will determine whether a reaction is endothermic or exothermic.

Figure 18-9

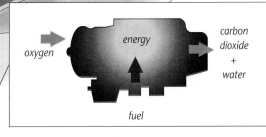

Exothermic reaction The process of burning jet fuel is an exothermic reaction—a reaction that releases energy. Do any substances used for fuels produce an endothermic reaction? Why?

(jet fuel) + oxygen → carbon dioxide + water + energy

DESIGN YOUR OWN
INVESTIGATION

18-2 Exothermic or Endothermic?

Preparation

Purpose To determine changes in energy during chemical reactions.

Process Skills observing and inferring, comparing and contrasting, recognizing cause and effect, forming operational definitions

Time Required 1 class period

Materials See reduced student text.

Possible Hypotheses Students may hypothesize that an increase in temperature indicates an exothermic reaction, and a decrease in temperature indicates an endothermic reaction.

The Experiment

Troubleshooting Students need to measure the temperature of a test tube containing hydrogen peroxide alone to see if the temperature changes over time.

Process Reinforcement Check science journal entries to see that students all observed an increase in temperature with time for these reactions. From their data, students should be able to infer that the reactions were exothermic.

Possible Procedure Put on goggles and lab aprons. Add approximately 5 mL of hydrogen peroxide to one test tube. Take the temperature of the hydrogen peroxide, and record it in the data table. Add a small piece of liver to the test tube and measure the temperature of the tube every 30 seconds for 6 minutes or so. For accuracy, repeat this test two more times. Then repeat the procedure with a piece of raw potato.

Teaching Strategies

Discussion Talk about what students would expect to happen if the reactions are exothermic: *Give off heat to the environment and feel warm* If endothermic: *Take up heat and feel cold*

Exothermic or Endothermic?

An endothermic reaction absorbs energy from the environment. An exothermic reaction releases energy into the environment. In the preceding section, the chemical reactions involved in digesting food were described as exothermic. In this Investigation, you will design an experiment that involves exothermic reactions.

Preparation

Problem
How can you prove that a reaction is exothermic or endothermic?

Form a Hypothesis
Write a hypothesis about how to determine whether a chemical reaction is endothermic or exothermic.

Objectives
• Design an experiment to test the hypothesis.

• Measure the energy released during a chemical reaction.

Materials
8 test tubes and rack
3% hydrogen peroxide solution
raw liver
raw potato
thermometer
stopwatch
25-mL graduated cylinder

Safety

Wear a lab apron and goggles at all times. Be careful when handling glass thermometers. Test tubes containing the hydrogen peroxide should be placed and kept in racks. **CAUTION:** *Hydrogen peroxide can irritate skin and eyes and damage clothing.*

Program Resources

Activity Masters, pp. 77–78, Design Your Own Investigation 18-2

How It Works, p. 19, How Does Photographic Film Work? [L2]

Making Connections: Across the Curriculum, p. 39, Pyrotechnics—The Art of Fireworks [L2]

Meeting Individual Needs

Visually Impaired Provide magnifying glasses for visually impaired students to facilitate their observing the reactions as well as their reading the thermometers.

DESIGN YOUR OWN
INVESTIGATION

Plan the Experiment

1 As a group, examine the materials, and design a procedure to test your hypothesis. How will you measure results?

2 To measure the heat given off or taken in during a reaction, what must you know? How many measurements should you take during the experiment?

3 Each experiment needs to be done several times to obtain accurate data. Each time the experiment is carried out is called a trial. Use the average of all the trial results as your data for supporting or disproving your hypothesis.

 **4** *In your Science Journal* or as a database, make a data table similar to the one shown below.

Check the Plan

1 Before you carry out this Investigation, make certain that your teacher approves your plan and the setup of your equipment.

2 How often will you take measurements? Is it important to keep track of time between measurements? Record data immediately in the data table.

3 Carry out your investigation.

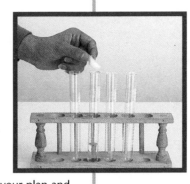

Going Further

Make a hypothesis about the relationship between how much fat a food contains and how much energy it will release. Design an experiment to test your hypothesis.

Analyze and Conclude

1. Infer Did a chemical reaction take place? What evidence do you have for your answer?

2. Analyze Was your hypothesis supported? Was this an exothermic reaction? Use your data to explain your answer.

3. Control Variables What were the variables in this experiment?

4. Conclude What was the source of the energy released during the experiment?

5. Predict What would change if you had used smaller pieces of liver and potato?

Sample data

Data and Observations

Trial	Starting Temperature	Temperature after Adding Liver/Potato Minutes					
1		Students should see an overall					
2		increase in temperature due to the					
3		release of thermal energy.					
4		The reactions are exothermic.					
Total							
Average							

Meeting Individual Needs

✔ Assessment

The graph will show an increase in temperature over time or the rate of each reaction for the two reactions. The graph for the liver will have a steeper slope. L1

Figure 18-10 **What endothermic and exothermic reactions occur in each of these tasks?** *Exothermic—burning of match and gas, digestion and respiration. Endothermic—cooking the egg.*

3 ASSESS

Check for Understanding

Have students identify the reaction in the Apply question as either endothermic or exothermic and explain their reasoning. *Exothermic; as the burning continues, it produces energy.*

Reteach

Analogy Have students visualize an endothermic reaction as "uphill" because it needs to be "pushed" (provided with energy) all the way. An exothermic reaction is "downhill" because, once it begins, it rolls on its own. L1

Extension

Activity Have students devise their own energy graphs of processes such as: (a) the pulling and releasing of the rubber band in the Explore activity on page 563; (b) a sparkler being lit. L3

4 CLOSE

Have partners write a story that involves at least three chemical reactions. Then direct them to make a chart tracing each energy change as it happens.
COOP LEARN L1

Tracing Energy Changes

SKILLBUILDER

Making and Using Graphs
Make a graph of the data you collected in the Investigation. Plot the temperature against time, beginning with the starting temperature. Plot the points for the liver in one color and the potato in a different color. For each set of data, connect the dots using straight lines. What do these lines show? If you need help, refer to the **Skill Handbook** on page 640.

In the reactions you just observed, the chemical energy in hydrogen peroxide was released when the liver and the potato were added to it. The chemical energy was changed to thermal energy, as evidenced by the increase in the temperature over time.

Using information from this chapter,

Figure 18-10 traces the energy changes and identifies the endothermic and exothermic reactions that may occur on a camping trip. In addition to those shown in the figure, what other chemical reactions might occur on the trip?

Make a list of things that you do and things around you that involve chemical reactions. Try to trace the changes in chemical energy for each item on your list.

Figure 18-10

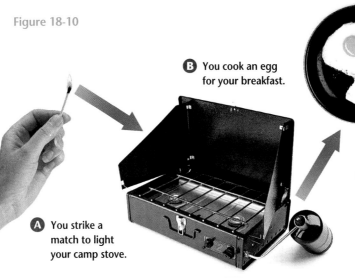

B You cook an egg for your breakfast.

A You strike a match to light your camp stove.

C You eat the egg and start off on a long hike. What endothermic and exothermic reactions occur in each of these tasks?

check your UNDERSTANDING

1. Describe the role of energy in the chemical reactions involved in baking a cake. Are the reactions endothermic or exothermic?
2. In writing a word equation for burning wood, where would you place thermal energy?
3. **Apply** Sometimes during thunderstorms, lightning may strike a tree and set it on fire. As you know, burning is a chemical reaction. Tell where the energy comes from to begin this reaction.

---

check your UNDERSTANDING

1. Thermal energy from the oven must be provided to start the reaction. Chemical energy is changed to mechanical energy to make the cake rise. Because the cake would stop baking if the oven were turned off, this must be an endothermic reaction.

2. over the arrow to show thermal energy is needed to start the reaction and on the right side to show that thermal energy is a product of the reaction

3. from the electrical energy in the lightning

Speeding Up and Slowing Down Reactions

Speeding Up Reactions

Baking a cake involves reactions that require energy in order to take place. Some reactions, such as the breaking down of table salt, would never take place without a great deal of energy. Is there a way to speed up a chemical reaction without using lots of energy?

Section Objectives

- Describe how a catalyst affects a chemical reaction.
- Explain how to control a chemical reaction with an inhibitor.

Key Terms

catalyst
inhibitor

Planning the Lesson

Refer to the Chapter Organizer on pages 550A-D.

Concepts Developed

This section focuses on the rates of chemical reactions. Catalysts and inhibitors are introduced.

Find Out! ACTIVITY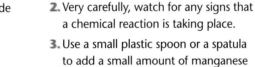

Can a chemical reaction be made to go faster without adding energy?

Remember that hydrogen peroxide breaks down very slowly when exposed to light.

What To Do

1. Wearing goggles and an apron, pour about 5 mL of 3-percent hydrogen peroxide into each of two test tubes.

2. Very carefully, watch for any signs that a chemical reaction is taking place.

3. Use a small plastic spoon or a spatula to add a small amount of manganese dioxide to one of the test tubes. Record your observations *in your Journal.*

4. Light a wooden splint and blow it out.

5. Place the glowing splint just inside the mouth of the tube without the manganese dioxide. What happens?

6. Repeat with the test tube containing the manganese dioxide. What happens?

7. Heat this tube in a beaker of boiling water until no liquid is left. What do you see?

Conclude and Apply

1. Although oxygen doesn't burn, it must be present for burning to take place. How do you know oxygen was produced in this reaction?

2. What can you infer about the role of manganese dioxide in this reaction?

571

Find Out!

Can a chemical reaction be made to go faster without adding energy?

Time needed   20 minutes

Materials   10 mL hydrogen peroxide, 2 test tubes and rack, manganese dioxide, spatula or spoon, goggles, apron, wooden splint, water, beaker, hot plate

Thinking Processes   observing and inferring; recognizing cause and effect; separating constants, variables, and controls

Purpose   To investigate how a catalyst speeds up a reaction.

Teaching the Activity

Safety   Be sure students wear goggles and aprons and avoid contact with chemicals.

Science Journal   Have students describe what a catalyst is, using their own words. Then tell how a catalyst affected this reaction. **L1**

Expected Outcomes

The manganese dioxide will greatly speed up the reaction, causing oxygen to collect in the test tube, but it will not be used up itself. Students should conclude that the catalyst helps the reaction along without being permanently changed by the reaction.

Conclude and Apply

1. The splint relights.

2. It serves as a catalyst.

✔ Assessment

Performance   Have pairs of students design and perform their own experiment to prove that the catalyst is, in fact, what is speeding up the reaction. For example, students might set up a control, performing the same experiment with and without the catalyst. Use the Performance Task Assessment List for Designing an Experiment in **PASC,** p. 23.

L2   **COOP LEARN**

Bellringer

 Before presenting the lesson, display **Section Focus Transparency 56** on an overhead projector. Assign the accompanying **Focus Activity** worksheet. L1

LEP

Demonstration To observe the effect of a catalyst upon a chemical reaction, try this demonstration.

Materials needed are 25 g potassium sodium tartrate, a 600-mL beaker, very hot water, 100 mL 6% hydrogen peroxide, 3 g cobalt chloride, a thermometer, a stirring rod, heat source, goggles, and an apron.

Dissolve the potassium sodium tartrate in 300 mL of very hot water. Then add the hydrogen peroxide and heat to 85°C. Finally, add the catalyst, cobalt chloride, at which time a reaction will occur. **CAUTION:** *Beaker may overflow. Avoid contact with chemicals.* Ask students to describe in their journals what they observe and name the reactants and products involved in the reaction.

2 TEACH

Tying to Previous Knowledge

This section develops directly from the previous section, which addressed energy requirements for reactions. Even exothermic reactions were said to need start-up energy to occur. Catalysts reduce the amount of start-up energy required. Inhibitors increase the amount of start-up energy needed.

Student Text Question

Manganese dioxide is a catalyst in the breaking down of hydrogen peroxide. How do you know? *It is still present after the reaction, but makes it go faster.*

DID YOU KNOW?

Because of enzyme catalysts, the chemical reactions in your body take place more than a million times faster than they would otherwise.

The manganese dioxide in the Find Out activity helped the reaction take place and made the reaction go faster. However, the manganese dioxide was not permanently changed as a result of the reaction. The word equation for this reaction is:

manganese dioxide
hydrogen peroxide → water + oxygen

Any substance that speeds up a chemical reaction without being permanently changed is called a **catalyst**. Manganese dioxide is a catalyst in the breaking down of hydrogen peroxide. How do you know?

Figure 18-11

Enzyme catalysts make possible the many chemical reactions that keep you alive.

Ⓐ **Amylase**, the enzyme found in saliva, helps break down starch into sugar.

Ⓑ **Pepsin**, in the presence of the acid in your stomach, starts the digestion of protein.

Ⓒ **Trypsin** helps to continue protein digestion in your small intestine.

Ⓓ **Lipase**, the enzyme that is produced in your pancreas, is secreted into your small intestine, where it breaks down fats.

572 Chapter 18 Chemical Reactions

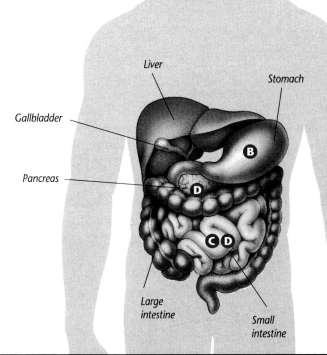

Salivary glands
Liver
Stomach
Gallbladder
Pancreas
Large intestine
Small intestine

■ **Enzymes**

Catalysts are also involved in many of the chemical reactions that take place in organisms. These catalysts are protein substances called enzymes. **Figure 18-11** summarizes your body's production of enzymes and what these enzymes do. The digestion of food, for example, depends on enzymes. In this way, enzyme catalysts make possible the many chemical reactions that keep you alive.

Program Resources

Study Guide, p. 62
Multicultural Connections, p. 40, Jabir ibn Hayyan–Muslim Alchemist L1
Science Discovery Activities, 18-3
Section Focus Transparency 56

Multicultural Perspectives

Who Drinks Milk?
Discussion As people grow older, their digestive systems often lose the ability to manufacture lactase, the enzyme that catalyzes the reaction that breaks down milk sugar, or lactose. This condition is especially common in Asia and in Africa. Have students think about why some people with lactose intolerance can eat yogurt. *Yogurt has enzymes.*

Figure 18-12

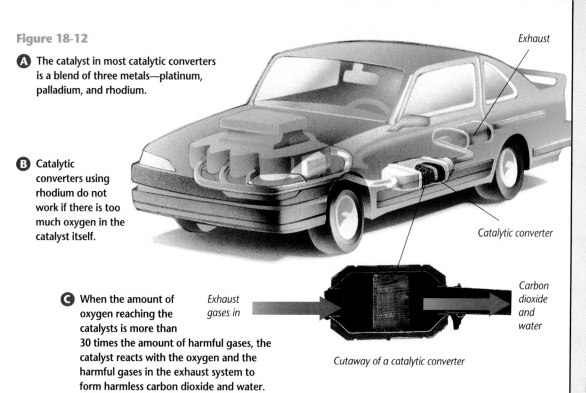

Ⓐ The catalyst in most catalytic converters is a blend of three metals—platinum, palladium, and rhodium.

Ⓑ Catalytic converters using rhodium do not work if there is too much oxygen in the catalyst itself.

Ⓒ When the amount of oxygen reaching the catalysts is more than 30 times the amount of harmful gases, the catalyst reacts with the oxygen and the harmful gases in the exhaust system to form harmless carbon dioxide and water.

Exhaust

Catalytic converter

Exhaust gases in

Carbon dioxide and water

Cutaway of a catalytic converter

■ Catalytic Converters

Air pollution caused by automotive exhaust gases is a problem. This pollution is reduced by using catalysts to speed up the chemical reaction between oxygen and the harmful substances produced by cars. The reaction takes place inside a device called a catalytic converter.

In one type of converter, shown in **Figure 18-12**, air and exhaust gases are passed through a bed of small beads that are coated with the catalysts platinum, rhodium, and palladium. The harmful gases, under the influence of the catalyst, combine with oxygen to form harmless carbon dioxide and water.

How Do We Know?

The Role of Chlorophyll

One of the ways we know that the catalyst chlorophyll is necessary for photosynthesis is by studying organisms that lack chlorophyll. Nongreen plants and fungi get their nourishment by living on food produced by other organisms. When separated from their host organisms, nongreen plants soon die.

From experimental evidence, we know that chlorophyll alone cannot produce the photosynthesis reaction. When chlorophyll extracted from green plants is mixed with water and carbon dioxide and exposed to light, photosynthesis does not take place. The process is quite complex and not completely understood.

Theme Connection Catalysts and inhibitors are examples of agents that impose change and stability. Catalysts and inhibitors slow down or speed up reaction patterns without changing the arrangement between the reactants themselves, much the way playing a record at the "wrong" speed changes the tempo of musical notes but not the relationship among the notes themselves.

Inquiry Question Why do you think an automobile with a catalytic converter must use only lead-free fuel? *The lead would coat the catalysts and make them ineffective.*

Visual Learning

Figure 18-11 Use the illustration to review the parts of the digestive system. Have students locate the relevant part of the digestive system, as a volunteer reads aloud the captions.

Across the Curriculum

Math

In one person's stomach it took 1.5 minutes to break down a sample of food with enzyme A present. Assuming this enzyme speeds up the reaction by a factor of 1 000 000, how long in years and months would it take a person without enzyme A to digest the same sample? *about 2 years, 10 months* L2

SCIENCE JOURNAL

Have students write definitions of catalysts and inhibitors in their journals. L1

ENRICHMENT

Research Have students find out from your state's air control board about the levels of pollutants from motor vehicle exhaust present in your area over a given period of time. Have students graph the values and explain any trends in the data. L2

Student Text Question

How do you think these paints help keep rust from forming? *The paints keep iron surfaces away from oxygen that may react with them.*

3 ASSESS

Check for Understanding

Have students answer questions 1 and 2 of Check Your Understanding. As an extension of the Apply question, direct small groups of students to set up cleaning tests in which they handwash equally dirty pieces of cloth in a regular detergent and in one that has enzyme boosters. Have students share their findings with the class. L1

Reteach

Activity Ask students to visualize inhibitors as "coatings" that prevent reactants from getting together so they can react. Conversely, catalysts can be seen as "matchmakers" that help reactants get together so they can react more easily. L1

Extension

Analogy Have students think of examples of catalysts and inhibitors in real-world settings. For example, an unselfish point guard on a basketball team and good music at a party are catalysts. A "ball-hog" point guard can *inhibit* a basketball team, and inappropriate music can inhibit social interaction at a party. L3

4 CLOSE

Have students name two catalysts and two inhibitors and tell how each controls the rate of a reaction. L1

Slowing Down Reactions

Figure 18-13

Ⓐ Bread slice 1

Bread slice 2

Ⓐ The slice of bread on the left is made of flour, salt, and yeast. The one on the right is made from the same ingredients and a food preservative. Both slices are exposed to the air.

Ⓑ Bread slice 1

Bread slice 2

Ⓑ After a few days, the slice without the preservative shows signs of mold. Preservatives inhibit the growth of mold on food.

Ⓒ Bread slice 1

Bread slice 2

Ⓒ A few more days later and the slice without a preservative is almost totally covered with mold. Although the preservative inhibited mold growth in the other slice for a time, it too shows signs of mold.

Have you ever wanted to make a sandwich and discovered that the bread had molded? Or maybe you have wondered why some foods spoil more quickly than do others. Sometimes ingredients are added to food to slow down the rate at which foods combine with oxygen. Oxidation is a major cause of spoilage in foods.

Any substance that slows down a chemical reaction is called an **inhibitor**. Most preservatives that are added to foods are inhibitors. Foods containing inhibitors will spoil eventually, but inhibitors greatly decrease the rate at which this occurs. Some antibiotics also act as inhibitors by slowing down the action of those enzymes in the body that tend to help bacteria grow. Some paints are advertised as being rust inhibitors. How do you think these paints help keep rust from forming?

check your UNDERSTANDING

1. You want to slow down the following reaction:

iron + oxygen → iron oxide

Which would act as an inhibitor, oil or water?

2. When a rusty iron nail is added to a test tube containing hydrogen peroxide, bubbles quickly form. The nail undergoes no obvious change. What conclusion can you draw about the iron oxide (rust) that is on the nail?

3. **Apply** Enzymes are sometimes used in detergents to boost their cleaning power. What do you think that means?

check your UNDERSTANDING

1. oil
2. It acts as a catalyst in breaking down hydrogen peroxide.
3. The enzymes act as catalysts to help the detergent remove stains.

Science and Society

What's In Your Food Besides Food?

INGREDIENTS: SELECTED U.S. PEA-NUTS, DEXTROSE, PARTIALLY HYDRO-GENATED VEGETABLE OIL (TO PREVENT SEPARATION), SALT, SUGAR.

NO CHOLESTEROL

0 48001 27036 7

NET WT. 12 OZ. / 340g

When George Washington Carver first made peanut butter, he probably gathered the nuts, roasted them, mashed them, and spread them on bread—all within a day or two. Now the peanut butter may spend weeks on a grocer's shelf and months in your cupboard.

The peanut butter can readily be changed by reactions with oxygen in the air. To prevent chemical changes, food manufacturers add chemicals, called food additives, to food.

Food Additives

Salt was one of the first additives and was used to preserve meats and fish. Natural additives, such as salt and ascorbic acid (vitamin C), are still used today. The food industry, however, depends on hundreds of different chemicals to keep your food edible for a longer time.

There are also additives to improve the appearance or texture of food. Iodine in salt and vitamin D in milk are added to prevent diseases.

All these chemicals are regulated by the Food and Drug Administration (FDA). Occasionally, the FDA has approved additives that were later shown to cause cancer. There are additives that cause reactions ranging from mild to life-threatening for certain individuals. One is the flavor enhancer monosodium glutamate, known as MSG.

*inter*NET CONNECTION

A list of accepted food additives and frequently asked questions is available from the FDA on the World Wide Web. Find out more about an additive in one of your favorite foods. Are there any risks to consuming it? In what other foods is it used?

A **food chemist** works to develop additives. Food chemists study chemistry, biology, and biochemistry.

*inter*NET CONNECTION

Visit the website of the Food and Drug Administration at
FDA Headquarters
http://www.cfsan.fda.gov/list.html

Purpose

Health CONNECTION

The Health Connection extends the concept of a catalyst presented in Section 18-4 by discussing caffeine, a stimulant found in many foods and beverages.

Content Background

Caffeine is the most widely used legal drug in the U.S. It stimulates the brain and heart but has numerous health effects, some serious, which are being studied. Americans get more caffeine from soft drinks than from any other source. The average person drinks 40 gallons per year, creating a $2.5 billion industry. Two-thirds of all soft drinks contain caffeine, much of it from decaffeinated coffee! Coffee is the second-biggest caffeine source, and America's second-largest import after petroleum. We drink 165 billion cups of coffee yearly, supporting a $4.5 billion industry. Tea and cocoa drinks are next in popularity. Twenty million vending machines sell caffeinated drinks. Some over-the-counter pain relievers, cold remedies, and diet pills contain caffeine. (Data from Frances Sheridan Goulart, *The Caffeine Book,* Dodd, Mead & Co., 1984)

Teaching Strategy

Hold a class discussion on the effects of caffeine-containing products. What effects do students believe they have observed in themselves and others?

Activity

For the What Do You Think activity, encourage students not only to record their own caffeine consumption, but to have their family members record their caffeine consumption as well. Then have them look for a pattern of consumption in their families, such as "all cola drinkers" or "all tea drinkers." Find out if any family members have been

Health CONNECTION

The story goes that a monk in an Arabian monastery centuries ago observed goats in the field nibbling the berries from a coffee plant. The goats ran and played all night long without ever seeming to be tired. The monk decided to try the berries himself. He gathered some, added them to boiling water, and found that the resultant brew, which contained caffeine, did indeed make him more alert.

Effects of Caffeine

Whether that story is true or not, caffeine has been used for centuries as a stimulant. Caffeine is a combination of carbon, hydrogen, nitrogen, and oxygen. When caffeine is taken into the body, chemical changes take place that turn caffeine into other chemicals that interfere with your body's natural substances and stimulate the nerves.

Alleged results can be inability to sleep, irregular heartbeat, nervousness, and high cholesterol.

Sources of Caffeine

Many products besides coffee contain caffeine. Tea, soft drinks, chocolate, cocoa, and several over-the-counter drugs contain fairly large amounts of caffeine. It is also hard to know how much caffeine you take into your body. For example, a cup of coffee can contain 29 milligrams of caffeine or 176 milligrams of caffeine, or

Are You Too Awake?

Average Amounts of Caffeine	
Coffee	80 mg-120 mg
Hot cocoa	1-8 mg
Tea	40 mg
Soft drinks	40 mg

any amount in between. The amount of caffeine depends in part on the way the coffee is prepared. Some researchers say that people can take up to 600 milligrams of caffeine per day without any harm being done. Others say that even smaller amounts can do damage.

USING MATH

The chart shows the average amounts of caffeine in a cup of four drinks. Keep a record of how much caffeine you take in a day from these sources. Would you be better off with less?

warned by their doctors to stay away from caffeine. Have they been able to do it? L1

Going Further ||||||►

Caffeine use is so prevalent in our culture that its health effects, which remain under study, could be enormous. Have students locate recent magazine articles on the health effects of caffeine; the different amounts of caffeine in different brands of soda, tea, and coffee; and the marketing of decaffeinated coffees, teas, and sodas. Ask each student to read one article, write a brief summary, and share his or her findings with the class. Use the Performance Task Assessment List for Writing in Science in **PASC,** p. 87 L1.

NATIONAL GEOGRAPHIC
SCIFACTS

NATIONAL GEOGRAPHIC
SCIFACTS

How strong is spider silk?

The intricate, fragile-looking threads in spider webs can outperform steel for strength and surpass nylon for durability. These features of spider silk have intrigued scientists for years. The spiders swing from the threads and use them to make cocoons for eggs and to make webs for capturing prey.

Recently University of Wisconsin researchers used natural spider silk as a material for sewing up laboratory animals that had undergone surgery. Surprisingly, the animals' immune systems did not reject the silk.

How do spiders produce this amazing material? To form silk threads, a spider forces liquid silk through numerous tiny holes in organs called spinnerets, located under the spider's abdomen. Scientists discovered that the material's molecules align lengthwise as they dry, forming a strong, lightweight fiber. In effect, the silk becomes a liquid crystal before it hardens. All of this happens with the help of water vapor in the air.

When researchers attempt to produce a fiber comparable in strength and durability to spider silk, they have to process the synthetic material at very high pressure in the presence of caustic sulfuric acid.

Scientists hope to discover a simpler process similar to the one spiders use.

WEB WEAVERS

All spiders spin silk, but web weavers such as the Black Widow are the superspinners. These aerial artists can weave silk that is sticky or smooth, in fine strands or thick sheets.

Spinnerets

About 600 species of spiders are native to the state of Connecticut. More than 30 000 spider species have been identified worldwide, and scientists suspect that number represents only about one-fourth of the total.

Black widow
Latrodectus mactans

High-elevation arachnid
A salticid spider was taken up Mount Everest to an elevation of 6705 meters, a record height for survival in animals.

Science Journal

Research the types of spiders common in your area. Describe, draw, and name each one *in your Science Journal*. In addition, find out how spider silk differs from silk produced by silk worms.

Science Journal

Answers will vary, depending on the area in which the students live. Like spiders, silkworms also have spinnerets from which liquid silk is emitted. The silk hardens as it comes into contact with the air. Unlike spiders, however, silkworms can be domesticated to produce a high-quality silk that is used to make fabric.

How strong is spider silk?
Purpose
Many plants and animals produce substances that may be helpful to humans. Thus, scientists often attempt to reproduce these substances artificially through chemical reactions. These SciFacts, which discuss a scientific attempt to produce a material similar to spider silk, relate to the discussion on chemical reactions in Section 18-1.

Content Background
Spiders produce different types of webs, but all are designed to trap food. Orb webs are circular with spokes branching out to the ends of the circle. Some orb-web spiders weave special designs into their webs, which may help hide the spider as it waits for its prey. Other spiders weave sheet webs close to the ground. Lines of silk are woven above the sheet web to better trap insects. Funnel webs are wide at one end and narrow at the other. The spider waits for its food in the narrow part of the funnel web.

Activity
Take students to the playground, park, or other grassy area to search for spiderwebs. Have them identify as many different types of webs as possible. Then have students produce sketches of the various webs they examined.

Teens in SCIENCE

Purpose

Teens in Science extends the discussion of chemical reactions inside people in Section 18-1 by describing an observed effect of radiation upon some human cells.

Content Background

X rays are electromagnetic waves (like light waves) that are extremely short and have very high energy. X rays are used to make images of the body's interior because they pass through tissues to varying degrees. However, their high energy can damage tissue and promote growth of cancer cells, so as much of the body as possible is shielded when X rays are being taken. A shield like the dentist's apron is made of lead, which absorbs X rays and won't let them pass through. It is especially important to shield the reproductive organs from X rays, as this can result in damage to offspring.

Teaching Strategy

Ask students to name ways in which X rays are used. These include medical diagnosis, killing cancer cells, detectors at airports, inspection of metals for stress cracks, detecting art forgeries, and studying crystal structures.

Demonstration

Borrow an X-ray film from a radiologist (or find a reproduction in a book) and show it to the class. Have students carefully study the image and explain that they are seeing a negative. Dark areas received more X rays that passed through softer tissues, whereas white areas (bone) received very few X rays because X rays couldn't easily pass through bone.

Teens in SCIENCE

Cellular Fun

Have you ever wondered exactly why your doctor or dentist uses certain types of equipment?

Edie Shin, a 17-year-old high school senior in Orland Park, Florida, did more than wonder. "I noticed that my dentist always covered me with a lead apron before she took X rays of my mouth. I decided to find out why she takes such great precautions with X rays. In my chemistry class, we'd been talking about the effects of radiation on human cells. I wanted to find out just how harmful X rays can be."

Irradiating Cells

A local hospital gave Edie a supply of human cells that had come from the throat culture of a healthy 50-year-old man. Edie used the hospital's equipment to expose the throat cells to a minimum of 50 RADS of radiation every week for nearly a year. Each exposure equaled nearly 250 times the amount of radiation that you are exposed to during regular dental X rays.

Edie soon discovered that the cells she had irradiated looked very different. "Many cells died over the course of the experiment. But the cells that survived grew as much as ten times their normal size. Some had as many as six nuclei."

Results of Radiation

Edie came to the conclusion that when precautions such as using the lead apron are taken, normal dental X rays do not harm human cells. However, when human cells are exposed to high levels of radiation over a long period of time, a series of chemical reactions occur in the cellular water of normal cells. This can result in mutations.

At first, Edie wasn't sure what was going to happen. She found that, to her, the best thing about science was that the investigation and questioning were themselves rewards.

You Try It!

Make a list of any medical or dental equipment that you would like to understand better.

Choose one item from your list. Telephone your doctor's or dentist's office and ask him or her to explain why this equipment is used.

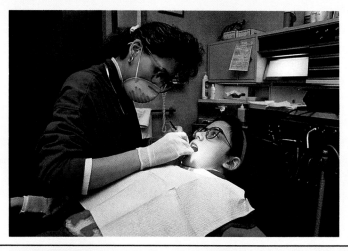

Going Further ▹

Medicine today has several ways of looking inside the body in addition to X rays. These include MRI, CAT scans, ultrasound, and PET scans (positron emission tomography). Have student teams select one of these methods to research. Then have them share their findings with the class in the form of posters. Use the Performance Task Assessment List for Poster in **PASC**, p. 73.

L2　COOP LEARN

Science Journal

Review the statements below about the big ideas presented in this chapter, and think about the questions. Then, re-read your answers to the Did You Ever Wonder questions at the beginning of the chapter. *In your Science Journal*, write a paragraph about how your understanding of the big ideas in the chapter has changed.

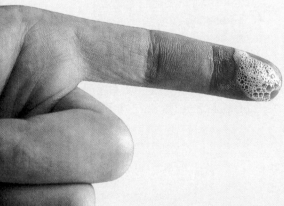

1 A chemical reaction involves the changing of substances into other substances. *How does this change differ from a physical change?*

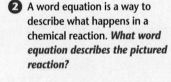

2 A word equation is a way to describe what happens in a chemical reaction. *What word equation describes the pictured reaction?*

3 Every chemical reaction involves changes in energy. Energy may be absorbed, as in an endothermic reaction. Or energy may be released, as in an exothermic reaction. *Is the pictured reaction endothermic or exothermic?*

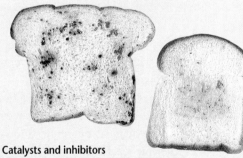

4 Catalysts and inhibitors help control the rate at which some chemical reactions take place. Catalysts and inhibitors themselves remain chemically unchanged at the end of the reaction. *Does the figure show the effect of a catalyst or does it show the effect of an inhibitor?*

Science Journal

Did you ever wonder...
• Cars today have catalytic converters coated with catalysts. Leaded gas would coat the catalysts and make them ineffective. (p. 573)
• Some are inhibitors, which slow down chemical reactions so foods stay fresh. (p. 574)
• Thermal energy is produced by a chemical reaction when food is digested and when body fat burns during respiration. (p. 567)
• Iron combines with oxygen to form iron oxide, or rust. (p. 556)

Project

Sunlight is the source of energy that promotes photosynthesis in plants. Have students test the effects of this energy source versus artificial light by giving two potted bean plants the same amount of light every day but providing one with sunlight and the other with electric light. Ask students to graph the growth. In other tests, students can vary the amount of light. Have students determine optimum light conditions for health and growth and write a brief report about their experiment. **P** **L1**

Direct students' attention to the diagrams. Use the diagrams to review the main ideas presented in the chapter. Have volunteers read each statement and question aloud. Then have students, working in groups of two or three, answer the questions.

Teaching Strategy

Write each of the following terms on an index card:
  physical change
  chemical reaction
  endothermic reaction
  exothermic reaction
  catalyst
  inhibitor
Write the definition of each term on an index card. Write an example of each on an index card. Put all the cards in a box and let pairs of students take turns matching each term with its definition and example.

Answers to Questions

1. A physical change alters the form of a substance but does not change the identity of the matter.

2. hydrogen peroxide $\longrightarrow$ water + oxygen

3. exothermic

4. an inhibitor

GLENCOE TECHNOLOGY

 **MindJogger Videoquiz**

Chapter 18 Have students work in groups as they play the Videoquiz game to review key chapter concepts.

Using Key Science Terms

1. product
2. exothermic reaction
3. inhibitor
4. Reactants are the starting substances in a chemical reaction; products are those substances formed as a result of the reaction.
5. Catalysts speed up a chemical reaction, while inhibitors slow down a chemical reaction.

Understanding Ideas

1. oxygen
2. Oxygen is needed to combine with food to produce carbon dioxide, water, and energy.
3. Exothermic reactions release energy. Endothermic reactions absorb thermal energy from the environment. The warmth of the body is evidence.
4. An inhibitor slows down a chemical reaction. Some inhibitors slow the action of enzymes that promote the growth of bacteria.
5. The dark glass will protect the contents from light, which acts as a catalyst and speeds up certain reactions that can harm the contents.

Developing Skills

1. Concept maps should show that chemical reactions have rates controlled by catalysts and inhibitors; its components are reactants and products, and energy changes are classified as endothermic or exothermic.
2. Some objects may have remained dry; others may have been treated with a rust inhibitor.
3. Carbon dioxide was added by blowing into the water. Bromothymol blue reacts to carbon dioxide by turning yellow.
4. Not enough activation energy had been supplied by the toaster for the reaction to proceed on its own.

Using Key Science Terms

catalyst   inhibitor
chemical reaction   product
endothermic reaction   reactant
exothermic reaction

For each term below, identify which key science term from the list is opposite in meaning.

1. reactant
2. endothermic reaction
3. catalyst

Use your knowledge of the key science terms to answer these questions.

4. Describe how reactants and products are a part of a chemical reaction.
5. Describe how catalysts and inhibitors affect a chemical reaction.

Understanding Ideas

Answer the following questions in your Journal using complete sentences.

1. What is the common reactant in the chemical reactions that take place when an apple turns brown, tools rust, or newspaper turns yellow?
2. Explain why it is necessary to have oxygen to digest food.
3. Differentiate an endothermic reaction and an exothermic reaction. What evidence do you have that respiration is an exothermic reaction?
4. What is an inhibitor? How might an inhibitor be useful in medicine?
5. Why are some products kept in dark glass containers?

Developing Skills

Use your understanding of the concepts developed in the chapter to answer each of the following questions.

1. **Concept Mapping** Create a concept map of chemical reactions using the following terms: *catalyst, chemical reactions, components, endothermic, energy, exothermic, inhibitor, product, rate, reactant.*
2. **Observing and Inferring** Continue to think about the Explore activity on page 551. Why do some iron products rust? Why don't other iron products rust? Make a list of factors that may have caused some products to rust faster than others.
3. **Recognizing Cause and Effect** Repeat the Find Out activity on page 554. Using two jars with one cup of distilled water in each, put the nozzle of a bicycle pump in one jar of water and gently pump air through it for 10 strokes. In the other jar, use the straw to gently blow air into the water for 10 seconds. Add 10 drops of bromothymol blue solution to each jar. What do you observe? What caused the color change?
4. **Sequencing** You know that burning is an exothermic reaction. If this is true, why doesn't toast continue to get darker after it is removed from the toaster?
5. **Interpreting Data** Two identical pieces of apple were left out in the air. One piece was first dipped in lemon juice. Based on the results in the table on the following page, do you think lemon juice

5. The lemon juice is an inhibitor; it prevented the apple from reacting with oxygen.

is a catalyst, or is it an inhibitor? Explain your reasoning.

Time	With Lemon Juice	Without Lemon Juice
10 min	no browning	edges brown
20 min	no browning	more browning
30 min	no browning	surface brown

Critical Thinking

In your Journal, *answer each of the following questions.*

1. How do catalytic converters reduce air pollution?
2. Food eaten has chemical potential energy. Explain why some athletes eat food high in carbohydrates before an event.
3. When water is broken down into hydrogen and oxygen, how do the products differ from the reactants?

4. What two factors might be responsible for bananas ripening more slowly if they are kept in the refrigerator?
5. If your doctor prescribed an inhibitor, how would it affect how your nerves transmit impulses across the synapse?

Problem Solving

Read the following problem and discuss your answers in a brief paragraph.

You are a highway engineer assigned to maintain your company's equipment in good condition.

1. How can you use your knowledge of chemical reactions to design a maintenance program that will protect the equipment from rust?
2. Your work site is near the sea. What other factors in the environment might you need to consider?

CONNECTING IDEAS

Discuss each of the following in a brief paragraph.

1. **Theme—Systems and Interactions** In photosynthesis, is sunlight a catalyst or a reactant?
2. **Theme—Systems and Interactions** When matter changes chemically, it is neither created nor destroyed; only its form is changed. Identify some chemical changes that you notice from day to day.
3. **Theme—Energy** Explain how gasoline has chemical potential energy.
4. **A Closer Look** Write a word equation for the burning of propane to form water and carbon dioxide. Be sure to indicate whether heat is needed for the reaction to proceed or if the reaction gives off heat.
5. **Life Science Connection** What is the word equation for the chemical reaction that changes sugar to energy in the cell? Is the reaction exothermic or endothermic? Explain your answer.

Critical Thinking

1. Catalytic converters get rid of most of the harmful exhaust and unburned fuel by speeding up their oxidation.
2. Food is stored energy, and eating more carbohydrates before an event provides more energy to use.
3. The reactant is a liquid compound made of two elements; the products are two single element gases.
4. The cold slows down reactions that require heat to occur. Also, the refrigeration might prevent certain gases needed for ripening from reacting with the bananas.
5. The inhibitor would make it more difficult for the impulses to be transmitted, because it would slow down the impulses.

Problem Solving

1. Keep all metal surfaces painted with rust-proof paint. Store equipment in a shed at night, keeping exposure to water at a minimum.
2. salt and moisture present

Connecting Ideas

1. catalyst
2. Answers will vary but might include rusting of metal and respiration.
3. Gasoline is a fossil fuel and thus contains stored energy. This energy is released when the fuel is burned in a car's engine.
4. propane + oxygen $\xrightarrow{heat}$ water + carbon dioxide + energy
5. sugar + oxygen $\xrightarrow{enzyme}$ carbon dioxide + water + energy; exothermic, because it gives off energy.

✔ Assessment

Portfolio Review the portfolio options that are provided throughout the chapter. Encourage students to select one product that demonstrates their best work for the chapter. Have students explain what they learned and why they chose this example for placement into their portfolios.

Additional portfolio options can be found in the following **Teacher Classroom Resources:**

Multicultural Connections, pp. 39, 40

Making Connections: Technology and Society, p. 39
Concept Mapping, p. 26
Critical Thinking/Problem Solving, p. 26
Take Home Activities, p. 27
Performance Assessment P

Chapter Organizer

SECTION	OBJECTIVES	ACTIVITIES & FEATURES
Chapter Opener		**Explore!**, p. 583
19-1 Traffic In and Out of Cells (4 sessions, 1.5 blocks)	1. **Describe** the function of the cell membrane. 2. **Explain** how materials move in and out of cells. 3. **Compare and contrast** osmosis and diffusion. **National Content Standards: (5-8) UCP2, UCP4, A1-2, B2-3, C1, C3, D1, E2**	**Find Out!**, p. 584 **Explore!**, p. 586 **Find Out!**, p. 588 **Skillbuilder:** p. 589 **Design Your Own Investigation 19-1:** pp. 590–591 **A Closer Look**, pp. 592–593
19-2 Why Cells Need Food (2 sessions, 1 block)	1. **Explain** the importance of energy to cells. 2. **Describe** the process of respiration in terms of its products and reactants. 3. **Relate** the number of mitochondria in different types of cells to their levels of activity. **National Content Standards: (5-8) UCP2, A1, B3, C1, C3**	**Find Out!**, p. 596 **Investigate 19-2:** pp. 598–599 **Chemistry Connection**, pp. 600–601 **Science and Society**, p. 607 **Health Connection**, p. 608 **How It Works**, p. 609
19-3 Special Cells with Special Jobs (2 sessions, 1 block)	1. **Explain** the differences between types of cells and how their differences are related to cell functions. 2. **Identify** the levels of organization in life-forms from cell to tissue, to organ, to organ system, to organism. **National Content Standards: (5-8) UCP1, A1, C1, F1, F4-5**	

ACTIVITY MATERIALS

EXPLORE!

p. 583 scissors, green onion, 2 containers, distilled water, saturated salt water
p. 586* tea bag, clear glass, hot water

INVESTIGATE!

pp. 598–599 2 rubber stoppers with plastic tubing inserted, 2 test tubes, 20 mL 25% sucrose solution, metric ruler, 2 flasks, yeast cubes, glass-marking pencil, watch or clock, tap water, graduated 10-mL cylinder

DESIGN YOUR OWN INVESTIGATION

pp. 590–591 2 raw eggs, 250-mL glass jars with lids(2), 400 mL white vinegar, 200 mL distilled water, 200 mL syrup, graduated cylinder, balance, paper towels, wax pencil, goggles

FIND OUT!

p. 584* cheesecloth, sand and gravel, funnel, stirring rod, 2 glass jars, water
p. 588 dialysis membrane bag, 1/4 cup cooked rice, 200-mL beaker, iodine solution
p. 596* 2 clear glass jars, 2 thermometers, 2 large balls of cotton, 50 dry kidney beans, 50 kidney bean seeds that have been soaked overnight

KEY TO TEACHING STRATEGIES

The following designations will help you decide which activities are appropriate for your students.

- **L1** Basic activities for all students
- **L2** Activities for average to above-average students
- **L3** Challenging activities for above-average students
- **LEP** Limited English Proficiency activities
- **COOP LEARN** Cooperative Learning activities for small group work
- **P** Student products that can be placed into a best-work portfolio
- Activities and resources recommended for block schedules

Need Materials? Call Science Kit (1-800-828-7777).

[00:00] OUT OF TIME? We recommend that students do the activities with an asterisk.

Chapter 19 How Cells Do Their Jobs

TEACHER CLASSROOM RESOURCES

Student Masters	Transparencies
Study Guide, p. 63 **Critical Thinking/Problem Solving**, p. 5 **Activity Masters**, Design Your Own Investigation 19-1, pp. 79–80 **Science Discovery Activities**, **19-1** **Multicultural Connections**, p. 41 **Laboratory Manual**, pp. 113–114, Diffusion	**Teaching Transparency 37**, Movement of Molecules **Teaching Transparency 38**, Plasmolysis and Equilibrium **Section Focus Transparency 57**
Study Guide, p. 64 **Concept Mapping**, p. 27 **Making Connections: Across the Curriculum**, p. 41 **Take Home Activities**, p. 28 **Activity Masters**, Investigate 19-2, pp. 81–82 **Critical Thinking/Problem Solving**, p. 5 **Science Discovery Activities**, **19-2** **Laboratory Manual**, pp. 115–116, Cell Respiration **Multicultural Connections**, p. 42	**Section Focus Transparency 58**
Study Guide, p. 65 **Making Connections: Integrating Sciences**, p. 41 **Making Connections: Technology & Society**, p. 41 **Critical Thinking/Problem Solving**, p. 27 **Science Discovery Activities**, **19-3**	**Section Focus Transparency 59**

ASSESSMENT RESOURCES	TEACHING & TECHNOLOGY
Review and Assessment, pp. 113–118 **Performance Assessment**, Ch. 19 **PASC*** **MindJogger Videoquiz** **Alternate Assessment in the Science Classroom** **Computer Test Bank**	**Spanish Resources** **Cooperative Learning Resource Guide** **Lab and Safety Skills** **Science Interactions, Course 2, CD-ROM** **Computer Competency Activities**

*Performance Assessment in the Science Classroom

NATIONAL GEOGRAPHIC TEACHER'S CORNER

Index to National Geographic Magazine	National Geographic Society Products Available From Glencoe	Additional National Geographic Society Products
The following articles may be used for research relating to this chapter: • "The Awesome Worlds Within a Cell," by Rick Gore, September 1976.	To order the following products for use with this chapter, contact your local Glencoe sales representative or call Glencoe at 1-800-334-7344: • *STV: The Cell* (Videodisc) • *Newton's Apple Life Sciences* (Videodisc)	To order the following products for use with this chapter, call the National Geographic Society at 1-800-368-2728: • *Everyday Science Explained* (Book) • *Discovering the Cell* (Video) • *Photosynthesis: Life Energy* (Video)

Teacher Classroom Resources

These are key components of the classroom resources package.

TEACHING AIDS

Section Focus Transparencies

57 SECTION FOCUS TRANSPARENCY — Section 19-1

YOU CAN'T LIVE ON JUNK FOOD
Potato chips may not be a wise desert dessert.

THE FAR SIDE — By GARY LARSON

"Uh-oh, I've got a feeling I shouldn't have been munching on these things for the last mile."

1. Why aren't potato chips a good choice in this situation?
2. What other foods may not be smart to eat in the desert? Why?

L1

58 SECTION FOCUS TRANSPARENCY — Section 19-2

CONSUMER'S DELIGHT
Some survivors don't need anyone. Some need help to get by. In this picture, some of the living things shown can make their own food. They are called producers. Other living things depend on these producers for their survival. They are called consumers.

1. Name the producers and consumers in this picture. Which producers do the consumers need to survive?
2. Are you a consumer or a producer? Explain your answer.
3. If all plants died, could any animals survive? Why? What if all animals died? Could any plants survive? Explain.

L1

59 SECTION FOCUS TRANSPARENCY — Section 19-3

THE RIGHT TOOL FOR THE RIGHT JOB
In kitchens across our country, food preparation is made easier by having the proper utensils available. Handy gadgets are made with particular uses in mind. Here are some examples.

1. How do these kitchen utensils make food preparation easier?
2. How might our body cells be adapted for their special assignments?

L1

Teaching Transparencies

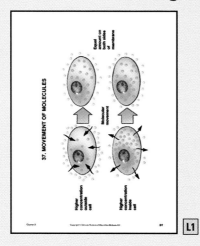

37. MOVEMENT OF MOLECULES

L1

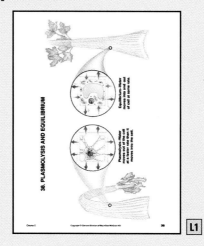

38. PLASMOLYSIS AND EQUILIBRIUM

L1

HANDS-ON LEARNING

Science Discovery Activity*

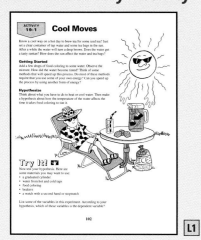

ACTIVITY 19-1 — **Cool Moves**

Know a cool way on a hot day to brew tea for some iced tea? Just set a clear container of tap water and some tea bags in the sun. After a while the water will turn a deep brown. Does the water get a tasty suntan? How does the sun affect the water and tea bags?

Getting Started
Add a few drops of food coloring to some water. Observe the mixture. How did the water become tinted? Think of some methods that will speed up this process. Do most of these methods require that you use some of your own energy? Can you speed up the process by using another form of energy?

Hypothesize
Think about what you have to do to heat or cool water. Then make a hypothesis about how the temperature of the water affects the time it takes food coloring to tint it.

Try It!
Now test your hypothesis. Here are some materials you may want to use:
• a graduated cylinder
• water from hot and cold taps
• food coloring
• beakers
• a watch with a second hand or stopwatch

List some of the variables in this experiment. According to your hypothesis, which of these variables is the dependent variable?

102

L1

Laboratory Manual*

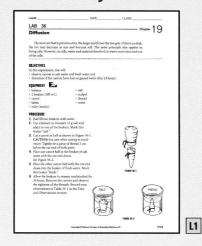

NAME _____ DATE _____ CLASS _____

LAB **36** — Chapter **19**
Diffusion

The more air that is put into a tire, the larger and firmer the tire gets. If there is a leak, the tire may decrease in size and become soft. The same principle also applies to living cells. However, in cells, water and material dissolved in water move into and out of the cells.

OBJECTIVES
In this experiment, you will
• observe carrots in salt water and fresh water and
• determine if the carrots have lost or gained water after 24 hours.

EQUIPMENT
• balance   • salt
• 2 beakers (500 mL)   • scalpel
• carrot   • thread
• labels   • water
• ruler (metric)

PROCEDURE
1. Half fill two beakers with water.
2. Use a balance to measure 15 g salt and add it to one of the beakers. Mark this beaker "salt."
3. Cut a carrot in half as shown in Figure 36-1. CAUTION: Use care when cutting to avoid injury. Tightly tie a piece of thread 2 cm below the cut end of both parts.
4. Place one carrot half in the beaker of salt water with the cut end down. See Figure 36-2.
5. Place the other carrot half with the cut end down into the beaker of fresh water. Mark this beaker "fresh."
6. Allow the beakers to remain undisturbed for 24 hours. Remove the carrots and observe the tightness of the threads. Record your observations in Table 36-1 in the Data and Observations section.

FIGURE 36-1

SALT   FRESH

FIGURE 36-2

113

L1

Take Home Activity

Chapter 19

RESPIRATION

How can you see the products created by the respiration of your body's billions of cells? When you breathe, you supply your cells with the oxygen they need to carry out their jobs. The products of cell respiration are eventually given off and carried away by the circulatory system. Where do these products go? You probably know that one of the products, carbon dioxide, is present in the air you exhale. Energy, another product of respiration, is given off as heat. You notice it easily after you have been exercising. But what about the third product of respiration—water?

You need: a small mirror

• Breathe onto the mirror.

• What happens?
• Where does this mist come from?
• Where do you suppose it goes when it leaves the mirror?
• What is this form of water called and why can't we see it?
• Can you think of other examples of water droplets in the air?
• Plants carry out respiration also. The byproducts of plant respiration are the same as those from humans. Have you ever observed a closed terrarium containing plants? What indirect evidence of respiration did you see?

28

L1

*There may be more than one activity for this chapter.

Chapter 19 How Cells Do Their Jobs

REVIEW AND REINFORCEMENT

Study Guide*

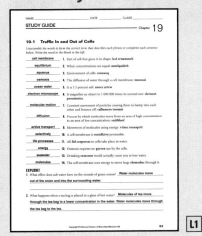

Concept Mapping

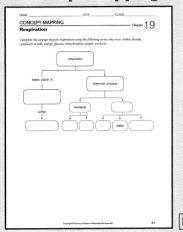

Critical Thinking/ Problem Solving

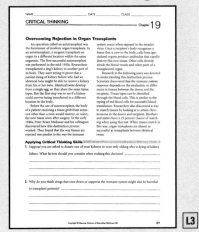

ENRICHMENT AND APPLICATION

Integrating Sciences

Across the Curriculum

Technology and Society

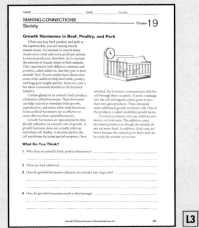

Multicultural Connection**

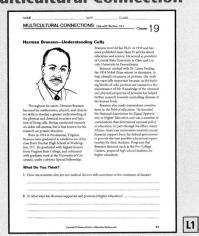

ASSESSMENT

Performance Assessment

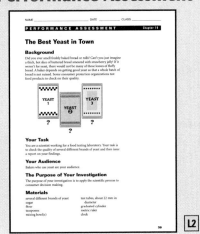

Review and Assessment

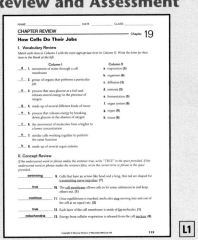

How Cells Do Their Jobs

HOW CELLS DO THEIR JOBS

THEME DEVELOPMENT

The themes that are supported by this chapter are systems and interactions and energy. Interactions among molecules on either side of a cell membrane cause materials to move into and out of the cell. Various levels of organization exist in life-forms; the well-being of an organism is dependent on the interactions of these levels.

CHAPTER OVERVIEW

In this chapter, students explore the nature of the cell membrane, learn the importance of energy to the cell, and explore differences in types of cells.

Tying to Previous Knowledge

Have students draw on their knowledge of cell structures to make a drawing of a plant cell and list the structures that do work. Explain to students that they will discover just how these structures obtain the energy they need to do their jobs.

INTRODUCING THE CHAPTER

Encourage students who swam in ocean water to describe the water's taste. Develop the idea that water-dwelling creatures are adapted to either a freshwater or saltwater environment.

Did you ever wonder...

✓ **Why rice gets soft when it is cooked?**

✓ **Why you get hot when you run fast?**

✓ **How you can make a cup of tea without stirring?**

Science Journal

Before you begin to study about how cells do their jobs, think about these questions and answer them *in your Science Journal*. When you finish the chapter, compare your journal write-up with what you have learned.

Answers are on page 610.

Have you ever been swimming in the ocean? If so, you know that ocean water is quite different from water found in a lake or river. It's salty! Ocean water is about 3.5 percent salt, while freshwater bodies such as lakes may contain less than 0.005 percent salt.

You may know that it is deadly to place saltwater fish in a freshwater tank and freshwater fish in a saltwater tank. Why can some organisms only live in salt water? What happens to cells in different environments?

In Chapter 17, you learned that organisms are made of cells. Most cells are "on duty" 24 hours a day, every day, taking in nutrients and giving off products. How can cells do all the things they do? This chapter will help you find out.

▶ *In the activity on the next page, explore how salt affects living things.*

582

| 00:00 | **OUT OF TIME?** |

If time does not permit teaching the entire chapter, use the Chapter Overview on this page, Reviewing Main Ideas at the end of the chapter, and the Chapter 19 audiocassette to point out the main ideas of the chapter.

Learning Styles		
Kinesthetic	Find Out, pp. 588, 596; Design Your Own Investigation, pp. 590-591
Visual-Spatial	Explore, pp. 583, 586; Find Out, pp. 584, 588; Visual Learning, pp. 585, 587, 589, 595, 597, 601, 603; Activity, p. 589; Demonstration, p. 593
Interpersonal	Activity, pp. 585, 603; Sharing Research, p. 589; Multicultural Perspectives, p. 604
Intrapersonal	Enrichment, pp. 586, 595; A Closer Look, pp. 592-593; Science at Home, p. 610
Logical-Mathematical	Investigate, pp. 598-599; Activity, p. 600; Chemistry Connection, pp. 600-601
LS **Linguistic**	Discussion, p. 604; Across the Curriculum, pp. 595, 604; Visual Learning, p. 605; Enrichment, p. 605

Explore! ACTIVITY

How does salt affect living things?

What To Do

1. Use a pair of scissors to cut a 6-cm piece from the green end of a green onion.

2. Cut one end of the section into thin strands.

3. Dip the cut end of the onion into a container of distilled water. Wait about four minutes and watch what happens. Record your observations *in your Journal*.

4. Take the onion and put it in a container of salt water.

5. *In your Journal* tell what happened to the strands of green onion? If the onion changed shape, what might have caused the change?

Explore!

How does salt affect living things?

Time needed   10 minutes

Materials   scissors, green onion, 2 containers, distilled water, saturated salt water

Thinking Processes observing and inferring, comparing and contrasting

Purpose   To observe the effects of movement of molecules within living tissue.

Preparation   Label the containers "distilled water" and "salt water."

Teaching the Activity

Point out that the concentration of water molecules is greater in distilled water.

Science Journal Suggest that students make sketches in their journals of the way the onion strands look after sitting in salt water and in distilled water. **L2**

Expected Outcome

Students will observe the onion strands spread out in distilled water and return to nearly normal shape in salt water.

Answer to Question

5. The onion strands spread out in distilled water and returned to nearly normal shape in salt water. The changes were due to the movement of water molecules.

✔ Assessment

Performance Explain that chefs sometimes salt foods such as eggplants or cucumbers and let them sit for a while before using them in recipes. Ask students to design an experiment to show the effect of salting these foods. Then have students write a brief explanation of a chef's purpose in doing this. *Salting draws water out of these foods and makes them less soggy.* Use the Performance Task Assessment List for Designing an Experiment in **PASC**, p. 23.

ASSESSMENT PLANNER

PORTFOLIO
Refer to page 612 for suggested items that students might select for their portfolios.

PERFORMANCE ASSESSMENT
Process, pp. 586, 596
Skillbuilder, p. 589
Explore! Activities, pp. 583, 586
Find Out! Activities, pp. 584, 588, 596
Investigate, pp. 590–591, 598–599

CONTENT ASSESSMENT
Oral, p. 588
Check Your Understanding, pp. 593, 602, 606
Reviewing Main Ideas, p. 610
Chapter Review, pp. 611–612

GROUP ASSESSMENT
Opportunities for group assessment occur with Cooperative Learning Strategies.

19-1

PREPARATION

Planning the Lesson

Refer to the Chapter Organizer on pages 582A–D.

Concepts Developed

In this section, students investigate cell processes and gain an understanding of the nature of the cell membrane.

1 MOTIVATE

Bellringer

 Before presenting the lesson, display **Section Focus Transparency 57** on an overhead projector. Assign the accompanying **Focus Activity** worksheet. L1

LEP

Find Out!

What substances can pass through a barrier?

Time needed 20 minutes

Materials double layer of cheesecloth, small amount of sand and gravel, funnel, stirring rod, two glass jars, water

Thinking Process observing and inferring

Purpose To make a model of a selectively permeable material.

Preparation Gauze can be substituted for the cheesecloth.

Teaching the Activity

Discussion Have students predict what will occur when they pour the mixture through the funnel. Ask students to explain the basis for their predictions. **LEP**

Science Journal After stu-

19-1 Traffic In and Out of Cells

Section Objectives

- Describe the function of the cell membrane.
- Explain how materials move in and out of cells.
- Compare and contrast osmosis and diffusion.

Key Terms

diffusion
osmosis

The Cell Membrane

You have now observed that onion strands soaked in salt water appear different from those soaked in distilled water. Somehow the salt in the salt water affected the cells. Did the salt coat the cells? Did it enter the cells? Did the water enter the cells? What caused the onion strands to change as they did in the presence of salt? To answer that question, you'll need to think back to Chapter 17, where you learned that all cells are covered by a thin cell membrane. The cell membrane gives the cell its shape. In order to live, a cell must obtain certain materials from its environment and release other materials. Given that a cell has a membrane, how is it possible for these materials to enter and leave the cell?

Find Out! ACTIVITY

What substances can pass through a barrier?

What To Do

1. Obtain a double layer of cheesecloth, a small amount of sand and gravel, a funnel, a stirring rod, and two glass jars.

2. Put the sand and gravel in one jar and add enough water to cover them by about 1 cm. Stir the mixture thoroughly.

3. Place the layers of cheesecloth inside the funnel.

4. While holding the funnel over an empty jar, pour your mixture through. Record your observations in your Journal.

5. Wait several minutes and remove the cheesecloth from the funnel. Inspect the cheesecloth's contents. Record your findings *in your Journal.*

Conclude and Apply

Describe what happened. What was the job of the cheesecloth?

---

dents have recorded their findings in their journals, have them suggest ways that a barrier such as cheesecloth might be used by cells. L1

Expected Outcome

Students should observe that only certain substances passed through the cheesecloth.

Conclude and Apply

Students will find larger particles of sand and gravel in the cheesecloth. The cheesecloth fil-

tered larger particles from the mixture.

✔ Assessment

Content Ask students to think of ways that filters or barriers are used in everyday life. Have students draw cartoon panels to illustrate their responses. Use the Performance Task Assessment List for Cartoon/Comic Book in **PASC,** p. 61.

The cell membrane

Even if you use a simple light microscope, such as the one in your classroom, you will not be able to see the details of a cell membrane's structure. How then do we know about the structure of the cell membrane?

One time or another,

you've probably experienced the thrill of pulling and pushing metal objects with the invisible force supplied by a magnet. In the 1920s, scientists discovered that beams of tiny particles called electrons could be pushed or pulled with a very powerful magnet. Years later, researchers put this knowledge

to use by building a new type of microscope called the electron microscope.

This microscope is much more powerful than a light microscope and can magnify an object over 1 000 000 times its normal size. Using it, we are able to see the microscopic structures of a cell membrane.

Activity To help students observe diffusion, use this activity.

Give small groups of students beakers of water at room temperature. Have them add 1 tsp of instant coffee or powdered soft drink and observe what happens to the water. *It changed color.* **What does this indicate?** *The powder molecules move through the water via diffusion—moving from high to low concentration.*
COOP LEARN L1

2 TEACH

Tying to Previous Knowledge

Have students recall from Chapter 17 how an electron microscope differs from the light microscopes used in the classroom.

Suggestions for further reading about the electron microscope are *Window on the Unknown* by Corinne Jackson, *Through the Microscope* by Ron Taylor, and *Atoms and Cells* by Lionel Bender.

Figure 19-1

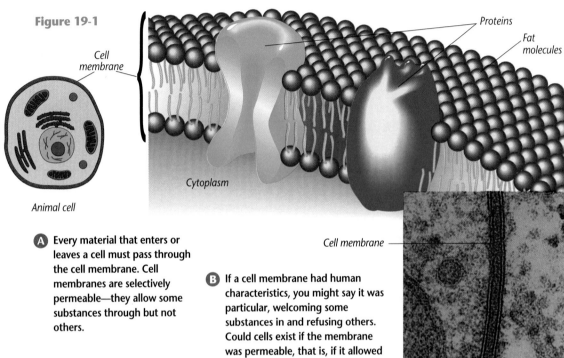

Cell membrane

Proteins

Fat molecules

Cytoplasm

Animal cell

Cell membrane

Electron microscope view

A Every material that enters or leaves a cell must pass through the cell membrane. Cell membranes are selectively permeable—they allow some substances through but not others.

B If a cell membrane had human characteristics, you might say it was particular, welcoming some substances in and refusing others. Could cells exist if the membrane was permeable, that is, if it allowed all substances to pass through?

Visual Learning

Figure 19-1 Explain that the cell membrane is flexible and that the large protein molecules move around much like icebergs in the ocean. Show a tray with a layer of red plastic balls with a few yellow balls representing protein molecules. As students run their hands over the balls, the yellow balls will change position. **Could cells exist if their membranes were permeable, that is, if they allowed all substances to pass through?** *No; if cell membranes were permeable, all of the cell structures and vital fluids could pass out of the cell.*
L1

■ Modeling the Cell Membrane

You have just seen that all of the sand passed through the cheesecloth, while most of the gravel was held back. The thin cell membrane that covers every cell, shown in **Figure 19-1**, works in a similar way to the cheesecloth.

The cell membrane permits certain molecules to pass in or out, depending on the type and size of the molecules. Water, oxygen, and carbon dioxide molecules pass through the

19-1 Traffic In and Out of Cells **585**

Meeting Individual Needs

Visually Impaired Have visually impaired students compare the texture of the sand and gravel prior to use in the Find Out activity. At the conclusion of the activity, have students feel the cheesecloth that lined the funnel. Encourage students to use these tactile observations to explain what occurred.

ENRICHMENT

Discussion Challenge students to describe ways to alter the materials used in the Find Out activity so that all of the gravel and sand will pass through the cheesecloth. You may also have students redesign the activity so all of the sand and gravel would be filtered from the mixture. Allow interested students to test their redesigned activities. L2

Explore!

How do tea bags work?

Time needed 20 minutes

Materials tea bag, clear glass, hot water

Thinking Process observing and inferring

Purpose To observe the results of molecular motion.

Preparation You may extend this activity by having students place another tea bag in a large jar of hot water.

Teaching the Activity

Discussion As students observe the color change in the water, ask them to note the point at which the water stops getting darker. Develop the idea that there is a limit to the amount of tea that will be taken up by the water. **LEP**

Science Journal Have students write their findings in their journals. **L1**

Expected Outcome

Due to prior knowledge, students will likely predict that the tea will leave the bag. However, they will need to apply their understanding of molecular motion to explain how this occurred.

Answers to Questions

2. The water changes from clear to the color of the tea. Molecules of tea pass through the bag into the surrounding water.

✔ Assessment

Process Ask students to predict what would happen if they placed a tea bag in cold water. *It takes longer for the tea to spread through cold water.* Then have pairs of students design and carry out an experiment to determine the effect of cold and hot water on the tea bag. Use the Performance Task Assessment List for Designing an Experiment in **PASC,** p. 23. **COOP LEARN**

---

cell membrane easily while other substances, such as sugars or sodium, are stopped or slowed down. Entry of

Figure 19-2

As molecules move, they bump into one another. The collision of two molecules causes both to change directions and move away from each other.

these kinds of substances may require energy on the part of the cell membrane to get them in or out of the cell, or may make use of special molecules in the cell membrane itself.

All matter is made up of molecules that are constantly moving. A diagram of molecular motion is shown in **Figure 19-2.** By bumping into each other and bouncing off, molecules move from an area where they are crowded together to places where there are fewer of them. Molecular motion like this occurs in the air and in the cytoplasm of your cells.

Explore! ACTIVITY

How do tea bags work?

Would you believe that a cup of tea can help you understand something about molecular motion? Try this activity.

What To Do

1. Place a wet tea bag in a clear glass of hot water. Without stirring, carefully watch where the tea color first appears. Wait two minutes, then describe the glass *in your Journal.*

2. Make an observation every 2 minutes for 10 minutes. What change in water color do you notice? How do you think the tea color got into the water?

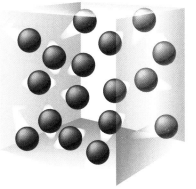

In the Explore activity, you made a model that demonstrated molecular movement. The bag containing tea represented a cell membrane. When you placed the tea bag in the hot water, molecules that make up tea moved from inside the bag out into the water. Likewise, the water moved

through the bag to the tea leaves. Why did these molecules move? As you learned in Chapter 6, the process of convection helped the tea move from the tea bag and throughout the glass because the hot water was in motion. The movement of molecules was also involved.

586 Chapter 19 How Cells Do Their Jobs

⎛Program Resources⎞

Study Guide, p. 63
Laboratory Manual, pp. 113–114, Diffusion **L1**
Science Discovery Activities, 19-1
Section Focus Transparency 57

ENRICHMENT

Research Have students investigate and report to the class on techniques of electron microscopy. Students may wish to include diagrams with their presentations. Use the Performance Task Assessment List for Oral Presentation in **PASC,** p. 71. **COOP LEARN** **L3**

Movement of Molecules

Molecules are in constant motion, causing many substances to move in and out of cells. This constant movement plays a role in changing the concentration of materials inside and outside the cells. **Figure 19-3** shows how the state of equilibrium is achieved when the concentrations of molecules inside and outside of the cell are not the same. When the concentrations become equal inside and outside of the cell, the molecules continue back and forth through the membrane at an equal rate in each direction.

■ Diffusion

The process by which the constant motion of molecules causes movement from an area of high concentration to an area of low concentration is called **diffusion**. In Chapter 14, you learned about the behavior of gas molecules, and in Chapter 16, you

DID YOU KNOW?

The word *cell* comes from the Latin word *cella*, which means small room.

Figure 19-3

Some molecules move from areas of high concentration to areas of low concentration, even when those areas are on opposite sides of a cell membrane.

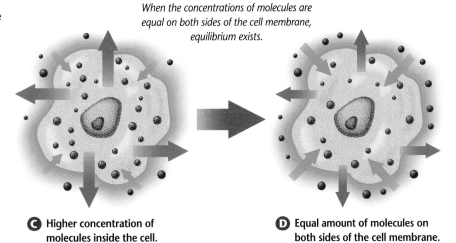

A Higher concentration of molecules outside the cell.

B Equal amount of molecules on both sides of the cell membrane.

When the concentrations of molecules are equal on both sides of the cell membrane, equilibrium exists.

C Higher concentration of molecules inside the cell.

D Equal amount of molecules on both sides of the cell membrane.

19-1   Traffic In and Out of Cells   **587**

Program Resources

Critical Thinking/Problem Solving, p. 5, Flex Your Brain
Teaching Transparency 37 L1
Teaching Transparency 38
Activity Masters, pp. 79–80, Design Your Own Investigation 19-1

learned how carbon dioxide and oxygen molecules are exchanged in the lungs by the process of diffusion. Whether diffusion involves gas molecules through air, or water molecules through a cell membrane, it occurs without the use of energy on the part of the cell.

Figure 19-4

When ink enters water, the ink molecules diffuse from areas of high concentration to areas of low concentration. The water molecules also move and become mixed with the ink molecules.

In time, the ink molecules—evenly dispersed with the water molecules—make the whole solution appear ink-colored.

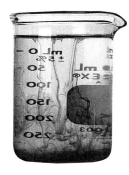

Find Out! ACTIVITY

How does diffusion occur?

What To Do

You can observe diffusion in the following activity.

1. With your teacher's help, prepare a dialysis membrane bag according to your teacher's directions.
2. Place about 1/4 cup of cooked rice in the bag.
3. Place the bag into a 200-mL beaker of iodine solution.
4. Wait five minutes and look at the rice in the bag.

Conclude and Apply

1. *In your Journal*, describe any changes you see in the rice.
2. What evidence do you have that iodine molecules diffused into the bag?

In the Find Out activity, you observed the diffusion of iodine molecules through the bag into the rice. A change in the color of the rice gives you evidence that diffusion has occurred. What would happen if you soaked uncooked rice in a pot of water overnight? Diffusion would also occur. When the dry rice grains, containing no water molecules, are placed in a container filled with water, water molecules will pass from the area of high concentration to the area of low concentration. The fact that the rice grains plump up, get soft, and change color gives you evidence that water

Figure 19-5

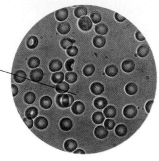

A Blood cells gained water molecules and bulged.

B Blood cells lost water molecules and shriveled.

has diffused into the grains. How does the amount of water in the pot before and after soaking indicate diffusion has occurred?

■ Osmosis: Diffusion of Water

Most cells live in an aqueous environment. In other words, they are bathed by fluids that are mostly water. This constant presence of water is important. All life processes in cells take place in water. If a cell does not receive an adequate supply of water, it will die because it cannot carry out its life processes.

Recall the Explore activity at the beginning of this chapter. You observed the onion section spread out like a fan in the distilled water. The change in the onion section was due to diffusion of water into the onion cells. When the onion was in the

distilled water, there was a higher concentration of water molecules outside the onion cells than inside. As a result, water molecules diffused into the onion cells. A similar experiment was performed on the blood cells in **Figure 19-5**. When you placed the onion in salt water, the cells lost water. This time, water molecules moved out of the onion cells and into the surrounding water.

The diffusion of water through a cell membrane is called **osmosis**. This is illustrated in **Figure 19-6**. In the following Investigate, you will observe and measure osmosis.

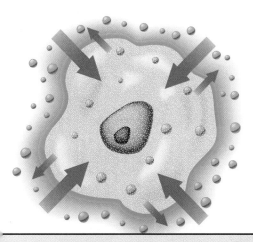

Figure 19-6

In osmosis, water molecules diffuse from an area of high concentration to an area of low concentration through a selectively permeable membrane. Eventually equilibrium is reached, and then the number of water molecules moving into the cell becomes equal to the number moving out of the cell. Osmosis requires no energy use by the cell.

SKILLBUILDER

Observing and Inferring

Make a chart of the observations you made during the last Find Out activity. Head one column *Before* and head a second column *After*. Make a third column headed Inference. Explain how the three columns are related. If you need help, refer to the **Skill Handbook** on page 642.

19-1 Traffic In and Out of Cells **589**

SKILLBUILDER

Inferences noted in the third column should describe what caused the changes noted in columns one and two. [L1]

Student Text Question
How does the amount of water in the pot before and after indicate that diffusion has occurred? *There is less water in the pot after the rice has been soaked, which indicates that water has diffused into the rice.*

Activity To demonstrate how temperature affects the rate of diffusion of molecules, have students try this activity. Have students prepare a beaker of cold water and a beaker with an equal amount of hot tap water. Have students add one fizzy antacid tablet to each beaker *at the same time*, and record how long it takes for each tablet to dissolve. Ask students to explain their observations. *The antacid tablet dissolved faster in the hot tap water. Students may conclude that hot water provides more energy for faster movement of molecules.*
COOP LEARN [L1]

Visual Learning

Figure 19-5 Ask students to recall the Explore activity, "How does salt affect living things," on page 583. **Assume that you could observe the onion cells under a microscope when placed in distilled water. Would the cells most likely resemble the blood cells in Figure A or B? Why?** *Figure A, water molecules would move into the onion cells from high to low concentration.*

ENRICHMENT

Sharing Research Discuss with students the importance of the sharing of research and discoveries among members of the scientific community. The compound microscopes students will use this year reflect the research and discoveries of different scientists working in various parts of the world. A Dutch spectacle-maker, Zacharias Janssen, discovered the principle of the compound microscope; a Dutch scientist, Anton van Leeuwenhoek, made microscope lenses that could magnify up to 270 times; and an American optical instrument manufacturer, C.A. Spencer, made the first American microscope.

19-1 Osmosis in a Cell

Preparation

Purpose To study osmosis by measuring how liquid diffuses through an egg membrane.

Process Skills making and using tables, observing and inferring, measuring in SI, comparing and contrasting, interpreting data

Time Required This lab should be started early in the week because it requires work on three days. On the day before the investigation—15 minutes to read the lab and prepare the eggs; on lab day—30 minutes to design and set up the experiment; on the following day—20 minutes to record data and answer questions.

Materials See reduced student text.

Possible Hypotheses Students may hypothesize that in one part of the experiment, water will move out of the egg and into syrup. In the other part, they may hypothesize that the distilled water will move into the egg.

The Experiment

Process Reinforcement Be sure students set up their data tables correctly. To reinforce student ability to interpret data, review how they will design the tables and record the original and final mass of the egg and the original and final volume of the liquid.

Possible Procedures On the day before, place one egg in each of two jars and add enough vinegar to cover the eggs. After 24 hours, pour off the vinegar. Observe and record what happened. Carefully remove the eggs from the jars, rinse them with water, and dry them with a paper towel. At the beginning of the actual experiment, determine the mass of the dry eggs. The mass should be recorded in the data table under original mass. Return each egg to a clean dry jar, add exactly 200 mL of distilled water to one jar and label it, and add exactly 200 mL of corn syrup to the

other jar and label it. After 24 hours, carefully remove the eggs and dry them, record the mass of the eggs, and then measure and record the remaining liquid in the beakers.

Teaching Strategies

Troubleshooting If eggs are left in vinegar longer than 24 hours, they will swell and become difficult to handle. Caution students to handle the eggs carefully.

Osmosis in a Cell

Osmosis occurs across cell membranes in plants and animals. A chicken egg can be used in the study of osmosis because osmosis can occur across the egg membrane after the shell is removed.

Preparation

Problem
How can osmosis be measured?

Form a Hypothesis
Design an experiment using eggs without shells to show osmosis. When these eggs are put in different environments, predict what will happen.

Objectives
- Model osmosis in cells using eggs.
- Measure changes related to osmosis in the eggs.

Materials
raw eggs (2)
white vinegar
small, clean beakers or jars (2)
lids
graduated cylinder
pan balance
label or wax pencil
200 mL distilled water
200 mL corn syrup

Safety Precautions

Wear goggles when pouring liquids.

Data and Observations				Sample da
Egg in Solution	Mass of Egg		Volume of L	
	Original	Final	Original	
Distilled Water (A)	mass will	increase	200 mL	1⁹
Syrup (B)	mass will	decrease	200 mL	2

Meeting Individual Needs

Learning Disabled Before beginning the Investigate activity, have students make a checklist of the procedures they will follow. For some students, you may prefer to prepare and duplicate forms listing these items:

- Obtain all equipment and materials needed.
- List the steps of the experiment.
- Make tables for collecting data. L1

Plan the Experiment

1 The day before the experiment, prepare two eggs. To remove the hard shells, cover the eggs in white vinegar and soak covered for 24 hours. *In your Science Journal*, record what happened to them. Then carefully remove the eggs from the jars, rinse them with water, and dry with a paper towel.

2 You now have two raw eggs without shells to test your hypothesis. Using the materials provided, design a step-by-step procedure to test your hypothesis about how osmosis occurs in different environments. You will need at least 24 hours to see the results of the osmosis experiment.

3 Make sure you label the beakers containing the eggs.

4 Measure the eggs. Record the data in a table similar to the one on page 590.

Check the Plan

1 What is the variable in your experiment? What is the constant? Have you kept all conditions the same except for the factor you varied?

2 What do you expect to change? The color of the egg? The mass of the egg? The volume of liquid?

3 Before you carry out your experiment, make certain your teacher approves your plan.

4 Carry out the investigation. Make observations and measure changes. Record your data.

Analyze and Conclude

1. Observe What did the vinegar do to the egg shell?

2. Compare and Contrast Compare what happened to the mass of the egg in distilled water to what happened to the mass of egg in syrup.

3. Compare and Contrast Compare what happened to the volume of distilled water to what happened to the volume of syrup.

4. Infer Infer from your data whether a substance passed through the egg membrane. What evidence do you have that supports your inference?

5. Conclude From your investigation, make a general statement about how osmosis occurs.

6. Calculate Determine how much one egg increased in mass and how much the other decreased. Do the same for the volume of the liquids. How are these numbers related?

Going Further

If you put a freshwater fish in salt water or a saltwater fish in fresh water, the fish will die. Use your knowledge of osmosis to explain why.

Discussion Discuss that some molecules cross the membrane around the egg and others do not. Water molecules pass through but not molecules of corn syrup. Remind students that osmosis is the diffusion of water molecules through a selectively permeable membrane from an area of higher water concentration to an area of lower water concentration. Students should discuss where the concentration of water molecules is greater in each experiment.

Science Journal Have students record their data in their science journals.

Expected Outcome

Students will observe that the mass of the eggs and the volumes of distilled water and corn syrup changed. Students should conclude that osmosis took place in both beakers. Water flowed into the egg in the distilled water and out of the egg in corn syrup.

Analyze and Conclude

1. dissolved the eggshell

2. The mass of the egg in the distilled water increased. The mass of the egg in syrup decreased.

3. The volume of distilled water remaining in the beaker decreased. The volume of syrup in the beaker increased.

4. The change in the mass of the eggs and in the volume of the liquids showed that a substance—water—crossed the egg membrane.

5. The direction in which osmosis occurs depends on the relative concentration of water on each side of the membrane. Water molecules move across a membrane from a greater concentration of water molecules to a lesser concentration.

6. Answers will vary, but the figures should show that as mass goes up in the egg in one beaker, the volume of liquid left in the beaker goes down. As the mass of the egg in one beaker goes down, the volume of liquid left in the beaker goes up.

✔ Assessment

Content Have groups of students identify examples of osmosis that occur in the human body. *Responses may include water moving from the blood into cells, water moving from the digestive tract into the blood, and water leaving the blood in the kidneys.* Use the Performance Task Assessment List for Group Work in **PASC**, p. 97.

COOP LEARN

Going Further ⫸

If a saltwater fish is placed in fresh water, there is a higher concentration of water molecules outside the fish. The fish will gain water and die. If the freshwater fish is placed in salt water, it will lose water and die.

Purpose

A Closer Look reinforces students' understanding of the nature of the cell membrane. This excursion also shows how protein molecules actively transport substances across the cell membrane.

Content Background

According to the fluid mosaic model, the cell membrane is primarily made up of two layers of lipids with some proteins and carbohydrate molecules scattered throughout. Free-moving proteins embedded in the lipid layers act as tubes through which substances move into or out of the cell. Other protein molecules act like pumps to push materials through the membrane. Carbohydrate molecules are located on the surface of the membrane. These molecules are attached to either proteins or lipids that cut through the membrane.

Teaching Strategy

Have pairs of students read the selection together. Challenge students to make a three-dimensional model that illustrates the fluid mosaic model. **COOP LEARN** L2

Discussion

Students often have the misconception that the cell membrane is porous, like a strainer. Ask students to explain how such a membrane structure would be harmful to the cell. *It might allow vital materials to leak out.*

What Do You Think?

Insufficient protein molecules are present in the cell membrane. As a result, glucose has no passageways to move through, so it builds up

DID YOU KNOW?

Contact lens solution must contain the same amount of dissolved salt in water as that found in the tears that bathe your eyes. Unequal concentrations would cause diffusion of water into or out of the cells on the surface of your eye.

As you observed in the last activity, osmosis occurred in the eggs. For the same reason that the eggs gained or lost water, and the blood cells bulged or shriveled, you can now explain why people who get their water from the ocean remove the salt before drinking it. If you were stranded on a deserted island, why wouldn't you drink seawater? As you guessed, drinking seawater would actually cause you to lose water.

■ Active Transport

While water can diffuse through a cell membrane, certain materials require energy to pass through it. Think back to the Find Out activity with the cheesecloth. Some of the larger pieces of gravel were too large to pass through the holes in the cheesecloth. You would have to use energy from your body to force the large pieces through the small holes.

In the same way, the cell membrane uses energy to move large molecules, such as sugar molecules, through it. If materials require energy to move through the cell membrane, active transport occurs. You will learn more about how the cell membrane uses energy to move substances in the "A Closer Look" article below.

In this section, you have examined diffusion and osmosis, two important

---

The Cell Membrane

The key to understanding how the cell membrane works lies in its structure. Scientists now know that the cell membrane not only acts as a filter, but is also a very active structure.

Structure

The cell membrane looks somewhat like a double-layered cake. Each layer is made up of a sheet of fat molecules. Larger protein molecules, which play a key role in the working of the membrane, are embedded in these layers. This model, shown on page 593, is known as the fluid mosaic model.

Some substances, such as glucose, move across the membrane by diffusing through channels made by tube-shaped protein molecules.

Active Transport

Sometimes cells require nutrients that are not in higher concentration in their environment. If the concentration of a substance outside a cell is lower than inside, the cell will lose that substance to the outside by diffusion. Cell membranes need to allow certain molecules to move in the reverse direction— from areas of low to high concentration. This type of movement requires energy and is known as active transport.

---

in the blood.

Going Further ▌▌▌▌▌▶

Have student groups place a few raisins in a beaker of water and let them stand overnight. Have students explain the process observed. *Most students will notice that the raisins absorb water and identify the process as passive.* Use the Performance Task Assessment List for Making Observations and Inferences in **PASC**, p. 17.

COOP LEARN L2

processes that depend on the movement of molecules. Both diffusion and osmosis cause materials to move through the cell membrane without the help of energy, while active transport requires energy. Through these processes, living cells obtain the substances they need as well as eliminate other materials they produce. In the next section, you will learn about other life processes that occur in living cells.

check your UNDERSTANDING

1. Name two functions of the cell membrane.
2. How are osmosis and diffusion alike? How are they different?
3. A bottle of ammonia is left open in the back of a classroom. What causes the odor of ammonia to be detected in the front of the room after only several minutes?
4. **Apply** A plant cell is surrounded by a particular substance. If the plant cell itself contains a larger concentration of this same substance than is present outside the plant cell, in which direction would you expect diffusion of the substance to occur? Explain your answer.

In active transport, molecules called carrier proteins attach to the molecules of the substance to be transported. Energy released by the cell is transferred to the carrier protein, which then changes shape. This change in shape moves the molecules into the cell.

What Do You Think?

Based on the fluid mosaic model of a cell membrane, find out what is different about the cell membrane of a diabetic.

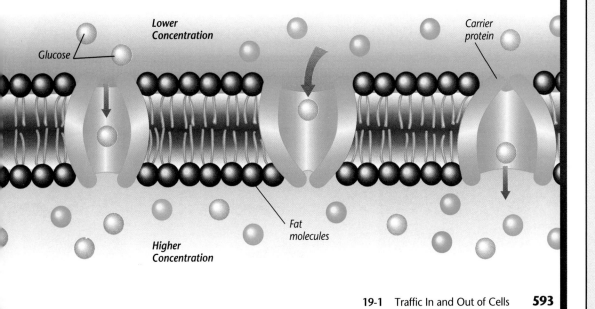

Lower Concentration

Glucose

Carrier protein

Fat molecules

Higher Concentration

19-1 Traffic In and Out of Cells **593**

check your UNDERSTANDING

1. The cell membrane allows certain substances to enter and leave the cell.
2. Both processes are types of transport that occur due to molecular motion. Diffusion can occur with any gas or liquid. Osmosis is the diffusion of water through a membrane.
3. Due to molecular motion, the ammonia gas diffuses through the air from the area of high concentration (the bottle) to an area of lower concentration (the room).
4. The molecules of the substance will move from the area of greater concentration (inside cell) to the area of lesser concentration (cell surroundings).

3 ASSESS

Check for Understanding

To help students compare concentrations inside and outside the cell, ask a volunteer to make a diagram on the chalkboard. Use Figures 19-3 and 19-6 as a model. L1

Reteach

Demonstration Place several carrot strips in beakers of distilled water and salt water. Ask students to observe and explain what occurs. *The carrot strips in distilled water (or tap water) became firm and crisp due to the movement of water into the cells. The carrot strips in salt water become soft due to the movement of water from the cells.* L1

Extension

Inference Ask students to explain why fresh fruits and vegetables in produce markets are sprinkled with water. *to keep water in the environment so it will not diffuse out of the fruits and vegetables, thus making the fruits and vegetables plump and fresh looking.*

4 CLOSE

Activity

Have students make a network tree concept map to compare movement by osmosis and diffusion. Begin with the phrase *Transport through membranes.* L1

PREPARATION

Planning the Lesson

Refer to the Chapter Organizer on pages 582A–D.

Concepts Developed

In the previous section, students explored the nature of the cell membrane and processes by which materials pass through it. In this section, students explore another cell process—cellular respiration. They discover where it occurs and explain how food and oxygen are related to it. Students also discover the role of energy within a cell.

1 MOTIVATE

Bellringer

 Before presenting the lesson, display **Section Focus Transparency 58** on an overhead projector. Assign the accompanying **Focus Activity** worksheet. [L1]

LEP

Discussion Display pictures of persons using energy—playing basketball, football, and other sports; doing aerobics; jogging, gardening; and so on. Discuss the source of their energy (food from green plants, which had captured the sun's energy). **LEP** [L1]

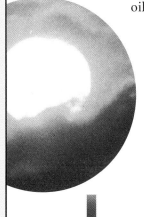

Section Objectives

- Explain the importance of energy to cells.
- Describe the process of respiration in terms of its products and reactants.
- Relate the number of mitochondria in different types of cells to their levels of activity.

Key Terms

respiration
fermentation

19-2 Why Cells Need Food

Cells and Energy

You may recall that energy can be found in many forms. If you have ever been to the ocean or a large lake on a windy day, you may have seen the motion of waves. This is an example of mechanical energy. Wet clothes hanging on a clothesline eventually dry due to thermal energy. Televisions and video games run on electrical energy.

When you think about the production of energy, you usually think of power plants. Power plants use energy-containing materials, such as oil, coal, water, or trash to produce the electrical energy that people use in their everyday activities.

Your cells, and the cells of other living organisms, also run on energy. Cells use this energy to carry out their life activities. At this very moment, your brain cells are using energy to allow you to read the words on this page. Muscle cells in your fingers are using energy to enable your fingers to turn the pages! Where does this energy come from? The original source of energy for the activities of living things is the sun. Think back to your earlier studies of plants and animals. **Figure 19-7** illustrates one path of energy transfer.

Figure 19-7

A Green plants convert light energy from the sun into sugars through the process of photosynthesis. The sugar produced contains a form of chemical energy. This same chemical energy is passed on to you through the food chain.

B A food chain is the feeding relationship that transfers energy through a community of producers, herbivores, and carnivores. In the food chain shown, which is the producer?

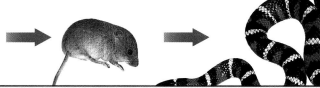

594

┌─── **Program Resources** ───┐

Study Guide, p. 64
Concept Mapping, p. 27, Respiration [L1]
Making Connections: Across the Curriculum, p. 41, Cholesterol in Your Diet [L2]
Science Discovery Activities, 19-2
Integrated Science Videodisc, Photosynthesis and Cellular Respiration
Section Focus Transparency 58

┌─── **Meeting Individual Needs** ───┐

Behaviorally Disordered For students who need a quick, hands-on demonstration of diffusion, place two or three drops of vanilla extract in a balloon. Blow up and tie the ends of this balloon and two others. Have students observe the balloons, and note any differences in them. They will be able to smell the vanilla outside the balloon as diffusion takes place. [L1]

■ Respiration

Each of your cells changes chemical energy to other forms of energy through processes such as respiration. **Respiration** is a chemical process in which glucose molecules are broken down to release energy. In many ways, a cell can be compared to a power plant. Each cell requires fuel to convert energy. A summary of this chemical process of respiration is given in **Figure 19-8**.

Figure 19-8

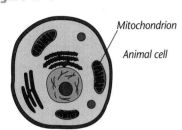

Mitochondrion

Animal cell

Organisms that depend on oxygen carry out respiration. Your brain cells, kidney cells, skin cells, and the cells in your big toe are using energy released during respiration. So are the leaves on the trees in the local park.

How can you tell if your body cells are producing energy? Feel your own forehead. It feels warm, doesn't it? What you are feeling is the heat energy produced as a product of the thousands of respiration reactions occurring in your body.

But what about plants? Plants don't feel warm. Do plants carry out respiration? In the following activity, you can prove to yourself that plants carry out respiration, and that this process converts one form of energy to another.

A Glucose is one fuel human cells use. Glucose molecules are broken apart into simpler molecules. These then enter the mitochondrion, where they combine with oxygen molecules to form water and carbon dioxide.

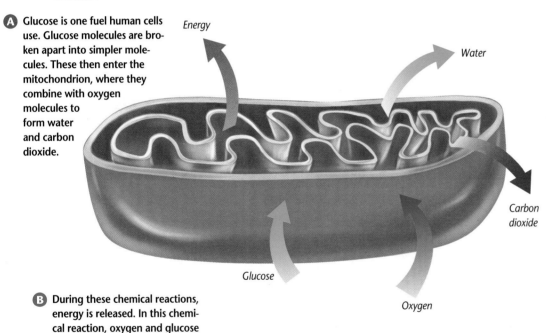

Energy

Water

Carbon dioxide

Glucose

Oxygen

B During these chemical reactions, energy is released. In this chemical reaction, oxygen and glucose are the reactants. What are the products?

19-2 Why Cells Need Food **595**

Tying to Previous Knowledge
Review the structures and functions of chloroplasts and mitochondria studied in Section 17-2.

Theme Connection The themes that this section supports are systems and interactions, and energy. A food chain shows how energy is transferred through a community by the feeding relationships of its producers, herbivores, and carnivores. All of the living members of a community need to obtain energy in this manner in order to carry out life processes.

Across the Curriculum

Daily Life
Ask students to list changes that occur in their bodies before, during, and after performing a strenuous physical activity such as playing basketball. Discuss how these changes provide evidence that respiration is occurring in their bodies. **L2**

Visual Learning

Figure 19-7 **In the food chain shown, which is the producer?** *the plant*
Figure 19-8 Have students use the figure to identify the initial materials that go into and come out of the process of respiration. *Oxygen and glucose go in; water and carbon dioxide come out.* Pose the question asked in Figure 19-8B: **In this chemical reaction, oxygen and glucose are the reactants. What are the products?** *water, carbon dioxide, and energy*

Flex Your Brain Use the Flex Your Brain activity to have students explore FOOD AND ENERGY. L1

Content Background

Emphasize that although photosynthesis takes place only in plants and certain other organisms, respiration takes place in all living things.

Find Out!

Does respiration release energy?

Time needed 3 hours

Materials two clear glass jars, two thermometers, 2 large balls of cotton, 50 dry kidney beans, 50 kidney beans that have been soaked overnight

Thinking Processes measuring in SI, recognizing cause and effect

Purpose To measure the thermal energy released during respiration.

Preparation Obtain cotton batting in a drugstore. Be sure students place 50 beans in a container of water to soak overnight.

Teaching the Activity

Troubleshooting Be sure students securely plug the top of each jar with a large wad of cotton.

Science Journal Suggest that students make a table to record their temperature readings. Have them write their observations and conclusions in their journals. L1

Expected Outcome

Students will observe that the temperature of the soaked beans rises. Students will infer that this change is due to the release of thermal energy during respiration.

Conclude and Apply
1. no
2. Respiration releases energy. When water was added to the

Find Out! ACTIVITY

Does respiration release energy?

Thermal energy is a form of energy released as a product of respiration. In this activity, you will actually measure some of the energy released by respiration in beans!

What To Do

1. Obtain the following materials: two clear glass jars, two thermometers, a large ball of cotton, 50 dry kidney beans, and 50 kidney beans that have soaked overnight.
2. Carefully place the soaked beans in one glass jar and label it.
3. Place 50 dry beans in the other glass jar.
4. Put a thermometer in each jar and seal with a large wad of cotton.
5. Take a temperature reading in each jar every half hour for three hours and record your observations *in your Journal*.

Conclude and Apply

1. Did the two different bottles have the same temperatures each time you took a reading?
2. What conclusions can you draw between your observations and your knowledge of respiration?

You saw evidence that the soaked beans released energy by respiration. You were able to measure the release of stored energy in the soaked beans by comparing the temperatures of the two treatments. Cells use the energy released by respiration in a variety of ways. Nerve cells need energy to transmit messages through the body. Plant cells need energy to form beautiful and complex flowers.

Although every living cell uses energy, the amount of energy one cell needs may differ from the amount of energy needed by a different cell. How does one cell have more energy available to it than another cell? One

beans, they began to respire much more rapidly than unsoaked seeds.

✔ Assessment

Process Have students graph their data (temperature readings) for each of their two bottles and explain how the graph supports their conclusions. Use the Performance Task Assessment List for Graph from Data in **PASC,** p. 39. P

Program Resources

Laboratory Manual, pp. 115–116, Cell Respiration L1
Critical Thinking/Problem Solving, p. 5, Flex Your Brain
Take Home Activities, p. 28, Respiration L1

answer lies in the number of mitochondria found within the cell.

The Role of Mitochondria

As you recall, mitochondria are the sites of respiration. Not all cells contain the same number of mitochondria, however. Why would there be more mitochondria present in the cytoplasm of brain cells than there are in the cytoplasm of skin cells? **Figure 19-9** addresses this question.

You have learned that energy is one product of respiration. Cells, like most factories, produce waste. The waste products of respiration are water and carbon dioxide. They are released from your body when you are exhaling. Nearly all organisms give off carbon dioxide as a result of respiration. However, the rate at which carbon dioxide is given off differs from organism to organism. In the activity that follows, you will investigate how temperature affects the rate at which respiration occurs.

Figure 19-9

The number of mitochondria per cell varies from fewer than 100 to several thousand. The greater the activity and energy use of a cell, the greater the number of mitochondria the cell contains.

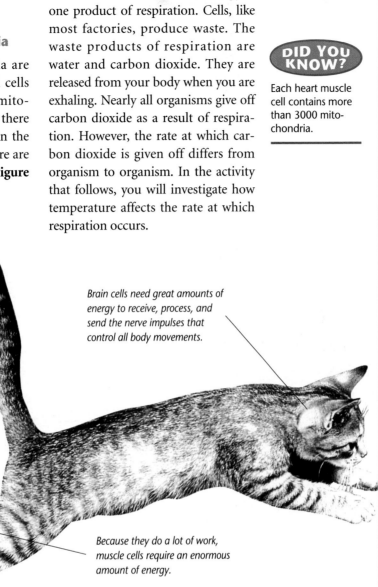

Brain cells need great amounts of energy to receive, process, and send the nerve impulses that control all body movements.

A The amount and type of work a cell has to do determine how much energy it uses.

Because they do a lot of work, muscle cells require an enormous amount of energy.

B Which type of cells in this cat—muscle, brain, or skin cells—can you infer have the least number of mitochondria?

When compared to other cells, the energy needs of skin cells are low.

Content Background

Typical cells contain from 300 to 1000 mitochondria, depending on their level of activity. Mitochondria have their own DNA, RNA, and ribosomes. They are produced only by the division of existing mitochondria. A cell cannot make new mitochondria from raw materials.

Visual Learning

Figure 19-9 As students study the figure, emphasize that cell activities require energy, most of which is supplied through the functioning of mitochondria. To reinforce students' understanding, ask these questions. **Do you think that brain cells contain many mitochondria?** *Yes, because they are very active and require large amounts of energy.* **Why might some muscle cells, which are always undergoing some movement, have greater numbers of mitochondria than other cells that are less active?** *Active cells require more energy.* Then have students answer the question posed in 19-9B: **Which type of cells in this cat—muscle, brain, or skin cells—can you infer have the least number of mitochondria?** *Skin cells require the least energy and therefore have the fewest number of mitochondria.*

19-2 Why Cells Need Food **597**

19-2 Respiration and Temperature

Planning the Activity

Time needed 40 minutes

Materials See student text activity.

Process Skills making and using tables, measuring in SI, observing and inferring

Purpose To observe the effects of temperature on respiration.

Preparation Make the 25 percent sucrose solution prior to the activity. Dissolve 150 g sugar in 450 mL water. Yeast cakes can be purchased at a grocery store. Use a razor blade or sharp knife to cut yeast cakes in half. One-half yeast cake is sufficient for each test tube.

Teaching the Activity

Have students work in groups of two or three. **COOP LEARN**

Discussion Before students perform the activity, you may wish to have them review the reactants and products of respiration by examining Figure 19-8. The cause of the rise of liquid within the tube is the result of increasing pressure inside the test tube due to the production of carbon dioxide gas. The faster the process of respiration, the more carbon dioxide produced, resulting in greater pressure and therefore a higher column of liquid. L2

Troubleshooting Be sure students position the glass tubing so that the end lies below the surface of the liquid.

Process Reinforcement Ask students what they might look for if they were trying to determine whether or not respiration was occurring in the cells of an organism. *Because carbon dioxide is a waste product of respiration, the presence of carbon dioxide could be evidence of respiration.*

Possible Procedures Students may wish to work in pairs, with one member of each pair taking responsibility for measuring the height of the liquid in either the warm or the cold tube. Use a stopwatch or kitchen

Respiration and Temperature

In this activity, you will observe evidence of respiration in yeast cells. You will also relate temperature to the rate of a reaction.

Problem

How is respiration influenced by temperature?

Materials

2 rubber stoppers with plastic tubing inserted	tap water
	2 test tubes
	metric ruler
20 mL 25% sucrose solution	2 flasks
	yeast cubes
	watch or clock
	10-mL graduated cylinder
	glass-marking pencil

Safety Precautions

What To Do

1 Copy the data table *into your Journal.*

2 Pour 10 mL of the sucrose solution into each test tube.

3 Place the yeast into the test tubes and mix well.

4 Insert a rubber stopper into each test tube. The end of the plastic tube inside the test tube should be below the surface of the liquid. Use the figure as a guide.

5 a. Add enough cold water to a flask to reach a height of 3 cm. Label this flask "cold" (see photo **B**).
b. Label the second flask "warm" and repeat Step 5a using warm water.

6 Carefully place one test tube in each flask (see photo **C**).

7 *Measure* the height of the liquid in each plastic tube

Meeting Individual Needs

Physically Challenged Help physically challenged students perform activities such as the Investigate by making the following modifications.

• Assign a partner to each physically challenged student.

• If necessary, have the partner premeasure samples required for the experiment. If possible, have partners make measurements to-

gether.

• Have the physically handicapped student be responsible for all observations in the experiment.

• Have partners interpret their data as a team. **COOP LEARN**

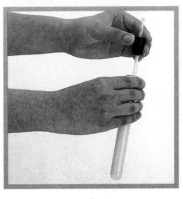

A   **B**   **C**

to the nearest millimeter. Position the ruler so that the 0.0 mm mark lines up with the bottom of the flask, as in the figure. Record your measurements in the table under "Starting Height."

8 Take measurements every 5 minutes for 20 minutes.

9 *Calculate* and record the total distance the yeast-water mixture moved. If the last reading was lower than the starting height, subtract this from your total.

Sample data

Data and Observations						
	Starting Height	5 Mins.	10 Mins.	15 Mins.	20 Mins.	Total Distance
Cold		Measurements should show a greater rise in				
Warm		the liquid of the test tube placed in warm water.				

Analyzing

1. *In your Journal,* describe what gas is released by yeast cells as they carry out respiration.

2. Which tube showed the greater rise in the height of the liquid? In which tube was more gas produced?

Concluding and Applying

3. Do you think that temperature has an effect on the rate of respiration? What evidence do you have to support your statement?

4. ~~Going Further~~ What can you *infer* about the rate of respiration in a fish swimming in cold water compared to the same fish swimming in warm water?

(**Program Resources**)

Activity Masters, pp. 81–82, Investigate 19-2

---

timer to measure 5-minute intervals. Call out "Time" so that students can all measure the height of the liquids in each tube at the same time.

Possible Hypotheses
Students may hypothesize that respiration and other life processes speed up in warm temperatures and slow down in cool temperatures.

Science Journal Have students write their findings in their journals.

Expected Outcome
The height of the liquid in the warm-water tube is higher than that in the cool-water tube.

Answers to Analyzing/Concluding and Applying
1. carbon dioxide

2. The warm tube showed the greatest rise. More gas was produced in the warm tube.

3. Yes; warm water made yeast cells respire faster.

4. Going Further: Cells of fish swimming in warm water would carry on respiration at a faster rate than if the fish were in cold water.

✔ Assessment

Performance Have groups of students analyze their results and then redesign the activity to show how much greater the temperature must be to make the yeast cells respire faster. Use the Performance Task Assessment List for Assessing a Whole Experiment/Planning the Next Experiment in **PASC,** p. 33.
COOP LEARN

GLENCOE TECHNOLOGY

 Software
Computer Competency Activities
Chapter 19

As you saw in the Investigate, a relationship exists between temperature and the rate at which respiration occurs. As temperature increases, so does the amount of respiration that occurs in an organism. This is true for your body as well as yeast cells. Think about the last time you exercised. Strenuous physical activities such as running or swimming require a lot of energy. If your arm and leg muscles

Figure 19-10

In order for the lion's leg muscles to get the energy they need, fast and numerous respiration reactions must occur. These reactions, as well as the muscle activity, release heat, which raises the lion's body temperature. The increased body temperature helps maintain the rate of respiration needed.

Chemistry CONNECTION

Does Mother Nature's Math Add Up?

When a cow eats grass, the cells in its body obtain energy from the grass through the process of respiration.

Respiration

During this process, glucose molecules in the grass are broken down to release energy. The cow will use this energy to

grow, produce a calf, and make milk. Respiration uses oxygen that has been delivered by the circulatory system to produce carbon dioxide and water, in addition to energy.

Respiration is a series of chemical reactions. In all chemical reactions, atoms are neither created nor destroyed. How can we be sure that Mother Nature's math adds up?

Chemical Equation of Respiration

The equation for the chemical reaction of respiration is illustrated on the next page.

The equation can also be written in such a way that we

Chemistry CONNECTION

Purpose

The Chemistry Connection explores the chemical equation for cellular respiration, discussed in Section 19-2. This excursion also shows that matter can never be created or destroyed, only rearranged.

Content Background

In theory, chemical reactions are reversible. But, in nature, many reactions do not run in reverse. For example, oxygen gas and hydrogen gas burn to form water vapor: $O_2 + 2H_2 \rightarrow 2H_2O$. During this spontaneous reaction, energy is released. However, water does not spontaneously decompose to form oxygen gas and hydrogen gas: $2H_2O \rightarrow O_2 + 2H_2$. This decomposition reaction requires an energy source.

Teaching Strategy

To reinforce students' understanding of chemical equations, display the following equations to the class. Have students identify types and number of atoms on either side of the arrow:

are to get the energy they need, fast and numerous respiration reactions must occur. These reactions, as well as the muscle activity, release heat, which raises your body temperature. Your increased body temperature maintains the rate of respiration to continually supply your muscles with energy. Due to this cycle, your body gets the energy it needs so you can reach the finish line!

Figure 19-11

This lion has just stopped running and is panting. As the lion pants, water from its mouth evaporates. The evaporation process reduces body temperature. As the lion stands and pants, does its rate of respiration increase or decrease? Why?

Visual Learning

Figure 19-11 Direct students' attention to the photo and have a volunteer read the caption aloud. **As the lion stands and pants, does its rate of respiration increase or decrease? Why?** *The rate of respiration goes down, because the body no longer requires a high level of energy and the panting cools the body, which also reduces the respiration rate.*

can count up the atoms.

To find out if the math adds up, you only need to count up the atoms on both sides of the arrow.

Let's start on the left side of the arrow. Each carbon atom is represented by the letter C.

How many carbon atoms (C) are there?

How many hydrogen atoms (H) do you see?

How many oxygen atoms (O)?

How many atoms total are on the left side of the arrow?

Now let's look at the right side of the arrow. How many carbon atoms do you see? How many hydrogen atoms? How many oxygen atoms? How many atoms total?

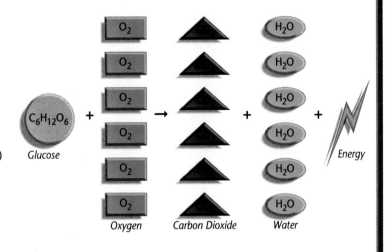

Glucose + → + Energy

Oxygen   Carbon Dioxide   Water

USING MATH

1. Is the sum of each kind of atom (C, H, O) on the left side of the arrow equal to the sum on the right side?

2. Does the total sum of atoms on the left side of the arrow equal the total sum of the atoms on the right side?

3. Were any atoms created or destroyed during respiration?

19-2 Why Cells Need Food **601**

$2Na + 2H_2O \rightarrow 2NaOH + H_2$

$Mg + 2HCl \rightarrow H_2 + MgCl_2$

Be sure students understand that coefficients indicate number of molecules and subscripts indicate number of atoms. [L2]

Activity

Challenge students to create their own shorthand system for writing the chemical equation for respiration. (For example, students may wish to use geometric symbols to represent elements.) Have students dis-

play their equations to the class. Discuss the problems that could arise if every scientist used his or her own system for representing chemical reactions.

USING MATH

1. yes
2. yes

3. no

Going Further ⫸

Students can work in small groups to create a skit that demonstrates how the atoms in the reactants of cellular respiration are rearranged to create new products. Use the Performance Task Assessment List for Skit in **PASC,** p. 75. **COOP LEARN** [L2]

Check for Understanding

To assess students' understanding of this section, use the following questions to begin a discussion. **Why must bacteria obtain energy through fermentation?** *Because bacteria lack mitochondria, many of them use fermentation instead of respiration.* **How are fermentation and respiration alike?** *Both processes can be used by cells to release energy. Both processes also release carbon dioxide.*

Reteach

Classify Display an aquarium in the classroom or show a picture of an aquarium. Have students identify those organisms in an aquarium in which photosynthesis takes place and those in which respiration takes place. [L1]

Extension

Inference Challenge students to explain how the job of chloroplasts and mitochondria are alike. *Energy changes form in both locations.* [L2]

4 CLOSE

Activity

Have students make a table that compares and contrasts photosynthesis, respiration, and fermentation. You may wish to start them off by writing the following headings on the chalkboard: *Process, Kind of Cell, Reactants, and Products.* [LEP] [L1]

Figure 19-12

A At the start of the bread-making process, bakers mix flour, water, salt, and yeast. They often also add some sugar or certain enzymes that convert some of the starch in the flour into sugar.

B During fermentation, the yeast cells use sugars as food and give off alcohol and carbon dioxide gas as waste chemicals. Heat makes the trapped gas expand and stretch the dough. The bread rises.

C As the bread bakes, the alcohol evaporates. The bubbles of carbon dioxide, however, stay trapped. The spaces that the bubbles form give bread its light texture.

■ Fermentation

But do all cells undergo respiration? Sometimes during periods of strenuous activity, muscle cells run low on oxygen. The muscles begin to tire. You begin to breathe harder and faster in an effort to supply the needed oxygen. But your body has another method to continue supplying smaller amounts of energy. When oxygen levels are low, the muscle cells begin to release energy from glucose by fermentation.

Fermentation is a process that releases energy by breaking down glucose without the use of oxygen. Yeast and some bacteria also use fermentation to release energy. **Figure 19-12** demonstrates one use of fermentation. Certain cells, such as some kinds of bacteria, lack mitochondria and therefore, cannot obtain their energy through respiration.

In this section, you learned that living organisms have mechanisms for supplying themselves with the energy they require. Different types of cells have different energy needs. In the next section, you will see how these cells are organized in the body so that they can complete their jobs.

check your UNDERSTANDING

1. Why do cells need energy?
2. Name three products of respiration.
3. What two reactants are needed for cell respiration?
4. How does fermentation differ from respiration?

5. **Apply** After examining a muscle cell from your lower jaw and a skin cell under a microscope, you find that the jaw muscle cell contains more mitochondria than the skin cell. What can you infer about the energy requirements of the jaw muscle cell?

check your UNDERSTANDING

1. Cells need energy to carry out their life processes.
2. energy, carbon dioxide, water
3. glucose and oxygen
4. Fermentation takes place without the use of oxygen. Glucose is converted to carbon dioxide and alcohol without oxygen. Respiration requires oxygen.
5. Jaw muscle cells must do more work than skin cells.

Special Cells with Special Jobs

A Variety of Cells

If you've ever used a tool kit, you know that there is nothing better than having the right tool for the right job. Some tools have flat and heavy parts for banging things, like a hammer. Others, such as screwdrivers, have long, thin parts for fitting into thin, tight places. Still others, such as a saw, have specially shaped teeth for cutting things. There is a definite relationship between the size and shape of a tool and its job, isn't there? The same is true of cells. Like the tools in a tool kit, there is a relationship between the structure of a cell and its parts, and how it functions in the body. **Figure 19-13** demonstrates this relationship.

Figure 19-13

Your body contains many different kinds of cells, each with its own unique shape and job.

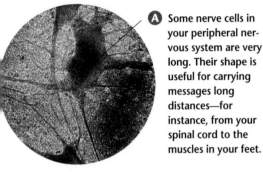

Ⓐ Some nerve cells in your peripheral nervous system are very long. Their shape is useful for carrying messages long distances—for instance, from your spinal cord to the muscles in your feet.

Nerve cell

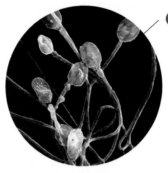

Ⓑ Sperm cells have an arrowlike head and a long, thin tail. Their shape helps them swim through body fluids.

Sperm cell

Ⓒ Some plant stem cells are long and hollow like drinking straws. Their shape helps them transport water and nutrients from the roots to the rest of the plant.

Plant cell

Ⓓ White blood cells, which surround and destroy harmful bacteria in your body, change into many different shapes while performing their job. How do you think these traits help a white blood cell do its job?

White blood cell

19-3   Special Cells with Special Jobs   **603**

2 TEACH

Tying to Previous Knowledge

Students should recall that one-celled organisms carry out all life functions. Cells in more complex organisms depend on other cells to help them carry out life functions.

Theme Connection The theme of systems and interactions is developed in this section. The ability of an organism to carry out its life processes is dependent on the successful interaction of its cells, tissues, organs, and organ systems.

Discussion Explain to students that organization of parts, with each doing a specific job, is known as division of labor, which is characteristic of tissues, organs, and systems. Have students consider whether division of labor is an efficient means of doing work. Ask them to identify the disadvantages of this process, such as what occurs when one part stops functioning properly. Challenge students to identify examples in which division of labor is used to accomplish a task. *Answers may include factory work, various departments in a company, the personnel structure of a school.*

Across the Curriculum

History/Language Arts

Have students research and write a report about the work of Aristotle, Galen, or Vesalius on animal organ systems. Students can present their reports in written or oral formats. L2

Connect to...

Earth Science

Some of the oldest fossil organisms were composed of one–celled bacteria and algae. These organisms resemble modern bacteria and blue-green algae in size and shape. Draw a time line showing how many years ago they appeared on Earth, when Earth was formed, and the present.

Figure 19-14

The cells of an organism—plant or animal—are arranged in levels of organization, much like letters of an alphabet are arranged in levels of organization to create a story. In a story, the levels are letters, words, sentences, paragraphs, and story.

Levels of Organization

The amoeba is a one-celled organism. The entire organism is composed of just one cell! The animal, plant, and fungus organisms that you are familiar with are all many-celled organisms. They contain more than one cell. One-celled organisms, however, still carry out the same life processes that occur in many-celled organisms such as your dog, your cat, and you.

Unlike one-celled organisms, the cells of many-celled organisms usually cannot function by themselves. Like parts in a machine, cells in your body work together to function effectively.

They are arranged in levels of organization. Each level is more organized than the one before.

What is meant by the phrase "levels of organization?" Your textbook shows different levels of organization. Letters of the alphabet are grouped together to form words. Words are grouped together in sentences. Many sentences arranged together are called paragraphs, and a group of paragraphs are organized into a story.

Cells in your body, like letters of the alphabet, are the most basic level of organization. Similar types of cells,

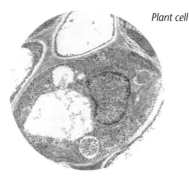
Plant cell

A *Cells* are the most basic level of organization. Specific types of cells have specific jobs.

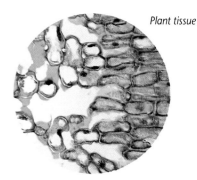
Plant tissue

B *Tissues* Similar types of cells, working together to perform the same function, make up tissues.

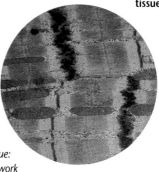

Animal cells and tissue: cardiac muscle cells work together to form heart tissue

604   Chapter 19   How Cells Do Their Jobs

Connect to . . .

Earth Science

Answer: appearance of oldest fossil organisms 3.2-3.5 billion years ago; formation of Earth 4.6 billion years ago

---

Program Resources

Making Connections: Integrating Sciences, p. 41 Industrial Chemicals and Wildlife L2

Making Connections: Technology & Society, p. 41, Growth Hormones in Beef, Poultry and Pork L3

Section Focus Transparency 59

Multicultural Perspectives

Dividing Labor

Explain that one thing that makes cultures different is the way labor is divided. In some societies, jobs are handed down through generations. In other societies, women farm and men weave. Have teams of students debate the advantages and disadvantages of dividing labor according to birth, gender, or other attributes. L2 working together to perform the same functions, are called **tissues**, like the words made up of letters.

Individual types of tissues usually are found with other types of tissue. An **organ** is a structure in the body made up of several different types of tissue that all work together to do a particular job, much like sentences made up of words. **Figure 19-14** shows the organization from cell to tissue to organ in a plant and the circulatory system in a person. The next level of organization in many-celled organisms like yourself is the organ system. **Organ systems** are simply groups of organs working together to perform a particular job, like the paragraphs that are groups of sentences.

The highest level of organization is an organism, like a bean plant, a frog, or even you. An **organism** is made up of several organ systems that work together, like paragraphs organized into a story. Each system plays a vital role in keeping the organism functioning normally. Can you think of other organisms? Think back to the organization of this book. Would the book provide as much information if there were no paragraphs or sentences? Just as a book needs parts to work together, so do living, many-celled organisms. Do you think you could be reading this book without a functioning nervous system?

Visual Learning

Figure 19-14 Continuing the analogy of levels of organization in creating a story in which the cells are the letters of the alphabet, have students tell what the tissues, organs, and organ systems represent. Then ask the question posed in Figure 19-14D: **Organ systems are groups of organs working together to perform a particular job. What is the next level of organization?** *organisms*

Inquiry Question In Chapter 7, you studied the skeletal and muscular systems. Describe how these two systems interact to allow movement. *Contraction and relaxation of muscles cause motion of the skeletal system.*

Plant organ: tomato plant leaf

Plant organ system: transport system— roots, stems, leaves

C *Organs* The third level of organization is organs— various types of tissue, all working together to perform a particular function.

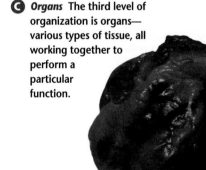

Animal organ: heart

D *Organ systems* are the next level of organization. Organ systems are groups of organs working together to perform a particular job. What is the next level of organization?

Animal organ system: circulatory system

19-3   Special Cells with Special Jobs   **605**

Check for Understanding

1. Provide models of an ear, a heart, a flower, and a leaf. Have students discuss the kinds of tissues that make up each organ. L2

2. Assign questions 1, 2, and the Apply question under Check Your Understanding. As students discuss their answers to the Apply question, provide a model of an earthworm or a diagram from a biology textbook to help them identify organs and systems in the earthworm. L1

Reteach

Question Ask students what organs work with the human nervous system to help a body respond to its environment. *eye, ear, and other sense organs* L1

Extension

Classify Have students write the name of a particular organ and organ system on two index cards. Collect the cards and shuffle. Students take turns removing a card from the deck and identifying whether it names an organ or an organ system. The cards could also be used for a classification game in which students group the organ cards according to organ systems. L3

4 CLOSE

Activity

Have students work in small groups to draw a diagram which shows the levels of organization in living things. **COOP LEARN** L1
P

In this chapter, you have seen that the structure of the cell membrane maintains the proper concentrations of molecules inside and outside the cell, so the cell can perform an important task: to provide usable forms of energy for life! You also observed that even though all cells in the body use energy, there are different types of cells for different jobs, and similar cells work together as a team to complete these jobs. **Figure 19-15** shows this same type of organization within a plant. Finally, you learned that all of the cells, tissues, organs, and organ systems in your body work together to form a whole, living organism.

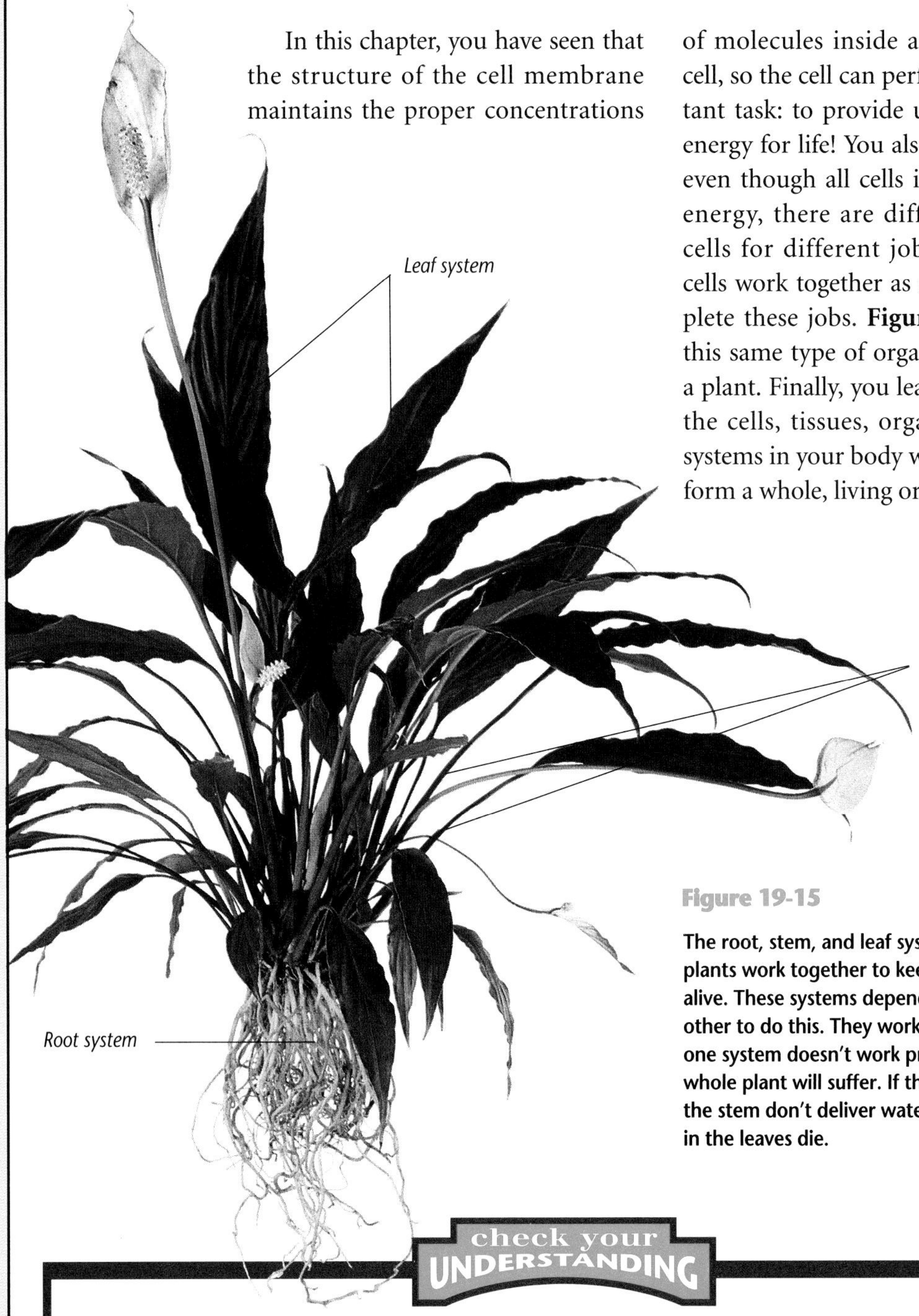

Leaf system

Stem system

Root system

Figure 19-15

The root, stem, and leaf systems of plants work together to keep the plant alive. These systems depend on each other to do this. They work as a team. If one system doesn't work properly, the whole plant will suffer. If the vessels in the stem don't deliver water, the tissues in the leaves die.

check your UNDERSTANDING

1. Give an example of two cells that differ in size and shape. Explain how their differences are related to cell function.

2. Explain why levels of organization are not found in an amoeba.

3. Apply What four levels of organization do you think can be found in an earthworm?

check your UNDERSTANDING

1. Possible answers include: white blood cells, shaped to engulf bacteria; plant stem cells, tube-shaped for carrying water; nerve cells, shaped for transmitting impulses.

2. Since an amoeba is one-celled, no levels of organization are needed. The one cell performs all its life functions.

3. cell, tissue, organ, organ system

Science and Society

Cryogenics: Frozen and Hopeful

No longer just in science-fiction movies—you can now buy the treatment. Your body can be frozen in order to be revived later. It's called cryonic suspension.

No one so far has been removed from cryonic suspension and revived. Only hamsters and dogs have been kept frozen for brief periods of time and revived. The success rate is, however, improving.

But when human beings are cryonically suspended, by law, they have to have died first. That means their revival will have to wait until medical science can reverse death, if ever, as well as cure whatever disease the person died of. If they died of old age, their cryonic suspension will continue until medical science can both revive and rejuvenate them.

How Does It Work?

Minutes after death, ideally, the patient's body is connected to a heart-lung machine. This enables oxygenated blood to continue bathing body tissues—especially the brain. The patient's temperature is lowered rapidly. When the body reaches the cryonics lab, the body's blood is replaced by a nontoxic chemical solution that acts like antifreeze. It replaces water in the tissues so no ice crystals rupture cell membranes. Once prepared, the body is cooled to the temperature of dry ice and then immersed in a vat of liquid nitrogen at -196°C. At this temperature, decay is essentially nonexistent. The body will remain unchanged for hundreds of years.

Questions, Anyone?

Note that we've been saying the body. Should that be the patient? Does a dead person whose body is now preserved cryonically have rights? Do heirs inherit property and other goods, or can the person store them for future need?

Will the person's memories awaken with the body? How much time are we talking about here—hundreds of years? Will revived people be able to make a life for themselves?

What Do You Think?

Write a reaction paper on cryonic suspension. Discuss one of the questions above or one of your own.

Going Further ⁞⁞⁞⁞⁞➤

Have students research cryogenics in magazine or newspaper indexes. Ask them to find articles about people who have chosen to be cryonically suspended. What reasons did people have for choosing this option? Students can use the information to write their own newspaper articles about a "frozen person." Use the Performance Task Assessment List for Newspaper Article in **PASC,** p. 69. **P**

Science and Society

Purpose

In Section 19-2, students learned that there is a relationship between temperature and the rate of respiration. As temperature rises, so does the rate of respiration. Increased respiration produces the energy the cells need and also increases waste products that must be excreted. This Science and Society section extends this relationship by examining technologies to lower the temperature of cells so much that respiration stops.

Content Background

Although scientists cannot cool bodies to absolute zero, the point when all molecular motion stops, cooling to about one-millionth of a degree above absolute zero is possible.

Teaching Strategy

Demonstrate the link between temperature and the rate of molecular motion by boiling and freezing water. Explain that ice is solid because water molecules move very slowly at the freezing temperature. Molecules move faster in the liquid state, while the rolling motion of boiling water and steam are created by very fast-moving molecules. Explain that molecules in the cells of living tissues also slow down a great deal at low temperatures.

Debate

Have teams of students choose one of the questions from the reading and debate its pros and cons.

What Do You Think?

Answers will vary depending on the question students choose and their own attitudes toward life and death.

Health CONNECTION

Purpose

The Health Connection examines the positive effects exercise has on cellular respiration, a topic introduced in Section 19-2. This excursion also describes how exercise improves the efficiency of the circulatory and respiratory systems.

Content Background

The breakdown of glucose involves both an anaerobic phase and an aerobic phase. In the anaerobic phase of cellular respiration, a molecule of glucose is broken down into two molecules of pyruvic acid. This phase, called glycolysis, occurs in the cytoplasm of the cell. During glycolysis, energy is released and captured to form two molecules of ATP. During the aerobic phase of cellular respiration, the molecules of pyruvic acid are further broken down in the presence of oxygen. This phase, known as the Krebs cycle, occurs in the mitochondria of the cell. As a result of the series of chemical reactions that occur during the Krebs cycle, carbon dioxide and water form and a great deal of energy is stored as ATP.

Teaching Strategy

Aerobic exercises are activities designed to promote the supply and use of oxygen in the body. Ask volunteers to describe and/or demonstrate any aerobic exercises with which they are familiar. L1

Activity

Ask students to make a list of five activities they have completed during the past 24 hours. Then have students use reference texts to determine the number of calories used in each activity. L2

Health CONNECTION

Shaping Up: You Can't Do It Overnight

Your life is full of activity—running for a bus, dancing, playing sports. You couldn't do any of these activities if the cells of your body didn't release energy.

Energy

Your cells release energy through the process of respiration, which takes place in the presence of oxygen. During activities, your cells receive plenty of oxygen

through the working of your respiratory and circulatory systems.

Activities such as sprinting or lifting weights require so much energy that your heart and lungs cannot work fast enough to provide your muscle cells with enough oxygen. When this occurs, your muscle cells release energy from glucose without oxygen through fermentation.

In people who are out of shape, strenuous activities don't have to last for very long before fermentation takes over in muscle cells. During fermentation, your muscles begin to tire, because of the build-up of a substance in your muscles. It takes time for the circulatory system to carry away this product. Even the next day, you may feel as if

your muscles can barely move. Fermentation doesn't produce as much energy as respiration and your body quickly tires out.

Benefits of Exercise

Exercise can strengthen the muscle in the heart's left ventricle so more blood can be pumped to your cells per heartbeat. Regular exercise can also strengthen the muscles of respiration so more air can be moved through the lungs per breath.

Exercise can also improve the oxygen-carrying properties of blood cells and increase blood volume in the body.

You Try It!

Getting into shape does not happen overnight. To maintain fitness, exercise must be performed on a regular basis. Your physician, physical education teacher, or a fitness professional can help you design the right fitness program for you.

Going Further ⫸

Have small groups of students work together to design a five-minute program of aerobic exercises appropriate for their age group. Encourage students to interview physical education specialists and/or fitness professionals for advice on the types of activities that should be included in the program. Each group should then demonstrate their exercise program to the class. Use the Performance Task Assessment List for Group Work in **PASC,** p. 97. COOP LEARN

The Artificial Kidney Machine

HOW IT WORKS

Every 40 minutes, your entire blood supply circulates through your kidneys. The kidneys cleanse about 180 liters of blood every 24 hours, filtering waste products and water from your blood. As they do this, your kidneys produce about 2 liters of urine a day. When kidneys are diseased or damaged, water and nitrogen waste products can collect in the blood and cause a variety of unhealthy or even life-threatening conditions.

What is Hemodialysis?

An artificial-kidney machine duplicates some of the kidney's functions by removing waste products and excess water. During the process, called hemodialysis, a patient's blood circulates through a filter.

The actual filtering device is a hemodialyzer. A hemodialyzer resembles a tube, like the tube inside a roll of paper towels. The tube is clear, with a thick bundle of white, hairy-looking material inside. This material is the filtering membrane, and it filters substances very much like the cell membranes of kidney cells.

How Does a Hemodialyzer Work?

The filtering membrane separates the hemodialyzer into two compartments. Blood from the patient's artery flows through one compartment, while a cleansing fluid flows through the second compartment.

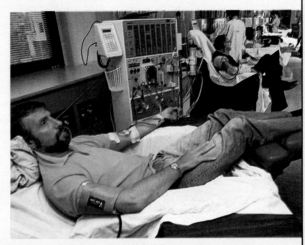

As blood circulates through the tubing, the blood with a high concentration of waste molecules diffuses through the membrane into the cleansing fluid. Here the concentration of waste molecules is very low. Fresh cleansing fluid is constantly added to the second compartment so that waste molecules continue to diffuse. The freshly cleansed blood is then returned to the patient.

A single treatment may take from two to four hour's and usually needs to be repeated every two to three days. Although not a cure, it is a treatment that allows patients to live longer.

Science Journal

The dialyzer and the cell membrane have much in common. What characteristics do they have in common? What processes occur in both the dialyzer and the cell membrane?

Science Journal

1. Both the dialyzer and the cell membrane are selectively permeable materials through which substances move into and out of the blood.
2. diffusion and osmosis

Going Further ⟩⟩⟩⟩

Have students use reference materials to learn more about the symptoms of acute glomerulonephritis, a condition that occurs mostly in children. Students can make posters about the disease. Use the Performance Task Assessment List for Poster in **PASC,** p. 73. L2

Purpose

How It Works reinforces Section 19-2 by explaining how an artificial kidney machine filters waste products of cellular respiration from the blood. This excursion also focuses on interactions and systems by describing how various body systems work together to benefit the whole organism.

Content Background

A common cause of kidney damage is the condition glomerulonephritis. It involves a swelling of the filtering units of the kidneys, which reduces the production of urine. Most cases of glomerulonephritis follow a *Streptococcus* infection. In some people, the infection causes the body to become allergic to the tissues of the filtering units of the kidneys or glomeruli. The tissues become damaged and cannot filter wastes from the blood. The blood of people suffering from glomerulonephritis must be filtered artificially through hemodialysis.

Teaching Strategy Develop the idea that the artificial kidney machine is an example of diffusion at work. Ask students why the cleansing fluid must be constantly changed. *to keep the urea concentration low*

Activity

Have students write a short paragraph that describes how their lives might differ if they suffered from kidney disease and required hemodialysis on a regular basis. L2

Have students look at the four diagrams on this page. Direct them to read the statements and answer the questions to review the main ideas in this chapter.

Teaching Strategy

Divide the class into four groups. Assign each group one of the diagrams. Each group should prepare a skit that depicts what is occurring in their diagram. For example, in diagram two, group members could represent molecules that pass through two students holding hands to represent the cell membrane. Students should continue moving through the cell membrane until there are equal numbers of students on either side of the structure. Have groups perform their skits for the class as a means of reviewing the chapter.

Answers to Questions

1. The cell membrane is a specialized structure that controls what enters and leaves a cell.

2. Osmosis is a form of diffusion involving water molecules only. It takes place when water diffuses across a cell membrane.

3. Such organisms release energy through fermentation.

4. These cells are organized into tissues, organs, and organ systems.

Science Journal

Review the statements below about the big ideas presented in this chapter, and answer the questions. Then, re-read your answers to the Did You Ever Wonder questions at the beginning of the chapter. *In your Science Journal,* write a paragraph about how your understanding of the big ideas in the chapter has changed.

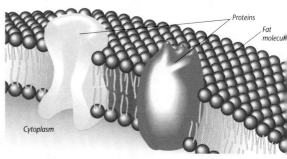

Proteins
Fat molecule
Cytoplasm

1 Cells carry out life processes with the help of the cell membrane. *What is the major function of the cell membrane?*

2 Some materials move through the cell membrane by the process of diffusion, in which molecules move from an area of high concentration to an area of lower concentration until equilibrium is reached. *What is osmosis?*

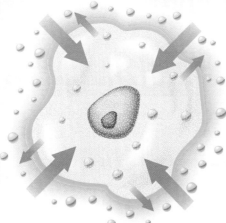

3 In the cells of most organisms, energy is released from glucose in the presence of oxygen by the process of respiration. *How do organisms release energy without oxygen?*

4 Most many-celled organisms are not just a collection of individual cells working by themselves. *Into what levels are these cells organized?*

Science at Home

Have students determine the effect salt has on a variety of vegetables such as cucumbers, carrots, and potato slices. Students may determine their own procedures and the amount of salt to be applied. Students may apply the salt directly or in a solution. Be sure that students record their procedures. Encourage students to record their observations in a table to share with the class. **P** **L2**

Science Journal

Did you ever wonder...
• Water enters the rice by diffusion, causing it to get soft. (pp. 588–589)
• Heat is produced by the chemical reactions that occur in cells as energy is released from food. (p. 595)
• Placing a tea bag in hot water causes molecules from the tea leaves to diffuse throughout the water. (p. 586)

Using Key Science Terms

diffusion

fermentation

organ

organisms

organ systems

osmosis

respiration

tissues

1. What are two ways molecules can reach equilibrium? How do these two ways differ?

2. How are the processes of fermentation and respiration alike? How are they different?

3. In what order do organ, organism, organ system, and tissue appear in the levels of organization? Briefly explain each term.

Understanding Ideas

Answer the following questions in your Journal using complete sentences.

1. How does respiration help an organism survive?

2. Would you expect to find more mitochondria in a more active cell or a less active cell? Why?

3. Do molecules stop moving through a cell membrane once equilibrium is reached? Explain your answer.

4. Name three types of tissue that are found in your body.

5. Why does your body get warm as you exercise?

6. In which direction do molecules flow during diffusion?

Developing Skills

Use your understanding of the concepts developed in this chapter to answer each of the following questions.

1. **Concept Mapping** Create a concept map of cell processes using the following processes: diffusion, osmosis, active transport, respiration, fermentation. Use additional terms as necessary.

2. **Observing and Inferring** Repeat the Explore activity on page 586 using a tea bag in a clear glass of room temperature water. Make an observation of the glass every two minutes for 20 minutes. How do your observations compare with those you made with the tea bag in hot water? What can you infer about the effect of hot water in this investigation?

3. **Predicting** Repeat the Find Out activity on page 596, adding water to the jar of dry beans. Predict what will happen to your temperature readings. What do the results of your temperature readings show?

Using Key Science Terms

1. Diffusion and osmosis. Diffusion is the movement of molecules from an area of high concentration to an area of low concentration. Osmosis is the diffusion of water.

2. Both are chemical processes that release energy. Fermentation breaks down glucose into alcohol and carbon dioxide without the use of oxygen. In respiration, oxygen is needed to combine with glucose to release energy.

3. The order is tissue, organ, organ system, organism. Tissues–similar types of cells working together to perform the same function; organ–several types of tissues that work together; organ system–a group of organs that work together; organism–made up of several organ systems that work together.

Understanding Ideas

1. Through respiration, an organism receives the energy it needs to carry out life processes.

2. More mitochondria in an active cell because it would have greater energy needs.

3. Molecules continue to move back and forth and at the same rate, maintaining equilibrium.

4. Possible answers include nerve tissue, skin tissue, blood vessel tissue, and muscle tissue.

5. The respiration rate increases, which raises body temperature.

6. Molecules move from the place where there are many of them to a place where there are fewer.

Developing Skills

1. Cell processes include molecular movement and require energy. Molecular movement can be passive (diffusion and osmosis) or active transport. Energy is obtained through respiration or fermentation.

2. The water changed color more slowly. The hot water causes the molecules of tea to move more quickly from the bag

into the water.

3. The temperature in the jar of dry beans should rise after water is added, showing that respiration is taking place.

Chapter 19 How Cells Do Their Jobs **611**

Critical Thinking

1. Results would be similar, but might not be as apparent.

2. Water molecules move from a place where there are many water molecules (soil) to a place where there are fewer water molecules (plant roots).

3. The salt causes water to move out of the plant's cells. This may result in damage or death to the plant.

4. They remain in the body fluids that surround the cells.

5. Respiration slows down and energy is not released.

Problem Solving

1. The liquid is water that came from vegetable cells. Salt caused the plant cells to lose water through the process of osmosis.

2. The lettuce wilted because water diffused out of the cells to a place of lower concentration. Loss of water in lettuce cells caused the cells to shrink and the lettuce to wilt.

Connecting Ideas

1. Cells without mitochondria generally obtain their energy through fermentation.

2. Because a muscle cell is more active than a skin cell, the energy requirements of the muscle cell are greater.

3. Cells obtain these materials from food and their environment.

4. Muscle cells produce energy anaerobically by fermentation, without oxygen. When this happens, muscles begin to tire and become stiff.

5. You would need to visit a dialysis center two or three times a week for three to four hours. You would lose six to twelve hours every week. You might not be able to work. You would incur a heavy financial burden. You would also need to monitor your diet.

Critical Thinking

In your Journal, *answer each of the following questions.*

1. In the first Investigate, you observed osmosis through an egg membrane. Do you think you would have obtained the same results if the shells had not been dissolved by the vinegar? Explain your answer.

2. Applying your knowledge of osmosis, explain how plant roots obtain water from the soil.

3. In snowy states, salt is used to melt ice on the roads. Explain what happens to many roadside plants as a result.

4. What do you think happens to substances that are not allowed to pass into your cells?

5. Why might a person feel very tired and weak after skipping several meals?

Problem Solving

Read the following problem and discuss your answers in a brief paragraph.

Chris made a salad of lettuce, tomatoes, carrots, and cucumbers. He seasoned the damp salad with herbs, salt, and pepper. Then he placed it in the refrigerator for a couple of hours.

When Chris returned, he took the salad from the refrigerator. The lettuce had wilted and the other vegetables were limp. He noticed that there was a liquid in the bottom of the bowl.

1. Where did the liquid in the bottom of the salad bowl come from?

2. Why did the lettuce wilt after having been left in the refrigerator for a couple of hours?

CONNECTING IDEAS

Discuss each of the following in a brief paragraph.

1. Theme—Systems and Interactions Why don't some cells release energy through respiration? What other process do they use to get energy?

2. Theme—Energy How do you think the energy requirements of a muscle cell compare to that of a skin cell? Explain your answer.

3. Theme—Systems and Interactions Describe how cells get the materials they need. Where do the nutrients, water, and oxygen come from?

4. Health Connection Sometimes when you exercise, your body is not able to provide enough oxygen to your muscle cells. What do you think happens then?

5. Science and Society How would your life change if you had to depend on hemodialysis?

✔ Assessment

Portfolio Review the portfolio options that are provided throughout the chapter. Encourage students to select one product that demonstrates their best work for the chapter. Have students explain what they learned and why they chose this example for placement into their portfolios.

Additional portfolio options can be found in the following **Teacher Classroom Resources:**

Multicultural Connections, pp. 41, 42

Making Connections: Technology and Society, p. 41

Concept Mapping, p. 27

Critical Thinking/Problem Solving, p. 27

Making Connections: Integrating Sciences, p. 41

Take Home Activities, p. 28

Making Connections: Across the Curriculum, p. 41

Laboratory Manual, pp. 113–116

Performance Assessment P

Life at the Cellular Level

In this unit, you investigated how a microscope can be used to observe cells. You learned that chemical reactions occur in living and nonliving matter.

Try the exercises and activity that follow—they will challenge you to use and apply some of the ideas you learned in this unit.

CONNECTING IDEAS

1. Find a way to calculate the magnification of a hand lens and a bifocal lens. Use the magnifying power you calculate to determine how much larger a single cell would appear using both of these tools.

2. Remember the biosphere experiment you set up at the beginning of this book? What were some of the important factors you took into consideration when deciding what to place into your biosphere? How would you explain the outcome of your experiment using what you have learned about respiration and photosynthesis?

  ACTIVITY

Where do you get your energy?

In the body, stored food is chemically combined with oxygen and energy is given off. How much energy is in a peanut? Let's burn one and see. Remember, a rise in temperature indicates energy is being given off. Although energy is required to start the burning reaction, energy does not have to be supplied to keep it going. The releasing of the chemical energy in the peanut keeps the reaction going.

What To Do

1. Get a ring stand; a ring; a utility clamp; a cork; a long straight pin; a small, empty can; aluminum foil; a thermometer; a peanut; water; and matches.

2. Set up your equipment as shown in the photograph. Place the cork and peanut 2 cm from the bottom of the can.

3. Pour 100 mL of water into the small can and record the water temperature.

4. Set fire to the peanut. **CAUTION:** *Use care around an open flame.*

5. What happened to the temperature of the water? Was energy present in the peanut?

Life at the Cellular Level

THEME DEVELOPMENT

Two themes supported in this unit were systems and interactions and energy. Students have learned about cells. They have seen how parts of the cell interact to perform specific cellular activities and how the cell releases energy to carry out those functions.

Connections to Other Units

The ideas developed in this unit relate to the discussion of the circulatory system in Unit 1, the skeletal system in Unit 2, and the respiratory system in Unit 4. This unit shows students how these systems can function.

Connecting Ideas
Answers

1. Students can trace the outline of a hand lens or bifocal lens onto a piece of graph paper, ideally one with a small scale, such as 1 cm squares. Count the number of squares within the outline. Hold the lens up and focus on the graph paper. Recount the number of squares now seen. Divide the original number of squares into the number counted with the hand lens to get the magnification of the hand lens.

2. Student answers will vary based on their original ideas.

Exploring Further
Where do you get your energy?

Purpose To demonstrate that food contains energy.

Background Information Nutritionists scientifically analyze the energy content of foods by combusting a food sample in a calorimeter—a device for accurately measuring the energy released. Using this data calorie charts are prepared.

Materials See student text.

Troubleshooting Keep a container of water handy in case of any fire emergency.

Answers to Questions

Students should observe that the water temperature rises. The energy that increases the temperature comes from the peanut, not the match. To demonstrate this repeat the experiment using the match in place of the peanut.

Assessment

Process Have students repeat the experiment using a marshmallow and compare the change in water temperature. Students can infer which food sample released more energy. Use the Performance Task Assessment List for Carrying Out a Strategy/Collecting the Data in **PASC**, p. 25. **COOP LEARN**

APPENDICES

Table of Contents

International System of Units

The International System (SI) of Measurement is accepted as the standard for measurement throughout most of the world. Three base units in SI are the meter, kilogram, and second. Frequently used SI units are listed below.

Table A-1: Frequently used SI Units	
Length	1 millimeter (mm) = 1000 micrometers (μm)
	1 centimeter (cm) = 10 millimeters (mm)
	1 meter (m) = 100 centimeters (cm)
	1 kilometer (km) = 1000 meters (m)
	1 light-year = 9 460 000 000 000 kilometers (km)
Area	1 square meter (m^2) = 10 000 square centimeters (cm^2)
	1 square kilometer (km^2) = 1 000 000 square meters (m^2)
Volume	1 milliliter (mL) = 1 cubic centimeter (cm^3)
	1 liter (L) = 1000 milliliters (mL)
Mass	1 gram (g) = 1000 milligrams (mg)
	1 kilogram (kg) = 1000 grams (g)
	1 metric ton (g) = 1000 kilo grams (kg)
Time	1 s = 1 second

Temperature measurements in SI are often made in degrees Celsius. Celsius temperature is a supplementary unit derived from the base unit kelvin. The Celsius scale (°C) has 100 equal graduations between the freezing temperature (0°C) and the boiling temperature of water (100°C). The following relationship exists between the Celsius and kelvin temperature scales:

$$K = °C + 273$$

Several other supplementary SI units are listed below.

Table A-2: Supplementary SI Units			
Measurement	Unit	Symbol	Expressed in Base Units
Energy	Joule	J	$kg \cdot m^2/s^2$ or $N \cdot m$
Force	Newton	N	$kg \cdot m/s^2$
Power	Watt	W	$kg \cdot m^2/s^3$ or J/s
Pressure	Pascal	Pa	$kg/(m \cdot s^2)$ or N/m^2

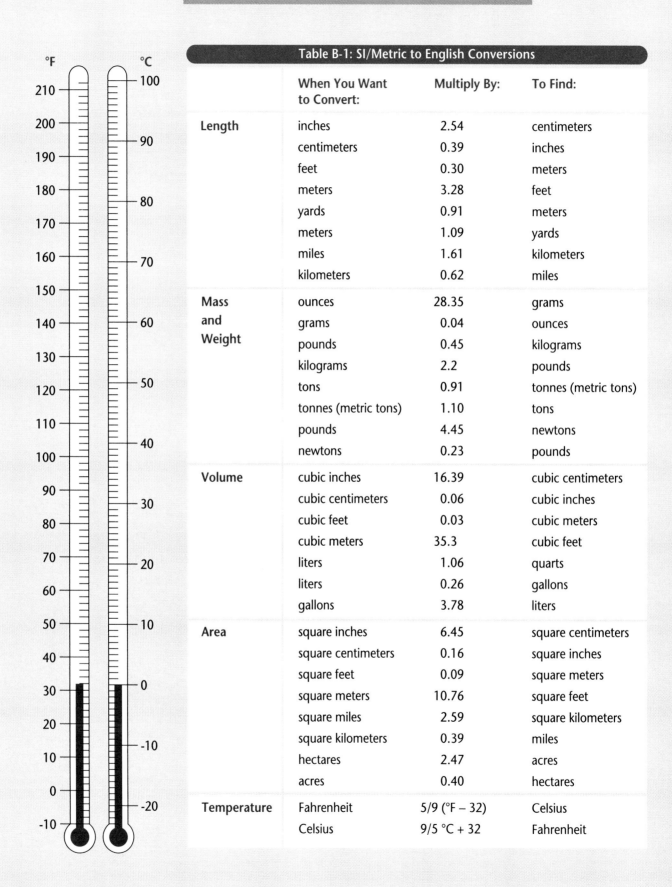

Table B-1: SI/Metric to English Conversions

	When You Want to Convert:	Multiply By:	To Find:
Length	inches	2.54	centimeters
	centimeters	0.39	inches
	feet	0.30	meters
	meters	3.28	feet
	yards	0.91	meters
	meters	1.09	yards
	miles	1.61	kilometers
	kilometers	0.62	miles
Mass and Weight	ounces	28.35	grams
	grams	0.04	ounces
	pounds	0.45	kilograms
	kilograms	2.2	pounds
	tons	0.91	tonnes (metric tons)
	tonnes (metric tons)	1.10	tons
	pounds	4.45	newtons
	newtons	0.23	pounds
Volume	cubic inches	16.39	cubic centimeters
	cubic centimeters	0.06	cubic inches
	cubic feet	0.03	cubic meters
	cubic meters	35.3	cubic feet
	liters	1.06	quarts
	liters	0.26	gallons
	gallons	3.78	liters
Area	square inches	6.45	square centimeters
	square centimeters	0.16	square inches
	square feet	0.09	square meters
	square meters	10.76	square feet
	square miles	2.59	square kilometers
	square kilometers	0.39	miles
	hectares	2.47	acres
	acres	0.40	hectares
Temperature	Fahrenheit	5/9 (°F − 32)	Celsius
	Celsius	9/5 °C + 32	Fahrenheit

APPENDIX C

Safety in the Science Classroom

1. Always obtain your teacher's permission to begin an investigation.
2. Study the procedure. If you have questions, ask your teacher. Understand any safety symbols shown on the page.
3. Use the safety equipment provided for you. Goggles and a safety apron should be worn when any investigation calls for using chemicals.
4. Always slant test tubes away from yourself and others when heating them.
5. Never eat or drink in the lab, and never use lab glassware as food or drink containers. Never inhale chemicals. Do not taste any substances or draw any material into a tube with your mouth.
6. If you spill any chemical, wash it off immediately with water. Report the spill immediately to your teacher.
7. Know the location and proper use of the fire extinguisher, safety shower, fire blanket, first aid kit, and fire alarm.
8. Keep materials away from flames. Tie back hair and loose clothing.
9. If a fire should break out in the classroom, or if your clothing should catch fire, smother it with the fire blanket or a coat, or get under a safety shower. NEVER RUN.
10. Report any accident or injury, no matter how small, to your teacher.

Follow these procedures as you clean up your work area.

1. Turn off the water and gas. Disconnect electrical devices.
2. Return all materials to their proper places.
3. Dispose of chemicals and other materials as directed by your teacher. Place broken glass and solid substances in the proper containers. Never discard materials in the sink.
4. Clean your work area.
5. Wash your hands thoroughly after working in the laboratory.

Table C-1: First Aid	
Injury	Safe Response
Burns	Apply cold water. Call your teacher immediately.
Cuts and bruises	Stop any bleeding by applying direct pressure. Cover cuts with a clean dressing. Apply cold compresses to bruises. Call your teacher immediately.
Fainting	Leave the person lying down. Loosen any tight clothing and keep crowds away. Call your teacher immediately.
Foreign matter in eye	Flush with plenty of water. Use eyewash bottle or fountain. Call your teacher immediately.
Poisoning	Note the suspected poisoning agent and call your teacher immediately.
Any spills on skin	Flush with large amounts of water or use safety shower. Call your teacher immediately.

APPENDIX D

Safety Symbols

These safety symbols are used to indicate possible hazards in the activities. Each activity has appropriate hazard indicators.

DISPOSAL ALERT
This symbol appears when care must be taken to dispose of materials properly.

ANIMAL SAFETY
This symbol appears whenever live animals are studied and the safety of the animals and the students must be ensured.

BIOLOGICAL HAZARD
This symbol appears when there is danger involving bacteria, fungi, or protists.

RADIOACTIVE SAFETY
This symbol appears when radioactive materials are used.

OPEN FLAME ALERT
This symbol appears when use of an open flame could cause a fire or an explosion.

CLOTHING PROTECTION SAFETY
This symbol appears when substances used could stain or burn clothing.

THERMAL SAFETY
This symbol appears as a reminder to use caution when handling hot objects.

FIRE SAFETY
This symbol appears when care should be taken around open flames.

SHARP OBJECT SAFETY
This symbol appears when a danger of cuts or punctures caused by the use of sharp objects exists.

EXPLOSION SAFETY
This symbol appears when the misuse of chemicals could cause an explosion.

FUME SAFETY
This symbol appears when chemicals or chemical reactions could cause dangerous fumes.

EYE SAFETY
This symbol appears when a danger to the eyes exists. Safety goggles should be worn when this symbol appears.

ELECTRICAL SAFETY
This symbol appears when care should be taken when using electrical equipment.

POISON SAFETY
This symbol appears when poisonous substances are used.

SKIN PROTECTION SAFETY
This symbol appears when use of caustic chemicals might irritate the skin or when contact with microorganisms might transmit infection.

CHEMICAL SAFETY
This symbol appears when chemicals used can cause burns or are poisonous if absorbed through the skin.

Care and Use of a Microscope

Coarse Adjustment *Focuses the image under low power*

Fine Adjustment *Sharpens the image under high and low magnification*

Arm *Supports the body tube*

Low-power objective *Contains the lens with low-power magnification*

Stage clips *Hold the microscope slide in place*

Base *Provides support for the microscope*

Eyepiece *Contains a magnifying lens you look through*

Body tube *Connects the eyepiece to the revolving nosepiece*

Revolving nosepiece *Holds and turns the objectives into viewing position*

High-power objective *Contains the lens with the highest magnification*

Stage *Platform used to support the microscope slide*

Diaphragm *Regulates the amount of light entering the body tube*

Light source *Allows light to reflect upward through the diaphragm, the specimen, and the lenses*

Care of a Microscope

1. Always carry the microscope holding the arm with one hand and supporting the base with the other hand.
2. Don't touch the lenses with your finger.
3. Never lower the coarse adjustment knob when looking through the eyepiece lens.
4. Always focus first with the low-power objective.
5. Don't use the coarse adjustment knob when the high-power objective is in place.
6. Store the microscope covered.

Using a Microscope

1. Place the microscope on a flat surface that is clear of objects. The arm should be toward you.
2. Look through the eyepiece. Adjust the diaphragm so that light comes through the opening in the stage.
3. Place a slide on the stage so that the specimen is in the field of view. Hold it firmly in place by using the stage clips.
4. Always focus first with the coarse adjustment and the low-power objective lens. Once the object is in focus on low power, turn the nosepiece until the high-power objective is in place. Use ONLY the fine adjustment to focus with the high-power objective lens.

Making a Wet Mount Slide

1. Carefully place the item you want to look at in the center of a clean glass slide. Make sure the sample is thin enough for light to pass through.
2. Use a dropper to place one or two drops of water on the sample.
3. Hold a clean coverslip by the edges and place it at one edge of the drop of water. Slowly lower the coverslip onto the drop of water until it lies flat.
4. If you have too much water or a lot of air bubbles, touch the edge of a paper towel to the edge of the coverslip to draw off extra water and force air out.

Animal Cell

Refer to this diagram of an animal cell as you read cell parts and their jobs.

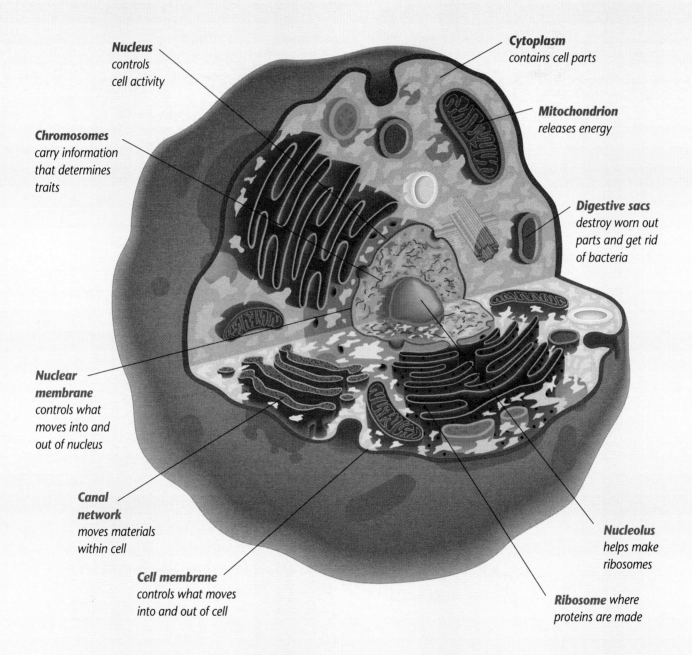

Nucleus controls cell activity

Cytoplasm contains cell parts

Mitochondrion releases energy

Chromosomes carry information that determines traits

Digestive sacs destroy worn out parts and get rid of bacteria

Nuclear membrane controls what moves into and out of nucleus

Canal network moves materials within cell

Cell membrane controls what moves into and out of cell

Nucleolus helps make ribosomes

Ribosome where proteins are made

Plant Cell

Refer to this diagram of a plant cell as you read cell parts and their jobs.

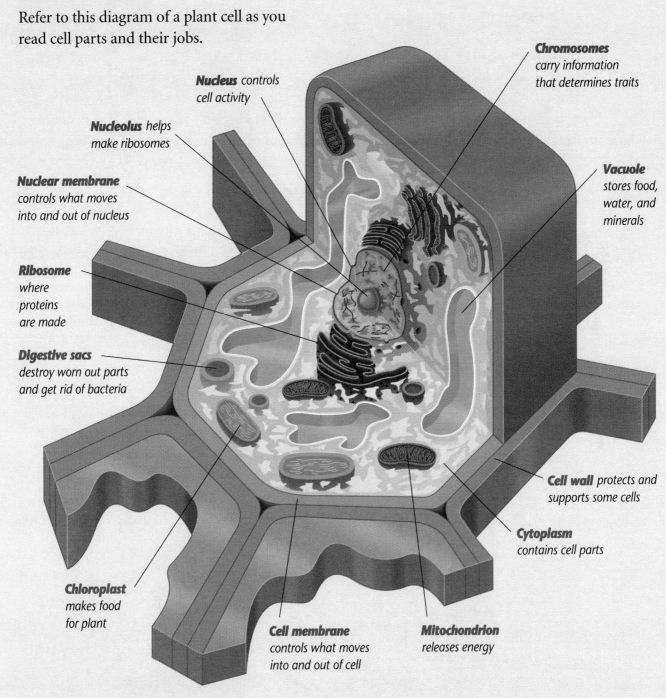

Nucleus controls cell activity

Nucleolus helps make ribosomes

Nuclear membrane controls what moves into and out of nucleus

Ribosome where proteins are made

Digestive sacs destroy worn out parts and get rid of bacteria

Chromosomes carry information that determines traits

Vacuole stores food, water, and minerals

Cell wall protects and supports some cells

Cytoplasm contains cell parts

Chloroplast makes food for plant

Cell membrane controls what moves into and out of cell

Mitochondrion releases energy

Diversity of Life: Classification of Living Organisms

Scientists use a five-kingdom system for the classification of organisms. In this system, there is one kingdom of organisms, Kingdom Monera, which contains organisms that do not have a nucleus and lack specialized structures in the cytoplasm of their cells. The members of the other four kingdoms each have cells that contain a nucleus and structures in the cytoplasm that are surrounded by membranes. These kingdoms are Kingdom Protista, Kingdom Fungi, the Plant Kingdom, and the Animal Kingdom.

Kingdom Monera

Phylum Cyanobacteria one-celled prokaryotes; make their own food, contain chlorophyll, some species form colonies, most are blue-green

Bacteria one-celled prokaryotes; most absorb food from their surroundings, some are photosynthetic; many are parasites; round, spiral, or rod shaped

Kingdom Protista

Phylum Euglenophyta one-celled; can photosynthesize or take in food; most have one flagellum

Phylum Crysophyta most are one-celled; make their own food through photosynthesis; golden-brown pigments mask chlorophyll; diatoms

Phylum Pyrrophyta one-celled; make their own food through photosynthesis; contain red pigments and have two flagella; dinoflagellates

Phylum Chlorophyta one-celled, many-celled, or colonies; contain chlorophyll and make their own food; live on land, in fresh water or salt water; green algae

Phylum Rhodophyta most are many-celled and photosynthetic; contain red pigments; most live in deep saltwater environments; red algae

Phylum Phaeophyta most are many-celled and photosynthetic; contain brown pigments; most live in saltwater environments; brown algae

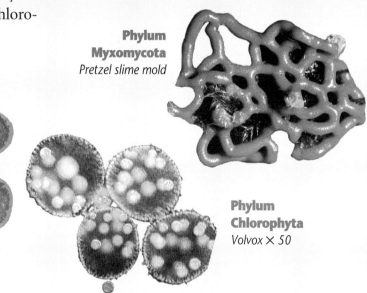

Phylum Myxomycota
Pretzel slime mold

Bacteria
Clostridium botulinum
× 13 960

Phylum Chlorophyta
Volvox × 50

Phylum Sarcodina one-celled; take in food; move by means of pseudopods; free-living or parasitic; sarcodines

Phylum Mastigophora one-celled; take in food; have two or more flagella; free-living or parasitic; flagellates

Phylum Ciliophora one-celled; take in food; have large numbers of cilia; ciliates

Phylum Sporozoa one-celled; take in food; no means of movement; parasites in animals; sporozoans

Phylum Myxomycota, Phylum Acrasiomycota one- or many-celled; absorb food; change form during life cycle; cellular and plasmodial slime molds

Kingdom Fungi

Phylum Zygomycota many-celled; absorb food; spores are produced in sporangia; zygote fungi

Phylum Ascomycota one- and many-celled; absorb food; spores produced in asci; sac fungi; yeast

Phylum Ascomycota
Yeast × 7800

Phylum Basidiomycota many-celled; absorb food; spores produced in basidia; club fungi

Phylum Deuteromycota members with unknown reproductive structures; imperfect fungi

Lichens organism formed by symbiotic relationship between an ascomycote or a basidiomycote and a green alga or a cyanobacterium

Plant Kingdom

Spore Plants

Division Bryophyta nonvascular plants that reproduce by spores produced in capsules; many-celled; green; grow in moist land environments; mosses and liverworts

Division Lycophyta many-celled vascular plants; spores produced in cones; live on land; are photosynthetic; club mosses

Division Sphenophyta vascular plants with ribbed and jointed stems; scalelike leaves; spores produced in cones; horsetails

Division Pterophyta vascular plants with feathery leaves called fronds; spores produced in clusters of sporangia called sori; live on land or in water; ferns

Lichens
Old Man's Beard lichen

Division Bryophyta
Liverwort

Seed Plants

Division Ginkgophyta deciduous gymnosperms; only one living species called the maidenhair tree; fan-shaped leaves with branching veins; reproduces with seeds; ginkgos

Division Cycadophyta palmlike gymnosperms; large compound leaves; produce seeds in cones; cycads

Division Coniferophyta deciduous or evergreen gymnosperms; trees or shrubs; needlelike or scalelike leaves; seeds produced in cones; conifers

Division Gnetophyta shrubs or woody vines; seeds produced in cones; division contains only three genera; gnetum

Division Anthophyta dominant group of plants; ovules protected at fertilization by an ovary; sperm carried to ovules by pollen tube; produce flowers and seeds in fruits; flowering plants

Animal Kingdom

Phylum Porifera aquatic organisms that lack true tissues and organs; they are asymmetrical and sessile; sponges

Phylum Cnidaria radially symmetrical organisms with a digestive cavity with one opening; most have tentacles armed with stinging cells; live in aquatic environments singly or in colonies; includes jellyfish, corals, hydra, and sea anemones

Phylum Platyhelminthes bilaterally symmetrical worms with flattened bodies; digestive system has one opening; parasitic and free-living species; flatworms

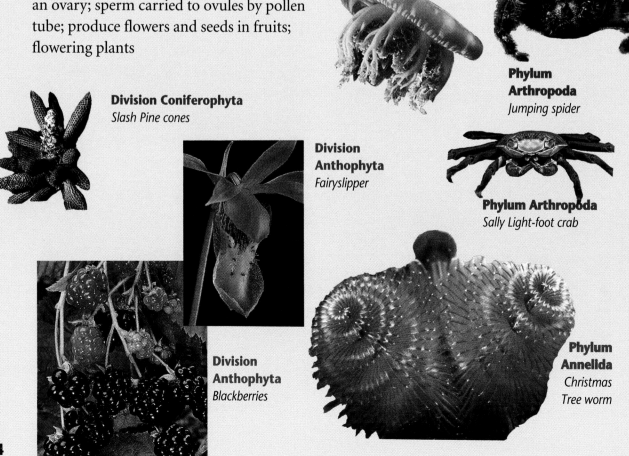

Phylum Cnidaria
Jellyfish

Phylum Arthropoda
Jumping spider

Division Coniferophyta
Slash Pine cones

Division Anthophyta
Fairyslipper

Phylum Arthropoda
Sally Light-foot crab

Division Anthophyta
Blackberries

Phylum Annelida
Christmas Tree worm

Phylum Nematoda round bilaterally symmetrical body; digestive system with two openings; some free-living forms but mostly parasitic; roundworms

Phylum Mollusca soft-bodied animals, many with a hard shell; a mantle covers the soft body; aquatic and terrestrial species; includes clams, snails, squid, and octopuses

Phylum Annelida bilaterally symmetrical worms with round segmented bodies; terrestrial and aquatic species; includes earthworms, leeches, and marine polychaetes

Phylum Arthropoda very large phylum of organisms that have segmented bodies with pairs of jointed appendages, and a hard exoskeleton; terrestrial and aquatic species; includes insects, crustaceans, spiders, and horseshoe crabs

Phylum Echinodermata saltwater organisms with spiny or leathery skin; water-vascular system with tube feet; radial symmetry; includes starfish, sand dollars, and sea urchins

Phylum Chordata organisms with internal skeletons, specialized body systems, and paired appendages; all at some time have a notochord, dorsal nerve cord, gill slits, and a tail; include fish, amphibians, reptiles, birds, and mammals

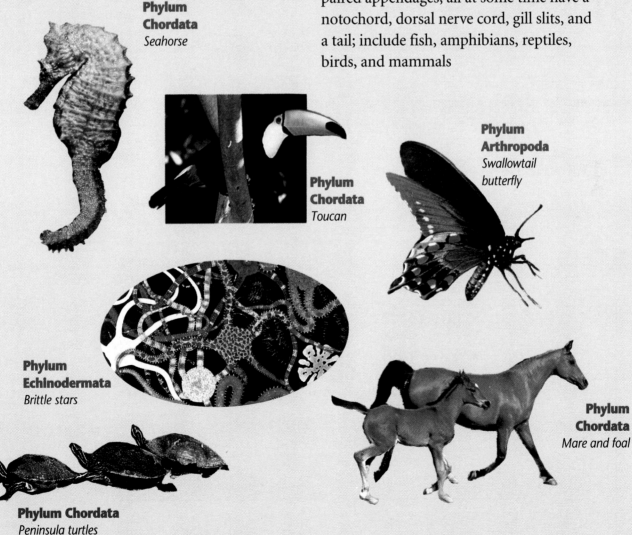

Phylum Chordata
Seahorse

Phylum Chordata
Toucan

Phylum Arthropoda
Swallowtail butterfly

Phylum Echinodermata
Brittle stars

Phylum Chordata
Mare and foal

Phylum Chordata
Peninsula turtles

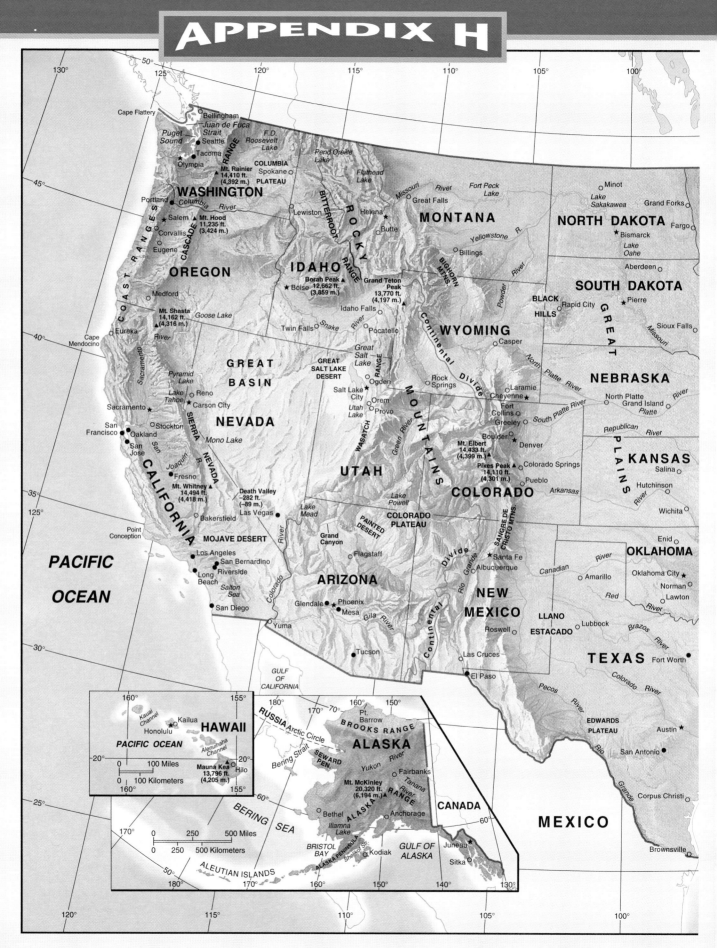

UNITED STATES

- ⊛ National capital
- ★ State capital
- ● Major city
- ○ Other city
- —— International boundary
- ---- State boundary

0 — 150 — 300 Miles
0 — 150 — 300 Kilometers

Projection: Albers Equal Area

APPENDIX I

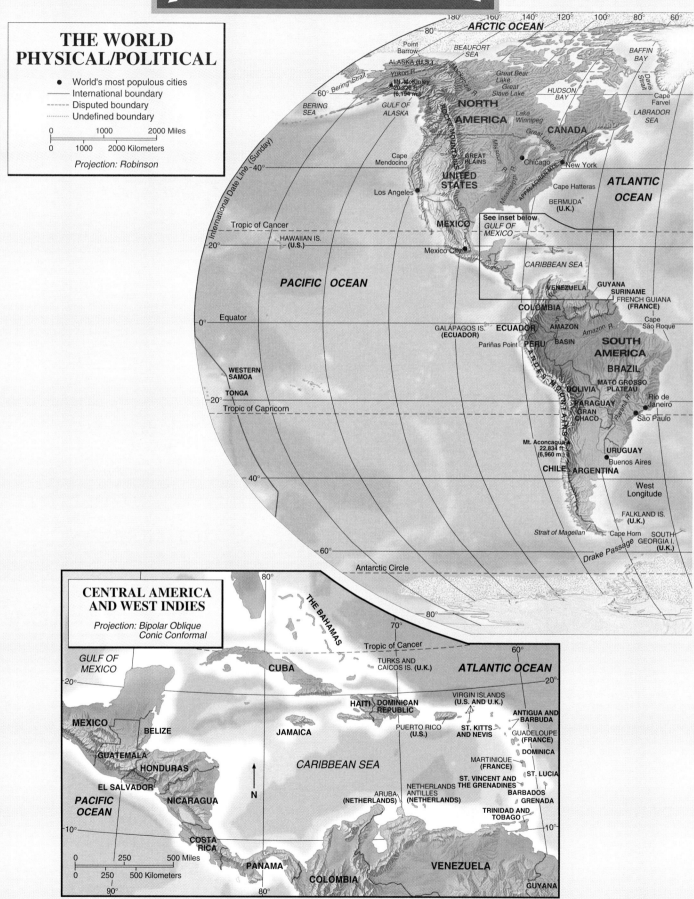

THE WORLD PHYSICAL/POLITICAL

- ● World's most populous cities
- —— International boundary
- ----- Disputed boundary
- ········· Undefined boundary

| 0 | 1000 | 2000 Miles |
| 0 | 1000 | 2000 Kilometers |

Projection: Robinson

CENTRAL AMERICA AND WEST INDIES

Projection: Bipolar Oblique Conic Conformal

| 0 | 250 | 500 Miles |
| 0 | 250 | 500 Kilometers |

628

ARCTIC OCEAN

COMMONWEALTH OF INDEPENDENT STATES

1	ARMENIA	6	KYRGYZSTAN
2	AZERBAIJAN	7	MOLDOVA
3	BELARUS	8	RUSSIA
4	GEORGIA	9	TAJIKISTAN
5	KAZAKSTAN	10	TURKMENISTAN
		11	UKRAINE
		12	UZBEKISTAN

40° 20° 0° 20° 40° 60° 80° 100° 120° 140° 160° 180°

KALAALLIT NUNAAT (GREENLAND) (DENMARK)

SVALBARD IS. (NORWAY)
FRANZ JOSEF IS. (RUSSIA)
Cape Zelaniya
KARA SEA
LAPTEV SEA
EAST SIBERIAN SEA

GREENLAND SEA
NORWEGIAN SEA
JAN MAYEN (NORWAY)
North Cape
BARENTS SEA
VERKHOYANSK RANGE

Denmark Strait
ICELAND
Arctic Circle
See Inset below
URAL MOUNTAINS
SIBERIA
CENTRAL SIBERIAN PLATEAU

FAROE IS. (DENMARK)
NORTH SEA
Lake Ladoga
WEST SIBERIAN PLAIN
Ob R.
Yenisey R.
Lena R.
ASIA
YABLONOVI RANGE
SEA OF OKHOTSK

EUROPE
NORTH EUROPEAN PLAIN
Volga R.
RUSSIA
Lake Baikal
Cape Lopatka
KURIL IS. (RUSSIA)

ALPS
Danube
Mt. Elbrus 18,510 ft. (5,642 m.)
CASPIAN DEPRESSION
KAZAKSTAN
ALTAI MTNS.
MONGOLIA
GOBI
Changchun
Shenyang
NORTH KOREA
SEA OF JAPAN
JAPAN

Cape Finisterre
BLACK SEA
GEORGIA
ARMENIA
ARAL SEA
CASPIAN SEA
UZBEKISTAN
KYRGYZSTAN
TIANSHAN
Beijing
Tianjin
Seoul
SOUTH KOREA
Tokyo

AZORES IS. (PORTUGAL)
TURKEY
TURKMENISTAN
TAJIKISTAN
TAKLIMAKAN
CHINA
Mt. Everest 29,028 ft. (8,848 m.)
Chongqing
Wuhan
EAST CHINA SEA

MEDITERRANEAN SEA
LEBANON
SYRIA
AZERBAIJAN
AFGHANISTAN
HIMALAYAS
BHUTAN
Chang Jiang (Yangtze R.)
Shanghai
TAIWAN
Tropic of Cancer

MOROCCO
TUNISIA
ISRAEL
JORDAN
IRAQ
IRAN
PLATEAU OF IRAN
PAKISTAN
NEPAL
Delhi
Ganges R.
BANGLADESH
Calcutta
HONG KONG
MACAO (PORTUGAL)

CANARY IS. (SPAIN)
W. SAHARA (MOROCCO)
ALGERIA
LIBYA
EGYPT
QATTARA DEPRESSION
Cairo
KUWAIT
QATAR
BAHRAIN
SAUDI ARABIA
UNITED ARAB EMIRATES
INDIA
Mumbai (Bombay)
MYANMAR
LAOS
VIETNAM
SOUTH CHINA SEA
Manila
MARSHALL ISLANDS

Cape Blanc
SAHARA
MAURITANIA
MALI
NIGER
CHAD
SUDAN
YEMEN
OMAN
ARABIAN SEA
BAY OF BENGAL
THAILAND
PHILIPPINES
GUAM (U.S.)

CAPE VERDE
SENEGAL
GAMBIA
GUINEA-BISSAU
GUINEA
BURKINA FASO
NIGERIA
BENIN
AFRICA
CENTRAL AFRICAN REP.
ERITREA
DJIBOUTI
ETHIOPIA
Cape Asir
CAMBODIA
BRUNEI
MALAYSIA
FEDERATED STATES OF MICRONESIA

SIERRA LEONE
LIBERIA
CÔTE D'IVOIRE
GHANA
TOGO
CAMEROON
SÃO TOMÉ AND PRÍNCIPE
Cape Comorin
SRI LANKA
SINGAPORE
PALAU

EQUATORIAL GUINEA
GABON
CONGO
RWANDA
CONGO (ZAIRE) BASIN
UGANDA
KENYA
Lake Victoria
SOMALIA
MALDIVES
Equator
KIRIBATI

Zaire
ZAIRE
BURUNDI
Kilimanjaro 19,340 ft. (5,895 m.)
TANZANIA
SEYCHELLES
INDONESIA
PAPUA NEW GUINEA
SOLOMON ISLANDS
NAURU

ANGOLA
MALAWI
ZAMBIA
MOZAMBIQUE
COMOROS
Jakarta
INDIAN OCEAN
Cape York
CORAL SEA
TUVALU

NAMIBIA
ZIMBABWE
BOTSWANA
MADAGASCAR
MAURITIUS
RÉUNION (FRANCE)
Tropic of Capricorn
WESTERN PLATEAU
AUSTRALIA
GREAT DIVIDING RANGE
NEW CALEDONIA (FRANCE)
VANUATU
FIJI

ATLANTIC OCEAN
SOUTH AFRICA
SWAZILAND
LESOTHO
Mozambique Channel
Mt. Kosciusko 7,310 ft. (2,228 m.)
TASMAN SEA
NEW CALEDONIA (FRANCE)

Cape of Good Hope
East Longitude
Prime Meridian
N
KERGUELEN IS. (FRANCE)
NEW ZEALAND
International Date Line (Monday)

60°
Antarctic Circle
ANTARCTICA
80°

20° 0° 20° 40° 60° 80° 100° 120° 140° 160° 180°

EUROPE

Projection: Azimuthal Equal Area

FINLAND
NORWAY
SWEDEN
St. Petersburg
ESTONIA
LATVIA
Moscow
RUSSIA

IRELAND
UNITED KINGDOM
DENMARK
LITHUANIA
RUSSIA
BELARUS

London
NETHERLANDS
GERMANY
POLAND
UKRAINE

ATLANTIC OCEAN
BELGIUM
LUXEMBOURG
CZECH REPUBLIC
SLOVAKIA
MOLDOVA

Paris
FRANCE
SWITZERLAND
AUSTRIA
HUNGARY
ROMANIA

SLOVENIA
CROATIA
BOSNIA HERZEGOVINA
SERBIA
YUGOSLAVIA

PORTUGAL
SPAIN
ITALY
MONTENEGRO
ALBANIA
MACEDONIA
BULGARIA
BLACK SEA
GEORGIA

GIBRALTAR (U.K.)
MEDITERRANEAN SEA
GREECE
TURKEY

TUNISIA
MALTA
CYPRUS
SYRIA
LEBANON

0 250 500 Miles
0 250 500 Kilometers

629

APPENDIX J

PERIODIC TABLE OF THE ELEMENTS

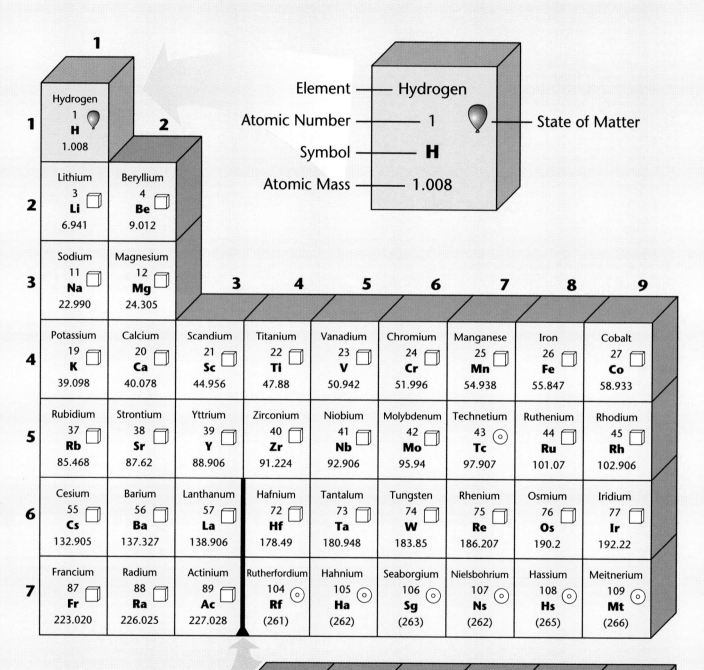

Lanthanide Series

Actinide Series

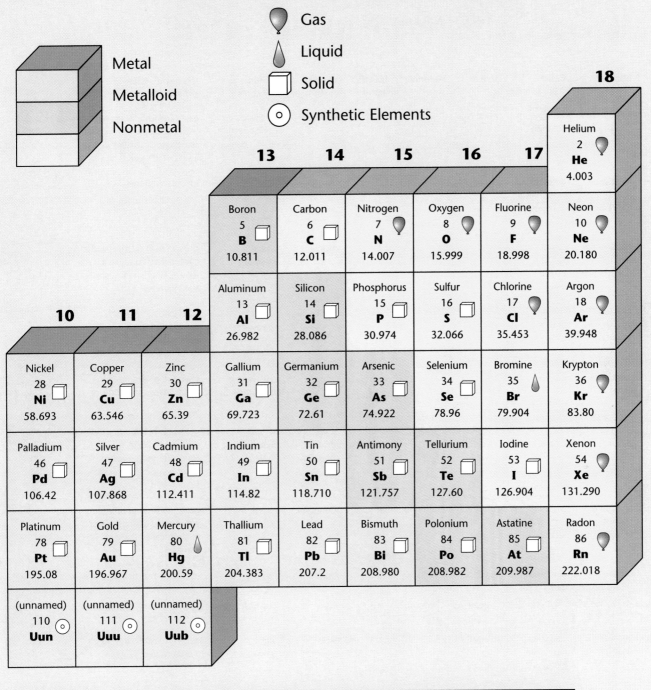

Gas

Liquid

Solid

⊙ Synthetic Elements

Metal

Metalloid

Nonmetal

18

Helium
2
He
4.003

13 **14** **15** **16** **17**

| Boron 5 **B** 10.811 | Carbon 6 **C** 12.011 | Nitrogen 7 **N** 14.007 | Oxygen 8 **O** 15.999 | Fluorine 9 **F** 18.998 | Neon 10 **Ne** 20.180 |

| Aluminum 13 **Al** 26.982 | Silicon 14 **Si** 28.086 | Phosphorus 15 **P** 30.974 | Sulfur 16 **S** 32.066 | Chlorine 17 **Cl** 35.453 | Argon 18 **Ar** 39.948 |

10 **11** **12**

| Nickel 28 **Ni** 58.693 | Copper 29 **Cu** 63.546 | Zinc 30 **Zn** 65.39 | Gallium 31 **Ga** 69.723 | Germanium 32 **Ge** 72.61 | Arsenic 33 **As** 74.922 | Selenium 34 **Se** 78.96 | Bromine 35 **Br** 79.904 | Krypton 36 **Kr** 83.80 |

| Palladium 46 **Pd** 106.42 | Silver 47 **Ag** 107.868 | Cadmium 48 **Cd** 112.411 | Indium 49 **In** 114.82 | Tin 50 **Sn** 118.710 | Antimony 51 **Sb** 121.757 | Tellurium 52 **Te** 127.60 | Iodine 53 **I** 126.904 | Xenon 54 **Xe** 131.290 |

| Platinum 78 **Pt** 195.08 | Gold 79 **Au** 196.967 | Mercury 80 **Hg** 200.59 | Thallium 81 **Tl** 204.383 | Lead 82 **Pb** 207.2 | Bismuth 83 **Bi** 208.980 | Polonium 84 **Po** 208.982 | Astatine 85 **At** 209.987 | Radon 86 **Rn** 222.018 |

| (unnamed) 110 ⊙ **Uun** | (unnamed) 111 ⊙ **Uuu** | (unnamed) 112 ⊙ **Uub** |

| Gadolinium 64 **Gd** 157.25 | Terbium 65 **Tb** 158.925 | Dysprosium 66 **Dy** 162.50 | Holmium 67 **Ho** 164.930 | Erbium 68 **Er** 167.26 | Thulium 69 **Tm** 168.934 | Ytterbium 70 **Yb** 173.04 | Lutetium 71 **Lu** 174.967 |

| Curium 96 ⊙ **Cm** 247.070 | Berkelium 97 ⊙ **Bk** 247.070 | Californium 98 ⊙ **Cf** 251.080 | Einsteinium 99 ⊙ **Es** 252.083 | Fermium 100 ⊙ **Fm** 257.095 | Mendelevium 101 ⊙ **Md** 258.099 | Nobelium 102 ⊙ **No** 259.101 | Lawrencium 103 ⊙ **Lr** 260.105 |

APPENDIX K

Minerals with Nonmetallic Luster

Mineral (formula)	Color	Streak	Hardness	Specific gravity	Crystal system	Breakage pattern	Uses and other properties
talc ($Mg_3Si_4O_{10}(OH)_2$)	white, greenish	white	1	2.8	monoclinic	cleavage in one direction	easily cut with fingernail; used for talcum powder; soapstone; is used in paper and for tabletops
apatite (Ca_5 (F,Cl, OH)$(PO_4)_3$)	green, blue, violet, brown, colorless, yellow	colorless	5	3.2	hexagonal	basal cleavage	used as a fertilizer
kaolinite ($Al_2Si_2O_5(OH)_4$)	white, red, reddish brown, black	white	2	2.6	triclinic	basal cleavage	clays; used in ceramics and in china dishes; common in most soils; often microscopic-sized particles
gypsum ($CaSO_4 \cdot 2H_2O$)	colorless, gray, white, brown	white	2	2.3	monoclinic	basal cleavage	used extensively in the preparation of plaster of paris, alabaster, and dry wall for building construction
sphalerite (ZnS)	brown	pale yellow	3.5-4	4	cubic	cleavage in six directions	main ore of zinc; used in paints, dyes, and medicine
sulfur (S)	yellow	yellow to white	2	2.0	ortho-rhombic	conchoidal fracture	used in medicine, fungicides for plants, vulcanization of rubber, production of sulfuric acid
muscovite ($KAl_2(Al_3Si_3O_{10})(OH)_2$)	white, light gray, yellow, rose, green	colorless	2.5	2.8	monoclinic	basal cleavage	occurs in large flexible plates; used as an insulator in electrical equipment, lubricant
biotite ($K(Mg,Fe)_3(AlSi_3O_{10})(OH)_2$)	black to dark brown	colorless	2.5	2.8-3.4	monoclinic	basal cleavage	occurs in large flexible plates
halite ($NaCl$)	colorless, red, white, blue	colorless	2.5	2.1	cubic	cubic cleavage	salt; very soluble in water; a preservative
calcite ($CaCO_3$)	colorless, white, pale blue	colorless, white	3	2.7	hexagonal	cleavage in three directions	fizzes when HCl is added; used in cements and other building materials
dolomite ($CaMg(CO_3)_2$)	colorless, white, pink, green, gray, black	white	3.5-4	2.8	hexagonal	cleavage in three directions	concrete and cement, used as an ornamental building stone

Mineral (formula)	Color	Streak	Hardness	Specific gravity	Crystal system	Breakage pattern	Uses and other properties
fluorite (CaF_2)	colorless, white, blue, green, red, yellow, purple	colorless	4	3-3.2	cubic	cleavage	used in the manufacture of optical equipment; glows under ultraviolet light
limonite (hydrous iron oxides)	yellow, brown, black	yellow, brown	5.5	2.7-4.3	–	conchoidal fracture	source of iron; weathers easily, coloring matter of soils
hornblende $(Ca,Na)_{2-3}$ $(Mg,Fe,Al,)_5$ $Si_6(Si,Al)_2$ $O_{22}(OH)_2$	green to black	gray to white	5-6	3.4	monoclinic	cleavage in two directions	will transmit light on thin edges; 6-sided cross section
feldspar (orthoclase) $KAlSi_3O_8$	colorless, white to gray, green and yellow	colorless	6	2.5	monoclinic	two cleavage planes meet at 90° angle	insoluble in acids; used in the manufacture of porcelain
feldspar (plagioclase) $(NaAlSi_3O_8)$ $(CaAl_2Si_2O_8)$	gray, green, white, pink	colorless	6	2.5	triclinic	two cleavage planes meet at 86° angle	used in ceramics; striations present on some faces
augite $((Ca,Na)$ (Mg,Fe,Al) $(Al, Si)_2 O_6)$	black	colorless	6	3.3	monoclinic	2-directional cleavage	square or 8-sided cross section
olivine $((Mg,Fe)_2$ $SiO_4)$	olive green	colorless	6.5	3.5	ortho-rhombic	conchoidal fracture	gemstones, refractory sand
quartz (SiO_2)	colorless, various colors	colorless	7	2.6	hexagonal	conchoidal fracture	used in glass manufacture, electronic equipment, radios, computers, watches, gemstones
garnet $(Mg,Fe,Ca)_3$ $(Al_2Si_3O_{12})$	deep yellow-red, green, black	colorless	7.5	3.5	cubic	conchoidal fracture	used in jewelry, also used as an abrasive
topaz (Al_2SiO_4) $(F,OH)_2)$	white, pink, yellow, pale blue, colorless	colorless	8	3.5	ortho-rhombic	basal cleavage	valuable gemstone
corundum (Al_2O_3)	colorless, blue, brown, green, white, pink,	colorless	9	4.0	hexagonal	fracture	gemstones; ruby is red, sapphire is blue; industrial abrasive

Minerals with Metallic Luster

Mineral (formula)	Color	Streak	Hardness	Specific gravity	Crystal system	Breakage pattern	Uses and other properties
graphite (C)	black to gray	black to gray	1-2	2.3	hexagonal	basal cleavage (scales)	pencil lead, lubricants for locks, rods to control some small nuclear reactions, battery poles
silver (Ag)	silvery white, tarnishes to black	light gray to silver	2.5	10-12	cubic	hackly	coins, fillings for teeth, jewelry, silver plate, wires; malleable and ductile
galena (Pbs)	gray	gray to black	2.5	7.5	cubic	cubic cleavage perfect	source of lead, used in pipes, shields for X rays, fishing equipment sinkers
gold (Au)	pale to golden yellow	yellow	2.5-3	19.3	cubic	hackly	jewelry, money, gold leaf, fillings for teeth, medicines; does not tarnish
bornite (Cu₅FeS₄)	bronze, tarnishes to dark blue, purple	gray-black	3	4.9-5.4	tetragonal	uneven fracture	source of copper; called "peacock ore" because of the purple shine when it tarnishes
copper (Cu)	copper red	copper red	3	8.5-9	cubic	hackly	coins, pipes, gutters, wire, cooking utensils, jewelry, decorative plaques; malleable and ductile
chalcopyrite (CuFeS₂)	brassy to golden yellow	greenish black	3.5-4	4.2	tetragonal	uneven fracture	main ore of copper
chromite (FeCr₂O₄)	black or brown	brown to black	5.5	4.6	cubic	irregular fracture	ore of chromium, stainless steel, metallurgical bricks
pyrrhotite (FeS)	bronze	gray-black	4	4.6	hexagonal	uneven fracture	often found with pentlandite, an ore of nickel; may be magnetic
hematite (specular) (Fe₂O₃)	black or reddish brown	red or reddish brown	6	5.3	hexagonal	irregular fracture	source of iron; roasted in a blast furnace, converted to "pig" iron, made into steel
magnetite (Fe₃O₄)	black	black	6	5.2	cubic	conchoidal fracture	source of iron, naturally magnetic, called lodestone
pyrite (FeS₂)	light, brassy yellow	greenish black	6.5	5.0	cubic	uneven fracture	source of iron, "fool's gold," alters to limonite

SKILL HANDBOOK

Table of Contents

Organizing Information

▶ Classifying

You may not realize it, but you make things orderly in the world around you. If you hang your shirts together in the closet, if your socks take up a particular corner of a dresser drawer, or if your favorite CDs are stacked together, you have used the skill of classifying.

Classifying is the process of sorting objects or events into groups based on common features. When classifying, first observe the objects or events to be classified. Then, select one feature that is shared by most members in the group but not by all. Place those members that share the feature into a subgroup. You can classify members into smaller and smaller subgroups based on characteristics.

How would you classify a collection of CDs? You might classify those you like to dance to in one subgroup and CDs you like to listen to in the next column, as in the diagram. The CDs you like to dance to could be subdivided into a rap subgroup and a rock subgroup. Note that for each feature selected, each CD only fits into one subgroup. Keep select-

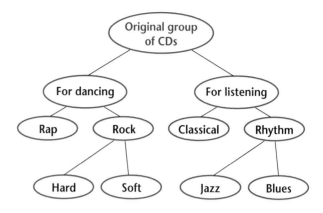

ing features until all the CDs are classified. The diagram above shows one possible classification.

Remember, when you classify, you are grouping objects or events for a purpose. Keep your purpose in mind as you select the features to form groups and subgroups.

▶ Sequencing

A sequence is an arrangement of things or events in a particular order. A sequence with which you are most familiar is the use of alphabetical order. Another example of sequence would be the steps in a recipe. Think about baking chocolate chip cookies. Steps in the recipe have to be followed in order for the cookies to turn out right.

When you are asked to sequence objects or events within a group, figure out what comes first, then think about what should come second. Continue to choose objects or events until all of the objects you started out with are in order. Then, go back over the sequence to make sure each thing or event in your sequence logically leads to the next.

▶ Concept Mapping

If you were taking an automobile trip, you would probably take along a road map. The road map shows your location, your destination, and other places along the way. By looking at the map and finding where you are, you can begin to understand where you are in relation to other locations on the map.

A concept map is similar to a road map. But, a concept map shows relationships among ideas (or concepts) rather than places. A concept map is a diagram that visually shows how concepts are related. Because the concept map shows relationships among ideas, it can make the meanings of ideas and terms clear, and help you understand better what you are studying.

Network Tree   Look at the concept map about Protists. This is called a network tree. Notice how some words are circled while others are written across connecting lines. The circled words are science concepts. The lines in the map show related concepts. The words written on the lines describe the relationships between concepts.

Network Tree

When you are asked to construct a network tree, write down the topic and list the major concepts related to that topic on a piece of paper. Then look at your list and begin to put them in order from general to specific. Branch the related concepts from the major concept and describe the relationships on the lines. Continue to write the more specific concepts. Write the relationships between the concepts on the lines until all concepts are mapped. Examine the concept map for relationships that cross branches, and add them to the concept map.

Events Chain   An events chain is another type of concept map. An events chain map, such as the one on the effects of gravity, is used to describe ideas in order. In science, an

Events Chain

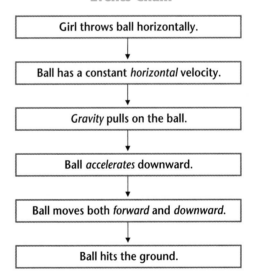

events chain can be used to describe a sequence of events, the steps in a procedure, or the stages of a process.

When making an events chain, first find the one event that starts the chain. This event is called the initiating event. Then, find the next event in the chain and continue until you reach an outcome. Suppose you are asked to describe what happens when someone throws a ball horizontally. An events chain map describing the steps might look like the one on page 637. Notice that connecting words are not necessary in an events chain.

Cycle Map   A cycle concept map is a special type of events chain map. In a cycle concept map, the series of events does not produce a

Cycle Map

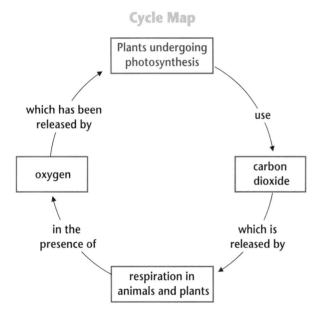

final outcome. Instead, the last event in the chain relates back to the initiating event. Look at the cycle map for photosynthesis.

As in the events chain map, you first decide on an initiating event and then list each event in order. Since there is no outcome and the last event relates back to the initiating event, the cycle repeats itself. Look at the cycle map of insect metamorphosis.

Spider Map   A fourth type of concept map is the spider map. This is a map that you can use for brainstorming. Once you have a central idea, you may find you have a jumble of ideas that relate to it, but are not necessarily clearly related to each other. By writing these

Spider Map

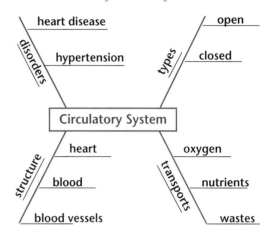

ideas outside the main concept, you may begin to separate and group unrelated terms so that they become more useful.

There is usually not one correct way to create a concept map. As you construct one type of map, you may discover other ways to construct the map that show the relationships between concepts in a better way. If you do discover what you think is a better way to create a concept map, go ahead and use the new way. Overall, concept maps are useful for breaking a big concept down into smaller parts, making learning easier.

▶ Making and Using Tables

Browse through your textbook, and you will notice tables in the text and in the activities. In a table, data or information is arranged in such a way that makes it easier for you to understand. Activity tables help organize the data you collect during an activity so that results can be interpreted more easily.

Parts of a Table   Most tables have a title. At a glance, the title tells you what the table is about. A table is divided into columns and rows. The first column lists items to be compared. In the table shown to the right, different magnitudes of force are being compared. The row across the top lists the specific characteristics being compared. Within the grid of the table, the collected data is recorded. Look at the features of the table in the next column.

What is the title of this table? The title is "Earthquake Magnitude." What is being compared? The distance away from the epicenter that tremors are felt and the average number of earthquakes expected per year are being compared for different magnitudes on the Richter scale.

Using Tables   What is the average number of earthquakes expected per year for an earthquake with a magnitude of 5.5 at the focus? Locate the column labeled "Average number expected per year" and the row "5.0 to 5.9." The data in the box where the column and row intersect is the answer. Did you answer "800"? What is the distance away from the epicenter for an earthquake with a

Earthquake Magnitude		
Magnitude at Focus	Distance from Epicenters that Tremors are Felt	Average Number Expected Per Year
1.0 to 3.9	24 km	>100 000
4.0 to 4.9	48 km	6200
5.0 to 5.9	112 km	800
6.0 to 6.9	200 km	120
7.0 to 7.9	400 km	20
8.0 to 8.9	720 km	<1

magnitude of 8.1? If you answered "720 km," you understand how to use the parts of a table.

Making Tables   To make a table, list the items to be compared down in columns and the characteristics to be compared across in rows. Make a table and record the data comparing the mass of recycled materials collected by a class. On Monday, students turned in 4 kg of paper, 2 kg of aluminum, and 0.5 kg of plastic. On Wednesday, they turned in 3.5 kg of paper, 1.5 kg of aluminum, and 0.5 kg of plastic. On Friday, the totals were 3 kg of paper, 1 kg of aluminum, and 1.5 kg of plastic. If your table looks like the one shown below, you are able to make tables to organize data.

Recycled Materials			
Day of Week	Paper (kg)	Aluminum (kg)	Plastic (kg)
Mon.	4	2	0.5
Wed.	3.5	1.5	0.5
Fri.	3	1	1.5

▶ Making and Using Graphs

After scientists organize data in tables, they may display the data in a graph. A graph is a diagram that shows how variables compare. A graph makes interpretation and analysis of data easier. There are three basic types of graphs used in science—the line graph, the bar graph, and the pie graph.

Line Graphs A line graph is used to show the relationship between two variables. The variables being compared go on two axes of the graph. The independent variable always goes on the horizontal axis, called the *x*-axis. The dependent variable always goes on the vertical axis, called the *y*-axis.

Suppose a school started a peer study program with a class of students to see how science grades were affected.

Average Grades of Students in Study Program

Grading Period	Average Science Grade
First	81
Second	85
Third	86
Fourth	89

You could make a graph of the grades of students in the program over the four grading periods of the school year. The grading period is the independent variable and is placed on the *x*-axis of your graph. The average grade of the students in the program is the dependent variable and would go on the *y*-axis.

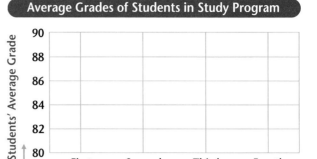

After drawing your axes, you would label each axis with a scale. The *x*-axis simply lists the four grading periods. To make a scale of grades on the *y*-axis, you must look at the data values. Since the lowest grade was 81 and the highest was 89, you know that you will have to start numbering at least at 81 and go through 89. You decide to start numbering at 80 and number by twos through 90.

Next, plot the data points. The first pair of data you want to plot is the first grading period and 81. Locate "First" on the *x*-axis and locate "81" on the *y*-axis. Where an imaginary vertical line from the *x*-axis and an imaginary horizontal line from the *y*-axis would meet, place the first data point. Place the other data points the same way. After all the points are plotted, connect them with straight lines.

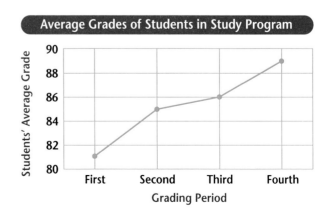

Bar Graphs Bar graphs are similar to line graphs. They compare data that do not continuously change. In a bar graph, vertical bars show the relationships among data.

To make a bar graph, set up the *x*-axis and *y*-axis as you did for the line graph. The data is plotted by drawing vertical bars from the *x*-axis up to a point where the *y*-axis would meet the bar if it were extended.

Look at the bar graph comparing the masses lifted by an electromagnet with different numbers of dry cell batteries. The *x*-axis is the number of dry cell batteries, and the *y*-axis is the mass lifted.

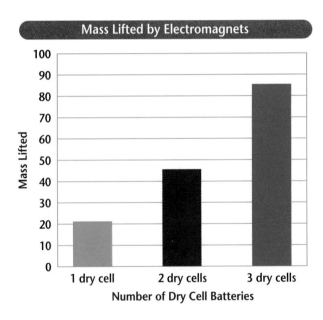

Mass Lifted by Electromagnets

Pie Graphs A pie graph uses a circle divided into sections to display data. Each section represents part of the whole. All the sections together equal 100 percent.

Suppose you wanted to make a pie graph to show the number of seeds that germinated in a package. You would have to count the total number of seeds and the number of seeds that germinated out of the total.

You find that there are 143 seeds in the package. This represents 100 percent, the whole pie.

You plant the seeds, and 129 seeds germinate. The seeds that germinated will make up one section of the pie graph, and the seeds that did not germinate will make up the remaining section.

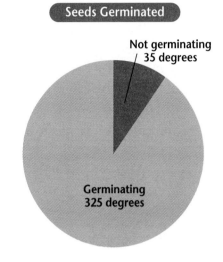

Seeds Germinated

To find out how much of the pie each section should take, divide the number of seeds in each section by the total number of seeds. Then multiply your answer by 360, the number of degrees in a circle, and round to the nearest whole number. The section of the pie graph in degrees that represents the seeds germinated is figured below.

$$\frac{129}{143} \times 360 = 324.75 \text{ or } 325 \text{ degrees}$$

Plot this group on the pie graph using a compass and a protractor. Use the compass to draw a circle. Then, draw a straight line from the center to the edge of the circle. Place your protractor on this line and use it to mark a point on the edge of the circle at 325 degrees. Connect this point with a straight line to the center of the circle. This is the section for the group of seeds that germinated. The other section represents the group of 14 seeds that did not germinate. Label the sections of your graph and title the graph.

Processing Information Critically

▶ Observing and Inferring

Imagine that you have just finished a volleyball game. At home, you open the refrigerator and see a jug of orange juice on the back of the top shelf. The jug feels cold as you grasp it. Then you drink the juice, smell the oranges, and enjoy the tart taste in your mouth.

As you imagined yourself in the story, you used your senses to make observations. You used your sense of sight to find the jug in the refrigerator, your sense of touch when you felt the coldness of the jug, your sense of hearing to listen as the liquid filled the glass, and your senses of smell and taste to enjoy the odor and tartness of the juice. The basis of all scientific investigation is observation.

Scientists try to make careful and accurate observations. When possible, they use instruments, such as microscopes. Other instruments, such as thermometers or a pan balance, measure observations.

Measurements provide numerical data that can be checked and repeated.

When you make observations in science, you'll find it helpful to examine the entire object or situation first. Then, look carefully for details. Write down everything you see before using other senses to make additional observations.

Scientists often make inferences based on their observations. An inference is an attempt to explain or interpret observations or to say what caused what you observed. For example, if you observed a CLOSED sign in a store window around noon, you might infer the owner is taking a lunch break. But, it's also possible that the owner has a doctor's appointment or has taken the day off to go fishing. The only way to be sure your inference is correct is to investigate further.

When making an inference, be certain to use accurate data and observations. Analyze all of the data that you've collected. Then, based on everything you know, explain or interpret what you've observed.

▶ Comparing and Contrasting

Observations can be analyzed by noting the similarities and differences between two or more objects or events that you observe. When you look at objects or events to see how they are similar, you are comparing them. Contrasting is looking for differences in similar objects or events.

Suppose you were asked to compare and contrast the planets Venus and Earth. You would start by looking at what is known about these planets. Then, make two columns on a piece of paper and list ways the planets are similar in one column and ways

Comparison of Venus and Earth		
Properties	Earth	Venus
Diameter (km)	12 742	12 112
Average density (g/cm³)	5.5	5.3
Percentage of sunlight reflected	39	76
Daytime surface temperature	300	750
Number of satellites	1	0

they are different in the other.

Similarities you might point out are that both planets are similar in size, shape, and mass. Differences include Venus having a hotter surface temperature and a dense, cloudy atmosphere that reflects more sunlight than Earth. Also, Venus lacks a moon.

▶ Recognizing Cause and Effect

Have you ever watched something happen and then made a suggestion as to why or how it happened? If so, you have observed and inferred. The event is an effect, and the reason for the event is the cause.

Suppose that every time your teacher fed the fish in a classroom aquarium, she or he tapped the food container on the edge of the aquarium. Then, one day your teacher just happened to tap the edge of the aquarium with a pencil while making a point about an ecology lesson. You observed the fish swim to the surface of the aquarium to feed. What is the effect, and what would you infer to be the cause? The effect is the fish swimming to the surface of the aquarium. You might infer the cause to be the teacher tapping on the edge of the aquarium. In determining cause and effect, you have made a logical inference based on your observations.

Perhaps the fish swam to the surface because they reacted to the teacher's waving hand or for some other reason. When scientists are unsure of the cause of a certain event, they design controlled experiments to determine what causes the event. Although you have made a logical conclusion about the behavior of the fish, you would have to perform an experiment to be certain that it was the tapping that caused the effect you observed.

▶ Measuring in SI

The metric system is a system of measurement developed by a group of scientists in 1795. It helps scientists avoid problems by providing standard measurements that all scientists around the world can understand. A modern form of the metric system, called the International System, or SI, was adopted for worldwide use in 1960.

Metric Prefixes			
Prefix	Symbol	Meaning	
kilo-	k	1000	thousand
hecto-	h	100	hundred
deka-	da	10	ten
deci-	d	0.1	tenth
centi-	c	0.01	hundredth
milli-	m	0.001	thousandth

The metric system is convenient because unit sizes vary by multiples of 10. When changing from smaller units to larger units, divide by 10. When changing from larger units to smaller, you multiply by 10. For example, to convert millimeters to centimeters, divide the millimeters by 10. To convert 30 millimeters to centimeters, divide 30 by 10 (30 millimeters equals 3 centimeters).

Prefixes are used to name units. Look at the table for some common metric prefixes and their meanings. Do you see how the prefix *kilo-* attached to the unit *gram* is *kilogram*, or 1000 grams? The prefix *deci-* attached to the unit *meter* is *decimeter*, or one-tenth (0.1) of a meter.

Length   You have probably measured lengths or distances many times. The meter is the SI unit used to measure length. A baseball bat is about one meter long. When measuring smaller lengths, the meter is divided into smaller units called centimeters and millimeters. A centimeter is one-hundredth (0.01) of a meter, which is about the size of the width of the fingernail on your ring finger. A millimeter is one-thousandth of a meter (0.001), about the thickness of a dime.

Most metric rulers have lines indicating centimeters and millimeters. The centimeter lines are the longer, numbered lines, and the shorter lines are millimeter lines. When using a metric ruler, line up the 0 centimeter mark with the end of the object being measured, and read the number of the unit where the object ends.

Surface Area   Units of length are also used to measure surface area. The standard unit of area is the square meter (m^2). A square that's one meter long on each

side has a surface area of one square meter. Similarly, a square centimeter (cm^2) is one centimeter long on each side. The surface area of an object is determined by multiplying the length times the width.

Volume   The volume of a rectangular solid is also calculated using units of length. The cubic meter (m³) is the standard SI unit of volume. A cubic meter is a cube one meter on each side. You can determine the volume of rectangular solids by multiplying length times width times height.

Liquid Volume   During science activities, you will measure liquids using beakers and graduated cylinders marked in milliliters. A graduated cylinder is a cylindrical container marked with lines from bottom to top.

Liquid volume is measured using a unit called a liter. A liter has the volume of 1000 cubic centimeters. Since the prefix *milli-* means thousandth (0.001), a milliliter equals one cubic centimeter. One milliliter of liquid would completely fill a cube measuring one centimeter on each side.

Mass   Scientists use balances to find the mass of objects in grams. You will use a beam balance similar to the one illustrated. Notice that on one side of the balance is a pan and on the other side is a set of beams. Each beam has an object of a known mass called a *rider* that slides on the beam.

Before you find the mass of an object, set the balance to zero by sliding all the riders back to the zero point. Check the pointer on the right to make sure it swings an equal distance above and below the zero point on the scale. If the swing is unequal, find and turn the adjusting screw until you have an equal swing.

Place an object on the pan. Slide the rider with the largest mass along its beam until the pointer drops below zero. Then move it back one notch. Repeat the process on each beam until the pointer swings an equal distance above and below the zero point. Add the masses on each beam to find the mass of the object.

You should never place a hot object or pour chemicals directly on the pan. Instead, find the mass of a clean beaker or a glass jar. Place the dry or liquid chemicals in the container. Then find the combined mass of the container and the chemicals. Calculate the mass of the chemicals by subtracting the mass of the empty container from the combined mass.

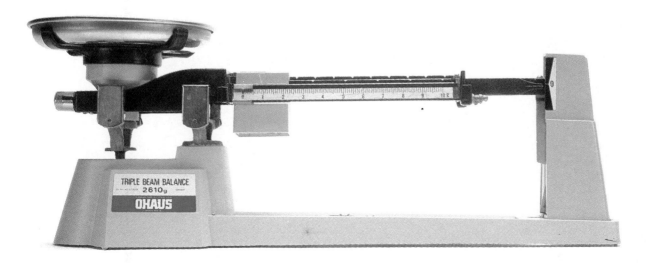

Practicing Scientific Methods

You might say that the work of a scientist is to solve problems. But when you decide how to dress on a particular day, you are doing problem solving, too. You may observe what the weather looks like through a window. You may go outside and see if what you are wearing is warm or cool enough.

Scientists use an orderly approach to learn new information and to solve problems. The methods scientists may use include observing, forming a hypothesis, testing a hypothesis, separating and controlling variables, and interpreting data.

▶ Observing

You observe all the time. Anytime you smell wood burning, touch a pet, see

lightning, taste food, or hear your favorite music, you are observing. Observation gives you information about events or things. Scientists try to observe as much as possible about the things and events they study so that they can know that what they say about their observations is reliable.

Some observations describe something using only words. These observations are called qualitative observations. If you were making qualitative observations of a dog, you might use words such as furry, brown, short-haired, or short-eared.

Other observations describe how much of something there is. These are quantitative observations and use numbers as well as words in the description. Tools or equipment are used to measure the characteristic being described. Quantitative observations of a dog might include a mass of 459 g, a height of 27 cm, ear length of 14 mm, and an age of 283 days.

▶ Using Observations to Form a Hypothesis

Suppose you want to make a perfect score on a spelling test. Begin by thinking of several ways to accomplish this. Base these possibilities on past observations. If you put each of these possibilities into sentence form, using the words if and then, you can form a hypothesis. All of the following are hypotheses you might consider to explain how you could score 100 percent on your test:

> If the test is easy, then I will get a perfect score.

> If I am intelligent, then I will get a perfect score.

Scientists make hypotheses that they can test to explain the observations they have made. Perhaps a scientist has observed that plants that receive fertilizer grow taller than plants that do not. A scientist may form a hypothesis that says: If plants are fertilized, then their growth will increase.

▶ Designing an Experiment to Test a Hypothesis

Once you state a hypothesis, you probably want to find out whether or not it explains an event or an observation. This requires a test. A hypothesis must be something you can test. To test a hypothesis, you design and carry out an experiment. Experiments involve planning and materials. Let's figure out how to conduct an experiment to test the hypothesis stated before about the effects of fertilizer on plants.

First, you need to write out a procedure. A procedure is the plan that you follow in your experiment. A procedure tells you what materials to use and how to use them. In this experiment, your plan may involve using ten bean plants that are each 15 cm tall (to begin with) in two groups, Groups A and B. You will water the five bean plants in Group A with 200 mL of plain water and no fertilizer twice a week for three weeks. You will treat the five bean plants in Group B with 200 mL of fertilizer solution twice a week for three weeks.

You will need to measure all the plants in both groups at the beginning of the experiment and again at the end of the three-week period. These measurements will be the data that you record in a table. A sample table has been done for you. Look at the data in the table for this experiment. From the data, you can draw a conclusion and make a statement about your results. If the conclusion you draw from the data supports your hypothesis, then you can say that your hypothesis is

Growing Bean Plants		
Plants	Treatment	Height 3 Weeks Later
Group A	no fertilizer added to soil	17 cm
Group B	3 g fertilizer added to soil	31 cm

reliable. Reliable means that you can trust your conclusion. If it did not support your hypothesis, then you would have to make new observations and state a new hypothesis, one that you could also test.

▶ Separating and Controlling Variables

In the experiment with the bean plants, you made everything the same except for treating one group (Group B) with fertilizer. In any experiment, it is important to keep everything the same, except for the item you are testing. In the experiment, you kept the type of plants, their beginning heights, the soil, the frequency with which you watered them, and the amount of water or fertilizer all the same, or constant. By doing so, you made sure that at the end of three weeks, any change you saw was the result of whether or not the plants had been fertilized. The only thing that you changed, or varied, was the use of fertilizer. In an experiment, the one factor that you change (in this case, the fertilizer), is called the independent variable. The factor that changes (in this case, growth) as a result of the independent variable is called the dependent variable. Always make sure that there is only one independent variable. If you allow more than one, you will not know what causes any change you observe in the dependent variable.

Many experiments also have a control, a treatment that you can compare with the results of your test groups. In this case, Group A was the control because it was not treated with fertilizer. Group B was the test group. At the end of three weeks, you were able to compare Group A with Group B and draw a conclusion.

▶ Interpreting Data

The word *interpret* means to explain the meaning of something. Information, or data, needs to mean something. Look at the problem originally being explored and find out what the data shows. Perhaps you are looking at a table from an experiment designed to test the hypothesis: If plants are fertilized, then their growth will increase. Look back to the table showing the results of the bean plant experiment.

Identify the control group and the test group so you can see whether or not the variable has had an effect. In this example, Group A was the control and Group B was the test group. Now you need to check differences between the control and test groups. These differences may be qualitative or quantitative. A qualitative difference would be if the leaf colors of plants in Groups A and B were different. A quantitative difference would be the difference in numbers of centimeters of height among the plants in each group. Group B was in fact taller than Group A after three weeks.

If there are differences, the variable being tested may have had an effect. If there is no difference between the control and the test groups, the variable being tested apparently had no effect. From the data table in this experiment on page 647, it appears that fertilizer does have an effect on plant growth.

▶ What is Data?

In the experiment described on these pages, measurements have been taken so that at the end of the experiment, you had something concrete to interpret. You had numbers to work with. Not every experiment that you do will give you data in the form of numbers. Sometimes, data will be in the form of a description. At the end of a chemistry experiment, you might have noted that one solution turned yellow when treated with a particular chemical, and another remained clear, like water, when treated with the same chemical. Data therefore, is stated in different forms for different types of scientific experiments.

▶ Are All Experiments Alike?

Keep in mind as you perform experiments in science, that not every experiment makes use of all of the parts that have been described on these pages. For some, it may be difficult to design an experiment that will always have a control. Other experiments are complex enough that it may be hard to have only one dependent variable. Real scientists encounter many variations in the methods that they use when they perform experiments. The skills in this handbook are here for you to use and practice. In real situations, their uses will vary.

Representing and Applying Data

▶ Interpreting Scientific Illustrations

As you read this textbook, you will see many drawings, diagrams, and photographs. Illustrations help you to understand what you read. Some illustrations are included to help you understand an idea that you can't see easily by yourself. For instance, we can't see atoms, but we can look at a diagram of an atom and that helps us to understand some things about atoms. Seeing something often helps you remember more easily. The text may describe the surface of Jupiter in detail, but seeing a photograph of Jupiter may help you to remember that it has cloud bands. Illustrations also provide examples that clarify difficult concepts or give additional information about the topic you are studying. Maps, for example, help you to locate places that may be described in the text.

Captions and Labels Most illustrations have captions. A caption is a comment that identifies or explains the illustration. Diagrams, such as the one of the feather, often have labels that identify parts of the item shown or the order of steps in a process.

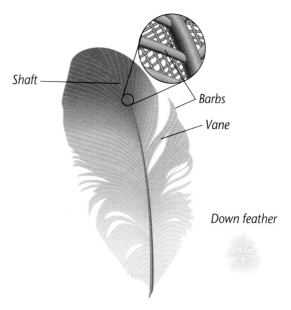

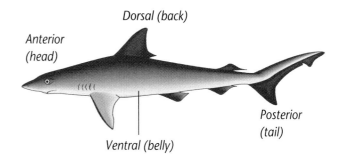

Shaft

Barbs

Vane

Down feather

Contour feather

Learning with Illustrations An illustration of an organism shows that organism from a particular view or orientation. In order to understand the illustration, you may need to identify the front (anterior) end, tail (posterior) end, the underside (ventral), and the back (dorsal) side of the organism shown.

Dorsal (back)

Anterior (head)

Posterior (tail)

Ventral (belly)

You might also check for symmetry. Look at the illustration on the following page. A shark has bilateral symmetry. This means that drawing an imaginary line through the center of the animal from the anterior to posterior end forms two mirror images.

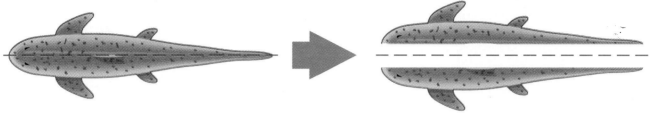

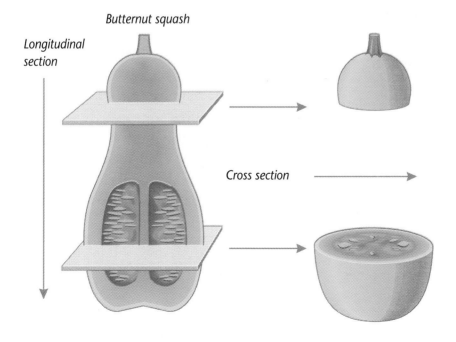

Bilateral symmetry

Two sides exactly alike

Radial symmetry is the arrangement of similar parts around a central point. An object or organism such as a hydra can be divided anywhere through the center into similar parts.

Some organisms and objects cannot be divided into two similar parts. If an organism or object cannot be divided, it is asymmetrical. Study the sponge. Regardless of how you try to divide a sponge, you cannot divide it into two parts that look alike.

Some illustrations enable you to see the inside of an organism or object. These illustrations are called sections.

Look at all illustrations carefully. Read captions and labels so that you understand exactly what the illustration is showing you.

Butternut squash

Longitudinal section

Cross section

▶ Making Models

Have you ever worked on a model car or plane or rocket? These models look, and sometimes work, just like the real thing, but they are usually much smaller than the real thing. In science, models are used to help simplify large processes or structures that may be difficult to understand. Your understanding of a structure or process is enhanced when you work with materials to make a model that shows the basic features of the structure or process.

In order to make a model, you first have to get a basic idea about the structure or process involved. You decide to make a model to show the differences in size of arteries, veins, and capillaries. First, read about these structures. All three are hollow tubes. Arteries are round and thick. Veins are flat and have thinner walls than arteries. Capillaries are very small.

Now, decide what you can use for your model. Common materials are often best and cheapest to work with when making models. Different

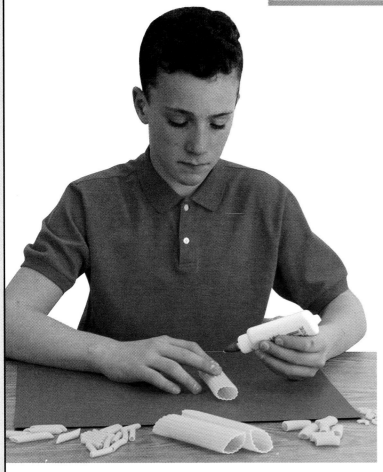

Predicting

When you apply a hypothesis, or general explanation, to a specific situation, you predict something about that situation. First, you must identify which hypothesis fits the situation you are considering. People use prediction to make everyday decisions. Based on previous observations and experiences, you may form a hypothesis that if it is wintertime, then temperatures will be lower. From past experience in your area, temperatures are lowest in February. You may then use this hypothesis to predict specific temperatures and weather for the month of February in advance. Someone could use these predictions to plan to set aside more money for heating bills during that month.

Sampling and Estimating

When working with large populations of organisms, scientists usually cannot observe or study every organism in the population. Instead, they use a sample or a portion of the population. Sampling is taking a small portion of organisms of a population for research. By making careful observations or manipulating variables with a portion of a group, information is discovered and conclusions are drawn that might then be applied to the whole population.

Scientific work also involves estimating. Estimating is making a judgment about the size of something or the number of something without actually measuring or counting every member of a population.

kinds and sizes of pasta might work for these models. Different sizes of rubber tubing might do just as well. Cut and glue the different noodles or tubing onto thick paper so the openings can be seen. Then label each. Now you have a simple, easy–to–understand model showing the differences in size of arteries, veins, and capillaries.

What other scientific ideas might a model help you to understand? A model of a molecule can be made from gumdrops (using different colors for the different elements present) and toothpicks (to show different chemical bonds). A working model of a volcano can be made from clay, a small amount of baking soda, vinegar, and a bottle cap. Other models can be devised on a computer.

Suppose you are trying to determine the effect of a specific nutrient on the growth of water lilies. It would be impossible to test the entire population of water lilies, so you would select part of the population for your experiment. Through careful experimentation and observation on a sample of the population, you could generalize the effect of the chemical on the entire population.

Here is a more familiar example. Have you ever tried to guess how many beans were in a sealed jar? If you did, you were estimating. What if you knew the jar of beans held one liter (1000 mL)? If you knew that 30 beans would fit in a 100–milliliter jar, how many beans would you estimate to be in the one–liter jar? If you said about 300 beans, your estimate would be close to the actual number of beans.

Scientists use a similar process to estimate populations of organisms from bacteria to buffalo. Scientists count the actual number of organisms in a small sample and then estimate the number of organisms in a larger area. For example, if a scientist wanted to count the number of microorganisms in a petri dish, a microscope could be used to count the number of organisms in a one square millimeter sample. To determine the total population of the culture, the number of organisms in the square millimeter sample is multiplied by the total number of millimeters in the culture.

GLOSSARY

This glossary defines each key term that appears in **bold type** in the text. It also indicates the chapter number and page number where you will find the word used.

abyssal plain: flat plains on the ocean bottom; created by sediments from the continental shelves and slopes filling in the hills and valleys on the ocean floor. (Chap. 12, p. 378)

action force: the force you exert when you push on something; every action force has an equal and opposite reaction force that occurs at exactly the same time—for example, as you walk on the ground (action force), the ground pushes back at you (reaction force). (Chap. 1, p. 37)

alveoli: thin-walled balloon-like sacs within your lungs; each alveolus is surrounded by capillaries; oxygen diffuses from the alveoli into the capillaries and then into your red blood cells; at the same time, carbon dioxide diffuses from the capillaries into the alveoli and is exhaled from the lungs. (Chap. 16, p. 489)

artery: thick, muscular vessel that transports blood away from the heart; the right ventricle pumps oxygen-poor blood toward the lungs through the arteries, and the left ventricle pumps oxygen-rich blood toward the body cells through the arteries. (Chap. 3, p. 86)

asthma: lung disorder in which the bronchial tubes become constricted quickly; results in reduced flow of air to the lungs, shortness of breath, wheezing, and coughing; often an allergic or stress reaction. (Chap. 16, p. 502)

atherosclerosis: condition that results when arteries supplying oxygen and nutrients to the heart become clogged with fatty deposits and calcium buildup; if the clogging continues, the person may have a heart attack. (Chap. 3, p. 102)

atmosphere: layer of gases hundreds of kilometers thick that surrounds Earth; divided into the troposphere, stratosphere, mesophere, and thermosphere; made up of a mixture of solids, liquids, and gases such as oxygen, nitrogen, methane, and carbon dioxide. (Chap. 15, p. 456)

atom: smallest particle of an element; different atoms are needed to "build" various substances—for example, you'd need to chemically combine two atoms of the element oxygen and one atom of the element hydrogen to make one molecule of the compound water. (Chap. 14, p. 441)

atomic theory of matter: theory that all matter consists of atoms, all atoms of an element are identical, different elements have different atoms, and atoms maintain their properties in chemical reactions. (Chap. 14, p. 439)

balanced forces: forces whose actions cancel each other; for example, when equal upward and downward forces are exerted on the same object, the object does not accelerate. (Chap. 1, p. 35)

blood pressure: pressure created by the force of blood flowing against the inner walls of arteries, veins, and capillaries; in a healthy young person, normal systolic blood pressure is about 110-120 mm Hg and normal diastolic blood pressure is about 70-80 mm Hg. (Chap. 3, p. 95)

bone marrow: gel-like substance filling spaces within the bones; red marrow makes new blood cells and is found in spongy bone; yellow marrow is found in the long parts of bones and contains fat and some blood-cell producing components. (Chap. 7, p. 215)

Boyle's law: explains the relationship between pressure and the volume of a gas; states that the volume of a certain amount of gas is inversely proportional to the pressure, if the temperature remains constant. (Chap. 14, p. 432)

brain stem: part of the brain that controls involuntary body activities such as digestion, breathing, and the beating of your heart; all nerve impulses sent to and from the brain travel through the brain stem. (Chap. 8, p. 250)

buoyant force: upward force exerted by water equal to the weight of water the object is displacing. (Chap. 1, p. 44)

capillary: microscopic, thin-walled vessel through which exchanges between blood and tissues take place; form an extensive network connecting your arteries to your veins. (Chap. 3, p. 86)

cardiac muscle: muscle found only in the heart; pumps blood through the heart and to the cells of the body by rhythmically relaxing and contracting (Chap. 7, p. 226)

cartilage: rubbery material which first forms the skeletal system; later replaced by bone; the end of your nose and external parts of your ears are made of flexible cartilage; cartilage also helps reduce bone-on-bone friction at joints. (Chap. 7, p. 220)

catalyst: substance that speeds up a chemical reaction; examples include the body's enzyme catalysts and catalytic converters that convert harmful automotive gases to harmless carbon dioxide and water. (Chap. 18, p. 572)

cell: basic unit of structure and function in all living things; your body contains more than 10 trillion cells; each different type of cell has a specific job to do. (Chap. 17, p. 517)

cell membrane: flexible, double-layered, outer covering of most cells; surrounds the cell's cytoplasm; oxygen and food enter and water and other products leave the cell through the cell membrane. (Chap. 17, p. 528)

cell wall: adaptation of plants, fungi, and bacteria; the cell wall is the outermost rim of the cell, outside the cell membrane; provides support and protection for the cell. (Chap. 17, p. 532)

cellular respiration: process in which your body gets energy to operate; oxygen from the respiratory system combines with nutrients stored in your cells and releases energy, carbon dioxide, and water. (Chap. 16, p. 497)

cerebellum: part of the brain that coordinates speed and timing of muscle action and maintains balance; sends messages to your cerebrum, which processes the messages, then sends out impulses along motor nerves to activate the muscles needed to carry out specific actions. (Chap. 8, p. 246)

cerebrum: largest part of the brain; carries out complex functions such as memory, speech, and thought; processes messages from your cerebellum and directs muscles to carry out movements; receives nerve impulses and changes them into such sensations as sound and taste and smell. (Chap. 8, p. 245)

Charles' law: explains the relationship between temperature and the volume of a gas; states that gases increase or decrease their volume as the temperature rises and falls, provided pressure and amount of gas are held constant. (Chap. 14, p. 435)

chemical reaction: process in which substances are changed into other substances; involves changes in energy, which may be absorbed or released; for example, the chemical reactions of digestion release some of the chemical energy in food, which can be used for work by the body. (Chap. 18, p. 553)

chloroplasts: structures in the cytoplasm of green plant cells; contain chlorophyll, a green pigment that lets plants produce their own food by converting light energy into chemical energy in a sugar called glucose. (Chap. 17, p. 533)

chromosomes: threadlike structures found in the nucleus of plant and animal cells; made up of proteins and DNA, the molecules that control the cell's activities, including its reproduction; chromosomes can be seen only when a cell is dividing. (Chap. 17, p. 531)

cilia: tiny hairlike structures lining your nasal passages, trachea, and lungs; cilia beat in an upward direction, causing inhaled particles caught in mucus to move to the throat, where they are swallowed and broken down by stomach acids. (Chap. 16, p. 501)

cleavage: ability of a mineral to break along a smooth flat surface as a result of the orderly arrangement of the atoms making up its crystals; mica and barite have perfect cleavage. (Chap. 10, p. 311)

coinage metals: copper, silver, and gold, which have been used because they are malleable and easily stamped into coins; however, because the price of these metals has increased, gold and silver are no longer used in U.S. coins, and some U.S. pennies now contain zinc. (Chap. 9, p. 280)

compact bone: hard, thick, outer bone layer in which compounds of calcium and phosphorus are concentrated; protects internal organs, supplies support and a place for muscle attachment; contains blood vessels, bone cells, nerves, elastic fibers, and some spaces filled with bone marrow. (Chap. 7, p. 215)

compound machine: machines, such as fishing rods and bicycles, that are a combination of simple machines; compound machines make it possible to accomplish tasks that simple machines alone can't do. (Chap. 5, p. 167)

conduction: transfer of heat by two objects in physical contact; because wood is a poorer conductor of heat than metal, the handles of many pots and pans are made of wood. (Chap. 6, p. 191)

continental shelf: first part of the ocean floor after you leave the shore; formed of flat, gently slanted sections that extend into the ocean, then dip steeply down, forming a continental slope. (Chap. 12, p. 378)

convection: heat carried by a moving fluid; convection is a major cause of winds; many ocean currents result from convection. (Chap. 6, p. 192)

cystic fibrosis: lung disease that is passed genetically from parents to children; thick mucus blocks air passages of the respiratory system, causing lung damage, wheezing, and frequent lung infections. (Chap. 16, p. 502)

cytoplasm: gel-like, constantly moving material found inside the cell membrane of both plants and animals; most of the cell's life processes take place in the cytoplasm. (Chap. 17, p. 529)

D

diaphragm: thin sheet of muscle under your lungs; when you inhale, your diaphragm contracts, helping to increase the size of your chest cavity; when you exhale, your diaphragm relaxes, helping to reduce the size of your chest cavity. (Chap. 16, p. 492)

diffusion: process by which molecules move from an area of higher concentration to an area of lower concentration until equilibrium is reached; diffusion occurs without the use of energy by the cell. (Chap. 19, p. 587)

ductile: ability of an element to be pulled into a wire without breaking; most malleable metals are also ductile; copper, which is very ductile, is also an excellent conductor of electrical signals and is used to make electrical wire. (Chap. 9, p. 272)

E

effort force: the force applied to a machine; for example, when you are in a rowboat and use an oar, you are exerting an effort force on the oar. (Chap. 5, p. 150)

emphysema: lung disease in which the air passageways or alveoli are no longer able to expand or contract; the lungs become scarred and less oxygen goes into the bloodstream, so the heart must work harder, often resulting in heart problems as well. (Chap. 16, p. 503)

endothermic reaction: chemical reaction is which energy is absorbed as the reaction continues; takes place, for example, during the refining of aluminum from bauxite ore as electrical energy is absorbed and aluminum oxide separates into oxygen and aluminum. (Chap. 18, p. 566)

epicenter: point on Earth's surface directly above the focus; when primary and secondary seismic waves reach the epicenter, they generate surface waves that

travel outward and cause great damage. (Chap. 2, p. 62)

exothermic reaction: chemical reaction in which energy is released; takes place, for example, during the explosion of fireworks, during which the chemical potential energy in the fireworks is released and changed into sound, light, and thermal energy. (Chap. 18, p. 567)

extrusive: igneous rock formed by lava on Earth's surface; extrusive igneous rock cools in air and moisture, forming minerals quickly that are much smaller than those in intrusive rock; rhyolite and basalt are examples. (Chap. 11, p. 333)

F

fault: fracture within Earth along which rock movement occurs; faults are caused by compression forces, which produce reverse faults, tension forces, which produce normal faults, or shearing forces, which produce slip-strike faults; earthquakes occur as a result of all three types of faults. (Chap. 2, p. 56)

fermentation: energy-releasing process that breaks down glucose without the use of oxygen; muscle cells release energy by fermentation when oxygen levels are low; bakers use fermentation by yeast to make bread rise. (Chap. 19, p. 602)

focus: point deep beneath Earth's surface where the sudden movement along a fault releases the energy that causes an earthquake. (Chap. 2, p. 61)

foliated: texture of metamorphic rocks whose mineral grains have been flattened and lined up in parallel bands; foliated rocks, such as slate, are easily separated along their foliation layers. (Chap. 11, p. 342)

force: a push or pull; more force is needed to move a large object than a small object; force equals mass times acceleration ($F = ma$); two types of forces are friction and gravity. (Chap. 1, p. 28)

fossil fuel: coal, natural gas, and oil formed millions of years ago from the remains of once-living organisms; contains chemical potential energy and when burned produces thermal energy; fossil fuels are nonrenewable natural resources. (Chap. 13, p. 397)

fracture: ability of a mineral to split into pieces with rough, jagged, or curved surfaces; quartz is a mineral with fracture. (Chap. 10, p. 311)

GLOSSARY

gem: valuable, rare mineral; examples include diamonds, sapphires, rubies, and emeralds; skilled stonecutters cut and grind facets into gemstones to bring out their beauty. (Chap. 10, p. 306)

generator: any machine that can convert kinetic energy to electrical energy; for example, a simple generator spins a coil of wire through a magnetic field, which causes the electrons in the wire to move, producing electricity. (Chap. 13, p. 394)

gills: main respiratory structures of most fish; when water passes over the gill filaments, oxygen diffuses from the water into the capillaries; at the same time, carbon dioxide moves from the capillaries into the water. (Chap. 16, p. 486)

hardness: property that can help identify a mineral; a measure of how easily a specific mineral can be scratched; the Mohs scale lists 10 minerals in order of hardness from 1 to 10. (Chap. 10, p. 310)

heat: energy in transit; heat is energy transferred from an object of higher temperature to an object of lower temperature by conduction, convection, or radiation. (Chap. 6, p. 190)

heat engine: inefficient means of doing work because most of the heat produced is transferred, wasting thermal energy; a car engine is an example of a heat engine. (Chap. 6, p. 201)

hemoglobin: oxygen-binding, iron-containing substance found in red blood cells; because of hemoglobin, red blood cells can transport oxygen from the lungs throughout your body and carry some carbon dioxide back to the alveoli. (Chap. 16, p. 494)

hormone: chemical messenger made by the endocrine glands, which are ductless and empty directly into the bloodstream; hormones are produced in one part of your body and cause change in another part of your body. (Chap. 8, p. 256)

hydroelectric: power plant that uses water power to generate electricity; hydroelectric energy is a clean, plentiful, renewable alternative to fossil fuels but ecosystems may be destroyed in power plant construction. (Chap. 13, p. 408)

hypertension (high blood pressure): disorder of the circulatory system that can result in organ damage; a known cause is atherosclerosis; hypertension places extra stress on the heart, which has to beat faster to get oxygen to tissues. (Chap. 3, p. 104)

igneous rock: forms when lava or magma cools and crystals grow together, creating solid igneous rock; two types of igneous rock are extrusive and intrusive. (Chap. 11, p. 332)

inclined plane: ramp or slope that reduces the force needed to lift an object; a more gradual inclined plane increases the distance but reduces the effort force needed to move an object; inclined planes are in everyday use—for example, ramps make buildings accessible to people in wheelchairs. (Chap. 5, p. 154)

inertia: tendency of an object to resist changes in motion; objects with large mass have great inertia and resist acceleration; as an object's inertia increases, the force needed to accelerate that object also increases. (Chap. 1, p. 23)

inhibitor: substance that slows down a chemical reaction; examples include most preservatives added to food, which slow the spoilage rate, and some antibiotics, which slow enzymatic body reactions that help bacteria grow. (Chap. 18, p. 574)

intrusive: igneous rock formed slowly from magma deep beneath Earth's crust; generally contains large, visible minerals; examples are granite and gabbro. (Chap. 11, p. 332)

joints: places where two or more bones of the skeleton join; movable joints allow actions such as bending your legs; immovable joints do not allow motion and form the bones of the skull; examples of movable joints are gliding joints and hinge joints. (Chap. 7, p. 222)

kinetic energy: the energy of motion; for example, as a roller coaster car rushes downhill, its kinetic energy increases; the amount of kinetic energy possessed by an object is never greater than the amount of work done on that object. (Chap. 4, p. 122)

L

law of conservation of energy: energy can change from one form to another, but it cannot be created or destroyed. (Chap. 4, p. 138)

lever: a bar that turns or pivots on a fulcrum; a lever is a simple machine that can exert a large resistance force over a short distance but only a small force over a long distance. (Chap. 5, p. 151)

ligaments: strong bands of tissue that hold bones together; injury to ligaments can result in sprains, often at the knees, ankles, and fingers. (Chap. 7, p. 224)

longshore current: flow of ocean water that approaches parallel to the shoreline; has enough mechanical energy to erode the shoreline and carry away tons of loose sediments. (Chap. 12, p. 364)

lung: the main organ of your respiratory system; exchange of oxygen and carbon dioxide takes place in your lungs in structures called alveoli; air moves in and out of the lungs through the bronchial tubes, which branch from the lower end of the trachea. (Chap. 16, p. 489)

lung cancer: cancer that causes the most deaths among men and women in the United States; inhaling tar from cigarette smoke is the largest contributing factor to lung cancer; cancer cells grow faster than normal cells and soon normal cells weaken and are no longer able to keep lung tissue healthy. (Chap. 16, p. 504)

M

malleable: ability of an element to be hammered or pressed into a thin sheet without breaking; most metals are malleable; gold, silver, and platinum are often used in making jewelry because they are so malleable. (Chap. 9, p. 272)

mechanical advantage (MA): the MA of a machine can be found by dividing the resistance force by the effort force; the larger the MA, the more helpful the machine. (Chap. 5, p. 159)

metalloids: elements with properties of both metals and nonmetals; boron and silicon are the two most common metalloids; boron and its compounds are used to make cleaning products and

rocket fuel; silicon is a semiconductor and is important in the computer and electronics industry. (Chap. 9, p. 290)

metals: groups of elements often having several properties in common, such as luster, hardness, ductility, malleability, and conductivity; examples are nickel, iron, and gold; examples of metals found in the body are calcium and iron. (Chap. 9, p. 272)

metamorphic rock: forms when heat or pressure or both are applied to any kind of rock; for example, the metamorphic rock slate forms when heat and pressure compact the minerals in the sedimentary rock shale. (Chap. 11, p. 340)

mid-ocean ridges: chains of underwater mountains alongside rift zones; often contain active volcanoes. (Chap. 12, p. 383)

mineral: inorganic, naturally occurring, crystalline structure with a unique chemical composition; a mineral can be an element or compound; examples are graphite, diamonds, and halite. (Chap. 10, p. 303)

mitochondria: power plants in the cell's cytoplasm in which energy is released for organisms to do their work; food molecules are broken down in the mitochondria and converted to forms the cell can use. (Chap. 17, p. 531)

mitosis: reproductive process of cells in which the cell nucleus divides so each new cell will have the same type and number of chromosomes as the parent cell; the original cell becomes two identical cells through mitosis. (Chap. 17, p. 538)

N

neuron: basic unit of the nervous system made up of a cell body, branching dendrites that carry electrical impulses to the cell body, and an axon that carries these messages away from the cell body; neurons are grouped in bundles called nerves. (Chap. 8, p. 242)

newton: standard unit of force, which is abbreviated N; 1 newton of force equals the amount of force needed to accelerate a 1-kg mass at 1 m/s^2. (Chap. 1, p. 35)

nonfoliated: texture of metamorphic rocks whose mineral grains have been changed, rearranged, or combined (but did not form parallel bands); such as marble (Chap. 11, p. 342)

nonmetals: most common nonmetals are gases, such as oxygen and nitrogen: solid nonmetals, such as sulfur, are dull, brittle, and poor conductors of heat and electricity; your body is composed mostly of non-metallic compounds, as is most living material. (Chap. 9, p. 282)

nonrenewable resource: any natural resource that people use up more quickly than it can be replaced by natural processes, or may never be replaced, or can be replaced only over a very long period of time; fossil fuels are nonrenewable resources. (Chap. 13, p. 403)

nucleus: membrane-enclosed command center in the cytoplasm of plant and animal cells; has chromosomes that contain complex chemical information that directs all the cell's activities, including its reproduction. (Chap. 17, p. 530)

organ: body structure made up of various types of tissue that all work together to perform a specific function; for example, your heart is an organ. (Chap. 19, p. 605)

organ systems: groups of organs that work together to perform a specific function; for example, your circulatory system is an organ system. (Chap. 19, p. 605)

organism: a living thing; organisms may be one-celled or many-celled. (Chap. 19, p. 605)

osmosis: diffusion of water through a semi-permeable cell membrane from an area of higher concentration to an area of lower concentration until equilibrium is reached; osmosis occurs without the use of energy by the cell. (Chap. 19, p. 589)

ozone: gas layer in the stratosphere that absorbs some of the harmful rays of the sun, helping to protect Earth from harmful radiation that can cause sunburn and skin cancer. (Chap. 15, p. 462)

pollution: any substance with the potential to damage the environment; pollution can be caused by such factors as burning of fossil fuels, oil spills, and industrial and medical wastes. (Chap. 12, p. 371)

potential energy: the stored energy of position; gravitational potential energy is stored when work lifts an object against the force of gravity; elastic potential energy is stored when work stretches or twists objects such as rubber bands or diving boards. (Chap. 4, p. 123)

power: power is the work done divided by the time it took to do it; the formula can be written as power = work/time. (Chap. 5, p. 168)

pressure: the force or weight acting on each unit of area; can be calculated by dividing the weight of an object by the surface area the object occupies; in contrast, water exerts pressure in all directions, and calculating water pressure depends on the depth of the water. (Chap. 1, p. 40)

prevailing westerlies: global wind system between 30° and 60° latitude; these winds are deflected from their north or south path by Earth's rotation; northern prevailing westerlies are responsible for much of the movement of weather in the United States and Canada. (Chap. 15, p. 474)

products: the new substances formed by any chemical reaction; citric acid + ammonium hydroxide → ammonium citrate + water is a word equation in which the products are ammonium citrate and water. (Chap. 18, p. 557)

pulley: a simple machine composed of a wheel with a rope or chain passing over it; a single fixed pulley changes the direction of the force applied to an object, but not the amount of the force; a single movable pulley increases the effect of the effort force and lessens the amount of effort you have to exert to get the job done, but does not change the direction of the force. (Chap. 5, p. 153)

pulse: rhythmic expansion and contraction of your arteries each time your heart beats; you can feel your pulse in your carotid artery and measure your heart rate. (Chap. 3, p. 91)

radiation: heat traveling across a space; does not need intermediate matter to transfer thermal energy; examples are heat from the sun and candles. (Chap. 6, p. 197)

reactants: substances you start with in any chemical reaction; citric acid + ammonium hydroxide → ammonium citrate + water is a word equation in which the reactants are citric acid and ammonium hydroxide. (Chap. 18, p. 557)

reaction force: the force that pushes back on you when you push on something; action-reaction forces occur in pairs, exert equal and opposite forces, and happen at exactly the same moment—for example, when you lean lightly against a parked car (action force), the car pushes lightly back at you (reaction force). (Chap. 1, p. 37)

reflex: automatic body response to an environmental stimuli that may harm you; your nervous system protects you by the whole reflex sequence—for example, instantly pulling your hand away from a pot of boiling water before you can even think about how hot it is. (Chap. 8, p. 254)

renewable resources: any natural resource that can be replaced by natural processes in less than a century; examples are trees and crops. (Chap. 13, p. 402)

resistance force: the force applied by a machine; for example, when you are in a rowboat and use an oar, you are exerting an effort force on the oar, which exerts a resistance force on the water and pushes the boat through the water. (Chap. 5, p. 150)

respiration: series of chemical reactions by which energy is released from glucose in the presence of oxygen; in your body, heat is a waste product of the thousands of respirations taking place in the mitochondria of your cells. (Chap. 19, p. 595)

rift zone: ocean region where the seafloor is spreading; magma oozes up from Earth's interior in rift zones; rift zones may become a future site for mining precious metals. (Chap. 12, p. 379)

rock: a mixture of one or more minerals or a mixture of mineraloids, glass, or organic particles. (Chap. 11, p. 331)

rock cycle: process by which the materials of Earth change to form different kinds of rocks; all rocks are in a constant state of change. (Chap. 11, p. 350)

screw: this simple machine is an inclined plane wound around a post; the ridges spiraling around a screw are called threads; these threads change the screwdriver's turning force into a downward force that helps lift the wood up around the screw. (Chap. 5, p. 156)

sedimentary rock: formed when any type of rock and plant and animal remains are weathered into sediments and then recombined to form rock by compaction, cementation, or precipitating out of solution; examples are limestone and sandstone. (Chap. 11, p. 343)

seismic waves: waves generated by earthquakes; primary and secondary seismic waves originate at the earthquake's focus, usually travel through Earth's

interior, and can be measured by a seismograph; surface seismic waves travel from the epicenter of the earthquake and cause the greatest damage. (Chap. 2, p. 61)

skeletal muscles: muscles that move bones; skeletal muscles work in pairs—when one skeletal muscle contracts, its partner relaxes—the biceps and triceps are examples; you can control the movement of these muscles. (Chap. 7, p. 226)

smog: type of air pollution caused by the burning of fossil fuels; harms people, making them more susceptible to lung disease and heart trouble; harms plants, preventing them from absorbing the carbon dioxide they need. (Chap. 15, p. 461)

smooth muscle: moving food through your digestive system is an example of work done by smooth muscles; smooth muscle is found in many places inside your body, but you do not control its actions. (Chap. 7, p. 226)

solar cell: converts sunlight into electricity; solar energy is clean, efficient, and renewable; however, solar energy is available only when there is sunlight, and solar cells are expensive to manufacture. (Chap. 13, p. 413)

spinal cord: a long cord that runs the length of the backbone and acts as a connection between the brain and nerves of the body; the spinal cord is protected by vertebrae and cartilage discs; your brain and spinal cord make up the central nervous system. (Chap. 8, p. 250)

spongy bone: inner bone layer containing many openings filled with bone marrow and tiny hard bone spikes made of minerals; found toward the ends of many compact bones; helps keep the skeleton lightweight and functions as shock absorber. (Chap. 7, p. 215)

streak: color of a mineral obtained when it is broken up and powdered or when a piece of the mineral is rubbed across a streak plate; a mineral's color can provide clues to its identity. (Chap. 10, p. 308)

synapse: a tiny space between neurons; electrical impulses from the axon of a neuron cause transmitting chemicals to cross the synapse and stimulate electrical impulses in the next neuron. (Chap. 8, p. 243)

target tissue: the specific tissue affected by the action of a hormone; for example, puberty occurs because the pituitary gland, an endocrine gland, releases

GLOSSARY

hormones that stimulate the sex organs in both males and females to produce sex hormones. (Chap. 8, p. 257)

tendons: elastic, strong tissue bands that attach skeletal muscles to bones at movable joints. (Chap. 7, p. 226)

thermal equilibrium: reached when two objects are in physical contact and the temperature of one is the same as the temperature of the other; if two objects in physical contact have different temperatures, transfer of thermal energy will continue until they are in thermal equilibrium. (Chap. 6, p. 186)

tissue: group of similar types of cells that work together to perform the same function. (Chap. 19, p. 605)

trachea: an air-conducting tube; a grasshopper has thousands of tracheal tubes that move air directly into all the cells of its body; in humans, the trachea is the windpipe, a sturdy tube supported by rings of cartilage. (Chap. 16, p. 485)

trade winds: global wind system; created by air sinking and returning to the equator; airplane pilots often ride the warm, steady trade winds to conserve fuel and increase speed. (Chap. 15, p. 473)

troposphere: atmospheric layer closest to the ground; we live in the troposphere; contains 75 percent of all atmospheric gases in addition to dust, ice, and water; clouds form and weather occurs in this layer. (Chap. 15, p. 462)

veins: elastic vessels that transport blood back to the heart; carbon dioxide-rich blood is returned to the right atrium by veins; oxygen-rich blood is returned to the left atrium by veins; veins contain one-way valves to prevent blood from flowing backward. (Chap. 3, p. 86)

vents: openings in Earth's surface through which volcanic material erupts; different types of volcanoes result from such factors as the number of vents and the type and temperature of the lava. (Chap. 2, p. 67)

watt (W): unit of power found by dividing the work done, in joules (J), by the number of seconds it took to do the work. (Chap. 5, p. 169)

wedge: inclined plane with a sharp, thin edge that cuts through a variety of materials; examples of this simple machine are knives, axes, and chisels. (Chap. 5, p. 157)

weight: the force with which an object is pulled by Earth's gravitational force; calculated by multiplying acceleration due to gravity times the object's mass. (Chap. 1, p. 36)

wheel and axle: a simple machine composed of a small wheel attached to the middle of a bigger wheel; a wheel and axle always rotate together; less effort force is needed to turn a large wheel than to turn a small wheel. (Chap. 5, p. 152)

work: the transfer of energy through both force and motion, with the force in the direction of the motion; work = force × distance ($W = F \times d$); unit for work can also be expressed as the newton • meter (N • m), which is also called a joule (J); 1 J is the work done when the force of 1 N acts through the distance of 1 m. (Chap. 4, p. 119)

Glossary   **661**

SPANISH GLOSSARY

This glossary defines each key term that appears in **bold type** in the text. It also indicates the chapter number and page number where you will find the word used.

A

abyssal plain/planicie abismal: superficie plana en las profundidades del océano la cual ha sido formada por los sedimentos acarreados por las corrientes marinas (Cap. 12, pág. 378)

action force/fuerza de acción: fuerza que resulta cuando empujas o halas algo (Cap. 1, pág. 37)

alveoli/alvéolos: manojos de pequeños saquitos con paredes delgadas localizados en las extremidades más angostas de los bronquiolos (Cap. 16, pág. 489)

arteries/arterias: vasos sanguíneos que llevan la sangre desde el corazón a las demás partes del cuerpo (Cap. 3, pág. 86)

asthma/asma: trastorno pulmonar caracterizado por falta de aliento, sibilancias o tos (Cap. 16, pág. 502)

atherosclerosis/aterosclerosis: condición en la cual se obstruyen las arterias coronarias debido a la acumulación de depósitos grasos y de calcio (Cap. 3, pág. 102)

atmosphere/atmósfera: capa de gases con un espesor de cientos de kilómetros que rodea la Tierra (Cap. 15, pág. 456)

atom/átomo: la partícula más pequeña de un elemento (Cap. 14, pág. 441)

atomic theory of matter/la teoría atómica de la materia: teoría que asevera que la materia está compuesta de pequeñas partículas llamadas átomos (Cap. 14, pág. 439)

B

balanced forces/fuerzas equilibradas: fuerzas que se anulan entre sí al actuar sobre un objeto (Cap. 1, pág. 35)

blood pressure/presión sanguínea: presión ejercida por la sangre contra las paredes internas de los vasos sanguíneos (Cap. 3, pág. 95)

bone marrow/médula ósea: sustancia gelatinosa que se halla dentro de los huesos en donde se forman las nuevas células sanguíneas (Cap. 7, pág. 215)

Boyle's law/ley de Boyle: dice que el volumen de cierta cantidad de gas es inversamente proporcional a la presión, si la temperatura permanece constante (Cap. 14, pág. 432)

brainstem/bulbo raquídeo: la parte del encéfalo que lo conecta a la médula espinal (Cap. 8, pág. 250)

buoyant force/fuerza boyante: fuerza ascendente que los fluidos ejercen sobre todos los objetos (Cap. 1, pág. 44)

C

capillaries/capilares: los vasos sanguíneos más finos que, en forma de red, conectan las arterias con las venas (Cap. 3, pág. 86)

cardiac muscle/músculo cardíaco: tipo de músculo que forma las paredes del corazón, el cual bombea la sangre a través del mismo y del resto del cuerpo (Cap. 7, pág. 226)

cartilage/cartílago: materia blanda y elástica, menos dura que el hueso (Cap. 7, pág. 220)

catalyst/catalizador: sustancia que acelera una reacción química, sin ser alterada permanentemente ella misma (Cap. 18, pág. 572)

cell/célula: la unidad básica de la vida en todos los organismos vivos (Cap. 17, pág. 517)

cell membrane/membrana celular: estructura flexible que forma el límite externo de la célula (Cap. 17, pág. 528)

cell wall/pared celular: estructura rígida ubicada fuera de la membrana celular, la cual da apoyo y protege la célula (Cap. 17, pág. 532)

cellular respiration/respiración celular: proceso en el cual el oxígeno se combina con los nutrimentos almacenados en las células para liberar energía, dióxido de carbono y agua (Cap. 16, pág. 497)

cerebellum/cerebelo: parte del encéfalo que coordina las acciones de todos tus músculos y mantiene el equilibrio (Cap. 8, pág. 246)

cerebrum/cerebro: la parte más grande del encéfalo, la cual interpreta los impulsos que le llegan de los nervios desde las diferentes partes del cuerpo (Cap. 8, pág. 245)

Charles' law/ley de Charles: dice que el volumen de los gases aumenta o disminuye a medida que la temperatura sube y baja, siempre y cuando se mantengan constantes la presión y la cantidad de gas (Cap. 14, pág. 435)

chemical reaction/reacción química: proceso bien definido que resulta en la formación de nuevas sustancias que tienen propiedades diferentes de las de las sustancias originales (Cap. 18, pág. 553)

chloroplast/cloroplasto: pequeña estructura que contiene clorofila, un pigmento verde que permite que las plantas fabriquen su propio alimento (Cap. 17, pág. 533)

chromosome/cromosoma: estructura filamentosa compuesta de proteínas y DNA; molécula que controla las actividades celulares (Cap. 17, pág. 531)

cilia/cilios: estructuras filamentosas pequeñísimas que cubren el revestimiento húmedo de la tráquea y de los pulmones (Cap. 16, pág. 501)

cleavage/crucero: la propiedad de ciertos minerales de romperse a lo largo de superficies lisas y planas (Cap. 10, pág. 311)

coinage metals/metales de acuñación: metales que se usan en la fabricación de monedas (Cap. 9, pág. 280)

compact bone/hueso compacto: capa externa y gruesa que contiene grandes concentraciones de compuestos de calcio y fósforo, además de materias vivas y fibras elásticas (Cap. 7, pág. 215)

compound machine/máquina compuesta: es una combinación de máquinas simples, la cual permite realizar tareas que una máquina simple sola no puede realizar (Cap. 5, pág. 167)

conduction/conducción: proceso por el cual el calor se mueve a través de un material o de un material a otro (Cap. 6, pág. 191)

continental shelf/plataforma continental: primera área del océano que uno se encuentra al moverse mar adentro (Cap. 12, pág. 378)

convection/convección: transferencia de energía por el movimiento de un medio que transporta calor (Cap. 6, pág. 192)

cystic fibrosis/fibrosis cística: trastorno que afecta el sistema respiratorio al obstruir las vías respiratorias (Cap. 16, pág. 502)

cytoplasm/citoplasma: material gelatinoso que se encuentra en el interior de la membrana celular (Cap. 17, pág. 529)

diaphragm/diafragma: lámina muscular fina ubicada debajo de los pulmones (Cap. 16, pág. 492)

diffusion/difusión: proceso mediante el cual el movimiento constante de moléculas ocasiona el movimiento desde una región de mayor concentración hasta otra de menor concentración (Cap. 19, pág. 587)

ductile/dúctil: propiedad que poseen muchos metales para poder estirarse en forma de alambre sin romperse (Cap. 9, pág. 272)

effort force/fuerza de esfuerzo: fuerza que uno ejerce sobre una máquina (Cap. 5, pág. 150)

emphysema/enfisema: enfermedad que ocurre cuando las vías respiratorias o los alvéolos pierden la capacidad de expandirse y de contraerse (Cap. 16, pág. 503)

endothermic reaction/reacción endotérmica: reacción química en la cual se absorbe energía a medida que continúa la reacción (Cap. 18, pág. 566)

epicenter/epicentro: punto en la superficie terrestre que se encuentra directamente sobre el foco en un terremoto (Cap. 2, pág. 62)

exothermic reaction/reacción exotérmica: reacción química en la cual se libera energía (Cap. 18, pág. 567)

extrusive/extrusiva: roca ígnea que se forma de la lava que se enfría sobre la superficie terrestre (Cap. 11, pág. 333)

F

fault/falla: hendidura dentro de la Tierra donde ocurre un desplazamiento de rocas (Cap. 2, pág. 56)

fermentation/fermentación: proceso que libera energía al descomponer la glucosa, en ausencia del oxígeno (Cap. 19, pág. 602)

focus/foco: punto en el interior de la Tierra donde se originan las ondas sísmicas (Cap. 2, pág. 61)

foliated/foliada: roca metamórfica que posee una textura de granos aplanados y alineados en bandas paralelas (Cap. 11, pág. 342)

force/fuerza: un empujón o un halón (Cap. 1, pág. 28)

fossil fuel/combustible fósil: restos de plantas o animales antiguos que podemos quemar hoy en día para producir energía térmica (Cap. 13, pág. 397)

fracture/fractura: se dice que un mineral presenta una fractura cuando al romperse forma superficies curvas, ásperas o dentadas (Cap. 10, pág. 311)

G

gem/gema: mineral valioso que es raro o muy difícil de obtener (Cap. 10, pág. 306)

generator/generador: cualquier máquina que convierte la energía cinética en energía eléctrica (Cap. 13, pág. 394)

gills/agallas: órganos respiratorios de algunos animales acuáticos a través de los cuales se extrae el oxígeno del agua (Cap. 16, pág. 486)

H

hardness/dureza: medida que se usa para determinar la facilidad o dificultad con que se puede rayar un mineral (Cap. 10, pág. 310)

heat/calor: energía que se transfiere de un objeto con mayor temperatura a uno de menor temperatura (Cap. 6, pág. 190)

heat engine/motor térmico: motor que usa combustible para producir energía térmica para realizar trabajo (Cap. 6, pág. 201)

hemoglobin/hemoglobina: sustancia de los glóbulos rojos sanguíneos que contiene hierro, el cual se enlaza fácilmente con el oxígeno (Cap. 16, pág. 494)

hormone/hormona: sustancia química fabricada por una glándula sin conductos en una parte del cuerpo, la cual produce cambios en otra parte del cuerpo. (Cap. 8, pág. 256)

hydroelectric/hidroeléctrico: que usa agua para generar electricidad (Cap. 13, pág. 408)

hypertension/hipertensión: trastorno del sistema circulatorio en el cual la presión sanguínea es más alta de lo normal. También llamada presión alta (Cap. 3, pág. 104)

I

igneous rock/roca ígnea: roca que se forma al enfriarse un material derretido (Cap. 11, pág. 332)

inclined plane/plano inclinado: una rampa o un plano inclinado que reduce la fuerza que necesitas ejercer para levantar un objeto (Cap. 5, pág. 154)

inertia/inercia: tendencia a resistir cambios en el movimiento (Cap. 1, pág. 23)

inhibitor/inhibidor: cualquier sustancia que aminora una reacción química (Cap. 18, pág. 574)

intrusive/intrusiva: roca ígnea que se forma del magma que se enfría debajo de la superficie terrestre (Cap. 11, pág. 332)

J

joint/articulación: área en el esqueleto donde dos o más huesos se unen (Cap. 7, pág. 222)

K

kinetic energy/energía cinética: energía de movimiento (Cap. 4, pág. 122)

L

law of conservation of energy/ley de conservación de la energía: ley que dice que la energía no se puede crear ni destruir (Cap. 4, pág. 138)

lever/palanca: una barra que da vueltas o gira sobre un punto llamado fulcro (Cap. 5, pág. 151)

ligament/ligamento: Banda fuerte de tejido que mantiene unidos a los huesos (Cap. 7, pág. 224)

longshore current/corriente de costa: flujo de agua del océano que corre cerca de la costa y paralelo a ella (Cap. 12, pág. 364)

lung cancer/cáncer del pulmón: ocurre cuando se dañan los cilios y los pulmones no se pueden defender contra las enfermedades (Cap. 16, pág. 504)

lungs/pulmones: ubicados en la cavidad torácica, son los órganos principales del sistema respiratorio (Cap. 16, pág. 489)

M

malleable/maleable: propiedad de algunos metales de poder martillarse o extenderse en planchas o láminas sin romperse (Cap. 9, pág. 272)

mechanical advantage/ventaja mecánica: la ventaja mecánica de una máquina te deja saber cuantas veces se multiplica la fuerza de esfuerzo (Cap. 5, pág. 159)

metal/metal: uno de los grupos de elementos más importantes, generalmente son sólidos brillantes, maleables, dúctiles que conducen el calor y la electricidad (Cap. 9, pág. 272)

metalloide/metaloide: elemento que posee características tanto de los metales como de los no metales (Cap. 9, pág. 290)

metamorphic rock/roca metamórfica: roca que se forma cuando el calor o la presión o ambos cambian la roca (Cap. 11, pág. 340)

mid-ocean ridge/dorsal medioceánica: cadena de montañas submarinas que se extiende a lo largo de la zona de cuencas profundas (Cap. 12, pág. 383)

mineral/mineral: estructura cristalina inanimada que se da en forma natural y que posee una composición definida. (Cap. 10, pág. 303)

mitochondria/mitocondria: son las centrales eléctricas de una célula; estructuras en donde se descomponen las moléculas de alimento y la energía química se convierte en formas que la célula puede usar (Cap. 17, pág. 531)

mitosis/mitosis: proceso mediante el cual el núcleo de una célula se divide para producir dos núcleos, cada uno con el mismo tipo y número de cromosomas que tenía la célula original (Cap. 17, pág. 538)

neuron/neurona: célula nerviosa que es la unidad básica del sistema nervioso (Cap. 8, pág. 242)

newton/newton: unidad de fuerza cuya abreviatura es N (Cap. 1, pág. 35)

nonfoliated/no foliada: roca metamórfica en que los granos minerales cambian, se combinan o se ordenan de manera diferente pero no forman bandas visibles (Cap. 11, pág. 342)

nonmetal/no metal: elemento sólido quebradizo y mal conductor de calor y de electricidad; los no metales también pueden ser gaseosos (Cap. 9, pág. 282)

nonrenewable resource/recurso no renovable: recurso que la gente usa mucho más rápidamente de lo que la naturaleza tarda en reemplazar (Cap. 13, pág. 403)

nucleus/núcleo: centro de comando de la célula que dirige todas las actividades celulares (Cap. 17, pág. 530)

organ/órgano: estructura del cuerpo formada por varios tipos diferentes de tejido que funcionan juntos para desempeñar una función específica (Cap. 19, pág. 605)

organ system/sistema de órganos: grupo de órganos que funcionan juntos para desempeñar una función específica (Cap. 19, pág. 605)

organism/organismo: está compuesto de un conjunto de órganos que funcionan juntos (Cap. 19, pág. 605)

osmosis/osmosis: difusión del agua a través de una membrana celular (Cap. 19, pág. 589)

ozone/ozono: gas que absorbe parte de la radiación dañina que proviene del Sol (Cap. 15, pág. 462)

pollution/contaminación: materiales o efectos indeseados o dañinos en el ambiente (Cap. 12, pág. 371)

potential energy/energía potencial: energía almacenada debido a la posición de un objeto (Cap. 4, pág. 123)

power/potencia: cantidad de trabajo realizado dividida entre el intervalo de tiempo (Cap. 5, pág. 168)

pressure/presión: peso o fuerza que actúa sobre cada unidad de superficie (Cap. 1, pág. 40)

prevailing westerlies/predominio vientos ponientes: vientos situados entre las latitudes de 30° y 60° (Cap. 15, pág. 474)

product/producto: sustancia nueva que se forma en una reacción química (Cap. 18, pág. 557)

pulley/polea: una rueda con una cuerda o cadena que pasa sobre la cadena (Cap. 5, pág. 153)

pulse/pulso: la expansión y contracción rítmica de una arteria (Cap. 3, pág. 91)

radiation/radiación: transferencia de energía térmica a través del espacio (Cap. 6, pág. 197)

reactant/reactivo: sustancia con la que comienzas una reacción química (Cap. 18, pág. 557)

reaction force/fuerza de reacción: fuerza que resulta en oposición directa a la fuerza de acción (Cap. 1, pág. 37)

reflex/reflejo: respuesta automática del cuerpo a un estímulo potencialmente peligroso (Cap. 8, pág. 254)

renewable resource/recurso renovable: recurso natural que se puede reemplazar, por medio de procesos naturales, en menos de 100 años (Cap. 13, pág. 402)

resistance force/fuerza de resistencia: fuerza aplicada por una máquina (Cap. 5, pág. 150)

respiration/respiración: proceso químico en el cual se descomponen las moléculas de glucosa con el fin de liberar energía (Cap. 19, pág. 595)

rift zone/zona de cuencas profundas: regiones en donde el suelo marino se está expandiendo (Cap. 12, pág. 379)

rock/roca: mezcla de uno o más minerales, de mineraloides, de vidrio o de partículas orgánicas (Cap. 11, pág. 331)

rock cycle/ciclo de las rocas: proceso mediante el cual las materias terrestres cambian para formar diferentes clases de rocas (Cap. 11, pág. 350)

screw/tornillo: un plano inclinado enrollado alrededor de un poste (Cap. 5, pág. 156)

sedimentary rock/roca sedimentaria: roca que se forma de sedimetos asentados que se vuelven a combinar para formar una roca sólida (Cap. 11, pág. 343)

seismic wave/onda sísmica: onda ocasionada por los terremotos (Cap. 2, pág. 61)

skeletal muscle/músculo del esqueleto: músculo que mueve los huesos (Cap. 7, pág. 226)

smog/smog: tipo de contaminación del aire que aparece como una niebla de humo (Cap. 15, pág. 461)

smooth muscle/músculo liso: músculo que se encuentra en muchas partes dentro de tu cuerpo, tales como el estómago y los intestinos. (Cap. 7, pág. 226)

solar cell/célula solar: célula que convierte la luz solar en electricidad (Cap. 13, pág. 413)

spinal cord/médula espinal: conducto largo que se extiende desde el bulbo raquídeo hasta la parte inferior de la espalda. (Cap. 8, pág. 250)

spongy bone/hueso esponjoso: parte del hueso que se parece a una esponja porque contiene muchos orificios pequeñísimos; se encuentra en los extremos de muchos huesos (Cap. 7, pág. 215)

streak/veta: es el color del mineral cuando se desmenuza y se pulveriza (Cap. 10, pág. 308)

synapse/sinapsis: espacio pequeño que se encuentra entre las neuronas y a través del cual se transmiten los impulsos nerviosos (Cap. 8, pág. 243)

target tissue/tejido asignado: tejido específico que es afectado por una hormona (Cap. 8, pág. 257)

tendon/tendón: banda fuerte y elástica de tejido (Cap. 7, pág. 226)

thermal equilibrium/equilibrio térmico: dos objetos que están en contacto y poseen la misma temperatura (Cap. 6, pág. 186)

tissue/tejido: tipos de células similares que funcionan en conjunto para desempeñar la misma función (Cap. 19, pág. 605)

trachea/tráquea: pasaje por el cual el aire entra y sale del cuerpo (Cap. 16, pág. 485)

trade winds/vientos alisios: vientos causados por el aire descendiente que regresa al ecuador (Cap. 15, pág. 473)

troposphere/troposfera: zona inferior de la atmósfera que es la capa más cercana al suelo (Cap. 15, pág. 462)

veins/venas: vasos sanguíneos que transportan la sangre de regreso al corazón desde los pulmones o el cuerpo (Cap. 3, pág. 86)

vent/chimenea: abertura por donde el magma fluye a la superficie terrestre como lava (Cap. 2, pág. 67)

wedge/cuña: un plano inclinado el cual utiliza el extremo afilado y angosto para cortar a través de materiales (Cap. 5, pág. 157)

weight/peso: fuerza de gravitación ejercida sobre ti o cualquier otro objeto (Cap. 1, pág. 36)

wheel and axle/rueda y eje: una rueda pequeña pegada al centro de una más grande (Cap. 5, pág. 152)

work/trabajo: energía que se transfiere a través de la fuerza y del movimiento (Cap. 4, pág. 119)

INDEX

The Index for *Science Interactions* will help you locate major topics in the book quickly and easily. Each entry in the Index is followed by the numbers of the pages on which the entry is discussed. A page number given in **boldface type** indicates the page on which that entry is defined. A page number given in *italic type* indicates a page on which the entry is used in an illustration or photograph. The abbreviation *act.* indicates a page on which the entry is used in an activity.

Credits

Illustrations

Jonathan Banchick 28, 31; **George Bucktell** 413; **John Edwards** 15, 172, 364-365, 378, 379, 394-395, 401, 408-409, 417, 573; **Chris Forsey/Morgan-Cain & Associates** 58, 59, 62-63, 66, 69, 70, 79, (b) 192-193, 318, 319, 325, 332-333, 350-351, (b) 387, 397; **Nancy Heim/158 Street Design Group** 89; **Tonya Hines** (b) 87, (b) 90, (t) 99, 100, 101, 102, (r) 110, 572; **Tom Kennedy/Romark Illustrations** 38, (r) 49, (t) 140; **Ruth Krabach** (b) 245, 247, (tl) 250; **Gina Lapurga** (bl) 250, (r) 255, (c) 262; **Ortelius Design** 626-627, 628-629; **Felipe Passalacqua** 222, 223, 514, 515; **Bill Pitzer, National Geographic Society** 78, 259, 356, 507, 577; **Precision Graphics** 22-23, 27, 43, (l) 49, 82, 85, (t) 87, (t) 90, 93, 95, (l,c) 110, 124-125, 131, 138-139, (b) 140, 143, 145, 150-151, 154, 159, 160, 161, (b) 177, 187, 196-197, 198, 200, 201, 203, 207, 241, (t) 245, (r) 250, 257, 258, (tr, br) 262, 264, 276, 277, 285, 303, 340, 343, 344, 352, 357, 374, 392-393, 402-403, 405, 406, 418, 426, 432, 435, 437, 441, 444-445, 451, 463, 471, 472, (l) 479, 481, 484, 485, 487, 488, (t) 489, 493, 494, 501, 503, 508, 521, 526, 527, 528, 529, 530, 531, 532, 533, 538, 539, 547, 549, 562, 566-567, (l) 585, 586, 587, 589, (t) 595, 601, 605, (l,br) 610, 620 621, 630-631, 644, 650, 651; **Rolin Graphics Inc.** 47, (r) 213, 215, 224, 226, 227, 231, 235, 242-243, (l) 255, (tl,bl) 262, (t) 387; **Doug Schneider** 648; **Jim Shough** 155, 156, (t) 177, (l) 213; **John Walter & Associates** 52, 78, (b) 99, 121, 142, 183, (t) 193, 403, 468-469, 473, 474, 475, (r) 479, 492, (r) 585, 593, (b) 595, (t,bc) 610.

Photographs

Mark Thayer Studio 3 (t), 34, 40 (cr,r), 41, 42 (t), 49 (tr), 53 (r), 60, 116 (t), 128, 130 (b), 131 (t), 132 (l), 135 (b,c,tl,tr), 148, 149 (t), 151, 152 (t,l,r,), 153 (t), 156 (bl,br), 160, 161, 164 (r), 165 (l,r), 166 (br), 174 (t), 180, 186 (t,bl), 191 (t), 211 (t), 218 (b), 241, 245 (t,b),268-269, 268 (t), 269 (t), 273 (b), 277 (t), 278 (t), 279 (t), 280 (t,b), 281, 289 (b), 292, 299, 302 (tc), 303 (t), 308-309, 309 (t), 311 (7,8,9,10), 313 (l), 314 (r), 334 (c), 337, 338 (bl), 352 (t,bl,br), 353, 359 (l,r), 366 (tl,bl), 367 (l), 424 (l), 425 (t), 431 (r), 438 (l), 441, 446, 457, 485, 499 (l,c,r), 514 (t), 516, 517, 519 (t), 521 (t), 523 (b), 550 (b), 553 (tl,tr,bl), 561 (b), 564 (tcl), 565 (tc), 570 (l,lc), 574 (b), 582 (b), 588 (t), 602 (c); **RMIP/Richard Haynes** 2, 4, 5, 6, 7, 9, 10, 11 (l,r,c,b,), 14, 16, 19, 21 (b), 24, 25 (l,r), 32, 33, 36, 39, 44, 58 (b), 64, 65, 72, 76, 83, 86, 92, 96, 103 (t), 113, 117 (b), 119 (bl,bc,br), 123 (t), 126, 129 (b), 132 (r), 136, 137 (l,r), 149 (b), 153 (b), 157 (t,inset), 162, 163 (l,r), 164 (l), 166 (t), 170, 171 (tl), 179, 181 (b), 182, 183 (t,c,b), 184, 185 (l,r), 187, 189, 190, 194, 211 (b), 212, 216, 217 (l,r), 218 (t), 219, 221, 226 (tr), 227, 239 (t,b), 244 (t,c,b,), 248, 252, 253 (l,r), 254, 256, 257, 258, 265, 269 (c), 271 (b), 272 (b), 274, 275 (l,r), 278 (bl), 282, 286, 287 (l,r), 301 (b), 302 (t), 303 (b), 304, 305 (r), 308 (b), 310 (#2), 312, 316, 317 (l,c,r), 328 (t), 329 (b), 336, 341 (t), 344, 346 (t), 348, 361, 363 (t), 368, 373 (t), 377 (t), 380, 391 (br), 393, 398, 399 (l,c,r), 407, 410, 421, 425 (b), 428, 430, 431 (l), 434, 436, 440, 442, 443 (l,r), 455 (t), 458, 459 (l,r), 466, 467 (l,r), 470 (t), 483, 490, 491, 495, 497 (b), 498, 502, 511, 519 (b), 520, 522, 523 (l,c,r), 525, 530 (r), 540, 541 (l,r), 551 (t,b),

552, 553 (br), 557, 558 (l), 560, 561 (l,r), 564 (tl), 565 (b), 568, 569 (l,r), 570 (r,rc), 576, 583 (b), 584, 586, 588 (b), 596, 598, 599 (l,c,r); **Cover** (bk) Nicholas Devore/Photographers Aspen, (tl) NASA/Photo Researchers, (tr) Glencoe file, (c) Chip Clarke, (b) Norbert Wu; **16** (b) Superstock; **17** NASA; **18-19** National Geographic Journeys (c.) National Geographic Society/Vadim Gippenreiter-REI Adventures; **20** (b) Tom Sobolik/Black Star, (t)Scott McKiernan/Black Star; **21** (t) Studiohio, **26** Ralph Brunke; **29** Kenji Kerins; **30** (b) NASA, (t) Color-Pic,; **31** Superstock; **35** Mary Evans Picture Library; **37** Helen Marcus/Photo Researchers; **40** (l) Carl Purcell, (c,lc) David Stoecklein/The Stock Market, **42** (b) Stephen Frink/the WaterHouse; **45** Chris Sorensen; **46** Robert Frerck/Odyssey/Frerck/Chicago; **48** Studiohio; **49** (b) David Stoecklein/The Stock Market, (tr) Chris Sorensen Photography; **51** (l) Fredrik D. Bodin; **52-53** Francois Gohier/Photo Researchers; **52** Superstock; **54** Kenji Kerins; **55** (b) Tom Bean, (c) N.R. Rowan/Stock Boston, (t) Dell R. Fouts/Visuals Unlimited; **56** Herman Kokojan/Black Star; **57,** (b) David Young-Wolff/PhotoEdit, (t) Andrew Rafkind/Tony Stone Inter.; **58** (t) Greg Vaughn/Tom Stack & Assoc.; **59** (l) Greg Vaughn/Tom Stack and Assoc., (r) Kevin Schafer/Tom Stack and Assoc.; **63** Jose Fernandez/Woodfin Camp & Assoc.; **67** Ken Ferguson; **68** (b) Steve Lissau, (t) Gregory G. Dimijian/Photo Researchers; **70** (b) David Cavagnaro/Visuals Unlimited, (c) Krafft/Explorer/Science Source/Photo Researchers, Inc., (t) Nancy L. Cushing/Visuals Unlimited; **71** (b) Gary Braasch/Woodfin Camp & Assoc., (c) Paul Bierman/Visuals Unlimited, (t) Forest W. Buchanan/Visuals Unlimited; **74** (t) Sigfred C. Balatan/Black Star, (b) Robert Frerck/Odyssey Productions, Chicago; **75** James A. Sugar/Black Star; **77** Comstock Inc./Boyd Norton; **79** G. Dimijian/Photo Researchers; **82-83** Baron Wolman; **84** Doug Martin; **89** SIU Biomed Com/Custom Medical Stock; **91** Studiohio; **93** Doug Martin; **94** Kenji Kerins; **96-97** Doug Martin; **98** Steinmark/Custom Medical Stock Photo; **100** (b) Stephen Dalton/Photo Researchers, (t) Carl Roessler/Tony Stone Inter.; **101** Walter E. Harvey/Photo Researchers; **102** Cabisco/Visuals Unlimited; **103** (l)W.Ober/Visuals Unlimited; **103** (r) Sloop/Visuals Unlimited; **104** Paul Barton/The Stock Market; **105 106** Ed Kashi; **107** (t) Stock Montage, Inc., (b) Dr. Lesley Kohman/SUNY Syracuse/Stock Montage; **109** Courtesy of Tamika Walker; **110** (t) Carl Roessler/Tony Stone Inter., (b) Sloop/Visuals Unlimited; **114-115** National Geographic Journeys (c.) National Geographic Society/Bruce Curtis-Peter Arnold, Inc.; **116-117** Comstock/David MacTavish; **117** (tr) David Madison/Duomo; **118** Studiohio; **119** (t) International Stock Photo; **120** Ralph Brunke; **121** Comstock/Denver Bryan; **122** (l) Mark Burnett, (r) Tom Branch/Science Source/Photo Researchers; **123** (b,c) Tom Bean; **125** (t) Erika Klass; **127** (c) Rhoda Sidney/Stock Boston; **130** (tc) Loren Winters/Joe Strunk; **130** (t) Al Tielemans/Duomo; **133** Ken Regan/Time; **134** Ken Ferguson; **135** (tl) Sports Illustrated for Kids, October 1993; **138-139** NASA; **141** (l) Chris Sorensen, (r) Paul J. Sutton/Duomo; **144** (b) NBS Archives/Courtesy AIP Niels Bohr Library, (t) North Wind Picture Archives; **145** (b) Rhoda Sidney/Stock Boston; **147** Bob